Mechanical Properties of Materials

SOLID MECHANICS AND ITS APPLICATIONS

Volume 190

Aims and Scope of the Series

The fundamental questions arising in mechanics are: *Why?*, *How?*, and *How much?* The aim of this series is to provide lucid accounts written by authoritative researchers giving vision and insight in answering these questions on the subject of mechanics as it relates to solids.

The scope of the series covers the entire spectrum of solid mechanics. Thus it includes the foundation of mechanics; variational formulations; computational mechanics; statics, kinematics and dynamics of rigid and elastic bodies: vibrations of solids and structures; dynamical systems and chaos; the theories of elasticity, plasticity and viscoelasticity; composite materials; rods, beams, shells and membranes; structural control and stability; soils, rocks and geomechanics; fracture; tribology; experimental mechanics; biomechanics and machine design.

The median level of presentation is the first year graduate student. Some texts are monographs defining the current state of the field; others are accessible to final year undergraduates; but essentially the emphasis is on readability and clarity.

For further volumes:
http://www.springer.com/series/6557

Joshua Pelleg

Mechanical Properties of Materials

Joshua Pelleg
Materials Engineering
Ben Gurion University of the Negev
Ben Gurion Street
Beer Sheva
Israel

ISSN 0925-0042
ISBN 978-94-007-9442-9 ISBN 978-94-007-4342-7 (eBook)
DOI 10.1007/978-94-007-4342-7
Springer Dordrecht Heidelberg New York London

Softcover reprint of the hardcover 1st edition 2013

Printed on acid-free paper

Springer is part of Springer Science+Business Media (www.springer.com)

To my wife Ada, children Deenah, Ruth and Asher and their families

Preface

As the title implies, the purpose of this textbook is to present a different approach to the teaching of Materials Science and Engineering than the one that is now commonly used. In earlier times, studies of materials were usually referred to as "metallurgy". Currently most textbooks still begin by introducing the student to basic concepts, accompanied by laboratory exercises relevant to each concept. These exercises are usually quite self-explanatory and have no special prerequisite – except for some common sense – to perform them. In the preparation of this new text, the author has been influenced by two guiding principles. The first is that the student should begin by acquiring some degree of knowledge of and experience with performing mechanical tests. Thus, Chap. 1 places its emphasis on developing this skill. This lays the groundwork for the student to begin performing laboratory tests simultaneously and in parallel with their studies of new concepts throughout the course. Only after this chapter do we deal with basic concepts such as dislocations, plastic deformation, etc.

The second guiding principle was the assumption that students exposed to a course of mechanical behavior of materials in its various forms are if not at a junior then at least at a sophomore level and sometimes even at a senior level. Thus they should already have been exposed to the basic concepts of mechanical and structural fundamentals. Consequently, in this text elements of the theory of elasticity and plasticity are not covered, but rather the student is referred to appropriate books or other publications. Inherent in this assumption is the belief that students at this level are already familiar with the concepts of strain and stress tensors, principal stresses, normal stress, and to the description of stress at a point. The same approach is applied to other basic structural principles, and the elements of crystallography, assuming familiarity of students with the fundamentals of materials science and engineering materials.

Chapter 1 sets the framework of mechanical testing, excluding investigation of dynamic stresses (fatigue) and the effect of temperature on applying static load on materials (creep). These are considered separately. Chapter 2 introduces the basic concept of dislocations that are needed to explain various observations of mechanical behavior. Dislocations are essential in this regard and the chapter is devoted to describing this concept, their motion and generation. In Chap. 3, plastic deformation is discussed in terms of dislocation motion; slip and slip planes are introduced and their association with acting shear stress is considered. Strengthening mechanisms such as strain (work) hardening are treated in Chap. 4. Chapter 5 is completely devoted to creep, the effect of temperature on materials which are loaded statically and which can lead to failure either by not being able to carry the applied load or by stress rupture. The concept of cyclic or repeated stress and its effect on material response in terms of fatigue failure is discussed in Chap. 6. Fracture, both brittle and ductile, is considered in Chap. 7. It discusses fracture occurring in static loading such as in simple tension, in creep as influenced by the effect of time and temperature and fracture resulting from the application of cyclic stresses. Finally, Chap. 8 is devoted to the mechanical behavior of small-size specimens where the dimensions are in the micron and nano scales.

The book is intended as a text for materials engineering students in the junior or possibly senior years of their studies. Students in other related disciplines such as metallurgy or mechanical engineering can benefit from such a text. The scope of the book makes it appropriate as a reference for graduate students as well and for practicing engineers in industry who could use such a textbook to refresh their knowledge in this field. In particular, when practicing engineers are required to supplement or update their understanding in the field of mechanical properties of materials, such a text can be invaluable in achieving these requirements.

I believe that there is sufficient material in this book for teaching at least a three credit course per semester, but it can be extended for a second semester as practiced in my university. In most universities where materials, metallurgy and related engineering courses are being taught, this is about the level of credit hours devoted to the subject of mechanical properties. The content of this textbook – except for Chap. 8 dealing with specimens of small dimensions – is not new and it can be found in many good text and reference books, or in research papers published in journals. The book is based on my lectures given while I was teaching this course at the Ben Gurion University of the Negev for junior undergraduate students, taught as a three credit course for two semesters.

Finally I would like to express my gratitude to all publishers and authors for permission to use and reproduce some of their illustrations and microstructures.

Beer Sheva, Israel Joshua Pelleg

Contents

Chapter 1
Mechanical Testing of Materials

1.1 Introduction

Of the many properties of materials the mechanical ones are of great significance, since they deal with the principal phenomena regarding stability under force. Deformation under applied forces and the fracture of materials depend on their structure. The macroscopic responses of materials to the acting forces may result in their changing shape or even disintegrating, if these forces are sufficiently large. Interatomic forces must be overcome by external forces in order to cause shape changes in a material, which may eventually lead to its separation into two or more parts, depending on the atomic forces which resist any structural change, either in shape or dimension. The overall macroscopic behavior and the changes occurring in materials are inspected, tested experimentally and described in terms of the acting force per unit area, namely stress and the displacement per unit distance or strain. In a perfectly ideal material, free of lattice defects (which, in reality, does not exist, except in the form of whiskers), tremendous forces are required to cause the above changes. Real crystals contain various defects; lattice defects, such as dislocations, are responsible for the ease of deformation, which may often be observed in functioning machine elements. Mechanical engineers are expected to prevent this from occurring. To this end, laboratory tests must be performed in order to realize the practical potential of a given material of interest. In this chapter, a discussion of the mechanical behavior of materials on a macroscopic scale is described as observed during laboratory experiments intended to forecast the actual performance in real service. (An understanding of their real behavior will be the subject of Chap. 2.) The observed behavior, as revealed by the various laboratory tests, will be considered in terms of dislocations, providing a basic conceptual framework for the mechanical properties of materials.

This chapter presents the most commonly used mechanical tests for characterizing materials – indicating the responses of materials to applied static or dynamic forces. Such applied forces may act at certain temperatures, causing the acceleration of failure. Industrial tests are usually performed at room temperature, unless some

J. Pelleg, *Mechanical Properties of Materials*, Solid Mechanics and Its Applications 190, DOI 10.1007/978-94-007-4342-7_1,

specific high-temperature property is of interest. Practical knowledge of the forming of metals and of the specific applications of materials in certain industries is based, for the most part, on the laboratory tests listed in this chapter. In a material in use, stresses inevitably arise which can render it useless for practical purposes, often leading to failure. One has to know the limitations of a material in use. Thus, for example, the load-carrying capacity of a material is critical for the evaluation of the stress that can be safely applied without causing breakage. Similarly, material deformation must be prevented by more than a specified, allowable permanent dimensional change when in use. Critical questions often arise, the answers to which may be readily provided by the laboratory tests found in this chapter.

In this chapter, it is assumed that a material is a homogeneous continuum consisting of an assembly of molecules and that the interaction between them may be ignored, since the continuum is large enough so that an average property can be assigned to the material tested. This approach serves rather well in the cases considered in this chapter. Each of the tests indicated in this part of the book are manifestations of the continuum model, since no detailed atomistic understanding is required in order to perform the stress-strain tests necessary to learn about some material of interest.

1.2 The Tension Test

One of the useful and simple tests for load-carrying capacity is the tension test, which relates stress and strain. Of the two fundamental stresses, referred to as 'normal stress' and 'shear stress', in this chapter we define 'normal stress', which is applicable to an axially loaded rod or to a specimen having another shape.

In a body, every acting force can be resolved into one component acting normally on a chosen area and two components acting tangentially on that area. For the force P acting normally on a small area ΔA of the specimen, an internal opposing stress balances the applied load. For a cylindrical free specimen, e.g., a rod-like shape, we can write:

$$\sigma = \lim_{A \to 0} \frac{\Delta P}{\Delta A} = \frac{dP}{dA} \tag{1.1}$$

or:

$$\mathrm{dP} = \sigma \mathrm{dA} \tag{1.2}$$

Integrate (1.2) as shown in (1.3):

$$\mathrm{P} = \int \sigma dA \tag{1.3}$$

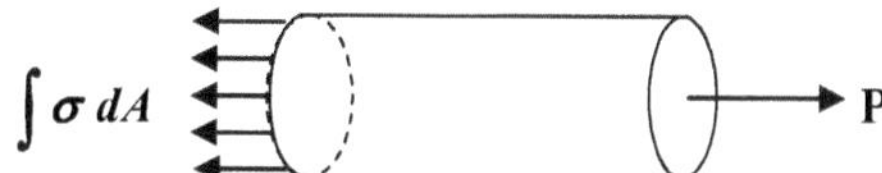

Fig. 1.1 Schematic drawing of uniaxial loading of a free body

Assuming that the stress σ is distributed uniformly over the cross-section means that it acts at a constant level and thus (1.3) can be written as:

$$P = \sigma \int dA = \sigma A \tag{1.4}$$

or

$$\sigma = \frac{P}{A}$$

Equation 1.4 provides the basic relation between normal stress and the load applied in tension. Figure 1.1 illustrates a simple free body cylindrical bar upon which an axial average load is acting. Usually, the stress is not acting uniformly over the cross-section of a specimen unless it is deformed, i.e., elongates uniformly, which is not always the case. In Sects. 1.2.7 and 1.2.8, the concept of instability and necking will be discussed. For the present discussion, 'stress' will refer to average stress.

In Fig. 1.1, a cylindrical rod is shown with an acting external load P, balanced by a resisting force expressed in Eq. 1.4. In practice, a tensile specimen does not have to be cylindrical. Often, flat rectangular specimens are used, depending on the cost and convenience of machinability, which may prove problematic for some materials. Furthermore, flat specimens are often used for sheet and plate stock. Specimens are machined according to specifications conforming to established standards of shape and dimension – to make them grippable in the jaws of the tensile machines. Regardless, if the specimens are round or rectangular, a 2″ (5.08 cm) gage length of reduced diameter (width) is machined in them to force the deformation to occur within the thinner section, where ultimately also the fracture occurs. A variety of tensile machines are available for tensile testing; some types are illustrated in Figs. 1.2 and 1.3. The tensile testing results of the rectangular specimens used in our laboratory are illustrated in Fig. 1.4 for 1020 and 1070 steels, 2024 aluminum alloy and for polyethylene. Note the yield elongation for the 1020 steel, showing also the upper yield point.

1.2.1 Elastic Deformation and the Relations Between Stress and Strain

At low loads, the deformation produced by applying a continuously increasing tensile load uniaxially is elastic, as shown in Fig. 1.5. Such behavior characterizes completely brittle materials, such as glass, ceramics, etc. In rubber-like materials,

Fig. 1.2 Instron dynamic tensile machine with a 15 ton load cell. Model 134

the stress/strain relation is not linear, as seen in Fig. 1.6. Figure 1.7 shows a portion of a stress/strain curve of metals in the elastic region, extending slightly also to the plastic region, indicating a transition from one mode of deformation to another.

Unloading the specimen, loaded only in the elastic region, restores the original size of the specimen and no permanent change can be observed in its dimensions. Thus, one can emphasize the elastic property as a reversible feature of the deformation process.

This property, in most materials where truly elastic behavior is observed, can be represented by the well-known Hooke's Law, which relates strain as a linear function of load (or stress). The relation between the applied stress and strain can be expressed by

$$\sigma = \mathrm{Ee} \tag{1.5}$$

Here e is the average linear strain, which relates the change in dimension of the specimen to its original length. In simple terms, the linear strain can be expressed as:

$$\boldsymbol{e} = \frac{\boldsymbol{l} - \boldsymbol{l}_0}{\boldsymbol{l}_0} = \frac{\Delta \boldsymbol{l}}{\boldsymbol{l}_0} = \frac{\boldsymbol{l}}{\boldsymbol{l}_0} - \mathbf{1} \tag{1.6}$$

with Δl representing the axial change in length as a result of deformation and l_0 is the gage length of the specimen. Thus, e is the ratio of the change in the gage

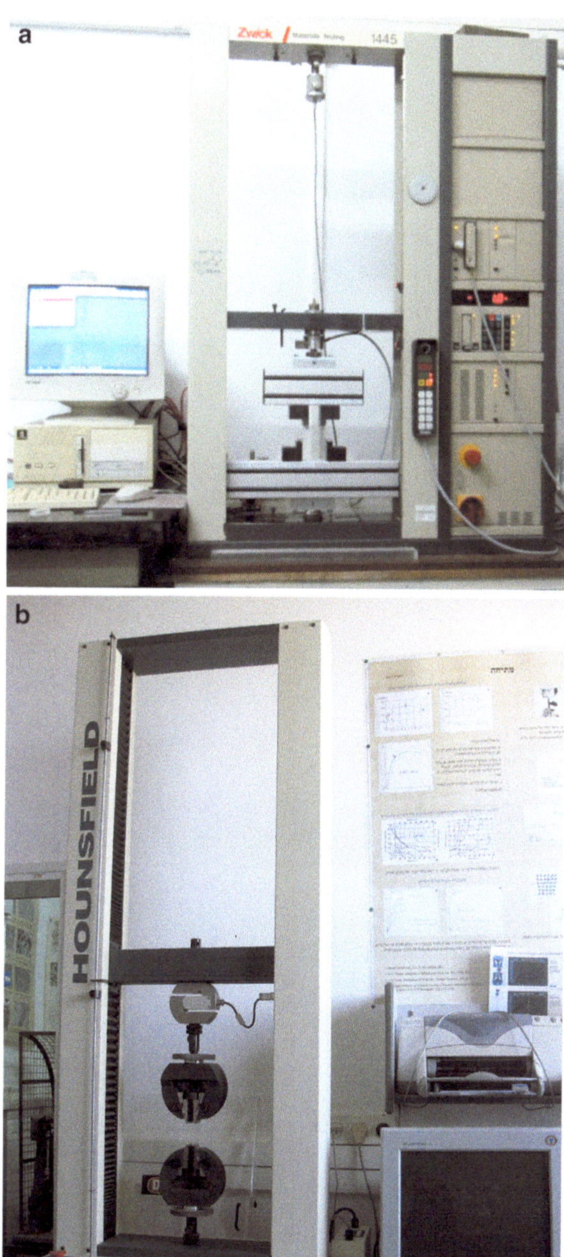

Fig. 1.3 (**a**) Zwick tensile machine with two load cells of 1 ton and 1 kg, respectively. Model 1445; (**b**) Hounsfield table tensile tester. Model H25 KT

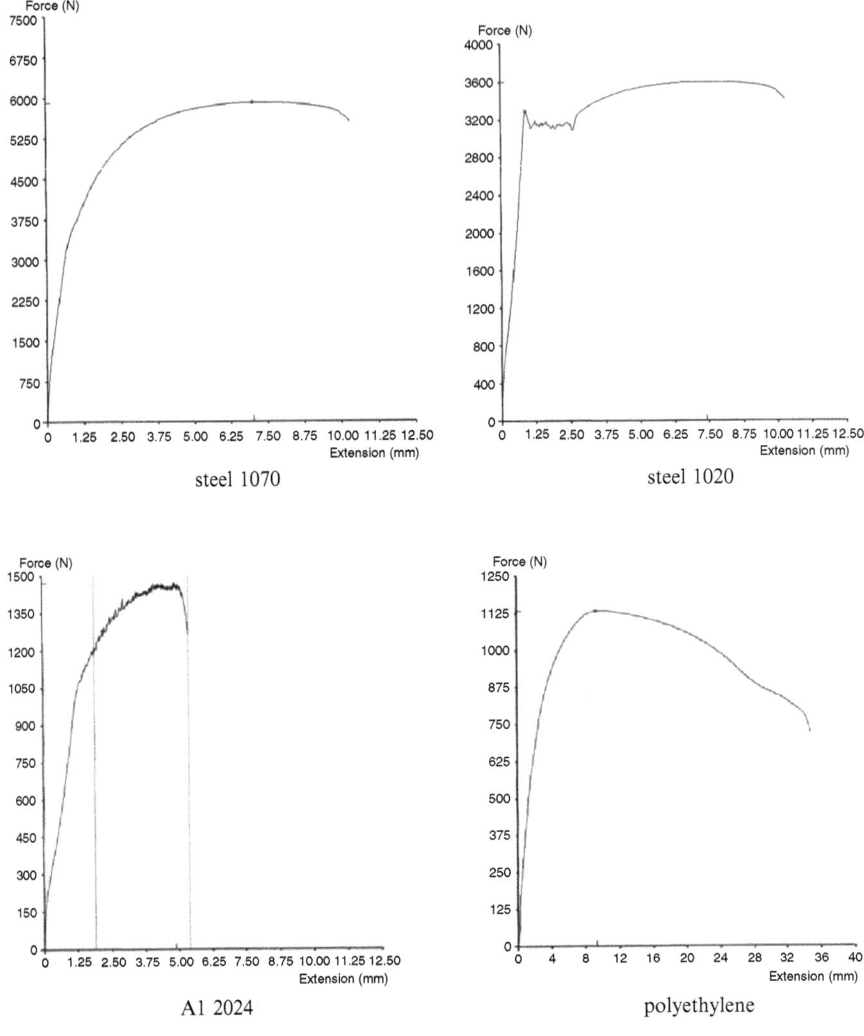

Fig. 1.4 Results of tension tests of various materials

length to the original gage length. Δl is often referred to as 'the deformation'. The last expression on the right of Eq. 1.6 is useful (as will be seen later) to express the elongation occurring during deformation.

Clearly, Eq. 1.6 may be expressed in accordance with the definition of the linear strain as:

$$e = \int_{l_0}^{l} \frac{dl}{l_0} = \frac{l - l_0}{l_0} \tag{1.6a}$$

Fig. 1.5 Schematic stress/strain line in the elastic region; $\sigma = Ee$

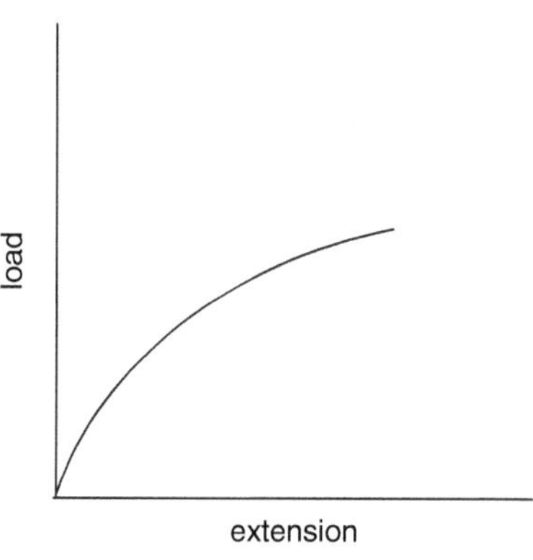

Fig. 1.6 Stress/strain relation in a rubber-like material

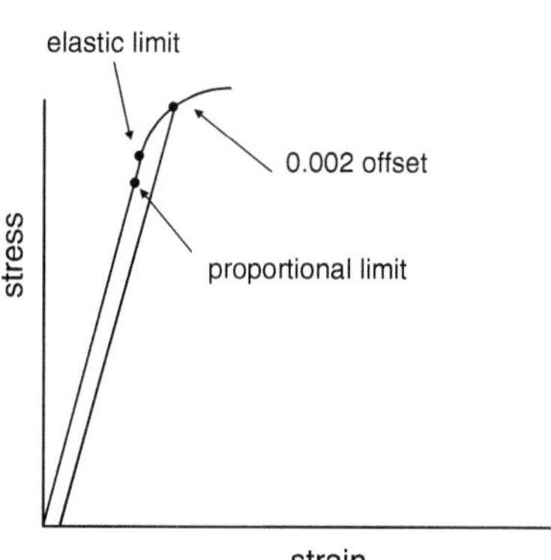

Fig. 1.7 The stress/strain curve extended somewhat into the plastic region to show the 0.002 offset yield

Linearity applies to most solids, except rubber-like materials, in which deformation is reversible, but the stress/strain relation is not linear. E is a constant of proportionality, known as 'Young's Modulus' or the 'elastic modulus', which varies to a large extent for different materials. The strain in Eq. 1.6 is called 'the engineering strain' and it is valid for small strains.

A different and useful concept for defining strain, when deformation is considered in more practical terms, is associated with the instantaneous change occurring in a specimen's length while a force is acting on it. Unlike cases of engineering strain (defined above), where reference was made to the constant gage length of the specimen, in the present case, reference is made to changes in the linear dimension at each instant of the test. If dl is the amount by which the length l changes, a strain can be defined similarly to Eq. 1.6 as:

$$\varepsilon = \frac{dl}{l} \tag{1.7}$$

Integrating Eq. 1.7, one gets:

$$\varepsilon = \int_{l_1}^{l_2} \frac{dl}{l} = \ln\frac{l_2}{l_1} \tag{1.8}$$

The integration is performed between two arbitrary values of l, lengths l_1 and l_2. Clearly, when l is the instantaneous length, and consideration is given to the original length, the integration limits should be changed, resulting in:

$$\varepsilon = \int_{l_0}^{l_i} \frac{dl}{l} = \ln\frac{l_i}{l_0} \tag{1.9}$$

ε is known as the natural, true or logarithmic strain. Equations 1.8 and 1.9 can be visualized basically as resulting from the summation of the small changes between the two values as given below:

$$\varepsilon = \sum\left(\frac{l_1 - l_0}{l_0} + \frac{l_2 - l_1}{l_1} + \frac{l_3 - l_2}{l_3} + \cdots\right) \tag{1.10}$$

It is often required to alternate between these two definitions of the strain. This can easily be performed, as shown below. Using Eq. 1.6, it is possible to write:

$$e = \frac{l}{l_0} - 1 \tag{1.6a}$$

or:

$$(e + 1) = \frac{l}{l_0} \tag{1.6b}$$

and with Eq. 1.9 written as:

$$\varepsilon = \ln\frac{l}{l_0} \tag{1.9a}$$

one obtains:

$$\varepsilon = \ln(e + 1) \tag{1.11}$$

It should be mentioned that for small strains ~0.1 engineering and true strains are the same as seen below. Expand Eq. 1.11 in series:

$$\boldsymbol{\varepsilon} = \ln(1 + e) = e - \frac{e^2}{2} + \frac{e^3}{3} - \cdots \tag{1.11a}$$

Again, for small strains, the terms where the exponents of e are larger than 1 can be neglected, resulting in:

$$\boldsymbol{\varepsilon} \cong \boldsymbol{e} \tag{1.11b}$$

When, for example, a load is acting on an elastic body in x direction, it elongates not only in the direction of the acting load, but contracts laterally, as well. Thus, contraction must occur in the transverse y and z directions. Empirically, it was observed that the transverse stains are constant fractions of the longitudinal extension. The ratio of the lateral contractive strain to the axial strain is called Poisson's ratio, denoted by ν and expressed as:

$$\nu = -\frac{\textit{lateral strain}}{\textit{axial strain}} = -\frac{\varepsilon_y}{\varepsilon_x} = -\frac{\varepsilon_z}{\varepsilon_x} \tag{1.12}$$

Therefore, for loading along the x axis, the other components are $\varepsilon_y = -\nu\varepsilon_x$ and $\varepsilon_z = -\nu\varepsilon_x$. In the case of a normal load, this fraction is ~0.33 and its value depends on the type of solid being tested. It might be in the range of 0.25 for a perfect elastic and isotropic solid.

Now, it is possible to express Eq. 1.5 in the light of the transverse stresses present, recalling that $\boldsymbol{\varepsilon} \cong \boldsymbol{e}$ (Eq. 1.11b) as:

$$\varepsilon_y = \varepsilon_z = -\nu\varepsilon_x = -\frac{\nu\sigma_x}{E} \tag{1.13}$$

Obviously, the negative sign indicates that the other two strains, namely ε_y and ε_z, experience a contraction while the test bar is being extended.

It is of interest to know what happens when three normal stresses are acting simultaneously on a test specimen. The simplest way to understand this is by visualizing a cube of unit dimensions while the stresses involved are σ_x, σ_y and σ_z. This is a three-dimensional state-of-stress problem. Given that this is in the elastic region and that the acting stresses are small, the assumption can be made that the normal stress, e.g. σ_x, does not produce shear strains on the appropriate planes. (There is discussion on shear strains and shear stress below in Sect. 1.2.9.) Using the principle of superposition and recalling that the stress components act in the respective directions and use appropriate strains, expressed in terms of Poisson's ratio, the effects of the three individual uniaxial loadings may be summed up as:

$$\varepsilon_x = \frac{\sigma_x}{E} - \frac{\nu\sigma_y}{E} - \frac{\nu\sigma_z}{E} = \frac{\sigma_x - \nu(\sigma_y + \sigma_z)}{E}$$

$$\varepsilon_y = \frac{\sigma_y}{E} - \frac{\nu\sigma_z}{E} - \frac{\nu\sigma_x}{E} = \frac{\sigma_y - \nu(\sigma_z + \sigma_x)}{E} \tag{1.14}$$

It can be seen, by eliminating E from two appropriate equations, that:

$$\frac{\varepsilon_x}{\sigma_x - \nu(\sigma_y + \sigma_z)} = \frac{\varepsilon_y}{\sigma_y - \nu(\sigma_z + \sigma_x)} = \frac{\varepsilon_z}{\sigma_z - \nu(\sigma_x + \sigma_y)} \tag{1.14a}$$

Equation 1.14 is the generalized Hooke's Law for isotropic materials. It is easy to remember the relations shown in (1.14), since they are in succession–x, y and z. If one of the normal strains in one of the directions, e.g., the z direction, is zero, then strain ε_z can be reduced, resulting in a strain expressed as:

$$\boldsymbol{\varepsilon_z = 0 = \frac{\sigma_z - \nu(\sigma_x + \sigma_y)}{E}} \tag{1.15a}$$

making the stress component σ_z:

$$\boldsymbol{\sigma_z = \nu(\sigma_x + \sigma_y)} \tag{1.15b}$$

Similar relations can be obtained for the strains in the x or y directions, when the appropriate normal strain in the respective directions is zero. Thus, it is possible to write the following relations for ε_x and ε_y:

$$\boldsymbol{\varepsilon_x = 0 = \frac{\sigma_x - \nu(\sigma_y + \sigma_z)}{E}} \tag{1.15c}$$

$$\boldsymbol{\sigma_x = \nu(\sigma_y + \sigma_z)} \tag{1.15d}$$

and:

$$\boldsymbol{\varepsilon_y = 0 = \frac{\sigma_y - \nu(\sigma_z + \sigma_x)}{E}} \tag{1.15e}$$

$$\boldsymbol{\sigma_y = \nu(\sigma_z + \sigma_x)} \tag{1.15f}$$

When the applied load produces a normal stress in one of the three directions that can be assumed to be zero, plain stress condition prevails. In thin films, for example, the z dimension is small compared to the other directions; the approximation that the normal stress is $\sigma_z = 0$ is a good one.

In this case, the equations for the plain stress can be obtained from Eq. 1.14. The strains are:

$$\varepsilon_x = \frac{\sigma_x - \nu\sigma_y}{E}$$
$$\varepsilon_y = \frac{\sigma_y - \nu\sigma_x}{E}$$
$$\varepsilon_z = -\frac{\nu(\sigma_x + \sigma_y)}{E} \tag{1.16}$$

In terms of the planar stresses, one can easily infer the following expressions from Eq. 1.16:

$$\sigma_x = \frac{E(\varepsilon_x + \nu\varepsilon_y)}{1 - \nu^2}$$
$$\sigma_y = \frac{E(\varepsilon_y + \nu\varepsilon_x)}{1 - \nu^2} \tag{1.17}$$

However, despite the assumption that in our thin film $\sigma_z = 0$, $\varepsilon_z \neq 0$ as seen from Eq. 1.15b. It is also easy to get the plain strain conditions when $\varepsilon_z = 0$. Useful relations can be obtained when Eq. 1.15b is substituted in Eq. 1.14 giving:

$$\varepsilon_x = \frac{\sigma_x(1 - \nu^2) - \nu\sigma_y(1 + \nu)}{E}$$
$$\varepsilon_y = \frac{\sigma_y(1 - \nu^2) - \nu\sigma_x(1 + \nu)}{E} \tag{1.18}$$

Even though the strain $\varepsilon_z = 0$, the stress is not as seen in Eq. 1.15b.

An additional constant for an elastic and isotropic material should be mentioned, namely the bulk or volumetric modulus designated by K. (The shear modulus will be discussed later in the chapter dealing with shear deformation.) It can be defined as the ratio of the hydrostatic pressure to the volume change of an elemental cube, which is the result of the acting pressure, p. This pressure is acting on all six faces of an elementary cube. Clearly, the pressure is the negative of the stresses. Thus, by definition, K is:

$$K = \frac{3\sigma}{\varepsilon_v} = \frac{-3p}{\varepsilon_v} \tag{1.19}$$

where ε_v is the volume strain. One can visualize the bulk modulus concept in the following manner – for small linear strains, it is possible to write for the volume strain:

$$\varepsilon_v = \varepsilon_x + \varepsilon_y + \varepsilon_z \tag{1.20}$$

By expressing these strains in terms of Eq. 1.14, for ε_v one obtains:

$$\varepsilon_v = \frac{(1-2\nu)}{E}(\sigma_x + \sigma_y + \sigma_z) \tag{1.21}$$

As mentioned above, the pressure is the negative of the stress and in hydrostatic pressure the acting stresses are equal; therefore, one can write Eq. 1.21 for the stresses acting on the elementary cube in terms of an average stress σ_a as:

$$\varepsilon_v = \frac{3\sigma_a(1-2\nu)}{E} \tag{1.22}$$

where:

$$\sigma_a = \frac{1}{3}(\sigma_x + \sigma_y + \sigma_z) \tag{1.23}$$

Equation 1.22 can also be expressed as:

$$\varepsilon_v = \frac{-3p(1-2\nu)}{E} \tag{1.22a}$$

From the definition of the bulk modulus K, it is possible to write:

$$K = \frac{E}{3(1-2\nu)} \tag{1.24}$$

This can be obtained by substituting for ε_v from Eq. 1.22 into Eq. 1.19. Alternatively, Eq. 1.24 can also be expressed in terms of the pressure as:

$$K = -\frac{p}{\varepsilon_a} \tag{1.25}$$

This can be obtained from Eq. 1.19 by defining an average strain ε_a similarly to the average stress in Eq. 1.23 as:

$$\varepsilon_a = \frac{1}{3}(\varepsilon_x + \varepsilon_y + \varepsilon_z) \tag{1.26}$$

and by substituting the average strain defined in Eq. 1.26 for ε_v. The compressibility factor β is defined as the inverse of the hydrostatic pressure $\frac{1}{\beta}$ and is related to the bulk modulus. Any expressions for K, namely, Eqs. 1.19, 1.24 and 1.25 can be equated to $\frac{1}{\beta}$

1.2.2 The Elastic and Proportional Limits

Materials that show a reversible deformation in response to an applied load acting on the external surfaces exhibit elastic behavior. The objective is to determine the transition from the elastic to the plastic behavior by evaluating the 'yield stress' (often called the 'yield point'), where the yielding of a test specimen starts. To do this, a load is applied to a test specimen, usually as a tensile test. The load required to produce extension at a constant rate is gradually increased. The early stages of the tension test are used to evaluate the yield strength. The plots obtained by applying tension are expressed as stress vs. strain, as shown in Fig. 1.7. Three methods are commonly used to define the transition from elastic to plastic deformation when a sample is extended by the application of a load. In Fig. 1.7, these three methods are marked to indicate the limit of the validity of Hooke's Law. Two of the three methods are the proportional and elastic limits. The 'proportional limit' is defined as the stress at which the stress/strain curve first deviates from a straight line. It can also be defined as the largest stress at which stress/strain relation is still linear. Below this value, the slope, namely the ratio of stress to strain, is constant (see Fig. 1.7). The material is said to behave according to Hooke's law. This is rarely used in specifications as the yield point of the material, since the deviation from linearity starts quite gradually and, therefore, inaccuracies or even controversies arise regarding the exact location of where the stress/strain line begins to curve. The 'elastic limit' is defined as the stress at which plastic deformation begins; in other words, it represents the largest load that a material can tolerate without noticeable or even measurable permanent change. This is tedious to evaluate and also cannot be determined precisely from the stress/strain curve. Its determination involves repeated, successive loading and complete unloading for the observation of the first permanent plastic set. As in the case of the proportional limit, its value, so determined, might raise controversy as to the credibility of the elastic limit. The elastic limit [henceforth: EL] is always greater than the proportional limit [henceforth: PL], i.e. EL > PL. The transition from the elastic to the plastic regions, which determines the yield point, was quite difficult to determine in the past, even when strain gages were used. Previously, the load and cross head displacement were recorded directly as a graph by means of a pen-plotter. At present, very accurate electric strain gages are in use and the measuring equipment is more sensitive. Nevertheless, it is still difficult to accurately determine these two definitions for the transition from elastic to plastic behavior. Note that these points almost coincide in the case of ductile materials. The need for a practical determination of yielding in a material resulted in a method for its evaluation – a technique known as the 'offset yield strength', also called 'proof stress'. Offset yield strength represents the practical yield strength for engineering applications. For its evaluation, the early stages of a tension test are used to evaluate the 'yield strength', which is defined as the stress at which a predetermined amount of permanent deformation occurs. To find the yield strength, so defined, a predetermined amount of permanent strain is set along the strain axis. A straight line is drawn parallel to the linear portion of the

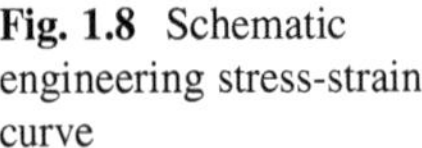

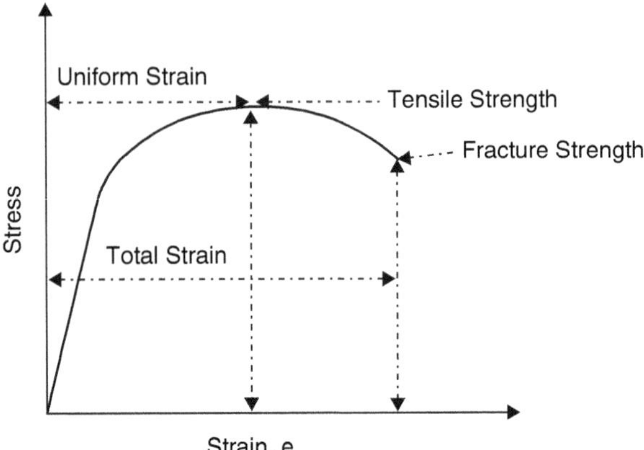

Fig. 1.8 Schematic engineering stress-strain curve

stress/strain curve. The point of intersection of this line and the stress/strain curve is projected on the stress axis, thus determining the yield stress. When the yield strength is reported, the amount of offset used in its determination should be noted as 'yield strength (at 0.2% offset)'. This is important, since often a 0.5% offset is used for very ductile material. Occasionally, a 0.1% offset is used. In Fig. 1.7, a 0.2% offset is shown.

Not all materials behave linearly at the early stages of tension. Elastic materials may be either linear or nonlinear. Rubber is known not to behave linearly. Figure 1.6 shows a nonlinear elastic behavior. Despite the non-linearity, unloading such a rubber-like material in the elastic region follows the same path as the loading. Many of the rubber-like materials and polymers behave in an inelastic fashion, which is a known concept in 'viscoelasticity' Viscoelastic solids exhibit both elastic and viscous behavior when loaded. Such material deformation is time dependent. (Creep shows viscoelastic behavior and will be discussed below in the chapter on creep.)

Basically, materials can be ductile and brittle. Ductile materials can withstand relatively large stress within the elastic region before yielding, whereas this is not the case in brittle materials, which usually break suddenly in the elastic region, often even below the proportional limit.

1.2.3 Plastic Deformation

A typical stress/strain curve is shown in Fig. 1.8. It is produced by applying a continuously increasing tensile load uniaxially to produce deformation in the specimen tested. This illustration is an example of an engineering stress/strain curve. In this tension test, a specimen is subjected to a continuously increasing uniaxial load during which simultaneous measurements are made on the extension. From these measurements, the so-called 'engineering stress/strain curve' is constructed. In a tensile test, a sample is extended at constant rate and the load needed to maintain this is measured. The stress (σ) (calculated from the load) and strain (e) (calculated from the extension) can either be plotted as 'nominal stress' against 'nominal strain' or as 'true stress' against 'true strain'. Nominal stress and strain are other expressions for the engineering curve indicated above. In this case, the stress is the ratio of the applied load to the original area between the two gage marks of the specimen, i.e., Eq. 1.4 is modified as:

$$\sigma = \frac{P}{A_0} \tag{1.27}$$

Similarly, a graph of the applied load vs. the extension will have the same character as the stress/strain relation shown in Fig. 1.8. The maximum stress (or load) is referred to as the 'tensile strength' and more often as the 'ultimate tensile strength' [henceforth: UTS]:

$$\sigma_{UTS} = \frac{P_{max}}{A_0} \tag{1.28}$$

Up to this point, the stain is uniform. However, beyond this point, an accelerated change in dimension (area) occurs, which is called 'necking' (to be discussed at a later stage). When the change in area is such that the specimen cannot sustain the applied load, fracture occurs. The stress level at which this occurs is known the 'fracture strength' (stress) and the strain to fracture ratio is the 'fracture strain':

$$\sigma_f = \frac{P_f}{A_0} \tag{1.29}$$

The uniform and non-uniform strains together comprise the total strain to fracture. Most metallic materials exhibit tension behavior as indicated in Fig. 1.8. Some body- centered cubic metals [henceforth: BCC], such as mild steel, do show a different transition from the elastic to the plastic regions. The elastic region is terminated by a sudden drop in stress (i.e., load). At this stage, deformation continues without an increase in the load level. Plastic deformation sets in and the horizontal section, usually jagged, where the extension continues, is often referred to as the 'yield plateau'. Beyond this plateau, a load increase is required for further deformation. Figure 1.4 illustrates the character of the tensile test for mild steel (1,020). The stress at the location where the sudden load drop occurs is called the 'upper yield stress' and the plateau level is called the 'lower yield stress' (the terms 'yield strength' or 'yield point' are also very often used). Note that in metals showing a yield drop it is much easier to define the transition to the plastic region. The crystal LiF also shows a sudden drop in stress value during deformation. At this point, it is worth mentioning that a material exhibiting absolute plasticity will indicate a stress/strain relation according to Fig. 1.9:

Fig. 1.9 Absolute plasticity

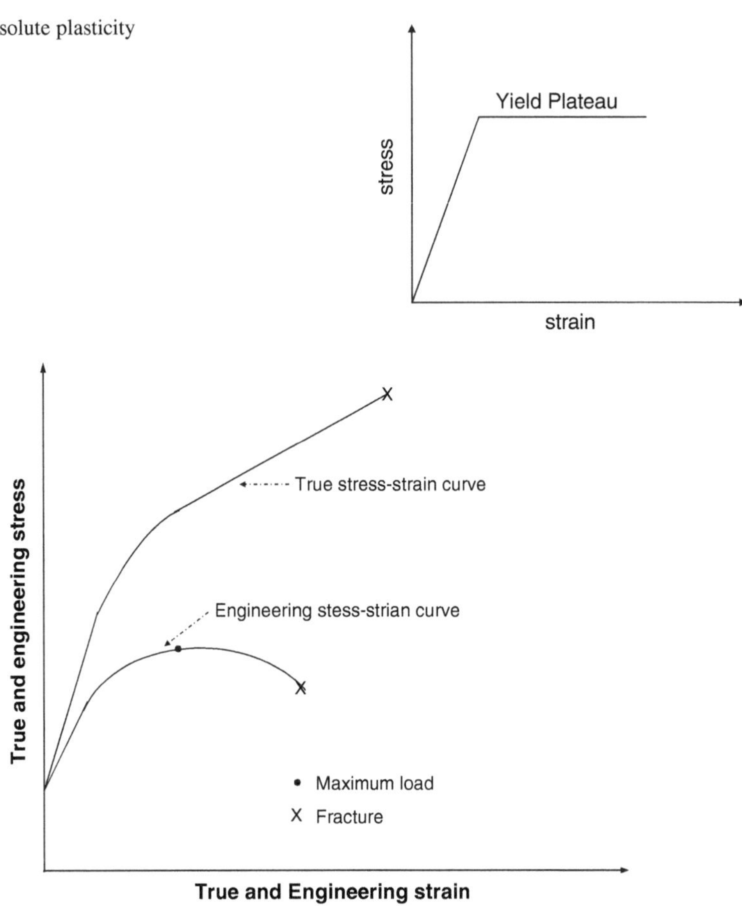

Fig. 1.10 Comparison of true stress/true strain and engineering stress/strain curves

1.2.4 *The True Stress/Strain Relation*

The engineering stress/strain curve seen earlier indicates a stress drop beyond the UTS, which means that no increase in load is essential for continued deformation. However, this is not the case, as seen from the true nature of a tensile test. A true stress/strain relation behaves quite differently than an engineering stress/strain curve. When true stress and true strain (natural strain, logarithmic strain) are related, an increase in load up to the point of fracture of the material is observed. Schematically, the engineering and the true stress/strain curves are compared in Fig. 1.10. In this figure, engineering strain is evaluated by reducing the area (as discussed in a separate section below), rather than by its extension.

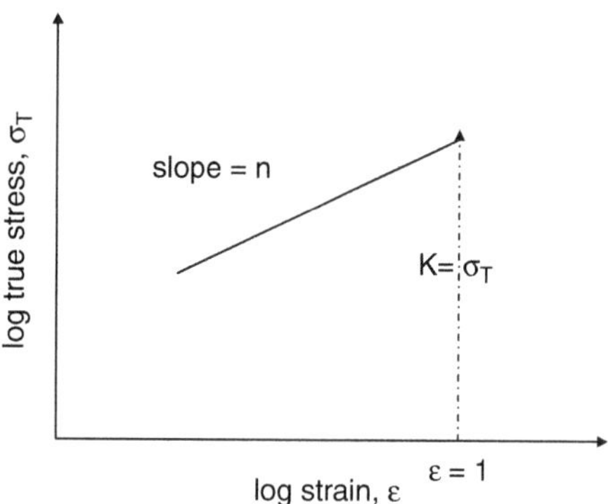

Fig. 1.11 Plot of log-log true stress/true strain

When an area is expressed in terms of the instantaneous dimensions of a specimen, i.e., its real area, true stress is given by Eq. 1.4. The true strain was defined in Eq. 1.9 and was expressed in terms of engineering strain in Eq. 1.11, appearing below:

$$\sigma_T = \frac{P}{A} \tag{1.4}$$

$$\varepsilon = \ln\frac{l_i}{l_0} \tag{1.9}$$

$$\varepsilon = \ln(e + 1) \tag{1.11}$$

No function relating true stress to true strain is known to exist for ductile materials, except for empirical relations. One such relation in the uniform plastic deformation region is the well-known expression, given as a power function attributed to Hollomon:

$$\sigma = K\varepsilon^n \tag{1.30}$$

Here, K is the strength coefficient (not the bulk modulus defined earlier) and n is the strain hardening exponent. Both parameters are material constants. K can be evaluated when $\varepsilon = 1$ and is the stress at this strain. Expressing this function as a log-log plot, a straight line between σ_T and ε is obtained, where the slope of this curve is the strain hardening exponent. Such a curve is shown schematically in Fig. 1.11. The value of n can be in the range $n = 0$ and $n = 1$. For perfectly ductile solids, the slope is zero with $n = 0$ and for elastic solids, $n = 1$. It was experimentally observed that Fig. 1.11 is applicable to a wide range of metals. Nevertheless, deviations from Eq. 1.30 were frequently observed.

The $\sigma_T - \varepsilon$ relation for various n values can be expressed schematically as shown in Fig. 1.12 for $n = 0$, $n = 1$ and $n = 1/2$.

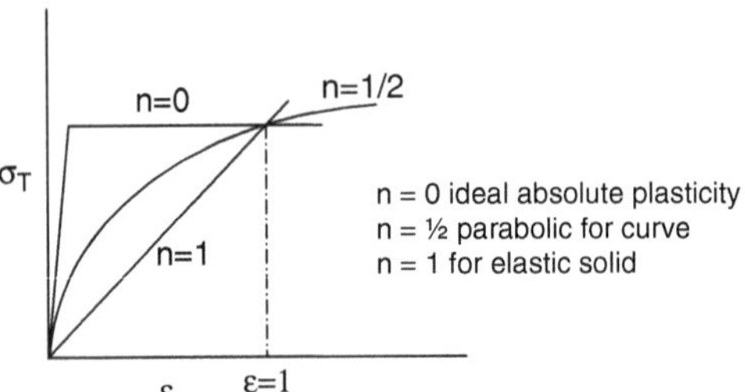

Fig. 1.12 $\sigma_T - \varepsilon$ curves for various n values

One can define a 'strain rate hardening exponent' as the change in slope (n). Since a log-log plot was used (Fig. 1.11) to define n, it is possible to write an expression for the rate of strain hardening in terms of $\frac{d\sigma}{d\varepsilon}$ as follows. Take logarithms on both sides of Eq. 1.30 and then use the derivative Eq. 1.31 obtained:

$$n = slope = \frac{d \ln \sigma}{d \ln \varepsilon} = \frac{\varepsilon d\sigma}{\sigma d\varepsilon}$$

$$\frac{d\sigma}{d\varepsilon} = n\frac{\sigma}{\varepsilon} \tag{1.31}$$

1.2.5 Elongation

'Elongation', e, is one of the measures of ductility. First rewrite the engineering strain Eq. 1.6 as:

$$e = \frac{l - l_0}{l_0} = \frac{\Delta l}{l_0} = \frac{l}{l_0} - 1 \tag{1.6}$$

Then multiplying it by hundred, to express the extension at any stage as per cent of change in the initial length, to provide a measure of the ductility of a specimen. This is commonly expressed as elongation. Thus, Eq. 1.6 becomes:

$$e = \frac{l - l_0}{l_0} \times 100 = 100\left(\frac{l}{l_0} - 1\right) \tag{1.32}$$

Elongation can then be specified at various stages of a tension test, such as e at necking or e at fracture with an appropriate index, such as e_n or e_f. The value of elongation, though a subjective property of the tested material, is a widely used (qualitative) property. It provides information on the material's ductility – to what extent can it be deformed without fracture. Furthermore, one can obtain an indication of how much a material will flow plastically, an important factor in metal forming. As mentioned earlier, l_0 and l refer to the 'gage length', the standard distance between two fixed points and its change at any stage of deformation.

1.2.6 The Reduction of Area

Another method for expressing ductility is the reduction of area, which has the merit of being an important parameter in the metal forming of sheet metal and, in particular, in deep-drawing operations, used during the fabrication process in the production of cups and other items. Measuring the cross-sectional area at the gage and its change at a particular instant during the test allows for the evaluation of ductility in terms of the percentage of area reduction:

$$q = \frac{A_0 - A}{A_0} \times 100 = \left(1 - \frac{A}{A_0}\right) 100 \tag{1.33}$$

Note that, unlike the definition of elongation Eq. 1.32, the ductility according to Eq. 1.33 uses the initial value of the area from which its final value is subtracted in the expression of the numerator. The reason for defining q in this way is the fact that it must be a positive number.

In plastic deformation the volume remains constant, as long as the test specimen remains cylindrical (or has the original shape), so:

$$\mathrm{V} = \mathrm{A_0 l_0} = \mathrm{Al} = \mathrm{A_i l_i} = \cdots \mathrm{A_f l_f} \tag{1.34}$$

A_i and l_i represent the instant area and length of the uniform elongation in the gage length, thus, it also applies to their final values at fracture, namely A_f and l_f. For a specimen that deforms uniformly over its entire length, i.e., up to UTS the nominal (engineering) strain, Eq. 1.6 will always be equal to the so called 'areal strain' (defined as $\frac{A_0 - A_{UTS}}{A_0}$). However, beyond UTS, where the deformation is non-uniform, the nominal and areal strains are not equal. Since, prior to fracture, most ductile metals neck down, the value of strain at fracture determined from the percentage of elongation is not reliable and will be less than that evaluated by the reduction of the area. From Eq. 1.34:

$$\frac{l}{l_0} = \frac{A_0}{A} \tag{1.34a}$$

and by substitution from Eq. 1.33, one obtains:

$$\frac{A_0}{A} = \frac{100}{100 - q} \tag{1.34b}$$

It is immediately clear that by Eqs. 1.6, 1.34a and 1.34b, the strain is related to the percentage of the reduction of the area:

$$e = \left(\frac{q}{100 - q}\right) 100 \tag{1.34c}$$

1.2.7 Necking

During the tensile testing of ductile material, a point is reached, usually when some sort of instability sets in, at which relatively large strains are concentrated locally and, as a result, a large decrease in cross-section occurs. The point at which instability sets in is usually a weaker point in the test specimen. All further deformation is concentrated at this point, resulting in an accelerated extension of the specimen, associated with thinning down at this region. Because of the analogy of the thinned-down cylindrical specimen to the human neck, this phenomenon occurring at the location where the fast deformation occurs is usually termed 'necking'. Equation 1.4 suggests that if the cross-sectional area decreases in an accelerated manner, the local stress greatly increases, becoming concentrated at the unstable location. We have seen earlier that, for deformation to occur, the load applied has to increase due to the work hardening (strain hardening) of the metal. However, the accelerated extension of the specimen (or its cross-section) is not associated with a substantial load increase for further deformation. The true stress/strain curve shows that work hardening does occur in the material and only the engineering stress/strain curve shows a drop in the load (stress) for further deformation. Thus, one could state that strain hardening of the material is less than the load-carrying capacity of the test specimen due to the accelerated decrease of the cross-section. Localized necking or thinning of the specimen quickly occurs until fracture sets in. Basically, the competition between strain hardening and the decrease in load-carrying capacity at some weak location determine the stability of a material during tensile deformation.

1.2.8 Instability in Tension

It has been indicated in the previous section that instability sets in at necking, which occurs at the maximum load in a ductile material tested by tension. Figure 1.9 shows the behavior of an absolutely plastic material, where plasticity sets in following yielding, since no strain hardening occurs in such a material. It becomes unstable immediately after yielding. In real metals, strain hardening occurs and, therefore, the load-bearing capacity of the materials increases. Thus, when strain hardening occurs, the load (or stress) has to be increased for further deformation (as seen in Fig. 1.10), for the true stress/strain curve, but when necking sets in, the change in load at the maximum (shown in Fig. 1.13) is $dP = 0$, which expresses instability. Up to this stage, a competition exists between strain hardening and the decrease in the cross-section of the material being tested. Beyond UTS, when necking begins, the decrease in the cross-section occurs at a faster rate than strain hardening (which is an indication of the load-bearing capacity of the material). The condition for this, as mentioned, is $dP = 0$. Rewrite Eq. 1.4, namely $P = \sigma A$, in terms of the derivative as:

$$dP = 0 = \sigma dA + A d\sigma \tag{1.35}$$

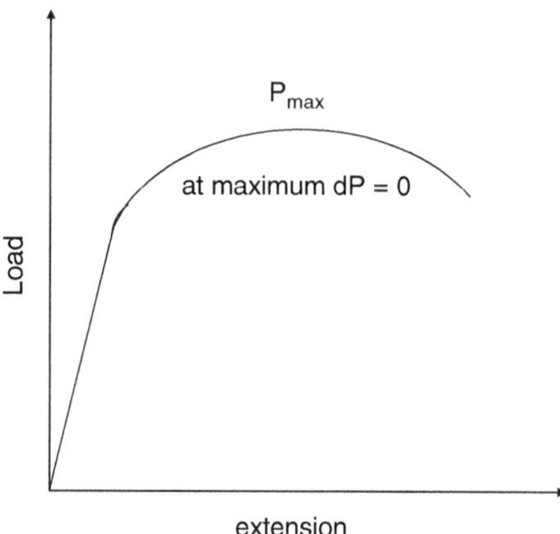

Fig. 1.13 Load extension; at P maximum dP = 0

and:

$$\frac{d\sigma}{\sigma} = -\frac{dA}{A} \tag{1.35a}$$

Due to the constancy of volume for ductile materials, according to Eq. 1.34:

$$\mathrm{dV} = 0$$

or:

$$\mathrm{dV} = 0 = \mathrm{d}\,(\mathrm{Al}) = \mathrm{ldA} + \mathrm{Adl} \tag{1.34a}$$

$$-\mathrm{dA/A} = \mathrm{dl/l} \tag{1.34b}$$

Equation 1.34b, according to Eq. 1.7, where $\frac{dl}{l} = \varepsilon$ can be written as:

$$\frac{dl}{l} = -\frac{dA}{A} = d\varepsilon \tag{1.36}$$

From Eqs. 1.35a and 1.36 one gets:

$$\frac{d\sigma}{d\varepsilon} = \sigma \tag{1.37}$$

Thus, the point of necking at the maximum load can be obtained from a $\sigma_T - \varepsilon$ relation by finding the point on the curve where the subtangent is equal to 1. Alternatively, it can be obtained by observing the location where the strain hardening equals the stress Eq. 1.37. This relation is shown in Fig. 1.14. Two curves

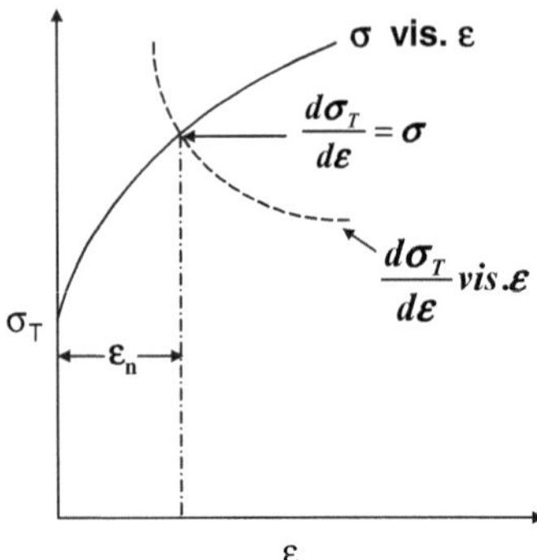

Fig. 1.14 The intersection of the two curves is the strain where necking sets in

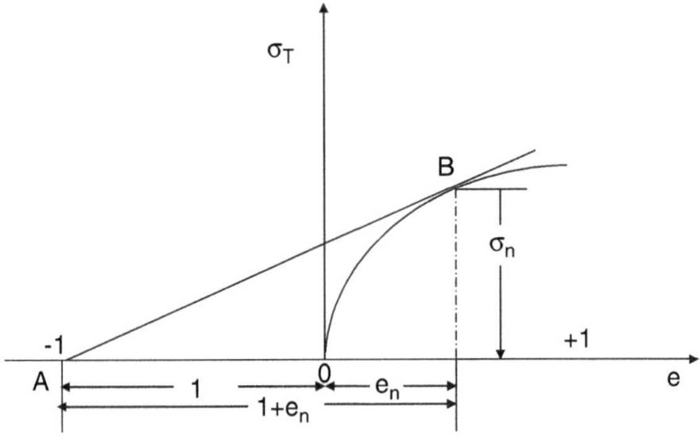

Fig. 1.15 Necking strain and true stress at necking, σ_n (Considére's construction)

intersect each other – one is the $\sigma_T - \varepsilon$ and the second is the $\frac{d\sigma_T}{d\varepsilon} - \boldsymbol{\varepsilon}$. Note that, for the sake of emphasizing that the stress is the true stress, the index T was used. The symbol ε_n in the Figure represents the necking strain. It is also possible to construct a curve of σ_T and engineering strain e, based on the following relation:

$$\sigma_T = \frac{d\sigma_T}{d\varepsilon} = \frac{d\sigma_T}{de}\frac{de}{d\varepsilon} = \frac{d\sigma_T}{de}\frac{\frac{dl}{l_0}}{\frac{dl}{l}} = \frac{d\sigma_T}{de}\frac{l}{l_0} = \frac{d\sigma_T}{de}(\mathbf{1}+e) \tag{1.38}$$

or:

$$\frac{d\sigma_T}{de} = \frac{\sigma}{(\mathbf{1}+e)} \tag{1.38a}$$

On the basis of this relation, a curve can be constructed (as mentioned above) with an abscissa of $e_u + 1$ and an ordinate, σ_T. A point on the abscissa, A, is obtained as a negative strain and, if a tangent is drawn to the point where it intersects the $\sigma_T - e$ curve, the maximum load is obtained as true stress at necking, σ_n, since the slope at this point is $\frac{\sigma}{(1+e)}$. The construction based on this relation is known as 'Considére's construction' and is shown in Fig. 1.15.

At this point, it is worth mentioning that there is a relation between σ_T and engineering stress σ_E, as follows:

$$\sigma_T = \frac{P}{A_i}\frac{A_0}{A_0} = \frac{P}{A_0}\frac{A_0}{A_i} = \sigma_E\frac{A_0}{A_i} = \sigma_E\frac{l_i}{l_0} \tag{1.39}$$

Equation 1.31 was given above as:

$$\frac{d\sigma}{d\varepsilon} = n\frac{\sigma}{\varepsilon} \tag{1.31}$$

and in Eq. 1.37 as:

$$\frac{d\sigma}{d\varepsilon} = \sigma \tag{1.37}$$

Therefore, from these two relations one can derive:

$$\sigma = n\frac{\sigma}{\varepsilon} \tag{1.38a}$$

or:

$$\mathbf{n} = \boldsymbol{\varepsilon} \tag{1.38c}$$

Thus, necking (which represents instability) occurs when $\varepsilon = n$ or it may be obtained simply from Eqs. 1.30 and 1.37 as follows:

$$\sigma = K\varepsilon^n$$

$$\frac{d\sigma}{d\varepsilon} = nK\varepsilon^{n-1} = \sigma = K\varepsilon^n$$

$$n = \varepsilon \tag{1.40}$$

1.2.9 *The Shear Stress and Shear Strain*

'Shear stress' is defined as the stress applied parallel to a face of a material, as opposed to normal stress, which is applied perpendicularly. Only shear stress can induce plastic deformation in a metal. Work done on a body by such a deformation remains stored in the material. Atomic planes slide over each other and, thus, permanently change the relative positions of the atoms. There is a critical shear stress for plastic deformation. An elementary cube removed from a structural material upon which forces expressed as stresses are acting is shown in Fig. 1.16. Here, the stresses acting normally to the plane and those acting within the plane

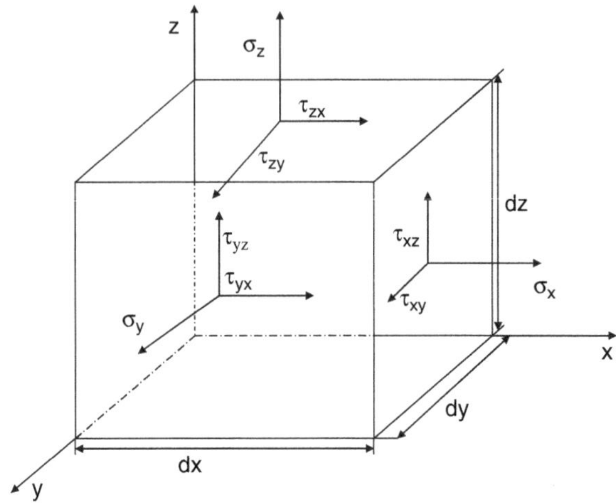

Fig. 1.16 An elementary cube removed from a structural material upon which (stress) forces are acting is shown

are illustrated. The stresses acting within the planes of the cube are shear stresses, also known as 'tangential stresses'. They are shown parallel to the axes, but are generally not parallel to them. For convenience sake, shear stress τ is resolved to a couple of components that are parallel to the coordinates and acting within the plane. To simplify the concept in a one dimensional element, the shearing stress is defined as the load acting along the area of the cross- section

$$\tau = \frac{\boldsymbol{load}}{\boldsymbol{area}} = \frac{\boldsymbol{P}}{\boldsymbol{A}}$$

As a result of applying the load, the square changes into a parallelogram under the strain produced as a consequence of deformation. This strain, known as 'shear strain', distorts the element, which can be written as:

$$\gamma = \frac{x}{l} = \tan\phi$$

For small elastic strains $\tan\phi \approx \phi$, but this is not applicable for plastic strains. The elastic shear modulus, G, is the ratio of the shear stress to the shear strain or:

$$\mathbf{G} = \tau/\gamma$$

This is related to the modulus of elasticity and to Poisson's ratio as:

$$\boldsymbol{G} = \frac{\boldsymbol{E}}{\mathbf{2(1+\nu)}}$$

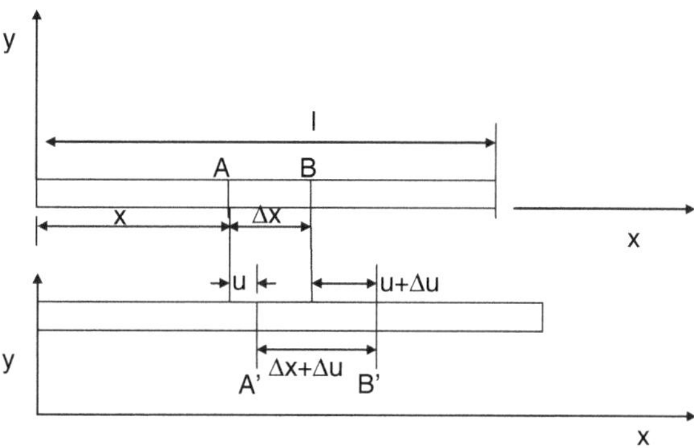

Fig. 1.17 A one-dimensional deformation of a bar

1.2.10 The Elastic Strain Energy

In Sect. 1.2.1, mention was made of the deformation of a unit cube. An elementary cube, removed from a structural material under stress, is shown in Fig. 1.16. Although the role of shear stresses has not yet been discussed here, they are shown together with the normal stresses. During the application of an increasing elastic force, work is done to deform the cube, which is stored in the material as elastic energy. Clearly, this energy is recovered on release of the applied force. Only the normal strain components are responsible for volume change in elastically deformed bodies. In this case, only normal stress is being considered with the load acting uniaxially along the x axis. An elemental volume change can be written as:

$$\mathrm{dV} = \mathrm{dxdydz} \tag{1.41}$$

A one-dimensional bar, upon which a force is acting and causes displacements in the body, is seen in Fig. 1.17 before and after application of that force. When applying a force on the bar, displacement, u, occurs. Point A will be displaced by u and move to position A′. Point B is displaced to position B′, thus moving a distance of u + Δu. The motion of the bar is considered as that of a rigid body and, thus, it elongates by Δu.

The strain energy can be written as:

$$dU = \frac{1}{2}\sigma_x e_x dV \tag{1.42}$$

Qualitatively, one can understand this relation from the fact that the work done on a body in the elastic region is the hatched area under the curve, seen in Fig. 1.18, whereas the product of σ_x and e_x represents the rectangular area. Basically, only half of the rectangle is the work applied to cause the deformation. Equation 1.42 can be obtained in the following way. The force is proportional to the displacement u:

$$\mathrm{P} = \mathrm{ku} \tag{1.43}$$

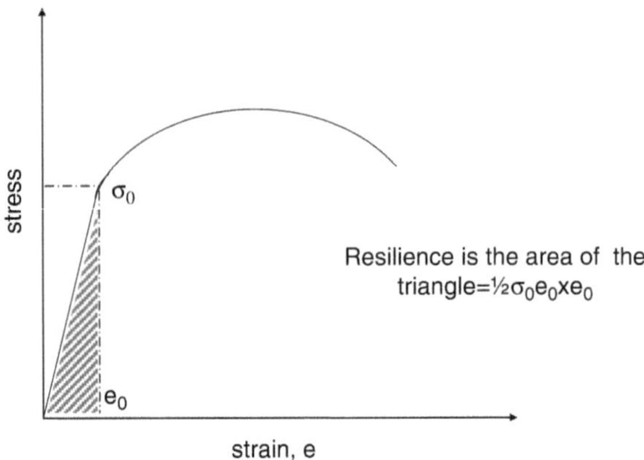

Fig. 1.18 Resilience; the hatched area is the energy stored on loading to σ_0

Recalling that the elemental work is given by (a scalar product):

$$\mathrm{dW} = \mathrm{P} \cdot \mathrm{du} \tag{1.44}$$

Therefore,

$$W = \int_0^u P \cdot du = \int_0^u kudu = \frac{ku^2}{2} = \frac{Pu}{2} \tag{1.45}$$

The change in external work is equal to the change in strain energy or $\Delta W = \Delta U$. The stress is given as $\sigma = P/A_0$ and the strain is given as $e = u/l_0$; thus, Eq. 1.44 can be written as:

$$dU = dW = Pdu = \sigma A_0 l_0 de \tag{1.46}$$

$$U = A_0 l_0 \int_0^e \sigma de = V_0 \int_0^e Eede = V_0 \frac{Ee^2}{2} = V_0 \frac{E}{2}\left(\frac{\sigma}{E}\right)^2 = V_0 \frac{1}{2}\frac{\sigma^2}{E} \tag{1.47}$$

For the elastic strain energy per unit volume:

$$U = \frac{\sigma^2}{2E} \tag{1.48}$$

or:

$$U = \frac{\sigma}{2} e \tag{1.48a}$$

Clearly, Eq. 1.42 is related to Eq. 1.48, as can be seen as follows. The change in strain energy per unit volume, dU/dV, can be integrated as:

$$U = \frac{1}{2}\int_V \sigma_x e_x dV = \frac{1}{2}\int_A \sigma_x dydz \frac{1}{2}\int_0^l e_x dx \tag{1.49}$$

where the first integral expresses the stress acting normally on the area dydz, which gives, after integration:

$$\int_A dydz = A$$

Then, with $\sigma_x = P/A$ and $e_x = u/l$, one can write

$$U = \frac{1}{2}A\frac{P}{A}\int_0^l \frac{u}{l} = \frac{Pu}{2} \tag{1.50}$$

This can also be written as:

$$U = \frac{Pu}{2} = \frac{1}{2}\sigma A e_x l = \frac{1}{2}\sigma e_x V \tag{1.50a}$$

since Al = V:

Thus, the strain energy per unit volume is:

$$U = \frac{1}{2}\sigma e_x = \frac{1}{2}Ee_x^2 = \frac{1}{2}\frac{\sigma^2}{E} \tag{1.50b}$$

Equations 1.50b and 1.48 are the same.

1.2.11 Resilience

During the deformation of a material in the elastic region, energy is absorbed. On unloading the test specimen, this energy is recovered. The 'resilience' of a material is its ability to absorb energy when it is deformed elastically up to its yield point and to recover it upon unloading. Resilience is designated by U_R, which is the resilience modulus given by:

$$U_R = \frac{1}{2}\sigma_0 e_0 = \frac{1}{2}\frac{{\sigma_0}^2}{E} \tag{1.51}$$

The yield stress is σ_0 and the corresponding yield strain is e_0. This is represented by the area under the curve in the elastic region in the stress/strain diagram. In Fig. 1.18, the resilience is shown graphically as a triangular area under the curve (cross-hatched) and given as $\frac{1}{2}\sigma_0 e_0$. Since the yield stress can be defined by the proportional or elastic limits or by the proof stress, the area under the curve may vary slightly. The second expression of Eq. 1.51 can be obtained by substituting for e_0 from Eq. 1.5. Note that Eq. 1.51 is basically the same as found in Eqs. 1.48 and 1.50b. The difference being that the resilience is related to the elastic strain energy at the yield point. In order to prevent the permanent deformation of some material in service, its resilience must be high, which means it must have a high yield stress and a low modulus of elasticity. Mechanical springs are an example of such materials. Thus, resilience is the maximum energy per unit volume that can be elastically stored.

1.2.12 Toughness

Another definition, similar to that of resilience, has also been defined in the plastic region. The term 'toughness' is used to represent the ability of a material to absorb energy in the plastic region and is expressed by the area under the stress/strain curve. Basically, it may be defined as the resistance to fracture of a deformed material, expressed as the amount of energy per volume that a material can absorb in the plastic region before breaking. Thus, by this definition, it is the integral, namely:

$$\boldsymbol{U_T = area\,under\,the\,curve} = \int_0^{e_f} \boldsymbol{\sigma de} \tag{1.52}$$

Toughness represents energy per unit volume to fracture. In the above, clearly the dimensions are energy per unit volume, i.e., $\left[\frac{kg}{m^2} \times \frac{m}{m} = \frac{kg.m}{m^3}\right]$. In the SI system, toughness is expressed in units of joules per cubic meter (J/m^3). Toughness and strength are related, but not necessarily the same, in particular when brittle and ductile materials are considered. In brittle materials, the fracture stress is about equal to the yield stress and e_f is very small (usually almost equivalent to the elastic strain at σ_f). e_f has a relatively smaller value than in ductile materials, giving $U_T \sim 2/3\sigma_f e_f$, since it is assumed that the stress/strain curve is a parabola. In ductile materials, on the other hand, toughness is approximated by $U_T \sim \sigma_f \varepsilon_f$. Equation 1.52 may also be used to express resilience, but the upper limit of integration should be changed up to the yield point only.

The above can be spelled out as follows. Let the ductile material be expressed in terms of true $\sigma - \varepsilon$. Equation 1.52 is then given as:

$$\boldsymbol{U_T} = \int_0^{\varepsilon_f} \boldsymbol{\sigma d\varepsilon} \tag{1.52a}$$

Substituting the flow stress from Eq. 1.30 for σ, one can write:

$$U_T = K \int_0^{\varepsilon_f} \varepsilon^n d\varepsilon = \frac{K \varepsilon_f^{\,n+1}}{n+1} = \frac{\sigma_f \varepsilon_f}{n+1} \cong \sigma_f \varepsilon_f \tag{1.52b}$$

It was mentioned in Sect. 1.2.4 that, in perfect ductile material, n = 1. Furthermore, it was assumed that in brittle materials the σ-e is a parabolic function, e.g., $\sigma \sim Ke^{1/2}$ and, thus, Eq. 1.52a is:

$$U_T = \int_0^{e_f} \sigma de = \int_0^{e_f} K(e)^{1/2} de = \frac{2}{3} K e_f^{\,1/2} e_f = \frac{2}{3} \sigma_f e_f \tag{1.52c}$$

Strength and toughness are related. Whereas strength indicates how much force a material can support before breaking, toughness indicates how much energy a material can absorb before fracture. A material may be strong and tough if it fractures by withstanding a high force and exhibiting high strain, but brittle materials that have high strength (~equal to its yield stress) may be strong, but not tough, since their strain value is limited. As such, toughness means a good combination of strength and ductility. A material having high strength and high ductility is ideal for high toughness, whereas a material with low strength and high ductility or one with high stress and low ductility is inferior for applications where toughness is a crucial structural requirement.

Toughness will be further discussed in paragraph 1.6 below and it will be shown that there are several variables that greatly influence the toughness of a material: strain rate (rate of loading), temperature and notch effect (i.e., triaxial state of stress).

1.2.13 Fracture Stress

Fracture can be either brittle or ductile. In brittle fracture, no observable plastic deformation occurs before fracture. Fracture is often called 'cleavage' and, in crystalline materials, the respective crystalline planes are called 'cleavage planes'. The bonding between such planes is generally weak. The cracks formed in brittle failure propagate normally to the applied stress (in tension, when a tensile stress is applied). In ductile fracture, extensive plastic deformation takes place before fracture. Many ductile metals, especially materials with high purity, can sustain very large deformation of 50–100% or more strain before fracture sets in. The stages in ductile fracture are void formation, coalescence (at this stage a crack forms), crack propagation and fracture. In well-behaved ductile materials, the fractured specimens are characterized by cup cone-shaped surfaces. Loading, environmental conditions and purity affect plastic deformation and, thus, the fracture of a specimen. Some materials are neither completely brittle nor completely ductile in their behavior and exhibit partially brittle and partially ductile characteristics. Surfaces showing the structural appearance of fractures are discussed below in Sect. 1.5. Such materials can occasionally show either brittle or ductile behavior, depending on those factors which influence plastic deformation. Grain boundaries have an important influence on fracture and, accordingly, brittle fracture (cleavage) can be transgranular, intergranular or intragranular, depending whether the grain boundaries are stronger or weaker than the grain itself.

Certain types of structures may sometimes exhibit characteristic fracture behaviors. The most commonly known example is that of the refractory body-centered cubic (BCC, nonferrous) metals, which usually show cleavage fracture. Face-centered cubic (FCC) metals are quite ductile and tough when external factors do not adversely influence their behavior.

1.3 Compression Stress

1.3.1 Introduction

'Compressive stress' is the opposite of tension and, thus, is indicated by negative values of the relevant parameters. It is reasonable to assume that Eq. 1.4 holds also for simple compression, when the compressive stress is also distributed uniformly over the cross-section and is given as force per unit area. The stresses acting in a body produce deformation. The deformation of a body causes the displacement of each point within it, relative to its prior position when in its non-deformed state. Contrary to tension, compressive stress, when applied to a tested specimen, usually leads to its shortening. Thus, strain has a negative sign. Unlike the affect of tension on an atomic level, the atoms are forced together. In general, the compressive strengths of materials are usually greater than their tensile strengths. Compressive stresses are involved in metalworking applications when forming various shapes by the plastic deformation of rods, sheet material, tubes, etc., either directly or as a stress component of the overall applied force. Therefore, compression tests are usually applied to obtain information on the quality of materials intended for fabrication.

1.3.2 The Compression of Brittle Materials

The stress/strain relation in brittle materials, under uniaxial compression, resembles that of elastic materials during tension testing. Glass, rocks, cast iron, etc. are such materials. However, during compression tests, such materials show higher strength than during tension testing and also greater strains before fracture. Some

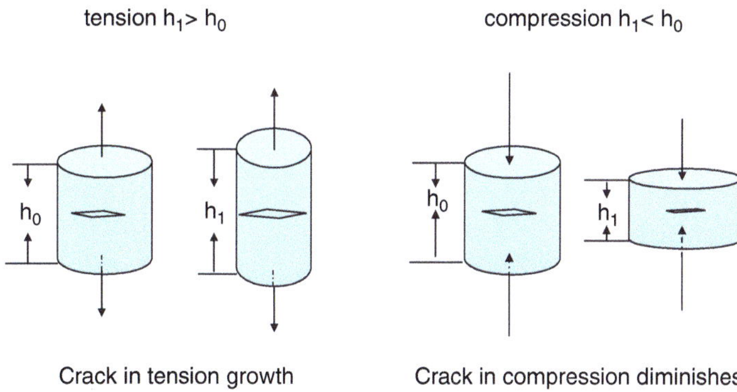

Fig. 1.19 The effect of the type of test on cracks

of these materials, which fracture in the elastic region under tension, may even show deviation from the line (often straight lines) characterizing their elastic, tensile behavior during compression testing. Nevertheless, many brittle materials do not exhibit yielding before failure, because fracture sets in earlier. A comparative list of the fracture strengths during the compression and tension testing of brittle materials is found in the book by Polakowski and Rippling. Note again that some materials show both ductile and brittle behavior, depending on the state or composition of the material being tested. Thus, some materials that have not undergone heat treatment, unlike heat-treated ones, may show such dual behavior. Steel is a typical example of dual behavior. For instance, it is known that high C steel shows brittle behavior at low temperatures and is ductile at high temperatures. The carbon content in steel also changes its behavior, making it either brittle or ductile, depending on the concentration (high carbon content embrittles steel). Concrete can given high compressive strength, e.g., in excess of 50 MPa. Even a very weak material, such as sandstone, may have a compressive strength of 5 or 10 MPa. Ceramics, glass and fiber-epoxy matrix composites typically have much higher compressive than tensile strength.

Micro-cracks are known to weaken materials under tension, since the cracks or other flaws open and propagate while the specimen is being tested. Contrary to this behavior, cracks in brittle materials strengthen the specimen being tested as the cracks close under the applied pressure. Schematically, this can be visualized in Fig. 1.19, where specimens with cracks are compared under tension and compression.

1.3.3 The Compression of Ductile Materials

Unlike the situation during tension, in which a specimen being tested usually gets longer, during compression, materials shorten. As stated above, materials being compressed can withstand much larger stresses than those under tension.

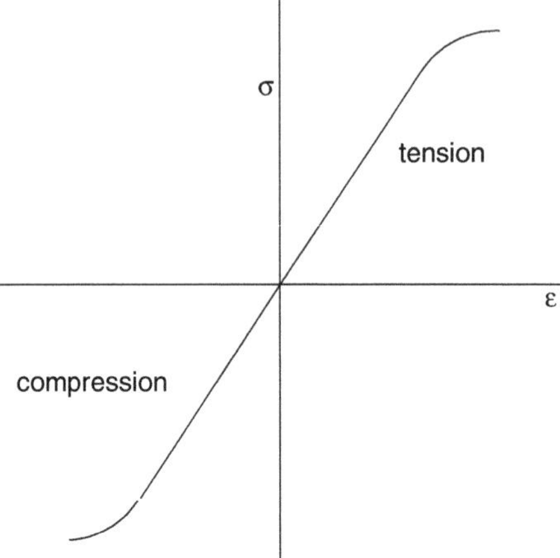

Fig. 1.20 Schematic tension and compression

The compression tests of ductile materials are similar to the tension tests, almost identical, as seen in Fig. 1.20. When the load is applied uniaxially at the edges of the specimen, frictional forces arise which oppose the lateral spread of the test piece. Work must be done to oppose this frictional force, manifested by the consumption of more energy during the compression process. This frictional force is not constant over the entire cross-section of the specimen, but rather is minimal at the center and maximal at the edges. As a consequence, barreling occurs in the specimen. This is equivalent but opposite to the necking observed in tension. The difference is clear; during tension, the specimen extends by thinning to a needle-like point leading to fracture, while barreling occurs when a load applied by compression is not homogeneous. In Fig. 1.21, the compression of a ductile material is shown schematically, both when the specimen shortened uniformly (homogeneous compression) and when barreling occurred, because the material did not deform uniformly. In homogeneous compression, h_0 is the specimen height before deformation and h_1 after compression. The specimens are loaded axially up to failure.

Compression-test specimens are simpler in shape than those used in tension-testing, since they do not require threads for gripping. Compression tests are sometimes used for testing nonstandard specimens and for the component testing of machine elements, in cases where the specimens for tension testing are difficult to machine. Following yielding, failure of ductile materials will occur, but brittle materials will crush when their limits of compressive strength are reached.

All the relations used in tension are applicable to compression, bearing in mind that the deformation is compressive. Thus, as in Eq. 1.4, compressive stress is given below as:

$$\sigma = \frac{P}{A} \tag{1.53}$$

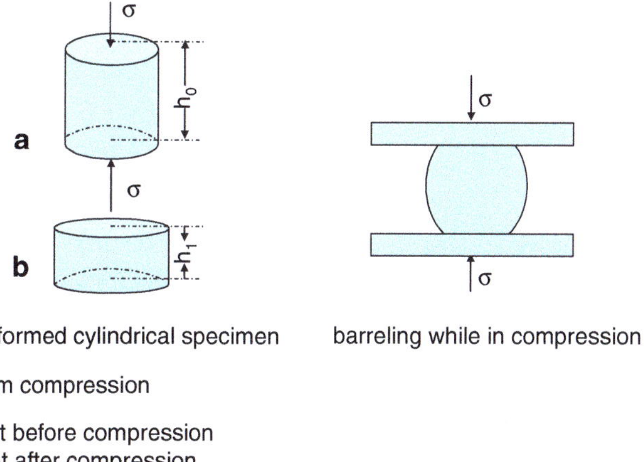

Fig. 1.21 Homogeneous and non-homogeneous compression; barreling is shown

As in tension, A (the area) will determine if the stress is expressed in terms of engineering or true stress. All equations from a tension test, i.e., Eqs. 1.5, 1.6, 1.7, 1.8, 1.9, 1.10, 1.11a, and 1.11b apply to compression, as well. Thus, for example, the ultimate compressive strength [henceforth: UCS] is the maximum compressive load divided by the original specimen cross-sectional area.

Recalling the negative definition of strain in compression, Eq. 1.9 for example:

$$\varepsilon = \int_{l_0}^{l_f} \frac{dl}{l} = \ln\frac{l_f}{l_0} \tag{1.9}$$

$$-\varepsilon = \ln\frac{l_f}{l_0} \tag{1.54}$$

or:

$$\varepsilon = \ln\frac{l_0}{l_f} \tag{1.54a}$$

A typical compression stress/strain curve is shown in Fig. 1.22. Note the difference between this diagram and that of a tension test. Whereas, in a tension test, a maximum, the UTS, is reached, followed by a decrease in stress until fracture, during compression, even in an engineering stress/strain curve, the stress continues to rise until fracture sets in. Clearly, this behavior is associated with the lack of necking in compression tests. In the schematic diagram, note that an elastic linear region is present where Hooke's Law applies, while Young's Modulus of elasticity during compression is $\sigma = E\varepsilon$. Although nonstandard specimens can be used in compression testing, the choice of which is dictated by the specific

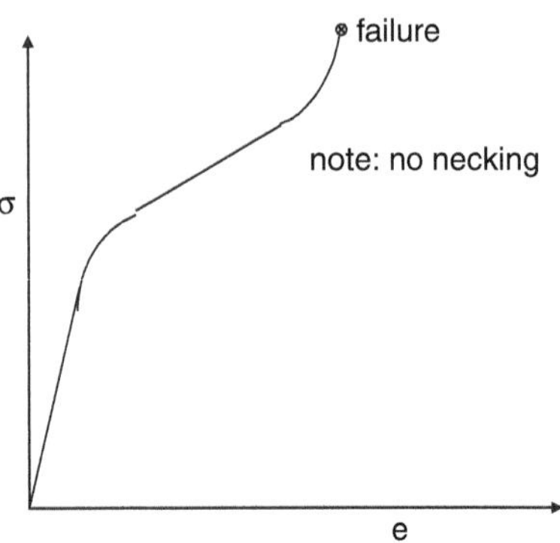

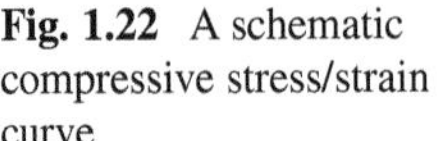
Fig. 1.22 A schematic compressive stress/strain curve

circumstances (the costs of materials to be tested, the feasibility of taking proper specimens from machine elements, etc.), the consequences to the test of applying improper specimens should be born in mind. Thus, for homogeneous compression, the ratio of specimen length, h, to its diameter, d, should be less than two-three, i.e., $h/d < 2.0–3.0$ and friction at the contact areas between the platens and the specimens should be eliminated. This can be achieved by confining the ends by rigid rings or introducing recesses in the anvils. Specimens having $h/d > 2–3$ should be avoided, to prevent buckling or shearing stresses which may arise during a compression test. The end surfaces of the specimens should be flat, in order to obtain an even distribution of the load. Samples with concentric end-grooves, to retain lubricant, are used to minimize barreling during compression. This way, a frictionless condition can be maintained between the platens and sample end faces. Barreling is the result of friction between the specimen end faces and the platen, which often also produces cracks in the specimen. In long structural elements, such as columns or bars, an increase of compressive force P leads to failure due to buckling at lower stress than the compressive strength.

1.3.4 The Effect of Hydrostatic Pressure on Compression

The beneficial effect of hydrostatic (isostatic) pressure on the workability of metals has been clearly shown by Bridgman in the early work on the enhancement of tensile ductility of various steels when subjected to high levels of hydrostatic pressure. As mentioned above, compressive stresses are involved in most metalworking and, hence, the interest in promoting enhancement of compressive ductility by the application of hydrostatic pressure, in particular for forming techniques in difficult-to-work alloys, such as gamma titanium aluminides.

Usually during a compression test, secondary stresses, either shear or tensile, arise and cause fracturing in the specimen. It is possible, if not to overcome, to reduce the effect of the tension acting in a transverse direction of the test specimen by superimposing a hydrostatic component on the applied compressive stress. Basically, hydrostatic pressure can be applied to any uniaxial deformation, i.e., by tension or by compression, but in this section the interest is on its effect on compressive deformation. The effect is on the elastic/plastic regions and on fracture, but other parameters, such as Young's Modulus, may also be affected. The type of material is also important in this regard. Usually, the major interest is to extend the elastic and plastic regions and to increase the fracture strength, particularly in metallic materials showing brittle behavior (e.g., cast iron) and in plastics, rocks and elements of soil mechanics. Extending deformation means to postpone all aspects of fracture. This implies suppressing the nucleation stage of cracks or other flaws and reducing their propagation by superimposing hydrostatic pressure. For example, experiments on the compressive formability of 7,075 aluminum alloy rings under hydrostatic pressure (see Liu, et al.) show that, at 100 MPa, the maximal compressive strain is increased by 32.42%. Fracture occurs by crack initiation at the outer wall, propagating to the middle of the inner wall along the direction of the maximum shear stress. When compressing the ring specimen without hydrostatic pressure, radial and tangential stresses emerge that cause fracture at a lower strain. Strokov and Barabanov have studied the effects of various hydrostatic pressures on graphite-based composite materials and they found that the elastic modulus, compressive strength and strain at fracture all increased. Compressive behavior is significantly affected by hydrostatic pressure. The maximum stress and fracture strain increased substantially as the pressure increased. The elastic modulus also increased continuously with increasing pressure. For the tested materials, they suggested a relation between the compressive pressure, σ_z, and the relevant parameters as:

$$\sigma_z = \frac{(D_c^2 - D_{sp}^2)P_h - D_c^2 P_e}{D_{sp}^2} \; [MPa] \tag{1.55}$$

D_c and D_{sp} are the loading chamber and the specimen diameters, respectively; P_e is pressure in the loading chamber at the specimen end, (in MPa) and P_h is outer hydrostatic pressure on the specimen in (MPa).

In a manner similar to tension, the UCS is defined as the maximum compressive load divided by the original specimen's cross-sectional area. However, brittle materials will fail suddenly after their ultimate strength is exceeded. These brittle materials have much greater compression strength than tensile strength, as mentioned above. This is the reason why brittle materials are mostly tested by compression.

To conclude this section on the effect of hydrostatic pressure, one may quote from the work of Tirosh and Abraham: "Von Karman (1911) carried out compression tests on marble and sandstone with superimposed hydrostatic fluid pressure. His successor Boker (1914) performed similar tests, but in tension mode under hydrostatic fluid pressure, using marble and cast zinc. Their results were qualitatively the same: (i) the confining hydrostatic pressure, if high enough, leads to a substantial increase in the material compressive and tensile strength, (ii) the tensile strain at failure is significantly increased by hydrostatic pressure, and it is frequently associated with localized necking."

1.4 The Hardness Test

Now, in the twenty-first century, when materials of various sizes are in technological use, e.g., micromaterials and nanomaterials, the testing methods for their properties must suit their various scales. Methods for the hardness measurement of samples of various sizes are available and the scale for macro-, micro- or nano-indentations is defined according to the forces applied and the resulting impressions. Universal Mechanical Testers [henceforth: UMT] are available for micro- and nanohardness measurements with a large selection of force sensors. The range of micro- and nanohardness measurements are beyond the scope of this section, but a large body of relevant information is found in the literature. Generally, hardness is a material property that indicates resistance to plastic deformation, usually by penetration. Hardness has a different significance for different professionals performing such tests, depending upon their field, e.g. it means something else to geologists than it does to mechanical or materials science engineers. These tests are indentation tests, mainly intended to provide information on the resistance of materials to deformation. The indenters and their shapes used for the indentation vary widely, leaving different impressions. The appropriate dimensions of these impressions on the surfaces are measured to obtain information of interest to those performing such tests. Resistance to plastic deformation is a significant piece of information for engineers, stemming from the relations of the hardness values to other parameters obtained by different testing machines, mainly to the flow stress obtained by tension or compression. The hardness of a material correlates directly with its strength, wear resistance and other properties. Therefore, hardness testing is widely used for material evaluation, because of its simplicity and low cost relative to direct measurements. Conversion charts, from hardness to other properties of interest, appear in the literature.

There are two main reasons why practicing engineers must be able to relate hardness data to tensile stress or yield stress: (a) hardness tests are faster than tension tests and no machining is required, except for surface preparation, and (b) expensive tensile machines are not readily available in every workshop or institution. Experimentally, a hard indenter is pressed into a flat surface of the material under a standard indenting force and the resulting indentation is measured in a manner dictated by the shape of the indenter.

Experiments have shown that the mean pressure under the indenter is $\sim 3 \times \sigma_0$. The large hydrostatic pressure under the indenter usually prevents brittle fracture and allows the material to yield (flow) plastically. To avoid side-effects during the indentation, such as severe mechanical wear or friction welding, the indenter should be much harder than the material being tested. Diamonds and sapphires have

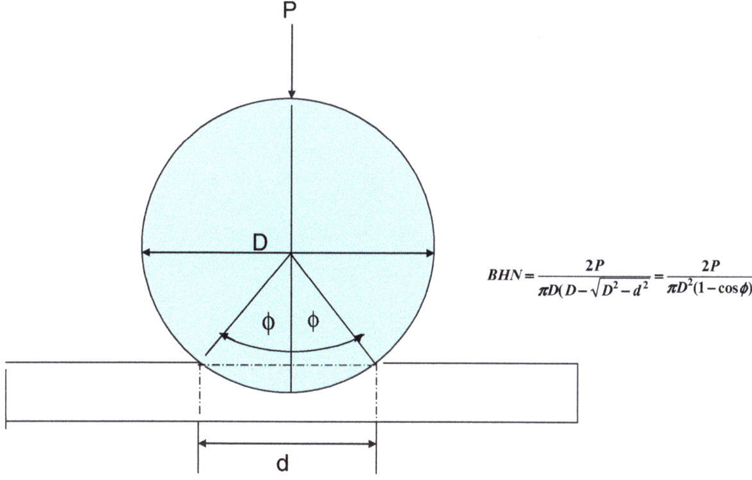

Fig. 1.23 Brinell hardness indentation on a plate showing the relevant parameters

low coefficients of friction against metals and, therefore, are useful as indenters. Below, the most commonly used indentation methods for hardness measurements are discussed.

1.4.1 Indentation by Spherical (Ball) Indenters

1.4.1.1 The Brinell Hardness Test

A test invented by Brinell in 1900 is the oldest hardness method in common use today. Brinell hardness indentation on a plate is shown in Fig. 1.23. The standard Brinell Hardness Test uses a 10 mm diameter tungsten carbide or steel spherical indenter to indent a metal surface under the static load (force) of 3,000 kg. The load is applied and maintained for 30 s and the diameter, d, of the projection of the impression is measured at the surface after removal of the ball (load) by a one ocular microscope at a magnification of 10×. An average of two readings of the diameter of the indentation trace at right angles are made. The Brinell hardness number [henceforth: BHN] is expressed by dividing the load applied by the surface area of the indentation, i.e. the spherical area, A_s, given in units of kg/mm^2 (or other SI units). The area is expressed in terms of the two diameters, namely that of the indenter ball, D, and the measured average diameter as shown below:

$$BHN = \frac{P}{A_s} = \frac{P}{(\pi D/2)(D - \sqrt{D^2 - d^2})} = \frac{P}{\pi D t} \tag{1.56}$$

t is the depth of indentation.

The above equation is obtained as a result of the following consideration; A_s is the difference between the half sphere (of the indenter) and the surface area of the segment bounded by two planes of diameters D (indenter) and d (impression) or:

$$\textit{the segment} = \frac{\pi D}{2}(D^2 - d^2)^{1/2}$$

$$\textit{the area of the halph sphere of the indenter} = \frac{\pi D^2}{2}$$

$$\textit{the difference} = \frac{\pi D^2}{2} - \frac{\pi D}{2}(D^2 - d^2)^{1/2} = \frac{\pi D}{2}\left[D - \sqrt{(D^2 - d^2)}\right] \tag{1.56a}$$

This is the denominator of Eq. 1.56.

Another relation can be derived from Eq. 1.56 as follows. Consider the illustration in Fig. 1.23, where the diameters D and d together with the angles ϕ are seen. Diameter d can be expressed as:

$$\mathrm{d} = \mathrm{D}\sin\phi$$

$$\sin^2\phi = 1 - \cos^2\phi \tag{1.56b}$$

substituting for d^2 in Eq. 1.56a, an expression for the BHN is obtained as:

$$BHN = \frac{P}{(\pi D^2/2)(1 - \cos\phi)} \tag{1.57}$$

For soft metals, the load is reduced to 500 kg to avoid very deep penetration. In order to maintain a constant value for the BHN, the ratio of load to ball-diameter should be kept constant and, if the load is varied, the ball-diameter should also vary according to the following relation:

$$\frac{P_1}{{D_1}^2} = \frac{P_2}{{D_2}^2} = \cdots \frac{P_i}{{D_i}^2} \tag{1.58}$$

It should be emphasized that the surface on which indentation is to be performed should be free of any scale or other foreign material and relatively smooth. No other surface preparation is required. The diameter of the indentation left in the test material is measured with a low-powered microscope. The use of a BHN table can simplify the determination of the BHN; take an average measured diameter and find the corresponding BHN (see the Table in the Appendix, for example in

Kehl with other hardness conversion information). This Table is based on Eq. 1.56. A BHN shows the test conditions and is often recorded as "85 HB 10/500/30," which means that a BHN of 85 was obtained using a 10 mm diameter hardened steel indenter with a 500 kg load applied for a period of 30 s. For testing of extremely hard metals, a tungsten-carbide ball is used, instead the usual steel ball. Compared to the other hardness test methods, the Brinell ball makes the deepest and widest indentation, so the test averages the hardness over a wider amount of material, which will more accurately account for multiple grain structures and any irregularities in the uniformity of the material. This method is the best for achieving the bulk or macro-hardness of a material, particularly those materials with heterogeneous structures. The Brinell Test is frequently used to determine the hardness of forgings and castings that have a grain structure too coarse for Rockwell or Vickers testing. Therefore, Brinell Tests are frequently done on large parts. By varying the test force and ball size, nearly all metals can be tested using a Brinell Test. Brinell values are considered load independent, as long as the ball size/load relationship is the same as indicated in Eq. 1.58.

The BHN can be converted into UTS, although the relationship is dependent on the material under consideration and, therefore, determined empirically. In the U.S.A., Brinell testing is typically done on iron and steel castings using a 3,000 kg test force and a 10 mm diameter carbide ball. Aluminum and other softer alloys are frequently tested using a 500 kg test force and a 10 or 5 mm carbide ball. Therefore, the typical range of Brinell testing in the U.S.A. is 500–3,000 kg with 5 or 10 mm carbide balls. In Europe, Brinell testing is done using a much wider range of forces and ball sizes. It is common in Europe to perform Brinell tests on small parts using a 1 mm carbide ball and a test force as low as 1 kg. These low load tests are commonly referred to as 'baby Brinell Tests'.

Because of the wide-spread use of the Brinell Hardness Tests, many empirical relations were established between BHN and other mechanical properties of interest. One of the most interesting properties of metals with industrial applications, for both the purchase of materials and their specifications, are the tensile properties. Some of these are listed below.

The BHN is approximately related to tensile strength. For plain carbon steels up to $\sim$1% C, rolled and annealed, the relation constants have been quite accurately determined and are expressed as:

$$\text{BHN} \times \text{Constant} = \text{UTS (lb per sq. in.)}$$

The constants below 175 BHN are 515 and 504 for transverse to rolling direction and rolling direction, respectively. Above 175 BHN these are 489 and 461, respectively. Thus, plain carbon steel, having a BHN of 310 transverse to the rolling direction, will have UTS in accordance with the above relation and using the proper constant:

$$489 \times 310 = 151{,}590\ \text{lb/in}^2$$

In Brinell Hardness Tests, polishing of the surface is not required, but it should be smooth and free from surface defects, such as oxide, scale, etc. In addition, a proper thickness of the specimen to be tested is important to eliminate what is known as the 'anvil effect'. Any effect of the supporting anvil on a specimen being tested will contribute its hardness to the measured value, making the measurement erratic when the specimen is thin. In general, the thickness of the specimen to be tested should be ~10 times the penetration depth of the impression. Moreover, specimen edge effect should be eliminated. Thus, the minimum distance for the test should be at least 2.5 times the diameter of the impression.

1.4.1.2 The Meyer Hardness Test

Meyer's definition of hardness is more rational than the one proposed by Brinell, even if their tests are similar, since both use ball indenters. The main difference is in the area considered. In the Meyer Hardness Test, the entire projected area of impression is measured and not just the surface area of an indentation, as indicated in the equation below. The mean pressure between the surface of the indenter and the indentation is equal to the load divided by the projected area of the indentation. Meyer suggested that this mean pressure should be taken as the measure of hardness:

$$M = \frac{P}{\pi (d/2)^2} = \frac{4P}{\pi d^2} \tag{1.59}$$

M stands for Meyer hardness, d is the diameter of the projected indentation, while P has the same meaning as in the Brinell Test, namely load. As in Brinell hardness, Meyer hardness measures the load per area; thus, a mean pressure is calculated which is more basic than the one used by Brinell. An advantage of the Meyer Test is that it is less sensitive to the applied load than the Brinell Hardness Test. For cold-worked metals, the Meyer hardness measured is relatively constant and independent of load (see Fig. 1.24), whereas, in Brinell Hardness Tests, the BHN decreases with higher loads. For soft or annealed metals, the Meyer hardness increases continuously with load, due to strain hardening produced by the indentation (discussed below), which is contrary to the behavior of the BHN. Figure 1.24 shows schematically the variation in hardness numbers for both BHN and M vs. the change in impression diameter. It can be seen that at a value of $d = D$, where d is the diameter of the projected area and D is the diameter of the indenter, the ratio BHN/M = 1/2 when the loads are the same, regardless if the metal tested is soft or cold-worked.

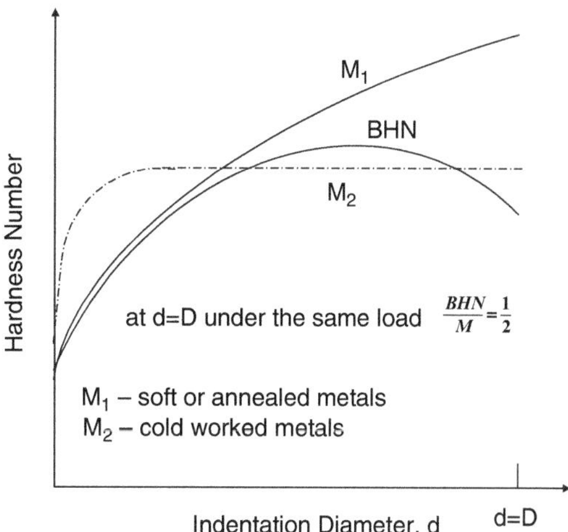

Fig. 1.24 Comparison of BHN and M test

This is shown below in Eq. 1.60

$$\begin{aligned}\frac{BHN}{M} &= \frac{\dfrac{P}{(\pi D/2)\left[D-\sqrt{(D^2-d^2)}\right]}}{4P/\pi d^2} \\ &= \frac{\pi d^2 P}{4P(\pi D/2)\left[D-\sqrt{(D^2-d^2)}\right]} = \frac{D}{2(D-0)} = \frac{1}{2}\end{aligned} \tag{1.60}$$

In this Figure, M_1 and M_2 refer to soft or annealed and cold-worked metals, respectively. Note that in the Meyer indentation of a soft (annealed) metal, the plastic region under the indenter work hardens. The hardness of a cold-worked metal, after an initial steep increase in hardness, actually remains constant during the entire test. This behavior of the Meyer indentation should be compared with that of the Brinell indentation, as seen in Fig. 1.24. The BHN reaches a maximum, then, with further increase in the size of the diameter, it decreases. The shape of the Meyer and BHN curves are similar to the $\sigma_T - \varepsilon$ and engineering $\sigma - e$ relations, respectively. On the basis of this similarity between the respective curves, empirical relations have been suggested between tensile and hardness properties which are fairly reasonable.

On the basis of the results shown in Fig. 1.24 for soft materials, Meyer has proposed an empirical relation between the load and the size of the indentation, usually known as Meyer's Law given below:

$$P = kd^m \tag{1.61}$$

P = pressure, k = property of the material tested, m = Meyer's Index, also known as 'Meyer's exponent', which is a material property, d = diameter of the indentation (chordal diameter). The parameters k and m are constants, but are also dependent on the size of the indenter. The similarity of this relation to the empirical expression for plastic flow (see Eq. 1.30) is clear. Here, as in Eq. 1.30, m is the strain hardening exponent. The constants of Eq. 1.61 can be evaluated from a log-log plot of P vs. d, the slope of the straight line being m and the value of P at d = 1 is from the intercept k:

$$\mathbf{ln}P = \mathbf{ln}k + m\,\mathbf{ln}d \tag{1.61a}$$

The magnitude of m is between 2 and 2.5. Fully annealed metals have an m value of about 2.5, while it is ca. 2 for fully strain-hardened metals. This parameter is roughly related to the strain-hardening coefficient in Eq. 1.30 for the true stress/true strain curve plus 2. Note, however, that below ca. d = 0.5 mm, the value of m can surpass 3. Because of this, Meyer's Law is often restricted to values of d greater than 0.5 mm and up to the diameter of the indenter.

Note that, as seen in Fig. 1.24, BHN deviates from the M hardness value at about its maximum. Up to this maximum, many metals tested indicate about the same values for BHN and M.

Experimentally, the indentation is measured from a small value to up to some greater value. Rather than using various size indenters to obtain d, increasing the value of the load will produce deeper indentations, yielding different values for the projected area. Using this approach makes it easier to obtain the parameters k and m.

Although Meyer hardness is a more fundamental measure of indentation hardness than Brinell's, it is, nonetheless, rarely used for practical hardness measurements. Hardness values from the Meyer test can be converted into BHNs and vice-versa. As a matter of fact, there are multidirectional conversion tables for all the various hardness numbers obtained by the different test methods.

1.4.1.3 Comments on the Ball Indentations

1. During the ball tests, when the load is removed, the impression remaining on the metal surface represents plastic deformation. It is impossible to measure the elastic range contribution to the indentation, because of the elastic springback.
2. Unless precautions are taken to keep P/D^2 constant during Brinell testing, as indicated in Eq. 1.58 (and which may be inconvenient), the BHN will generally vary with load. Over a range of loads, the BHN reaches a maximum at some intermediate load. This can be inferred from Fig. 1.24, where one sees that the change in BHN reaches a maximum and then decreases. BHN is f(P), so an increase in BHN means an increase in P. Therefore, it is not possible to cover the entire range of harnesses encountered in metals of industrial interest with a single load. To obtain the entire hardness range, it is essential that the load and ball-size ratio relationship be the same.
3. The main drawback of these tests is the need to optically measure the indent size, which is done with a low-power microscope (magnification of 10×).

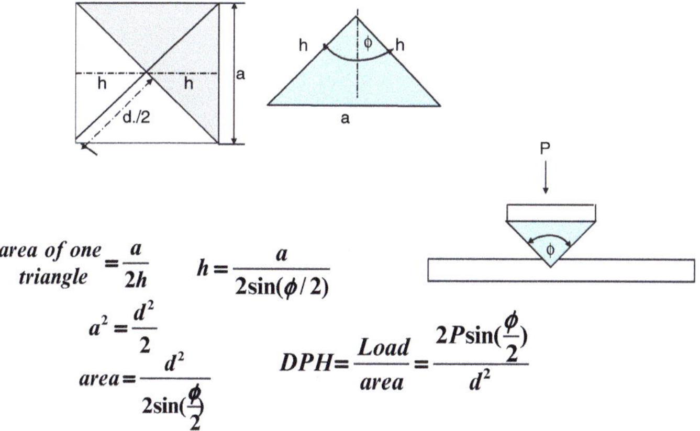

Fig. 1.25 Vicker's indentation

4. A wide range of loads and ball sizes can be used to cover all harnesses encountered in practice, from soft to hard metals, suiting every industrial application.
5. The test is nondestructive, thus parts or samples can normally be reused.
6. Brinell Test methods are standardized (ASTM).

1.4.2 Indentation by Pyramid and Cone Indenters

1.4.2.1 The Vickers Hardness Test

The Vickers Hardness Test uses a diamond pyramid with a square base for indentation, so it is also known as the Diamond Pyramid Hardness Test. The indenter has an angle of $\phi = 136°$ between two opposite faces. The results of this test are also known as 'diamond pyramid hardness' [henceforth: DPH] or the 'diamond pyramid number' [henceforth: DPN], defined as load divided by the surface area of the pyramid-shaped indentation (impression). The area is simple to evaluate from the geometry of the shape of the indentation, which requires measuring the diagonals and using the known angle between the two opposite faces. These two diagonals are measured on the screen of the Vickers Tester and their average is used in the DPH formula. The area of the sloping surface of the indentation is calculated as indicated in Fig. 1.25 and the steps are shown below for deriving the expression for DPH measurements:

$$\boldsymbol{DPH} = \frac{\mathbf{2}\boldsymbol{P}\ \mathbf{sin}\,(\boldsymbol{\phi}/\mathbf{2})}{\boldsymbol{d}^{\mathbf{2}}} = \frac{\mathbf{1.854}\boldsymbol{P}}{\boldsymbol{d}^{\mathbf{2}}} \tag{1.62}$$

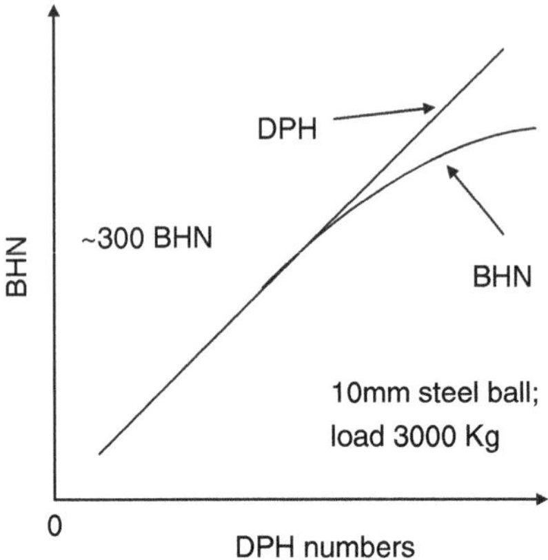

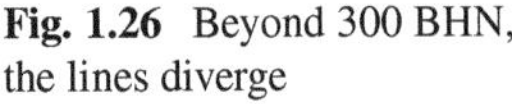
Fig. 1.26 Beyond 300 BHN, the lines diverge

The DPN can also be determined from tables after measuring the diagonals (and averaging the values). As in the tables for BHN evaluation without calculation, diagonal length values are compiled with the respective DPH numbers.

The 136° angle between the opposite faces of the indenter was chosen because of the similarity between BHN and DPN methods of hardness measurement. Due to this similarity, the DPNs are on the same hardness scale as the BHNs and they are about the same up to ~300 BHN. Only beyond this BHN value do the two curves shown in Fig. 1.26 deviate one from the other. Significant deviation between the two types of measurements occurs at high hardness values, particularly above ~600 BHN, due to the deformation of the indenter. The DPH curve is linear, as seen in schematic Fig. 1.26, because of the insignificant deformation of the diamond indenter compared to that of the Brinell.

The load varies from 1 to 120 kg and may be easily reset to the desired load at the push of a button. Thus, the Vickers Hardness Test can be applied to different materials having a very wide range of hardness values. The load is applied via the square-based pyramid indenter against the smooth, firmly-supported, flat surface of the test specimen. It is recommended to polish the specimen's surface, because a small impression is usually obtained in such a test.

The advantages of the Vickers test are:

1. Despite several different loading settings (in our automated laboratory the application and removal of loads is controlled automatically) almost identical hardness numbers are obtained on uniform materials. This is much better than arbitrarily changing the scale when using other hardness testing methods.
2. The Vickers Hardness Test yields more accurate diagonal readings.
3. Only one type of indenter is used for all types of metals and surface treatments. This test covers a wide range of hardness and, as a result, a continuous scale can be established.
4. Because of the wide load range, this test is adaptable for the testing of almost any metallic material, the softest and hardest ones.

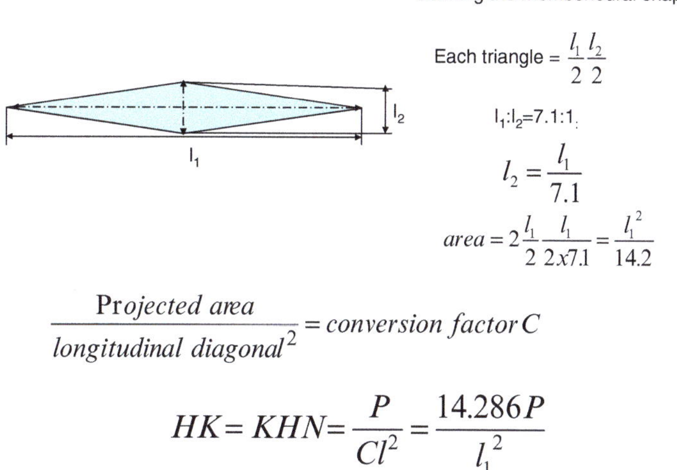

Fig. 1.27 Knoop Hardness Test

As is common in hardness specifications, the DPN is specified as shown in the following example – 356HV0.5, where 356 is the calculated hardness on the Vickers scale and 0.5 is the test force in kg. The thickness of the test piece should be about 1.5 times the diagonal length of the impression, but it will depend on the relative hardness of the material and the load applied.

The Vickers Test is a versatile hardness test, since it may be adapted for microhardness testing, as well as for a variety of materials, such as metals, ceramics and composites. It is useful for applications in which macrohardness measurements are unsuitable, e.g., testing very thin materials, such as coatings, carburized and decarburized steel, measuring individual microstructures within a larger matrix, or measuring the hardness gradients of a part along the cross-section. Another microhardness test is the Knoop Test (discussed below).

1.4.2.2 The Knoop Hardness Test – Microhardness

A Knoop diamond indenter may be used in a Tukon Hardness Tester (or in a Buehler Instrument Model 1600-6400). This test method was devised in 1939 by F. Knoop. The indentation produced by the Knoop Hardness [henceforth: HK] Test also applies a pyramid-shaped diamond indenter (developed by the National Bureau of Standards). The apical angles are 130° and 172°. Thus, a narrow rhombus-shaped indenter is used and the impression has a similar shape, as seen in Fig. 1.27.

The long diagonal is seven times longer than the short diagonal, i.e., d_{long}: $d_{short} = 7{:}1$. As a result, the length of the impression is approximately seven times the width and the depth is 1/30 of the length of the longer diameter. The loads are often less than 1 kg-force and even a value of 25 g may be used. This test has practical applications for the testing of thin films, coatings, carburized or decarburized layers in steels and for phase identification in microstructures. The latter use is a consequence of the HK indenter, used to sample minute grains of interest in some microstructure. In modern technology, where thin-films play an important role, microhardness testing has become crucial; hence, the major role played by the use of microindentation, such as the HK Test. During this test, the pyramid-shaped diamond indenter is pressed against a material, making a rhombohedral impression with one diagonal seven times longer than the other. Then, the Knoop hardness number [henceforth: KHN] is determined by the depth to which the Knoop indenter penetrates. The HK Test also allows for the hardness testing of brittle materials, such as glass and ceramics, since the indentation pressures are lower than in the Vickers Hardness Test. The Knoop indenter leaves an impression of ca. 0.01–0.1 mm in size. Because of the small size of the impression under load, hardness may only be calculated after measuring the length of the longest diameter with the aid of a calibrated microscope. HK is defined, as usual, by load (in kgf) per projected area (in mm^2) and is given as:

$$\boldsymbol{HK = KHN = \frac{P}{Cl_1^2} = \frac{14.286P}{l_1{}^2}} \tag{1.63}$$

l_1 is the long diagonal and C a conversion factor, which ideally is 7.028×10^{-2}, but this depends on the load. The derivation of the formula is given in Fig. 1.27. Again, the accepted way expressing a KHN is as follows, for example – 356HK0.5, where the first number and the letter are the measured hardness value, the second letter indicates HK and 0.5 is the load in kgf.

The measuring apparatus is preset to apply a 25 g load. The duration of the contact between the indenter and the specimen should be 10–15 s. The length of the long diagonal of the impression is measured with a high-powered microscope. This procedure is repeated until at least five impressions have been made at widely spaced locations. The KHN is then calculated. Test loads are in the range of 10–1,000 g. The samples are normally mounted and polished. One scale covers the entire hardness range.

1.4.3 Indentation by Cone (or Spherical) Indenters

1.4.3.1 The Rockwell Hardness Test

The Rockwell Hardness Test is the most used and versatile of the hardness tests. Usually the indenter is a cone, called the ‘Brale indenter’, but in some hardness scales spherical indenters are also used. Historically, the Rockwell Hardness Test was developed because the common Brinell Hardness Test was relatively slow, not very useful for testing fully-hardened steel and left a large impression after indentation. The requirement for a quicker test was originally related to the need to determine the effects of heat treatment in steels. Basically, the test is a differential depth hardness measurement.

The determination of the Rockwell hardness number (HR, once symbolized as RHN) of a material involves the application of a minor load of 10 kg, followed by a major load and notation of the depth of penetration in terms of hardness number value directly from a dial which is an integral part of the machine. The value representing hardness is an arbitrary number. The chief advantage of the Rockwell Hardness Test is its ability to display hardness values directly, thus eliminating operator error, such as often occur during measurements and calculations in other hardness measurement techniques.

This test is typically used in engineering fields, such as metallurgy. Its popularity arises from its speed, reliability and small area of indentation. To get a reliable reading, the thickness of the sample being tested should be about 10 times the depth of the indentation. This is essential in order to eliminate what is known the 'anvil effect'. Readings should also be taken from a flat, perpendicular surface, because round surfaces give lower readings. A correction factor may be used when hardness is measured on a round surface. Each Rockwell machine has a special gadget which holds non-flat, perpendicular surfaces and allows their measurement.

Several scales are used to cover all hardness encountered in various materials. The differences between these scales are due to the loads applied and the indenters used. The most commonly used scales are the "B" and "C" scales. Both of these scales express hardness as an arbitrary, dimensionless number. The "A" scale uses a load of 60 kgf and a 120° diamond cone indenter and is the only scale that is continuous over a wide range of material hardness, thus allowing continued use with only one fixed load and indenter combination. The "C" and "D" scales use the same indenter, but their loads are 150 and 100 kgf, respectively. The "B", "F" and "G" scales use a 1/16 in. diameter steel sphere with the respective loads of 100, 60 and 150 kgf. The "E" scale, on the other hand, uses a 1/8 in. diameter steel sphere and 100 kgf. Usually, the diamond cone (Brale) indenter is used for hard materials, such as tungsten carbide or hard steels. The other scales are used for various soft materials; the "B" scale, for example, is used for materials such as aluminum, brass, soft steels, etc. The minor load in all cases is 10 kgf. A hardness tester must realize the limits of his/her measurements, e.g., readings below 20HRC and above 100HRB are generally considered unreliable.

As mentioned above, during the Rockwell Hardness Test, a material is indented by means of a diamond-cone or a hardened steel-ball indenter. A preliminary minor load, usually 10 kgf, is applied first, causing initial penetration and holding the indenter in place. This also cleans the area, eradicating undesirable surface conditions, such as scaling. The dial is set to zero while the minor load is on and then the major load is applied. The major load is applied in addition to the preliminary

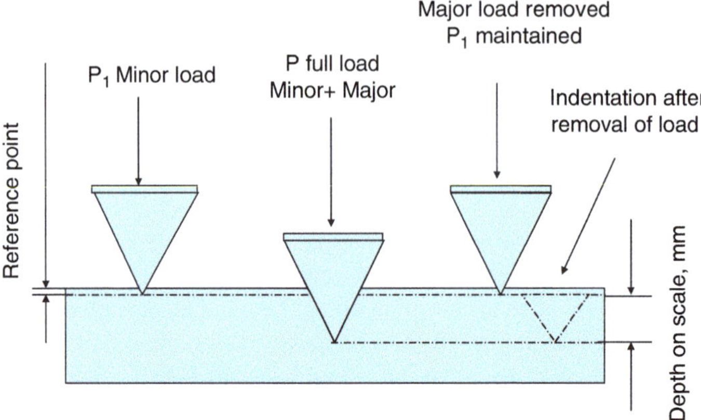

Fig. 1.28 The steps in a Rockwell test; application of minor and major loads are indicated

minor load. Upon removal of the major load, but while the preliminary minor load is maintained, the HR appearing on the dial is recorded, expressed as:

$$\mathrm{HR} = \mathrm{RHN} = \mathrm{C}-\mathrm{depth} \tag{1.64}$$

C is a constant dependent on the indenter and is 100 for the diamond-cone and 130 for the steel-ball indenter. The depth of penetration due to the major load is measured in units of mm. A currently used HR designation appears as 62HRC, indicating the hardness read from the dial, followed by HR (Hardness Rockwell), after which the specific Rockwell scale (e.g., C) is indicated. Thus, 62HRC indicates that the specimen has a hardness reading of 60 on the C scale (previously recorded as 62Rc and still often used as such). Another example of a hardness value on the A scale is 80HRA.

In summary, to start the test, the indenter is set into the sample near the minor load. The major load is then applied and held for some time. The force on the indenter is then decreased by removing the major load and leaving only the minor load. The HR is calculated from the depth of the permanent penetration of the indenter into the sample, i.e. the difference in indenter position before and after application of the major load. Figure 1.28 shows the steps of the test method. A test piece is shown with the minor load applied, thus setting the reference point. Then the major load makes an indentation, which is then released while maintaining the minor load. Then a reading is taken on the scale, indicating the depth of penetration. The remaining dashed indentation is shown without taking into account the small elastic springback.

Among the advantages of the Rockwell hardness method are the direct HR readout and the rapid testing time. Disadvantages are the arbitrary scales for the hardness measurements and possible anvil effects from the specimen support. Typical industrial applications are for the quality control of metals, particularly steel, before and after heat treatment. Material inspection, weld evaluation in the heating zone and failure analysis are some additional applications. Plastics can also be tested, since special test scales with load and indenter combinations are available for quality evaluation; for example, the "M" and "R" scales are used for polymers. Furthermore, hardness testing can be performed on thin samples using surface-sensitive measurement by means of the superficial Rockwell technique (discussed below). The appropriate hardness scales are "N" and "T" for metals and "W", "X" and "Y" for nonmetallic materials and soft coatings. Rockwell Test Blocks and Calibration Sets are occasionally used for calibration, to test the reliability of the measurements.

In order to test measurement reliability, the conditions of machined surfaces are generally satisfactory, but, for a high degree of accuracy, surface polishing is required. The penetration depth of the indenter depends on the hardness of the material and for reproducible results in thin-sheet metal, care should be taken regarding the effect of the supporting anvil, since the influence of a test is felt to a depth of $\sim$10 times the height of the impression. No hardness measurement on a thin section of a soft material is reliable if the mark of the indentation is visible on the reverse side of specimen. Since the Brinell and Rockwell tests are very often used, mutual conversions are of interest and appropriate conversion tables have been compiled. Empirical relations (accurate to within $\pm 10\%$) have been developed by the U.S. Bureau of Standards. Some known relations are presented below:

For the Rockwell range B 35–B 100

$$\text{BHN} = 7,300/\,(130 - \text{Rockwell B number})$$

For the Rockwell range C 20–C 40

$$\text{BHN} = 1,420,000/\,(100 - \text{Rockwell C number})$$

For Rockwell >C 40

$$\text{BHN} = 25{,}000/\,(100 - \text{Rockwell C number})$$

There is a second Rockwell tester, referred to as the 'Rockwell superficial hardness tester'. This machine works the same way as the standard Rockwell tester. It is used to test thin strips or lightly carburized surfaces, small parts, etc. The superficial tester uses a minor load of 3 kg and has the major load reduced to either 15 or 45 kg, depending on the indenter, which is one of those used with the common Rockwell scales. Using the 1/16 in. diameter, steel-ball indenter, a "T" is added (meaning thin-sheet testing) to the hardness designation. An example of superficial Rockwell hardness is 15T-22, which indicates that the superficial hardness is 22, with a load of 15 kg. If the 120 diamond cone were used instead, the "T" would be replaced with "N".

1.4.4 Comments on Hardness Tests

Mechanical property data, such as the UTS of bulk materials, are readily obtained from small samples by measuring hardness. Such data is also widely used for quality control in surface-treatment processes. However, for coatings and surface properties or the characterization of thin films, the macroindentation (discussed above) is inadequate. In fact, in the case of materials that have a fine microstructure, are multi-phase, non-homogeneous or tend to crack, macrohardness measurements will not identify the individual surface features. Microhardness measurements, as mentioned above, are appropriate and useful. Due to the small indentations, the hardness of the micro-constituents in a multi-phase structure can be evaluated.

Nanoindentation tests have become very important in modern technology and hardness measurements are performed by using minuscule indentation forces, on the order of 1 nN, and making precision measurements of the indention depth. There are various mechanisms and methods that have been designed to perform nanoindentation hardness tests. Modern nanoindentation machines are available to perform controlled tests and give reliable data. Nanoindentations are beyond the scope of this chapter.

1.5 The Torsion Test (Shear)

Not all deformation occurs by tension or compression. There is another way to induce stain in a specimen, namely by shear. In Sect. 1.2.9, shear stress, shear strain and the shear modulus were discussed. Pure shear stresses occur in specimens subjected to torsion. This type of deformation is not as popular as tension or compression tests, but it is of considerable importance for industry. A variety of machine elements operate under shear stress when exposed to rotational forces. Shafts, cylindrical rods, propellers, axels, etc. are some cases where shear-type deformation occurs. Analogous to tensile stress (force/unit area), shear stress τ is:

$$\tau = \frac{P}{A} \tag{1.65}$$

However, here the force is acting tangentially to the area. Figure 1.16 is shown below in two dimensions with the load P applied transversely to the square and the line length of the specimen remaining unchanged. This illustrates the shear strain described in Sect. 1.2.9 as (Fig. 1.29):

$$\gamma = \frac{x}{l} = \tan\phi \tag{1.66}$$

The shear modulus has been given above (Sect. 1.2.9) as:

$$\mathrm{G} = \tau/\gamma \tag{1.67}$$

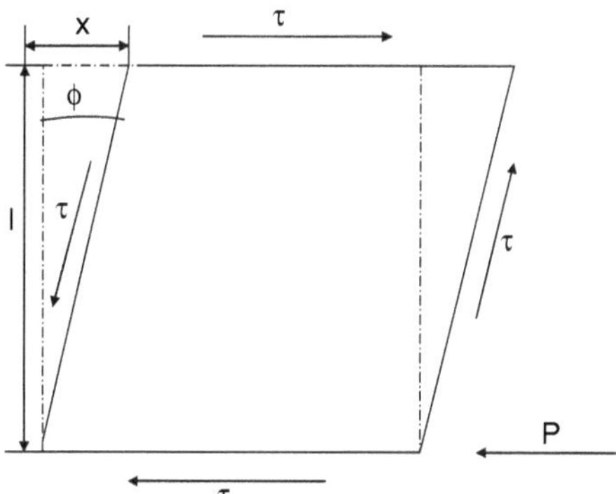

Fig. 1.29 Figure 1.16 in two dimensions showing the shear deformation of an elementary cube

and the relation between the elastic and the shear moduli as:

$$G = \frac{E}{2(1 + \nu)} \tag{1.68}$$

1.5.1 *Torsion in the Elastic Region*

Figure 1.30 is a schematic illustration of solid and hollow cylinders. Now, consider a solid cylinder. Torsion-testing machines are commercially available for tests which are performed by applying a twisting moment to a specimen. A specimen is mounted in a special chuck. Deformation, by twisting applied to the specimen, is measured by a device called a 'tropometer', which is an instrument for measuring the degree of torsion. The response of the material to a torsional force, i.e., the angle of twist and the torque, are recorded while the test is performed.

Assume that a cylinder of length L and radius r is twisted by an angle θ. A couple of stresses are applied by torsion about its axis. Also assume that the resulting strain is distributed uniformly. Figure 1.30c, d represent the cylinder from Fig. 1.30a in a different position and when it has been opened and spread out flat by shear forces. Figure 1.30d illustrates the outcome of a shear distortion. The arc length is a. For small angles, the arc length, AB, in Fig. 1.30c, d can be given as:

$$\mathrm{AB} = \mathrm{a} = \mathrm{r}\theta = \mathrm{L}\phi \tag{1.69}$$

Recalling (ref. Fig. 1.29) that tanϕ defines the shear strain γ, as in Eq. 1.66, i.e., for small angles:

$$\gamma = \tan\phi = \phi = \frac{a}{L} = \frac{r\theta}{L} \tag{1.69a}$$

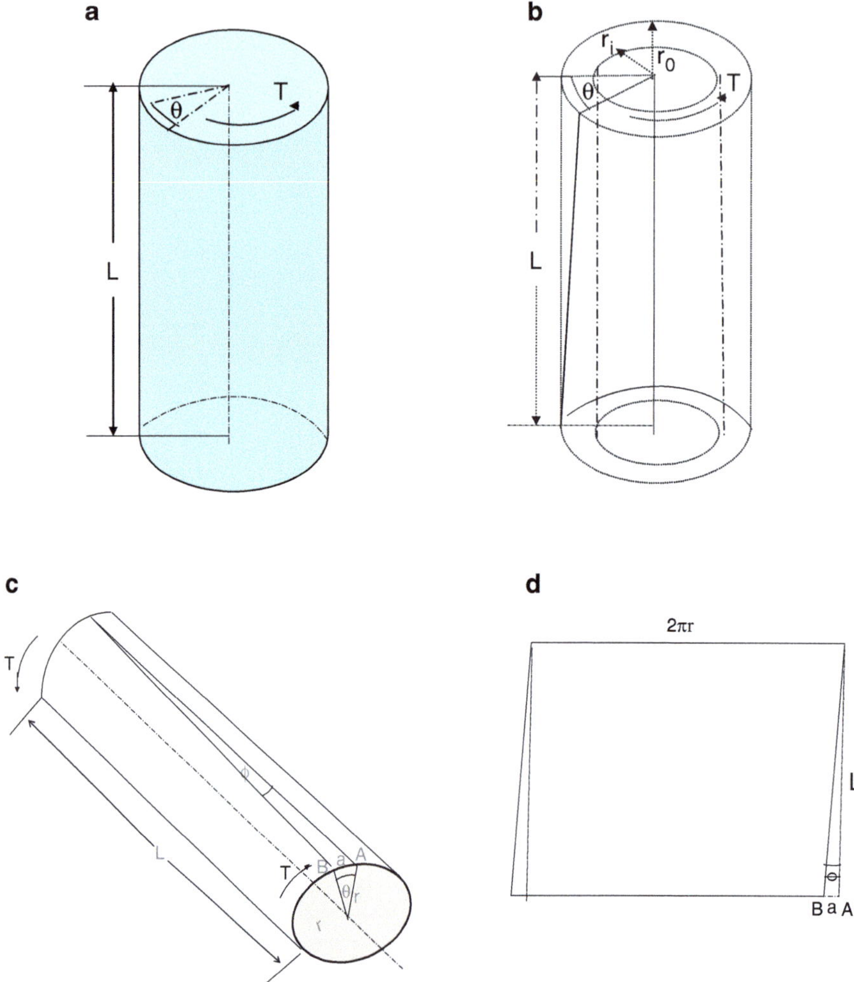

Fig. 1.30 Schematic illustration of cylinders under torsion: (**a**) solid cylinder, (**b**) hollow cylinder, (**c**) solid cylinder (shaft) redrawn for clarity, (**d**) the cylinder is opened and spread flat

Note that θ and γ are expressed in radians. However, the shear strain and stress define the shear modulus, G, and it is possible to express this relation by the use of Eq. 1.69a. Thus:

$$\tau = G\gamma = G\frac{r\theta}{L} \tag{1.70}$$

When torque is applied to a system, it is opposed by an internal resisting force. Torque arises due to a force couple acting at a plane normal to the axis of some structural element. The shear stress at the center of the bar is zero and increases with the radius on the application of the torque. The effect of the externally applied torque is such that the internal shear stress, resisting the twisting of the bar, is equal, but opposite, in sign. The force acting on an elemental surface area is $dF = \tau dA$, which is a vector normal to the radius along the end face dA. An elementary torque is the product of the radius and the force. Thus, an elementary internal torsional moment dT_i is:

$$dT_i = \tau r dA = \tau r(2\pi r dr) = 2\tau\pi r^2 dr \tag{1.71}$$

The total internal torque is obtained by integrating Eq. 1.71 from the center to the full radius which is equal to the applied torque T, resulting in

$$T = \int_0^r dT_i = \int_0^r 2\tau\pi r^2 dr \tag{1.72}$$

Substituting for τ from Eq. 1.70:

$$T = \int_0^r \frac{2G\theta\pi r^3}{L} dr = 2\frac{G\pi\theta}{L}\int_0^r r^3 dr = \frac{G\pi\theta r^4}{2L} \tag{1.73}$$

It is common to express the integral in Eq. 1.73 in terms of the diameter of the shaft D, making the expression:

$$T = \frac{G\theta}{L}\int_0^{\frac{D}{2}} 2\pi r^3 dr \tag{1.73a}$$

The term of the integral associated with the cross-section of the shaft is known as the 'polar moment of inertia' and is designated as J. Thus, rewriting Eq. 1.73a in terms of J, one gets:

$$\frac{T}{J} = \frac{G\theta}{L} \tag{1.74}$$

Note that G can be determined from the torque vs. the angle curve and derived as follows. Take the derivative of Eq. 1.74 to obtain the slope of the curve. Rearrange the resulting expression to obtain:

$$G = \frac{dT}{dQ}\left(\frac{L}{J}\right) \tag{1.74a}$$

This is important, since G (modulus of rigidity) may be directly determined, rather than by calculation from the modulus of elasticity described in Sect. 1.2.9 and given above as:

$$\boldsymbol{G} = \frac{\boldsymbol{E}}{\mathbf{2(1 + \nu)}} \tag{1.68}$$

E is determined by tensile tests.

Using Eq. 1.70 in conjunction with Eq. 1.74, an expression is obtained which is known as the 'torsion equation', namely:

$$\frac{T}{J} = \frac{\tau}{r} \tag{1.75}$$

Shear stress is related to the polar moment of inertia and the applied torque by:

$$\tau = \frac{T}{J} r \tag{1.76}$$

$$\frac{\tau}{r} = \frac{T}{J} = \frac{2\tau}{D} \tag{1.76a}$$

One realizes that J of Eq. 1.75 can be derived from the integral of either Eq. 1.73 or Eq. 1.73a for the torque and, with the use of Eq. 1.76a, a value for the circular section can be written as:

$$J = \frac{\pi D^4}{32} \tag{1.77}$$

The relation between the shear stress and the torque is obtained in terms of the diameter by the substitution of Eq. 1.77 into Eq. 1.76a:

$$\tau = \frac{16T}{\pi D^3} \tag{1.78}$$

Since the shear stress is maximal at the surface of the bar, Eq. 1.78 expresses the maximum shear stress.

In Fig. 1.30b, a hollow cylinder is shown with radius r_0 and inner radius r_i. Thus, the shell thickness is $(r_0 - r_i)$. One can express these radii in terms of diameters as $D_0/2$ and $D_i/2$. Integrating the integral of Eq. 1.73a for J within the integration limits of $D_0/2$ and $D_i/2$, a value of $\frac{\pi}{2 \times 16}(D_0 - D_i)$ is obtained. With this value of J and Eq. 1.76, when r is expressed in terms of the diameter (namely $D_0/2$), for the hollow cylinder one can write:

$$\tau = \frac{16TD_0}{\pi(D_0{}^4 - D_i{}^4)} \tag{1.79}$$

The use of a torsion test enables the determination of the modulus of shear by using Eq. 1.73 with the angle of twist, θ, and by applying a definite torque on a cylindrical bar of known dimensions (diameter and length).

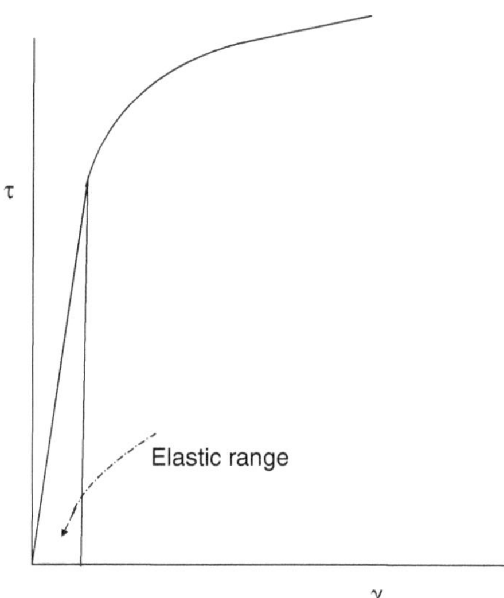

Fig. 1.31 Schematic shear stress/strain relation in the elastic and plastic regions

1.5.2 Torsion in the Plastic Region

Schematically, a torsion test expressing the relation between τ and γ is illustrated in a manner similar to that of a tension (or compression) test (Fig. 1.31). An experimental stress/strain relation in copper may be seen in Fig. 1.32 in terms of effective stress/strain, as well as in terms of shear stress/strain. This Figure compares tension and torsion tests of copper specimens which have undergone the same heat treatment and have the same grain size. Expressing the tension and torsion test results in terms of effective stress allows comparison between the two kinds of tests. The effective stresses and strains were given as $\sigma_e = \sigma$, $\varepsilon_e = \varepsilon$ for tension and $\sigma_e = \sqrt{3}\tau$, $\varepsilon_e = \gamma/\sqrt{3}$ for torsion, respectively.

During tension, a significant work hardening rate is observed, whereas during torsion, an initial increase in the work hardening rate is followed by slight increases at shear strains ranging from 120% to 330%, after which there are fewer additional increases in the work hardening rate than occur in tension. The dislocation structures of the specimens deformed by tension and torsion are presented in Figs. 1.33 and 1.34. In Fig. 1.33, the evolution of the dislocation structure under tension at various strains is presented. The specimens for investigation by means of transmission electron microscopy [henceforth: TEM] were cut transverse to the loading direction. Cell structure was obtained and, at higher values of reduction area (>40%), subgrain formation with sharper cell walls was observed until fracture set in. In Fig. 1.34, the dislocation structures in tension and torsion are compared. The dislocation cell structure during tension testing was obtained for specimens cut in the direction of the applied load, while, for the torsion-tested specimens, it was obtained parallel to the longitudinal axis.

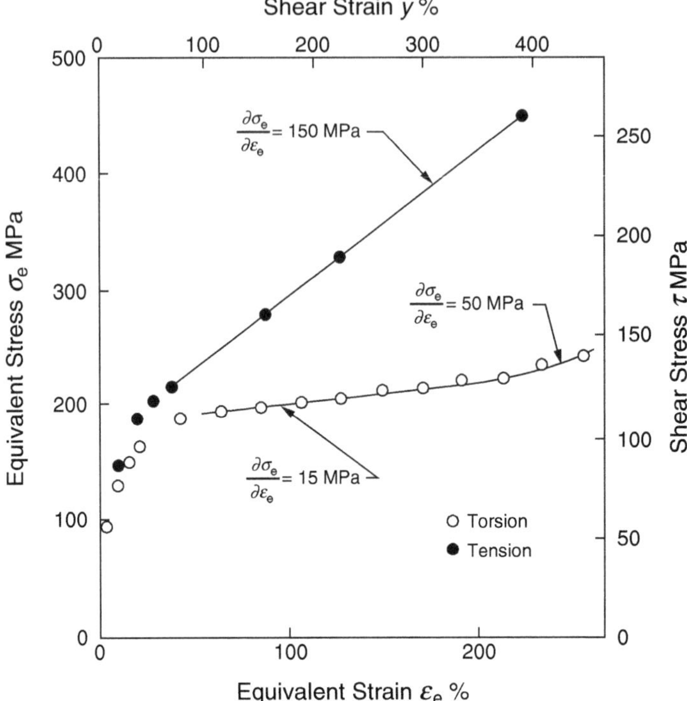

Fig. 1.32 Torsional stress-/strain curves for copper expressed as effective stress and effective strain, as well as shear stress and shear strain (Nabil Bassim and Liu1993. With the kind permission of Elsevier and Prof. Bassim)

It was observed that the work hardening rate in tension is much higher than in torsion. The dislocation structure in tension is characterized by sharp subgrain boundaries, while, in torsion, laminar dislocation structures are obtained, consisting of paired sheets evolved together with other paired sheets to form checker-board patterns. The resistance to the bowing out of free dislocation segments explains the difference in work hardening.

The relation between τ and γ, Eq. 1.69, is not linear beyond the elastic region and, on continued twisting of a cylindrical specimen, may be given for unit specimen length in terms of the radius:

$$\gamma = \mathrm{r}\acute{\theta} \tag{1.69b}$$

Thus,

$$dr = \frac{d\gamma}{\acute{\theta}} \tag{1.80}$$

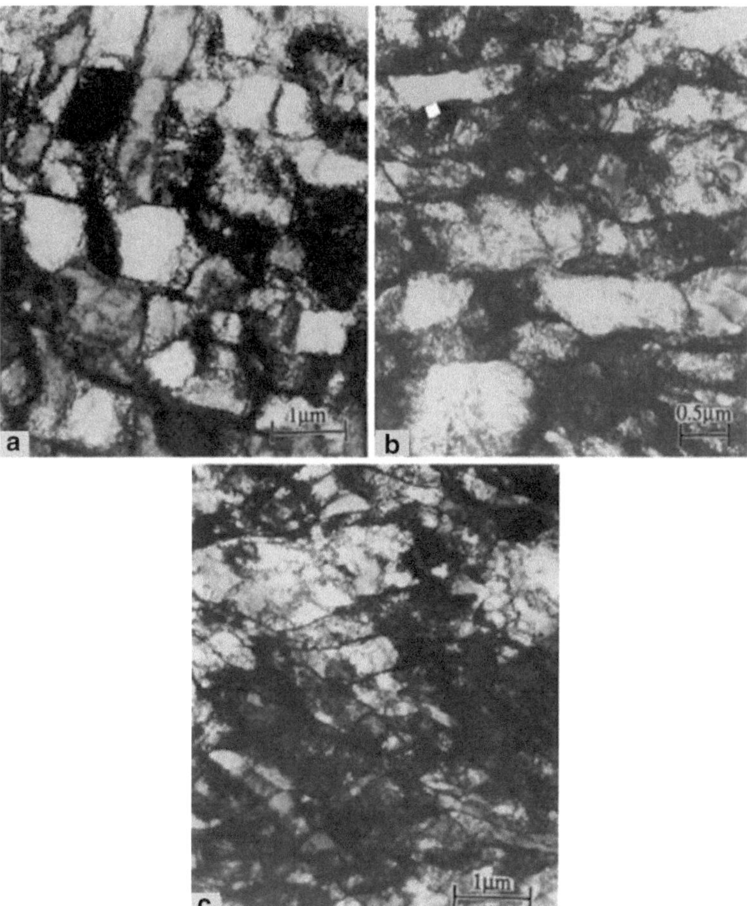

Fig. 1.33 Evolution of dislocation structure in tension at strains (**a**) 35%, (**b**) 96% and (**c**) 137% (Nabil Bassim and Liu 1993. With the kind permission of Elsevier and Prof. Bassim)

also from Eq. 1.69b:

$$r^2 = \frac{\gamma^2}{\acute{\theta}^2} \tag{1.69c}$$

For the moment (torque), T, Eq. 1.72 can be expressed by using the shear strain from Eq. 1.69a through Eqs. 1.69b and 1.69c as:

$$T = \int_0^{\gamma} 2\pi\tau \frac{\gamma^2}{\acute{\theta}^2} \frac{1}{\acute{\theta}} d\gamma = 2\pi \int_0^{\gamma} \tau \frac{\gamma^2}{\acute{\theta}^3} d\gamma \tag{1.81}$$

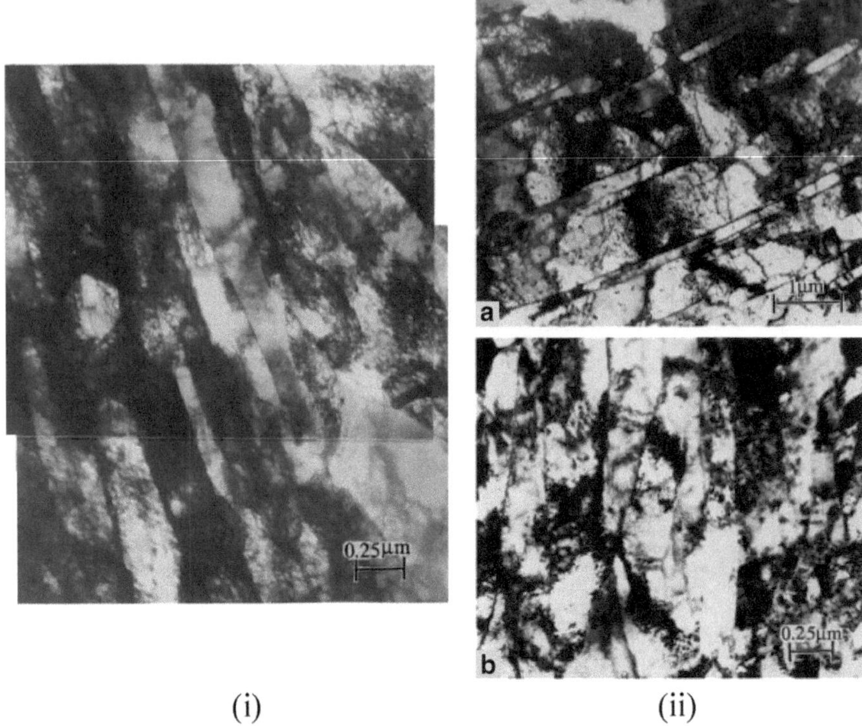

Fig. 1.34 Dislocation structure in copper (i) in tension at strain equal to 217%, (ii) in torsion at shear (*a*) 120% and (*b*) 440% (Nabil Bassim and Liu 1993. With the kind permission of Elsevier and Prof. Bassim)

From Eq. 1.69b one can write:

$$\frac{d\gamma}{d\acute{\theta}} = r \tag{1.82}$$

Now, one can express Eq. 1.81 in terms of the angle $\acute{\theta}$ by using Eq. 1.82 as:

$$T\acute{\theta}^3 = 2\pi \int_0^{\acute{\theta}} \tau r^2 \acute{\theta}^2 r d\acute{\theta} = 2\pi \int_0^{\acute{\theta}} \tau r^3 \acute{\theta}^2 d\acute{\theta} \tag{1.83}$$

Differentiating Eq. 1.83 with respect to $\acute{\theta}$, one can write:

$$\frac{d(T\acute{\theta}^3)}{d\acute{\theta}} = 2\pi\tau r^3 \acute{\theta}^2 \tag{1.84}$$

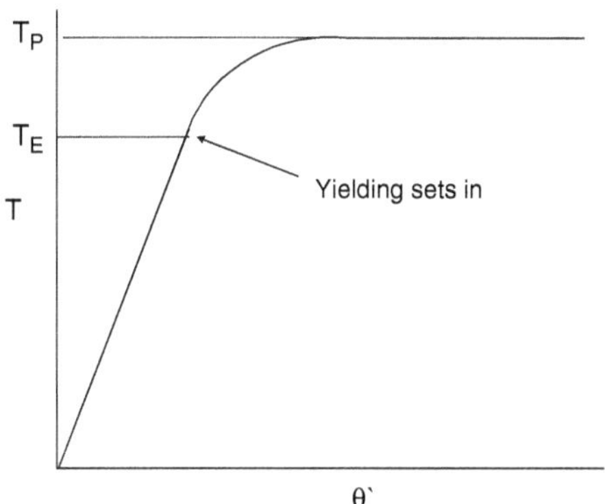

Fig. 1.35 A schematic illustration of torque vs. angle

$$\tau = \frac{d(T\acute{\theta}^3)}{d\acute{\theta}} \frac{1}{2\pi r^3 \acute{\theta}^2} \tag{1.85}$$

But a term from Eq. 1.85 can be written after the differentiation as:

$$\frac{d(T\acute{\theta}^3)}{d\acute{\theta}} \frac{1}{\acute{\theta}^2} = \acute{\theta} \frac{dT}{d\acute{\theta}} + 3T \tag{1.86}$$

Thus, based on Eqs. 1.85 and 1.86, one can give shear stress as:

$$\tau = \frac{1}{2\pi r^3} \left(3T + \acute{\theta} \frac{dT}{d\acute{\theta}} \right) \tag{1.87}$$

Shear stress is proportional to torque and the angle of torsion, as determined experimentally from the torque vs. the twist angle $\acute{\theta}$. The term $dT/d\acute{\theta}$ is graphically illustrated in Fig. 1.35.

As seen in the above illustration, the shear stress at various angles can be determined from the slope at some point providing $dT/d\acute{\theta}$ for evaluating Eq. 1.87 for a bar of known length and radius. Note that at the maximum torque (i.e., T_P) $dT/d\acute{\theta} = 0$ and, therefore, according to Eq. 1.87:

$$\tau_{\max} = \frac{3T}{2\pi r^3} \tag{1.87a}$$

Recall that the shear stress is maximal at the surface of the solid cylinder.

Figure 1.36 compares calculated and experimental torque with twist angle. Normalized torque vs. angle of twist for a hollow bar is shown. This Figure compares the values obtained by a numerical method with experimental results for a hollow bar made of an elastic, perfectly plastic material (Herakovich). T_E is the maximum elastic torque and θ_E is the maximum elastic twist angle.

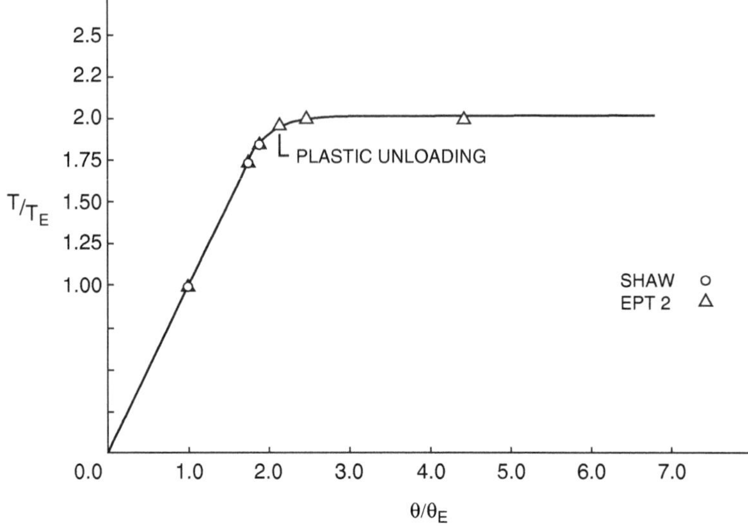

Fig. 1.36 Normalized torque vs. normalized angle of twist for a hollow tube. EPT means that a computer program EPT2 was applied in the evaluation of the curve (Herakovich and Hodge 1969. With the kind permission of Elsevier and Prof. Herakovich)

A graphic evaluation of Eq. 1.87 appears in many works (e.g., Bressan and Unfer). Figure 1.37 is a schematic illustration of this evaluation. In the Figure, the slope of the line, passing at point B, is shown. For small angles (the tangent of angle ≅ angle), one may write:

$$BD = CD\frac{dT}{d\acute{\theta}} = \acute{\theta}\frac{dT}{d\acute{\theta}} \tag{1.88}$$

At point B, AB = T. Inserting these values into Eq. 1.87, one gets:

$$\tau = \frac{1}{2\pi r^3}(3AB + \mathrm{B}D) \tag{1.89}$$

All the quantities are measurable and, for a given specimen, by applying a known torque at a twist angle θ, the shear stress can be determined.

The torsion/twist angle relation for work hardening material is assumed to be represented by a power law relation of the form:

$$T = T_0\acute{\theta}^n \tag{1.90}$$

This relation is similar to the tension/strain relation $\sigma = Ke^n$ given by Eq. 1.30 in Sect. 1.2.4. T_0 is a constant and n has the same meaning as in tension, i.e., the work hardening exponent. Differentiating Eq. 1.90 results in:

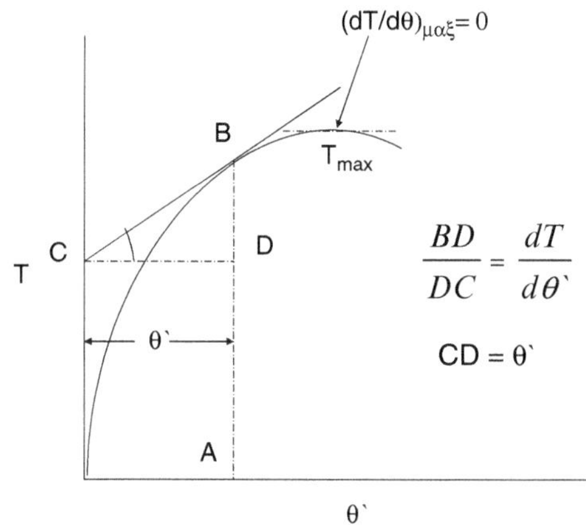

Fig. 1.37 Graphic illustration of Eq. 1.87 for the shear stress determination

$$\frac{dT}{d\acute{\theta}} = nT_0\acute{\theta}^{n-1} \tag{1.91}$$

Substituting for T and dT/dθ into Eq. 1.87, an expression for the shear stress is obtained in terms of the strain hardening exponent:

$$\tau = \frac{T}{2\pi r^3}(3+n) \tag{1.92}$$

Experiments indicate that the work hardening index, n, determined from torsion tests is not of the same magnitude as the one determined from tension tests (Bressan and Unfer).

In a derivation similar to that of Eq. 1.87, the stress acting on a tube (i.e., a hollow cylinder) may be obtained. Using the relation in Eq. 1.72 within the limits of r_0 and r_i (as shown in Fig. 1.30b), we can write:

$$T = \int_{r_i}^{r_0} 2\pi\tau(r)r^2 dr \tag{1.93}$$

Here, $\tau(r)$ is the shear stress at radius r; r_0 and r_i are the external and internal radii of the hollow cylinder. Assuming that a material is perfectly plastic beyond the elastic limit and that shear stress is uniform in the wall of the cylinder, the integration of Eq. 1.93 yields:

$$\mathrm{T} = \frac{2\pi\tau(\mathrm{r})}{3}\left[\mathrm{r_0}^3 - \mathrm{r_i}^3\right] \tag{1.94}$$

Which can then be written as:

$$3T = 2\pi(\tau_0 r_0{}^3 - \tau_i r_i{}^3) \tag{1.95}$$

In the multiplication with the outer and inner radii in Eq. 1.95, the appropriate signs were given for τ (r). Then, dividing this expression by $r_0{}^3$, we can write:

$$\frac{3T}{r_0{}^3} = 2\pi\left[\tau_0 - \tau_i\left(\frac{r_i}{r_0}\right)^3\right] \tag{1.96}$$

$$\tau_0 = \frac{3T}{2\pi r_0{}^3} + \left(\frac{r_i}{r_0}\right)^3 \tau_i \tag{1.97}$$

Eq. 1.97 for τ_0 represents the shear stress on the outer surface of the cylinder. For a solid cylinder, where $r_i = 0$, this equation is equivalent to Eq. 1.87a.

1.5.3 Axial Change in Torsion

Axial change in specimens can occur when large plastic strains are involved during torsion. Experiments applying torsion to various sized specimens, both free-end and fixed-end, were recorded by Wu, et al. in extruded and cast high-purity Al. A specially designed extensometer measured the strain in solid shafts or tubular specimens. Axial extension increases with shear strain in torsion of the free-end specimens. In the fixed-end specimens, axial change was affected by temperature, but almost none of the free-end specimens were affected within the investigated temperature range. It was claimed by Wu, et al. that, in the free-end torsion test of a solid shaft, the axial change is smaller than it is in tubular specimens; therefore, solid shafts are not useful for the study of the axial effect and the evaluation of its magnitude. The stated reason for this was that the axial elongation, being a function of shear strain, varies linearly along the radius of the solid shaft. Thus, the inner core of the shaft has less axial strain than the outer layer, which acts to restrict the axial elongation of the whole shaft, resulting in residual axial stresses in the specimen.

Figure 1.38 shows the axial change of various cast and extruded specimens as a function of shear strain. The torque vs. twist-angle curves were converted into axial strain vs. shear strain curves. As seen in Fig. 1.38, the axial extension increases with shear strain during the free-end torsion of extruded and cast aluminum. The size of the specimen is important. Also, the production method seems to have an influence on axial length vs. shear stress. Unlike the case of the cast specimens in the extruded aluminum, axial extension starts to increase at the early stage of the torsion test and is barely affected by temperature within the test range, as seen in Fig. 1.39.

Axial changes in specimen length due to the application of torsion have been observed in various materials. This phenomenon, the aforementioned 'axial effect', is of interest not only for research purposes but for practical reasons as well. It is a significant fact that, during torsional deformation in the large strain range, axial changes occur, the magnitude of which depend on several factors, such as

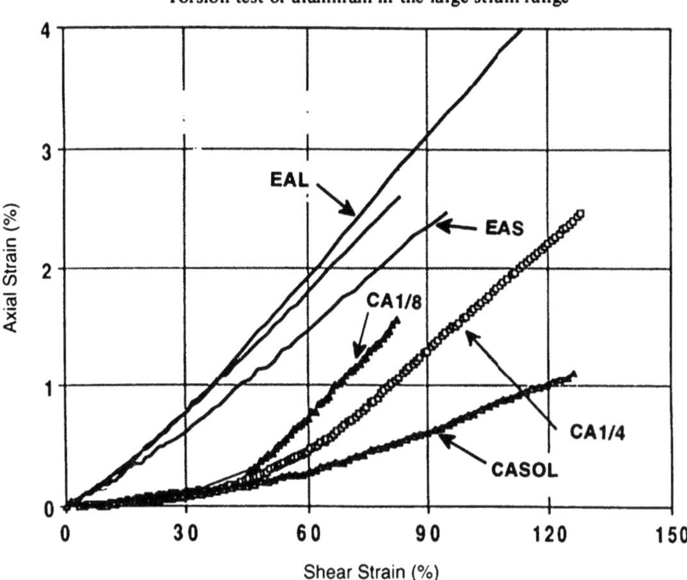

Fig. 1.38 Axial strain in free-end torsion for cast and extruded aluminum. EAL-extruded long specimen, EAS-extruded short specimen, CA1/8 and CA1/4 tubular specimens with 1/8 and 1/4 in. wall thicknesses respectively and CASOL-cast solid specimen. Wall thickness of the respective specimens is 1/4th of an inch (Wu et al. 1998. With the kind permission of Elsevier and Prof. Wu)

homogeneity, temperature, the type of loading (fixed-end or free-end), the geometry and the dimensions. Thus, deformation is not entirely the consequence of pure shear strain.

1.5.4 Fracture by Torsion Test

The inhomogeneous nature of plastic deformation under large strains affects the type of failure that occurs. In general, failure due to torsion is different from failure occurring due to tension. Thus, a ductile material fails in shear along some plane, where the shear stress is maximal. Further discussion on failure (fracture) during torsion is below, in Chap. 7, which deals with fracture in general.

In summary, one can quote the statement in Kocks and Stout: "Torsion tests are useful for two reasons. First, they can provide data to very large strains without changes in the specimen geometry and without any effects of friction at the grips. Second, these data are in plane-strain mode, which has occasionally been observed to exhibit systematic differences in behavior with respect to uniaxial tests." In sheet rolling, for example, plane strains are involved during deformation.

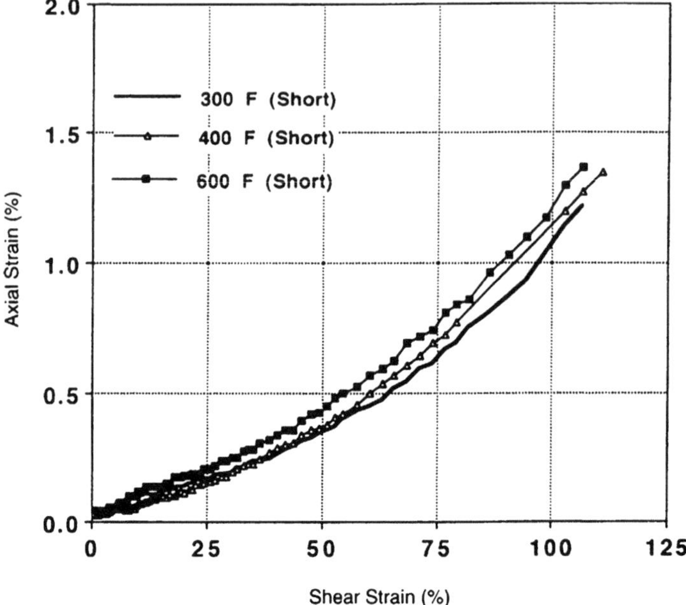

Fig. 1.39 Axial strain in free-end torsion for cast aluminum at the temperatures indicated (Wu et al.1998. With the kind permission of Elsevier and Prof. Wu)

1.6 The Impact Tests

Impact testing is a very popular and fast method for evaluating the fracture toughness of materials. One of the purposes of this method is to evaluate the energy absorbed by a standard specimen during tests at a very high strain rates. Depending on the type of specimen and the way it is held in the impact machine, two types of impact tests are known – Charpy and Izod, respectively. The former is common in the U.S.A., while the latter is mainly used in the U.K. and Europe. Although the Charpy Impact Test is known also as the 'Charpy V-Notch Test', because the specimen is usually notched, sometimes un-notched specimens are used to bypass the need for machining. Impact tests are performed as a function of temperature, to determine the ductile-to-brittle transition, which is temperature dependent. The sinking of a Liberty cargo ship during WWII accelerated the study of ductile/brittle behavior and this research turned up several parameters, some relevant to the remaining Liberty ships. Thus, the three prerequisites for ductile/brittle transition are:

(a) a low temperature
(b) a triaxial state of stress
(c) a high strain rate.

Conditions (a) and (b) existed in the Liberty ships and any of them could have caused failure in their hull construction, some of which broke in half without warning. In the North Atlantic Sea, they were exposed to low temperatures and most likely the catastrophic failure was the consequence of being in the ductile-to-brittle transition temperature range or even below it. Weld zones existed in those ships, zones known to act as stress concentrations, where triaxial states of stress commonly exist, augmenting the reduction of the resistance of the material and weakening or even damaging its fracture toughness. This is a classic example of the critical importance of exposing specimens to serious impact tests. At that time, high strain rates were not known to be present in those ships, but to get sufficient warning regarding the performance of structural materials – high strain rate is crucial in impact testing.

The effect of the notch is to embrittle the material and concentrate stresses at the root of the notch or in its vicinity. It also raises the elastic limit of the material in the notch area. When a crack forms at the root of the notch, the increased stress induces quick crack propagation. The capacity to absorb energy cannot be detected without a notch and many test specimens would simply bend without fracturing.

Figure 1.40 shows an impact tester used at the Ben-Gurion University of the Negev in Beer-Sheva, Israel. It is a Mohr & Federhaff AG (Manheim, Germany) impact testing machine with interchangeable pendulum and hammer. The striking energy is either 1.5 or 5 kgm, depending on the pendulum used. The dimensions of Charpy V-notch specimens, according to ASTM standards, are $55 \times 10 \times 10$ mm, having a notch machined 2 mm deep, with a 45° angle and a 0.25 mm radius along the base. The notch is midway on the specimen. A keyhole-shaped specimen is another variation of the test specimen in appearance and size. A keyhole notch (or 'U-notch') is 5 mm deep with a 1 mm radius at the base of the notch. The Charpy Test is performed by striking a suitable specimen with a hammer on a pendulum arm, while the specimen is safely held in a horizontal position and the blow occurs on the opposite side of the notch (behind the notch). Since the specimen is supported by its two edges, as a simple beam, the test may be considered to be a dynamic beam-breaking test. The energy absorbed by the specimen is recorded. The impact energy is calculated, based on the height to which the pendulum has risen after striking the specimen, as seen on the scale. Clearly, the actual test is performed after calibrating the machine for air friction, etc., without an engaged specimen. In practice, the pendulum is raised to a certain height, released and swings to another height on the scale, which is considered as the zero point of the impact machine. When the actual test is run, the difference in the heights is noted; i.e., the difference between the predetermined zero position of the machine and the height reached by the pendulum after fracturing the specimen. The impact machine measures the energy absorbed while breaking the specimen. Tough materials absorb much energy, while brittle materials tend to absorb very little energy prior to fracture.

The ductile-to-brittle transition is one of the major tests using the impact tester. This test is performed by applying the Charpy Test to identical specimens at different temperatures and then plotting the impact energy as a function of temperature to evaluate the ductile/brittle transition. This provides crucial information regarding

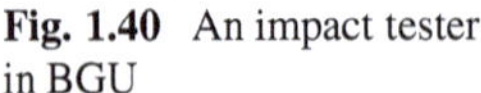

Fig. 1.40 An impact tester in BGU

the service temperatures at which it is safe to use certain materials. Impact resistance appears to decrease markedly with temperature. Since no thermocouple is used during the impact blow, and the temperature has to be recorded for plotting the results, it is customary to determine a few temperatures by immersing the test specimen in liquids of known temperatures or in those whose temperatures can be measured. The duration of immersion of a test sample for temperature equalization is determined, after which the specimen is quickly inserted in the holder of the impact tester and the test is performed. Commonly used temperatures are: water at room temperature, boiling water ($\sim$100°C), acetone with dry ice ($\sim$55°C), ice in water ($\sim$0°C), brine ($\sim$20°C), etc. An idealized plot is shown schematically in Fig. 1.41. In this Figure, known as the 'energy-temperature curve', the transition is recorded by energy absorption. This criterion for determining when a material becomes brittle is known as the 'energy transition mode'.

There are three ways to record the transition temperature: (a) average energy transition, which indicates the transition temperature in the middle, between the upper and lower shelves, as shown in the Figure. Note (as indicated in the figure) that, below the transition temperature, the failure is brittle, whereas, above it, the failure is ductile with relatively high energy absorbed; (b) the temperature at which

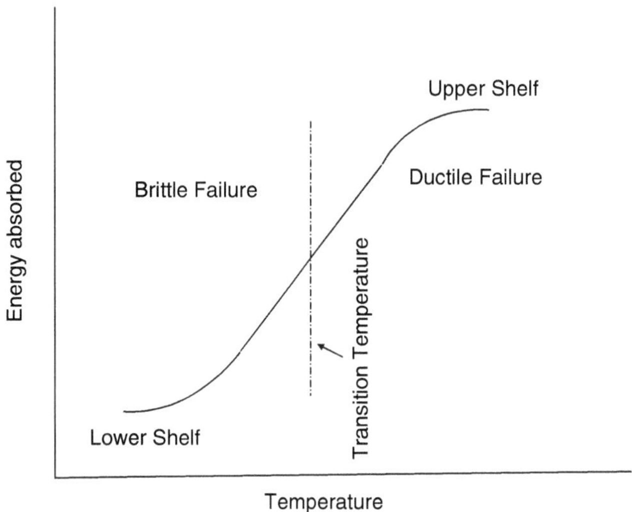

Fig. 1.41 Ideal transition temperature by energy absorbed mode

the impact energy is 40 ft-lb. In this method, the 40 ft-lb energy intersection with the curve provides the transition temperature; and (c) the 15 ft-lb impact energy level arbitrarily determined at a temperature of −40°F. This test confirms the quality of materials. The transition temperatures recorded by (a) and (b) are relatively reasonable criteria for accepting materials

In reality, materials may show fracture surfaces which are neither completely brittle nor completely ductile. By considering the fracture surface, one can determine the transition temperature by the fracture, i.e., the fracture transition mode (also shown in Fig. 1.42). Using this method, the appearance of the fresh fractured surface is examined, usually visually, to determine: (a) the lowest temperature at which the fracture is 100% fibrous; (b) the highest temperature at which the fracture shows 100% cleavage, becomes granular; and (c) the temperature at which the fractured surface is 50% granular and 50% fibrous. Thus, between temperatures (a) and (b), a transition-temperature range exists, in which the character of the fracture may be mixed**.** The transition temperature, when the fracture surface is 100% fibrous is often called 'fracture transition plastic' [henceforth: FTP]. This criterion actually corresponds to the upper shelf transition, shown in Fig. 1.42, and is quite a liberal definition of the transition temperature, since, above this temperature, materials are ductile. A severe, but well-defined, transition temperature corresponds to the lower shelf, indicated in Fig. 1.42. Such a transition is also known as the 'nil ductility transition temperature'. At this temperature, some steel breaks in a completely brittle manner, without any plastic deformation. Fractured surfaces, when inspected visually, also show lateral changes in the vicinity of the notch. A lack of contraction is an indication of a brittle fracture, while a ductile fracture may show ~1% lateral contraction. These different measurements of the same materials do not usually give the same transition temperatures.

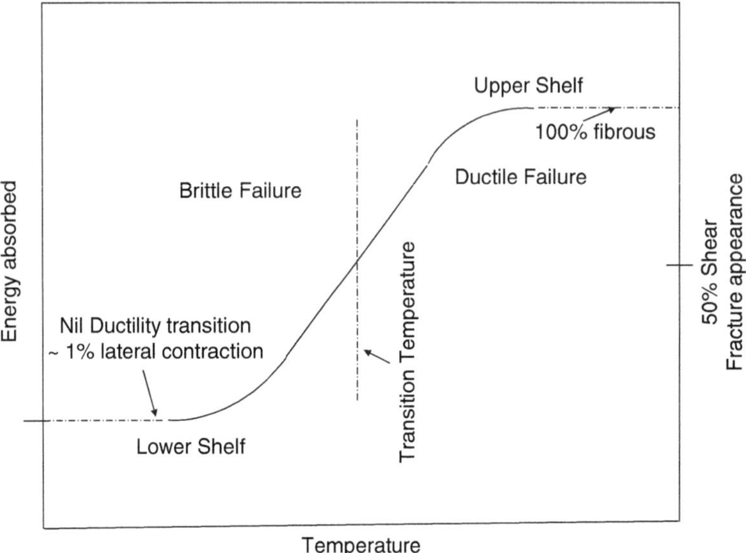

Fig. 1.42 Fracture transition temperature

In Izod impact testing, the specimen is clamped into the impact machine fixture at one end and is free at the other end. The notch is just at the edge of the supporting fixture and facing the direction of the striking pendulum. The length of the specimen is 75 mm and the notch is machined at 25 mm from the edge, but other details of the specimen size are the same as in the Charpy Test. In this test, the short side of the specimen, just below the notch, is clamped into position. Thus, the Izod Impact Test differs from the Charpy Impact Test in that the sample is held vertically, i.e., in a cantilevered beam configuration, as opposed to a three-point bending configuration.

The importance of impact tests is that they provide a critical measure of service life, showing the resistance of structural parts. This test also provides the designer with quick and easy quality control – with a means to check whether materials actually have the specific properties required to meet the service standards. An impact test may be used to compare materials in regard to their general toughness, but a major disadvantage is that all results are merely comparative and, thus, qualitative. These impact tests have many applications for testing construction materials to be used in making pressure vessels, bridges, buildings, ships, etc. Even if specimens are standardized, the test results of Izod and Charpy impacts are not directly comparable. The test results recorded are typically the averages of tests performed on five specimens. Clearly, impact testing involves the kinetic energy needed to initiate fracture and continue the fracture until the specimen breaks. However, the major disadvantage of impact testing is that it is impossible to differentiate between the energy consumed in crack initiation and the energy directly absorbed during crack propagation. Therefore, no direct indication of fracture toughness level K_f can be obtained. Brittle materials take little energy to start a crack, while ductile ones require high energy to initiate and propagate a crack. Many materials are capable of either ductile or brittle failure, depending upon the type of test, rate and temperature conditions. The ductile-to-brittle transition actually shifts according to these variables. ASTM dictates the standards for the test specimens and the way to perform the notched-bar impact test. SI units are commonly used as the standard to indicate the test results and, in many publications, ft-lb units are common. Often, impact strength is calculated by dividing impact energy by the thickness of the specimen, possibly given in J/m or ft-lb/in.

Instrumented impact testing became widely used during the past decade or so. In such tests, instrumented impact is continuously recorded with the load on the specimen as a function of time prior to fracture.

One should remember that impact energy is the sum of the overall factors involved in the fracture of a specimen and is, therefore, composed of three elements: (a) elastic strain energy; (b) plastic work following yielding and prior to fracture; and (c) work creating fractured surfaces, which can be identified with some surface energy. In brittle materials, factors (a) and (c) dominate, whereas, in ductile materials, the energy of impact fracture is dominated by the plastic work. It should also be recalled that the standardization of specimens is common practice, to allow for a comparison between different materials, since factors, such as specimen size and notch characteristics (depth and curvature), determine the results. Due to the importance of the notch characteristics, impact tests also indicate the notch sensitivity of the materials. Note that today similar tests are also applied to polymers, ceramics and composites.

In summary, tough materials absorb much of the energy when fractured, while brittle materials absorb very little energy. The impact test is a dynamic test in which a notched-test specimen is broken by a single blow in the impact machine and the energy absorbed, while breaking the specimen, is measured. Impact tests are mechanical tests used to determine the properties of materials and their performances – factors which are useful in engineering design and structural analysis. It is a **qualitative test** predominantly for comparison.

1.7 Anelasticity

1.7.1 Introduction

Contrary to the assumption made in the previous sections discussing strain (displacement) only as a function of stress, many observations of the behaviors of materials under loads have shown that this is not strictly true, since a time factor must also be taken into consideration. The behavior of most metals (as discussed above in Sect. 1.2) shows small time dependence of strain under load, while other materials, especially polymers, may show significant time dependence. In the plastic region, a similar phenomenon, creep, also represents time-dependent change under

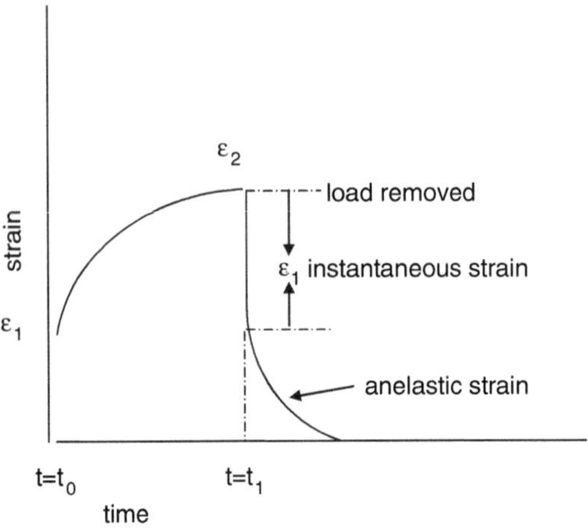

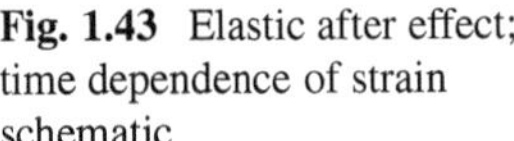
Fig. 1.43 Elastic after effect; time dependence of strain schematic

strain, which poses a great problem for engineers and machine-element designers. In metals, the effect of anelasticity is very small and is generally neglected.

1.7.2 *The Elastic After Effect*

An ‘elastic after effect’ is often observed by design engineers, especially in polymers (plastics), which are widely used in modern technology. In essence, if such a material is loaded resulting in a strain, ε_1 at a time t_0, at some time t_1, an additional strain will be produced as ε_2, which is a delayed elastic strain and occurs over a period of time without further increase in load. Over time, the rate of the elastic strain ε_2 diminishes and no further elastic deformation occurs. At this point, a sudden unloading of the test specimen results in the return of the strain to its previous value, ε_1, as shown in Fig. 1.43; this is then followed by a further slow decrease in strain to zero. The test specimen is restored asymptotically to its original dimension. Since thermal effects are involved during elastic deformation, depending on the strain rate, the temperature may be a significant parameter to be taken into account.

A material can be deformed so rapidly that its temperature cannot equalize with the environmental temperature, as in an adiabatic process with constant entropy and a decrease in temperature. In uniaxial tensile straining, the relation expressing such an adiabatic process is given by:

$$\left(\frac{\Delta T}{\Delta \varepsilon}\right)_S = -\frac{\alpha V_m E T}{c_v} \tag{1.98}$$

where V_m, α, E, T and c_v are the molar volume, the coefficient of linear expansion, the isothermal Young’s Modulus, the absolute temperature and the specific heat

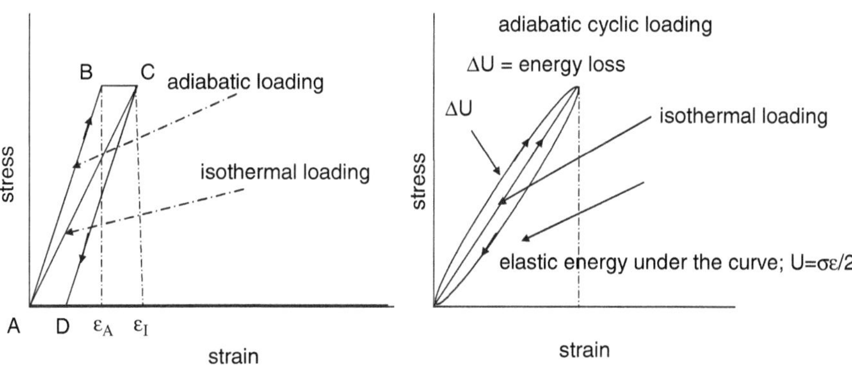

Fig. 1.44 (**a**) Adiabatic and isothermal loading; (**b**) hysteresis loop

at constant volume, respectively. The above equation can easily be derived from thermodynamic principles relating them to stress/strain. For uniaxial compressive strain, however, there is an increase in temperature and, in the above relation, a change in sign has to be added. This thermal effect in adiabatic straining is known as the 'thermoelastic effect'.

In most materials, α is positive, however, for rubber, it is negative; therefore, when rubber-like is being studied, the sign in Eq. 1.98 should be changed. Appendix I shows how the above relation may be obtained.

1.7.3 The Thermoelastic Effect

In metals, the thermoelastic effect is usually small, but, in polymeric materials, it can be of appreciable magnitude. During the continued loading and unloading of a material, a hysteresis loop may develop. This represents an energy loss and, if the loading and unloading process is performed cyclically, the loss of energy due to vibrations in a solid is known as 'damping'. In Fig. 1.44, this process is also shown in slow motion, illustrating the steps producing the hysteresis loop.

This Figure also shows that when the specimen is loaded at a slow rate, i.e., isothermally along the line AC, sufficient time is available for heat dissipation and, thus, the resulting strain is ε_I. Unloading will follow the same line from point C to zero (point A). When, however, straining is done rapidly in the elastic region (almost in an adiabatic manner) from A to point B, the strain obtained is represented by ε_A and the process is associated with a small temperature decrease. The specimen slowly heats up along the line BC and achieves a state of relaxation due to increasing strain. At this time, if the specimen is again unloaded rapidly along the line, CD, an adiabatic-like process occurs accompanied by heating. After some time has elapsed, thermal equilibration is reached with the surroundings and the specimen cools down, contracting to its original size along the line, DA. The closed loop, ABCD, is a

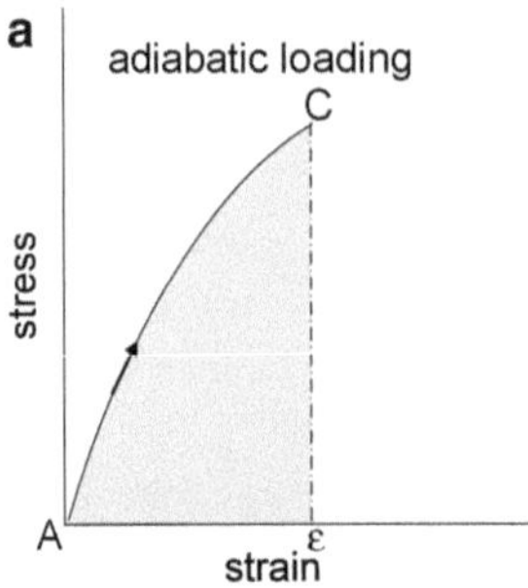

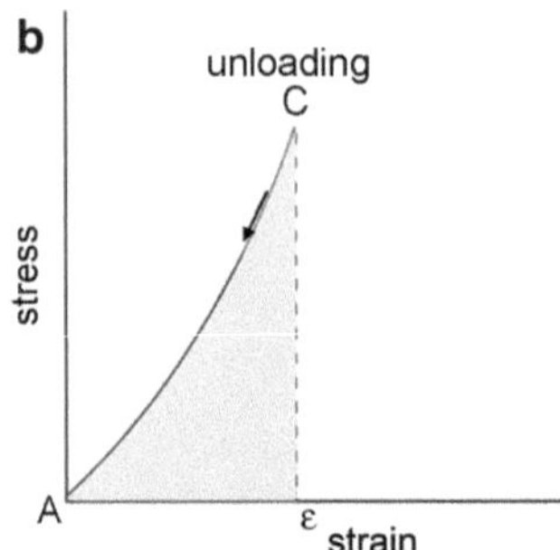

Fig. 1.45 Adiabatic loading and unloading

hysteresis loop and, when the loading and unloading is continuous, an elliptical-shaped loop results from the cyclic loading. Note that the strain difference between an isothermal and adiabatic loading is $(\varepsilon_I - \varepsilon_A)$.

The hysteresis loop in metals is small and it is a function of the frequency of loading and unloading. With continuous rapid vibrations (namely, loading and unloading), due to the energy loss represented by the loop, this phenomenon can be used for damping. As is clearly shown in the above Figure, the area under the curve represents the elastic energy, $U = \sigma\varepsilon/2$; thus, the loop represents a loss/damping of this energy. The separate curves, denoting the loading and unloading in Fig. 1.45 below, indicate that the closed loop is, indeed, an energy loss. These curves show that the areas under the lines represent work and, thus, the difference between them (the loop) is basically an elastic energy loss.

The work done can be expressed mathematically in terms of $\sigma - \varepsilon$ and, due to the phase lag in strain, its time dependence may be expressed as:

$$\sigma = \sigma_m \sin \omega t \tag{1.99}$$

$$\varepsilon = \varepsilon_m \sin(\omega t - \phi) \tag{1.100}$$

$$\frac{d\varepsilon}{d(\omega t)} = \varepsilon_m \cos(\omega t - \phi) \tag{1.101}$$

The work of the closed loop is:

$$\int \sigma d\varepsilon = \int_0^{2\pi} \sigma_m \varepsilon_m \sin \omega t \cos(\omega t - \phi) d(\omega t) = \pi \sigma_m \varepsilon_m \sin \phi \tag{1.102}$$

This is the work lost per cycle of the elliptical hysteresis loop, often referred to as 'mechanical hysteresis'. σ_m and ε_m are the maximal amplitudes of stress and strain (see the loop in Fig. 1.44b), ω and t are frequency and time, respectively. In this case, the phase angle, ϕ (representing the lag of strain behind stress), is independent of strain. The above may also be written as:

$$\frac{\pi\sigma_m^2}{E}\sin\phi = E\varepsilon_m^2\pi\sin\phi \tag{1.103}$$

If the hysteresis loop is narrow, Eq. 1.102 can be approximated by $1/2\sigma_m\varepsilon_m$. In Appendix II, equations for harmonic oscillations are given, which serve as the bases for the expressions found above and below.

1.7.4 Energy Losses/Hysteresis Loop

The energy loss in the elastic region indicated above, represented by the hysteresis loop, is only one aspect of the known phenomena associated with energy loss. Dissipation of energy can be induced by grain-boundary motion, dislocation motion, solute atom oscillations, such as C or N in steel or other BCC metals, stress-induced ordering, etc. Generally, the term ‘internal friction’ is used to describe these phenomena and several methods are discussed in the literature for measuring this effect that can provide important structural information.

1.7.5 Internal Friction

Internal friction is most generally defined as ‘the dissipation of mechanical energy inside a gaseous, liquid or solid medium’. In solids, it usually means energy loss connected with deviations from Hooke’s Law, manifested by some stress/strain hysteresis under cyclic loading and its conversion into heat. Figure 1.44 shows an energy-loss loop; proof that this loop represents energy loss may be seen in Fig. 1.45, where the two branches of loading and unloading and the areas under their curves are shown. Clearly, the difference between both curves indicates the energy loss, shown as the loop in Fig. 1.44b. The size of such a loop depends on the frequency of the loading and unloading in a cyclic deformation. In Sect. 1.7.4, other phenomena were mentioned in which energy loss was observed, but there are also some additional, external phenomena that should be mentioned, such as magnetic hysteresis. Energy loss is described in various ways, depending on the research objective and the experimental measuring method. In a single stress/strain cycle, the relative energy scattered can be determined from the difference in areas shown in Figs. 1.44 and 1.45, where ΔU is the energy loss and U corresponds to the total vibrational energy per cycle at the strain indicated. As stated above, the change in external work is equal to the change in strain energy or ΔW = ΔU. Damping of the vibration is expressed by:

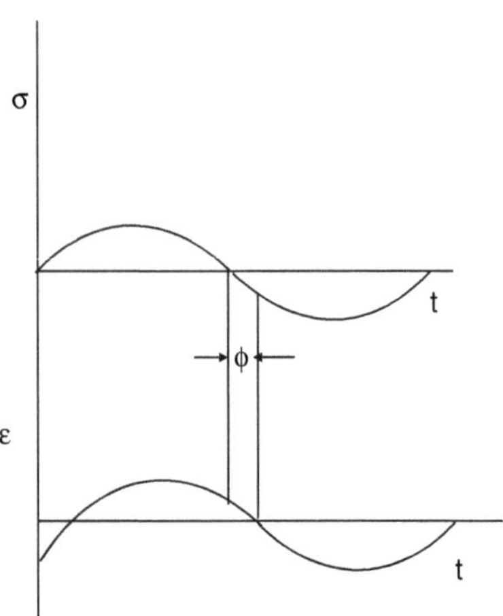

Fig. 1.46 ϕ indicates the phase lag between stress and strain

$$\boldsymbol{D} = \frac{\Delta \boldsymbol{W}}{\boldsymbol{W}} = \frac{\Delta \boldsymbol{U}}{\boldsymbol{U}} \tag{1.104}$$

where D is called the ‘specific damping capacity’.

The stress and strain are not in phase and the strain lags behind the stress. Angle ϕ indicates this lag between stress and strain (Fig. 1.46).The fractional energy loss from the above relation (Eq. 1.104) may also be expressed as:

$$\boldsymbol{D} = \frac{\pi \sigma_m \varepsilon_m \sin \phi}{\frac{1}{2}\sigma_m \varepsilon_m} = 2\pi \sin \phi \tag{1.105}$$

For small ϕ, sinϕ ≅ ϕ.

The area of the hysteresis loop is a function of the frequency of the loading and unloading. At very small frequencies, the cycle may be almost isothermal; the hysteresis loop is very small. If it is high, the loading and unloading paths are almost entirely adiabatic and, again, the hysteresis loop is very small. At some intermediate frequency, the hysteresis loop is at maximum. A schematic illustration of the influence of the frequency, f (angular frequency = ω = 2πf), on the energy loss is seen in Fig. 1.47.

In Fig. 1.44, the stages of energy loss during loading and unloading are indicated and the slopes of the pure adiabatic loading and the isothermal straining are different. The stress/strain ratio is steeper in adiabatic loading. Note that the slopes of the σ-ε, indicated in Fig. 1.47, are not the same, due to the effect of the frequencies. Thus, it is expected that the elastic moduli will also be different.

There is another common expression for damping, described by the use of two adjacent amplitudes of vibration. Since the square of the amplitude is proportional to the vibrational energy, D can be written as:

$$D = \frac{\Delta W}{W} = \frac{\Delta U}{U} = \frac{A_n^2 - A_{n+1}^2}{A_n^2} \tag{1.106}$$

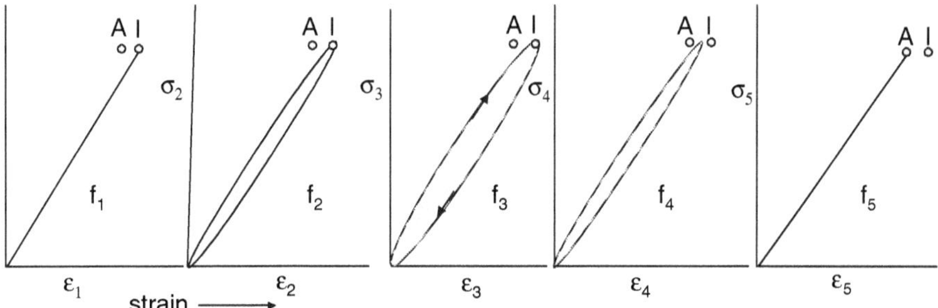

Fig. 1.47 The effect of frequency on the hysteresis loop. *A* and *I* represent the locations of strain for adiabatic and isothermal straining. Note that, at intermediate frequency f_3, the hysteresis loop is the largest

Referring to the first and the next amplitudes as A_1 and A_2 and re-expressing Eq. 1.106, it is possible to rewrite it as:

$$D = \frac{\Delta W}{W} = \frac{\Delta U}{U} = \frac{2\Delta A}{A} \tag{1.107}$$

since:

$$A_2^2 - A_1^2 = (A_1 - A_2)(A_1 + A_2) \tag{1.108}$$

and from Eqs. 1.106 and 1.108 as:

$$\frac{\Delta U}{U} = \frac{\Delta W}{W} = D = \frac{(A_1 - A_2)(A_1 + A_2)}{A_1^2} \tag{1.109}$$

The difference between A_1 and A_2 is small, thus $A_1 - A_2 = \Delta A$ and $A_1 + A_2 = 2A$. Rewriting Eq. 1.109 results in Eq. 1.107. Note that the above calculation is based on the decrease in the amplitude in a freely vibrating system.

A commonly used measure of internal friction is based on the logarithmic decrement of successive amplitudes of vibration (which is related to the decay of the vibrations):

$$\delta = \ln\frac{A_1}{A_2} \tag{1.110}$$

Using Eqs. 1.108 and 1.110 for δ, one can obtain:

$$\delta \approx \frac{1}{2}\frac{\Delta W}{W} = \frac{1}{2}D = \frac{\Delta A}{A} \tag{1.111}$$

According to Zener, there is yet another way to express internal friction when damping is small, by the use of the phase angle, ϕ, as:

$$\tan\phi = \frac{1}{\pi}\ln\frac{A_1}{A_2} = \frac{\delta}{\pi} \tag{1.112}$$

which can also be written as:

$$\phi = \frac{\delta}{\pi} = \frac{1}{2\pi}\frac{\Delta W}{W} = \frac{1}{2\pi}D = \frac{1}{2\pi}\frac{\Delta A}{A} \tag{1.113}$$

since, for small angles, $\tan\phi \approx \phi$. Figure 1.46 shows the phase lag between stress and strain.

Most of the literature credits the early work of Zener on anelasticity; it was he who coined the term ‘internal friction’ for related phenomena. The similarity between energy loss and the propagation of a vibrating wave has resulted in the use of equivalent mathematical expressions for internal friction phenomena, as well. Harmonic oscillation expressions are applied (see Appendix II).

One of the equations for the energy absorbed per cycle in terms of the fraction of the strain, “a” is:

$$a = \frac{\varepsilon_R - \varepsilon_U}{\varepsilon_R} = \frac{\varepsilon_R(1 - \varepsilon_U)}{\varepsilon_R} \tag{1.114}$$

In Fig. 1.44, $\varepsilon_1 = \varepsilon_u$ (unrelaxed strain) and $\varepsilon_2 = \varepsilon_R$ (relaxed strain). As mentioned earlier, these strains correspond to the adiabatic and isothermal strains, respectively (see Fig. 1.47). Time-dependent strain is often approximated as an exponential function of time. The time dependency of straining a specimen, in terms of unrelaxed and relaxed strains, as indicated in the previous equation, can, thus, be expressed by:

$$\varepsilon = \varepsilon_R\left[1 - a\exp\left(-\frac{t}{\tau}\right)\right] \tag{1.115}$$

and for unloading:

$$\varepsilon = a\varepsilon_R \exp - \left[\frac{(t - t_1)}{\tau}\right] \tag{1.116}$$

τ is known as the ‘relaxation time’, i.e., the time required for a time-dependent component strain to increase within 1/e of its final value. Similarly, upon unloading, it indicates the decrease in strain by 1/e of its initial value.

The mechanical loss (the energy absorbed per cycle) can also be given in terms of unrelaxed and relaxed elastic moduli:

$$\tan \delta = Q^{-1} = \frac{E_U - E_R}{E_{ave}} \frac{\omega \tau}{1 + (\omega \tau)^2} = \Delta \frac{\omega \tau}{1 + (\omega \tau)^2}$$

$$E_{ave} = \sqrt{E_U E_R} \tag{1.117}$$

E_U and E_R are the unrelaxed and relaxed elastic moduli, respectively. ω is 2π times the measured frequency and τ is the time constant for the relaxation process. Clearly, the condition of maximal loss is when $\omega\tau = 1$, giving $\tan\delta_{max} = \Delta/2$. The time constant, τ, is given by:

$$\tau = \tau \exp\left(\frac{U}{kT}\right) \tag{1.118}$$

Referring to the frequency effects mentioned earlier, at high frequencies, deformation occurs with very small heat evolution; since the process is adiabatic, the heat interchange is small (there is only a short time available for heat flow) and the modulus, E_U, is the unrelaxed one. Damping is small. At low frequencies, deformation takes place slowly at constant temperature (above it was indicated that, under these conditions, sufficient time is available for thermal equilibration with the surroundings and, thus, almost no temperature change is observed, i.e., the temperature is almost constant) and the stress/strain ratio provides the relaxed modulus, E_R. At intermediate frequencies, the energy loss is the highest. The ratio of the moduli is called the 'modulus defect':

$$\frac{E_U - E_R}{E_{ave}} \tag{1.119}$$

Various methods are used to express energy loss. In the case of damping in a vibrating system, which can be natural or externally induced, there is a peak in the resonance frequency of the specific contributor to internal friction. It is often customary to express internal friction phenomena as propagating or standing waves. The simplest way to describe the decay of a wave is by expressing the amplitude of the vibrations as a sinusoidal function. From an engineering point of view, hysteresis produces a thermal effect and also damps the vibrations. An evaluation of the energy scattering (energy loss) per cycle appears above.

To summarize this section, note the equality of various expressions as follows:

$$\tan \phi \approx \phi = \frac{1}{2\pi} \frac{\Delta W}{W} = \frac{\delta}{\pi} = Q^{-1} \tag{1.120}$$

The relations expressed above hold for low values of friction. Internal frictions at high values depend on the mechanism operating and must be considered separately in each case. Experimentally, it was found that, in a spring-mass system, the natural frequency is $\omega_0 = \sqrt{\frac{k}{m}}$. This is the frequency of vibration in the absence of externally applied force (see Appendix II). The force on the spring is $F = F_0 \cos \omega_0 t$.

If a frictional force (damping), proportional to the velocity, is also present, then the harmonic oscillator is described as a 'damped oscillator'. Depending on the friction coefficient, such a system can:

(a) oscillate with a frequency smaller than in the non-damped case and with an amplitude decreasing over time (*under damped* oscillator).
(b) decay exponentially to the equilibrium condition without oscillations (*over-damped* oscillator).

Appendix II can serve as a guide for further consideration of anelasticity.

Appendix I

One of Maxwell equations is given by (see any book on thermodynamics for example in Gaskell)

$$\left(\frac{\partial T}{\partial P}\right)_S = \left(\frac{\partial V}{\partial S}\right)_P \tag{a}$$

$$dS = \frac{1}{T}\left(\frac{\partial E}{\partial T}\right)_V dT + \frac{1}{T}\left[\left(\frac{\partial E}{\partial V}\right)_T + P\right] dV \tag{b}$$

$$C_P - C_V = \left[P + \left(\frac{\partial E}{\partial V}\right)_T\right]\left(\frac{\partial V}{\partial T}\right)_P \tag{c}$$

$$\left(\frac{\partial V}{\partial S}\right)_P = \frac{T}{\left(\frac{\partial E}{\partial T}\right)_V \left(\frac{\partial T}{\partial V}\right)_P + \left[\left(\frac{\partial E}{\partial V}\right)_T + P\right]} \tag{d}$$

$$\left(\frac{\partial T}{\partial V}\right)_P (C_P - C_V) = \left[P + \left(\frac{\partial E}{\partial V}\right)_T\right] \tag{e}$$

From (d) and (e)

$$\left(\frac{\partial V}{\partial S}\right)_P = \frac{T}{\left(\frac{\partial E}{\partial T}\right)_V \left(\frac{\partial T}{\partial V}\right)_P + (C_P - C_V)\left(\frac{\partial T}{\partial V}\right)_P} \tag{f}$$

$$\alpha = \frac{1}{V}\left(\frac{\partial V}{\partial T}\right)_P \tag{g}$$

Here α is the coefficient of thermal expansion

$$\frac{1}{\alpha V} = \left(\frac{\partial T}{\partial V}\right)_P \tag{h}$$

From (f) and (h)

$$\left(\frac{\partial V}{\partial S}\right)_P = \frac{\alpha V T}{\left[\left(\frac{\partial E}{\partial T}\right)_V + (C_P - C_V)\right]} \tag{i}$$

$$C_V = \left(\frac{\partial E}{\partial T}\right)_V \tag{j}$$

From (i) and (j)

$$\left(\frac{\partial V}{\partial S}\right)_P = \frac{\alpha V T}{C_P} \tag{k}$$

From (a) and (k)

$$\left(\frac{\partial T}{\partial P}\right)_S = \frac{\alpha V T}{C_P} \tag{l}$$

In the above expression all terms are positive therefore when an external pressure is applied the system experiences an increase in temperature, and therefore in tension the sign in (i) will be negative.

We can write also that

$$\left(\frac{\partial T}{\partial \sigma}\right)_S = -\frac{T\alpha V}{C_P} \tag{m}$$

For solids the difference between C_P and C_V is of the order of 1% at room temperature. The small difference can often be ignored. Thus $C_p \sim C_V$ and

$$\left(\frac{\partial T}{\partial \sigma}\right)_S \approx -\frac{T\alpha V}{C_V} \tag{n}$$

Recalling that $\sigma = E\varepsilon$

$$\left(\frac{\partial T}{\partial \sigma}\right)_S = \left(\frac{\partial T}{\partial (E\varepsilon)}\right)_S = \left(\frac{\partial T}{E\,\partial \varepsilon}\right)_S \tag{o}$$

From (m) and (n)

$$\left(\frac{\partial T}{\partial \varepsilon}\right)_S \approx -\frac{TEV}{C_V}\alpha \tag{p}$$

In most materials α is positive but for rubber it is negative, therefore when this material is under consideration the sign in Eq. o should be changed.

Appendix II

In harmonic vibrations of a system composed of mass m and a spring the displacement x according to Newton's law is expressed as

$$m\frac{d^2x}{dt^2} + kx = 0 \tag{1}$$

This is a consequence of the forces acting on the mass and spring respectively: From classical mechanics the force is

$$\mathrm{F} = \mathrm{ma} \tag{2}$$

and a is the acceleration given as

$$a = \frac{d^2x}{dt^2} \tag{3}$$

Thus the force F is given as

$$\mathrm{F} = \mathrm{m}\frac{d^2x}{dt^2} \tag{4}$$

And the restoring force is given by

$$\mathrm{F_s} = -\mathrm{kx} \tag{5}$$

where k is the spring constant. Summing the forces acting in the system the following expression is obtained

$$m\frac{d^2x}{dt^2} + k\frac{dx}{dt} = 0 \tag{6}$$

A solution in the form of

$$\mathrm{x} = \mathrm{A}\exp(\gamma \mathrm{t}) \tag{7}$$

can be used for Eq. 6. Taking the first and second derivatives of (7)

$$\dot{x} = A\gamma \exp(\gamma t) \text{ and } \ddot{x} = A\gamma^2 \exp(\gamma t) \tag{8}$$

Let us now substitute these values into Eq. 6 after dividing by m we obtain

$$\mathrm{A}\gamma^2 \exp(\gamma \mathrm{t}) + \mathrm{A}\frac{k}{m}\exp(\gamma t) = 0 \tag{9}$$

or

$$\gamma^2 + \frac{k}{m} = 0 \tag{10}$$

$$\gamma^2 = -\frac{k}{m} \text{ and } \gamma = \pm\sqrt{-\frac{k}{m}} \tag{11}$$

denote

$$\sqrt{\frac{k}{m}} = \omega_0 \tag{12}$$

ω_0 is known as the natural frequency of the spring-mass system expressing its natural vibration in the absence of an externally applied force.

$$\gamma = \pm i\omega_0 \tag{13}$$

where i is clearly $\sqrt{-1}$.

Substituting the two roots of γ into (7) results in

$$x = A\exp(\pm i\omega_0 t) \tag{14}$$

or

$$x = A_1\exp(i\omega_0 t) + A_2\exp(-i\omega_0 t) \tag{15}$$

Recaling that $\exp(\pm ix) = \cos x \pm i \sin x$ Eq. 15 can be rewritten as

$$x = A_1(\cos \omega_0 t + i \sin \omega_0 t) + A_2(\cos \omega_0 t - i \sin \omega_0 t) \tag{16}$$

This can also be changed to the form of

$$x = B \cos \omega_0 t + C \sin \omega_0 t \tag{17}$$

$$\text{with} \quad B = (A_1 + A_2) \text{ and } C = i\,(A_1 - A_2) \tag{18}$$

B and C can be determined from the initial conditions. Let's assume that $x = 0$ at $t = 0$, B must be equal also zero; thus x from Eq. 17 is given by

$$x = x_0 \sin \omega_0 t \tag{19}$$

x_0 is the amplitude of vibrations. When Eq. 12 is substituted into Eq. 19 we obtain

$$x = x_0 \sin\sqrt{\frac{k}{m}}t \tag{20}$$

and since the angular frequency $\omega = 2\pi f$ (29) can also be written as

$$x = x_0 \sin 2\pi ft \tag{20b}$$

When some frictional force is also acting exerting a linear damping force in the form of

$$F_f = -cv \tag{21}$$

v and c are the velocity and damping coefficient (viscous) respectively and

$$v = \frac{dx}{dt} \tag{22}$$

the following known expression is obtained by summing the forces acting on the mass:

$$ma = -kx + (-cv)$$

$$\text{or} \quad ma + cv + kx = 0 \tag{23}$$

$$m\frac{d^2x}{dt^2} + c\frac{dx}{dt} + kx = 0 \tag{24}$$

$$\ddot{x} + \frac{c}{m}\dot{x} + \frac{k}{m}x = 0 \tag{24a}$$

This differential equation can be solved by substituting

$$x = A\exp(\gamma t) \tag{25}$$

where γ is usually a complex number thus:

$$\dot{x} = A\gamma \exp(\gamma t) \text{ and } \ddot{x} = A\gamma^2 \exp(\gamma t) \tag{26}$$

Substituting (25) and (26) into (24a) we obtain a quadratic equation which can readily be solved

$$A\gamma^2 \exp(\gamma t) + A\frac{c}{m}\gamma \exp(\gamma t) + A\frac{k}{m}\exp(\gamma t) = 0 \tag{27}$$

Divide Eq. 27 by $A\exp(\gamma t)$ to get the following quadratic equation

$$\gamma^2 + \frac{c}{m}\gamma + \frac{k}{m} = 0 \tag{28}$$

The solution of this quadratic equation is

$$\lambda_{1,2} = -\frac{c}{2m} \pm \frac{\sqrt{\left(\frac{c}{m}\right)^2 - 4\frac{k}{m}}}{2} \tag{29}$$

Looking at the second term of the above relation, it can be written as:

$$\sqrt{\left(\frac{c}{2m}\right)^2 - \frac{k}{m}} \tag{30}$$

or

$$\gamma_{1,2} = -\frac{c}{2m} \pm i\sqrt{\left[\frac{k}{m} - \left(\frac{c}{2m}\right)^2\right]} \tag{31}$$

In the case of small damping $\left(\frac{c}{2m}\right)^2 \ll \frac{k}{m}$

Denoting $\sqrt{\left[\frac{k}{m}\right] - \left(\frac{c}{2m}\right)^2} = \omega$ and substituting it for γ into Eq. 25 the following expression is obtained

$$x = A\,\mathbf{exp}\left[\left(-\frac{c}{2m}\right)t \pm i\omega t\right] \tag{32}$$

Recalling that

$$\exp(\pm ix) = \cos x \pm i \sin x$$

Equation 32 after rearranging and expressing iω in terms of cosines and sines can be written as

$$x = \mathbf{exp}\left(-\frac{ct}{2m}\right)[B\cos\omega t + C\sin\omega t] \tag{33}$$

where $B = (A_1 + A_2)$ and $C = i(A_1 - A_2)$

This is a result of expressing Eq. 33 as

$$A_1\,\mathbf{exp}\left(-\frac{ct}{2m}\right)[\cos\omega t + i\sin\omega t] + A_2\,\mathbf{exp}\left(-\frac{ct}{2m}\right)[(\cos\omega t - i\sin\omega t)] \tag{34}$$

and distinguishing between the additions of the two terms of ± iω by giving two values for A.

$\sqrt{\left[\frac{k}{m}\right] - \left(\frac{c}{2m}\right)^2} = \omega$ is the damped natural frequency of the system.

In this case the amplitude of the oscillations decreases with time according to the exponential term in (35), i. e., $\mathbf{exp}\left(-\frac{ct}{2m}\right)$.

It is observed that from the radical in (30), i.e., (29) $\left(\frac{c}{2m}\right)^2 - \frac{k}{m} = 0$ or $\left(\frac{c}{2m}\right)^2 = \frac{k}{m}$ is obtained, and therefore $\boldsymbol{c} = \mathbf{2}\sqrt{\boldsymbol{km}} = 2\mathrm{m}\omega_0$ and with the definition of ω_0 as seen above $\boldsymbol{\omega_0} = \sqrt{\frac{k}{m}}$. where ω_0 is the natural circular frequency.

The hysteresis loop is formed because the strain cycle lags behind the stress cycle. If the stress and strain vary sinusoidally with time as

$$\sigma = \sigma_0 \sin \omega \mathrm{t} \text{ and} \tag{35}$$

$$\varepsilon = \varepsilon_0 \sin(\omega \mathrm{t} - \phi) \tag{36}$$

where ϕ is the lag, the loop is elliptic as shown in Fig. 1.46. The area of the loop is a measure of the energy loss per unit volume of the body per cycle due to the damping (which dissipates as heat).

References

P.W. Bridgman, *Studies in Large Plastic Flow and Fracture with Special Emphasis on the Effects of Hydrostatic Pressure* (McGraw-Hill, New York, 1952)

G.R. Buchanan, *Mechanics of Materials* (Holt, Reinhart and Winston, Inc., New York, 1988)

A.H. Cottrell, *The Mechanical Properties of Matter* (Wiley, New York, 1964)

G.E. Dieter, *Mechanical Metallurgy*, 2nd edn. (McGraw-Hill Inc., New York, 1976)

D.R. Gaskell, *Introduction to Metallurgical Thermodynamics* (McGraw Hill Book Company, New York, 1973)

G.L. Kehl, *Metallographic Laboratory Practice* (McGraw-Hill Book Company, Inc., New York, 1949)

A.L. Nadai, *Theory of Flow and Fracture of Solids*, vol. 1 (McGraw-Hill, New York, 1950)

A. Nowick, B. Berry, *Anelastic Relaxation in Crystalline Solids* (Academic, New York, 1972)

N.H. Polakowski, E.J. Rippling, *Strength and Structure of Engineering Materials* (Prentice Hall, Inc., Englewood, 1966)

D. Tabor, *The Hardness of Metals* (Oxford University Press, Oxford, 2000). ISBN 0198507763

C. Zener, *Elasticity and Anelasticity of Metals* (University of Chicago Press, Chicago, 1942)

Further References

J.D. Bressan, R.K. Unfer, J. Mater. Process. Technol. **179**, 23 (2006)

A. Considére, Ann. Ponts Chaussee **9**, 574 (1885)

L. Gang, W. Li-liang, Y. Shi, W. Zhong-ren, Trans. Nonferrous Met. Socc. China **16**, 1103 (2006)

C.T. Herakovich, P.H. Hodge Jr., Int. J. Mech. Sci. **11**, 53 (1969)

U.F. Kocks, M.G. Stout, Model. Simul. Mater. Sci. Eng. **7**, 675 (1999)

M. Nabil Bassim, C.D. Liu, Mater. Sci. Eng. **A164**, 170 (1993)

C.G. Peters, F. Knoop, Metals in thin layers – their microhardness. Met. Alloys **13**, 292 (1940)

V.I. Strokov, V.N. Barabanov, Strength Mater. **19**, 501 (1987)

J. Tirosh, O. Abraham, Mech. Mater. **39**, 760 (2007)

H.-C. Wu, Z. Xu, P.T. Wang, Int. J. Plast. **13**, 873 (1998)

Chapter 2
Introduction to Dislocations

2.1 Introduction

When exposed for the first time to materials engineering, students often ask: "Why study dislocations? Why can't we get an engineering diploma without making this effort?" Such questions did not cross the minds of Taylor, Orowan and Polanyi, the fathers of modern dislocation theory in 1934, when they independently suggested their novel theory of dislocations, following concepts developed by Volterra in 1905. Their basic insights were crucial for the development of the modern science of structure, for the understanding of structural properties and, in particular, for the essential concept of the deformation of materials. A dislocation, due to its extent, is considered to be a line defect and is one of the various types of defects found in materials which determine each and every property of a crystal. Briefly, materials are not perfect and contain a variety of defects. These defects, that determine the properties of a material, are:

(a) point defects – various vacancy and interstitial configurations and foreign atoms;
(b) line defects – the various configurations of dislocations;
(c) planar defects – grain boundaries and surfaces (internal and external surfaces), stacking faults, etc.; and
(d) volume defects – voids and precipitates.

By expanding this concept, a point defect might also be considered to be a volume defect, but of atomic dimensions. Thus, the volume of a single vacancy or interstitial defect has the dimensions of $\sim$ one atom. In general, defects disrupt the periodicity of a lattice.

The defects considered in this chapter are the dislocations that determine the mechanical properties of materials. When Taylor, Orowan and Polanyi published their ideas on deformation and strengthening mechanisms, no actual observations of dislocations per se were as yet available; their abstract theory was based on observations of the behavior of materials in response to acting forces. It is commonly said that: "Seeing is believing." A giant step forward, toward eliminating any

J. Pelleg, *Mechanical Properties of Materials*, Solid Mechanics and Its Applications 190, DOI 10.1007/978-94-007-4342-7_2,

doubt as to the existence (and necessity) of dislocations, was taken with the development of modern technology for their visual observation by means of TEM. Field ion microscopy [henceforth: FIM] and atom probe techniques (magnification: typically three million times and above) permit the observation of dislocations at much higher magnifications, on an atomic scale. Even using a good optical microscope (which is not appropriate for the direct observation of dislocations), their presence can be detected using the etch pits technique. Etching attacks the sample surface preferentially around those dislocations emerging from the surface, because such sites are more highly strained. (Crystal orientation can be determined by the geometrical shape of the etch pits associated with the dislocations). The various phenomena associated with the mechanical properties of materials are readily explained by dislocations, which shed light on all the topics mentioned in the chapters below. Thus, plastic deformation and microplasticity, strengthening mechanisms, fracture, creep and fatigue are all explained by the specific dynamics of dislocations.

Plastic deformation can be realized by the motion of dislocations. During the process, bonds break and re-bond. The energy required to break a single bond is far less than that required to break all the bonds of the atoms on an entire plane at once; this is what causes entire planes to slip. With an increase in the density of the dislocations, overlapping between the strain fields of adjacent dislocations gradually increases the resistance to further dislocation motion. This causes hardening of the material as deformation progresses. The effect is known as 'strain hardening'. Dislocation pile-ups at various obstacles having large strain fields coalesce and resemble small micro-cracks, which propagate under the effect of more applied stress. These cracks grow as a result of the action of normal stresses. At high stresses, creep is controlled by the movement of dislocations. The non-conservative motion of dislocations is also known as 'dislocation climb'. Fatigue is one reason for the failure of many structures and a prerequisite of fatigue is the initiation of a crack. Experimental studies have revealed that fatigue first starts with the accumulation of high-density dislocations.

Thus, seen from the crucial perspective of dislocation theory, it is possible to gain a better understanding of the properties of materials, especially those engineering materials discussed below. Continuing the response to: "Why must we learn about dislocations"? – it can be further stated:

(a) to reconcile between the theoretical and the observed strengths of crystals;
(b) to explain yielding and the associated phenomena;
(c) to explain crystal growth (not directly relevant to this book).

2.2 The Theoretical Strength of Crystals

By evaluating the theoretical shear stress required to initiate plastic deformation, Frenkel calculated that the theoretical strength of a crystal is approximately one-fifth of the elastic shear modulus. However, experimental values for the shear strength of

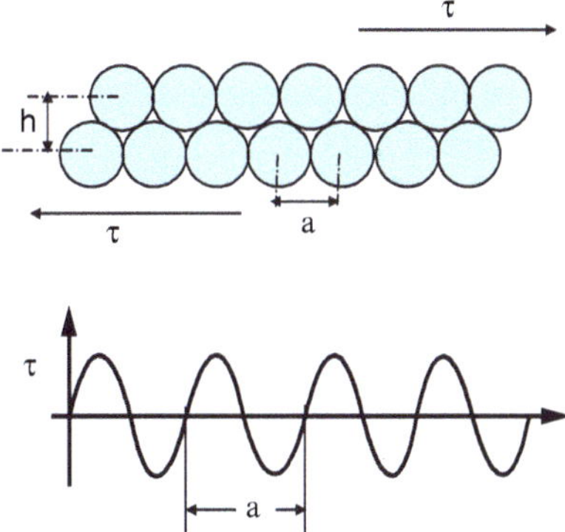

Fig. 2.1 Slip of one row of atoms over another by a distance a of the lattice spacing. A sinusoidal motion is assumed on the application of shear stress

metals were found to be several orders of magnitude smaller. The justification of this discrepancy was one of the major challenges in materials science at the beginning of 1930s. The presence of dislocations was the clue to understanding this discrepancy. The calculations of theoretical strength that appear below are performed in the light of Frenkel's approach, using a sinusoidal function for estimations. Frenkel calculated the critical amount of shear stress required to move adjacent atomic planes past one another, i.e., the energy per unit area involved in shearing two atomic layers from their equilibrium configuration. In Fig. 2.1, two rows of atoms in a close-packed structure are seen schematically, one of which is sheared over the other. The top row had to move a distance "a" from one atomic equilibrium position to another. As mentioned above, a sinusoidal function was suggested for the calculation of the theoretical strength, since an atom moves from one equilibrium site to another along its path of movement to shear an atomic row over its immediate neighbor below.

The applied shear stress, τ, produces a displacement, x, of "a", which is basically the inter-atomic distance, given as:

$$\tau = k \sin \frac{2\pi x}{a} \tag{2.1}$$

For small displacements, here, the value of the sine equals the value of $\frac{2\pi x}{a}$ and, thus Eq. 2.1 becomes:

$$\tau = k \frac{2\pi x}{a} \tag{2.2}$$

Expressing Hooke's Law in terms of the shear modulus and shear stress one obtains:

$$\tau = \frac{Gx}{h} \tag{2.3}$$

where h is the height between the two rows (or planes of atoms) and G is the shear modulus. x/h is obviously the shear strain, since x represents the displacement of the atoms from their equilibrium position. By equating Eqs. 2.2 and 2.3, a value can be obtained for k as:

$$k = \frac{Ga}{2\pi h} \tag{2.4}$$

Substituting this value of k into Eqs. 2.1, 2.5 can be written as:

$$\tau = \frac{Ga}{2\pi h} \sin\left(\frac{2\pi x}{a}\right) \tag{2.5}$$

When $\sin\frac{2\pi x}{a} = 1$, the shear stress is maximal; this occurs at x = a/4 (since $\sin(\pi/2) = 1$) and Eq. 2.5 becomes:

$$\tau_{max} = \tau_0 = \frac{Ga}{2\pi h} \tag{2.6}$$

At a/4, the lattice becomes unstable, i.e., it will yield at a value of τ_0, which is the critical shear stress. Since h ≅ a, the maximum shear stress for the instability of the lattice is about:

$$\tau_{max} = \tau_0 \sim \frac{G}{2\pi} \sim \frac{G}{6} \tag{2.7}$$

For metals, G is ~27–77 GPa (for W ~161). Thus, according to Eq. 2.7, $\tau_0 \sim 4.5$–13 GPa. The observed critical shear-stress value is ~0.0069 GPa and, as such are two to three orders of magnitude smaller than the theoretical value. In a more realistic, refined approach, without using a sinusoidal function, τ_{max} is ≈G/10–G/30. Even these refined values are ~2 orders of magnitude greater than the experimental values.

The inevitable conclusion is that real crystals must contain defects, such as the dislocations suggested by Taylor, Orowan and Polanyi, which reduce their mechanical strength or, more specifically, their resistance to slip when the applied stress reaches a critical value. The 1934 postulate showed that shear is possible at much lower stresses than in a perfect crystal.

2.3 Seeing (Dislocations) Is Believing

The success of the dislocation theory was outstanding, since it was developed without dislocations ever having been seen. The postulate made by Taylor, Orowan, and, regarding the necessity of the existence of dislocations, has since been confirmed in many ways. Many authors have published beautiful photos of dislocations in various configurations and many may be seen in the works of Amelinckx. Some of the techniques used will be briefly discussed below.

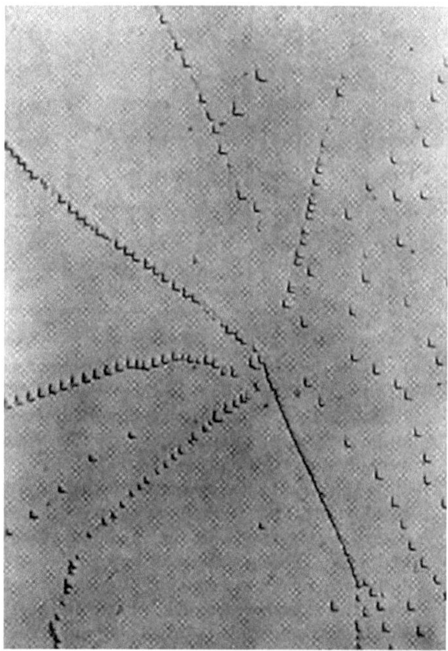

Fig. 2.2 Nb poly-crystal showing substructure in one of the grains ×850 (Pelleg 1969a)

2.3.1 Etch Pits

The earliest approach confirming that dislocations are present in crystals and that they move either conservatively (glide) or non-conservatively (climb or cross slip) used the etch pit technique. This technique involves etching a surface by means of an etching solution (acidic or basic) appropriate to the material under consideration. When a dislocation line emerges from the surface of a metallic material and intersects it, the local strain field existing around it increases the relative susceptibility of that material to acidic (or basic) etching and an etch pit of some geometrical shape forms. These etch pits can be counted, giving some idea of the density of the dislocations. If the material is strained (deformed) and repeatedly re-etched, a series of etch pits can be produced, effectively tracing the movement of the dislocation in question. The etch-pit method is applicable to a number of different research areas involving various types of dislocations. For example, one method for analyzing grain orientation is the etch-pit technique, where the geometry of the etch figures can be related to the grain orientation. Basic etch-pit shapes can have the geometry of a square, a rectangle, a triangle and a hexagon, but they may be distorted. Etch pits formed after etching may be related to dislocations, stacking faults or other defects, such as sub-boundaries (low-angle grain boundaries). Figures 2.2, 2.3, 2.4, 2.5, and 2.6 show etch-pits in single, bi-crystal and poly-crystal niobium. Figure 2.2 represents an Nb poly-crystal showing substructure in one of the grains at x850 by optical microscopy.

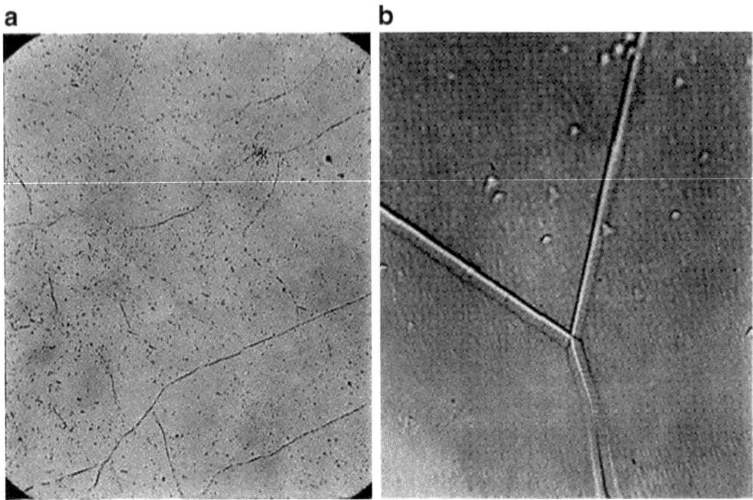

Fig. 2.3 Nb single crystals showing sub-grain boundaries: (**a**) ×60, (**b**) ×745 (Pelleg 1969b)

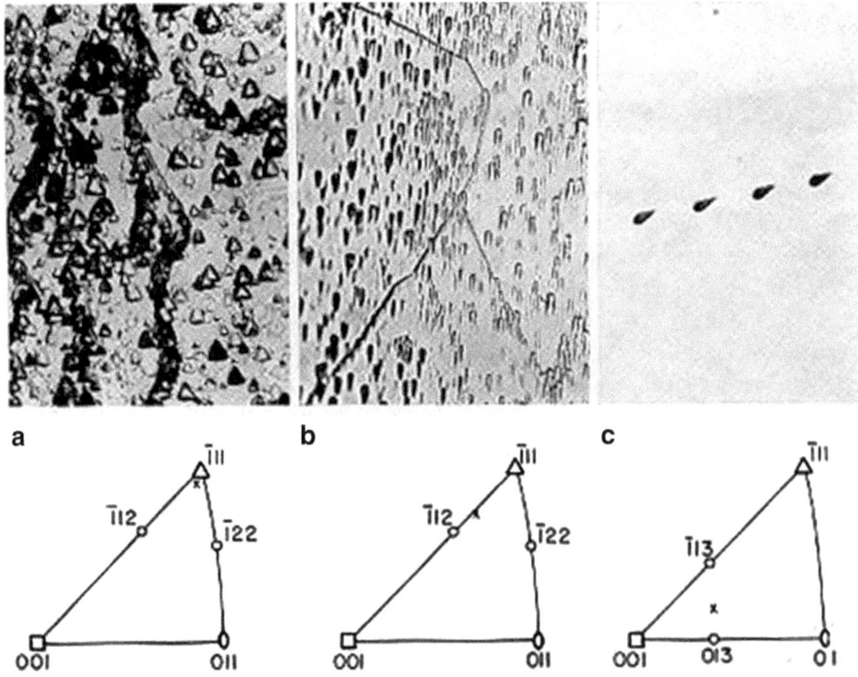

Fig. 2.4 Etch pits at the orientations indicated: (**a**) equilateral etch pits of as grown dislocations in Nb single crystals 2° from (111) – ×1,400; (**b**) distorted etch pit and sub-boundaries in a grain of bi-crystals 14° from (111) – ×760; (**c**) conical etch pits in a grain of bi-crystals 38° from (111) – ×1,200 (Pelleg 1969c)

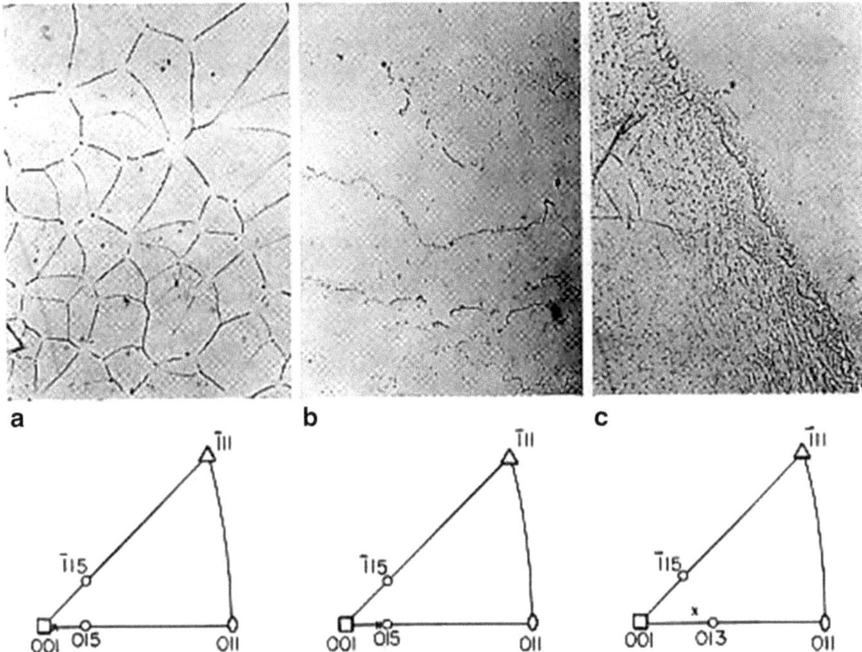

Fig. 2.5 Structures observed in the vicinity of (001) plane: (**a**) cellular substructure in a single crystal 3.5° from (001) ×230, (**b**) sub-boundaries in a grain of a tri-crystal 8° from (001) ×90; (**c**) sub-boundaries formed by pips in a grain of a bi-crystal 14° from (001) ×760 (Pelleg 1969c)

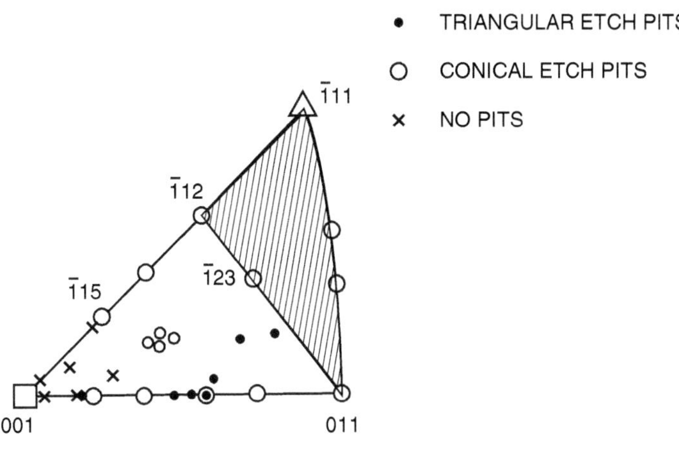

Fig. 2.6 High-density *triangular etch pits* are found in the *shaded region* of the stereographic triangle, but pits are absent in the *unshaded region*

Individual dislocations, having equilateral triangle-shaped etch pits, outline the sub-boundaries. In Fig. 2.3, sub-grain boundaries are outlined in Nb single crystals by arrays of dislocations revealed by etch pits. The etch pits in Nb, at the orientations indicated, are shown together with stereographic triangles in Fig. 2.4, presenting as grown dislocations (equilateral etch pits), distorted and conical-shaped etch pits and sub-boundaries. The cellular substructure and sub-boundaries in single-crystal Nb were distorted when a change in orientation occurred, depending on the angle of the dislocation relative to the surface. At certain orientations, the etch pits were low in density or absent. Poles within the shaded region of Fig. 2.6 responded favorably to the etchant. Etch pits were absent in the unshaded region observed when the orientation was in the vicinity of the (001) plane, as indicated in Fig. 2.5. All the etch pits formed on {111} planes in various materials were equilateral triangles; these observations also supported the basic ideas that had been previously put forward by dislocation theory. In particular, this method established a correlation between dislocation density and the stress required for deformation; this correlation is also fundamental to the understanding of other observed phenomena, such as strain hardening. The movement of dislocations was observed by repeated straining and etching, thus showing the locations of the etch pits before and after deformation, when re-etching the polished surface. The etch-pit technique is still useful and valid even though various new experimental techniques have been developed for the observation of individual dislocations, notably TEM.

2.3.2 *Transmission Electron Microscopy (TEM)*

Almost all the dislocations can be detected and studied by TEM. Thus, dislocation lines of either edge or screw orientation, partial dislocations, stacking faults, low-angle grain boundaries, cell formation, dislocation pile-ups at some obstacle, etc., can be readily explored. Such studies require the use of thin foils, transparent to electrons. Transmission through good, dislocation-free material is different than it is through a strained region containing dislocation. Dislocations may be observed due to the scattering of the electrons in the strained region around the dislocations. As the voltages increase (modern electron microscopes can operate at voltages in excess of 10^6 eV) the electron microscopes have larger penetration distances and can register greater detail in the defective region. The movement of dislocations, as a consequence of straining specimens, and the interactions between the dislocations can be directly observed. Today, modern TEM technology often includes straining devices which can even deform the specimens.

An illustration of dislocation walls and cell structures in pure Al, rolled for 5% and 15% deformations, is seen in Figs. 2.7, 2.8, and 2.9. Arrays of dislocations are also visible in the pure Al specimen. A reference specimen with a relatively low dislocation density is shown for comparison. Cell formation requires cross slip and, therefore, in order to observe such a feature by TEM at low strains, materials of high stacking fault energy (discussed later on) are preferred (Figs. 2.10, 2.11, and 2.12):

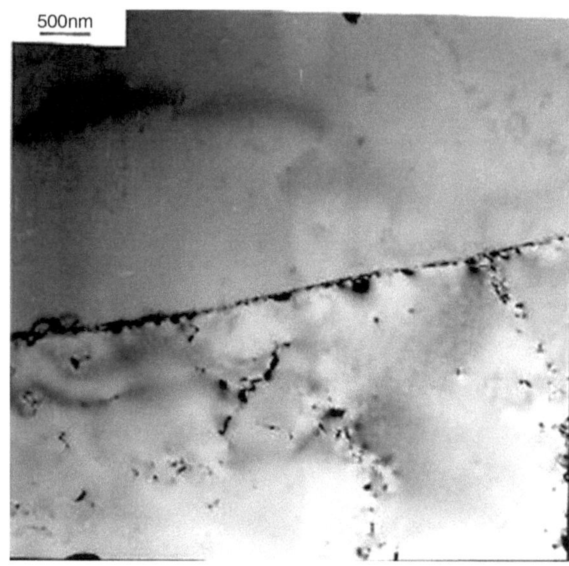

Fig. 2.7 Reference specimen of pure Al, annealed and unstrained without tilt. Relatively small amounts of dislocations are present. ×15 (Courtesy of Prof. Roni Shneck of Ben-Gurion University of the Negev)

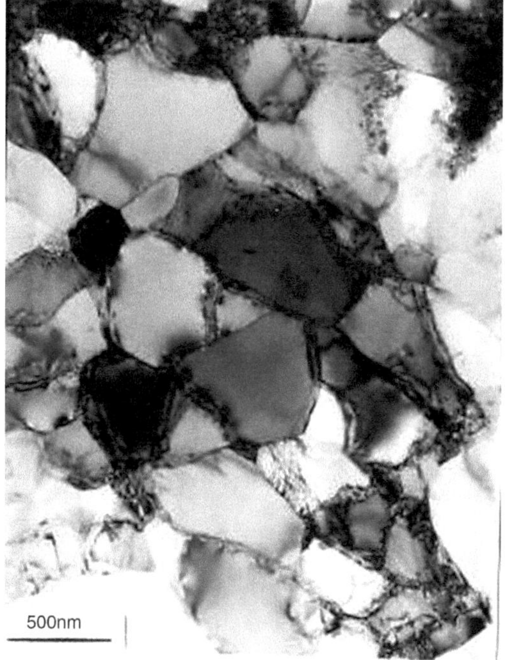

Fig. 2.8 Pure Al deformed 5%, rolled (in transverse direction) and annealed. Note the cell structure formed by dislocation walls (Courtesy of Prof. Roni Shneck of B.G.U.)

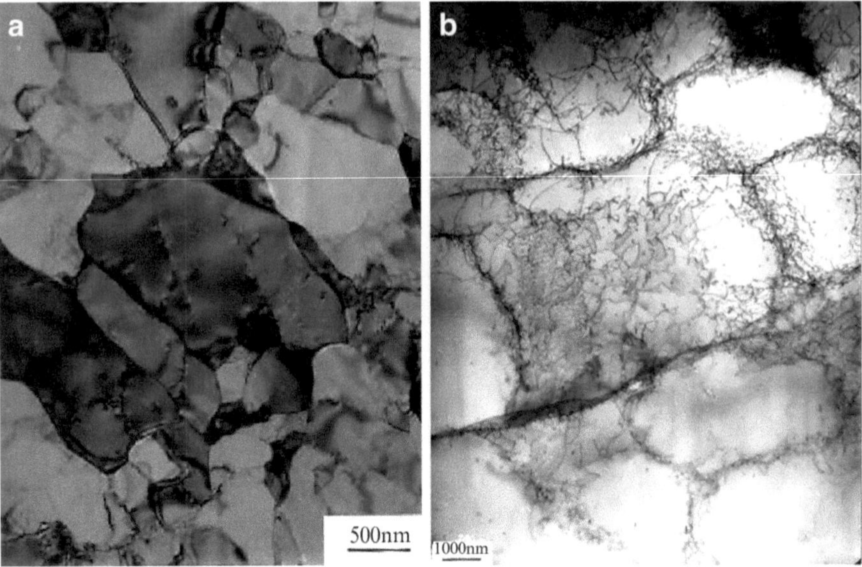

Fig. 2.9 (**a**) Like Fig. 2.8, but sectioned in a longitudinal direction; (**b**) dislocation cells and tangled dislocations in a cell (Courtesy of Prof. Roni Shneck of Ben-Gurion University of the Negev)

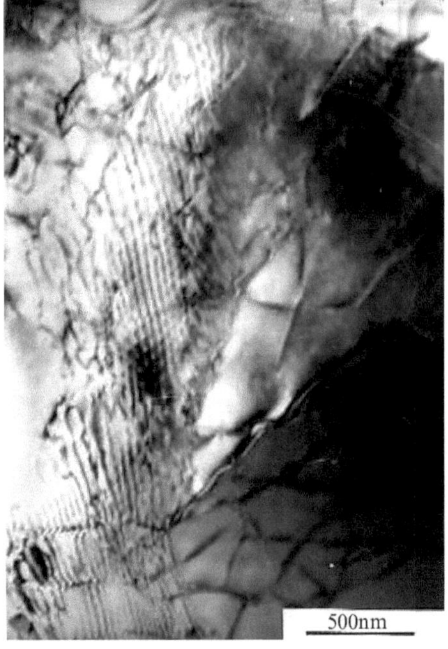

Fig. 2.10 Pure Al 15% deformation by transverse rolling. 30° tilt. ×37.5; Note the dislocation structure in the cell wall (Courtesy of Prof. Roni Shneck of B.G.U.)

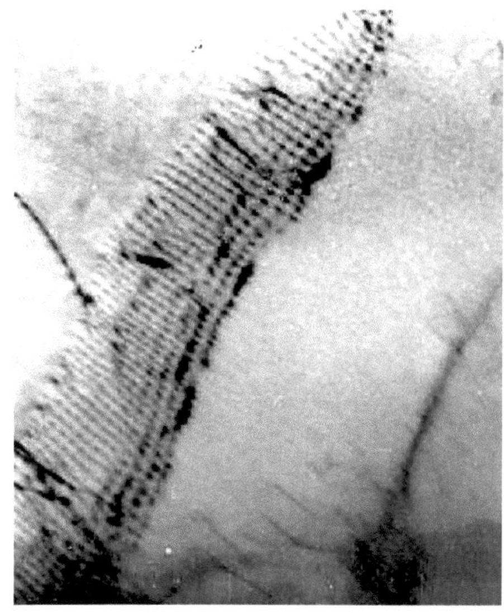

Fig. 2.11 Dislocation structure in a cell wall (Courtesy of Prof. Roni Shneck of B.G.U.)

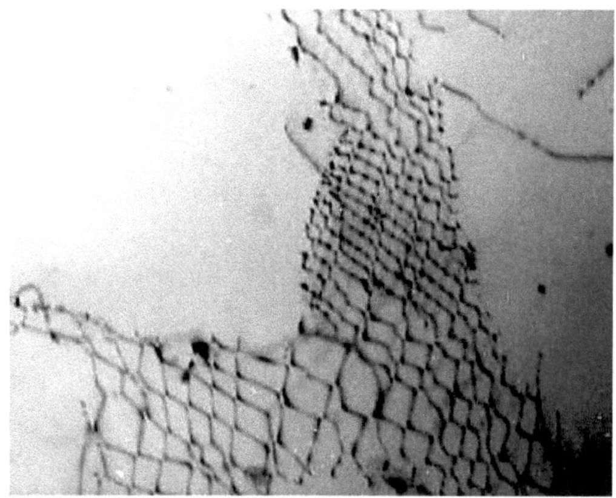

Fig. 2.12 Dislocation network in cell walls (Courtesy of Prof. Roni Shneck of B.G.U.)

2.3.3 Field Ion Microscopy (FIM)

By conventional TEM resolution on an atomic scale, is not feasible to observe, detect and examine the position of individual atoms or point defects. FIM (developed by Muller) allows for the study of structure on atomic scale at a resolution of 2.3 Å. By field evaporation, a layer of atoms can be peeled off in order to observe the successive layer of the remaining structure and to study the point-defect concentration and dislocations. Sharp tungsten tips, electro-polished to a hemispherical shape of ∼300 Å or less, and positively charged at high voltages in the kV range are used. After reaching a high vacuum, He atoms are bled into the system, which is then ionized and produces an image of the ions on a screen equipped with a channel plate. Characteristic image patterns are formed on the screen, which can also be photographically reproduced. Figures 2.13 and 2.14 illustrate, among other

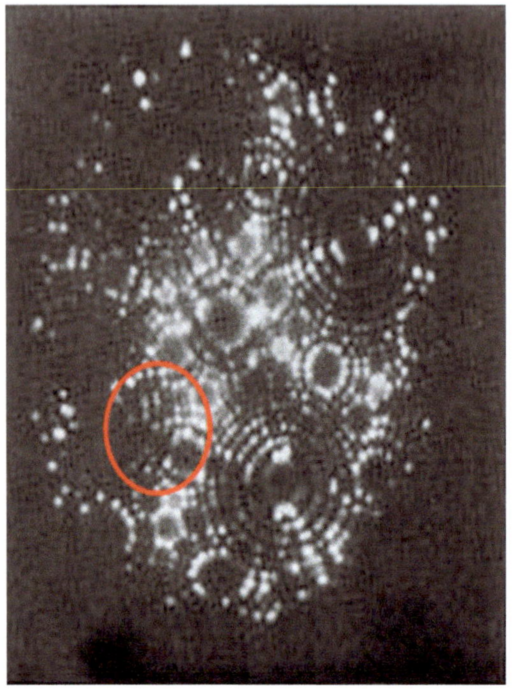

Fig. 2.13 Field ion micrograph of clean W tip outlining the location of a dislocation; <100> oriented tip; 7.48 kV and He pressure of 2.93×10^{-3} Pa (Pelleg 1992)

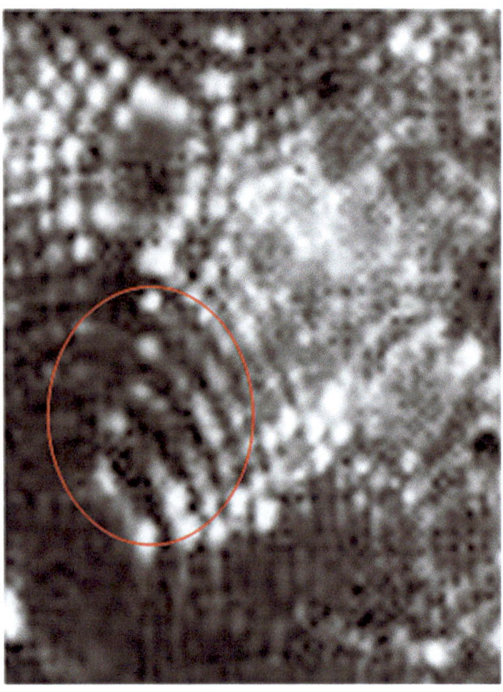

Fig. 2.14 FIM pattern after field evaporation at 9.76 kV. Dislocation is *outlined* in the micrograph after stripping Zr from the tip. 8.75 kV (Pelleg 1992)

defected regions, the dislocations outlined in red. In Fig. 2.13, a FIM micrograph is shown, oriented at <100>, obtained at a He pressure of 2.93×10^{-3} Pa and 7.48 kV. Figure 2.14 illustrates the presence of a dislocation in a field-evaporated zirconiated W tip at 8.75 kV. The field evaporation was performed at 9.76 kV leaving behind a clean W tip. The tip radii were $(0.5–1.0) \times 10^{-7}$ m.

Other techniques, such as X-ray diffraction (Lang's technique) and the decoration or Moiré fringes method are outside the scope of this book.

2.4 The Geometrical Characterization of Dislocations

Figure 2.15 illustrates a dislocation having an edge orientation. An 'edge dislocation' is a defect where an extra partial plane of atoms exists in the crystal, distorting nearby planes of atoms. This extra plane can be seen schematically in Fig. 2.15 and is marked as ABCD; its termination, the line CD, is the dislocation. An additional defect, a vacancy, is also shown occupying the volume of a missing atom. In Fig. 2.16, the front face of Fig. 2.15 is shown and the dislocation in the cubic crystal is marked. The symbol for the edge dislocation is the shape of the letter T, drawn either inversely or in its normal shape. The conventional symbol for a positive dislocation is the inverse symbol of the letter.

Basically, there are two types of dislocations, edge and screw. An intermediate dislocation exists between these, known as 'mixed dislocation', which is present in real materials and has the characteristics of both edge and screw. The mixed dislocations are the most common form in materials. It is more difficult to draw an illustration for a screw dislocation. Figure 2.17 is a schematic view of a screw dislocation. Screw dislocation was named for continuous, pivotal motion around a given point, but with each circumnavigation reaching a different plane than the

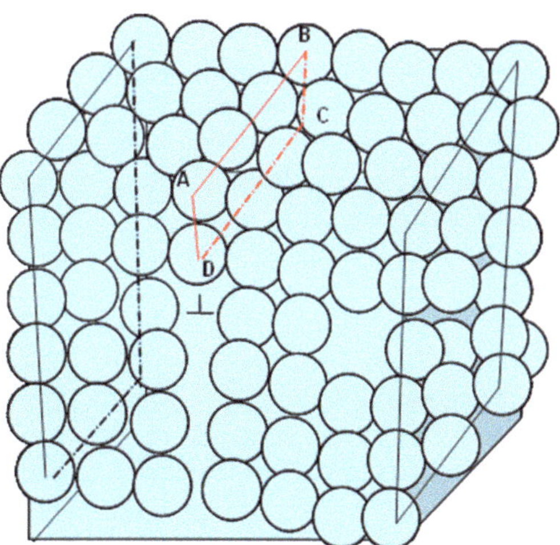

Fig. 2.15 Line DC defines an edge dislocation; it is the termination of an extra plane. The space of the missing atom is a vacancy

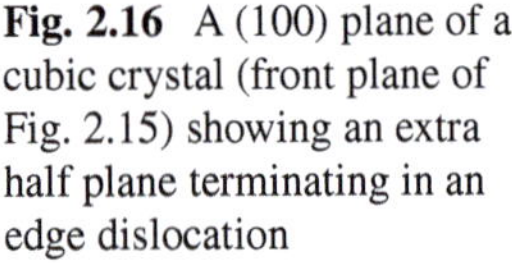

Fig. 2.16 A (100) plane of a cubic crystal (front plane of Fig. 2.15) showing an extra half plane terminating in an edge dislocation

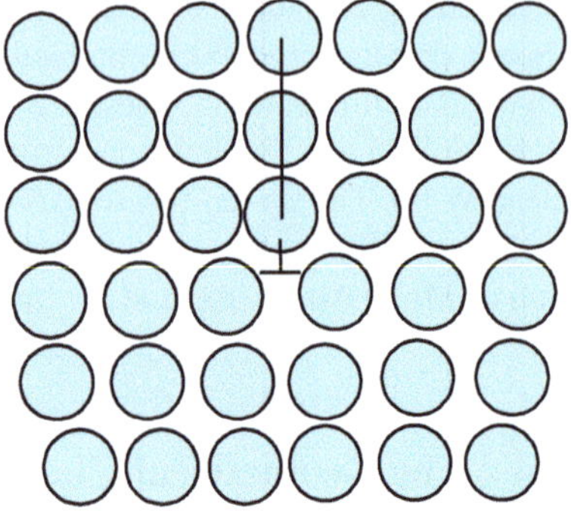

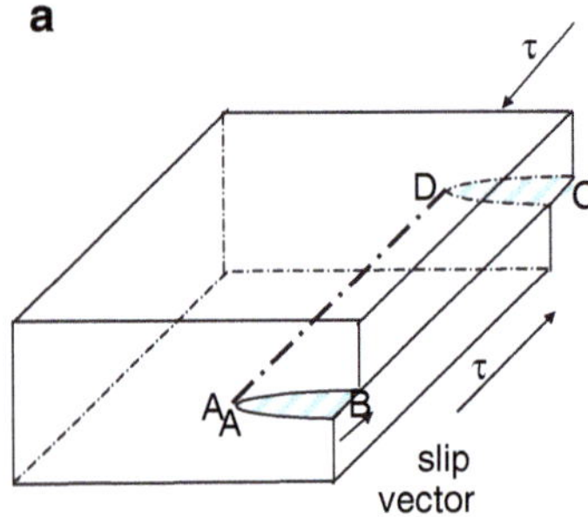

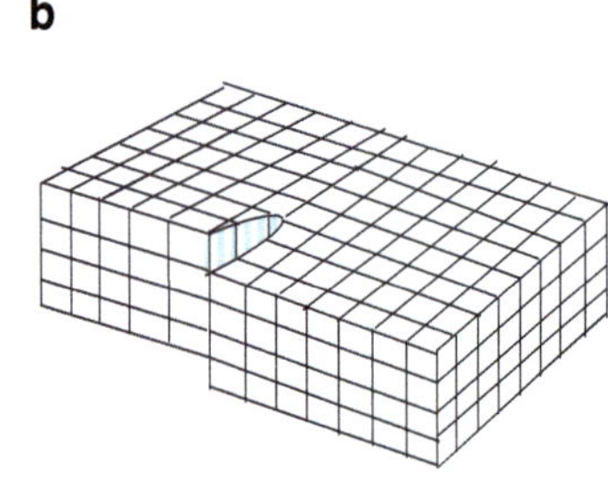

Fig. 2.17 A screw dislocation **AD**; (**a**) formed by slip by a pair of shear stresses; (**b**) the same but more detailed

previous one, not unlike the motion along a screw. With each turn of a screw, an elevation of one step is made. A good example of such motion is a modern parking structure, in which each level spirals up or down a floor. Figure 2.17a illustrates this feature better than Fig. 2.17b.

A dislocation is characterized by its 'Burgers vector', which is defined by what is known as the 'Burgers circuit'. A dislocation has two properties – a line direction and its Burgers vector. In an edge dislocation, the Burgers vector is perpendicular to the line direction, whereas, in screw dislocations, it is parallel.

To form a Burgers circuit, a starting point, S, is chosen in an undeformed plane of a crystal, often called a 'good region' of the crystal and, then, by a stepwise procedure moving from one atomic position to the next, a closed loop is formed in the clockwise direction (by convention) and the number of lattice steps in each direction is counted until the loop is closed. The finishing point, F, closes at point S. Now, a region of another crystal is chosen containing a dislocation and the same number of steps is taken in the same clockwise direction as in the previous case, to similarly form a loop. If the chosen region, indeed, contains a dislocation, point F, the final step taken, will not close the loop and the difference in magnitude between S and F is the Burgers vector. Thus, (A;a) in perfect dislocation-free crystals, the circuit closes at starting point S; (A;b) in crystals containing dislocations, the

b is Burgers vector. The stepwise circuit did not close the circuit. The excess step indicates the presence of a dislocation.

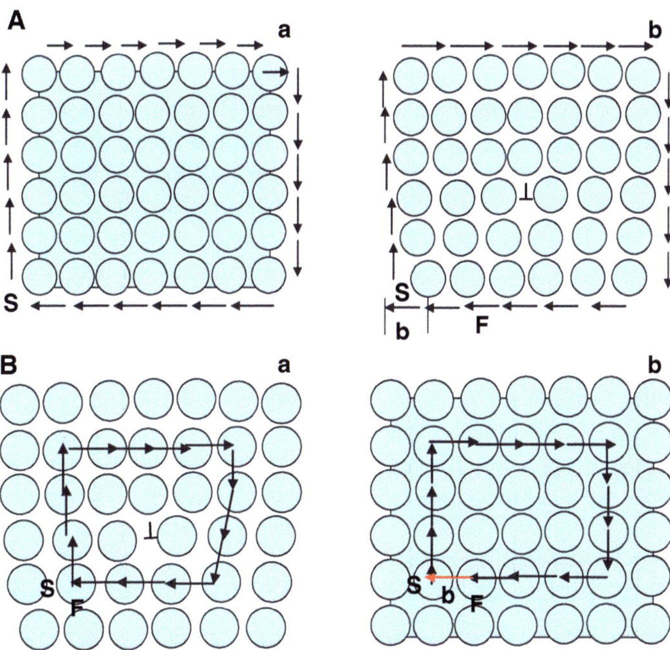

Fig. 2.18 (**A**) Burger circuits in a perfect crystal and in a faulted one. The steps of the circuit from one atomic position to the next have to be the same in both lattices: (*a*) the circuit in a dislocation-free crystal; (*b*) the circuit around a dislocation. (**B**) Burgers circuits to show the strength of the vector ***b*** as having the magnitude of the distance between the atoms of the lattice; (*a*) around a dislocation and (*b*) in the reference crystal

loop of the stepwise circuit is not terminated at starting point S. The number of steps required to reach closure indicates the number of dislocations present in the chosen region. Figure 2.18 is an illustration showing a Burgers circuit around one dislocation. The Burgers vector is perpendicular to the dislocation line and, thus, it is an edge dislocation, as indicated by its symbol. The Burgers vector, which is a unit slip vector, is supposed to be the same as an interatomic distance of the lattice. However, this is not so, as seen in Fig. 2.18B;a. In this case, of course, the closed circuit defining the Burgers vector was drawn around the dislocation. In the vicinity of this dislocation, the lattice is strained elastically, B;a. To show that the magnitude of **b** is indeed a lattice vector, the Burgers circuit should be performed differently, as indicated in Fig. 2.18B;a. The first step is to make a closed circuit in a crystal containing dislocation, starting at point S and finishing at F, the same lattice point. Then, exactly the same circuit is made in the reference crystal, counting the same number of steps in all directions. In this method of making the Burgers circuit, the position of F is separated from S by one lattice unit, which is the special vector needed to close the circuit in the reference crystal – by definition, this is the true Burgers vector **b**. According to this version of the Burgers vector, the true Burgers vector, **b**, appears in the reference crystal, rather than in the lattice containing the dislocation, as indicated in the first version with the closed circuit. Note that if the first method (Fig. 2.18A;b) made the circuit sufficiently farther away from the 'bad region', then the vector **b** obtained would be practically the same as the one obtained during by second method (Fig. 2.18B;b), since distortion decreases with distance.

It follows that the Burgers vector of a (perfect) dislocation is, by necessity, a lattice vector. The Burgers vector has a magnitude and orientation which shows the direction of slip. However, a significant controversy exists in the literature regarding the sign of the dislocation and the question still remains regarding the notation of the direction of slip in the Burgers vector. The thought in one school favors the method indicated in Fig. 2.18B;a, assuming that the direction of the dislocation line is into the drawing and the circuit is made clockwise. In this case, the dislocation is considered to be positive, as the dislocation symbol is drawn in the Figure. Another group of researchers in the dislocation field (including Burgers himself) takes an opposite approach. Thus, one must clearly define whether the circuit is drawn clockwise or counter clockwise around the dislocation. The vector is always the same, but its sign will be different and, therefore, consistency in describing a dislocation is important. The presentation of a Burgers circuit for the evaluation of **b** is called the 'RH Convention' (i.e., the right-handed convention, clockwise and positive).

A dislocation line cannot end inside a crystal, because the Burgers vector cannot become zero (it is constant) at some point along its line. It can terminate at a free surface, internal surface or interface (e.g., a grain boundary), form a closed loop inside the crystal or branch into other dislocations (forming a node). At the node (where a dislocation branches out), for all dislocations pointing out from it, Burgers-vector conservation must exist, analogously to Kirchoff's Law for the flow of electric current. This is Frank's rule of the conservation of the Burgers vector. Thus:

$$\mathbf{b}_{\mathrm{i}} + \mathbf{b}_2 + \mathbf{b}_3 = 0$$

and more generally:

$$\sum_{i}^{n} \mathbf{b}_{\mathrm{i}} = 0 \tag{2.8}$$

So, for three dislocations, namely the one which branches into two other dislocations, we can write that:

$$\mathbf{b}_1 = \mathbf{b}_2 + \mathbf{b}_3 \tag{2.8a}$$

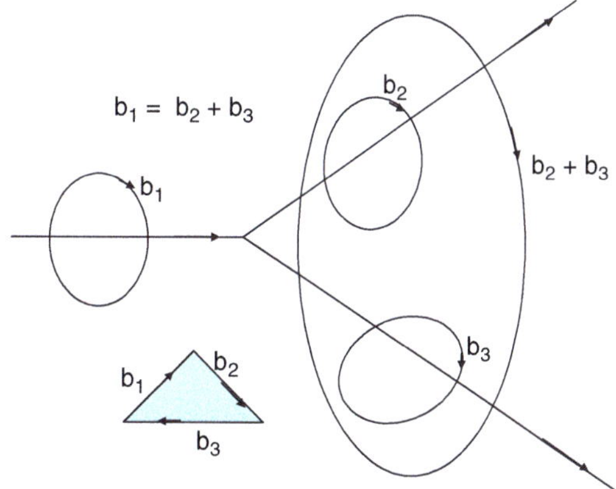

Fig. 2.19 Dislocation with Burgers vector $\mathbf{b_1}$ is branching off into two dislocations with vectors $\mathbf{b_2}$ and $\mathbf{b_3}$

Equation 2.8 can be understood by the schematic illustration shown in Fig. 2.19. Accordingly, since the Burgers circuit can be drawn around the dislocation line entering the node and the two lines leaving the node in the positive direction, it follows that the sum of the Burgers vectors provides all the dislocation-line vectors, t_i, according to Eq. 2.8. However, if the dislocation line branches out and not all of them are pointing away from the node, Eq. 2.8a is appropriate in this case. After the node, the dislocations can be enclosed by one single extended Burgers circuit, like the one drawn around the dislocation before the branching.

In summary (and since no agreement on the convention for **b** is available), in order to define the dislocation line, a unit vector **t** (which is the same as the translation vector in the crystal), is chosen, such that it is tangent to the dislocation line and the positive direction of the dislocation is taken in the positive direction of **t**. This helps to define the **b** vector. In Fig. 2.18a, view the dislocation line as running into the sheet of paper, and consider it to be positive; therefore, the Burgers vector **b** is defined by the RH convention. When reversing the direction of **t** (i.e., the dislocation line is running out of the drawing), the sign of **b** is **also** reversed. +On the basis of the above convention, a dislocation can be characterized by **t** and **b** as: (a) edge dislocation $\mathbf{b}\cdot\mathbf{t} = 0$ (i.e., the direction and Burgers vector are perpendicular); RH screw dislocation $\mathbf{b}\cdot\mathbf{t} = \mathrm{b}$ and left-handed (LH) screw dislocation $\mathbf{b}\cdot\mathbf{t} = -\mathrm{b}$ (in both directions the dislocation and Burgers vector are parallel).

2.5 The Formation of Dislocations

As a student, I first encountered the concept of dislocation as it was presented by my lecturer (Professor Polakowski) by means of a very simple, but instructive, method, as depicted in Figs. 2.15 and 2.16. Consider, for the sake of simplicity, the simple cubic structure shown in Fig. 2.16 and further assume that a slot has been cut into it by a knife of atomic scale; now a partial plane (not necessarily a half plane) of

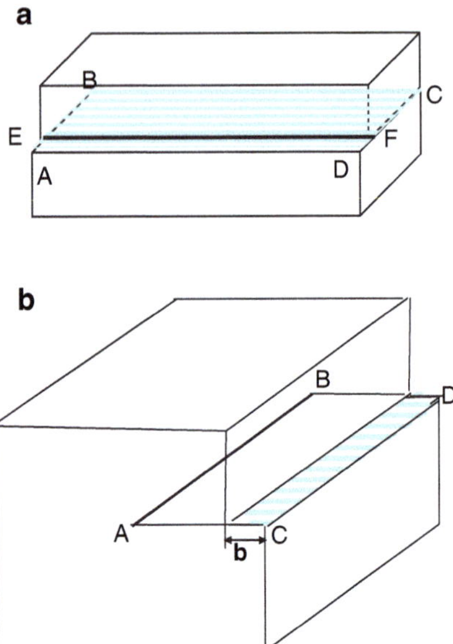

Fig. 2.20 (**a**) *ABCD* is the slip plane and *EF* is the dislocation line at the start of deformation. (**b**) Edge dislocation **AB** formed by shear displacement in the slip plane *ABCD* following a cut in the plane and gluing of the faces formed by the cut

atoms has been inserted into that slot and then glued together. The termination of the plane at its edge, as defined by the row of atoms, is a defect in the crystal of line dimensions and is an 'edge dislocation'. This imaginative method of producing a dislocation distorts the region around the line defect, as shown in Fig. 2.16. The above procedure can be further visualized by an analogy to a stack of cards. If half a card is inserted in a deck of cards, a defect is formed in the deck at the termination line of the half card. As mentioned previously, in real crystals, due to the straining caused by the extra half plane, elastic distortion arises around it and in its immediate vicinity extending out for several planes. The dislocation line itself has a high energy. The strain decreases over distance. Clearly, this method of producing dislocations is useful for understanding their character. In a similar manner, one can form an edge dislocation, as depicted in Fig. 2.20a, as follows. Part way into a plane of an elastic body, make a cut and apply a deforming shear stress on that plane, say of dimension **b**. Then glue the newly-formed faces (produced by the cut) together to reproduce a continuous solid. A dislocation line formed in this way is clearly the demarcation line or the boundary between the slipped and unslipped parts indicated in the figure. Here, the upper part illustrated slipped over the shaded slip plane, marked as ABCD, forming one atomic displacement **b**. In Fig. 2.20b, this slip has progressed over the shaded slip plane and a dislocation line has formed at AB. In a similar manner, a screw dislocation may be formed (see Fig. 2.17), but the slip vector, in that case, will be parallel to the dislocation line.

Dislocations in a real crystal are present in sufficient amounts, depending on the material and its history of production. The growth of a single crystal by itself is pictured as occurring by a process of screw dislocation (which is an integral necessity for growth per se). However, the dislocation concept of strain hardening, for example, requires the formation of a large number of dislocations and their multiplication. Plastic deformation can produce large numbers of dislocations, as will be discussed in a later section. During plastic deformation, the density of the dislocations increases very much, even to a level of $10^{10}/mm^2$, depending on the severity of the deformation. Annealed material has a density of $\sim10^6$–$10^7/mm^2$. Silicon wafers usually have a very low density of dislocations, $<100\ cm^{-2}$, while semi-insulating GaAs wafers have a density on the order of $\sim10^5\ cm^{-2}$.

Three mechanisms for dislocation formation in a polycrystalline material are considered here: (a) by homogeneous nucleation; (b) by induction at grain boundaries and interfaces in the lattice; and (c) by surface stress. In regard to homogeneous nucleation, it is very unlikely, since it requires high stresses of about the same level as the theoretical stress in a dislocation-free crystal. When considering grain boundaries in materials, they can produce dislocations which propagate into the grain; in particular, sites of irregularities and steps or ledges at the boundaries are dislocation sources in the early stages of plastic deformation. Lastly, stresses on surfaces are usually larger than the average stress within the lattice; the presence of irregularities, such as steps, can induce dislocation propagation into a crystal. Thus, dislocations nucleate from the surface.

In single crystals, dislocations are formed at the free surfaces. Suffice it to say that sites, such as precipitates, dispersed phases, oxides, etc., can all serve as sources for dislocation initiation and their generation. Note that an important concept for dislocation generation is the collapse of vacancies to form disc-shaped dislocations, as suggested by Nabarro.

2.6 The Motion of Dislocations

The importance of understanding the motion of dislocations is directly linked to the fact that the plastic deformation of crystalline materials is mostly carried out by either their conservative or non-conservative advance. When enough force is applied to a crystal structure (but the orders-of-magnitude are less than the theoretical strength in a dislocation-free crystal), the atoms at the edge of the extra plane (the dislocation line) passes through the planes of atoms, breaking and forming bonds with them until it runs out at the opposite side of the crystal, forming one unit step, as shown in Fig. 2.21 below. This procedure occurs in many discreet steps until the dislocation has moved through the entire lattice, leaving plastic deformation in its wake. The following question immediately arises: Why does this deformation require so much force (stated above)? The atoms around these dislocations are symmetrically located on opposite sides of the extra half plane. In Fig. 2.21, this is indicated by short dashes at a dislocation site, such that the forces acting on the dislocation, on its both sides, are equal and opposite in sign. The force opposing the motion of a dislocation is balanced by the force encouraging its motion. Thus, in a first approximation, there is no net force on the dislocation and the motive

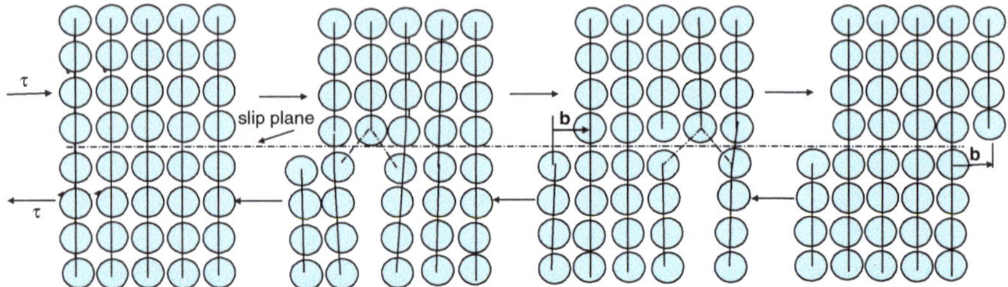

Fig. 2.21 The motion of a dislocation involves the breaking and reforming of a bond (*short dashes*) between an edge atom of the dislocation and its two neighbors, only one bond at a time, resulting in plastic deformation. A step is formed when the dislocation leaves the crystal

stress required is ~0. Nevertheless, the small force actually required to move a dislocation was explained by Peierls-Nabarro. 'Peierls-Nabarro stress', as it became to be known, is given as:

$$\tau_{PN} = \frac{2G}{(1-\nu)} \exp\left(-\frac{2\pi w}{b}\right) \tag{2.9}$$

Here, w is the width of the dislocation, given by:

$$w = \frac{a}{2\,(1-\nu)} \tag{2.10}$$

In Eq. 2.3, "h" was used for the distance between the interatomic planes "a". Equation 2.9 may be substituted for the width to obtain:

$$\tau_{PN} = \frac{2G}{(1-\nu)} \exp\left[-\frac{\pi a}{(1-\nu)b}\right] \tag{2.11}$$

The width of the dislocation, indicated in Eq. 2.9, defines the magnitude of the Peierls-Nabarro stress and is a measure of the degree of distortion that has occurred due to dislocation. It is basically the distance over which the dislocation causes disregistry; thus, it is the magnitude of the displacement of the atoms from their perfect-crystal positions. When "w" is several 'atomic spacings' (the distance between the atomic nuclei), then the dislocation is considered to be 'wide'; if it is on the order of one or two atomic spacings, it is 'narrow'. Dislocations glide easily in the case of wide dislocations, such as in close-packed crystal structures, where the atomic spacing is large. Hence, these materials are ductile and only small stresses are needed to produce large strains. Ceramics, for example, tend to have narrow "w", so that high stresses are required to move dislocations and the ceramics become hard and brittle as a result. Therefore, "w" is defined in terms of "a", the atomic spacing in Eq. 2.10. τ_{PN} varies exponentially with $\left(\frac{a}{b}\right)$, i.e., for large, inter-planar spacing, which corresponds to the distance between close-packed planes, the shear stress is a minimum on such slip planes. Thus, interatomic bonds across such slip planes are weak and the stress is small. Having a small **b** makes the ratio in the exponent larger, so it is understood why dislocations tend to move in the closest-packed direction. The shear stress is much smaller than the theoretical strength in close-packed planes. Actual calculations for an FCC lattice gives $\tau_{PN} \sim 10^{-3}G$, which can be obtained on a FCC (111) plane with $a = b\sqrt{\frac{2}{3}}$. Though this equation indicates the low stress required for dislocation motion, it is not accurate, since the structure at the core of the dislocation and the changes in energy during slip are not known.

The kind of dislocation motion described above largely explains, for example, why FCC metals can be ductile, as long as some obstacle does not retard their motion. Contrary to the idea of breaking all the atomic bonds at once between atoms on a plane (requiring high stress), to break the bonds between individual neighboring atoms at a dislocation site, the stress required is lower by orders. The motion of the dislocation line, depicted in Fig. 2.21, relates to an edge dislocation, but the motion of a screw dislocation is also the result of shear stress; the difference between these two kinds of dislocations is in the direction of their motion. Whereas the direction of motion of an edge dislocation is parallel to the applied stress, in screw dislocation, it is perpendicular. Nevertheless, the net plastic deformation of both the edge and screw dislocations is the same. It has been indicated above that, in general, most dislocations are of a mixed type, exhibiting both characteristics, i.e., at one part of the line there is an edge, while at the other, there is a screw orientation.

2.6.1 *Conservative Motion*

Only shear stress on a glide plane is of interest in regard to the movement of dislocations; this is so because only a shear component acting in the glide plane has an effect on the dislocation motion. A normal stress component acting on the glide plane will not contribute to this motion. Glide occurs in a slip plane without any change in the extra half-plane atomic configuration, namely no atoms are added or removed while the dislocation advances under the influence of stress. It does occur in preferential planes and directions, depending on the structure type. In conservative motion, whether caused by edge or screw dislocation the resultant step formed is the same, despite the difference in movement direction, as shown in Fig. 2.22 for both positive and negative edge dislocations under the same shear stresses.

Again, the directions of the motion of positive and negative dislocations are opposite, yet the end results will be similar as long as the screw dislocation is moving according to the same pair of shear stresses. Both the RH and LH screw dislocations produce the same steps under the same stress for edge dislocation, as seen in Fig. 2.22. LH screw dislocations move to the rear, while RH screw dislocations progress to the front. Students should practice schematically drawing the motions of RH and LH screw dislocations. Figure 2.23 is a schematic presentation of the motion of a positive-edge dislocation as shown in the upper illustrations in Fig. 2.22.

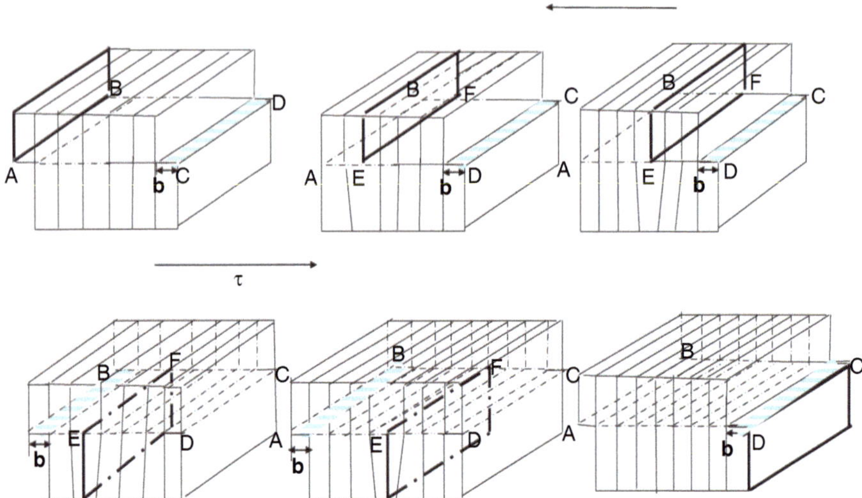

Fig. 2.22 Movements of positive and negative dislocations are shown under the same shear stresses. The movement directions of the dislocations are opposite, but the results of the shear are the same – a step is formed on the same crystal plane

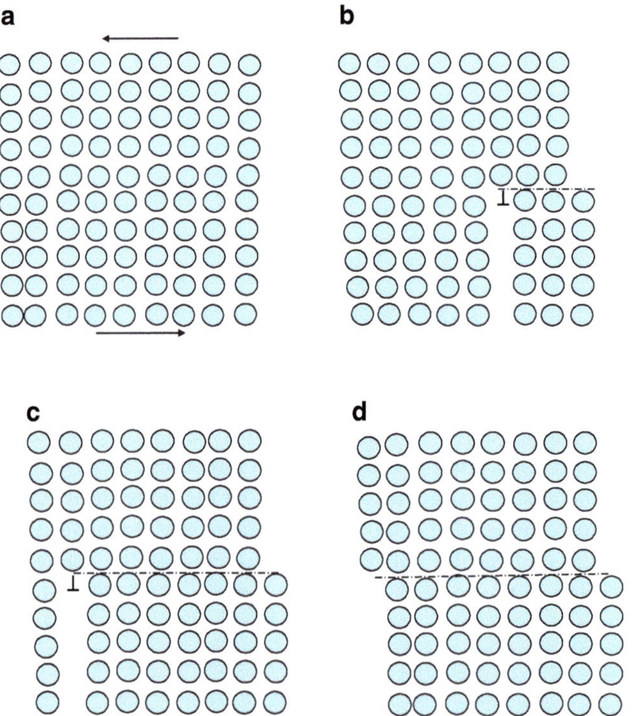

Fig. 2.23 Edge-dislocation motion on its slip plane under the influence of shear stress. The crystal deforms as seen through (**a**)–(**d**). When the dislocation has run out of the crystal (**d**), a step is formed having the slip vector's dimension

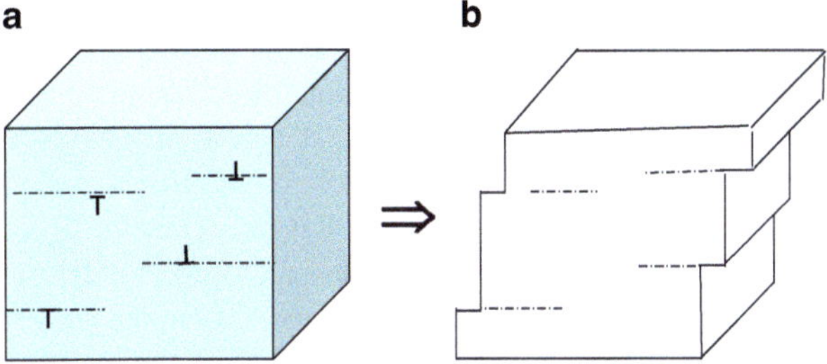

Fig. 2.24 Steps formed by shear stress caused by dislocations leaving their respective slip planes

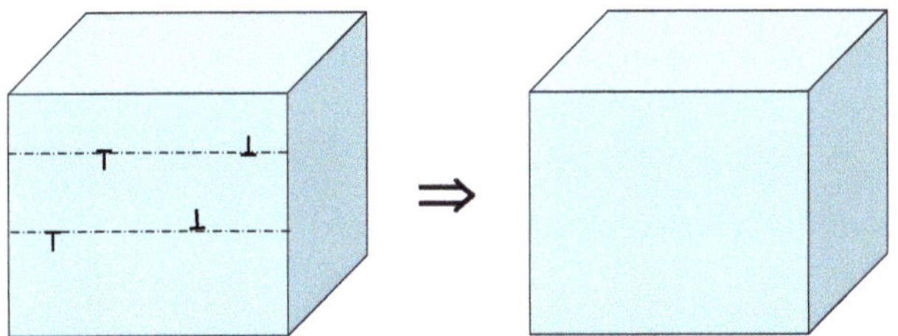

Fig. 2.25 Two sets of dislocations having opposite signs on the same slip plane. They annihilate each other, resulting in a perfect crystal

At a first glance, one might suggest how to make a perfect defect-free crystal. One would simply have to apply an appropriate shear and squeeze the dislocation out of the crystal, leaving behind a dislocation-free crystal, thus obtaining a high-strength material, as calculated in Sect. 2.2. Obviously, this naive thought might arise in the minds of students who have not yet learned all about dislocations. They are unaware that, during the deformation process, very many dislocations are formed, multiply and interact as the stress is applied and that their movement is often blocked by a host of possible obstacles. In fact, materials actually become stronger not because the dislocations have been squeezed out of them, but rather due to the application of stress and the resulting interactions between the dislocations formed.

When more than one set of dislocations leave a crystal, more steps are formed. To illustrate this effect, take the following Figure 2.24, having two positive and two negative edge dislocations in their respective slip planes, and apply a pair of shear stresses. The slip planes, shown (partially) in Fig. 2.24b, indicate the planes of (a) where the dislocations have moved forming the steps upon leaving the crystal. If the positive and negative dislocations were on the same planes, as shown in Fig. 2.25, they would annihilate each, other resulting in a perfect lattice. One has to realize that, above an edge dislocation, the atomic planes experience compression (because an extra half plane has been inserted), while, below it, the atoms are being pulled apart or tension exists. In other words, the strain fields above and below a dislocation line are compressive and tensile, respectively. The energy of a crystal is reduced by the reaction of the compressive and tensile strain fields, resulting in the cancellation of the pair of dislocations.

An edge dislocation is able to glide across a plane, containing the edge dislocation line and its Burgers vector. Thus, an edge dislocation is confined to movement within one plane, unlike the motion of screw dislocation. The glide plane is defined by its normal, given as $\mathbf{b} \times \mathbf{t}$. A screw dislocation does not have a defined glide plane, since $\mathbf{b} \times \mathbf{t} = 0$ and it has a cylindrical symmetry about its axis. Therefore, every plane passing through that axis (known as the 'zone axis') may become a glide plane through which a screw dislocation may move. Therefore, screw dislocation is not restricted to only one plane. Often, the step-by-step motion of edge dislocation, breaking and reforming each individual bond, is more easily explained by an analogy to the motion of a caterpillar, or to the removal of a wrinkle in a large carpet. The localized force employed by the special motion of a caterpillar or the force to move the carpet with the wrinkles into its correct place is much smaller than that required to pull an entire carpet straight.

2.6.2 Non-conservative Motion (Climb)

As indicated above, the motion of an edge dislocation is restricted to one plane – only the glide plane (or slip plane). The glide plane can be left by the dislocation only through the process of 'climb', illustrated in Fig. 2.26 for a positive- and a negative-edge dislocation. For climb to happen in a positive dislocation (the top illustrations of Fig. 2.26), vacancies, arriving by diffusion, must occur adjacent to the dislocation line by replacing one or more of its atoms. This is a thermally activated process, causing part of the extra plane, defining the dislocation, to become shorter. Since these vacancies never appear at the same time in sufficient amounts, only part of the dislocation climbs. The climb of a segment of a dislocation forms a 'jog' (i.e., a segment of a dislocation line having a vector normal to the glide plane). Jog formation is shown in Fig. 2.27. The dislocation segment that climbed (CD) and the jogs formed are heavily outlined in this figure. Thus, rather than an entire dislocation line shifting all at once, only segments of a dislocation climb. In addition to the jogs, 'kinks' are also formed during the process of climb; these are the segments of the dislocation line that have remained within the glide plane that did not climb. Jogs are immobile, unlike kinks, and cannot glide together with the dislocation segments that did not climb (i.e., kinks), since their direction of motion is in a plane normal to the glide plane and not within the glide plane itself. This is shown in Fig. 2.28. A kink is basically an edge dislocation and, if it is lying in the slip plane, it does not provide resistance to motion and may be considered a step in the dislocation line. A diffusion of vacancies (or interstitials) is required for the motion of jogs. Jogs and kinks can also be formed by dislocation interactions (discussed later on).

Screw dislocation cannot climb, since it does not have a specific, well-defined glide plane, as mentioned above. Every plane passing through the zone axis may be considered a glide plane; when in the motion, screw dislocation can pass from one glide plane to another, within the family of planes sharing a common zone axis.

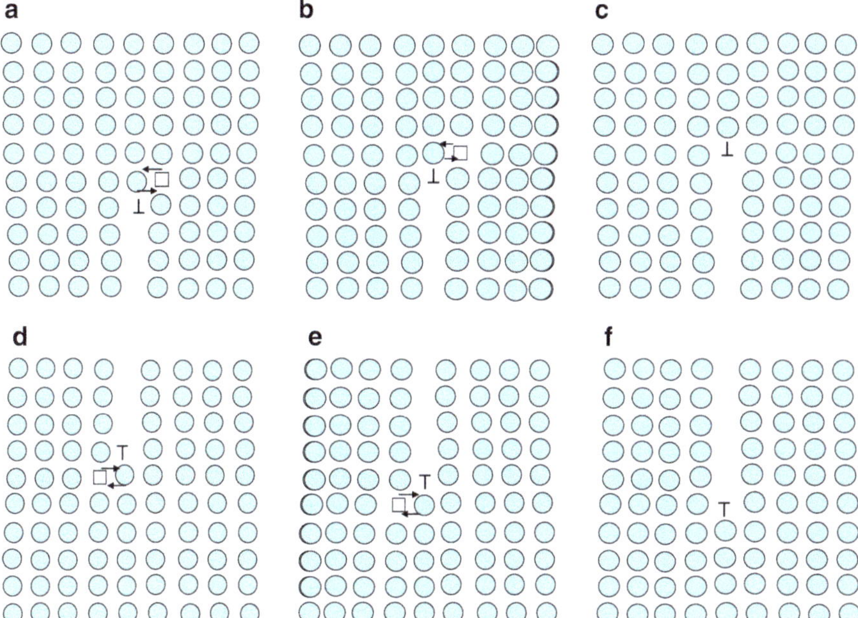

Fig. 2.26 Climb in positive and negative dislocations. When a point defect, such as a vacancy, reaches the vicinity of a dislocation, as shown in (**a**) it changes places with the indicated atom of the dislocation and climb occurs, resulting in (**b**) and (**c**) Dislocation climb means that a part of the dislocation has left its slip plane

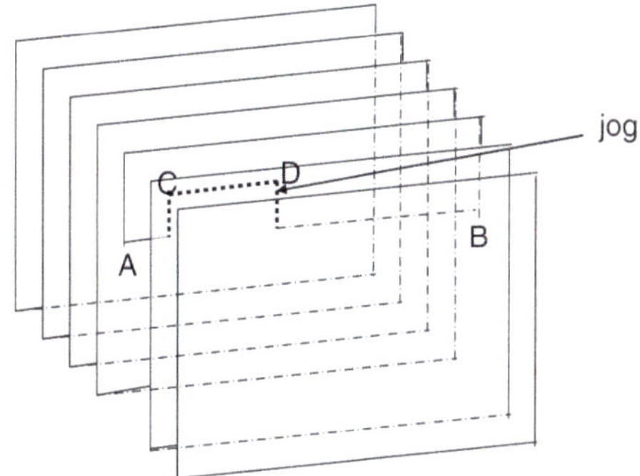

Fig. 2.27 Jog formation by climb

'Cross slip' is the term used for the process consisting of passing a screw dislocation from one plane to another plane with an equivalent glide plane to that of the original one. Thus, cross slip involves the initial slip plane and the cross-slip plane, where the glide continues. Usually, when a screw dislocation meets some obstacle in its glide, it can circumvent that obstacle by the process of cross slip. Many obstacles can impede dislocation movement, such as: other dislocations, precipitates, grain boundaries, etc.

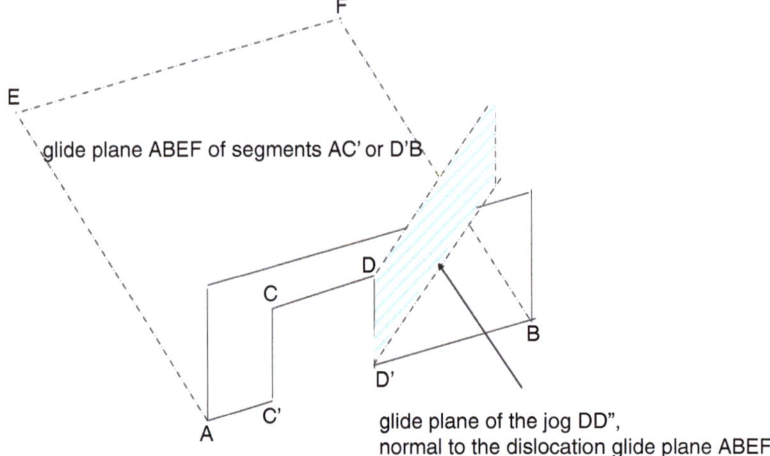

Fig. 2.28 One of the jogs *DD* is shown with its glide plane *shaded*; it is normal to the glide plane of the dislocation before its climb

To conclude Sect. 2.6 on dislocation motion, the following can be added. Dislocations that are able to move by pure slip are called 'glissiles' and dislocations which cannot glide without some thermal process, such as point defect diffusions, are 'sessiles'. Glissile dislocations play a role in plastic deformation, in which good elongation is of importance, since large strains can be achieved by the motion of many dislocations without considerable impedance to their motion. Sessile dislocations, on the other hand, are among the factors responsible for high strength, but lower strains. Thus far, nothing has been said about a very important factor that influences the motion of dislocations, namely the structure of materials, which will be discussed in a later section in regard to FCC metals.

2.7 The Energy of Dislocations

It was previously mentioned that dislocations have energy, since their formation requires work and the resultant energy is stored in the material. Chapter 1 discussed the application of stress to produce strain, introducing elastic strain energy caused by atoms displaced from their equilibrium position. Whenever atoms are displaced from their lattice equilibrium position, strain energy is introduced into the lattice. A dislocation must have a certain strain energy, since it distorts the lattice and displaces atoms from their lattice equilibrium position; therefore, a strain is introduced into the surroundings. A strain field, thus, exists at the site of the dislocation, affecting its motion. Bonds outside the radius of the dislocation line, say at a distance r_0, are elastically strained, but some of the energy lies within the radius, r_0, stored in what is known as the 'core of the dislocation'. Thus, the total energy per unit length of the dislocation line, E_{TE}, is composed of the core energy and the energy outside r_0, given by:

$$E_{TE} = E + E_{core} \tag{2.12}$$

where E represents the energy outside the dislocation core. Note that, in Chap. 1, U was used to symbolize energy, but, in keeping with the energy notation in this section, the term E will be used. For the present, consider only E (discussion on the contribution of the core follows at a later stage).

Figure 1.1 had illustrated an example where, as a result of extending a rod of length "l" and cross-section "A" by tension, the applied force had extended the specimen by dl and the introduced strain energy, dE, was simply the force applied times the change in length, dl:

$$dE = Fdl \tag{2.13}$$

According to Eq. 1.27:

$$F = \sigma A \tag{1.27}$$

and also according to Eq. 1.9, one can write:

$$d\varepsilon = \frac{dl}{l} \tag{1.9}$$

The constancy of the volume in Eq. 1.34 was presented and V = Al. Now, by simple substitution, one can write:

$$dE = \sigma V \frac{dl}{l} = \sigma V d\varepsilon \tag{2.14}$$

Due to the constancy of the volume, Eq. 2.14 can now be replaced by the energy per unit volume, dE′:

$$\frac{dE}{V} = dE' = \int_0^{\varepsilon} \sigma d\varepsilon \tag{2.15}$$

and, by applying Hooke's Law, the last term from Eq. 2.15 is obtained:

$$E' = \frac{1}{2}\sigma\varepsilon = \frac{1}{2}E\varepsilon^2 \tag{2.15a}$$

For shear stress and the shear modulus, Eq. 2.15a should be written as:

$$E' = 1/2\tau\gamma = 1/2G\gamma^2 \tag{2.15b}$$

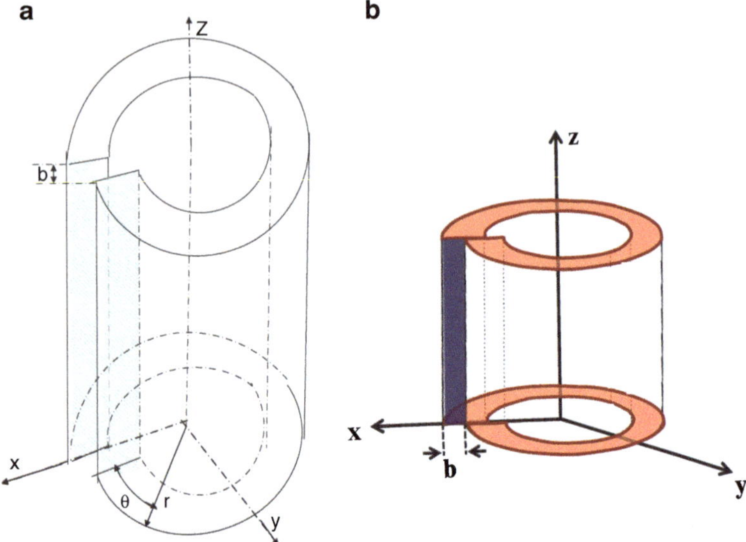

Fig. 2.29 Elastic distortion of a cylindrical ring (**a**) produced by screw dislocation; (**b**) formed by edge dislocation

The general relation from the principle of superposition is:

$$E' = \frac{1}{2}\left(\sigma_{xx}\varepsilon_{xx} + \sigma_{yy}\varepsilon_{yy} + \sigma_{zz}\varepsilon_{zz} + \tau_{xy}\gamma_{xy} + \tau_{xz}\gamma_{xz} + \tau_{yz}\gamma_{yz}\right) \tag{2.16}$$

But when a single strain component exists, the strain energy per volume is either $1/2E\varepsilon^2$ or $1/2G\gamma^2$. In Eq. 2.15a, E is Young's Modulus and not the energy.

2.7.1 *Screw Dislocation*

It is simpler to consider the case of screw dislocation, in which only a shear stress is active and results in shear strain. Consider a segment taken from a cylinder with length l and radius r, as shown in Fig. 2.29a (in polar coordinates), and which behaves like an elastic continuum. The strain in the shell of radius dr is presented below. A radial slit has been cut in the cylinder (not shown) parallel to the z axis and then the cut surfaces were displaced one over the other by shear deformation at amount **b** and glued together. This displacement by the shear is **b** pointing in the direction of z. Now, unfold the shell and spread it flat, for the sake of understanding the derivation. The unfolded cylindrical shell from Fig. 2.29a is shown in Fig. 2.30, indicating the distortion that has occurred due to the shear stress. The circumference is $2\pi r$ and the angle of the shear strain is $\gamma = b/2\pi r$. The displacement is in the z

Fig. 2.30 The segment of the cylinder seen in Fig. 2.29a in unfolded form

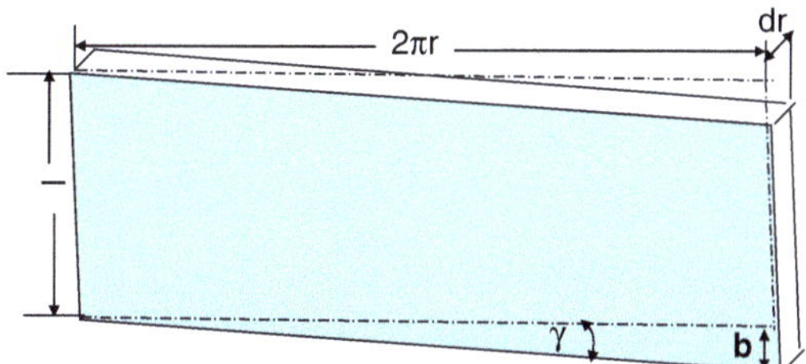

direction and the energy stored per unit volume between r and dr for this case may be written as $Gb^2ldr/4\pi r$. Then for r, between the limits r_0 and r after integrating Eq. 2.17, is obtained:

$$E' = \int_{r_0}^{r} \frac{Gb^2 l}{4\pi} \frac{dr}{r} = \frac{Gb^2 l}{4\pi} \ln \frac{r}{r_0} \tag{2.17}$$

For a unit dislocation line, $E'/l = E$, one can write the energy of a screw dislocation as:

$$E = \frac{Gb^2}{4\pi} \ln \frac{r}{r_0} \tag{2.18}$$

Equation 2.18 is the result of the following steps:

(a) the area under a stress/strain curve is the stain energy per unit volume (see Eq. 1.50b); in our case, for shear it is: $E = \frac{1}{2}\tau\gamma$;
(b) $\tau = G\gamma$;
(c) substituting for τ leads to $E = \frac{1}{2}G\gamma^2$;
(d) the volume of the thin shell (Fig. 2.29) is $dV = 2\pi rl$; therefore, the energy in the elemental volume is:
(e) $E' = \frac{1}{2}\frac{Gb^2}{(2\pi r)^2}2\pi rldr = \frac{Gb^2}{4\pi r}ldr$;
(f) integrating between the limits r_0 and r and taking the energy per unit line gives Eq. 2.18.

Since the above displacement was in the z direction, there are no displacements in the x and y directions. Thus, all the hydrostatic and shear components not act except for the shear stress in the z direction and its respective strain, as derived above. This strain is γ and, thus, the stress is $\tau = Gb/2\pi r$. Note that, in the above equation, the strain should actually be written as $\gamma_{z\theta} = \gamma_{\theta z}$ and the shear stress, accordingly, as $\tau_{z\theta} = \tau_{\theta z}$; but for simplicity sake, the indexes were dropped. The above is a consequence of the fact that, in screw dislocation, only two shear stresses are active in a radial plane in the z direction and in a plane perpendicular to the z axis and parallel to the radius, respectively for $\tau_{\theta z}$ and $\tau_{z\theta}$. The stress field has a radial symmetry and is independent of θ, as seen in Eq. 2.18. It might be indicated in the system under consideration (polar coordinates) that since the shear and, thus, the displacement were in the z direction, no strains exist in the x and y directions and the corresponding strains are zero, as are the stresses shown below:

$$\sigma_r = \sigma_\theta = \sigma_z = \tau_{r\theta} = \tau_{\theta r} = \tau_{zr} = \tau_{rz} = 0 \tag{2.19}$$

There is a limit to the validity of Eq. 2.18. For example, it is unlikely that when in the expression $\tau = Gb/2\pi r$, $r \to 0$ the stress is infinite since $\tau \to \infty$. Equation 2.18 is valid only outside a radius of ~5–10 Å. By using reasonable values for r and r_0, Eq. 2.18 can be approximated as:

$$E \approx Gb^2 \tag{2.20}$$

This can be obtained as follows. Assume that $r \sim 10^5 b$ (providing a dislocation density of $\sim 10^{10}$ dislocations/m^2). Since Hooke's Law applies only for small strains and not to the very high values that exist in the core of the dislocations, a value of b/4 has often been suggested for r_0. Substituting these values into Eqs. 2.18, 2.20 is obtained.

An estimated value of the core energy is then:

$$E_{core} \cong \pi(5b)^2 \left(\frac{G}{30}\right)\frac{1}{G} \approx \frac{Gb^2}{10} \tag{2.21}$$

This expression is obtained by the following assumptions:

(a) that the stress level in the vicinity of the core is $\tau = G/6$, as shown in Eq. 2.7;
(b) that the radius is approximately 5b;
(c) that the area is $\sim\pi\ (5b)^2$;
(d) that the elastic energy per unit volume, acting on the dislocation core, is $E' = 1/2G\gamma = 1/2\tau^2/G$ (Eq. 2.15b);
(e) by using the last term in (d) with the assumption in (a) and multiplying by the area of (c), Eq. 2.21 is obtained:

$$1/2\tau^2/G = \frac{1}{2}\frac{\tau^2}{G} = \pi(5b)^2\frac{1}{2}\left(\frac{G}{30}\right)^2\frac{1}{G} \approx \frac{G}{10}$$

The total energy, according to Eq. 2.12, can now be expressed by Eqs. 2.18 and 2.21, yielding for $E_{TE} \approx Gb^2 + Gb^2/10 \cong Gb^2\ (1 + 1/10)$. The core energy of 0.1, according to the estimated expressions, is small or even negligible in its contribution to the overall energy, so it is often neglected.

2.7.2 *Edge Dislocation*

The analysis of edge dislocation is somewhat more complicated, but it can be done in a manner similar to that of screw dislocation, again using a ring-shaped segment of a material. The edge dislocation line is also parallel to the z axis, as in the case of screw dislocation, but the Burgers vector is normal to the z axis, as shown in Fig. 2.29b. Thus, the **t** vector of the dislocation line points in the direction of the z axis. Since the line is in the z direction and its displacement is parallel to the x axis, i.e., the Burgers vector is normal to the z axis, there will be no strain (no displacement) in the z direction, but the stress will not be zero.

The energy of an edge dislocation per unit length of the dislocation line (stated without proof) is similar to the energy of a screw dislocation, but it is divided by (1 − ν), where ν is Poisson's ratio, as given by:

$$E_{edge} = \frac{Gb^2}{(1-\nu)4\pi} \ln \frac{r}{r_0} \tag{2.22}$$

As indicated above, a value of ~0.1Gb^2 is the contribution of the dislocation core, which is neglected. Equation 2.22 indicates that the energy of an edge dislocation is larger by 1/(1 − ν), i.e., 1.5 times greater than that of a screw dislocation having the same length "l".

It is known that, in materials, the dislocation line's direction and the Burgers vector are neither perpendicular nor parallel – being mixed dislocations, consisting of both screw and edge characteristics. Therefore, it is appropriate to sum up the dislocation energies given in Eqs. 2.18 and 2.22, while neglecting the contribution of the core. This may be expressed as:

$$E_{total} = \frac{Gb^2}{4\pi} \ln \frac{r}{r_0} + \frac{Gb^2}{(1-\nu)4\pi} \ln \frac{r}{r_0} = \frac{Gb^2}{4\pi} \ln \frac{r}{r_0} \left[1 + \frac{1}{(1-\nu)}\right] \tag{2.23}$$

In this summation, done in order to estimate the energies of mixed dislocations, the interactions between the edge and screw components are ignored. Note that, here, energy is given per unit length. Several comments are now in order:

(a) Energy is proportional to b^2; thus, a dislocation of large Burgers vector, e.g. of strength 2b, is not stable and is likely to decompose into two vectors, having a lower energy:

$$b^2 + b^2 = 2b^2 < (2b)^2$$

(b) The energy of dislocations, as seen in Eqs. 2.18, 2.22 and 2.23, is proportional to b^2; therefore, the most stable dislocations are those with minimum Burgers vectors, namely those with the closest-packed directions.
(c) Since $r_0 = 0$ is not reasonable, providing an infinite value for energy, it is customary in the continuum model to avoid this by introducing a hole along the dislocation line. Thus, r_0 is the radius of a hole with a diameter of about 10^{-7} cm.

Equation 2.23 may be presented in a different form. To this end, one needs a schematic drawing of a mixed dislocation in a slip plane, as shown in Fig. 2.31a.

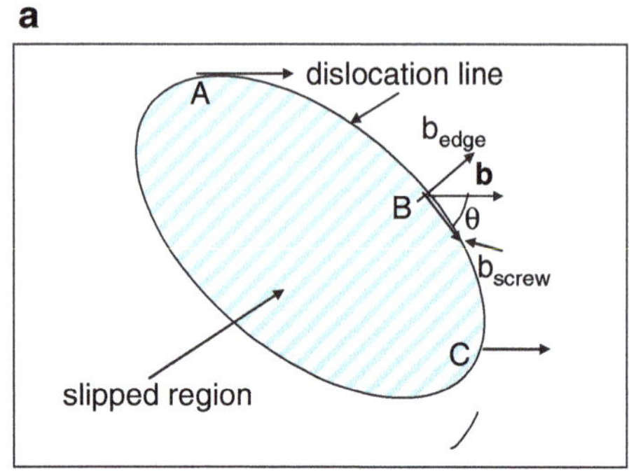

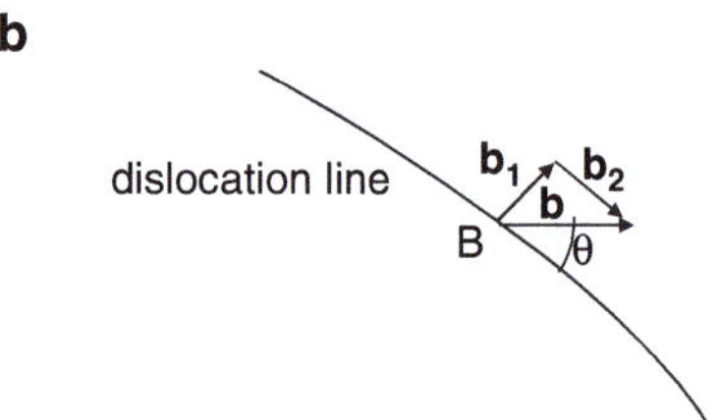

Fig. 2.31 A 'general dislocation line' is the boundary between the slipped and unslipped regions: (**a**) at point B, the Burgers vector **b** can be resolved into $\mathbf{b_{edge}}$ and $\mathbf{b_{screw}}$; (**b**) the dislocation at point B is shown (*magnified*) and **b** is resolved into edge and screw components

In this figure, the dislocation is in the shape of a closed line and, for a given displacement of the vector **b**, both the edge and screw orientations are shown. Some mixed characteristics can be seen, for example at point B. The Burgers vector is constant, pointing to the displacement of the line, but the dislocation line changes direction. At point A, the line is parallel to **b** and, therefore, it has the character of a screw dislocation, while, at point C, the segment of the dislocation line is perpendicular, like a pure edge. At point B, **b** can be resolved into two components with edge and screw orientations designated as b_{edge} and b_{screw} (Fig. 2.31b). In Figure 2.31a, the slipped region is shaded and the direction of the dislocation line at point B is shown. One can write for the edge and screw components:

$$\mathbf{b_{edge}} = \mathbf{b}\sin\theta$$

$$\mathbf{b_{screw}} = \mathbf{b}\cos\theta$$

Substituting these values in Eq. 2.23, to find the mixed dislocation at point B in Eq. 2.24, one gets:

$$E_{total} = \left[\frac{Gb^2\cos^2\theta}{4\pi} + \frac{Gb^2\sin^2\theta}{(1-\nu)4\pi}\right]\ln\frac{r}{r_0} = \frac{Gb^2}{4\pi(1-\nu)}\ln\frac{r}{r_0}\left[1-\nu\cos^2\theta\right] \tag{2.24}$$

Thus, the value of a mixed dislocation is between that of screw and edge dislocations.

2.8 Line Tension

The expressions for the energies above were given per unit length of dislocation. When not normalized for a unit length, the energy is proportional to the length of the dislocation:

$$\mathrm{E} \approx \mathrm{lGb}^2 \tag{2.25}$$

As a consequence, a dislocation will tend to decrease its length to a minimum to decrease its energy. A curved dislocation line has a tension, **T**, which acts along its line and can be expressed as a change in energy with its length

$$T = \frac{\partial E}{\partial l} \approx Gb^2 \tag{2.26}$$

The units of line tension in Eq. 2.26 are in energy per unit length.

A dislocation line can encounter various obstacles to its motion. In such a case, the dislocation becomes pinned by the obstacle and curves due to the forces exerted by the obstacle resisting its motion. Movement by the dislocation cannot occur under small forces. In Fig. 2.32, the bowing of a dislocation line is shown after it has been pinned at both ends, at points A and B, by obstacles, perhaps foreign particles in the material or some other precipitates. For the sake of better understanding this figure, consider Eq. 2.17, which may now be rewritten as:

$$T = \alpha Gb^2 = \frac{lGb^2}{4\pi} \ln \frac{r}{5b} \tag{2.27}$$

In Eq. 2.17, r_0 was replaced by 5b, assumed to represent the core of the dislocation. The line tension, T, is acting tangentially, to shorten the dislocation line and, thus, reduce the dislocation energy. The curving of the dislocation line segment, due to

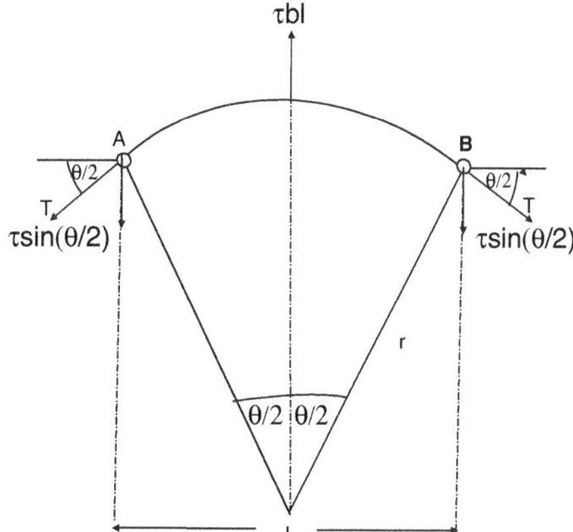

Fig. 2.32 The line tension of a dislocation when blocked by obstacles A and B

its normal force, τbl, as a result of the applied shear stress, is shown in the above figure. For the force equilibrium in the y direction, since it is balanced by the line tension, it is possible to write:

$$\tau bl = 2T \sin\frac{\theta}{2} \tag{2.28}$$

The coefficient 2 arises from the fact that the line tension acts at points A and B of the dislocation line segment. For small angles, sin (θ/2) = θ/2. Substituting for T from Eq. 2.27, τ can be expressed as:

$$\tau = \frac{\alpha G b \theta}{l} = \frac{\alpha}{l} G b \frac{l}{r} = \frac{\alpha G b}{r} \tag{2.29}$$

The last term in Eq. 2.29 is a consequence of substituting θ = l/r for θ/2 from Fig. 2.32. Note (from the first term in Eq. 2.29) that the stress required to motivate a dislocation being detained by some obstacle depends on its length l. For small values of l, the required force is higher.

Line tension in dislocations will be significant for the discussion (below) of dislocation multiplication sources, according to the ideas of Frank-Read.

2.9 The Stress Field of a Dislocation

It has been mentioned above that it is simpler to consider screw dislocations, since they possess a cylindrical symmetry and only shear stress acts when displacement occurs in the z direction. Elastic distortion occurs with the introduction of a dislocation into an isotropic material.

2.9.1 *Screw Dislocations*

Here is another case of a cylindrical ring cut from good material surrounding a screw dislocation, as shown in Fig. 2.29a. Recall that dislocations are defects in crystals and, therefore, are accompanied by elastic stress fields. Now, a radial cut is made parallel to the z axis (as shown in Fig. 2.29a), displaced by **b**, and the surfaces are joined together. The shear strain obtained is:

$$\gamma_{\theta z} = \gamma_{z\theta} = \frac{b}{2\pi r} \tag{2.30}$$

The stress field of a screw dislocation is pure shear. As indicated earlier, high strains exist in the core region and, therefore, Hooke's Law of elasticity does not apply there and, thus, will not be considered. The dislocation line is parallel to the z axis. There are no displacements in the x and y directions; the other stress components are zero:

$$\sigma_r = \sigma_\theta = \sigma_z = \tau_{r\theta} = \tau_{\theta r} = \tau_{rz} = \tau_{zr} = 0$$

Once again, the magnitude of the elastic stress is directly proportional to the shear modulus, G, and the Burgers vector, $\boldsymbol{b}$, and inversely proportional with increasing distance from the dislocation, given as:

$$\tau_{\theta z} = \tau_{z\vartheta} = \frac{Gb}{2\pi}\frac{1}{r} \tag{2.30a}$$

The stress field of a screw dislocation is manifested by two active stresses: $\tau_{\theta z}$ acting in radial planes parallel to the z axis and $\tau_{z\theta}$ acting in planes normal to the z axis. The stress has a long range, because it is inversely proportional to r and perpendicular to the radius. To get a feeling for this, take a distance of 10^4b, for 1/r the stress is $\sim 10^{-5}$G, which is about the yield stress of some crystals. The stress vanishes at infinity.

Generally, these displacements can be expressed as:

$$u = 0$$

$$v = 0$$

$$w = \frac{b}{2\pi}\arctan\frac{y}{x} = \frac{b}{2\pi}\theta \tag{2.31}$$

The above is obtained from the equilibrium equations of elasticity:

$$\gamma_{xy} = \frac{\partial u}{\partial y} + \frac{\partial v}{\partial x}$$

$$\gamma_{xz} = \frac{\partial w}{\partial x} + \frac{\partial u}{\partial z}$$

$$\gamma_{yz} = \frac{\partial w}{\partial y} + \frac{\partial v}{\partial z} \tag{2.32}$$

or, generally:

$$\gamma_{ij} = \frac{1}{2}\left(\frac{\partial u_i}{\partial x_j} + \frac{\partial u_j}{\partial x_i}\right) \tag{2.32a}$$

The expression for w in Eq. 2.31 is obtained as follows. While making a circuit around a dislocation line, the arctan changes by 2π and, thus, w changes as π.

In rectangular coordinates and by applying Eq. 2.32, while recalling that the displacements u and v = 0, one gets:

$$\gamma_{xz} = \frac{\partial w}{\partial x} + \frac{\partial u}{\partial z} = -\frac{b}{2\pi}\frac{y}{(x^2+y^2)} = \frac{b}{4\pi}\sin\frac{\theta}{r}$$

$$\gamma_{yz} = \frac{\partial w}{\partial y} + \frac{\partial v}{\partial z} = \frac{b}{2\pi}\frac{x}{(x^2+y^2)} = \frac{b}{4\pi}\cos\frac{\theta}{r}$$

$$\gamma_{xy} = \frac{\partial u}{\partial y} + \frac{\partial v}{\partial x} = 0 \tag{2.32b}$$

Deriving the displacement via these relations for the strain, the stress is given by multiplying with G. In rectangular coordinates, there are only two stress components:

$$\tau_{xz} = G\left(\frac{\partial w}{\partial x} + \frac{\partial u}{\partial z}\right) = -\frac{Gb}{2\pi}\frac{y}{(x^2+y^2)}$$

$$\tau_{yz} = G\left(\frac{\partial w}{\partial y} + \frac{\partial v}{\partial z}\right) = \frac{Gb}{2\pi}\frac{x}{(x^2+y^2)} \tag{2.32c}$$

Note that no normal stresses act and, of all the shear stresses, only the two above exist. All the other stresses are zero, as indicated below.

$$\sigma_x = \sigma_y = \sigma_z = \tau_{xy} = \tau_{yx} = 0$$

One can note that, when going from Cartesian to cylindrical coordinates (or vice versa), the relations indicated below may be used:

$$r^2 = x^2 + y^2$$

$$\tan\theta = y/x$$

$$z = z \tag{2.33}$$

The last term in Eq. 2.32c is a consequence of Fig. 2.29a, schematically representing all the relevant parameters of the cylinder, but not drawn for clarity.

2.9.2 Edge Dislocations

As shown in Fig. 2.29b, **b** is parallel to the x axis and vector **t** is in the z direction. The displacement is in the x direction; therefore, no strain in z direction exists. In cylindrical coordinates, one can write:

$$\sigma_r = \sigma_\theta = -D\frac{\sin\theta}{r}$$

$$\tau_{\theta r} = \tau_{r\theta} = D\frac{\cos\theta}{r}$$

$$\sigma_z = \nu\,(\sigma_r + \sigma_\theta) \tag{2.34}$$

where:

$$D = \frac{Gb}{2\pi\,(1-\nu)} \tag{2.34a}$$

and:

$$\tau_{rz} = \tau_{zr} = \tau_{z\theta} = \tau_{\theta z} = 0$$

In Cartesian coordinates, the above gets the form of:

$$\sigma_x = -\,Dy\frac{(3x^2 + y^2)}{(x^2 + y^2)^2}$$

$$\sigma_y = Dy\frac{(x^2 - y^2)}{(x^2 + y^2)^2}$$

$$\tau_{yx} = \tau_{xy} - Dx\frac{(x^2 - y^2)}{(x^2 + y^2)^2}$$

$$\sigma_z = \nu\left(\sigma_x + \sigma_y\right) \tag{2.35}$$

and:

$$\sigma_{xz} = \sigma_{zx} = \sigma_{yz} = \sigma_{zy}\tau_{xz} = \tau_{zx} = \tau_{yz} = \tau_{zy} = 0$$

Note that, although no displacement and strain evolves in the z direction, a stress is acting, as shown in Eq. 2.35. Clearly, dilatational (expanding) stresses are also active, in addition to the shear stress. The largest normal stress is σ_x, which acts parallel to the slip vector **b**. The slip plane, according to Fig. 2.29b, is along the plane passing through $y = 0$. The sign above the edge dislocation is negative, therefore, the stress is compressive and maximal just above the slip plane of the dislocation, whereas the stress is tensile, thus, positive (σ_x) and maximal just below the slip plane. The signs of the stress and strain reverse if the sign of the Burgers vector is reversed.

In cases of mixed dislocations, it is possible to add their components, namely the solutions of their screw and edge orientations.

For arbitrary orientations in an isotropic material, when stress depends on the orientation of the dislocation and the Burgers vector and the crystal axis are significant, stress fields and stress distributions around edge dislocations are shown and discussed in detail in Read's book.

2.10 The Forces Acting on Dislocations

The presence of forces acting on dislocation lines, balancing line tension, was discussed above in Sect. 2.8. When an external stress is applied to a dislocation it will tend to move. In general, dislocations in stress fields sense a force that tends to move them. The motion of a dislocation, as seen in Sect. 2.6 above, can either be a glide or a climb in cases of edge dislocations or cross slips in screw dislocations.

2.10.1 The Glide Forces

An external force acting on a dislocation may induce its glide; this effect may be seen in Fig. 2.33a. A slip plane, A, is shown, where a segment, dl, of a dislocation is depicted that moved a distance, dx. The ratio of the area swept out by the motion of the segment, dA = dldx to the total slip area, multiplied by **b**, gives the displacement, which can be written as:

$$displacement = \frac{dldx}{A}\mathbf{b} \tag{2.36}$$

The applied force is F = τA and, therefore, the work, dw, performed by the force is given by F x distance, namely:

$$dw = F\frac{dldx}{A}\mathbf{b} = A\tau\frac{dldx}{A}\mathbf{b} \tag{2.37}$$

If one defines the force on a unit length of dislocation as 'the work done when the unit length of the dislocation segment dl moves a unit distance dx', one gets:

$$F = \frac{dw}{dldx} = \frac{dw}{dA} = \tau\mathbf{b} \tag{2.38}$$

This force is normal to the dislocation line. Apparently, only a shear stress, acting on the dislocation in this plane, is necessary for its motion and only a force component, lying in the glide plane, is effective in this regard. A normal component of force or stress will not contribute to glide motion, since only components of force acting in the glide plane in the direction of the Burgers vector can move a dislocation. Practically speaking, the dislocation line never moves in total, because it may be anchored somewhere by an obstacle (as above in Sect. 2.8 on line tension, where the dislocation became curved).

The above relation might have also been obtained in the following way. The work, w, by moving the dislocation a distance, **b**, on slip area, A, is:

$$\mathrm{w} = \mathrm{A}\tau\mathbf{b} \tag{2.39}$$

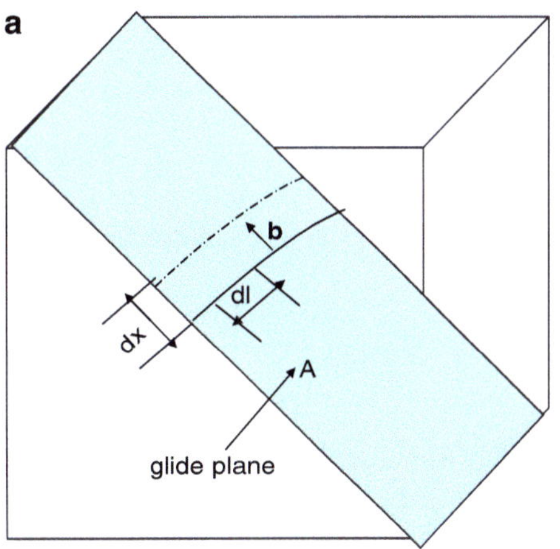

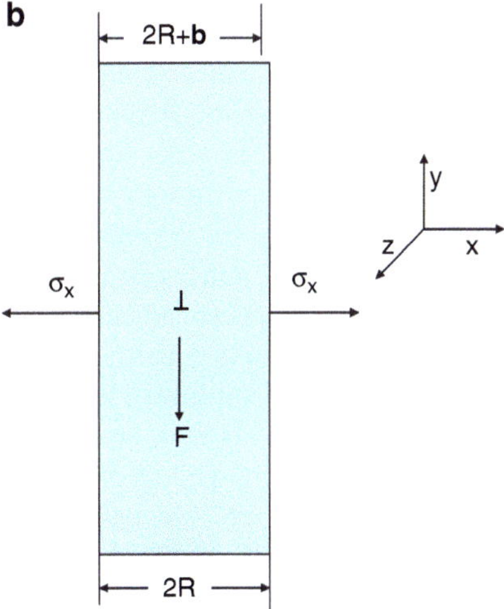

Fig. 2.33 (**a**) A segment of dislocation, dl, has moved a distance, dx, in the glide plane of area A as a result of a force acting on it perpendicularly. (**b**) This force is acting in the negative y direction, because tensile stress induces the negative climb of dislocation. Note that the width at the lower edge of the curve is less by **b** than the upper edge, since a dislocation is defined by the extra half plane

where $A\tau = F$. Incremental work, dw, done on an incremental area is dldx (the area swept out by the elemental dislocation line, dl, when moving an incremental distance, dx) may be related to the work moving the dislocation. Thus:

$$\frac{dw}{w} = \frac{dldx}{A} \tag{2.40}$$

Following all the prior relations, dw can be expressed as:

$$dw = A\tau b \frac{dldx}{A} = \tau bdldx \tag{2.41}$$

but dw = Fdx and, when acting on the dislocation segment, dl, dw = Fdxdl. Redefining the force acting on a unit dislocation as $F' = F/dl$, for Eq. 2.41 one can obtain:

$$F' = \tau b \tag{2.42}$$

F' is the same as F, given and defined in Eq. 2.38.

2.10.2 Climb

Climb was first discussed in Sect. 2.6.2 above. Figure 2.26 illustrated the climb process and Fig. 2.27 showed the jog formation resulting from climb. Climb is the movement of dislocations out of their glide planes as a result of interactions with vacancies (or interstitials, for negative climbs, i.e., the growth of an extra plane or part of it). Now, it is possible that a force, acting on an edge dislocation, will be directed perpendicularly to the slip plane, normal to its Burgers vector, allowing an edge dislocation to leave its slip plane. The driving force for dislocation climb is the movement of vacancies through a crystal lattice. If a vacancy moves next to the extra half plane of atoms that defines an edge dislocation, the atom in the half plane, closest to the vacancy, can trade places with that vacancy, thus causing climbing. Compressive force produces positive climb in the positive y direction**,** whereas tensile force induces negative climb. During positive climb, the crystal shrinks in the direction perpendicular to the extra atomic half plane, because atoms are being removed from it. Since a negative climb involves the addition of atoms to the half plane, the crystal grows in the direction perpendicular to that half plane. Therefore, compressive stress, in the direction perpendicular to the half plane, promotes positive climb, while tensile stress promotes negative climb. Figure 2.33a illustrates negative climb schematically. Assume that the climb occurred over a distance L in the y direction. The work done is then:

$$w = FL = \sigma_x \mathbf{b} L \tag{2.43}$$

and the work per unit distance of the tensile stress, $\sigma_{x,}$ on the crystal was given above in Eq. 2.42. Thus, the force per unit length is:

$$F' = \sigma_x \mathbf{b} \tag{2.39a}$$

Jogs are formed by climb. They are favorable sites for the absorption and emission of point defects. In thermal equilibrium, the atoms at jog sites are in dynamic equilibrium, and arrive and leave the jog at equal rates. If there is an increase in vacancies, for example, in the vicinity of a dislocation line above the thermal equilibrium value, the probability of atomic exchange at a jog with a vacancy increases, climb occurs and the extra plane (defining a dislocation) line shrinks. Therefore, excess vacancies promote the process of climb. Similarly, excess of interstitial atoms adds atoms to the existing jog, which causes it to grow. In summary, when atoms are removed from the extra plane, the crystal collapses locally, but when atoms are added, the atomic plane expands locally. As stated above, either compressive or tensile strains are induced during the climb process. These stresses work during climb and the direction of climb depends on the type of stress. In Fig. 2.33b, tensile stress is shown promoting negative climb. If, however, a compressive stress is applied, the process resembles an act of squeezing out the extra plane.

The formation of jogs by climb may be considered to be a thermally activated process, since point defects are involved. It is proper to talk about nucleation and the motion of jogs. An activation energy exists, expressed by:

$$n_j = n_0 \exp\left(-\frac{U_j}{kT}\right) \tag{2.44}$$

where n_0 is the number of atom sites per unit length of dislocation and U_j is the activation energy of jog formation. The activation energies for vacancy formation (for positive climb) E_f and motion E_m determine the activation energy E_{SD} of self-diffusion. Thus, the activation energy of climb, $U_{c,}$ can be expressed as :

$$U_c = U_j + U_f + U_m = U_j + U_{SD} \tag{2.45}$$

Excess point defects are also a driving force for climb.

2.11 The Forces Between Dislocations

A dislocation senses a force due to the presence of a stress field around it. The presence of more dislocations, each with its own surrounding stress field, means that all these forces influence one another. Consider the presence of two dislocations without the application of an external stress. The total force may be assumed to be the sum of the individual forces. In Sect. 2.7 above, it was indicated that a dislocation must have certain strain energy and, indeed, a strain field exists at the site of each dislocation, affecting its motion.

2.11.1 Screw Dislocations

Given two parallel dislocations with the same sign at sites 1 (the origin) and 2 (as shown schematically in Fig. 2.34a), the stress fields at these screw dislocations have a radial symmetry and their Burgers vectors are parallel to the z axis. Due to their radial symmetry, their stress is:

$$\tau_{z\theta} = \frac{Gb_1}{2\pi r} \tag{2.46}$$

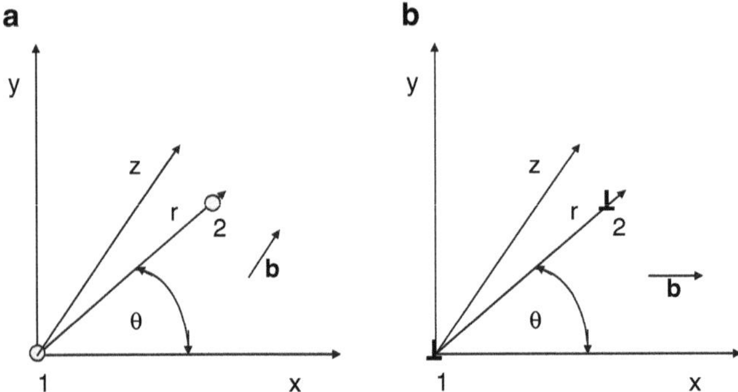

Fig. 2.34 Two same-signed, parallel dislocations at sites 1 and 2: (**a**) screw; (**b**) edge

There is only one force, because of the radial symmetry, F_r, which is a radial component. The first stress field is acting on dislocation 2 at a distance r from the first dislocation and, therefore, this force is:

$$F_r = \tau_{z\theta} b_2 = \frac{Gb_1 b_2}{2\pi r} \tag{2.46a}$$

These two same-signed, parallel dislocations repel each other, depending inversely on the distance, r, (how far apart they are from each other) and, therefore, they stay relatively distant on the same or on near slip planes.

In Cartesian coordinates, two forces exist, those of F_x and F_y, expressed as:

$$F_x = \frac{Gb_1 b_2 x}{2\pi(x^2 + y^2)} \tag{2.47}$$

$$F_y = \frac{Gb_1 b_2 y}{2\pi(x^2 + y^2)} \tag{2.48}$$

2.11.2 Edge Dislocations

Again, consider two parallel edge dislocations, as shown in Fig. 2.34b. They act in plane zx. Here, the dislocation line is parallel to the z axis. The first stress field is acting on dislocation 2, which is a parallel slip plane. The force has two components, F_x and F_y, in the x and y directions, as indicated by the subscripts. For the dislocation at the point of origin, these two components may be expressed as:

$$F_x = \tau_{xy} b_2 = \frac{Gb_1 b_2}{2\pi(1-\nu)} \frac{x(x^2 - y^2)}{(x^2 + y^2)^2} \tag{2.49}$$

F_x exerts a glide force on the second dislocation in the x direction. The dislocation at the origin (dislocation1) exerts a climb force component F_y on dislocation 2, shown in Fig. 2.34b in the y direction (a force component perpendicular to the glide plane).

$$F_y = -\sigma_1 b_2 = \frac{Gb_1 b_2}{2\pi(1-\nu)} \frac{y(3x^2+y^2)}{(x^2+y^2)^2} \tag{2.50}$$

Note that when dislocation 2 is above dislocation 1, i.e., when its coordinates are (0, y), they do not exert a glide force on each other. In polar coordinates, these forces are given as:

$$F_r = F_x \cos\theta + F_y \sin\theta = \frac{Gb_1 b_2}{2\pi(1-\nu)} \frac{1}{r} \tag{2.51}$$

$$F_\theta = F_y \cos\theta - F_x \sin\theta = \frac{Gb_1 b_2}{2\pi(1-\nu)} \frac{\sin 2\theta}{r} \tag{2.52}$$

Equations 2.46, 2.47, 2.48, 2.49, 2.50, 2.51, and 2.52 are based on the expressions in Sect. 2.9 above, specifically 2.9.1 and 2.9.2 for screw and edge dislocations, respectively. In Fig. 2.34a, b, the Burgers vectors are parallel to the z and x axes of the screw and edge dislocations, respectively. As mentioned in Sect. 2.11.1, there is a repulsive force between same-signed dislocations an attractive force between unlike dislocations. When $y = 0$, Eq. 2.49 reduces to:

$$F_x = \frac{Gb_1 b_2}{2\pi(1-\nu)} \frac{1}{x} \tag{2.53}$$

indicating that the force depends only on the distance between the dislocations. Also $F_y = 0$.

2.12 The Intersection of Dislocations

In earlier sections, the characteristics of single dislocations were considered. It was indicated that a dislocation moves quite freely in certain glide planes under applied shear stress. However, even well-annealed crystals contain many dislocations at a level of $\sim 10^4$. Therefore, it is quite probable that moving dislocations will encounter others, known as 'forest dislocations', which will hinder their freedom to glide. Even the most favored planes will contain dislocations and, thus, moving dislocations will interact with those others present in the material. The term 'dislocation intersection' refers to the interaction occurring between a moving dislocation and the others encountered while in motion. This section will simplify this concept by considering interactions between two dislocations.

When dislocations intersect jogs, kinks may form. Kinks are steps occurring in dislocation lines within the same slip plane, while jogs are steps in another

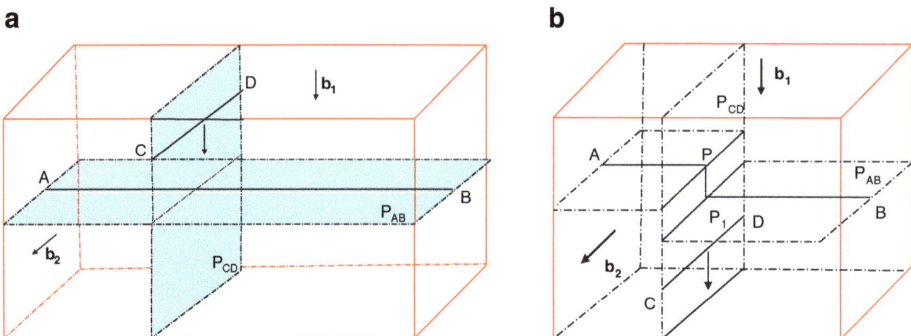

Fig. 2.35 The intersection of two edge dislocations: (**a**) before intersection; the dislocation in plane P_{CD} is moving in the direction shown; (**b**) after intersection, jog PP′ is formed on plane P_{AB}

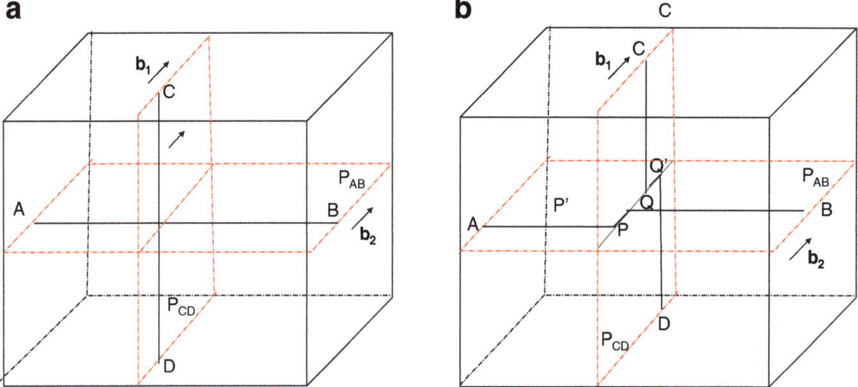

Fig. 2.36 Two orthogonal dislocations with parallel Burgers vectors: (**a**) before intersection; (**b**) after intersection

slip plane. Jogs may form during processes of thermal activation (as mentioned in Sect. 2.10.2 above), at some elevated temperature or at low temperatures by some intersection of dislocations. Of the many possibilities, the simplest case is when two edge dislocations, moving on two planes normal to each other, intersect. In Fig. 2.35, the intersection of two edge dislocations is shown before and after jog formation. A jog PP′ is formed after dislocation CD cuts through the AB dislocation line (Fig. 2.35b). The size of the step is that of Burgers vector $\mathbf{b_1}$, but it remains a part of dislocation AB. The segment is of edge character, since its Burgers vector is $\mathbf{b_2}$, that of the overall dislocation AB of which it is a part. Since the Burgers vector and the dislocation line APP′B are in the same glide plane, jog PP′ will not hinder the motion of the dislocation. There is an energy increase in dislocation AB by the length of $\mathbf{b}$ (i.e. $\mathbf{b_1}$), which can be expressed by $\alpha Gb^2\mathbf{b} = \alpha Gb^3$.

The intersection of two orthogonal dislocations with parallel Burgers vectors is illustrated in Fig. 2.36, before and after intersection. In this case, two kinks are

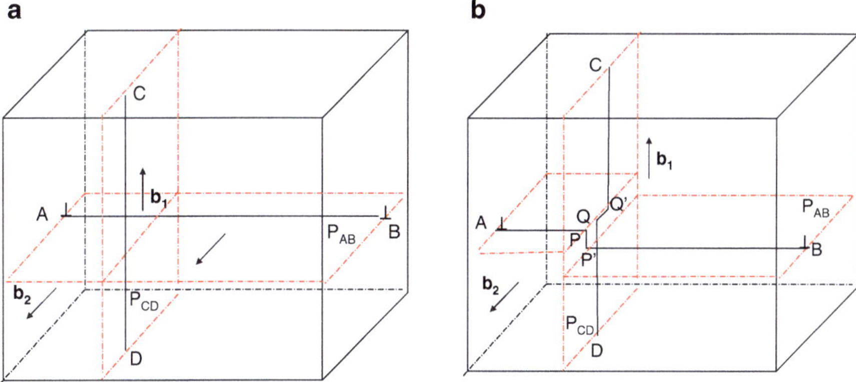

Fig. 2.37 The intersection of edge and screw dislocations: (**a**) before intersection; (**b**) after intersection

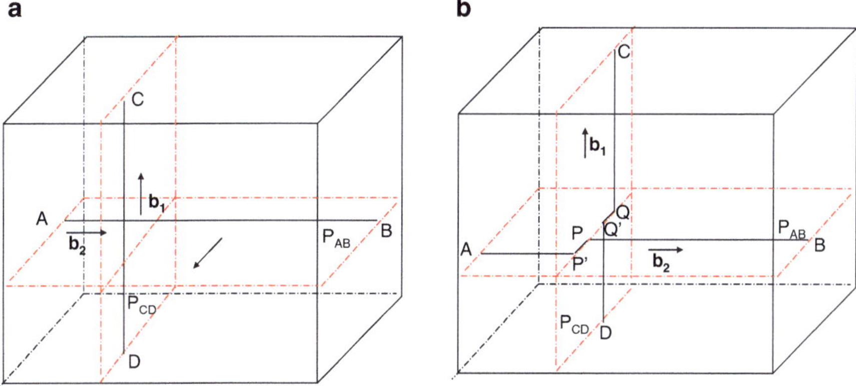

Fig. 2.38 The intersection of screw dislocations: (**a**) before intersection; (**b**) after intersection. Two steps are formed, both having an edge character

formed with the Burgers vectors, namely $\mathbf{b_1}$ and $\mathbf{b_2}$ for the AB and CD dislocations, respectively. Both these kinks have a screw orientation and an increase in energy occurred in each of their dislocation lines.

Other cases are the intersection between edge-screw and screw-screw dislocations, illustrated in Figs. 2.37 and 2.38, respectively. For drawing simplicity, the intersecting planes in the figures below are orthogonal, but they do not have to be. Figure 2.37 illustrates that, in the edge-screw case, a jog PP′ in the edge and a kink QQ′ in the screw dislocations, respectively, are formed. The jog is formed, because plane P_{AB}, where dislocation AB is moving, is a spiral surface, not unlike a ramp, so its motion, after cutting the screw dislocation, is not at the same level as it was at the beginning of its motion. The PP′ jog has an edge character and can move with dislocation AB in the P_{AB} plane, whereas QQ′ has to climb, since it cannot move up along the screw dislocation. The segments PP′ and QQ′ in Fig. 2.38 both have an

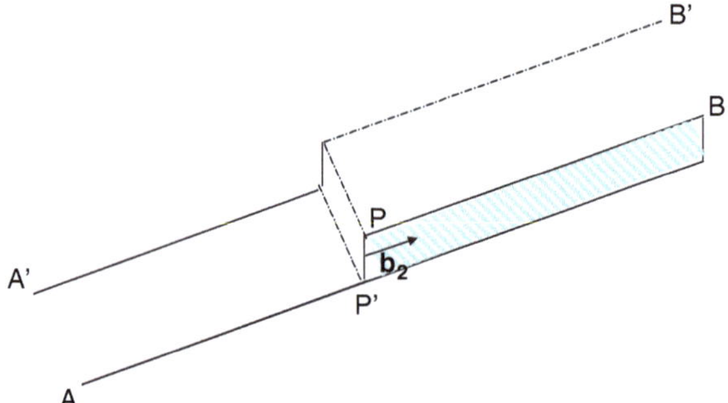

Fig. 2.39 The jog, PP′, and the screw dislocation line, APP′B, after intersection. This jog has an edge character and its glide plane is defined by the *shaded area*. This is the plane where the jog can glide. Movement of the screw dislocation to A′B′ can occur only by climb

edge character, since the respective Burgers vectors are perpendicular to them. The AB screw dislocation, after cutting through dislocation CD and moving into P_{CD} again (as described in Fig. 2.37 above), also concludes its motion at a level higher than where it began, thus forming a jog. The segment PP′ is actually a jog, but it is not easily drawn. However, by means of Fig. 2.39, the jog formation in Fig. 2.38 and its motion are better understood. Note that the shaded area defines the glide plane of the jog, which has an edge orientation and, thus, its motion is along the screw axis of AP′PB. The only motion of the screw dislocation to A′B′ is by climb.

Some comments in summation:

(a) the movement of dislocations generates steps as jogs and kinks;
(b) jogs severely influence the glide of dislocations;
(c) jogs and kinks can occur in any dislocation, not just edge dislocations, though they are often difficult to render graphically;
(d) As will be discussed in a next chapter, the role of jogs in strain hardening is of great importance, since no easy glide is possible.

2.13 Dislocation Multiplication

It was indicated in earlier sections that in a well-annealed crystal $\sim 10^4$ dislocations exist and, in a heavily deformed ones, that number might even reach $\sim 10^{10}$. Therefore, means to produce such large number of dislocations are required. Several sources for dislocations in materials were mentioned above in Sect. 2.5. Formation of a large number of dislocations is a prerequisite for understanding many of the mechanical behavior of materials, such as for examples large plastic strains. One of the most well-known concepts for generating a large number of dislocations at high speeds is by means of the Frank-Read mechanism, also known as the 'Frank-Read generator'. In essence, according to this method, one single dislocation is sufficient to produce a large number of them at velocities of deformation during deformation. Figure 2.40 is the graphic method used by most publications in the field to show how this is done. A well-documented Frank-Read dislocation source, in action in silicon was demonstrated by Dash, which is available in almost every communication on the subject and, thus, will not be reproduced here.

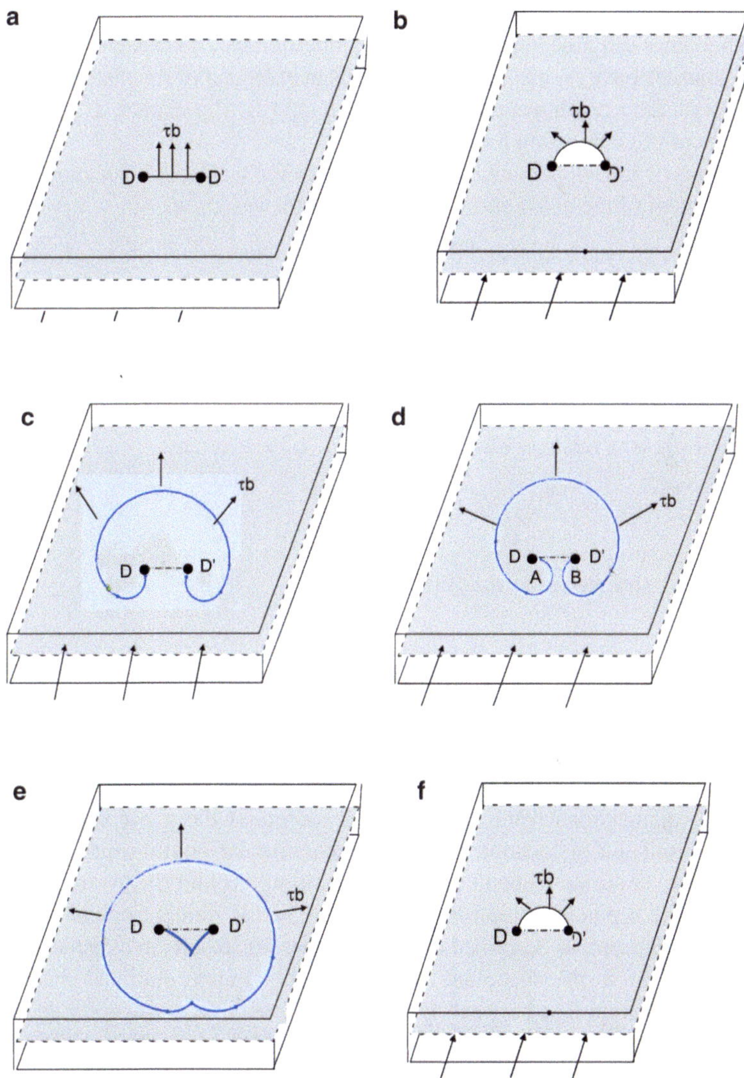

Fig. 2.40 Schematic multiplications of a single dislocation according to the concept of a Frank-Read source

The stepwise operation of a Frank-Read source on a slip plane is as follows:

(a) assume that a dislocation segment D-D′ is pinned at both ends, as seen in the illustration;
(b) an applied stress produces a glide force, τb, acting on this segment in its glide plane, causing it to curve maximally into a semicircle shape. The line tension balances the force on the dislocation;

(c) the stress, τ, is increased beyond a critical value (illustration b) and the dislocation becomes unstable, tending to expand indefinitely. The expanding dislocation loop doubles back on itself, forming spirals at each end of the dislocation;
(d) this dislocation continues to expand. In (c) and (d), slip occurs, as indicated by the area of the expanding loop;
(e) the spirals of the slipped area have joined, forming a closed dislocation loop and that segment of the dislocation is ready to start a new cycle.

The dislocation parts at A and B in illustration (d) have opposite signs and attract each other until the loop closes, as shown in illustration (e). The dislocation segment D-D′, which was left behind, repeats the same sequence as described above.

The force acting on the dislocation, causing it to curve, is balanced by the line tension (discussed above in Sect. 2.9). The line tension per unit length in that section is equal to the force acting normally on the dislocation segment. In equilibrium with the applied stress, a relation was derived by Eq. 2.29, given as:

$$\tau = \frac{\alpha G b}{r} \tag{2.54}$$

which can be expressed in terms of the radius, r, as:

$$r = \frac{a G b}{\tau} \tag{2.54a}$$

Beyond this radius, at the point of the applied stress, the dislocation becomes unstable, as previously mentioned, and the Frank-Read source, pinned at both ends, expands under a diminishing force. Equation 2.54 represents the stress required to generate dislocations from a Frank-Read source. If the shear stress increases any further and the dislocation passes the semicircular equilibrium state, it will spontaneously continue to bend and grow, spiraling around the A and B pinning points until the segments spiraling around the A and B pinning points collide and cancel each other out, as described earlier. This process results in a dislocation loop around D and D′ in the slip plane, which expands under continued shear stress. Also, the dislocation line between D and D′ can continue to generate dislocation loops in the manner described above. A Frank-Read loop can, thus, generate many dislocations in a crystal plane under applied stress. A single Frank-Read source may also operate when the line is pinned only at one end. Intersections with other dislocations, forming jogs which increase the length of the line, may even act as Frank-Read sources.

2.14 Partial Dislocations

Usually, there are preferred slip planes and directions in certain crystal systems. The combination of both the slip plane and the direction define the slip system. Slip planes are generally the closest packed planes in a structure. The slip direction of a slip plane is the one which has the highest linear atomic density. Up to this point, no

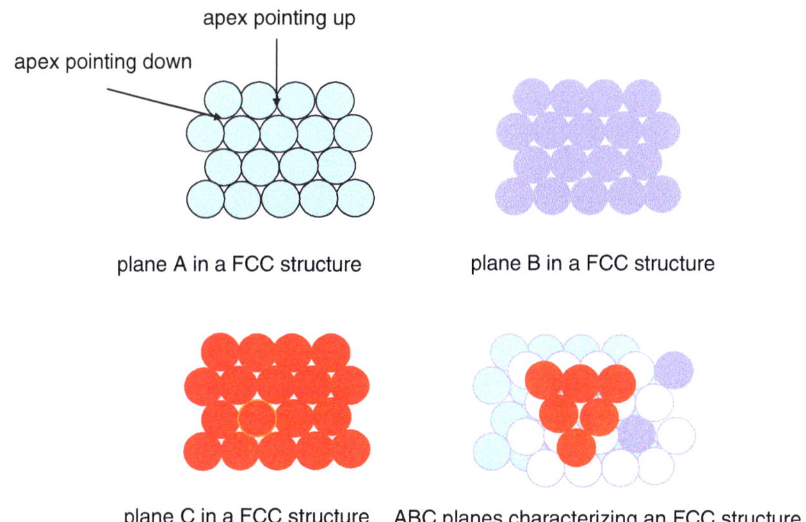

Fig. 2.41 Packed planes in an FCC metal. The sequence of the planes is A, B and C

specific consideration has been given to structure. However, when discussing partial dislocations, one cannot avoid the structural aspect. The most common approach is to consider FCC structures. However, before entering into the details, one must define the term 'partial dislocation'. Unlike the dislocation cases discussed thus far, in which the most characteristic feature was the Burgers vector, which is also a 'translation vector' of the lattice, in 'partial dislocation', the Burgers vector is not a translation vector of the lattice and partial dislocations are usually bordered by planar defects called 'stacking faults'. For the purpose of illustration, closely-packed planes make a good example. In FCC slip occurs on {111} planes in <110> directions. In Fig. 2.41, close-packed planes and their arrangement in an FCC structure are shown.

Note that the apexes of triangular-shaped voids have two orientations, with the apex pointing upward or downward. Plane B can occupy any of these voids. Assume that those with the apex pointing upwards are occupied by this plane. In the FCC structure, where the sequence of planes is A, B and C, plane C will occupy the voids with the apexes pointing down (in the considered case). The entire stacking is shown in the last illustration (ABC planes) such that all three planes may be observed. Thus, by necessity, plane C is only partially shown. An FCC unit cell and one of the {111} planes is depicted in Fig. 2.42 below.

2.14.1 Shockley Partial Dislocations

As indicated above, the translation vector in FCC structures is of type **a/2<110>**. Therefore, it should also be the Burgers vector. However, for energetic and

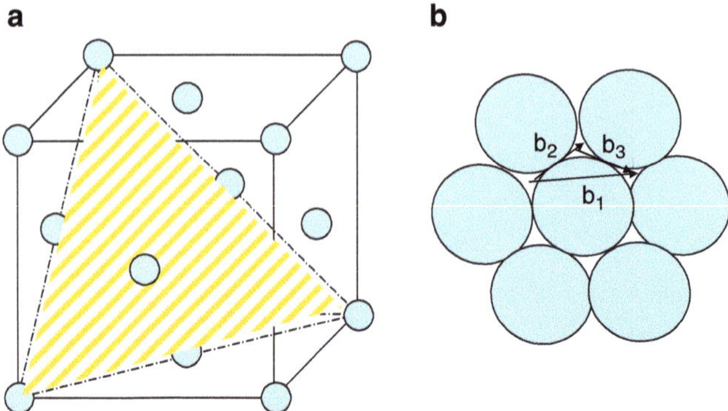

Fig. 2.42 FCC unit cell: (**a**) *shaded area* a {111} plane; (**b**) the *shaded* {111} plane showing the splitting of Burgers vector $\mathbf{b_1}$ into $\mathbf{b_2}$ and $\mathbf{b_3}$

topographic reasons, this vector splits into what is known as 'partial dislocations'. Two partials are formed and are termed 'Shockley partial dislocations' or simply 'Shockley partials'. First, consider the topological reason why a Burgers vector should split. In Fig. 2.42b, one sees the Burger vector, $\mathbf{b_1}$, which, as mentioned, is the translational vector. The motion would require surmounting the peak formed by the central atom. Rather than "taking the high road", as it were, an easier strategy is adopted by splitting into two steps and "going 'round the mountain", by passing through the valleys on either side, between each of the two neighboring atoms, as represented by the vectors $\mathbf{b_2}$ and $\mathbf{b_3}$. The crystallographic orientations of the partials (students should verify this as an exercise) are of the type **a/6<211>**.

As indicated in Sect. 2.7, the energy of a dislocation is proportional to b^2 in the form of:

$$E = \alpha G b^2 \tag{2.55}$$

α takes into account the type of dislocation. One can see from the following reaction that dislocation splitting is energetically favored. One such reaction is:

$$\frac{a}{2}[110] = \frac{a}{6}[211] + \frac{a}{6}[12\bar{1}] \tag{2.56}$$

Before looking at the energetic profile of this reaction, one has to verify its correctness according to the following steps:

	The un-dissociated vector	The partials
Component x	1/2	2/6 + 1/6 = 3/6 = 1/2
Component y	1/2	1/6 + 2/6 = 3/6 = 1/2
Component z	0	1/6 − 1/6 = 0

Thus, the reaction is proven correct and both sides of Eq. 2.56 are balanced. Next, one must check the energetic aspect. For this purpose, take the absolute values of

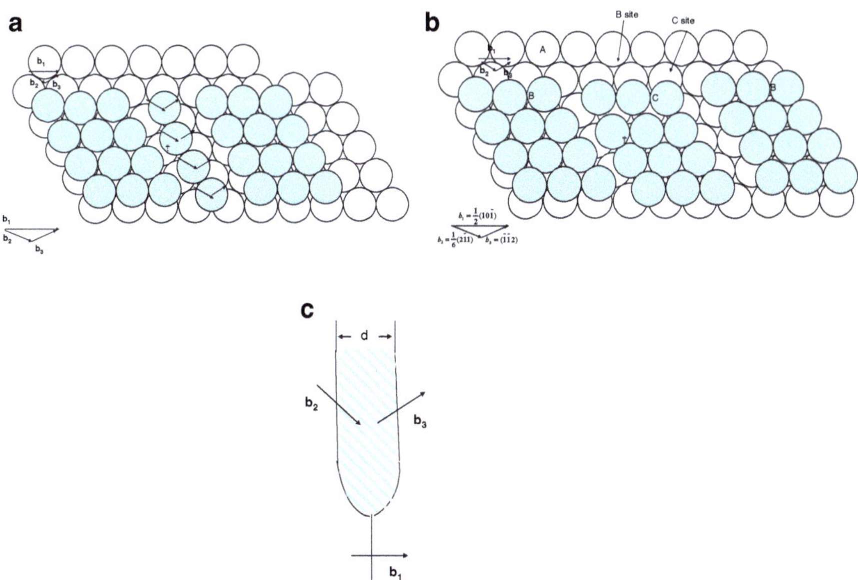

Fig. 2.43 Partial dislocations in a FCC structure: (**a**) the split motion of atoms from *site B* to *site C*; partial dislocation with Burgers vectors $\mathbf{b_2}$ and $\mathbf{b_3}$; (**b**) extended dislocation composed of the two partials and the stacking fault between them; (**c**) schematic presentation of a stacking fault

the indexes of the Burgers vectors. Since their energy is proportional to their squares (as seen in Eq. 2.55), for the left hand side of Eq. 2.56, one gets $a^2/2$ and, for the sum of the squares on the right hand side, $a^2/3$. Clearly, $a^2/2 > a^2/3$ and, thus, there is a decrease in energy, thus the splitting of the dislocation is also favored energetically.

In Fig. 2.43, the separation of perfect dislocation in FCC metals is illustrated. As indicated in Fig. 2.41, the planes are designated as A, B and C, and the same notation is also used for the voids between the atoms with up or down apexes. The reaction in Fig. 2.43 is:

$$\frac{1}{2}\left(10\bar{1}\right) = \frac{1}{6}\left(2\bar{1}1\right) + \frac{1}{6}\left(\bar{1}\,\bar{1}\,2\right)$$

This is another form of Eq. 2.56. In Fig. 2.43a, one row of atoms indicates the formation of the partial dislocations. Looking down on the Figure the dislocation on plane A, with $\mathbf{b_1}$ normal to it, is of edge orientation (it is the total dislocation). The single row of atoms in plane B is zig-zagged on both sides, which is characteristic of partial dislocation. The first movement of the row, by the amount of $\mathbf{b_2}$, shifts it to the zig-zag position, while the second movement, indicated by vector $\mathbf{b_3}$, brings the atoms to the final position, as would have occurred with a single displacement of $\mathbf{b_1}$. These two partial dislocations will repel each other to a point where a balance is reached between the elastic energy decrease, due to the splitting of the dislocation, and the increase of the stacking-fault energy. Stacking-fault energy varies widely from metal to metal, depending on the width of the fault. Thus, the width of the stacking fault in Cu is about 10 atomic spacings, whereas, in Al, it is only ~2 atomic separations. This means that the stacking-fault energy of Cu is low (~80) compared to that of Al (~200 mJ/m$^{2)}$. The combined defect of the partials and the stacking fault is called 'extended dislocation'. Schematically, the partials and the stacking fault are shown in Fig. 2.43c.

The width of a stacking fault is the consequence of the balance between the repulsive force between the two partial dislocations and the attractive force due to the stacking fault. When the stacking-fault energy is high, the splitting of the perfect dislocation into two partials is unlikely and glide in the material occurs only as a result of perfect dislocation glide. Lower stacking-fault energy will promote the formation of wider stacking faults and cross-slip or climb will be more difficult consequently. Thus, the mobility of the extended dislocation in materials with low stacking-fault energy decreases.

The usual way of schematically presenting the splitting of a perfect dislocation into Shockley partials with the fault between them is shown in Fig. 2.43c.

In Sect. 2.11, the force between two dislocations is given either by Eq. 2.53 or by Eq. 2.47. In analogy to Eq. 2.47, the repulsive force between the two partial dislocations, in terms of the distance, d, can be rewritten as:

$$F = \frac{Gb_2b_3}{2\pi}\frac{1}{d} \tag{2.57}$$

where d, in Eq. 2.57, stands for the thickness of the stacking-fault ribbon formed by the partial dislocations with vectors $\mathbf{b_2}$ and $\mathbf{b_3}$. The stacking-fault energy per unit area is γ and, when multiplied by the area, the energy of the fault is obtained. Thus, the force of the stacking fault (i.e., energy per distance) is equivalent to γ. By balancing these forces:

$$\gamma = \frac{Gb_2b_3}{2\pi}\frac{1}{d} \tag{2.58}$$

the equilibrium distance, d, of the separation between the partials can be obtained as:

$$\boldsymbol{d} = \frac{\boldsymbol{G}\boldsymbol{b_2}\cdot\boldsymbol{b_3}}{\boldsymbol{2\pi}}\frac{1}{\boldsymbol{\gamma}} \tag{2.58a}$$

The extended dislocation, discussed hitherto, consists of Shockley partials and a stacking fault, which can glide within its own glide plane; therefore, the accepted notation is glissile dislocation.

2.14.2 *Frank Partial Dislocations*

Partial dislocations can form not only by splitting a perfect dislocation, but also by inserting or partly removing a {111} plane. In Fig. 2.44a, the FCC stacking of slip

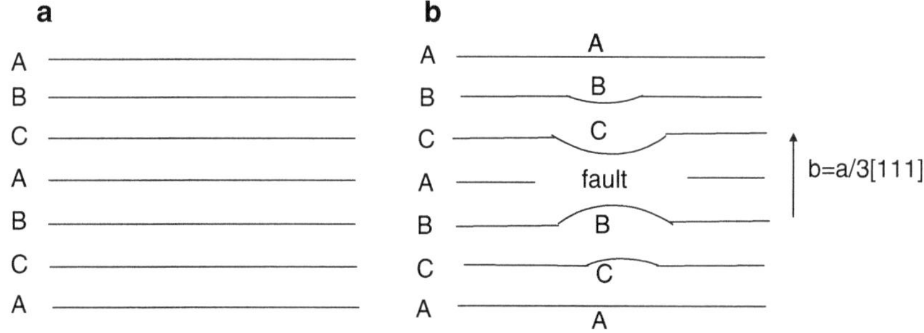

sequence in the missing part: A**BCBC**A is a stacking fault :

c

d

The sequence of stacking has been changed to A**BCBC**AC

Fig. 2.44 Stacking-fault formation: (**a**) the FCC stacking of slip planes before removal of a plane; (**b**) the stacking sequence of the {111} planes is modified at the region where the fault exists from the FCC to an HCP structure; (**c**) the sequence before the insertion of an additional plane (indicating the place of insertion); (**d**) the re-insertion of the plane produces a thin HCP structure

planes is illustrated schematically before the removal of a plane. The sequence of the stacking of the {111} planes is modified at the region where the fault exists from the FCC to a hexagonal close-packed (HCP) structure, as seen in Fig. 2.44b. In the center of this figure, in the region of the fault, the sequence is not ... ABCABC ..., but ... A**BCBC**A ... ; the underlined bold letters are characteristic of HCP packing. The Burgers vector indicated is in the <111> direction; its actual value is **b** = a/3<111>. The missing plane, indicated in Fig. 2.44b, may have been removed by the condensation and collapse of the vacancies, forming a disc-shaped layer. As seen in this Figure, this fault is bounded by an edge-oriented dislocation at each side. The Burgers vectors are normal to the {111} planes and are not the slip direction in FCC crystals. This is known as 'Frank sessile dislocation'. A Frank dislocation can also be formed by inserting part of a plane, which will also be a sessile dislocation, since it will not be able to glide in any of the ABC slip planes, as shown in Fig. 2.44c. A dislocation formed in this way can only move non-conservatively, which requires transport of either atoms or vacancies. It is worth noting that, when part of a plane is removed, the fault is described as 'intrinsic' and, when a partial plane is inserted,

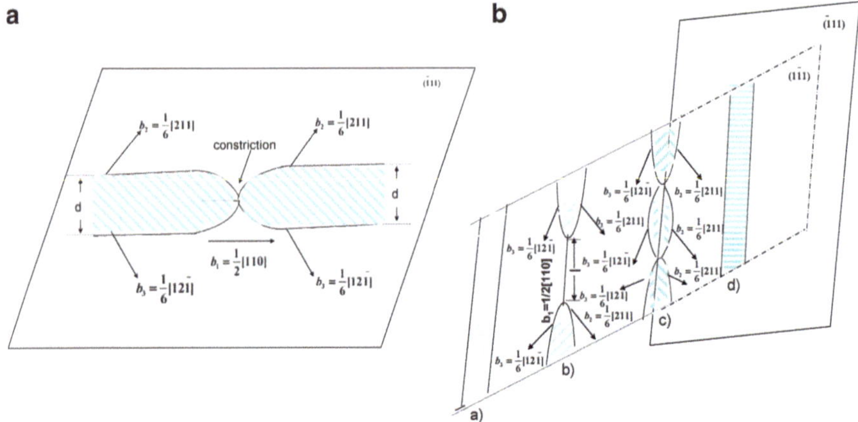

Fig. 2.45 The cross slip of a screw dislocation: (**a**) schematic constriction in an extended dislocation, making cross slip possible; (**b**) the stages of cross slip from one octahedral plane to an other in (*a*) a stacking fault before cross slip; (*b*) the partial dislocations of the extended dislocation combined along distance l; (*c*) the dislocation, after combination of the partials curves under the influence of stress, cross-slipping into the $\bar{1}11$ cross-slip plane; and (*d*) the extended screw dislocation and the stacking fault continue on from one slip plane into another

it is called an 'extrinsic-stacking fault'. In practice, both types of Frank partial dislocations may occur, either due to the condensation of excess vacancies, during crystal growth or by the condensation of interstitial atoms. They are also known as 'negative' or 'positive Frank partials', respectively.

2.14.3 The Cross Slip of Partial Dislocations

In earlier sections, it was stated that, in edge dislocations, the slip direction and the dislocation line define the slip system; however, in screw dislocations, the Burgers vector is parallel to the dislocation line and, thus, it may cross slip into planes belonging to the same form. The situation is different in cases of extended dislocations, where stacking fault influences cross slip. Figure 2.45 illustrates cross slip in a faulted crystal bounded by two partial dislocations, in accordance with ideas promoted by Seeger. An edge dislocation with its partials is able to move within its glide plane along with its faulted region-the extended dislocation-but it will not be able to move into another octahedral plane unless it climbs. A screw dislocation or a screw component will not have such a problem as long as the direction of slip and the Burgers vector are common to both {111} planes. However, cross slip can occur only if a 'constriction' (i.e., a joining of the partials) forms. Figure 2.45a shows such a constriction in a fault at the point at which the partials come together, allowing cross slip to occur. In Fig. 2.45b, the stages of cross slip are indicated: (a) the faulted region in its glide plane; (b) the partial dislocations reunited by an

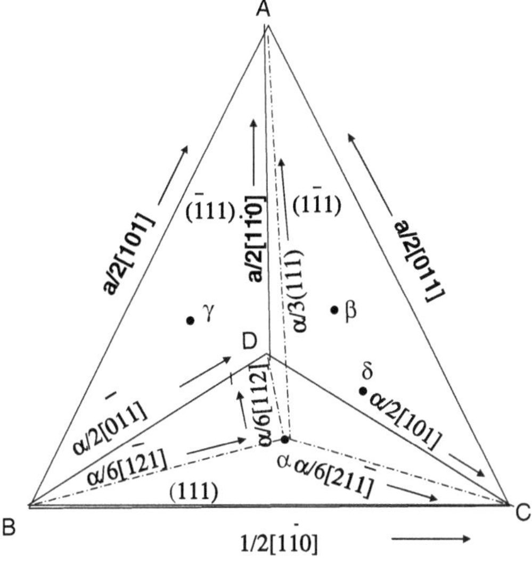

Fig. 2.46 A Thomson's tetrahedron aiding in visualizing dislocation reactions in FCC structures on {111} planes. Decomposition into Shockley and Frank partials is shown

applied stress over a length of l to form an un-extended dislocation, before it can cross slip from **(1$\bar{1}$0)** to another {111} plane (here, an **($\bar{1}$11)** plane is illustrated, which is a cross-slip plane); (c) the dislocation after the partials have combined under the influence of stress in the **($\bar{1}$11)** cross-slip plane bows out, after having first been anchored at two constrictions and then spread as a new extended dislocation; and (d) the extended screw dislocation and the stacking fault continue in another plane which is a cross-slip plane.

Cross slip depends on the width of the stacking fault. Wide ribbons require considerable force to bring the partials together for cross slip. In Al, for example (as indicated in Sect. 2.14.1) has high stacking-fault energy and, thus, due to its very thin ribbon, cross slip will occur very readily as compared to Cu. As a mater of fact, it is often stated that, in Al, the dislocations are relatively un-extended and, thus, frequent cross slip is observed. HCP structures behave quite similarly to FCC metals. In transition BCC crystals or in cubic ionic crystals (NaCl, MgO, etc.), with narrow stacking faults, cross slip occurs readily. Note that the process of cross slip occurs more readily at higher temperatures, since activation energy may be required for the removal of stacking faults, especially in cases in which the ribbon is broad.

2.14.4 The Thompson Tetrahedron

It is usually difficult to predict the outcome of dislocation reactions that will occur, especially when partial dislocations are involved. Without being exposed to elaborate crystallographic understanding of the subject, it is usual (and practical) to visualize possible reactions with the help of a Thompson tetrahedron. Figure 2.46

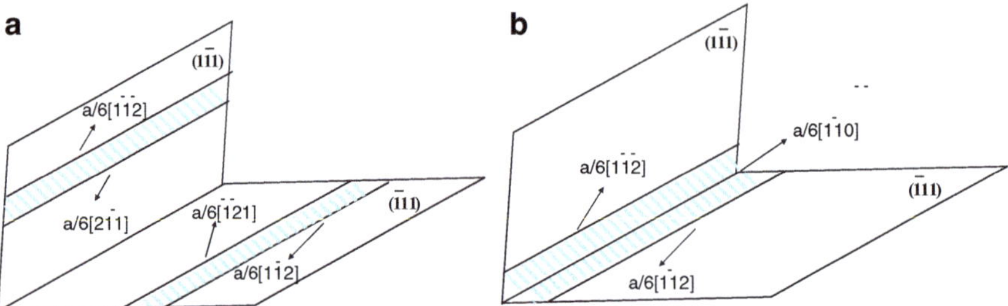

Fig. 2.47 A Lomer-Cottrell lock: (**a**) two partial dislocations before interaction, (**b**) after interaction of the leading partials forming a stair rod dislocation

presents a schematic diagram of such a commonly used tetrahedron, based on Thompson's concept. Four {111} planes form this tetrahedron, labeled as ABCD, where the external edges, such as AB, etc., indicate the <110> slip directions common in FCC structures before the dissociation into partials. The Burgers vector of the Shockley partial dislocations, a/6<112>, is obtained by the dissociation of a/2<110>. The Burgers vectors are indicated by lines connecting the corners to the centers of the faces, such as Aγ, Bγ, etc. Frank partial dislocations having a/3<111> Burgers vectors are represented by lines, such as Aα, Bβ, etc., which connect the corners of the triangles to the opposite triangle centers, labeled by Greek letters. Furthermore, stair- rod dislocations obtained by the intersection of two Shockley partials on two intersecting {111} planes are indicated as reactions between the lines connecting the centers of the triangles (i.e., {111} planes), for example:

$$\alpha\beta + \beta\gamma = \alpha\gamma$$

2.14.5 *Lomer-Cottrell Locks*

Dislocation gliding on intersecting {111} planes can form a series of obstacles known as 'Lomer-Cottrell barriers', preventing further glide. Since FCC crystal (and also in HCP and BCC structures) dislocations generally tend to dissociate into partials, causing 'extended dislocation faults', it is important to consider the case of Lomer-Cottrell barrier formation. Figure 2.47 illustrates the formation of such a barrier. In Fig. 2.47a, stacking faults, bounded by partial dislocations, are gliding on intersecting {111} planes before their interaction. In Fig. 2.47b, the result of the reactions between these partials is shown which forms a 'Lomer-Cottrell lock'. Here, partial leading and tracking dislocations, with a stacking fault

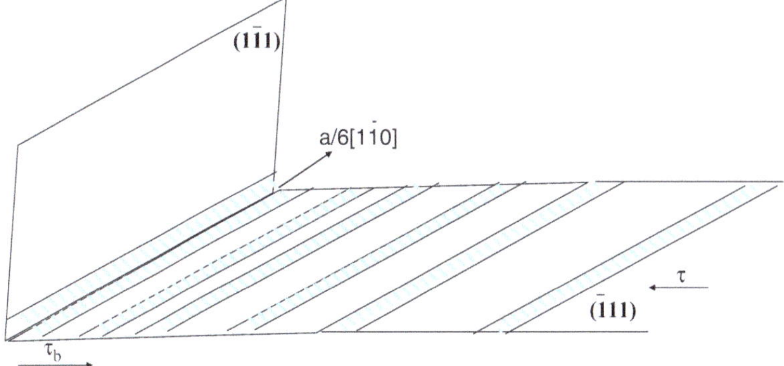

Fig. 2.48 Dislocation pile-ups (in the figure partials and their faults are shown) behind a Lomer-Cottrell lock, acting as obstacle to their movement

between them, glide on their respective planes. The leading partial dislocations on the intersecting planes have formed a new partial dislocation with Burgers vector $\boldsymbol{a/6[1\bar{1}0]}$ according to

$$a/6[\bar{1}\bar{2}1] + a/6[21\bar{1}] = a/6[1\bar{1}0]$$

This reaction is correct, as can be seen by checking the components of the Burgers vectors, and it is also favorable energetically. The consequence of the above reaction is the formation of a sessile dislocation, beyond which the trailing dislocations pile up. The Burgers vector of the newly formed partial dislocation, i.e. a/6<110>, as shown in the above reaction, is not the vector of the FCC lattice (but rather a/2<110>) located in plane {001}, which is not a slip plane in the FCC structure, so it cannot glide. As a matter of fact, all three partial dislocations – both the leading partials which entered into the reaction and the newly formed partial dislocation – form a very stable lock, which prevents the movement of other dislocations.

2.15 Dislocation Pile-Ups

Dislocations generated by a Frank-Read source quite often pile-up at various barriers, such as grain boundaries, precipitates, etc., and cannot glide easily. The later dislocations exert forces on the earlier ones. Pushing dislocations into a pile-up configuration requires stresses higher than those which act in the absence of pile-up dislocations. Figure 2.48 schematically illustrates a dislocation pile-up on a {111} plane, where movement is blocked by some barrier, e.g., a Lomer-Cottrell lock. High stress concentration is acting on the leading dislocations and, when the pile-up stress is greater than the theoretical shear stress, yielding is induced. A back stress acts on the obstacle, preventing further motion of the dislocations.

Near the obstacle, the density of the pile-ups is the greatest, but the distance between the individual dislocations increases as the same-signed dislocations repel each other. Since repelling forces are additive, they increase. One could say that a stress concentration exists at or close to the obstacle. The key factor in the motion of these dislocations is the first or leading dislocation in the vicinity of the obstacle. Assume that the leading dislocation has moved a distance, dx; all the trailing dislocations will move the same distance. The work done per unit length of dislocation is:

$$work = n\tau bdx \tag{2.59}$$

where n is the number of dislocations and τ is the applied stress. The leading dislocation, τ_i, works against the local stress (internal stress) of the obstacle. Thus, the work of the leading dislocation is:

$$\mathrm{work} = \tau_i \mathrm{bdx} \tag{2.59a}$$

At equilibrium, these equations should be equal and, thus, it is possible to derive the following from Eqs. 2.59 and 2.59a:

$$\mathrm{n}\tau = \tau_i{}^{\mathrm{b}} \tag{2.59b}$$

In other words, the internal stress at the head of the pile-up, composed of n dislocations, is n times greater than the applied stress. As seen in later chapters, stress concentration plays an important role both in strain hardening and in brittle fracture formation. The back stress of the pile-up, τ_b, acts on the source to create new dislocations. As long as:

$$\tau_b - \tau_a = \tau \tag{2.59c}$$

where τ_a is the stress required to operate a source, it will function to produce dislocations. Eshelby, Frank & Nabarro have calculated the number and distribution of dislocations that can pile-up in a slip plane of length L with acting shear stress τ as

$$n = \frac{\pi L\tau k}{Gb} \tag{2.60}$$

K = 1 for screw and $(1 - \nu)$ for edge dislocations.

A pile-up of n dislocations along a distance L may be considered to be a giant dislocation with Burgers vector nb. The breakdown of a barrier occurs by:

(a) slip on a new plane or if the material is polycrystalline in a new grain;
(b) climb;
(c) crack formation, due to the high stress.

2.16 Low (Small)-Angle Grain Boundaries

Materials are usually polycrystalline; but in order to study crystal properties without the complexities incorporated by grain boundaries, single crystals are more suitable subjects. However, if the difference between the grain orientations in a polycrystalline material is sufficiently small, then the problem of studying crystal properties becomes limited to the presentation of an array of dislocations. Figure 2.49a shows two crystals with a tilt misorientation angle of θ before joining. The two grains share a common axis about which both are rotated, defining the angle θ. By joining these two crystals to a bi-crystal, a dislocation wall is formed by an array of dislocations and each dislocation has the same sign and Burgers vector. The dislocations of the array line up one above the other to reduce the energy of the system, as shown in Fig. 2.49b. Each individual dislocation can be resolved up to a misorientation of ~10°, above which analysis of the grain boundary in terms of an array of dislocations is not possible. A simple relation can be obtained from Fig. 2.49b, connecting the distance between dislocations, the tilt angle and the Burgers vector (in the simple cubic bi-crystal), given as:

$$2\sin\frac{\theta}{2} \approx \theta = \frac{b}{D} \tag{2.61}$$

For θ = 1° and for **b** ~ 2.5 × 10^{-8}, the value of D ~ 140Å. In simple cubic materials, b = <100>. The larger the angle θ is, the closer the spacing between the dislocations

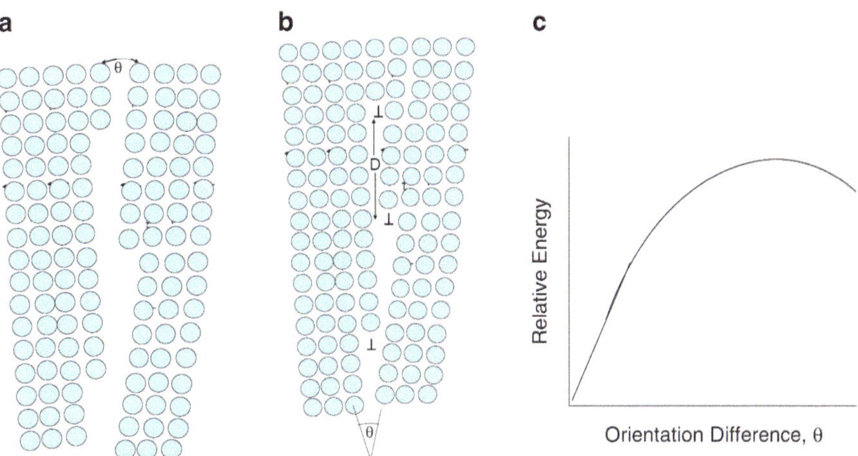

Fig. 2.49 A low-angle grain boundary: (**a**) two grains are shown with a common cube axis and their angular difference is θ; (**b**) the two grains are joined to form a bi-crystal. An array of edge dislocations is formed at the joint of the two grains with spacing *D* between them; and (**c**) relative, low-angle, grain-boundary energy

will be. Thus, for an angle of ~15° (the dislocation model is not valid anymore), the dislocations are at about a few atomic spacings one from another, at $\sim 9.5 \times 10^{-8}$, and it is impossible to distinguish between the individual dislocations.

This kind of presentation, of a bi-crystal as an array of edge dislocations, defines the low-angle grain boundary, which is also known as the 'simple tilt boundary'. Low-angle grain boundaries are not usually pure tilt boundaries, since the formation of the bi-crystal may occur via a process of screw dislocation and, thus, may include some characteristics of screw dislocation. Such a low-angle boundary is also known as a 'twist low-angle grain boundary', originating from the fact that the lattices of the grains are twisted when a screw dislocation component is present.

The energy of a small-angle grain boundary may be calculated on the basis of the dislocation model. Schematically, the relative energy, as a function of the angle of orientation difference, is shown in Fig. 2.49c. One can express this energy as:

$$\mathrm{E_b = E_0\theta(A - \ln\theta)} \tag{2.62}$$

This relation may be obtained as follows:

(a) In Sect. 2.7, Eq. 2.12 gives $\mathrm{E_{TE}}$, which includes the core energy:

$$\mathrm{E_{TE} = E + E_{core}} \tag{2.12}$$

(b) The core energy in Eq. 2.21 is given by:

$$E_{core} \cong \pi(5b)^2\left(\frac{G}{30}\right)\frac{1}{G} \approx \frac{Gb^2}{10} = \mathrm{CGb^2} \tag{2.21}$$

(c) In Eq. 2.22, the energy per unit length of dislocation is given by:

$$E_{edge} = \frac{Gb^2}{(1-\nu)4\pi}\ln\frac{r}{r_0} \tag{2.22}$$

(d) The energy of unit cell area is $\mathrm{E_{TE}} \times 1$ and $\mathrm{E_{TE}}$ is the energy per unit length of a dislocation passing or cutting through this area. Therefore, the energy per area of the boundary is $\mathrm{E_{TE}} \times 1/1 \times \mathrm{D}$;

(e) The energy per unit area of a boundary of length D is then:

$$\mathrm{E_b = (E_{TE} \times 1)\,/\,(1 \times D)} \tag{2.63}$$

(f) Substituting from Eqs. 2.12, 2.21 and 2.22 (using the proper subscripts):

$$E_b = \frac{Gb^2}{(1-\nu)4\pi}\ln\left(\frac{r}{r_0}\right)\frac{1}{D} + \frac{CGb^2}{D} \tag{2.64}$$

(g) Taking r = D (since outside this range the stress fields cancel each other out) and expressing D in terms of θ, from Eq. 2.61, one obtains:

$$E_b = \frac{Gb^2}{(1-\nu)4\pi}\ln\left(\frac{D}{r_0}\right)\frac{1}{D} + \frac{CGb^2}{D} = \frac{Gb^2}{(1-\nu)4\pi}\ln\left(\frac{b}{\theta r_0}\right)\frac{\theta}{b} + \frac{CGb^2\theta}{b} \tag{2.65}$$

(h) Arbitrarily taking $r_0 \sim b$, Eq. 2.65 can be written as:

$$E_b = \frac{Gb^2}{(1-\nu)4\pi}\ln\left(\frac{1}{\theta}\right)\frac{\theta}{b} + \frac{C\theta Gb^2}{b} \tag{2.66}$$

$$E_b = \frac{Gb\theta}{(1-\nu)4\pi}\left[\ln\left(\frac{1}{\theta}\right) + C(1-\nu)4\pi\right] \tag{2.67}$$

(i) Eq. 2.67 expresses the form of the low-angle grain boundary (just as well as does Eq. 2.62, commonly found in the literature) and:

$$E_0 = \frac{Gb}{(1-\nu)4\pi}, \quad A = C4\pi(1-\nu) \text{ and } \ln\frac{1}{\theta} = -\ln\theta \tag{2.68}$$

Note the shortcomings of this derivation: (a) it is valid only for real, small angles, as assumed to be the case when using Eq. 2.61; (b) the core energy is not known. The approximation in Eq. 2.21, at best, only provides an estimation; and (c) it is inherent in the derivation that A and E_0 are independent of θ.

Small-angle grain boundaries are illustrated above in Figs. 2.2 and 2.3, by means of the etch pits technique.

References

S. Amelinckx, *Dislocations and Mechanical Properties of Crystals* (Wiley, New York, 1957)

S. Amelinckx, *The Direct Observation of Dislocations.* Solid State Physics, Supplement 6 (Academic, New York, 1964)

W.C. Dash, *Dislocations and Mechanical Properties of Crystals* (Wiley, New York, 1957), p. 57

J.J. Gilman, W.G. Johnston, *Dislocations and Mechanical Properties of Crystals* (Wiley, New York, 1957)

J.P. Hirth, J. Lothe, *Theory of Dislocations* (McGraw-Hill, New York, 1968)

D. Hull, *Introduction to Dislocations* (Pergamon Press LTD, Oxford, 1969)

W.T. Read Jr., *Dislocations in Crystals* (McGraw-Hill Book Company, Inc., New York, 1953)

G. Schoeck, A. Seeger, *Report on the Conference "Defects in Crystalline Solids", Bristol 1954* (The Physical Society, London, 1955)

A. Seeger, Glide and work hardening in face centered cubic metals, in *Dislocations and Mechanical Properties of Crystals* (Wiley, New York, 1957)

A.R. Verma, *Crystal Growth and Dislocations* (Butterworth, London, 1953)

Further References

S. Amelinckx, W. Dekeyser, J. Appl. Phys. **29**, 1000 (1958)
W.D. Dash, J. Appl. Phys. **27**, 1193 (1956)
J.D. Eshelby, F.C. Frank, F.R.N. Nabarro, Philos. Mag. **42**, 351 (1951)
J. Frenkel, Z. Phys. **37**, 572 (1926)
A.R. Lang, J. Appl. Phys. **30**, 1748 (1959)
F.R.N. Nabarro, Proc. Phys. Soc. **58**, 699 (1947)
F.R.N. Nabarro, Proc. Phys. Soc. A **200**, 279 (1952)
R. Peierls, Proc. Phys. Soc. **52**, 34 (1949)
J. Pelleg, J. Less-Common Met. **17**, 319 (1969a)
J. Pelleg, Philos. Mag. **19**, 25 (1969b)
J. Pelleg, J. Less-Common Met. **17**, 130 (1969c)
J. Pelleg, R. Liu, Thin Solid Films **221**, 318 (1992)
W. Puschl, Prog. Mater. Sci. **47**, 415 (2002)
N. Thompson, Proc. Phys. Soc. Lond. **366**, 481 (1953)

Chapter 3
Plastic Deformation

3.1 Introduction

On the basis of the first two chapters, it is clear that deformation in materials encompasses the following:

(a) elastic deformation;
(b) plastic deformation;
(c) fracture.

'Plastic deformation' refers to effects involving time and temperature, such as creep or fatigue (which will be treated later in separate chapters). Fracture can be a consequence of deformation, resulting from either creep or fatigue, and static loading in the plastic region may also eventually lead to fracture. Brittle materials may fracture without any deformation in the elastic region. However, this section focuses on plastic deformation actually occurring during a short time while loading is in process. Plastic deformation predominantly occurs by slip, but 'twinning' is also a way to induce deformation in a material given the appropriate conditions for this to occur. This chapter considers the behavior of single crystals and also the respond of polycrystalline materials to applied forces. The significance of studying deformation in single crystals is due to the atomistic concept of deformation, rather than the continuum approach.

3.2 Critical Resolved Shear Stress (CRSS)

When stress exceeds a critical value, materials irreversibly change their dimensions by the process of plastic deformation. The stress applied may be tensile or compressive. Thus, a critical stress is necessary in order for plastic deformation to occur. Shear stress, acting in a specific plane, is responsible for plastic deformation. Basically, shear stress causes a change in the dimensions of a material, measured

J. Pelleg, *Mechanical Properties of Materials*, Solid Mechanics and Its Applications 190, DOI 10.1007/978-94-007-4342-7_3,

after the load has been removed from it. Plastic deformation is usually associated with a slip mechanism, but that is not the sole mechanism which may be involved. Clearly, twinning may occur simultaneously if no time element, temperature or cyclic stresses are involved. But before considering slip and the systems in which it occurs, it is worth exploring the details regarding the applied stress which initiates slip deformation. Slip is aided by dislocation movement.

Dislocations (discussed above in Chap. 2) may have a dual effect on the strength of materials. In the presence of a few dislocations, materials deform at a low-yield stress, τ_0, whereas, when the concentration of dislocations increases, materials become strong again. In the absence of dislocation, which can be considered as a reference point, the resistance to any change in the dimensions of a material is extremely large (close to the theoretical strength). Whiskers tested under tension approaches the theoretical strength of materials. Frenkel calculated a value for the yield stress τ_0:

$$\tau_0 \sim G/6 \tag{2.7}$$

In crystals, slip starts on specific planes and in definite crystallographic directions at some critical value, known as the ‘critical resolved shear stress’ (discussed below). In the following, the minimum stress required for deformation by initiating slip will be derived with the aid of Fig. 3.1. Figure 3.1b is a bar cut from a cubic single crystal, as shown in Fig. 3.1a. A tensile load, P, is applied to the bar (as illustrated), resulting in the stress:

$$\sigma = \frac{P}{A} \tag{3.1}$$

Slip occurs on the slip plane indicated by the dashed area and can be expressed as:

$$\text{slip area} = \frac{A}{\cos\phi} \tag{3.2}$$

It can be seen in Fig. 3.1b that ϕ is the angle between the normal to the slip planes and the tensile axis. The resolved force in the slip direction may be given as:

$$\text{resolved force in the slip direction} = \text{P}\cos\lambda \tag{3.3}$$

The resolved shear stress in the slip plane in the slip direction obtained from Eqs. 3.1 to 3.2 is:

$$\tau = \sigma\cos\phi\cos\lambda \tag{3.4}$$

One immediately observes, from Eq. 3.4, that the resolved shear stress is zero, either when $\phi = 90°$ or when $\lambda = 90°$. In the first case, the tensile axis is normal

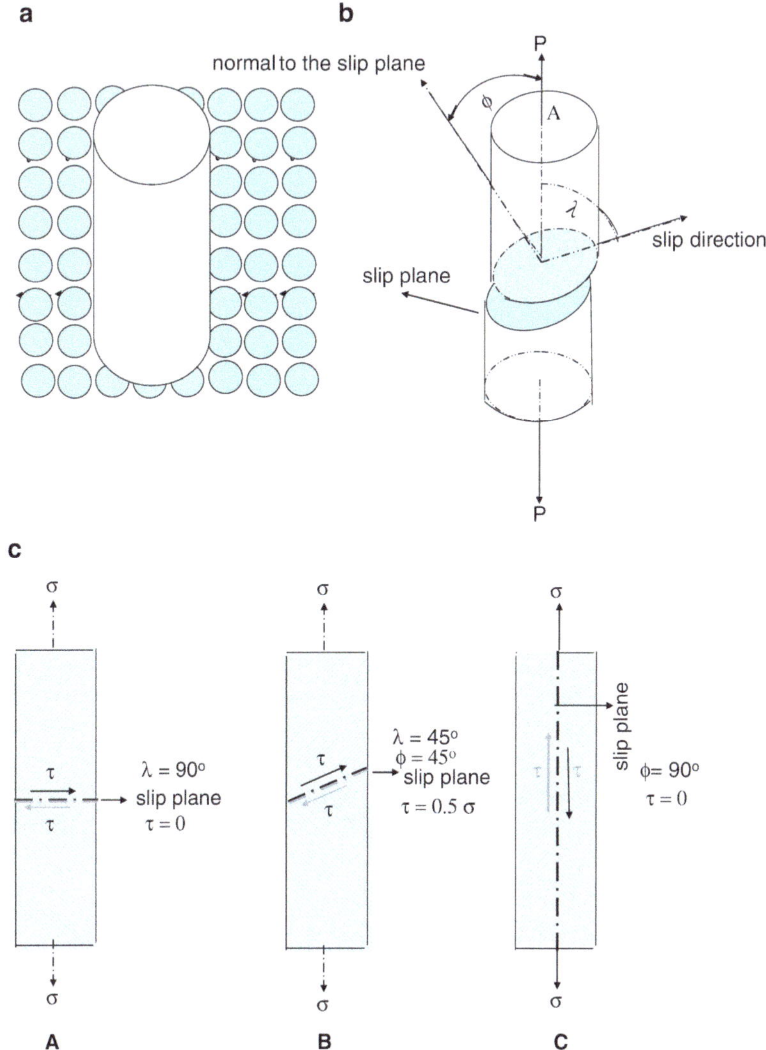

Fig. 3.1 (**a**) A *cube* showing the bar cut out of it; (**b**) the load applied the bar and the resolved shear stress. (**c**) The shear stress for some angles of Eq. 3.4: (**A**) $\lambda = 90°$, when the tensile axis is parallel to the slip plane, $\tau = 0$; (**B**) $\lambda = 45°$, $\phi = 45°$, $\tau = 0.5\sigma$, maximum shear stress; and (**C**) $\phi = 90°$, when the tensile axis is normal to the slip plane, $\tau = 0$

to the slip plane, whereas, in the second case, the tensile axis is parallel to the slip plane. Deformation by slip will not be expected when the tensile axis is parallel to the slip direction, because the shear stress is zero. The component of stress, normal to the slip plane, does not influence slip. The maximum shear stress is when $\phi = \lambda = 45°$.

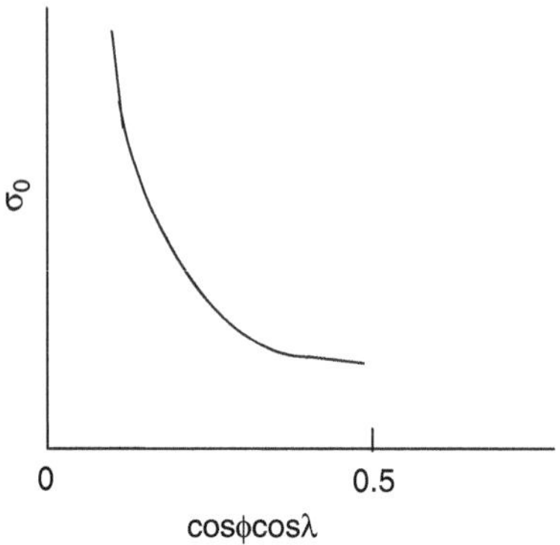

Fig. 3.2 The schematic variation of yield stress with orientation

Equation 3.4 gives the 'resolved shear stress'. The product in the equation is known as the 'Schmid factor' and determines whether the orientation is favorable for slip. The conditions for slip are given by Schmid's Law and the value of Eq. 3.4, often represented in the literature by τ_r, indicating the onset of plastic deformation, and called 'critical resolved shear stress' [henceforth: CRSS]. CRSS is a structure-sensitive property, since it is very dependent on impurities and the way the crystal was grown and handled.

On application of the tensile load, σ_0, the yield stress, varies widely, but, considering Eq. 3.4 in terms of yield, its value is a minimum when $\phi = \lambda = 45°$, thus giving for the product of the cosines (Schmid's factor) a value of 0.5. In this case, the maximum value attained for τ is $0.5\sigma_0$. Beyond 0.5, σ_0 increases again.

The early work of Schmid and Boas, who studied HCP structures, showed that, despite the wide variation in σ_0 with orientation, the CRSS is a constant of the material. According to them, the variation in tensile yield stress in Zn crystals is shown schematically by Fig. 3.2 by expressing Eq. 3.4 as:

$$\sigma_0 = \frac{\tau_0}{\cos\phi\cos\lambda} \tag{3.4a}$$

Schmid's Law has been experimentally proven for a large number of single crystals, but tests to substantiate this rule are usually more conveniently done on HCP crystals, due to the low number of slip systems; in FCC and BCC, where more slip systems are possible, the active slip plane is usually the one at which CRSS is first reached.

In summary, CRSS is the necessary component of shear stress, resolved in the direction of slip, which initiates slip in the crystal. It is a constant for a given crystal.

Complications may arise in polycrystalline materials, where each grain may be considered as a single crystal and has a different orientation than its neighbor. Taylor assumed that the strain of the grains (crystallites) and the deformation occur with the same value of CRSS on all available slip systems. According to Schmid's calculations for the yielding behavior of polycrystalline metals, the different initial CRSS values, due to the geometry of the slip systems, have no effect on the macroscopic yield surface of a randomly oriented polycrystalline material.

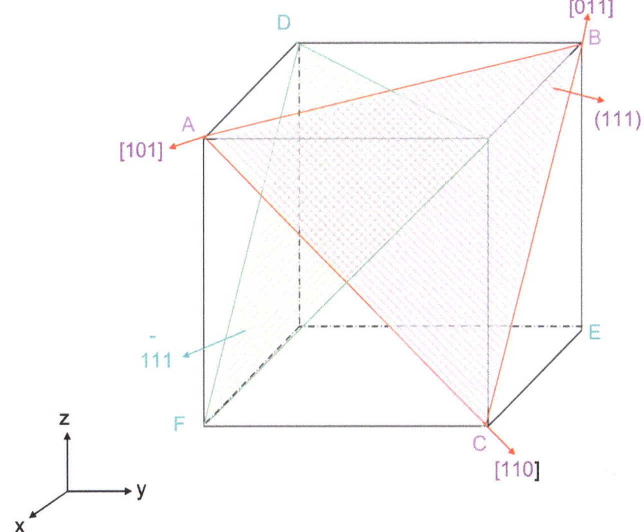

Fig. 3.3 An FCC structure showing two of the four {111} planes and the slip directions of the (111) plane shaded in *red color*

3.3 Slip

It was indicated several times in the previous chapters that plastic deformation occurs by slip on specific planes and in specific directions, depending on the structure. Slip plane and slip direction define the slip system of a material. One has to realize that: (a) the slip plane for slip is the most densely packed plane in a specific structure; (b) the slip direction is almost always the closely-packed direction of the atoms and; (c) of the several possible slip systems in a structure, the functional one will be the one with the largest CRSS. The most common structures, certainly in metals, are the FCC, BCC and HCP structures. A brief survey of various slip systems in each of these structures follows:

3.3.1 FCC Structures

In FCC structures, there are four {111} planes and three <110> directions; therefore, the number of slip systems is $4 \times 3 = 12$. Figure 3.3 illustrates an FCC structure with two of the four possible planes and the slip directions of the (111) plane.

At room temperature, the major source for plastic deformation by slip is dislocation motion through a crystal lattice. Crystal structure is not altered by plastic flow. In single crystals, deformation is anisotropic and is influenced by the active slip system. The motion of the dislocations involved in the slip process leaves linear

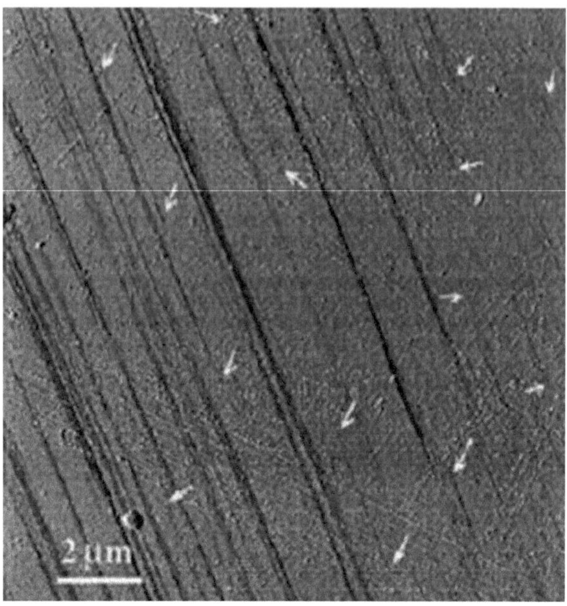

Fig. 3.4 An AFM image of a crystal surface showing *slip-line* traces in Ni_3Al. Short slip traces are indicated by *white arrows* (Coupeau et al. 1999. Courtesy of Elsevier Science Ltd.)

traces on the surface of the material, known as 'slip lines'. These may be observed under ordinary microscopes. Figure 3.4 illustrates such slip lines obtained by atomic force microscope (AFM) in Ni_3Al.

Slip progresses inside a crystal until the dislocation line reaches the end of the crystal, where it results in a visible step called a 'slip band'. A slip band is actually caused by the clustering of slip lines. Slip happens progressively, one step at a time, in such a manner that the crystal structure remains the same during the deformation. However, in cases of deformation by tension, a crystal lattice undergoes rotation in such a way that the slip direction rotates towards the tensile axis.

3.3.2 BCC Structures

Schematic Fig. 3.5 below illustrates the possible slip systems in BCC structures. Here, deformation and the choice of slip systems are more complicated than in FCC structures, since three planes may be active, but the directions are always <111>. For example, αFe can slip either on the {112} plane or on the {123} plane, in addition to the more common {110} slip plane. On both the {112} and {123} planes, the observed slip lines are often somewhat wavy. The active slip system largely depends on the material under deformation. Thus, the deformation of metals, such as W, Mo, αFe, etc., occur in the {110} – <111> system. BCC metals have six {110} planes and two <111> directions, giving altogether $6 \times 2 = 12$ possible slip systems. The aforementioned metals and also Na may slip during deformation also

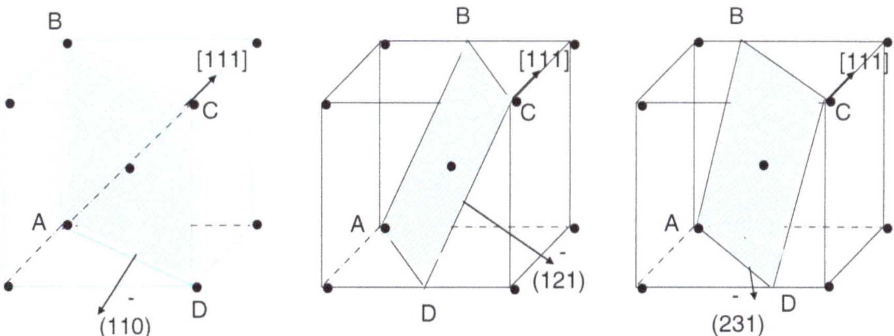

Fig. 3.5 A schematic illustration of slip systems in BCC structure. Note that the direction <111> does not continue the corresponding line of the side of the *rectangle*

on {112} planes. In such a case, there will be only one <111> direction, but there will still be 12 slip planes. Altogether, the number of slip systems is $12 \times 1 = 12$. Finally, the {123} plane may also be active under certain conditions. Besides the aforementioned αFe, K also shows such deformation. The number of slip systems in this case is $24 \times 1 = 24$. Thus, some BCC materials, such as αFe, can have up to 48 slip systems, as indicated above by summing up all the slip systems for the slip of αFe. Again, the multiplicity of slip systems is represented by wavy slip lines.

Usually in all the systems, slip occurs in the shortest direction (namely, the Burgers vector); however, unlike FCC structures, there are no truly close-packed planes in BCC crystal structure. It is believed that the activation energies for slip on {123} and {112} planes are not exactly identical to that of {110}, but sufficiently close to allow slip on these planes under certain conditions. This indicates that none of the BCC slip planes are truly close-packed; thus, BCC slip systems require additional energy (maybe in the form of heat) in order to function.

In other words, slip in BCC metals is temperature dependent. The close-packed planes in BCC are not really close packed and the planes in BCC with the highest atomic-packing density are on the {110} planes. These are known as 'primary slip planes', while the second set of planes upon which slip may occur are termed 'secondary slip planes'. More energy is required for slip on secondary planes than on primary planes. Compared to FCC crystals, more energy is required to cause slip in BCC crystals. Van der Walt, in his work on slip in BCC structures, even claims that {123} planes are rarely or never involved in slip, thus having no real secondary slip system.

3.3.3 HCP Structures

Figure 3.6 shows slip in single HCP crystals. Slip in HCP metals is much more limited than in FCC and BCC crystal structures when considering the main slip

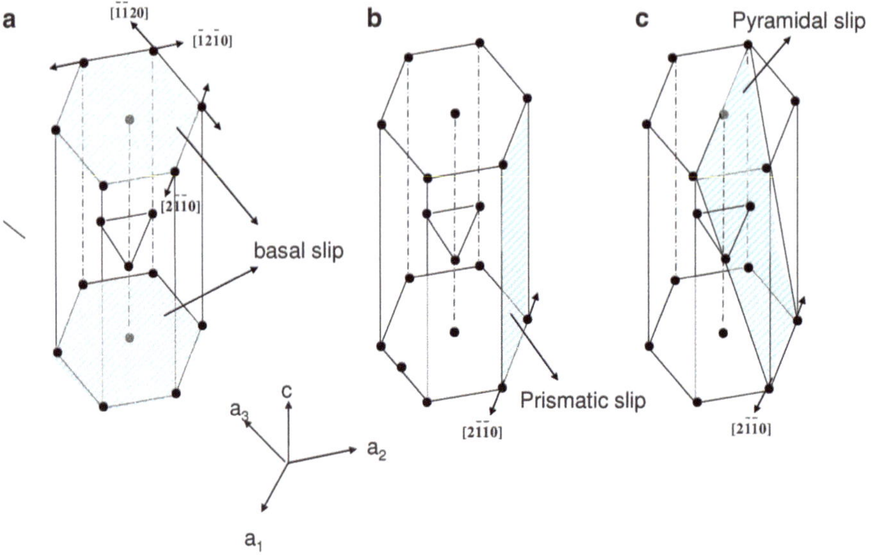

Fig. 3.6 A schematic illustration of slip systems in HCP: (**a**) basal slip; (**b**) prismatic slip; and (**c**) pyramidal slip

systems. This is because few active slip systems exist in HCP metals. As a result of the small number of slip systems, HCP metals are generally brittle. Nevertheless, the possible slip systems in HCP are presented here. The main and most common slip plane, the basal slip plane, is (0001) and its directions are <1120>. Thus, the number of slip systems is $1 \times 3 = 3$. Typical metals, in which slip occurs on the (0001) plane, are Zn, Cd and Mg. Prismatic and pyramidal slip are also likely in the prismatic and pyramidal planes {1010} and {1011}, respectively; thus, the number of prismatic slip systems is $3 \times 1 = 3$ and of the pyramidal ones, $6 \times 1 = 6$. As indicated above, the direction in all the slip systems is <1120>. Note that in Ti, for example, any of the aforementioned slip systems is possible. Of all the possible slip planes, only the one with the maximum resolved shear stress, having the crystallographic slip direction, will start to yield when deformed.

The (0001) slip plane has the same atomic packing as FCC crystals and is a true close-packed plane. The other systems (prismatic and pyramidal) function at high temperatures. Several researchers have observed that lattice parameters play an important role in determining the operative slip system. When the ratio of a to c (the lattice parameters) is $a/c = 1.632$, then ideal atomic packing exists, but this ratio may also be higher or lower. In Zn and Cd, $c/a = 1.856$ and in other metals the ideal ratio does not occur. In fact, when $c/a < 1.632$, the (0001) plane can no longer be considered the closest-packed plane. Cadmium, zinc, magnesium, titanium and beryllium have slip planes on {0001} and slip directions of <1120>. Thus, a total of three slip systems exist in these metals, depending on their orientation.

Unlike FCC and BCC crystals, HCP crystals have only one set of slip planes, {0001}, a family of densely-packed planes, as mentioned earlier. HCP crystals have secondary slip systems, like BCC crystals, and also exhibit twinning. They become brittle at low temperatures, while FCC crystals generally remain ductile and tough even at low temperatures. Even BCC crystals have more slip systems and, as such, more ways for dislocations to propagate. Thus, BCC crystals are more ductile than HCP crystals, which are more brittle. In BCC crystals, the 48 slip systems can more readily lead to dislocation pile-ups, for example, meaning high ductility coupled with higher strength, as observed in metals such as V, Cr, Fe, Mo, W and others.

3.4 The Slip in Polycrystalline Materials

As mentioned above, when a metal consists of a single crystal, it deforms anisotropically when stressed, depending on the orientation of the operative slip system. Polycrystalline materials consist of many grains, each with a different orientation. Yet, all of these grains have identical packing (structure), though each crystal has its principal slip planes or directions oriented differently from those of its neighbors. In each such grain slip lines (or bands) indicate deformation that has taken place. Figure 3.7 is a schematic illustration of a polycrystalline material as it might be seen under a microscope, before and after deformation, showing slip-line formation in some of its differently oriented grains (crystallites). This Figure is hypothetical, since it erroneously assumes that the grains do not change in shape or size. Such a structure (Fig. 3.7) might possibly occur only when the macroscopic deformation is very small.

In polycrystalline materials, slip planes and directions vary from one crystal (grain) to another; thus, CRSS varies from one crystal to another. The crystal with the largest CRSS yields first. If a grain is oriented unfavorably with respect to the applied-stress direction, its deformation is impeded and vice versa for

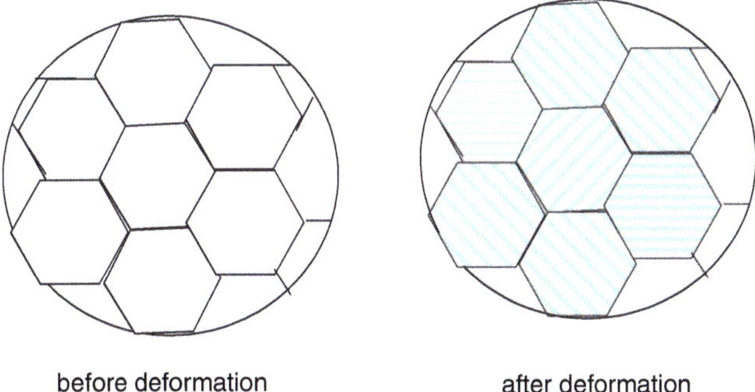

Fig. 3.7 A schematic illustration of a material before and after deformation. Note the non-realistic assumption that no change in grain shape or size has occurred

Fig. 3.8 Continuous lattice rotation due to slip in Zr deformed by rolling (Tenckhoff 1978. Courtesy of Springer Customer Service)

favorably-oriented grains. Grain boundaries are obstacles to slip, since the slip direction of a favorably-oriented crystal may change when it crosses a grain boundary. As a result, the strength of polycrystalline materials is higher than that of single-crystal materials.

Compared to single crystals on a macroscopic scale, polycrystallinity permits a metal to act as an 'isotropic body' when stressed, even though each grain would behave in an anisotropic manner if isolated. The plastic deformation of polycrystalline materials is, thus, determined by means of a compromise between the deformations within the individual grains (occurring by dislocation movement in response to externally applied stress). The movement of crystal dislocations allows the grains to deform in certain directions, as dictated by the restrictions of grain orientation. The change in crystallite shape is a consequence of the prerequisite that grains must adjoin during the deformation process, while the stress acting on them is in local equilibrium. These crucial requirements dictate a change in the shape of the grains under stress (unlike hypothetical Fig. 3.7 above), occurring, for example, during some processing operation, such as rolling. Furthermore, these shape-changes are also associated with the crystallographic lattice rotation of the crystallographic orientations of the grains with respect to each other. Usually, different textures evolve during plastic deformation, as a polycrystalline material changes shape.

The rotation of a deforming grain is caused by the external stresses exerted upon it by its neighbors. The resulting preferential alignment of crystalline grains (known as 'texture') reflects the deformation conditions, as well as the deformation mechanisms of the material. This may also cause a polycrystalline material to develop anisotropic (i.e., direction-dependent) physical properties, such as elasticity or electrical conductivity. Texture development generally occurs when the degree of deformation is quite large. The lattice rotation due to slip in Zr undergoing rolling may be seen in Fig. 3.8. Slip, which is, among other things, temperature dependent, occurs in various systems characterizing HCP structures, but twinning systems may also be activated. If the stress value required for slip is higher than the cohesion strength, fracture is likely to occur. In general, such a situation should be avoided. Good ductility is required for shaping operations (rolling, drawing, stamping, etc.), but, at the same time, plastic stability is important, so that a finished product does not flow and change its dimensional characteristics when exposed to acting stresses while in service.

Now, to return to the topics of microplasticity and macroplasticity, 'microplasticity' deals with materials on an atomic scale, describing dislocation movements, intersections and the response of single crystals to them. By scaling it up, one can predict the behavior and properties of materials in bulk. In general, by studying plastic deformation on the microplastic (atomic) level, one can better understand the macroplastic properties of materials on the continuum level (as was reflected in the mechanical tests described in Chap. 1). The yielding criteria in polycrystalline materials, to induce plastic deformation, is a compromise made by the slip mechanism of each grain as dictated by CRSS (Schmidt's law). However, one cannot take it for granted that polycrystalline materials exhibiting microplastic deformation at stresses well below the yield stress, will have all or a majority of their grains deformed plastically at the same yield-stress level. Rather, one should assume that, even at the macroscopic yield-stress level, not all the grains have undergone slip, either because of their orientation and/or because crystallites may exist where the CRSS for slip has not yet been achieved. If a grain is oriented unfavorably to the stress direction, its deformation is impeded.

3.5 Twinning

Plastic deformation may take place not only due to slip, but also by twinning or a combination of both. Whereas, in slip, atoms move a whole number of atomic spacing, in accordance with the Burgers vector of the specific structure, in deformation by twinning, atoms move a fractional atomic spacing, which leads to a rearrangement of the lattice structure. Furthermore, thin slip lines are usually observed (visually or under a microscope) on the specimen surfaces of polished test specimens. 'Twins' are generally represented as wide bands or broad lines. The slip lines can usually be removed by appropriate etching techniques; twins maintain their microscopic appearance even after relatively deep etching. However, the major difference is related to structural changes in the lattice induced by twinning, i.e., the stacking sequence changes. These changes are associated with the actual motion of the atoms. There is no change in lattice orientation during slip. Figure 3.9 schematically illustrates plastic deformation by slip and a combination of slip and twinning. As in the case of Fig. 3.7, this illustration is hypothetical, since, for the sake of convenience, no changes in the shape or size of the grains has been indicated after deformation.

Two types of 'twins' are common: 'annealing twins' and 'mechanical twins'. Mechanical twins, also known as 'deformation twins', are mostly found in HCP

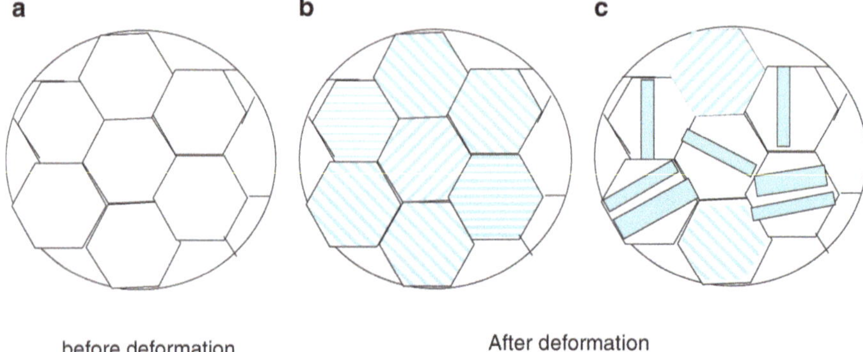

Fig. 3.9 Schematic hypothetical illustration (no change in shape is shown): (**a**) before deformation; (**b**) after deformation by slip only; (**c**) after deformation by slip and twinning. Note the twin bands in some crystallites

metals, such as Mg, Zn, Cd, Ti, etc. Annealing twins, as the name implies, are more frequent in FCC metals, such as Al, Cu, brass, etc. These FCC metals have previously undergone some type of metalworking, followed by heat treatment. Annealing twins are also called 'growth twins', since crystals can be grown from vapor, liquid or solid phases. Twins are formed because of a change in the normal growth mechanism.

Tentative experimental data show that certain twinning systems become operative at CRSS, which is analogous to Schmid's law for slip. Szczerba, et al., reported a CRSS law for twinning in Cu-8 at.% Al single crystals, which is analogous to Schmid's law for slip. Some criteria have to be met for twinning to occur, namely the ratio of the resolved shear stress, τ_{RSS}, to the critical stress of a twin system, τ_C, must be greater than that of any slip system, namely $\tau_{RSS}/\tau_C >$ CRSS for slip. Furthermore, the τ_{RSS} should be greater than the threshold (namely greater than some minimum) stress necessary for twinning to occur.

As in the case of slip, twinning occurs on well-defined planes and directions:

$$\begin{aligned}&\text{a) FCC Twin plane}: \{\mathbf{111}\} \quad \text{Twin direction}: <\mathbf{11\bar{2}}>\\&\text{b) BCC Twin plane}: \{\mathbf{112}\} \quad \text{Twin direction}: <\mathbf{11\bar{1}}>\\&\text{c) HCP Twin plane}: \{\mathbf{10\bar{1}2}\} \quad \text{Twin direction}: <\mathbf{10\bar{1}\bar{1}}>\end{aligned}$$

In HCP crystals, additional twinning systems may operate, depending on the conditions of deformation and heat treatment. Thus, Christian also indicates the following twin systems:

$$\begin{array}{lll}\text{Mg} & \{\mathbf{10\bar{1}1}\} & <\mathbf{10\bar{1}\bar{2}}>\\ \text{Zr, Ti} & \{\mathbf{11\bar{2}1}\} & <\mathbf{11\bar{2}\bar{6}}>\\ \text{Zr, Ti} & \{\mathbf{11\bar{2}2}\} & <\mathbf{11\bar{2}\bar{3}}>\\ \text{Body centered tetragonal Sn} & \{\mathbf{101}\} & <\mathbf{\bar{1}01}>\end{array}$$

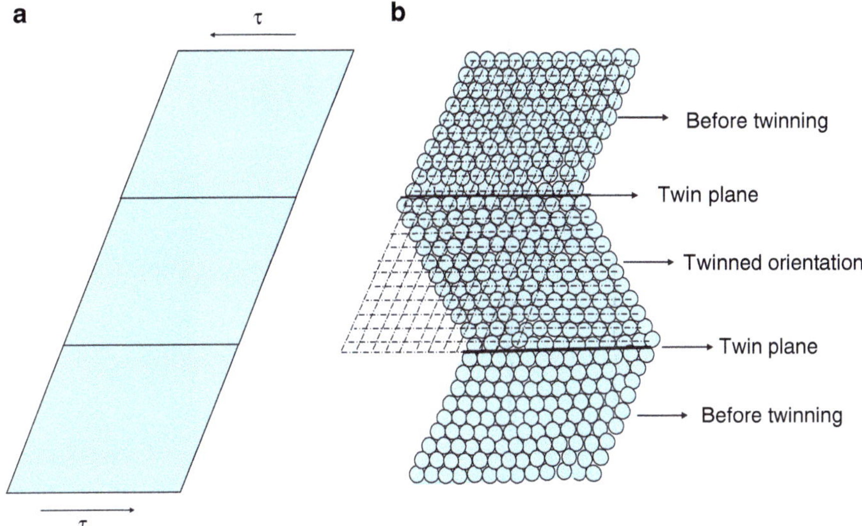

Fig. 3.10 Schematic illustration of a twinned region in the material: (**a**) before application of shear τ and (**b**) after twinning deformation. Note that the twinned region is a mirror image of the crystal part before twinning

In deformation by twinning, the atoms of each slip plane in some part of the lattice move different distances, causing half of the crystal lattice to become a mirror image of the other half. This mirror imaging is seen in Fig. 3.10. A very common feature, observed experimentally in recrystallized FCC structures like Cu, is the presence of twin bands in the grains with parallel sides, as shown in Fig. 3.9. The number of the bands increases with further grain growth in the specimen during heat treatment. The energy of a twin boundary is very low, compared to grain boundary energy, about in the range of low-angle grain boundaries. However, when the interfacial energy is larger than that of a grain boundary, annealing twins are rarely observed. Such is the case, for example, in Al, where the twin interfacial energy is higher than that of the grain boundary and annealing twins are rare. This recalls the fact that cross slip occurs frequently in the Al system (as mentioned in Sect. 2.14.1 above). Al, having a high stacking-fault energy, seldom dissociates into partial dislocations and, if it does, the fault is narrow and cross slip commences readily.

Deformation may often involve both twinning and slip. This is depicted in Fig. 3.11 for the $Ni_3Al(\gamma')$–$Ni_3Nb(\delta)$ system. Large deformations were produced by the formation of $\{211\}$ twins in δ, which generated γ' deformation bands with dominant slip systems on $\{100\}$ and $\{111\}$ to accommodate large twin shear.

High-temperature alloys are of interest for technological use and their mechanical properties offer attractive potential for elevated temperature applications. Their use often requires resistance to cyclic deformation. The role of cyclic deformation

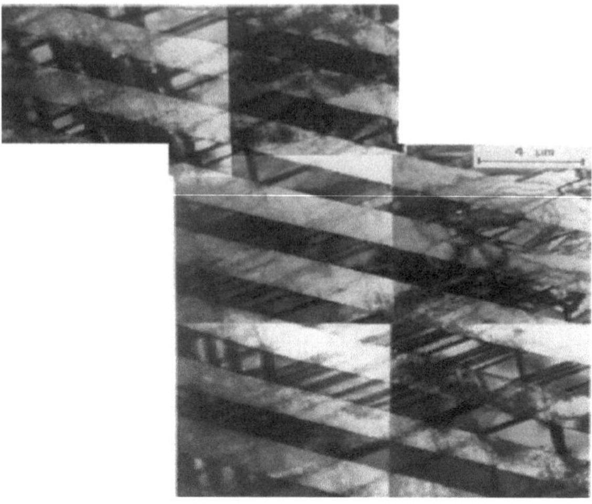

Fig. 3.11 Twins in δ and associated deformation bands in γ′ (Bhowal and Metzger 1978. Courtesy of Springer Customer Service)

on twinning in the directionally solidified γ(Ni)/γ/′(Ni_3Al)-δ(Ni_3Nb) alloy indicates that δ twins are formed during the tension part of the stress and, under the subsequent compression, untwining occurred. Under tension, the δ deforms by four variants of {211} twinning. However, {011} twins result when compression stresses are present. Plastic strain, during the compression part of a cycle, is attained through untwining of the {211} δ twins. The δ twins are seen in Fig. 3.12. Solute effects on deformation are also of importance. Solute effects on the deformation mode exist and have been observed in metastable beta titanium alloys. The transition from twinning to slip occurs due to the presence of the ω phase in aged Ti-V alloys. It was found that, when the ω phase was present in low to moderate volume fractions, the aged samples of Ti-V exhibited a strong tendency to form coarse deformation twins after prolonged aging. At high ω concentrations, slip formation is initiated and macroscopic embitterment sets in. These specimens were deformed by tension or by rolling approximately 3–5 pct. The orientation of the primary and secondary twins was of {112} planes and <111> directions, as seen in Fig. 3.13.

In Chap. 1, Fig. 1.4, a yield drop in the tension test of a 1020 steel was shown. A similar yield drop in tension occurs and a serrated curve is observed in specimens when twinning deformation occurs. In FCC materials, such as Cu alloys, during low-temperature deformation serrated curves occur at ~4.2 K. The serrated curve is characterized by a sudden fall in stress, followed by a rise in stress and then another fall, repeatedly during the mechanical twinning, which is the reason for their observation during tensile stress tests. Figure 3.14 shows such a serrated curve schematically.

In slip, dislocations move across the slip planes of a crystal, which remains the same single crystal, since the atoms move into equivalent crystallographic positions. The unit cell also remains unchanged. When partial dislocations move across the slip planes, the atoms on either side of the slip planes do not move into equivalent positions; the stacking sequence changes. A bi-crystal is formed in twin orientation.

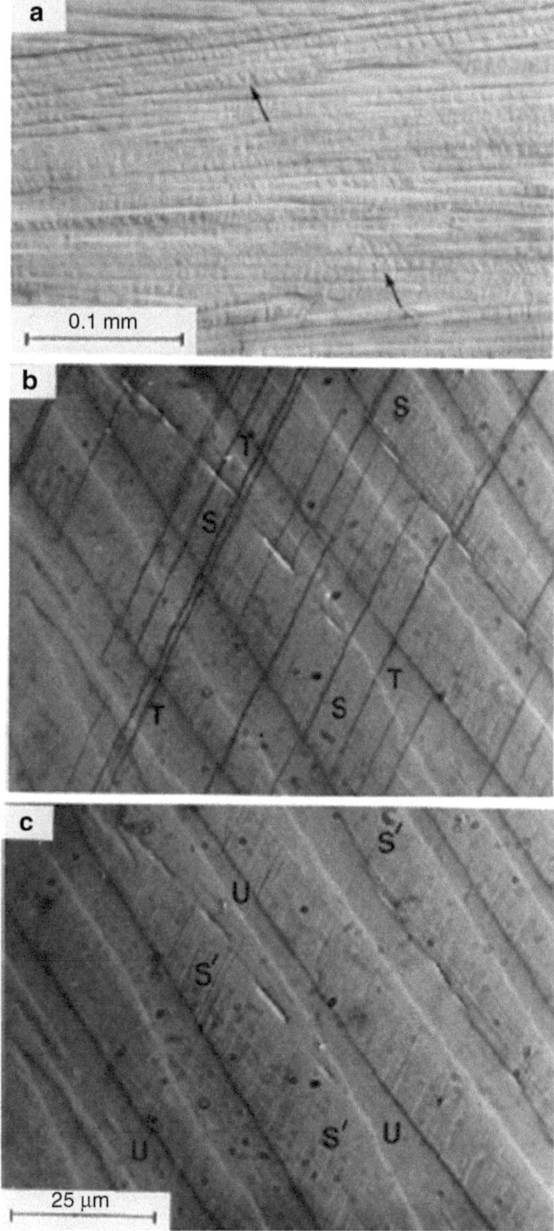

Fig. 3.12 Surface markings in a γ(Ni)/γ' (Ni_3Al)-δ(Ni_3Nb) alloy illustrating δ twins: (**a**) the presence of two variants of the δ twin is indicated by *arrows*; (**b**) the δ twins (*T*) under tension (magnification same as (**c**) *below*); (**c**) the disappearance of the δ twins of (**b**) following compression (Bhowal et al. 1978. Courtesy of Springer Customer Service)

Thus, partial dislocations and stacking faults are believed to be associated with twin formation. Stacking faults during recrystallization may cause twins to appear during annealing. A feature of many FCC metals is that both the twin and the parent matrix have the same crystal lattice and the same unit cell. Twinning produces a change in

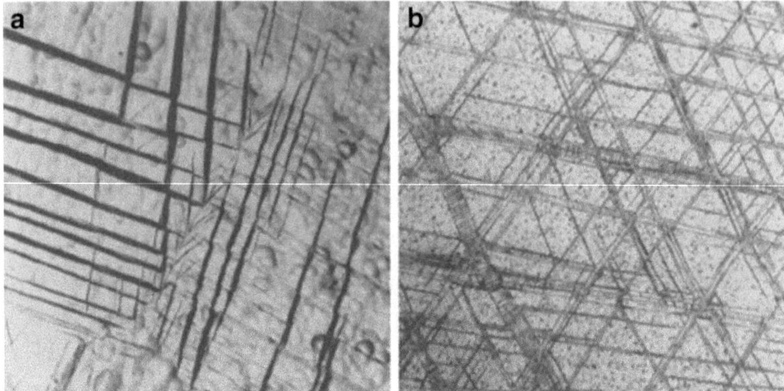

Fig. 3.13 Deformation twins in a tensile Ti-V alloy specimen showing a twin-to-slip transition in 20% V alloy deformed ~3–5% under tension: (**a**) pre-polished, as quenched contains ω with surface parallel to tensile axis; (**b**) a polished section of a tensile specimen perpendicular to gage length; aged 6 min at 400° (Paris et al. 1976. Courtesy of Springer Customer Service)

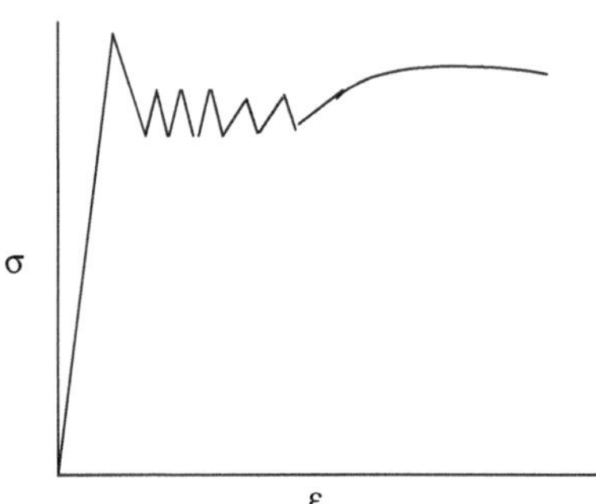

Fig. 3.14 A schematic illustration of the yield drop and serration in the load elongation curve

shape, but not in their crystal structure. The shear movement of a plane in the twin is proportional to its distance from the twin plane, thus producing the mirror image.

Serration is formed by the movement of the partial dislocations. This motion changes a part of the crystal into a twin orientation. It is believed that dislocations are involved in twinning, but the mechanism is not yet clear. The faulted region associated with partial dislocations (Frenkel) has mirror symmetry (Gilman). Partial dislocations are sometimes involved in twinning. The boundary, between the twinned an untwined portions of a crystal, constitutes a dislocation line with a Burgers vector **b**. Low stacking-fault energy provides wide ribbons; this is the case, for example, in Ag single crystals. Gubicza et al., have studied the deformation behavior in Ag and reported on the evolution of grain structure, dislocation and twin densities when Ag is processed to very high strains. It was concluded that the contribution of twinning to plasticity increases significantly with increasing strain in FCC metals, such as Ag, where the stacking-fault energy is very low and in accordance with theoretical models. They explained their observations as being a consequence of the very low stacking-fault energy, because the annihilation of dislocations is hindered by their high degree of dissociation into partials.

HCP and BCC metals are prone to show serrations during low-temperature deformation. Thus, the enhanced ductility of zirconium at ~4.2 may be attributed to the ability of this metal to twin readily at low temperatures (Briottet et al.). All stress/strain curves for the ~4.2 tests exhibited sudden, characteristic drops in the load. Transverse specimens showed a much higher incidence of twinning, greater strain hardening, more plastic deformation between serrations and fewer serrations than longitudinal specimens. However, these researchers concluded that twinning per se is probably not responsible for the discontinuous yielding. In BCC, serrations in the stress/strain curves occur at high strain rates, such as are obtained by impact and/or low temperatures. Twin bands obtained in BCC iron by high strain rates, such as impact, are known as 'Neumann bands'.

The morphological change of solids has been largely developed in the last years. The localization of plastic deformation in homogeneous materials may be associated with instabilities of stress/strain curves. This phenomenon can have very different aspects: the Portevin-Le Chatelier effect, Lüders bands, twinning, thermomechanical effects and avalanches of dislocations. During the past years, more and more researchers have doubted the ideas that twinning is the main contributor to the instability of the stress-strain curve at low temperatures and the contribution of dislocation is emphasized. The cooperative contribution of the various causes of instability must be appreciated. Nevertheless, this section focuses on twinning. Twinning is associated with the coordinated deformation of a large number of atoms, possibly leading to serrations in the deformation curves (giving a jagged appearance). Loud clicks are heard during the formation of twins, commonly known as 'tin cry'. This occurs because twin formation can be extremely rapid. The serration of the stress-strain curve is a sign of twin formation. Many investigators reported twin formation and serrated stress–strain curves associated with twinning. For further details on twinning, see the literature on the crystallography of deformation.

3.6 Yield Phenomena

3.6.1 Introduction

In Sect. 1.2.2, the elastic and proportional limits were discussed in regard to the transition from the elastic to the plastic deformation regions. In Fig. 1.7, these transition points were indicated together with a 0.2% offset yield point. In Fig. 1.4, the stress/strain relation under tension was shown by means of tests done to 1020 (low-carbon) steel, 1070 steel and a polymer (polyethylene), characterizing sharp yield points and otherwise ductile materials. Most structures are designed to act in

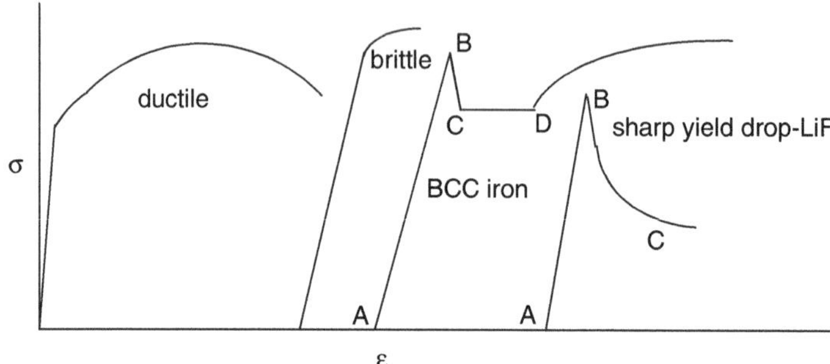

Fig. 3.15 Yielding in ductile materials, brittle materials, BCC iron and LiF. The C-D zone is the "yield elongation" region

the elastic region, since permanent change in a structural material must be avoided. Nonetheless, the plastic region is of great importance for the fabrication and shaping of structural materials requiring high ductility. Thus, the transition from the elastic to the plastic regions is of great importance. Consequently, extensive research was carried out and is available in the literature. A number of interesting phenomena were observed while characterizing yield and information was also provided on the behavior of materials during their transition into the plastic range. Some of the data on yield are considered below.

3.6.2 Sharp Yield

In metals, where the elastic–plastic transition is gradual, for practical purposes, the deviation from linearity may be taken to be the yield point (the 0.002 offset point). The most commonly known sharp yield was first observed in low carbon content BCC iron, also known commercially as 'mild steel'. In this case, sharp yield is followed by a sudden drop to a lower value, before further deformation takes place. In Fig. 3.15 such a yield drop in low-carbon steel can be seen. Deformation in the C-D region occurs without an increase in the stress level beyond a specific value, known as the 'lower yield point'. The highest stress in the elastic region is known as the 'upper yield point'. The C-D region is not smooth, but jagged. This kind of yield-point drop can be detected in what is known as a 'hard tensile machine', which is characterized by very small elastic distortion.

Note that N can also produce a sharp yield point followed by a sudden drop. In the theory put forth by Cottrell and Bilby, the yield in BCC iron was explained by the involvement of interstitial solutes, such as C or N, in the sharp yield. The nonhomogeneous deformation in the 'yield-elongation' region (the C-D zone) begins at a point of stress concentration, often at the grips, and propagates through the specimen as bands. The grips used for holding the specimens during tensile stress are sites of stress concentration. Beyond point D, deformation proceeds as stress increases with further straining, as shown in Fig. 3.15. The sharp-yield drop occurs not only in low C (or N) αFe single crystals, but also in polycrystalline iron, where the yield-elongation region is well developed. Cottrell has indicated that interstitial solid solutions, such as Cd, etc., may show similar yield-point drops, as seen in αFe, but less pronounced. Wain and Cottrell observed sharp yield points also in crystals of zinc containing N. Their experiments showed not only that yield points could be produced in zinc crystals, but also that this effect is associated with impurities in the metal, as in the case of yield in αFe. Johnson and Gilman, who studied LiF, revealed that sharp yield points also occur in this crystal. They reported that, in order to obtain a sharp yield drop, the necessary criteria are: (a) an increase in the number of moving dislocations and (b) a direct relation between the stress and the velocity of the dislocations. By knowing the strain rate, given as:

$$\dot{\boldsymbol{\varepsilon}} = \mathrm{nvb} \tag{3.5}$$

and the velocity of dislocation motion:

$$\mathrm{v} = \mathrm{k}\tau^{\mathrm{m}} \tag{3.6}$$

they were able to calculate stress-elongation curves for LiF showing sharp yield drops. By varying the exponent, m, in Eq. 3.6 and the density of the mobile dislocation ρ of Eq. 3.7, the magnitude of the yield drop could be changed. In the early stages of deformation, the density of mobile dislocations is given as:

$$\rho = \rho_0 + \mathrm{C}\varepsilon^{\alpha} \tag{3.7}$$

The exponent, m, is in the range 1–100. For a certain value of ρ, increasing m decreases the yield drop. The value of m for LiF is ~16.5 and for αFe, ~35. Equation 3.6 is an empirical relation with k being a constant. In Eq. 3.5, n is the number of moving dislocations/cm^2. Calculations were performed for $\varepsilon < 0.1$, namely in the early stages of the deformation. A dislocation-density evaluation is done either using the etch pits technique or by electron microscopy in the range of $\varepsilon < 0.1$. Equation 3.7 gives ρ_0 as the initial dislocation density, the constant $C = 10^8/cm^2$ and the constant α is $\sim 1 \pm 0.5$.

3.6.3 *Lüders Bands*

When a yield drop is observed in mild steel, surface marks usually develop at the point of stress concentration. These band-like markings are called 'Lüders bands'. Deformation in this range is non-uniform. Grips in the vicinity of specimen fillets are considered be locations of stress concentration. To best see these bands, the specimens should be reasonably well polished. Schematic Fig. 3.16 shows a specimen in which such bands may develop. Note that these bands should not be

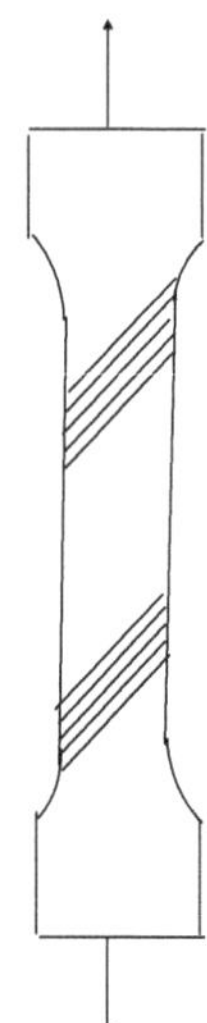

Fig. 3.16 Lüders bands at 45° from the tensile axis

considered as slip lines or slip bands. The propagation of the Lüders bands continues while the deformation is going on, until the entire specimen is covered by such bands. In industry, especially in the automobile industry, these bands are liable to appear, particularly on fenders, as veins called 'stretcher strains', that ruin the surface finish. The appearance of Lüders bands happens when the sharp yield drop occurs, meaning that the propagation of such bands occurs without the further need of stress; as long as the yield-point elongation continues, these bands will spread. Only after the yield-point elongation stops will the stress rise again. The appearance of Lüders bands is a result of dislocations having been pinned by C (or N), and the higher stress is required to free pinned dislocations that have been anchored by interstitial atoms. As will be detailed below, the avoidance of the formation of stretcher strains can be achieved by applying a small prestrain shortly before the fabrication, but somewhat beyond the lower yield point (e.g., by skin rolling) and only then starting the actual fabrication of a mild steel part (e.g., to be used in the automobile industry). After reloading, no Lüders bands appear, since the sharp-yield drop and the low yield-point elongation have been destroyed by the prestrain (or prefabrication straining). Adding Ti or similar elements, which react with either C or N or both, is another method for avoiding the appearance of stretcher strain marks.

3.6.4 Stain Aging

Schematic Fig. 3.17 illustrates strain aging in mild steel. The yield phenomenon indicated by a sharp yield-point drop is shown in region I. Following inhomogeneous

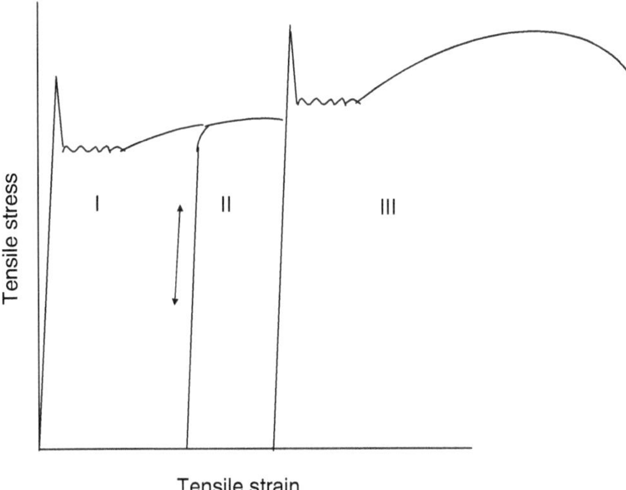

Fig. 3.17 Strain aging. In region *I*, a sharp yield drop is seen with Lüders bands formation. In region *II*, no yield drop is seen. The reappearance of a sharp yield drop in region *III* is also shown

deformation (represented by lower yield-point elongation) and after Lüders bands have formed and propagated, and passing it by some small deformation, if one interrupts the tensile test by unloading the specimen, one can observe an interesting behavior having practical applications in the automotive industry. If this specimen is reloaded immediately – or after the elapse of a short time – following unloading, the yield-point drop does not reappear and the Lüders bands do not form. A smooth transition from the elastic to the plastic regions is the result of the reloading, as shown in region II of Fig. 3.17. However, if a specimen has been unloaded and is left for a few days in the unloaded state, then the yield-point phenomenon reappears. This time, the upper yield point is higher than the one initially observed in that specimen, as shown in region III of Fig. 3.17. This is explained by the fact that the interstitial atoms C or N return to the dislocations and form what is called an 'atmosphere' around the dislocations re-pinning them. This re-pinning requires the time-dependent diffusion of C or N to the dislocations. This may take several days at room temperature, but, at higher temperatures, the increased diffusion rates of the interstitial C atoms shorten considerably the time for the yield point return. The involvement of the element of time in yield-point return also clarifies the use of the term 'aging' (strain aging), which is appropriate, since the process of straining also involves a temperature factor. That the new yield point is higher than the one initially observed is associated with the fact that some work hardening occurred in region I before unloading the specimen.

The C or N atoms occupy the octahedral interstitial, distorting the unit cell tetragonally (i.e., into a body-centered tetragonal structure) and a large volume expansion occurs.

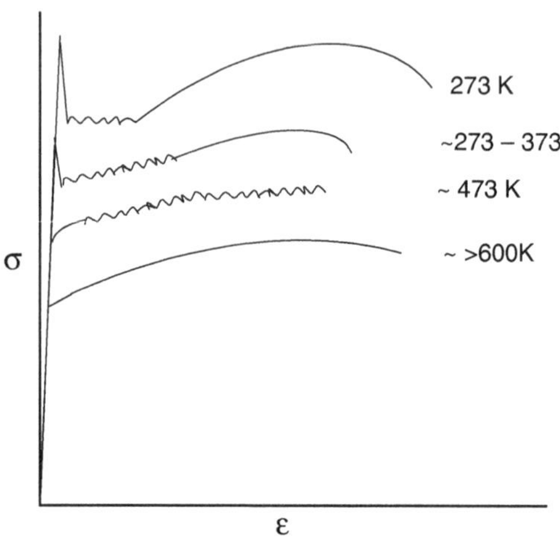

Fig. 3.18 The Portevin Le Chatelier effect. The influence of temperature on the appearance of a stress–strain curve under tension

3.6.4.1 The Portevin Le Chatelier Effect (PLE)

Another observed phenomenon, closely related to the yield-point phenomena associated with interstitial atoms, is that of temperature-dependent serrated curves. Serrated stress/strain curves are similar to those observed in twinned specimens, but their origin is different, as mentioned. In Fig. 3.18, a set of curves is shown along with the effect of temperature on their appearance. Such behavior, although observed initially in α Fe, is exhibited by several materials as they undergo plastic deformation, e.g., Al-Cu alloys (see, for example, Liang et al.), substitutional Al-2.5% Mg alloy (P. Barat et al.), etc. For the Portevin Le Chatelier effect [henceforth: PLE] to occur, solute atoms must segregate at the dislocation core. This requires sufficient mobility by diffusion of the segregated atoms. The local site of the dislocation core is energetically favorable, since it has space available to accommodate the solute atom which locks the dislocation, hindering its motion. A larger force (stress) is necessary to move the dislocation, as the cloud of solute atoms is dragged with it. At some stage, the dislocation eventually breaks away from the atmosphere of solute atoms, resulting in reduced drag stress for dislocation movement. Note that this process results in the formation of stretcher strains, mentioned above, which make the surface rough, prohibiting the use of that material in the automotive industry, unless remedied.

Many interstitial and substitutional alloys exhibit repeated yield-stress drops under tension, as influenced by strain rate and temperature. At sufficiently high temperatures and at a specific strain rate, these serrations gradually disappear, due to the relatively higher diffusion of the solute atoms, preventing the recapture of their atmosphere. The effect of temperature is also shown in Fig. 3.18.

Not only mild steels used in the automotive industry show Lüders bands as surface markings (see Fig. 3.19) or serrations resulting from the PLE effect, but other materials do as well. Al-Mg alloy sheets, potentially useful for automotive

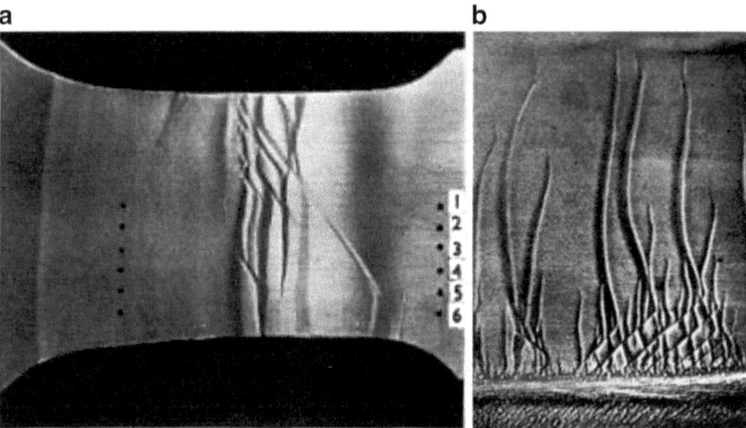

Fig. 3.19 Lüders markings on soft steel: (**a**) a tensile specimen; (**b**) a stamped sheet part (From Chadwick and Hooper 1953; B. B. Hundy 1953)

applications, may also show yield-point related phenomena, as seen in Fig. 3.20. In (a) the surface markings are illustrated and in (b) serrated curves are illustrated.

3.6.5 The Cottrell-Bilby Theory

Cottrell and Bilby did a theoretical study on yield-point phenomena (discussed in Sects. 3.6.2, 3.6.3, and 3.6.4). Their original concepts, regarding the dependence of the yield drop on impurities, discontinuous yielding and the strain aging associated with BCC iron, are based on dislocation locking by interstitial solutes (impurities). Various BCC metals and other structures may also show similar occurrences, but the first and most obvious encounter with yield phenomena occurred in α Fe. Cottrell and Bilby invented the concept of 'atmosphere', known ever since as 'Cottrell atmosphere', in order to explain how dislocations are anchored by interstitials, specifically C and N. These interstitial atoms settle down under the dislocation line to form an atmosphere. These atoms somewhat distort the lattice by forming a stress field around themselves and their own vicinity. This distortion may be reduced or relaxed once they settle down beneath the dislocation line. The preferential location of the C (N) atom below the dislocation is controlled by diffusion, which are time- and/or temperature dependent. Usually, after fabricating an iron billet (or some other shape), time goes by before it is used or tested. This elapsed time is sufficient for the C to diffuse into a dislocation and become mutually anchored there. Once an atom or atoms have diffused into a dislocation core, they will stay there. The dislocation has become pinned down and now, in order to tear such a dislocation away from its atmosphere, a somewhat higher force is required. This extra force is responsible for the upper yield point. After unpinning a dislocation, it is free to move at a lower

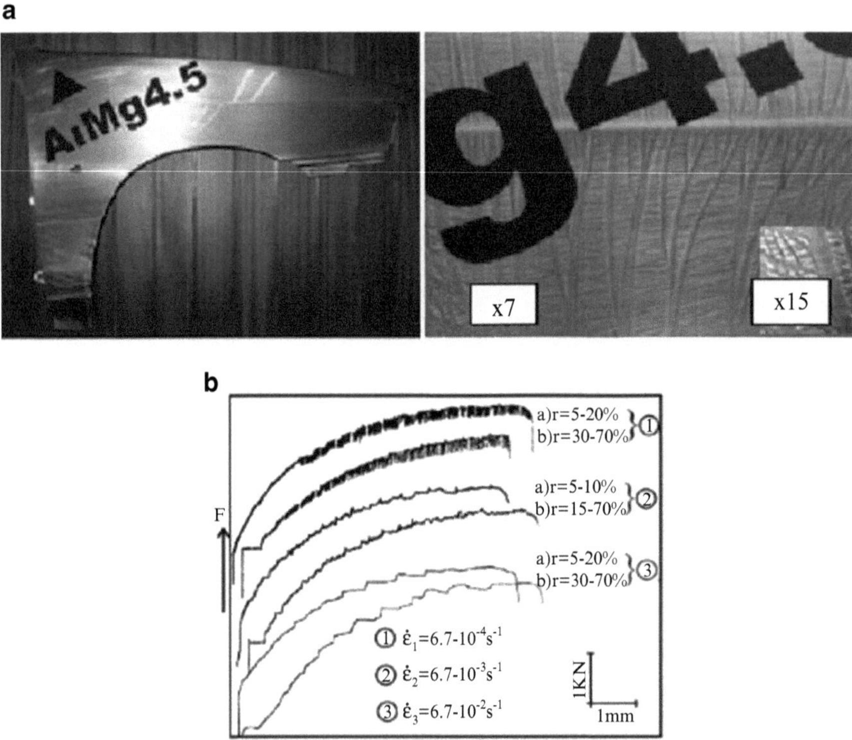

Fig. 3.20 Al-Mg alloy: (**a**) the surface appearance of a fender made of Al-Mg sheets; (**b**) load extension curves of AlMg6.5Mn sheets at different strain rates (Romahanji et al. 2004. Courtesy of Dr. Marija Korać, Technical Editor, metalurgija.Org., rs)

stress, producing the lower yield point. Once a dislocation is free to move, the material can continue to deform plastically in the usual way, i.e., the stress increase will produce some strain in accordance with the work-hardening concept.

Leaving a sample to age at room temperature for a few hours enables the carbon atoms to re-diffuse into the dislocation cores, resulting in the return of the upper yield point, as strain aging (region III in Fig. 3.17). Contrary to this, if a test specimen is pulled by tension immediately following unloading or within a short time afterwards, the yield point does not reoccur (region II, Fig. 3.17). The automotive industry is aware of this phenomenon and construction parts are shaped shortly after some prestrain is applied. Furthermore, as mentioned earlier, elements such as Ti, may be added to react with the interstitial atoms, such as N, in order to prevent the formation of Lüders bands. Cottrell atmospheres lead to the formation of Lüders bands, where the serration is believed to be the result of repeated yielding. Dislocations are repeatedly released from and recaptured by anchoring interstitial atoms.

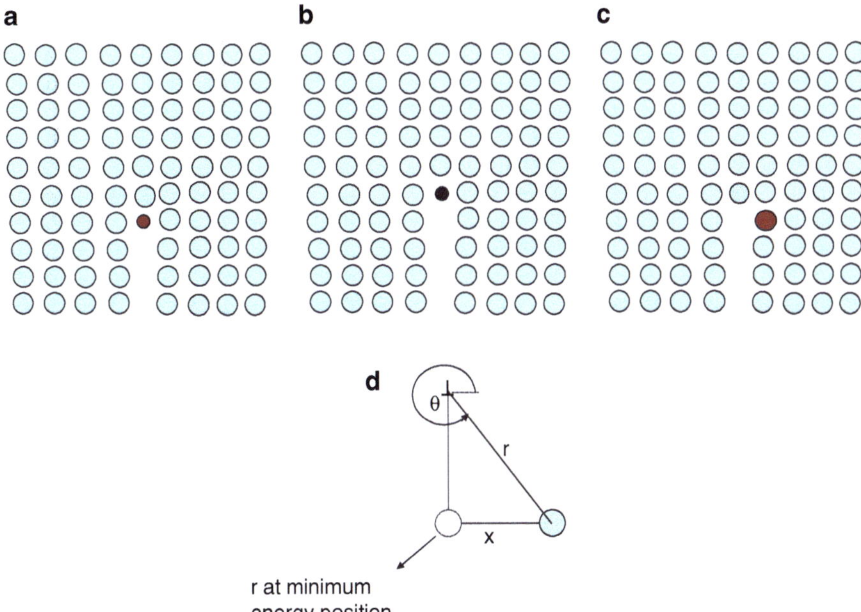

Fig. 3.21 Solutes in a matrix: (**a**) an interstitial solute under the dislocation; (**b**) a small substitutional solute replaces a matrix atom above the dislocation line; (**c**) a large substitutional atom is located near the extended site of the dislocation; and (**d**) moving a solute distance x from the minimum energy position

3.6.5.1 Elements of the Cottrell-Bilby Theory

How did Cottrell and Bilby derive their concept of yield-point phenomena that lock dislocations by means of impurities, such as C (or N) atoms? To better understand, consider Fig. 3.21a, where an interstitial impurity has diffused into a dislocation and settled underneath. These atoms are randomly distributed in the material before diffusion to the dislocation. A positive-edge dislocation is under tension just below it close to its core and under compression above it. In Fig. 3.22, the stress just above an edge dislocation is under pressure and below it under tension is indicated. Read's book provides a complete distribution of the stress field and the lattice strain.

When randomly distributed, N or C also has a stress field, which can be reduced (relieved) when it settles in the immediate vicinity and below the dislocation core. This location of the interstitial atom under the dislocation which is under dilation provides the lowest energy of the system. The C or N atoms, besides inducing volumetric expansion, also cause tetragonal distortion. This is a consequence of the fact that C atoms are located in interstitial sites at the center of the {001} planes (faces) and the [001] directions (edges), thus distorting the unit cell tetragonally in a [001] direction. This kind of distortion allows reactions with shear and hydrostatic stress fields to occur. Therefore, a C impurity will interact with both edge and screw dislocations.

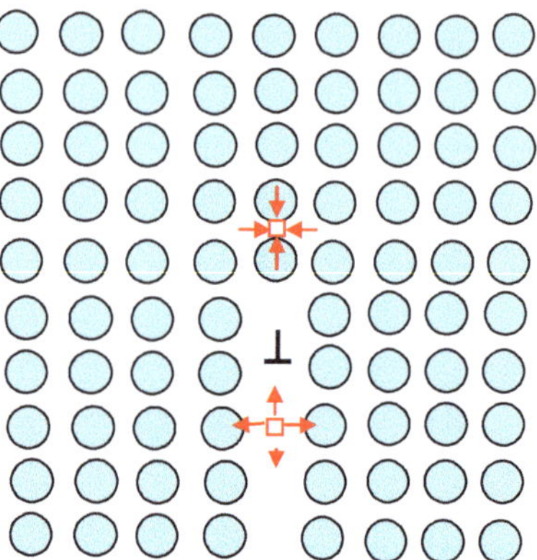

Fig. 3.22 Straining of the lattice *above* and *below* the edge dislocation is shown schematically

Consider, as in the model below, a system composed of hard balls having radii, r′. The radius of the hole where the atom settles is r_0. Mark the average radius, after the ball joins the space, as r_a. Now, express the average radius as:

$$r_a = r_0(1 + \varepsilon) \quad (3.8)$$

Equation 3.8 is a result of the following:

$$\varepsilon = \frac{r_a - r_0}{r_0} \quad (3.9)$$

$$\varepsilon r_0 = r_a - r_0 \quad (3.10)$$

$$r_a = r_0 + \varepsilon r_0 = r_0(1 + \varepsilon) \quad (3.11)$$

ε can be positive or negative. When the strain ε = 0, the distortion in the matrix is symmetric.

The energy of the interaction, U_i, is the work done against the local hydrostatic stress field and is given by:

$$U_i = p\Delta V \quad (3.12)$$

p is the hydrostatic pressure of the stress field and is given by:

$$p = -\frac{1}{3}(\sigma_x + \sigma_y + \sigma_z) \quad (3.13)$$

Obviously, the stresses indicated in Eq. 3.13 are the normal components of stress. Now, express ΔV with the help of Eqs. 3.8, 3.9, 3.10, and 3.11 as:

$$\Delta V = \frac{4\pi r_a^3}{3} - \frac{4\pi r_0^3}{3} = \frac{4\pi}{3}\left(r_a^3 - r_0^3\right) \quad (3.14)$$

$$r_a{}^3 = (\varepsilon r_0 + r_0)^3 \tag{3.15}$$

$$\Delta V = \frac{4\pi}{3}\left[(\varepsilon r_0 + r_0)^3 - r_0^3\right] = \frac{4\pi}{3}\left(\varepsilon^3 r_0^3 + 3\varepsilon^2 r_0^2 r_0 + 3\varepsilon r_0 r_0^2 + r_0^3 - r_0^3\right) \tag{3.16}$$

Neglecting terms of ε having high exponents, results in the following for ΔV:

$$\Delta V = \frac{4\pi}{3} 3\varepsilon r_0^3 = 4\pi\varepsilon r_0^3 \tag{3.17}$$

Substituting this value in Eq. 3.12 gives:

$$U_i = -\frac{4}{3}\pi\varepsilon r_0^3\left(\sigma_x + \sigma_y + \sigma_z\right) \tag{3.18}$$

The stress field of edge dislocations was previously given in Eq. 2.35 as:

$$\begin{aligned}
\sigma_x &= -Dy\frac{\left(3x^2 + y^2\right)}{\left(x^2 + y^2\right)^2} \\
\sigma_y &= Dy\frac{\left(x^2 - y^2\right)}{\left(x^2 + y^2\right)^2} \\
\tau_{yx} &= \tau_{xy} - Dx\frac{\left(x^2 - y^2\right)}{\left(x^2 + y^2\right)^2} \\
\sigma_z &= \nu\left(\sigma_x + \sigma_y\right)
\end{aligned} \tag{2.35}$$

Substituting the respective equation into Eq. 3.18, one obtains:

$$U_i = \frac{4(1+\nu)}{3(1-\nu)} Gb\varepsilon r_0{}^3\left(\frac{y}{x^2 + y^2}\right) \tag{3.19}$$

In polar coordinates, this is equal to:

$$U_i = \frac{4(1+\nu)}{3(1-\nu)} Gb\varepsilon r_0{}^3 \frac{\sin\theta}{r} \tag{3.20}$$

Recall the relations (from Eq. 2.33) between polar and Cartesian coordinates described above:

$$\begin{aligned}
&\mathrm{r}^2 = \mathrm{x}^2 + \mathrm{y}^2 \quad \text{or} \quad \mathrm{r} = \left(\mathrm{x}^2 + \mathrm{y}^2\right)^{1/2} \\
&\mathrm{x} = \mathrm{r}\cos\theta, \quad \mathrm{y} = \mathrm{r}\sin\theta \\
&\mathrm{tg}\theta = \mathrm{y}/\mathrm{x}
\end{aligned} \tag{2.33}$$

Thus, actually:

$$\sin\theta = \frac{y}{(x^2 + y^2)^{1/2}} \tag{3.21}$$

One can rewrite Eq. 3.20 as:

$$U_{(r,\theta)} = A\frac{\sin\theta}{r} \tag{3.22}$$

with A given as:

$$A = \frac{4\,(1+\nu)}{3\,(1-\nu)}Gb\varepsilon {r_0}^3 \tag{3.23}$$

has been calculated as A $\sim 3 \times 10^{-20}$ dyn cm^2.

According to Eq. 3.9, ε can be either positive or negative, depending on the value of r_a, which, in turn, depends on the radius of the impurity atoms if it is smaller or larger than the matrix atoms. The sign of $U_{(r,\,\theta)}$ is determined according the value of sinθ. Sinθ is positive in the range $\pi > \theta > 0$, namely in the upper part of the dislocation, which is under compressive stress. It will be negative just below the dislocation core, where tensile extension (dilation) exists. This is in the range $2\pi > \theta > \pi$. The maximum negative value is at $\theta = 3\pi/2$.

In lieu of Eq. 3.20, a somewhat different relation was approximated in the Cottrell-Bilby theory:

$$U = 4Gb\varepsilon {r_0}^3 \left(\frac{\sin\theta}{r}\right) \tag{3.24}$$

C or N can be accommodated under the dislocation line in the extended region, where $\theta = 3\pi/2$. The elastic theory does not hold in the dislocation core (as was indicated in Chap. 2). The calculations were done at a value of $r_0 = 2 \times 10^{-8}$ cm. For C, Cottrell calculates for $U_{max} \sim 1.0$ eV. For Zn dissolved in Cu, a value of $U_{max} \sim 1/8$ eV was calculated.

As mentioned above, the segregation of C atoms at a dislocation is possible when the atoms arrive there by diffusion, which is temperature-dependent. Room temperature is sufficient for C diffusion, but not for substitutional solute atoms. With the assumption that the concentration of C (or N) atoms is c, a binding energy can be calculated by:

$$c = c_0 \exp\left(\frac{U}{kT}\right) \tag{3.25}$$

c_0 is an average concentration and U is the elastic binding energy. Usually, a Maxwellian distribution is obtained at the vicinity of a dislocation, but, in the case of C or N, the elastic interaction is $U \gg kT$, so that the C atoms condense to form an atmosphere, i.e., a row of C atoms along the core of the dislocation line. Equation 3.25 allows for the calculation of a temperature at which this can occur, assuming that c = 1 and U = U_{max}.

For example, with the values of $c_0 = 10^{-5}$–10^{-6} and U_{max} ~0.5 eV, the calculated value of T ~700–750 K; therefore, at room temperature or below it, a strong yield-point occurs. Their theory explains the strong temperature dependence of the yield point. It increases with temperature decrease and, at 0 K, the atoms are fully condensed, forming the atmosphere. No thermal fluctuations help the atoms to escape from the atmosphere. Therefore, a high temperature is needed to tear the dislocation from the atmosphere formed by these C atoms. As the temperature increases, thermal fluctuations assist the applied stress to tear the dislocations away from the atmosphere.

3.6.5.2 The Stress to Unpin a Dislocation

The stress needed to move a dislocation away from a condensed line of C atoms may be obtained as follows:

$$U = 4Gb\varepsilon r_0^3\left(\frac{\sin\theta}{r}\right) = A\frac{\sin\theta}{r} \tag{3.24}$$

At a distance x from its minimum position, which is the initial location of the interstitial solute (shown in Fig. 3.21d above), the dislocation has an interaction energy given as:

$$U(x) = -A\frac{r_{max}}{(x^2 + r_{max}{}^2)} \tag{3.26}$$

The attractive force per unit length of dislocation is the derivative of Eq. 3.24, since, according to Eq. 2.13, energy and distance are related as:

$$\mathrm{dE} = \mathrm{Fdl} \tag{2.13}$$

Thus, the force needed to move a dislocation a distance x away from the atmosphere may be given by:

$$F_x = \frac{dU_x}{dx} = \frac{2Ar_{max}x}{(x^2 + r_{max}{}^2)^2} \tag{3.27}$$

The maximum value for x is obtained after equating the derivative to zero to obtain:

$$x = \frac{r_{max}}{\sqrt{3}} \tag{3.28}$$

and for F_{max}, after substitution for x:

$$F(x) = \left(\frac{3\sqrt{3A}}{8{r_{max}}^2}\right) = \frac{0.65A}{{r_{max}}^2} \tag{3.29}$$

From the above, one can obtain the critical tensile-yield stress, σ_{crit}, to unpin the dislocation in terms of b (force per area) as:

$$\sigma_{crit} = \frac{0.65}{{r_{max}}^2 b^2} \tag{3.30}$$

The above can also be expressed in terms of polar coordinates. The energy is maximal under the dislocation core, so for Eq. 3.23 one can write:

$$U_{max} = A\frac{\sin\theta}{r_{max}} \tag{3.31}$$

As indicated above (in Sect. 2.7), Eq. 2.13 relates energy and distance as:

$$dE = Fdl \tag{2.13}$$

thus:

$$\frac{dU}{dr} = F = -A\frac{\sin\theta}{{r_{max}}^2} \tag{3.32}$$

Force relates to area and stress as $F = \sigma A$, thus we can write for (3.32) in terms of the Burgers vector (lattice spacing)

$$\sigma_{crit} = \frac{A}{(b^2 {r_{max}}^2)} \tag{3.33}$$

since $\sin\theta = -1$ at $\theta = 3\pi/2$. Taking r_{max} just outside the core (elasticity does not hold in the core) as to be 2×10^{-8} and A $\sim 3 \times 10^{-20}$ dyn cm^2, the value of the breakaway stress $\sigma \sim 6 \times 10^{10}$ dyn/cm^2 may be calculated. The experimental value at 0 K is $\sigma \sim 1.25 \times 10^{10}$ dyn/cm^2.

When the dislocation is pulled apart from its atmosphere, glide can occur at a lower stress, as observed at the lower yield point in the yield elongation. When an external stress is applied, a dislocation is dragged with its atmosphere and starts to move. An opposing force tries to pull the dislocation back to its place. If the applied stress is not sufficiently large, the dislocation will not be able to free itself from its atmosphere, while will be dragged along with the moving dislocation. The movement of a dislocation, in such a case, is limited to the motion of the atoms forming its atmosphere. Since the velocity of the diffusional motion of the atoms is slower than that of the dislocation itself, such a process can be envisaged only in those cases where the rate of dislocation strain is low. Creep or micro-creep are such processes in which the rate of dislocation strain is $\sim 10^{-8}$ cm/s.

According to Cottrell, the maximum velocity of a dislocation, when it can still drag its atmosphere with it, is given by:

$$v = \frac{A}{r_{\max}} \frac{D}{kT} \tag{3.34}$$

In most of the cases, the rate of strain is high, the temperature is relatively low and plastic deformation occurs when a dislocation is freed from its atmosphere. With increasing applied stress, the upper yield point is reached. At this stress level, the force pulling the dislocation back to its atmosphere must be overcome. Once a dislocation has pulled away from its atmosphere, it moves into a relatively 'clean' region, free of interstitial atoms. The stress required to keep a dislocation in motion, after it has been freed from the atmosphere, is smaller and, therefore, the sudden drop in yield stress occurs. Following this event, a dislocation continues to glide, causing plastic deformation.

The relation existing between strain rate and dislocation velocity was, in essence, formulated by the Cottrell-Bilby theory and appears in many texts on the subject. When a specimen is tested by the application of tension, the strain rate,$\dot{\varepsilon}$, may generally be expressed as:

$$\dot{\varepsilon} = \rho b v \tag{3.35}$$

ρ is the moving dislocation density and v is its velocity. (Note that this equation is equivalent to Eq. 3.5). The density of dislocations increases with the increase of stress and velocity is a strong function of stress. Similar equations may be given for the strain rates at the upper and lower yield stresses:

$$\dot{\varepsilon} = \rho_U b v_U \tag{3.36}$$

$$\dot{\varepsilon} = \rho_L b v_L \tag{3.37}$$

The subscripts refer to the upper and lower yield stress cases. Since the strain rate is constant during tension, one can write:

$$\dot{\varepsilon} = \varepsilon_L = \varepsilon_U \tag{3.38}$$

Using the equality of strain rates in Eq. 3.38 and Eqs. 3.36 and 3.37, a ratio can be written as:

$$\frac{\rho_U}{\rho_L} = \frac{v_L}{v_U} \tag{3.39}$$

Generally, the relation between stress and velocity (as given in Eq. 3.6) is:

$$v = k\tau^m \tag{3.40}$$

τ is the shear stress acting on the dislocation and exponent m is the strain-rate sensitivity. The ratio of the upper to lower yield stresses may be derived from the previous two expressions, giving:

$$\frac{\tau_U}{\tau_L} = \left(\frac{\rho_L}{\rho_U}\right)^{\frac{1}{m}} = \left(\frac{v_U}{v_L}\right)^{\frac{1}{m}} \tag{3.41}$$

This last equation summarizes the general aspects of the upper/lower yield stress phenomenon. At upper yield stress, the mobile dislocation density is low, but the average dislocation velocity is high. At lower yield stress, there are many more mobile dislocations, but they are not moving as fast. Lüders bands are nucleated at the upper yield stress and propagate at the lower yield stress, with σ_L and v_L representing the deformation that is occurring in a narrow zone ahead of the advancing Lüders band. As mentioned in Sect. 3.6.2, increasing the value of m for a certain dislocation density, ρ, the yield-drop magnitude decreases, as evaluated by Johnson and Gilman for LiF.

When ρ_U is smaller than $10^3/cm^3$ and bearing in mind that the dislocation density in annealed iron is $\sim 10^6/cm^3$, the majority of dislocations are locked. 'Blue brittleness', as it is commonly known, is an undesirable effect observed in steel at the temperature range of 410–700 F, accompanied by low ductility and low resistance to impact. Blue brittleness occurring at relatively low temperatures is believed to be associated with strain aging.

Now, consider non-interstitial solutes, which can be smaller or larger than matrix atoms. In Fig. 3.21b, a substitutional solute, smaller than the matrix atoms, is shown. Such an atom replaces the solvent atom and settles down in the compressive region of the strain field. A larger substitutional atom, one larger than the matrix atoms (shown in Fig. 3.21c), is expected to be located at the dilated region in the vicinity of the dislocation. There is a difference between an interstitial and a substitutional solute. A solute atom, such as Zn in Cu (as mentioned above), distorts the lattice symmetrically; therefore, it can interact only with edge dislocations, but not with screw dislocations. Thus, the sharp yield drop, in the case of C in αFe, is quite emphasized; whereas, in the case of Zn in Cu, it is barely noticeable. Usually, upper and lower yield points are rarely seen in substitutional solid solutions and, even then, only under special conditions.

To summarize this section, it can be stated:

(a) the stress field of an edge dislocation has hydrostatic and shear components, whereas screw dislocations have only a shear component;
(b) a substitutional solute has a symmetrical stress field;
(c) an interstitial solute has a tetragonal distortion, in addition to volume expansion;
(d) tetragonal distortion allows interaction with both hydrostatic and shear stresses;

(e) tetragonal distortion allows interaction with edge and also screw dislocations in regard to the formation of an atmosphere;
(f) a substitutional solute, with its symmetrical stress field, can only react with a hydrostatic stress field and, therefore, it can interact only with edge dislocations. It will not interact with screw dislocations, that have no hydrostatic component;
(g) yield is more pronounced in Fe than by Zn in Cu;
(h) one can envisage dislocation pinning when a dislocation dissociates into two partials, at least one of which may have edge characteristics. Solutes that have only spherical symmetry can react with partial dislocations having edge characteristics and, indirectly, this way the screw dislocations can also be anchored.

3.7 The Bauschinger Effect (BE)

In general, dimensional changes occurring during plastic deformation are irreversible. Once a specimen has undergone straining into the plastic region, unloading will not restore its dimensions. However, the level of the stress applied to achieve deformation can be reduced, if the specimen is reloaded in the opposite direction. This was observed by Bauschinger as early as 1981. Figure 3.23 illustrates this phenomenon.

The change in the stress/strain characteristics, occurring when the direction of a deformation is reversed, happens in both single crystal and polycrystalline metals. This phenomenon was observed in various metals. Recognizing this effect is important for fabrication, since, quite often, a pre-stress is applied beyond the yield point into the plastic region. One of the first theories to explain the Bauschinger effect [henceforth: BE] dealt with dislocation pile-ups. In order to experimentally

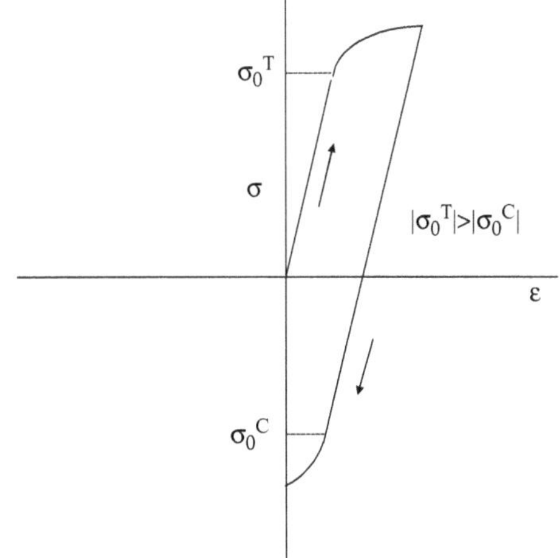

Fig. 3.23 Bauschinger effect (*BE*). A specimen under tension is pulled beyond its yield point into the plastic region and then reloaded under compression. The yield point under compression, σ_0^C, is lower than the yield point under tension, σ_0^T

study this effect, a system should be chosen having only a few slip systems, such as HCP structures. Zn is ideal for such a study, since, at proper orientation, slip will only occur in one slip system. This way, complications or effects from other slip systems are eliminated.

There are two reasons explaining lower yield stress in the reverse direction:

(a) The help during the reverse stress from the repelling forces of the dislocations in the pile up which arise on the loading in the opposite direction and
(b) due to the reversal of stress the dislocation sources are now acting in an opposite direction, generating and emitting the same kind of dislocations, but of opposite sign to those which were formed during tension. Opposite dislocations attract each other and tend to cancel each other out with the consequent softening of the crystal.

The BE is also observed in torsional and cyclic deformations. BE in polycrystalline material is discussed below.

3.8 The Effect of Impurity (Solute), Temperature and Orientation

The factors discussed in this section are those commonly thought to influence dislocation movement in crystals. Plastic deformation starts at the yield point of materials. Interstitial solutes, but not only interstitials, interact with dislocations to form a higher yield stress. In metals other than iron, the yield stresses (upper and lower) occur under special thermal conditions, usually at lower than room temperature. Whether the upper and lower yield points are observed or not, solutes have a strengthening effect. Once yielding occurs at a higher stress, the entire flow curve in ductile materials is affected, namely the entire curve expressing the stress-strain relation is moved upwards to higher stress levels. The effect of impurities depends, to a large extent, on the structure; thus, in BCC structures for example, a small addition of impurities elevates the yield point to such a level that embrittlement sets in at that level or slightly above it without any appreciable ductility (see Fig. 3.15 above). The behavior of such BCC metals can be compared to the ductile behavior mostly observed in FCC crystals. The previous sections dealt with the yield-point phenomena, which are consequences of interactions of such solutes with dislocations. Whatever the case may be, the effect of interstitial or substitutional atoms beyond yielding is associated with the hindrance of dislocation motion. Strengthening by solutes (including precipitates) is not considered in this section and attention will be given to the topic in the next chapter.

It is difficult to discuss the effects of solutes, temperature and orientation separately, because they are usually considered together in the experimental data. Thus, one can speak about the effect of temperature at a certain solute content or at a specific orientation. Nevertheless, an attempt will be made to discuss them

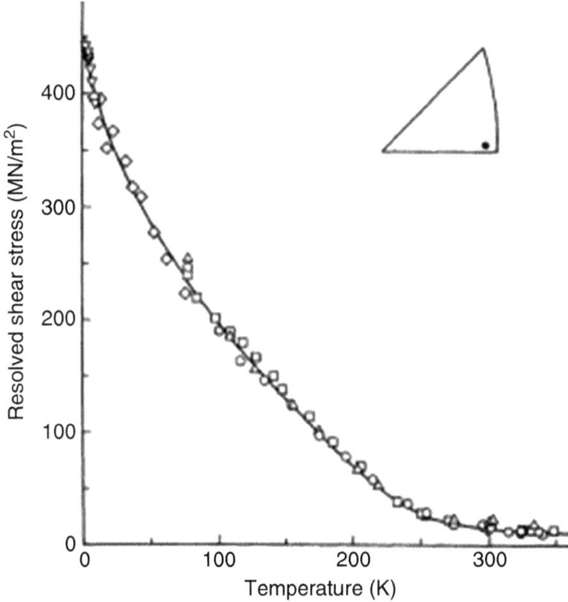

Fig. 3.24 The temperature dependence of the resolved yield stress in Nb (After Shin Takeuchi 1982)

separately (as was done above in regard to the effects caused by impurities). Now, regarding temperature, one must realize that this discussion is not about the temperatures that invoke phase changes in crystals.

At room temperature, deformation occurs predominantly by dislocation motion. Dislocations, their configurations and their interactions control most of the properties of crystals, among them their strength and ductility. In FCC metals, the force required to move dislocations is not strongly temperature-dependent and, therefore, the flow stress is not greatly influenced by temperature changes within reasonable limits (extreme temperature conditions are not studied here). Consequently, dislocation movement remains high even at low temperatures and the crystals remain relatively ductile. In contrast to FCC metals, the yield stress or CRSS of BCC single crystals is largely temperature-dependent, particularly at low temperatures. The temperature sensitivity of the yield stress of BCC crystals has been attributed to the presence of solutes, actually interstitial impurities, which are difficult to remove and require special purification techniques to reduce their amount. A very low level of such impurities must be reached in order to realize only the temperature effect without the contribution of impurities. Takeuchi, et al., studied high-purity single crystals of Nb under tension in the low-temperature range of 2–350 K in order to determine its plasticity. A sharp increase in the resolved yield stress was observed with decreasing temperature (shown in Fig. 3.24). Their results showed that normal slip took place only in a narrow region in the standard stereographic triangle near the $[110] - [\bar{1}11]$ boundary, as indicated in the Figure. They conclude that deformation occurs by normal slip at all temperatures.

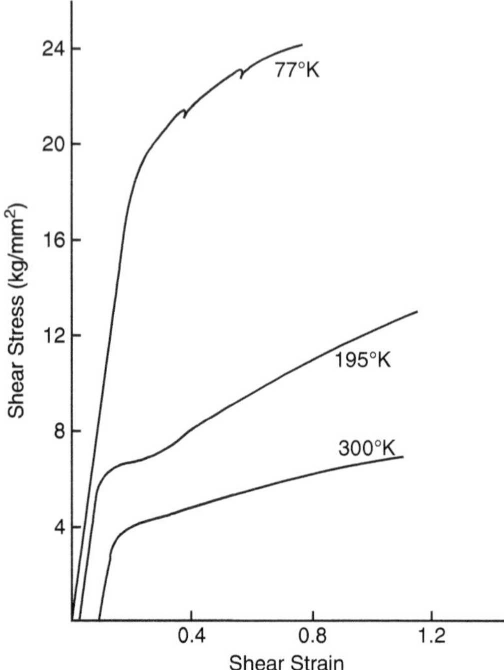

Fig. 3.25 Shear stress/shear strain curves for iron single crystals deformed by shear on the (112) plane (After Akinori Matsude 1977)

Matsude investigated Fe single crystals in shear deformation in the {110} <111> and {112} <111> slip systems at low temperatures. His results indicate that the resolved shear stress-shear strain curves are parabolic for all orientations tested in tension, but they are markedly different when tested by shear. They show the temperature effect, as seen in Fig. 3.25.

At high temperatures, other phenomena, such as recrystallization, grain growth, precipitation of a second phase, etc., can taint the temperature effect itself. Furthermore, other deformation modes might set in, such as creep. These subjects are not discussed here, but one can state, in general, that for ductile materials at higher than room temperature, the ductility properties increase at the expense of strength. Thus, yield stress will set in at lower stress levels, but their elongation or reduction area will increase. Toughness was previously discussed (Sect. 1.2.12) for ductile materials and the relation is (Fig. 3.25):

$$U_T \sim \sigma_f \varepsilon_f \tag{1.52b}$$

One can see that 'toughness' means a good combination of strength and ductility. Therefore, a moderate decrease in stress level with increased elongation, although somewhat inferior to high strength and high ductility, does not necessarily have to be at the expense of the overall room temperature properties. In any event, fabrications, such as rolling, drawing, etc., are performed at moderate temperatures and good elongation (reduction of area) is essential.

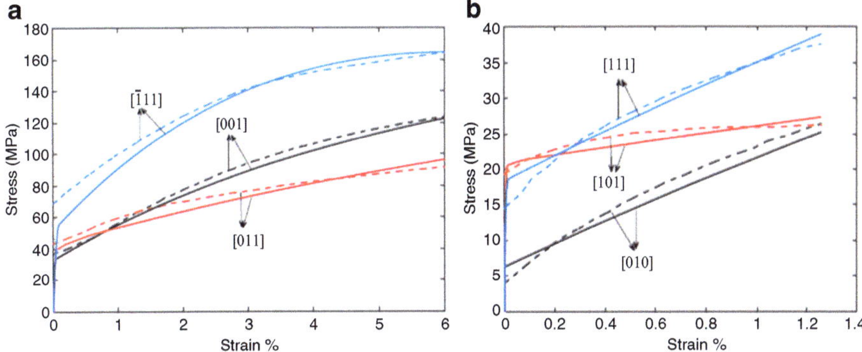

Fig. 3.26 (**a**) The tensile orientation dependence of [001], [011] and [$\bar{1}$11] oriented α-Fe single crystals. The *solid lines* are the simulation results and *dashes* represent experiments; (**b**) The orientation dependence of [010], [101] and [111] oriented Mo single crystals. The *solid lines* are the simulation results and *dashes* represent experiments (Reprinted with permission from Yalcinkaya et al. 2008, Copyright American Institute of Physics)

Experimentally, it was found that, of all the possible, equivalent slip systems, the one having the highest resolved shear stress will be the first to yield (slip). In Chap. 2, the slip systems of various structures were described. Usually, a standard stereographic triangle (see Cullity) is used to indicate the orientation of a single crystal. In FCC crystals, 12 systems are operating, i.e., four {111} planes and three <110> directions. To eliminate slip on other systems, such as the conjugate systems, the orientation of the crystal under consideration should be remote from the sides of the standard triangle. In particular, the [001] – [$\bar{1}$11] side should be avoided when studying slip, since, at such orientation or close to it, 3.4 equivalent slip systems may operate simultaneously. In HCP systems, in which the basal planes are similar to the FCC octahedral planes, one expects to have the easiest glide on such planes. However, the lattice parameters ratio, c/a, determines which system will be operative. The calculated ratio, on the basis of the hard sphere model, is 1.632. At lower ratios than the ideal, the (0001) plane loses its distinction of being the closest plane and slip on additional planes can occur. Zn, Cd, Mg, etc. slip on the (0001)-<11$\bar{2}$0> systems. In the Cd single crystal, for example, Uçar, et al. found that slip occurs on the basal plane in <11$\bar{2}$0> directions under tension in the 300–480 K temperature range, regardless of crystal orientation. However, above 350 K, slip systems depend on crystal orientation and temperature. Slip was also observed in other systems at some orientations. Thus, Ti_3AI single crystal pyramidal slip was observed by Minonishi et al., when the crystal was oriented so that the straining axis was close to the c-direction. At this orientation, slip on the basal and prismatic planes is inhibited. CRSS and the flow stress of BCC metals all depend on the orientation and temperature of the crystals. In the work of Yalcinkaya et al., pronounced influence was observed of the orientation on the yielding and hardening behavior of crystals in αFe and Mo single crystals, shown in Fig. 3.26.

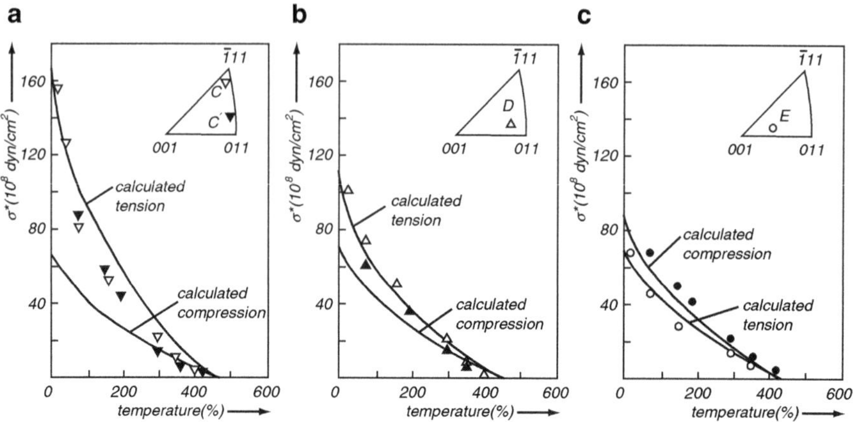

Fig. 3.27 Effect of orientation as a function of temperatures in Mo single crystals under compression: (**a**) Temperature dependence of the effective compression yield stress, σ*(▼) for orientation C′ and the effective tensile stress (∇) orientation C. (**b**) Temperature dependence of the effective compression yield stress σ*(▲) and the effective tensile stress (Δ) for orientation D; (**c**) Temperature dependence of the effective compression yield stress, σ*(•) and the effective tensile stress (○) for orientation E (Reprinted with permission from Liu et al. 1972, Copyright Wiley-VCH Verlag GmbH & Co. KGaA)

Here, experimental and calculated (by simulation) results are compared. The stress–strain relation is strongly orientation-dependent. Liu, et al. investigated the results of orientation on yield stress in Mo single crystals under compression and during slip behavior in the temperature range 77–550 K. Their results are reproduced in Fig. 3.27. The plastic behavior shows varying degrees of reversal effects in yield stress and slip band formation upon reversal of the sense of applied stress, depending on crystal orientation. Thus, one can state that orientation affects the yield and strength properties of single crystals and that the influence of temperature and impurity content cannot be ignored, since there is a mutual influence of these factors. The effect of orientation on single crystals helps to explain deformation in polycrystalline materials (discussed in the following section). Further discussion will clarify why grains (which are single crystals) must compromise, despite their individual desire to follow slip along the lines had they been single crystals.

3.9 Polygonization

Figure 3.28 is an illustration of the polygonization process in a crystal after small bending deformation. Dislocations of both signs are produced in their respective slip planes, depending on the structure, as seen in Fig. 3.28a. An excess of same-signed dislocations may be produced, as mentioned above, by bending. Early experiments were performed on Zn and Al crystals by Cahn. In Fig. 3.28b, dislocations

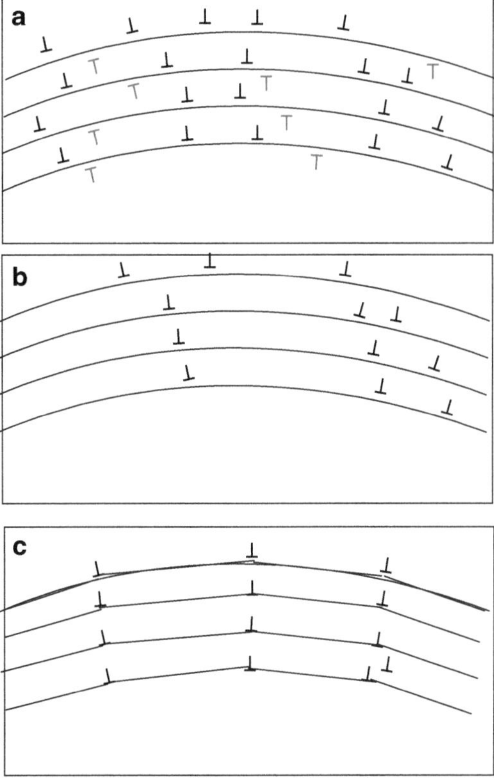

Fig. 3.28 Polygonization: (**a**) dislocations with both signs after deformation; note the excess dislocations of the same sign; (**b**) dislocation annihilation after annealing; (**c**) polygonization

with opposite signs attract and cancel each other out. Annealing at relatively low temperatures, below the recrystallization in the recovery range, provides the required energy for the rearrangement of the dislocations above each other into a dislocation wall. These dislocations all have the same sign and this configuration is a low-energy state of the crystal. Recovery results in the formation of a configuration, like a small-angle grain boundary, called 'polygonization'. Since the illustration below relates to edge dislocations, during the process of polygonization, the small-angle grain boundary obtained is similar to a tilt boundary. The alignment of the dislocations in the wall is shown in Fig. 3.28c.

In his work on bent silicon crystals at elevated temperatures, Vogel proved that the ease with which polygonization occurs depends on the orientation of the crystal. If a crystal is oriented so that only one slip system is operating in a given region, then an homogeneous aggregation of dislocations is produced, which readily climb to form dislocation walls; otherwise, polygonization may be inhibited. In Cahn's experiments, Zn slips mostly on the basal plane, thus, polygonization is easily observed, as explained by the orientation influence mentioned above.

3.10 Deformation in Polycrystalline Materials

Deformation in single crystals sheds light on the basic principles regarding the application of force to polycrystalline materials, which are aggregates of differently oriented grains, each one of which is a single crystal. Although properly oriented grains may behave as individual entities in regard to the type of deformation (slip, twinning or both), the response of the entire material is a compromise between the individual tendencies of the grains. The overriding requirement is to maintain the wholeness of the material and its continuity along the boundary. The neighboring grains must have some boundaries between them while joined by the effects of interatomic forces. These boundaries play an important role in the overall behavior of polycrystalline materials. All the factors mentioned earlier, such as solutes (impurities), temperature and orientation influence strongly the respond of the material to a force. Clearly, the structure of the materials themselves is a critical factor to be considered. Nevertheless, in polycrystalline materials, each grain deforms by slip in its respective slip system, but is assumed to have the same strain rate as the polycrystal. Such a slip must be a so-called 'multiple slip', for the grains to remain attached to their neighbors and maintain continuity. Thus, such a deformation is not regular or uniform as in single crystals.

In polycrystalline materials, slip planes and slip directions vary from one crystal to another. Thus, the resolved shear stress, τ, varies from one crystal to another. The crystal with the largest τ yields first. The presence of dislocations in each grain and their various configurations are basic to plastic deformation and it must be considered. Firstly, grain boundaries act as obstacles, retarding dislocation movement and hindering dislocation glide along the respective slip planes. Within grains, dislocation pile-ups may develop, because the grain boundaries act as obstacles to dislocation motion, as long as they can resist the force acting on the dislocation. Dislocations move along the same slip planes where the dislocation pile-ups act and, being same-signed, they repel each other. As the number of dislocations in a pile-up increases the stress on the obstacle (grain boundary, in our case) increases, as do the repelling forces between the individual dislocations in the pile-up. In fact, if there are n dislocations in a pile-up, the stress at the grain boundary will be n times the applied stress. Thus, regarding the strength of a grain boundary, only a local stress at the head of the pile-up, is sufficient to cause: (a) slip in an adjacent grain by propagating dislocation movement or (b) the grain boundary to collapse, with the possible consequence of a dislocation avalanche following the collapse of the boundary. Such an event may signal the initiation of fracture in some materials.

Another aspect is the grain size. As the grain size grows, it can accommodate more dislocations in a pile-up, which means that a lower applied stress is sufficient to act at the grain boundary. Following the Hall–Petch concept, Fig. 3.29 illustrates the effect of a stress required to propagate dislocations from grain to grain, i.e., to cause yielding in the second grain, given as:

$$\sigma_y = \sigma_0 + \frac{k_y}{\sqrt{d}} \tag{3.42}$$

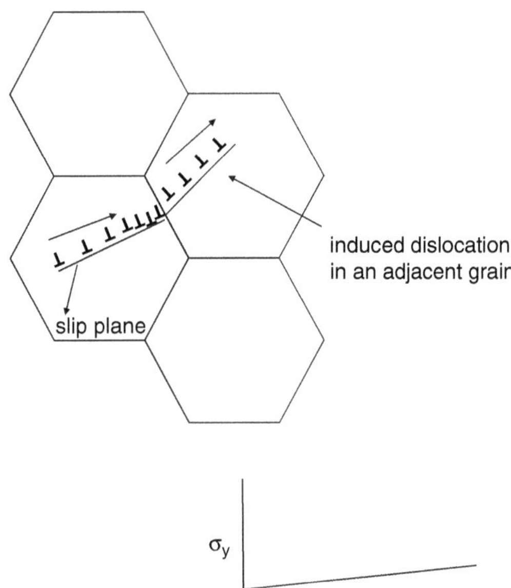

Fig. 3.29 A dislocation propagates from grain one to grain two, its adjacent neighbor

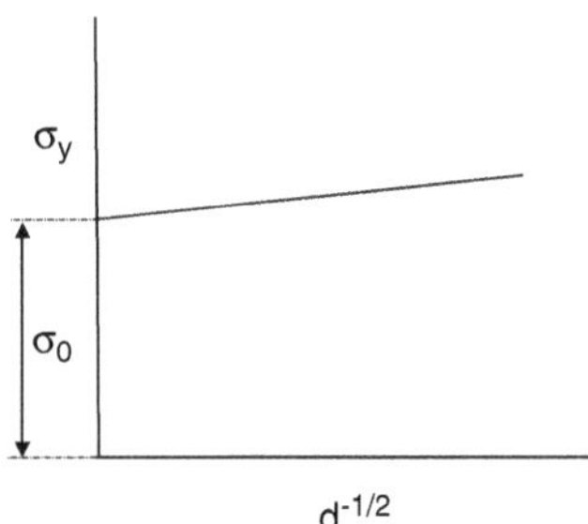

Fig. 3.30 A schematic plot, according to Eq. 3.42, showing the Hall–Petch relation. The intercept is σ_0 and the slope is k_y

σ_y is the lower yield stress, σ_0 is a materials constant for the starting stress for dislocation movement or the resistance of the lattice to dislocation motion (a frictional stress), k_y is a strength coefficient, which is a constant specific to each material, and d is the average grain diameter. The strength coefficient, k_y, is associated with the propagation of dislocations across grain boundaries and is responsible for their unpinning from their atmospheres. One can realize, from Fig. 3.29, that the smaller the grain, the smaller the number of dislocations in a pile-up; therefore, a higher applied yield stress will be required to produce a local stress at the head of the piled-up dislocation and vice versa. When the grain size is greater, more dislocations will be present in a pile-up, thus, requiring a lower applied stress to drive the dislocations across the grain boundary. Therefore, changes in grain size affect yield strength, due to the dislocations interacting with the grain boundaries as they move. Poly-crystals are stronger, due to geometric constrains, so they require more stress to yield. Strengthening by grains will be discussed in the next chapter.

The unpinning value of k_y may be evaluated from the slope of a plot, according to Eq. 3.42, i.e., the lower yield stress, σ_y, vs. grain size, $d^{-1/2}$. Also, the resistance to dislocation motion, σ_o, may be obtained from the intercept. Such a curve is shown schematically in Fig. 3.30.

This justifies the common practice of attempting to produce strong materials by creating structural parts with smaller grains that exhibit higher yield stress.

By changing grain size, one can influence dislocation movement and, consequently, yield strength. Two main reasons should be noted for the increase in yield stress as a result of smaller grain size: (a) as mentioned earlier, the number of boundaries increases when the grains are smaller, resulting in a smaller number of pile-ups in each small grain (boundaries hinder dislocation movement) and (b) the grain boundaries are also much more disordered than the grain itself, further preventing dislocations from moving from one grain to another on continuous slip planes. Impeding dislocation movement hinders the onset of plasticity due to the increased yield strength of the material. The higher the yield strength, the higher the applied stress needed to move a dislocation, in accordance with the inverse relation in the Hall–Petch equation 3.42.

The difference in orientation of two adjacent grains can be either small or large. In the first case, it is possible for the dislocation to overcome the obstacle and move into another grain. However, when there is a large difference in orientation between two adjacent grains, the dislocation may not necessarily move from one grain to another, but rather may create a new source of dislocation in the adjacent grain. Whatever the case may be, grain boundaries are obstacles to dislocation movement and, as a result, higher stress is then required to overcome the opposition to dislocation movement and this in turn strengthens the material.

The effect of grains and their boundaries is expected to be more pronounced in materials having HCP structure. Recall that, in HCP, there is usually one slip plane, as compared to other materials in which several equivalent slip planes might be operative. Therefore, in FCC or BCC structures, there is almost no grain which is not in the right orientation relative to the stress applied, so slip can occur at relatively low expense (i.e., a lower level of stress will have to be applied to attain deformation). In HCP, however, the grains are rarely in the appropriate orientation for slip to occur and, therefore, the cost is higher. Thus, one expects that the stress required for deformation in HCP will be higher relative to the single crystal case, whereas in FCC or BCC structures the difference will be less significant.

3.10.1 Preferred Orientation (Texture)

In polycrystalline materials, not all the grains are distributed randomly. A polycrystalline material in which the distribution of the grains is in perfectly random orientation shows isotropic properties. Such a polycrystalline material is said to be ‘texture free’. Ordinarily, materials are seldom in such an ideal state and some preferential orientation, known as ‘preferred orientation’, exists to various degrees. The mechanical (and also other physical) properties may be significantly influenced, particularly when the preferred orientation is strong. The degree of texture depends on the percentage of crystals (grains) having a preferred orientation, which, in turn, depends on the prior history of the fabrication of the material. Although the preferred orientation of an aggregate is not desirable, it is often practical to have a preferred orientation for certain applications. The most prominent use is

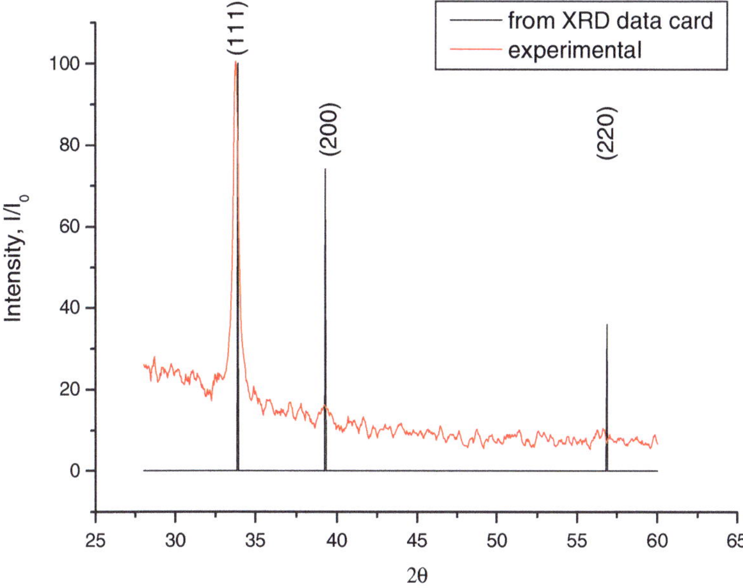

Fig. 3.31 Preferred orientation in ZrN; experimental and theoretical (XRD card 35–0753) diffractograms (Experimental data from Pelleg 2007)

for transformer steel sheet metal, where the cubic iron crystals (silicon steel) are expected to have <100> direction for magnetic applications. This direction is known as the 'easiest magnetization direction'. These ferrite steels are produced by rolling, which provides the desired texture.

Materials scientists have several methods for evaluating the degree of preferred orientation. The first and utmost among them is by X-ray diffraction [henceforth: XRD] techniques. Figure 3.31 illustrates a diffractogram of sputtered ZrN with a strong [111] preferred orientation, compared to a powder pattern from an XRD card (35–0753) with random orientation. Very small (200) and (220) peaks were also detected in the sputtered ZrN. The experimental intensity was normalized to the strongest (111) peak from the XRD card. Thin films usually have a preferred orientation. Preferred orientation is obtained by cold work, such as drawing, rolling, etc., each fabrication providing a characteristic preferred orientation. Such preferred orientations are called 'deformation textures'. More often, pole figures are used to study the preferred orientations resulting from cold work. The annealing of polycrystalline materials to induce recrystallization does not usually eliminate the texture of the material, although it may alter it. Annealing textures are related to deformation textures. In most FCC structures, the recrystallization (annealing) texture is (100)-[001]. Not only the fabrication of a bulk polycrystalline material causes texture formation, but textured structure can be obtained by growth or solidification, e.g., as in casting. Furthermore, in film deposition techniques, usually no random distribution of grains is obtained and a pronounced texture is observed. Epitaxial

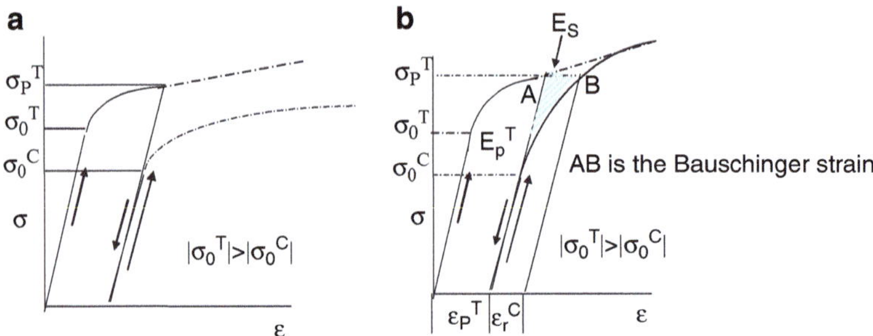

Fig. 3.32 (**a**) The Bauschinger effect. $\sigma_P{}^T$ is the prestrain under tension. $\sigma_0{}^T$ and $\sigma_0{}^C$ are the yield stresses under tension and compression, respectively and (**b**) AB is the Bauschinger strain, representing a value between two work hardening curves, i.e., between the reverse loading (here, under compression) and the prestrain value

films, often obtained by some deposition technique, are basically considered to be preferred orientations and their characteristic XRD peaks reflect the substrate orientation effect. Suffice it to say, the impurity content and the alloying and temperatures of fabrication, greatly influence the outcome of preferred orientation character.

3.10.2 The Bauschinger Effect (BE)

The Bauschinger effect occurs not only in single crystals, but is also observed in polycrystalline materials, as mentioned above. Basically, this effect is related to a reduction in the yield stress of a material after some prestrain by loading in one direction (say by tension), followed after unloading by reloading (say by compression) in the opposite direction. The test specimen may also be a polycrystalline material. Some aspect of orientation is involved in the BE, namely the opposite direction in reloading a specimen following a previous straining. In Fig. 3.32, the curves indicate the BE; in essence, Fig. 3.23 has been redrawn here to better illustrate the yield-stress decrease upon reloading in the opposite direction. Intuitively, one can relate yield stresses under compression to those found under tension to obtain a parameter which connects these two values, as given below and marked as σ_{0BE}. Equation 3.43 indicates the softening occurring in the case when a reversed deformation is performed after a straining in the opposite direction:

$$\sigma_{0BE} = \frac{\sigma_0{}^C}{\sigma_0{}^T} \tag{3.43}$$

In the literature (see, for example, Choteau, et al.), three other relations are given: (a) for the reduction of yield stress upon reloading; (b) for the strain shift between the two work hardening curves, seen in Fig. 3.32b, and; (c) for the energies stored and released on reloading. These relations are given as:

$$\beta_\sigma = \frac{\sigma_P^T - \sigma_0^C}{\sigma_P^T} \tag{3.44}$$

$$\beta_\varepsilon = \frac{\varepsilon_r^C}{\varepsilon_P^T} \tag{3.45}$$

$$\beta_E = \frac{E_s}{E_P^T} \tag{3.46}$$

In Fig. 3.32 (revised from Fig. 3.23), the first quadrant of the $\sigma-\varepsilon$ relation is redrawn. It can be deduced from Fig. 3.32b, for example, that the BE is a function of the prestrain during the first loading (in our example, under tension). Thus, $\sigma_P{}^T$, indicated in the above figures and in Eqs. 3.44, 3.45, and 3.46, represents the amount of prestrain during the first loading (under tension). Note that $\sigma_P{}^T$ represents yield stress when the deformation is reloaded a second time (after the specimen had been unloaded). The strain in reverse loading, $\varepsilon_r{}^C$ (the second loading under compression), is AB in Fig. 3.32b, which represents the Bauschinger strain. While Eq. 3.44 represents the reduction in yield stress, Eq. 3.45 is the strain shift after the reloading step, as related to the strain resulting from the prestrain. Finally, the energy parameter, β_E, is the area under the curve during tension, which represents energy. As noted in Chap. 1, energy is stored in deformations and is released when specimens are unloaded. The difference between the energies involved in the deformations' forward and reversed directions, in terms of stored energy, is given by Eq. 3.46. Note that the shaded area in Fig. 3.32 represents the energy saved during reverse loading.

At the basis of this phenomenon, strain hardening (discussed in the next chapter) is associated with the BE. The progress of deformations in one direction increases (e.g., due to tension), though, after tensile cold working, compressive yield stress actually becomes smaller. The greater the strain under tension is, the lower is the compressive yield strength. In polycrystalline materials, the grain orientations are different and, thus, their slip planes are also different. More favorable orientations of grains are found where the CRSS has already induced dislocation slip; less favorably-oriented grains remain in the elastic region while overall deformation goes on. By reversing the direction of deformation, on one hand, the grains still in the elastic region will tend to return to their original dimensions, thus assisting the applied reverse stress and resulting in a lower yield stress. On the other hand (as indicated in Sect. 3.7), the repelling forces of the piled-up dislocations and the reversal of the stress dislocation sources currently acting in the opposite direction greatly contribute to the lowering of the yield strength. During such a process, dislocation annihilations are likely to occur. It has been observed in many experiments that grain size influences the BE. It was also found that the larger the grain size, the smaller the BE. One can speculate that, due to the

Further References

P. Barat, A. Sarkar, Barat Arnab, Phys. Rev. B **74**, 245425 (2006)
J. Bauschinger, Zivilingur **27**, 289 (1881)
P.R. Bhowal, M. Metzger, Metall. Trans. **9A**, 1027 (1978)
P.R. Bhowal, K.M. Prewo, A.J. McEvily, Metall Trans. **7A**, 751 (1978)
J.F.W. Bishop, R. Hill, Philos. Mag. **42**, 414 (1951)
D.G. Brandon, Acta Met. **14**, 1479 (1966)
L. Briottet, A. Ambard, D. Guichard, Model. Simul. Mater. Sci. Eng. **9**, 259 (2001)
R.W. Cahn, J. Inst. Met. **76**, 121 (1949)
R.W. Cahn, Adv. Phys. **3**, 363 (1954)
R. Chadwick, W.H.L. Hooper, J. Inst. Met. **81**, 744 (1953)
M. Choteau, P. Quaegebeur, S. Degallaix, Mech. Mater. **37**, 1143 (2005)
A.H. Cottrell, B.A. Bilby, Proc. Phys. Soc. (Lond.) **A62**, 49 (1949a)
A.H. Cottrell, B.A. Bilby, Proc. Phys. Soc **A62**, 490 (1949b)
C. Coupeau et al., Scripta Mater. **41**, 945 (1999)
J. Gubicza, N.Q. Chinh, J.L. Lábár, Z. Hegedüs, T.G. Langdon, J. Mater. Sci. **44**, 1656 (2009)
E.O. Hall, Proc. Phys. Soc. B **64**, 747 (1951)
F. Heidelbach, Science **291**, 2330 (2001)
B.B. Hundy, Inst. Met. **81**, 744 (1953)
W.G. Johnson, J.J. Gilman, J. Appl. Phys. **33**, 2716 (1962)
M.L. Kronberg, F.H. Wilson, Trans. Am. Inst. Min. Metall. Eng. **185**, 601 (1949)
J.C.M. Li, J. Appl. Phys. **32**, 525 (1961)
S. Liang, Zhang Qingchuan, Jiang Huifeng, Front. Mater. Sci. China **1**, 173 (2007)
G.C. Liu, S.S. Lau, J.E. Dorn, Phys. Status Sol. (a) **11**, 645 (1972)
D.V. Lotsko, Yu.V. Milman, Journal De Physique IV, Colloque C7, supplément au Journal de Physique 111, **5** (novembre 1995)
L. Margulies, G. Winther, H.F. Poulsen, Science **291**, 2392 (2001)
A. Matsude, Trans. J. Inst. Met. **18**, 214 (1977)
Y. Minonishi, M. Legros, D. Caillard, Mater. Res. Soc. Symp. Proc. (Mater. Res. Soc.) **460**, 237 (1997)
H.G. Paris, B.G. LeFevre, E.A. Starke, Jr., Metall Trans. **7A**, 273 (1976)
J. Pelleg, A. Bibi, M. Sinder, Phys. B **393**, 292 (2007)
N.J. Petch, J. Iron Steel Inst. **173**, 25 (1953)
N.J. Petch, Acta Met. **12**, 59 (1964)
W.T. Read, W. Shockley, Phys. Rev. **78**, 275 (1950)
E. Romahanji, M. Popovič, D. Glšič, M. Stefanovič, M. Milovanovič, J. Metall. **10**, 5 (2004)
M.S. Szczerba, T. Bajor, T. Tokarski, Philos. Mag. **84**, 481 (2004)
N. Uçar, I. Karaman, B. Düzgün, Acta Met. Mater. **43**, 2103 (1995)
C.M. Van der Walt, Acta Met. **17**, 393 (1969)
F.L. Vogel Jr., Acta Met. **6**, 532 (1958)
S. Takeuchi, T. Hashimoto, K. Maeda, Trans. Jpn. Inst. Met. **23**, 60 (1982)
G.I. Taylor, J. Inst. Met. **62**, 307 (1938)
E. Tenckhoff, Metall. Mater. Trans. **9A**, 1401 (1978)
H.L. Wain, A.H. Cottrell, Proc. Phys. Soc. B **63**, 339 (1950)
T. Yalcinkaya, W.A.M. Brekelmans, M.G.D. Geers, Model. Simul. Mater. Sci. Eng. **16**, 085007 (2008)

Chapter 4
Strengthening Mechanisms

4.1 Introduction

The strengthening of materials is of great importance for engineering applications. Construction parts are designed not only to endure the anticipated forces, which are intentionally applied (those they are expected to withstand while in service), but also any sudden, short-duration forces that might cause catastrophic failure, if not taken into account. In order to avoid the probability of such failure, liberal safety factors are generally adopted by designers. Their approach is to strengthen materials beyond the magnitude which would be sufficient to prevent failure, even if a steady force was exerted during the entire period of their use. This extra strength value constitutes the safety factor required for ensuring the safe use of a construction part, even in the event that a sudden force of larger magnitude appeared during service. There are several mechanisms by which materials may be strengthened, listed below:

(i) Strain (or work) hardening;
(ii) Solid solution strengthening–

   a. interstitial atoms,
   b. substitution atoms;

(iii) Grain boundaries and grain size;
(iv) Second phase hardening (precipitates, other particles, and impurity atoms).

These subjects are discussed in this chapter. However, to understand the concept of hardening (strengthening), one has to realize that all the aspects of strengthening involve dislocation generation, dislocation movement and the interactions of dislocations with each other or with any of the above-listed entities that might hinder their free motion. To this end, strain hardening in single crystals will be the first topic of discussion on the basics of strengthening materials.

J. Pelleg, *Mechanical Properties of Materials*, Solid Mechanics and Its Applications 190, DOI 10.1007/978-94-007-4342-7_4,

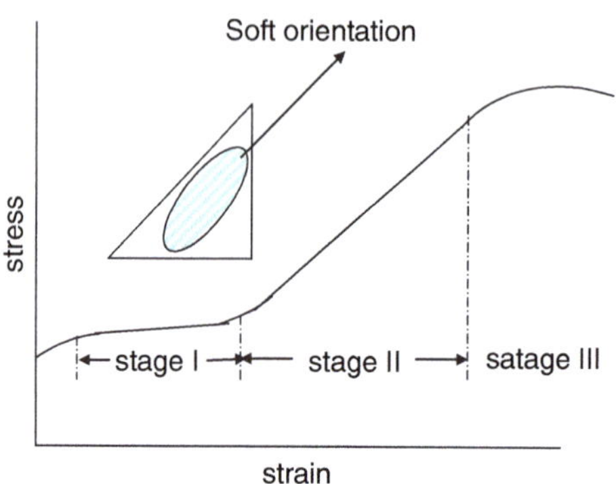

Fig. 4.1 The stress-strain curve of FCC single crystals. Three hardening stages occur when specimens have tensile axes in the *shaded area* of the standard *triangle*

4.2 Strain Hardening

The strengthening of a metal by means of plastic deformation is called 'strain hardening'. Plastic deformation is associated with the motion of a relatively large number of dislocations. Strengthening is the consequence of dislocation motion and interactions with other dislocations and features present in the material. The unhindered motion of dislocations dictates the capability of a crystal to deform plastically. It was stated in Sect. 4.1 that all the strengthening techniques listed above depend on hindering dislocation movement. In research studies done on FCC structures, in which the restriction of dislocation mobility was caused by other dislocations, three-stage strain hardening was observed, as presented in Fig. 4.1. This figure shows the generalized stress-strain curve for a single crystal. This type of three-stage hardening was also observed in other structures, as well, such as Ge, NaCl and in some BCC metals. In HCP single crystals, three-stage hardening is observed when the slip system is not prismatic. The stress-strain relation is divided into three stages: stage I or 'easy glide'; stage II or 'linear hardening' and; stage III, also known as the region of 'parabolic hardening' or 'dynamic recovery'. These three stages are characterized by the motion of the dislocations on different slip systems and their interactions, which determine the shape of Fig. 4.1. These stages correspond to different strain-hardening rates and to specific dislocation microstructures. The initial region, before stage I commences, is the transition from elastic to plastic deformation (and not relevant to this chapter).

Reports by Anongba indicate that also stages IV (see Haasen, for example) and V in the stress/strain curve of single crystals exist. Siethoff studied these stages in Si and Ge, while Nicklas and Mecking studied them in FCC metals. These stages appear after the dynamic recovery of stage III and have also been observed in heavily deformed FCC crystals.

There are too many independent factors that can influence the hardening stages in single crystals and, often, it is difficult to separate their influences, since they may operate simultaneously. Nevertheless, it is useful to consider them separately, as far as possible. Some of these factors are considered below.

4.2.1 Stage I

To reiterate from Chap. 3, Sect. 3.2 – when the CRSS has been reached, plastic deformation (i.e., slip) begins in the most favorable slip system. Dislocation movement is mainly unhindered. Free motion by slip defines stage I and is called the 'easy glide' of the stress-strain diagram. This commences immediately after yielding and has a low strain hardening rate, as indicated by the shallow slope seen in Fig. 4.1. This is a consequence of the unhindered motion of the dislocations. The length of the stage depends, as expected, on: crystal structure, orientation, temperature, purity, size and the absence of oxides on the external surface of the specimen. In general, the most obvious fact about plastic deformation by glide is the increase in dislocation density, observed microstructurally either by optical microscope (by means of the etch pits technique) or by TEM. Although the motion of the dislocations is relatively free, some hindering still occurs, as indicated by the slope in Fig. 4.1.

The slope of the linear hardening at this stage is shallow. The hardening rate is given as

$$\theta_I = \frac{d\tau}{d\gamma} \tag{4.1}$$

The dislocation density increases with strain upon stress application. The relation in stage I, between shear stress and dislocation density, is given by Mitchell as:

$$\tau = \frac{1}{2} G b \rho^{1/2} \tag{4.2}$$

where ρ is the dislocation density. The effects of some of the factors determining the shape and extent of strain at this stage are considered below.

4.2.1.1 Crystal Structure

The most intensively studied structure is that of FCC, where experimental observation lead to the three-stage hardening concept (illustrated in Fig. 4.1 above). Therefore, it is clear that the effects of other structures are considered relative to FCC. HCP crystals are expected to show distinct stage I hardening, because the basal plane, which is the most common slip plane, has basically the same character as the octahedral FCC faces. Indeed, most of the experimental work confirms this hypothesis and the (0001)–<1120> system is involved in stage I deformation. In stage I, it is expected that the slip lines will be long and straight, which has, indeed, been observed in many metals, e.g., see the deformation of Hf single crystals (Das and Mitchell). Twinning is very common in HCP structures and, therefore, pure slip deformation should be separated from the contribution of twinning, although, in much of the experimental work, twinning is also observed

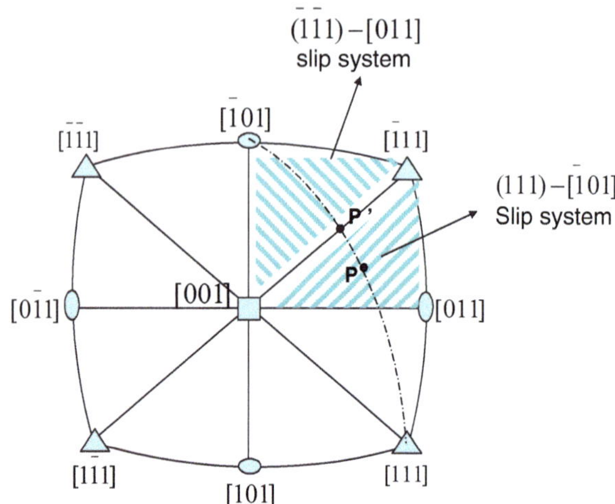

Fig. 4.2 A part of the standard projection of cubic crystals on (001). Two slip systems; the primary and the secondary (conjugate) are shown. Point *P* represents a tensile axis orientation

and discussed. It is convenient to study easy glide hardening in Zn and Cd, since these metals show well-developed basal glide, due to the large c/a ratio. Of the many reports in the literature, it is worth mentioning the work of Wielke on easy glide in these systems; he investigated this in the temperature range 4.2–400 K.

BCC metals also show easy glide at specific conditions of temperature and purity. Impurities are chronically present in BCC structures, unless very tedious and special purification techniques are applied; their presence masks the real hardening behavior of the pure material. Despite this difficulty, studies have been performed in very pure BCC systems, such as α Fe, Nb, Ta, etc. In Nb single crystals, Votava observed the development of an easy glide region for orientations inside the stereographic triangle. It has been reported that single crystals of Nb (Mitchell et al.) and Fe (Keh) can be made to deform in a manner similar to that of FCC metals, simply by controlling the orientation of the tensile axis. Similarly, Mo crystals (Guiu and Pratt) with the tensile axis orientated in a direction near the center of the unit stereographic triangle exhibits three work-hardening stages.

Thus, under the proper conditions, all structures can show easy glide.

4.2.1.2 Orientation

Before considering the orientation effect, now is the time to describe the concept of the 'standard stereographic triangle'. The stereographic projection of cubic structures will be considered and the projection of such crystals on plane (001) is shown in Fig. 4.2. For details of such projections, see, for example, Azaroff or Cullity.

In this Figure, eight triangles are seen in the standard [001] stereographic projection. Four poles of the {111} planes and four <011> directions are indicated. The dashed triangles represent two systems that can be operative, those of

(111) – [$\bar{1}$01] and ($\bar{1}\bar{1}$1) – [011]. Assume that P is the orientation of the tensile axis. One can follow the rotation of the tensile axis of the specimen during deformation on the large (dashed) circle connecting [$\bar{1}$01], P and [111]. At some stage of deformation, the tensile axis of the specimen reaches point P′ on boundary [001]–[$\bar{1}$11]. At this stage, the resolved shear stress on the primary (111)–[$\bar{1}$01] slip system is equal to that operating on the adjacent ($\bar{1}\bar{1}$1} – [011] slip system, known as the 'secondary' or 'conjugate slip system'. Now, deformation proceeds on both slip systems and duplex slip is obtained. One can identify section P′ – [$\bar{1}$01] with angle, λ, between the tensile axis and the slip direction, and section P′-[111] with the angle between the slip plane and the tensile axis in Schmid's relation (see Chap. 3, Sect. 3.2 and Eq. 3.4). It is relatively easy to determine the number of possible slip systems in the standard stereographic triangle by the number of its neighboring triangles. Specimens with tensile axes oriented at [001] can slip in eight slip systems, represented by the eight triangles shown in Fig. 4.2, consisting of four planes and two directions for each plane. At this orientation, the maximum resolved shear stress on the eight slip systems is equal. At the pole of [011] in the standard projection, four triangles are neighbors, but only two of them are shown in Fig. 4.2. Thus, [001] has two planes and two directions in each plane, altogether four slip systems. Again, the [001]–[011] boundary has two slip planes, but each of them has only one direction; therefore, there are two possible slip systems (namely direction × planes or 1 × 2 = 2). One immediately realizes that only two neighbor triangles are adjacent to this symmetry boundary. [$\bar{1}$11] is common to six triangles and, on the basis of the number of triangles, one can find that it has six slip systems, i.e., three planes and two directions in each plane. The [011] – [$\bar{1}$11] boundary has one plane with two directions; therefore, it has two equivalent slip systems, as expected, in accordance with the (two) triangles in its immediate vicinity. Finally, the [001] – [$\bar{1}$11] boundary, along the stereographic standard triangle, has two planes and one direction in each, since again only two triangles are neighboring each other, that of the primary and that of the conjugate slip systems. Thus, if the tensile axis is in high symmetry locations, where more than one slip system can be operative, the applied stress for deformation will be higher, contrary to the soft areas (e.g., the central part of the standard stereographic triangle, far away from the poles and boundaries; see the shaded area in Fig. 4.1), where mostly one slip system is operative.

Like the other factors mentioned earlier, the orientation of a single crystal has a profound effect on hardening. In all structures, if the orientation is not set for one slip system, easy glide may be completely absent. Stage I occurs only during single slip and the hardening rate is small compared to the other stages. Figure 4.3 shows the predicted curves for various orientations within reasonable agreement with the Cu single crystal experiment (see, Diehl and Wu, et al.). FCC single crystals possess 12 easy glide systems {111} <110> (as mentioned in Chap. 3, Sect. 3.3.1). Note that, as the tensile axis orientation approaches the [001] – [$\bar{1}$11] symmetry boundary, the length of easy glide decreases, whereas hardening rate increases. This is due to the contribution of secondary slip. At the high symmetry orientation of [$\bar{1}$11], easy glide is absent, due to multiple slip (orientation 5). Similarly, at the [001] orientation,

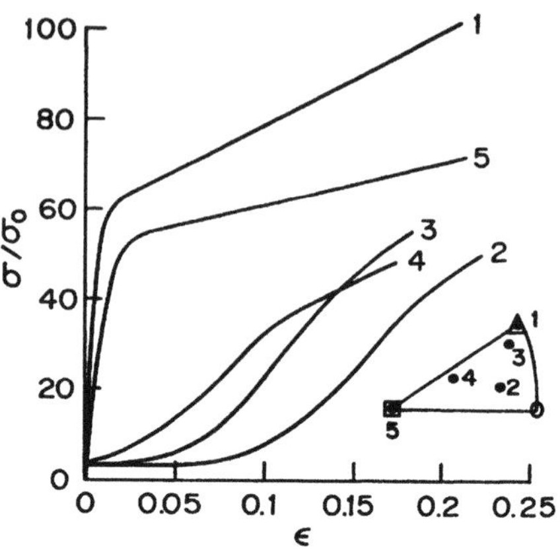

Fig. 4.3 Typical stress/strain curves for various orientations of the tensile axis in FCC crystals. Note the extent of the easy glide varying with orientation (Reproduced with permission from Bassani (1991). Courtesy of Professor Bassani)

where many slip systems are operative, no stage I is present. Orientation 2 (shown in the interior of the stereographic triangle in Fig. 4.3) has an easy glide with the greatest extent; it is in the region known as 'soft orientation'.

In HCP crystals, it is expected that strain hardening will be closely related to the strain hardening observed in FCC structures, especially when slip occurs on basal planes. In such cases, well-developed stage I hardening should be observed, the extent of which may be even larger than in FCC crystals. Zn and Cd show basal slip preferentially, since the c/a ratio is much higher than the ideal; therefore, it is reasonable to study strain hardening in these metals. In his work, Wielke showed the large easy glide in these metals at various temperatures, as indicated by the reproduced figure below (Fig. 4.4). Large easy glide is associated with slip in basal systems and work-hardening behavior closely resembles that of FCC crystals, as a result of the similarity between slip planes having the same character. The basal planes are exactly the same as the octahedral {111} FCC planes and slip occurs on the (0001) $<11\bar{2}0>$systems. Contrary to FCC metals, hexagonal crystals, oriented to slip in basal planes, particularly with a large c/a ratio, practically do not deform in secondary slip systems, because only one basal plane exists. If, however, the tensile specimen is oriented in such a way that other slip systems can contribute to the deformation, easy glide may be absent. Furthermore, one has to remember, when doing experiments with HCP crystals, that twinning may also contribute.

As in the other structures discussed above, in BCC crystals, the orientation of the tensile axis determines if three-stage hardening occurs, as well; it was observed in many BCC crystals, such as Mo, Nb, W, Ta, Fe, etc. (see, for example, Guiu and Pratt). The orientation of the tensile axis should be in the center of the unit stereographic triangle (soft orientation) for easy glide, as observed, for instance, in pure Nb. At such locations in the unit stereographic triangle, three work-hardening stages are exhibited. As in Nb single crystals, three-stage hardening was also observed in Ta, as shown in Fig. 4.5. In Fig. 4.5, various orientations are shown, only some with easy glide and a shallow slope (representing hardening at this stage).

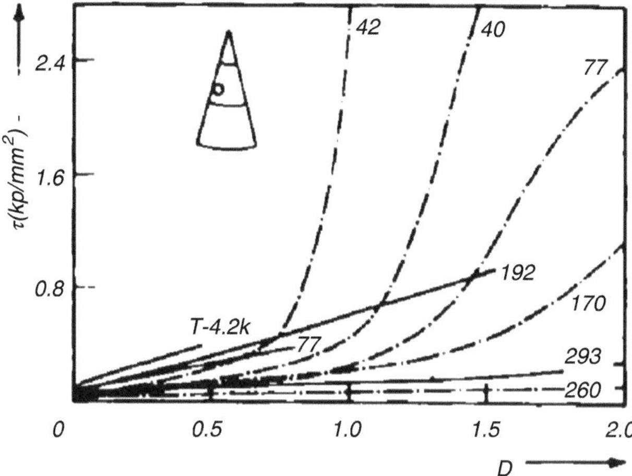

Fig. 4.4 Typical stress/strain curves of Zn (—) and Cd (-·-·-·) at various temperatures. Note the extent of easy glide, both in Cd and Zn (Reprinted from Wielke 1976. With permission of the author and Wiley-VCH Verlag GmbH & Co. KGaA)

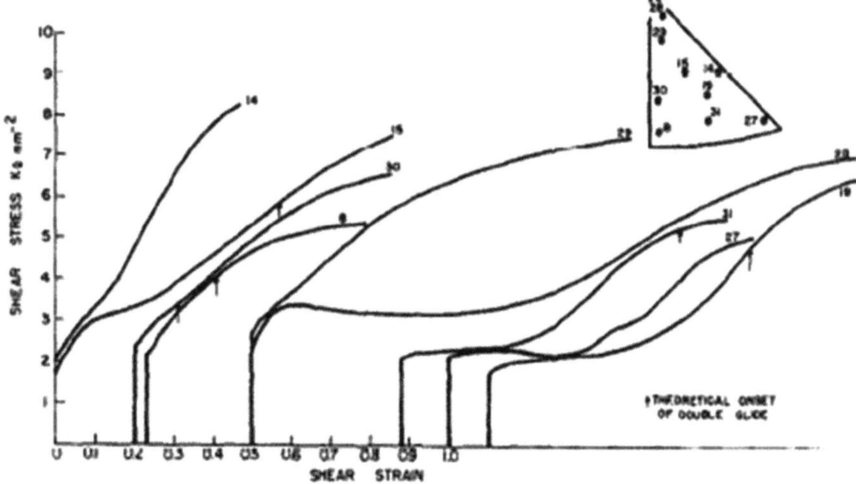

Fig. 4.5 Shear stress-shear strain curves of Ta single crystals deformed under tension at 373 K as a function of orientation, $\dot{\varepsilon} = 9.3 \times 10^{-4}\ s^{-1}$ (Reprinted from Mitchell and Spitzig 1965. With permission of the author and Elsevier Ltd.)

In other words, BCC structures can be made to deform in a manner similar to FCC materials by controlling the orientation of the tensile axis and, thus, easy glide can be observed. The crystallographic orientation of the tensile specimen axis is a major factor affecting the work hardening of BCC single crystals. It is worth mentioning that the conditions for single glide, necessary for the observation of stage I behavior, are more difficult to obtain in BCC metals than in the FCC crystals, which have well-defined slip systems on {111} <110>. As in other structures, slip on other systems may begin, suppressing easy glide, when the orientations are near symmetrical positions in the standard stereographic triangle.

4.2.1.3 The Effect of Temperature

Figure 4.6 is an example of the effect of temperature (and orientation) in Cu crystals which represents the temperature-dependent stress-strain behavior in FCC metals. It is evident from the experimental data that the effect of decreasing the temperature is to extend the strain-hardening stages and to prolong stage I. In general, it was observed that the extent of stage I increases with the decrease of deformation temperature.

Figure 4.4 shows the effect of temperature on the stress-strain curves for Cd and Zn. The extent of stage I, in both metals, increases with increasing temperature. At the same time, the strain-hardening rate decreases, as indicated by the slope of the lines in the figures. This is probably due to recovery processes at the higher temperatures, occurring at ~200 and 175 K in these metals. In general, it was observed that the extent of stage I increases with the decreasing deformation temperature, and not as indicated in Fig. 4.4, apparently due to the aforementioned recovery processes. In another publication, reproduced in Fig. 4.7 for Mg, the temperature-dependent critical shear stress, indeed shows the expected pattern, namely, the extent of stage I increases with decreasing temperature. The slip is on the (0001) $<11\bar{2}0>$ slip system. Two stages of deformation, indicated by A and B, can be seen for two temperatures. As the temperature decreases, the strain-hardening coefficient (the slope) also increases. As in other structures, this is the expected pattern, when no other influences exist.

Note again that if the tensile axis is at or close to a high-symmetry axis, as in experiment 305 (Fig. 4.6), it is not possible to see the temperature effect, since stage I is absent. The calculated stress-strain curves seen in Fig. 4.8 clearly show the effect of temperature on stage I strain hardening, in agreement with experimental observations. The pattern of increase in stage I hardening with decreasing temperature is expected in all structures (also shown below for BCC crystals). Figure 4.9 illustrates the effect of temperature on the shear stress-shear strain curve of Ta single crystal at a specific orientation. For more elucidatory observations, high-temperature curves are shifted on the strain axis.

4.2.1.4 Specimen Size

Experimental observations in polycrystalline materials indicate that smaller specimens are able to withstand higher applied stresses. Understanding this effect per se, even without its practical application, is crucial in light of current interest in microelectronic and nano technologies. Recently, Frick, et al. studied [111] Ni

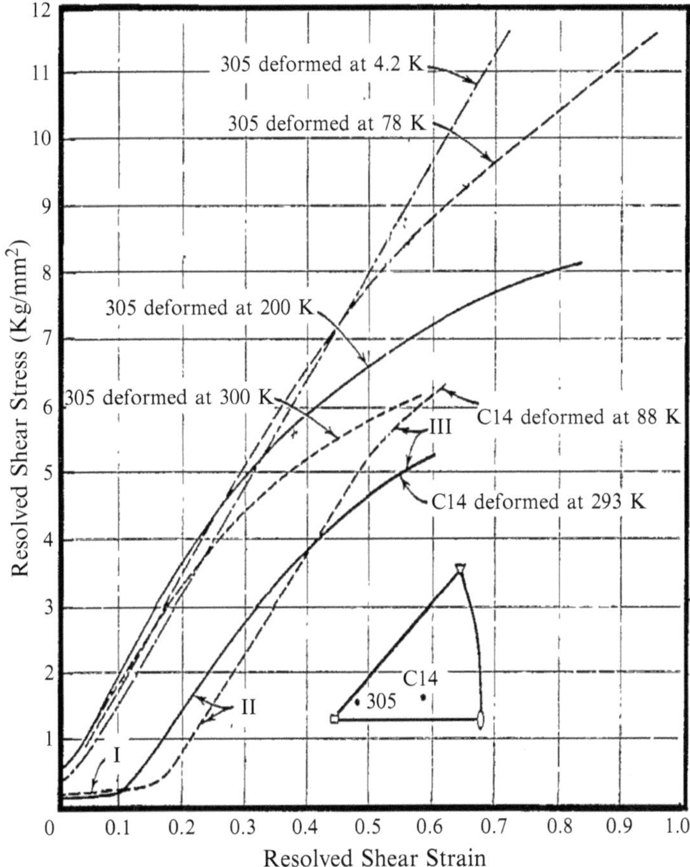

Fig. 4.6 The effect of temperature (and orientation) on Cu single crystals (Reproduced from Cuitino (1996). With permission of the author and Elsevier Ltd.)

crystals, observing a strong increase in strain hardening with decreasing diameter. The TEM results suggest that dislocation pile-up may play a significant role in strain hardening. It is believed that this mechanism is related to dislocation-dislocation interactions and that the required increased stresses are associated with the activation of dislocation sources with decreasing size. Their study describes the general size effect without specific consideration in regard to the stain-hardening stages.

4.2.1.5 Purity of the Specimen

It is commonly known from experiments in polycrystalline materials that the overall purity has a significant effect on the mechanical properties of materials. Thus, very high-purity material (e.g., as obtained after five zone-melting passes) has an appreciably lower yield stress than material which is less pure (e.g., after one zone-pass

Fig. 4.7 *Hardening curves* in Mg single crystals at 300 and 77 K at a deformation rate of $\sim = 8 \times 10^{-4}$ to 2×10^{-2} s^{-1} (Lavrent'ev et al. 1972. Courtesy of Springer Science)

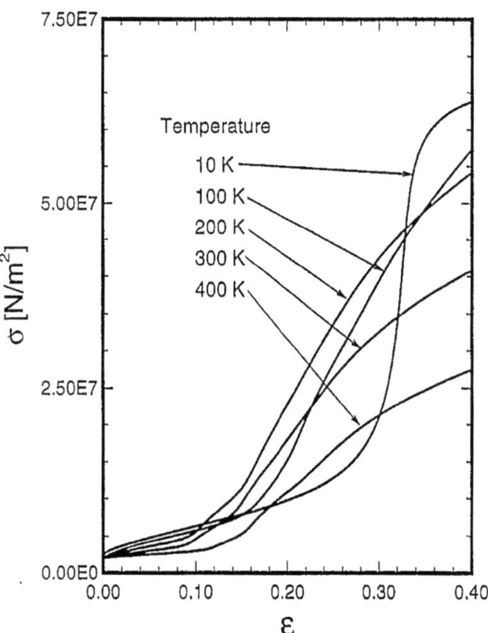

Fig. 4.8 Dependence of stress-strain behavior (calculated) on temperature (Reproduced from Cuitino (1996). With permission of the author and Elsevier Ltd.)

refinement) and, as a consequence, the entire flow stress is reduced. Following yield and flow stress reduction, much higher ductility is obtained. This is particularly true for BCC metals, which are very sensitive to impurities. One of the pronounced slip systems in BCC structures occurs on {110}-type planes in <111> directions. Thus, for example, the CRSS in W, deformed on the $(\bar{1}01) - [111]$ system, is reduced by 40,000 psi when tested at −196 K (see Koo) for a purity increase from one to five zone-melting passes. The strengthening effect of impurities is attributed to the elastic interaction with dislocations. Impurities may be distributed in the matrix as random substitution or interstitial solutes or may form some second-phase particles; this concept will be revisited in later sections regarding solid-solution strengthening.

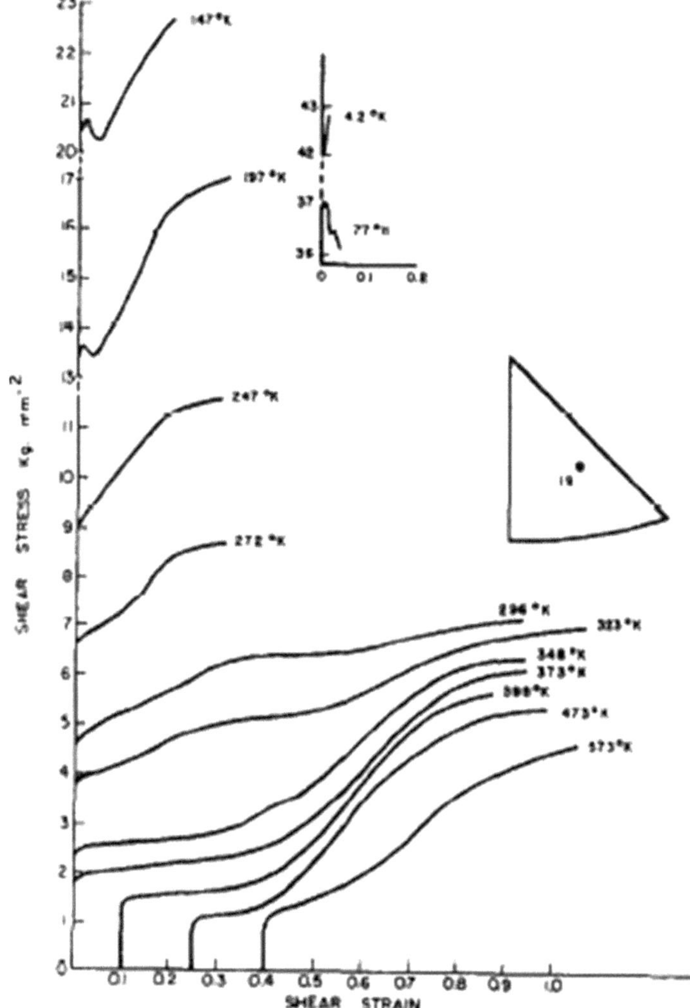

Fig. 4.9 The dependence of *shear stress/shear strain curves* on temperature in Ta single crystals at the orientation indicated. The *high temperature curves* are shifted on the strain axis (Reproduced from Mitchell and Spitzig (1965). With permission of the author and Elsevier Ltd.)

4.2.2 Stage II

Slip at stage I occurs on primary slip systems, as long as no impurity, temperature, strain rate or orientation effects change the pattern by introducing dislocation slip into other systems as well. As above, in stage I hardening, this section begins with a general discussion and general observations of stage II; later on, the various detailed effects will be considered. Stage II is an almost linear region of the stain-hardening curve, commencing after stage I, where strengthening increases rapidly with strain. As indicated earlier, more than one slip system is contributing.

The stress required for stage I hardening is usually stated as being proportional to the number (density) of dislocations or $\tau \propto \rho$; more specifically, it is given as:

$$\tau = \frac{1}{2} G b \rho^{1/2} \tag{4.2}$$

Kuhlmann-Wilsdorf and Comins gave a relation for the shear stress necessary for continued deformation (stage II) on the active glide system as:

$$\tau = \tau_0 + \alpha G b \rho^{1/2} \tag{4.3}$$

τ_0 is the frictional stress opposing dislocation motion and α is an empirical constant of $\sim 1/3$. This relation is linear, as indicated by the linear second-stage hardening shown in Fig. 4.1.

4.2.2.1 Crystal Structure

In all structures, three-stage strain hardening can be induced a priori. The tendency is to compare strain-hardening phenomena with those occurring in FCC single crystals (the most researched structures). Even in FCC structures, if the deformation axis (tension or compression) of the specimen tested is not in the shaded area (shown in Fig. 4.1) some of the stages may be absent. Thus, the appearance of stage II occurs during continuous deformation, immediately following stage I; though often stage II sets in at the beginning of straining, as a consequence of duplex or multiple slips occurring simultaneously in other slip systems. Therefore, deformation by slip is relatively easy when only one single system is operative, but is much more difficult when more slip systems operate simultaneously and the initial stress requirement is higher. In such cases, stage II sets in immediately after yielding. Experimental curves showing stage II hardening may be seen in Fig. 4.3, which represents FCC structures and agrees with the calculated curves. Here, stage II commences following yielding for lines 5 and 1 which are located at [001] and $[\bar{1}11]$, respectively. Directly obtaining stage II by experimentation in FCC structures, immediately after yielding or following a general deformation pass, as shown in Fig. 4.1, is relatively easy and can be achieved by selecting the proper orientation in the stereographic standard triangle.

In general, slip systems in HCP crystals are complicated by the fact that the basal (0001) plane is not always the preferred slip plane during deformation. Further twinning may be induced and contribute to deformation. Nevertheless, the considerations mentioned above for FCC are valid and slip in stage II occurs on basal planes, as shown in Fig. 4.10. The resolved shear stress-shear strain curves were constructed from the load elongation data, assuming slip on a $\{10\bar{1}0\}$ $<\bar{1}2\bar{1}0>$ system. Note

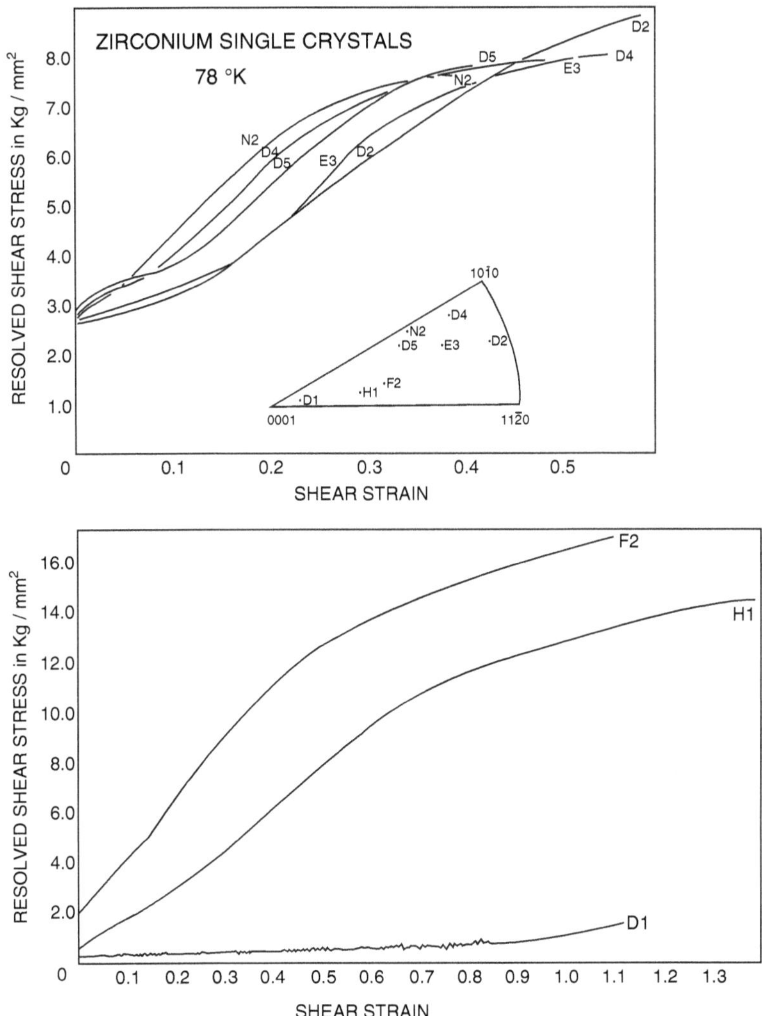

Fig. 4.10 Resolved *shear stress-shear strain curves* for the orientations indicated in the unit stereographic *triangle* for HCP crystals tested at 78 K (Reproduced from Akhtar and Teghtsoonian (1971). With permission of the author and Elsevier Ltd.)

that, at certain orientations of the tensile axis, stage I decreases and stage II becomes dominant. The conjugate slip system becomes operative at this stage.

In BCC crystals, various slip systems are possible, making it difficult to consider strain hardening in one type of slip system. In earlier experimental work, no three-stage hardening curves were found, but it was shown later that the same patterns of hardening exist in BCC materials as they do in FCC materials, but only under restricted conditions of purity, temperature and orientation. There is only a limited region at medium temperature in which all the hardening stages were observed under the same purity and orientation (Mitchell, et al.). As in the case of FCC, stage II can commence immediately after yielding or following stage I. Figure 4.5 above, for Ta, can represent other BCC structures, as well. Specimen 14, for example, shown oriented on the [001]–[111] boundary, has no stage I and has a high work-hardening rate in stage II. In this case, the orientation is very close to the boundary and double glide sets in. As indicated earlier, whenever more than one slip system is operative, hardening in stage II occurs and, often even the absence of stage I is realized (also indicated in Fig. 4.5 above).

4.2.2.2 Orientation

In all the structures considered above, it was strongly indicated that they exhibit three-stage hardening, provided they are oriented in such a manner that only one slip system is operative. However, just as stage I may be absent, stage II may also be absent, and deformation may begin and proceed from a stage III-type mechanism, also known as 'parabolic hardening'. The feature common to all these structures, essential for obtaining three-stage hardening, is that the tensile (or compression) axis is in the central part (soft region) of the standard triangle, where only one slip system is active (away from the symmetry boundaries or poles). To test deformation, in the light of the orientation influence on stage II hardening, one should start with an FCC structure, as shown above in the Fig. 4.3, with the deformation axis at poles [001] or $[\bar{1}11]$. Only stage II hardening is observed immediately after the elastic region. Equal slip on several systems occurs in the poles. Moving farther away from the boundaries or the poles, classic three-stage hardening is observed, but here, again, the extent of stage II and the rate of hardening, θ_2, are associated with the location of the tensile axis within the stereographic triangle. This can be seen for orientations 2, 3 and 4, of which orientation 2 (shown in the Fig. 4.3) is the most remote from the symmetrical sites (poles or boundaries) and shows the classic appearance of the theoretical three-stage hardening curve in the so-called 'single slip orientation'.

In HCP systems, definite orientation dependence was also observed in all the stages, shown, for example, in Fig. 4.10. Even a non-basal plane can show three-stage hardening. In Fig. 4.10, the deformation was adjusted for the prismatic slip system. Similar to the basal system during the prismatic slip, three slip systems (3×1) can be operative, i.e. three planes and one direction. Note that twinning can be easily induced in HCP systems and, for the study of pure shear deformation, this should be avoided. It was indicated (Akhtar and Teghtsoonian) that θ_2 exhibits a minimum near the $[11\bar{2}0]$ corner. Specimen D2 (in Fig. 4.10) is in the vicinity of the $[10\bar{1}0] - [11\bar{2}0]$ boundary and well-developed stage II hardening, between stages I and III, is observed. Further specimens, with deformation axes at locations N2, D4 and D5 in the vicinity of the symmetry boundary $[0001] - [10\bar{1}0]$ and along it, have about the same characteristics, with no change in the strain-hardening rate. For stage I, θ_1/G is typically 1.5×10^{-3} and θ_2/G is approximately five times higher. Similarly to FCC crystals, stage I is associated with single slip and stage II with conjugate slip.

In BCC metals, there is an optimum temperature for the observation of three-stage hardening, which is about 373 K in the case of Ta single crystals. Figure 4.5 shows the orientation dependence and it is seen that (analogous with FCC crystals) stage II is also affected by the location of the tensile axis. In these experiments, the stereographic unit triangle of the tensile specimens was [001]–[101]–[111]. As seen in specimen 19, which is remote from the corners and symmetry boundaries, found in the middle of the standard triangle, there is classic three-stage hardening with a long stage I, after which two slip systems start to operate simultaneously to produce stage II. The tensile axis of specimen 30, for example, along the symmetry boundary [001]–[101], enters immediately into a stage of double slip; thus, a stage II develops, which ends with the beginning of stage III. Specimen 14 has no stage I hardening, since it is located close to the [001]–[111] symmetry boundary and its strain-hardening rate, θ_2, is very high. A thorough discussion on the orientation dependence of BCC Ta is presented in the work of Mitchell and Spitzig. The three main conclusions that can be reached regarding the orientation dependence of strain hardening in all its stages are: (a) that whenever more than one slip system is operative, stage I may be entirely missing; (b) that the extent of stage II, its hardening rate and the stage at which it commences, depends on the orientation of the specimen deformation and; (c) stage II may also be entirely missing, such that the hardening curve starts immediately with stage III, having a similar parabolic shape as observed in polycrystalline materials.

4.2.2.3 The Effect of Temperature

The effects of temperature on HCP structures were seen in Fig. 4.4 for Cd and Zn single crystals. By focusing attention on Cd, one can see that by decreasing the temperature of deformation for a specific specimen's orientation and purity will increase the extent of stage II. At sufficiently high temperatures, stage II is absent and stage III-dynamic recovery-sets in. Another explanation of the decrease in stage II hardening with increasing temperature is that, after the onset of stage I, its extent increases with increasing temperature, contrary to observations in other HCP metals.

Experimental and calculated relations between temperature and stage II may be seen in Figs. 4.6 and 4.8, respectively for FCC metals. In Fig. 4.6, the effect of temperature is shown for two orientations and it is quite clear that decreasing the temperature increases the extent of stage II with almost no effect on θ_2. The decrease in stage II, in specimens with orientation 305, is a result of the fact that stage III sets in earlier at higher temperatures. In specimens having their deformation axes in site C14 orientation, a decrease in stage II has occurred; the appearance of stage III, and the contribution of stage I to the overall hardening curves, is associated with the shortening of stage II.

Figure 4.9 describes the temperature effect in BCC structures. For the given orientation over the temperature range 348–573 K, classic three-stage hardening is observed. In this temperature range, stage II (and also stage I) decreases at the expense of stage III, the extent of which increases with increasing temperature.

Similar observations were also made for Nb, which is quite similar in behavior to that of Ta. Below 348 K, a hump appeared in the transition from stage I to stage II, but this subject is out of the scope of this book.

4.2.2.4 Specimen Size

Experimental work has indicated the strong effect of specimen size on stress-strain curves. An extensive study by Suzuki, et al. on Cu single crystals of various dimensions shows that decreasing specimen size has a profound effect on the stress-strain relation. Easy glide increases with the decreasing radii of the specimen for each orientation investigated. Only a small effect, if any, was found on stage II. Figure 4.11 (reproduced from Suzuki, et al.) illustrates the effects of size when tensile axis orientations are at [011] and [001], respectively. Figure 4.12 distinctly shows that the decreased radius results in greatly increased slip in stage I (easy glide), but there is none in stage II. Figure 4.11b clearly indicates that, in some cases, orientation effects are more important than size effects, as seen for orientation in the immediate vicinity of [001] in specimens 136 and 96.

One can summarize the effect of size by stating that the size effect manifests itself mainly in the easy glide (stage I) region and produces almost no change in stage II. θ_1 and θ_2 do not seem to be strongly affected, since the slopes of their stress-strain curves are almost the same. One can understand size effect by the length of the dislocation slip path as it leaves the specimen to produce strain. One expects that a thin coating might prevent dislocations from leaving the specimen and, therefore, might affect the extent of dislocation glide.

4.2.2.5 The Purity of Specimens

There is experimental proof that purity has a significant influence on the mechanical properties of materials. This is especially true for BCC metals, which are very sensitive in regard to the purity level, particularly to interstitials. During their production, maintaining a reasonable purity level is very difficult and requires the application of special techniques, such as zone refining by several passes. High-purity BCC metals obtained by several zone-refining passes have appreciably lower CRSS for yield and flow stress than less pure crystals. See, for example, the work of Koo on high-purity W single crystals, investigated over a wide temperature range (from −196°C up to 29°C).

4.2.3 Stage III (Dynamic Recovery)

In general, stage III strain hardening is characterized by the continuous decrease of the strain-hardening rate, as inferred from the decreasing slope of the

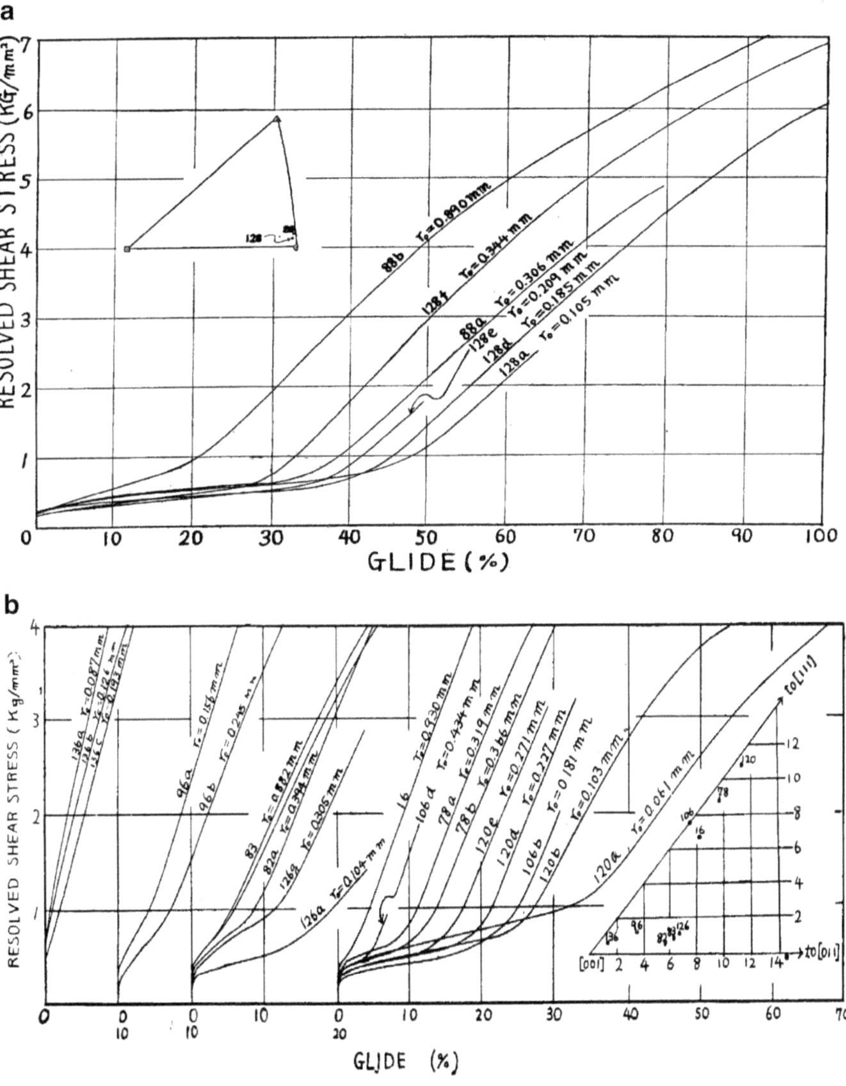

Fig. 4.11 *Shear stress/glide curves* of Cu crystals having various radii. The orientations are: (**a**) [011] and (**b**) [001] (Reproduced with the permission of JPSJ from Suzuki et al. 1956)

strain-hardening curve with increasing flow stress. The line in this region is parabolic and, thus, exhibits a low rate of strain hardening. The parabolic shape of stage III resembles the hardening in polycrystalline materials. Much published research states that this stage of hardening is temperature and strain-rate sensitive. The commonly used term 'dynamic recovery' refers to a relative stress reduction (softening) during deformation, contrary to stage II. The effect of dynamic recovery is not time-dependent, since it occurs while deformation is going on. Stage III continues up to the point of stress saturation.

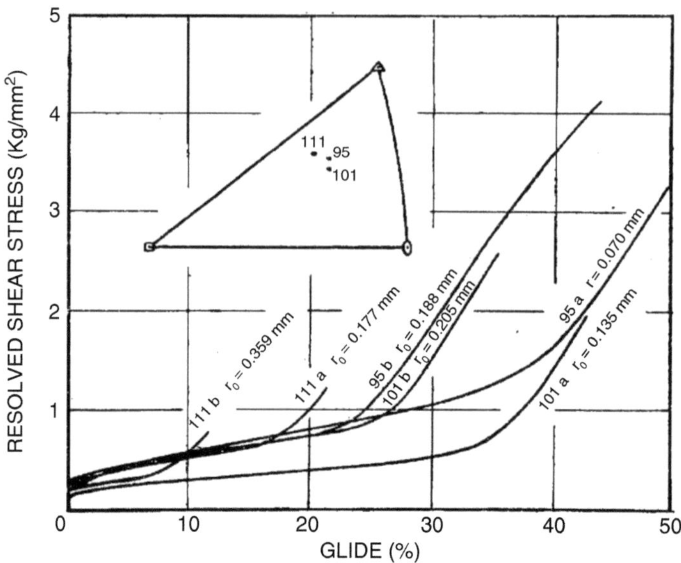

Fig. 4.12 *Shear stress/glide curves* of Cu crystals having various radii. The orientation is far from the corners (Reproduced with the permission of JPSJ from Suzuki et al. 1956)

4.2.3.1 The Effect of the Stacking Fault

Stage III is associated with the cross slip of screw dislocations (edge dislocations climb, but do not cross slip). Cross slip requires a constriction of the area (known as 'ribbon') between the partial dislocations, which, in turn, depends on the stacking-fault energy. As noted in Chap. 2 (Sect. 2.14.3 and Eqs. 2.58 and 2.58a), the wider the ribbon between the partials, the more difficult their constriction for cross slip. High-energy stacking faults have a narrow ribbon, if any, and their constriction is easy; therefore, cross slip occurs more easily and the consequence is less hardening. Al, for instance, is a material that exhibits high stacking-fault energy and, therefore, cross slip is made easier. Cross slip is seldom observed in Al, contrary to metals, such as Cu, that have much lower stacking-fault energy. It is clear that high stacking-fault materials show appreciably lower hardening, as manifested by the lower stress and strain required by Al to enter into this region.

4.2.3.2 The Effect of Temperature

Cross slip is thermally activated and enhanced at high temperatures. Thus, the stress-strain curve is affected by temperature, as shown in Fig. 4.13. Cuitino expressed the relation between stress and the density of an obstacle for dislocation glide as:

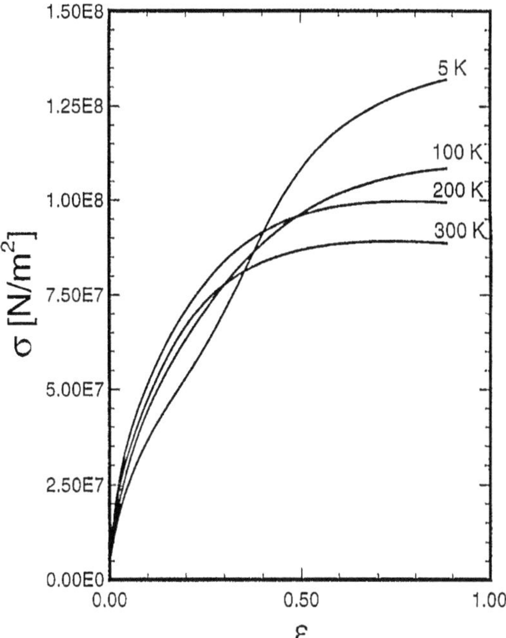

Fig. 4.13 The effect of temperature on the stress-strain relation of a Cu single crystal loaded in the [001] direction (Reproduced from Cuitino (1996). With permission of the author and Elsevier Ltd.)

$$\tau^{\gamma}{}_{III} = \alpha\,(T, d)\, b\mu\, (n^{\alpha}{}_{S})^{1/2} \tag{4.4}$$

In this equation, $\mu = G$, the shear modulus, $(n^{\alpha}{}_{S})$, represents the density of the obstacles and d is the distance between the partial dislocations. The resolved shear stress, $\tau^{\gamma}{}_{III}$, marks the inception of stage III. The higher the temperature, the lower the stress for the initiation of parabolic hardening, as indicated in Fig. 4.13. This is associated with the variation of α and temperature, as in Eq. 4.4. α has been computed and was found to decrease linearly with temperature up to a critical value, after which it remains constant (see Cuitino). For Cu at room temperature, $\alpha = 0.3$. The variation of α with temperature is important for the prediction of the CRSS, τ_{III}, for the onset of stage III.

Similar to the thermal effect on τ_{III} in Cu, Al single crystals also show the same pattern as seen in Fig. 4.14. The higher the deformation temperature, the lower the stress required for the initiation of stage III.

In summary of this section, note that high temperatures and high stacking-fault energies increase the rate of cross slip and reduce the strength of the obstacles opposing dislocation glide, whereas low temperatures and low stacking-fault energies reduce the rate of cross slip and, therefore, increase the strength of materials.

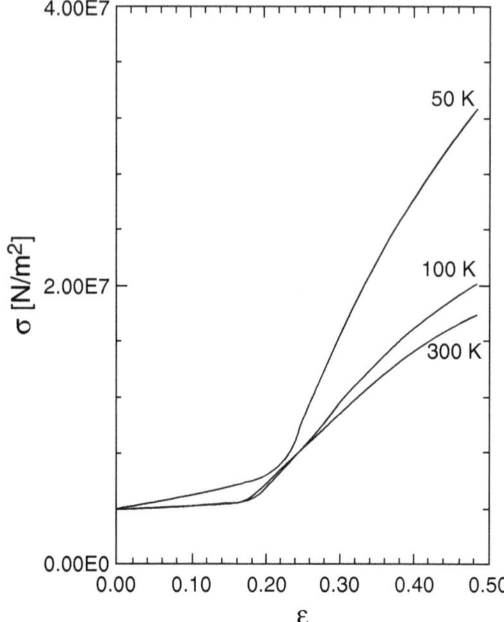

Fig. 4.14 *Stress strain curves* for Al single crystals at various temperatures (Reproduced from Cuitino (1996). With permission of the author and Elsevier Ltd.)

4.3 Microstructure

Deformation changes the structures of materials, but, here, our interest is focused on dislocation structures, as observed at various stages of strain hardening. Figures 4.15 and 4.16 represent the microstructural changes in Zr single crystals, representative of HCP crystals. HCP crystals are known to behave in a manner similar to that of FCC structures under appropriate conditions and also show a three-stage hardening process. It was observed in HCP crystals that the length of easy glide is orientation-dependent and that the hardening rate, θ_1, is lower when the tensile axis is located near the [0001] – $[10\bar{1}0]$ boundary, than it is for those crystals having a tensile axis near the [0001] – $[11\bar{2}0]$ boundary (see, e.g., Fig. 4.10). As mentioned earlier when the strain rate is normalized by the shear modulus one obtains a θ_1/G for stage I that is typically 1.5×10^{-3}; θ_2/G is approximately five times higher. An extensive stage III follows stage II. Stage III is characterized by a decreasing work-hardening rate under strain.

Note the length of the slip lines in (a), representing easy glide, and the slip markings in (b), when deformation occurs on two slip systems-the primary and the conjugate ones. The dislocation density in the un-deformed crystals is usually $\sim 10^6/cm^2$, which was also obtained in the Zr crystal by the etch pits technique. The dislocation structure, deformed at various strains, is shown in Fig. 4.16 below. This figure shows: (a) crystals deformed in stage I with a high density of dislocation loops; (b) the dislocation structure towards the end of stage II; (c) progressive

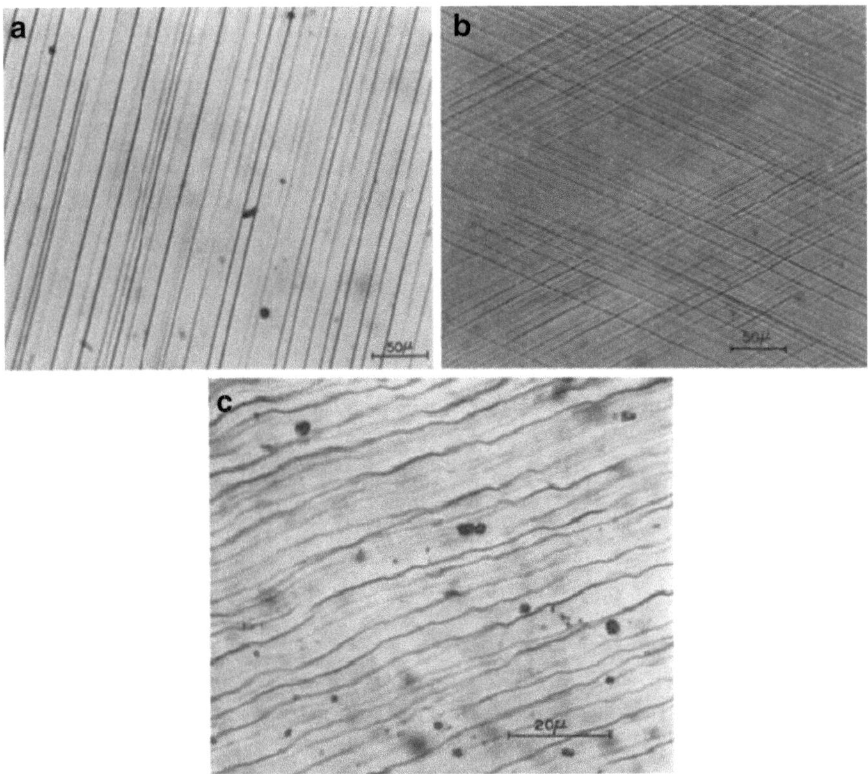

Fig. 4.15 *Slip markings* in Zr single crystals deformed at 78 K. (**a**) in easy glide $\gamma = 0.05$; (**b**) into stage II $\gamma = 0.22$, *conjugate slip markings* is seen; (**c**) stage III $\gamma = 0.5$ (Reproduced from Akhtar and Teghtsoonian (1971). With permission of the author and Elsevier Ltd.)

deformation in stage III with one-dimensional 'braids' developing along the trace of a $\bar{1}2\bar{1}0$ plane and; (d) the dislocation distribution in a crystal deformed nearly to fracture. The braids have finally broken into cell structures. The tangles form what is known as 'cell walls'; in these cells, low or zero density of dislocations can be seen. Previously, in Sect. 2.3.2, a cell structure was observed in Al, deformed by rolling and annealing (shown in Fig. 2.8). This illustrates typical cell structure in an FCC metal during stage III of the deformation curve. Similar cell structures were also observed by Püschl in cold-rolled Al. Cross slip characterizes stage III, which plays an important role in the re-arrangement and annihilation of screw dislocations. Thus, the decreasing rate of strain hardening at this stage is manifested by the action of cross slip.

Slip in FCC crystals is shown in Fig. 4.17. The slip lines in (a) represent steps in Cu at stage II. The slip steps appear to be quite homogeneously distributed. In (b), stage III slip bands are seen in the same crystal, which may often be resolved into a number of parallel steps (Fig. 4.18).

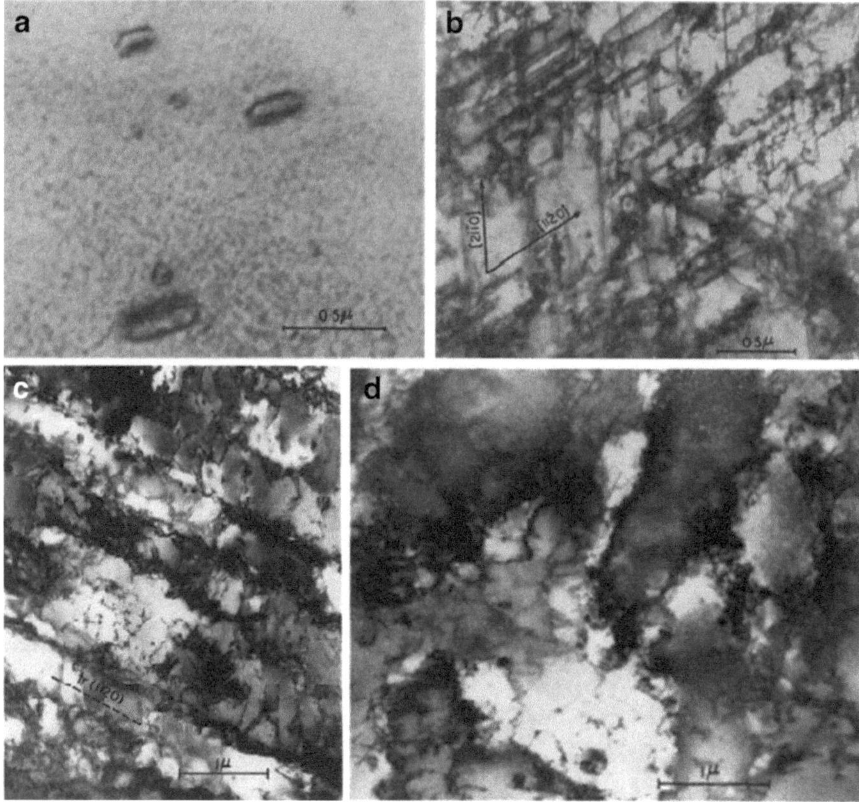

Fig. 4.16 Transmission electron micrographs: (**a**) easy glide, $\gamma = 0.05$; (**b**) Dislocation structure at the end of stage II $\gamma = 0.29$ just strained past τ_{III}; (**c**) Crystal deformed to $\gamma = 0.43$, dislocation braids formed along the tracer of [11$\bar{2}$0]plane; (**d**) Cell structure deformed to $\gamma = 0.58$ (Reproduced from Akhtar and Teghtsoonian (1971). With permission of the author and Elsevier Ltd.)

Experimental evidence for FCC crystals indicates that those with low energy are dissociated into partial dislocations and that the width of the stacking faults confined between them depends on their energy. High-energy crystals rarely dissociate into partials and the area between them is narrow. The association of partials in metals with wide separation requires high stresses for cross slip. Since cross slip is performed by screw dislocation, the interaction between such dislocations and with others, such as forest dislocations, must be involved in the strain-hardening stages. Forest dislocations act as obstacles to dislocation motion. Cross slip is thermally activated; thus, increasing both the temperature and the stacking-fault energy weakens the resistance of the obstacles to slip, helping cross slip to occur. On the contrary, decreasing temperature and stacking-fault energy suppresses cross slip and increases resistance of obstacles (e.g., forest dislocations) preventing slip. During stage I, the density of forest dislocations does not change greatly; as a matter of fact, it remains almost constant and, therefore, there is no great resistance to

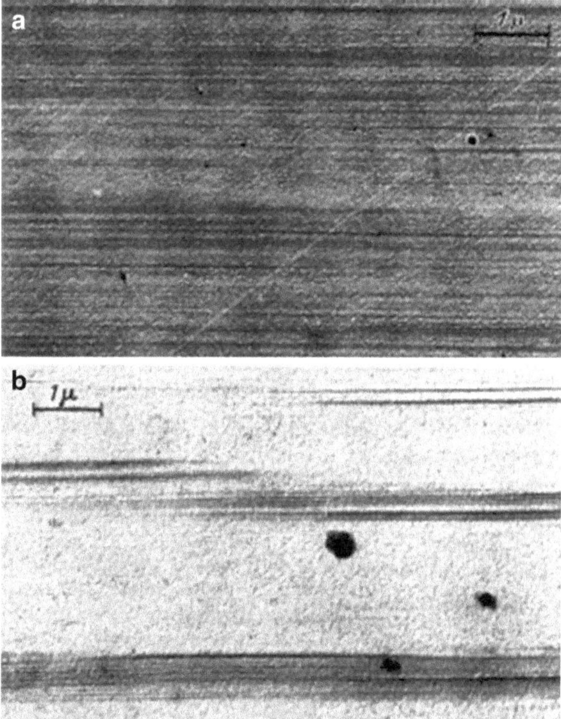

Fig. 4.17 *Slip* in Cu: (**a**) slip steps in stage II, (**b**) slip bands in the same crystal at stage III (From Kocks and Mecking (2003). With permission of the authors and Elsevier Ltd.)

dislocation motion. Hardening is mostly the result of dislocation motion through an array of obstacles, which is responsible for increasing strength. This is in accordance with the shallow slope of the first stage, indicating a low strain-hardening rate. Slip lines are long, due to the low resistance of the obstacles. At low temperatures, the degree of hardening is higher and the slope increases. With increased straining, more dislocations are formed by some multiplication mechanism, thus the density of obstacles increases. This multiplication occurs in secondary systems, occurring during stage II and resulting in high strain-rate hardening. Stage II hardening is a result of screw-screw interaction. Stage III is dominated by cross slip, which requires constriction of the partial dislocations, an essential part of the deformation process, in order to attain dynamic recovery. Theories of cross slip and the various concepts associated with this process are widely discussed in the work of Püschl.

4.4 Theories of Strain Hardening

Many theories have been suggested to explain the stages of strain hardening in crystals. Some of the early theories and later concepts on strain hardening are reviewed here from the perspective of the dislocation approach. Stages IV and V

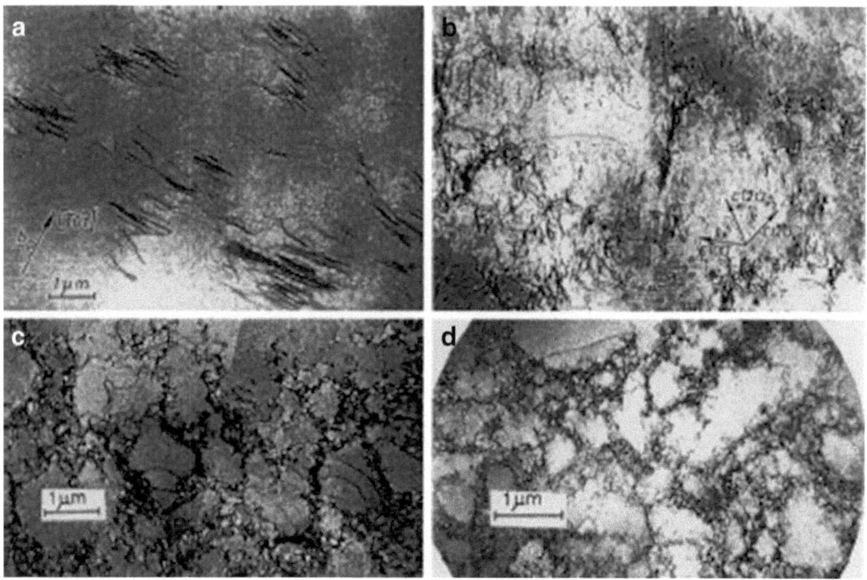

Fig. 4.18 Shows dislocation arrangements in Cu: (**a**) deformation was carried out into the end of stage I and slip occurred on one system, (**b**) deformation in early stage II also on one single slip system, low density dislocation randomly arranged, (**c** and **d**) dislocation structure at higher stress where multiple slips occurs due to a [001] orientation (From Kocks and Mecking (2003). With permission of the authors and Elsevier Ltd)

are not considered in the following. The present focus is on the three stages of strengthening:

1. Applying a sufficiently large strain above CRSS causes dislocations to move in their respective slip systems.
2. The applied stress must be sufficient to overcome Peierls-Nabarro forces and any other obstacles that may hinder dislocation movement.
3. Sufficient dislocations are generated by the sources for strain formation having the appropriate magnitudes.
4. Dislocation generation and the overcoming of the relatively small resistance of obstacles to motion results in small hardening and a shallow slope of the stress-strain curve. In this range, known as stage I, θ_1 is small. Note the influence of all the aforementioned factors.
5. Continued deformation activates other slip systems. Lattice rotation, due to gripping of the tensile specimens, is one of the factors inducing glide on additional slip systems (for a given orientation, purity, size, temperature, etc. of a specimen).
6. Large amounts of dislocations of other systems are anchored by various obstacles (for our purpose, by other dislocations), making dislocation movement more difficult. Increased hardening sets in.

7. Higher stresses are required for motion, partly because repulsive stress fields of dislocations have the same sign. This occurs in pile-ups, such as the Lomer-Cottrell ones. Dislocations, moving in the wake of earlier dislocations, are repelled, because individual dislocations within the same pile-up have the same sign.
8. Other obstacles may be caused by dislocation-dislocation interactions, resulting in jogs, tangles, etc. Cutting a dislocation-tree in a forest dislocation induces hardening, while other obstacles also contribute. Overcoming repulsive stress fields and the other obstacles requires ever-increasing stress for continued deformation.
9. As a consequence of the ever-increasing stress, a sharp slope in the stress-strain curve is obtained and θ_2 becomes much higher than hardening in stage I. This new region is called 'stage II'.
10. When τ is sufficiently large to enable continued dislocation motion and for cross-slip to occur, a mechanism exists by which obstacles can be by-passed. Low-energy stacking faults require constriction of the partials before cross slip can occur; this is more difficult than in materials with high stacking-fault energy.
11. As a consequence of avoiding obstacles by cross slip, strain-rate hardening during stage III is continually decreasing with straining.

4.4.1 *Stage I*

4.4.1.1 Taylor's Approach to Strain Hardening

Taylor's strain-hardening model has historical interest. Taylor was the first to realize that the experimental evidence on the deformation process must be related to the effect of dislocations. His most important assumptions follow:

(a) The density of dislocations increases with deformation (straining).
(b) Dislocations react elastically with other dislocations.
(c) Dislocations are anchored by other dislocations, since not all of them arrive to the specimen's surface.
(d) The average distance that a dislocation moves under an applied stress is L.

Therefore, strain is related to dislocation density according to:

$$\varepsilon = \rho L b \tag{4.5}$$

where ρ is the dislocation density, L the average distance that dislocations travel before they are stopped by other dislocations and b is the Burgers vector. A schematic illustration, according to Taylor, is shown in Fig. 4.19. Here, the shear stress for edge dislocations may be expressed as:

$$\tau = \frac{Gb}{8\pi}\frac{1}{l} = k\frac{Gb}{l} \tag{4.6}$$

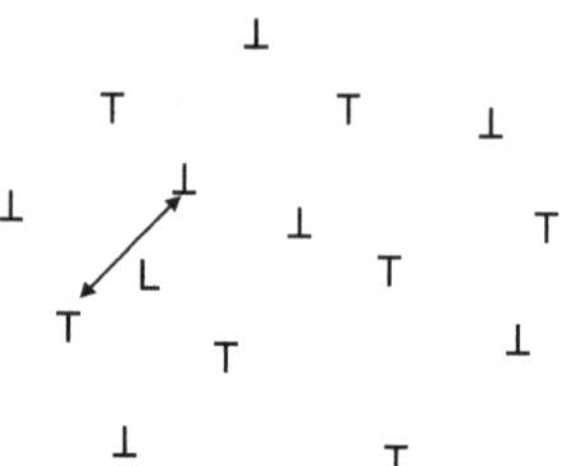

Fig. 4.19 An array of dislocations illustrated according to Taylor's concept of work hardening. The average separation distance is *L*

l is the average distance between dislocations, written as:

$$l = \rho^{-1/2} \tag{4.7}$$

After substitution into Eq. 4.6, a relation, in terms of dislocation density, is obtained for stress:

$$\tau = kGb\rho^{1/2} \tag{4.8}$$

Using Eqs. 4.5 and 4.8, stress can be expressed as:

$$\tau = kG\left(\frac{\varepsilon b}{L}\right)^{1/2} \tag{4.9}$$

Note that the effort made to obtain a parabolic relation was a consequence of the observed experimental stress-strain relation (behavior witnessed especially in the deformation of polycrystalline metals).

Taylor's theory is basically a two-dimensional model. Since sources emit dislocation rings, the strain in Eq. 4.5 can be expressed as:

$$\varepsilon = N\pi R^2 b \tag{4.10}$$

where πR^2 is the area of a ring emitted by the source and N is the number of such rings per unit volume. Thus, the average distance between a pair of dislocations, l, can be given as:

$$l = \left(\frac{1}{2}\pi RN\right)^{1/2} \tag{4.11}$$

Accordingly, an expression written in terms of the dislocation ring radius is:

$$\tau = const.\left(\frac{b}{R}\right)^{1/2} \varepsilon^{1/2} \tag{4.12}$$

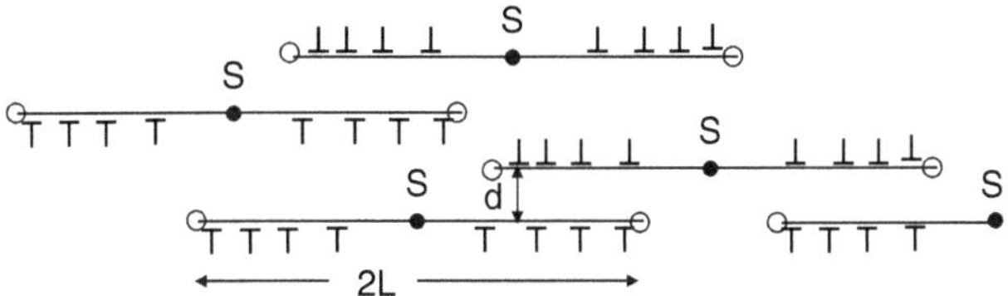

Fig. 4.20 A pile-up group on slip planes *d* distance apart, consisting of edge dislocations of opposite signs: ∘ represents obstacles; S, • are dislocation sources

One of the major shortcomings of this early and historically significant theory is that Taylor's model only considered single dislocations, whereas, in strain hardening, a large number of dislocations are involved. Nonetheless, some of Taylor's ideas were basic to later theories.

4.4.1.2 Mott's Model

Other theories were put forward taking into account the fact that not isolated, but many, dislocations are involved in the hardening process. One such significant and much-cited model was that of Mott. Mott rectified Taylor's model by adding certain other relevant considerations:

(a) Dislocation pile-ups, rather than individual dislocations, are involved.
(b) Sources generate dislocations on slip planes. Other dislocations, sessile in nature, impede the dislocation movement, acting as obstacles.
(c) High internal stress acts in front of the leading dislocation in a pile-up.
(d) This high stress activates sources in secondary slip systems.
(e) Dislocations in secondary slip systems may interact with those in primary systems.

At the obstacle, n, dislocations of the same sign pile-up to form a sort of 'super-dislocation' of strength nb, with n being the number of dislocations piled-up behind the obstacle. Figure 4.20 is a common schematic illustration of this model.

The stress may be expressed as:

$$\tau = \frac{G(nb)}{2\pi(Ld)^{1/2}} \tag{4.13}$$

with d being the distance between the active slip planes. The density of the super-dislocations is 1/(Ld) and, thus, the mean distance between them is $(Ld)^{1/2}$. An expression, like Eq. 4.5 for strain, is written in terms of the super-dislocations as:

$$\varepsilon = \rho\,(\mathrm{nb})\,\mathrm{L} \tag{4.14}$$

The density, ρ, is 1/(Ld), which can be substituted into Eq. 4.14, yielding:

$$\varepsilon = \frac{(nb)}{d} \tag{4.15}$$

From Eqs. 4.13 and 4.15, one obtains:

$$\tau = \frac{G}{2\pi}\left(\frac{\varepsilon nb}{L}\right)^{1/2} \tag{4.16}$$

This relation gives a parabolic law, like Taylor's Law, but one that is conceptually different and more in line with the many dislocation pile-up groups acting on various slip planes and interacting with each other.

Later theories are based on detailed experimental observations; however, the main efforts were directed at explaining the classic, three-stage stress-strain relation in regard to the deformation process in single crystals. Thus, there are separate theories for each stage and also some that cover the entire strain-hardening phenomenon.

4.4.1.3 Seeger's Theory for Stage I

Some of Seeger's assumptions regarding stage I are presented below:

(a) Long range interactions exist between dislocation loops.
(b) These loops are fairly widely spaced one from another.
(c) Dislocation loops occur in primary slip systems.

A small change in stress, δτ, will increase the number of loops by δn, which, in turn, produces a change in strain given as:

$$\delta\varepsilon = \mathrm{bNL}^2\delta\mathrm{n} \tag{4.17}$$

N is the number of sources per unit volume and L the distance that loops move. Each source emits n loops, which are generated at a stress τ. N can be defined in terms of d and L as:

$$N = \frac{1}{dL^2} \tag{4.18}$$

d is the distance between the slip planes, as indicated in Fig. 4.20. Substituting Eq. 4.18 in Eq. 4.17 results in:

$$\mathrm{d}\varepsilon = \frac{\mathrm{b}\delta\mathrm{n}}{\mathrm{d}} \tag{4.19}$$

The generation of δn new loops increases the back stress, τ_b, on the sources by:

$$\delta\tau_b = \frac{Gb\delta n}{2\pi L} \tag{4.20}$$

A stage is reached where the back stress is equal to the applied stress, i.e., $\tau = \tau_b$ and, consequently, the source will not emit any more dislocation loops. Substituting for δn from Eq. 4.19 into Eq. 4.20, one obtains:

$$\delta\tau_b = \delta\tau = \frac{Gd\delta\varepsilon}{2\pi L} \tag{4.21}$$

Thus,

$$\frac{\delta\tau}{\delta\varepsilon} = \theta_I = \frac{G}{2\pi}\frac{d}{L} \tag{4.22}$$

A more accurate calculation of θ_I gives:

$$\theta_I = \frac{8G}{9\pi}\left(\frac{d}{L}\right)^{3/4} \tag{4.23}$$

Note that Fig. 4.20 may be used to calculate strain hardening in stage I in accordance with Seeger's approach. Furthermore, Eq. 4.23 provides fairly close values to those of experimental observations for a number of metals, a level of $\sim 3 \times 10^{-4}$ G, when reasonable values are used, such as d = 300 Å and L ≈ 0.5 mm.

4.4.2 *Stage II*

It has been suggested that the factors responsible for hardening are:

(a) forest dislocations;
(b) Lomer-Cottrell locks; and
(c) jogs.

It is therefore possible to identify the stress with which one can activate dislocation sources and overcome any of the abovementioned factors. All the various theories suggested thus far emphasize one of the above factors and, thus, it may be stated that the stress required to overcome dislocation obstacles is the one capable of:

(a) overcoming the elastic stresses of the forest dislocations;
(b) overcoming long-range stresses of dislocations that are stuck in pile-ups behind Lomer-Cottrell locks; or
(c) overcoming the resistance of jogs to dislocation glide.

4.4.2.1 Seeger's Pile-Up Model

Experimental observations of stage II indicate that:

(a) Stage II hardening is linear, like stage I.
(b) Strengthening rate $\theta_{II} = \sim 2 \times 10^{-3}$ G.
(c) The slope of the stress-strain curve (the strengthening rate) is only slightly dependent on temperature.
(d) The real influence of temperature on the slope is not direct; however it indirectly effects the extent of stage II by determining when the strain-hardening curve enters stage III.
(e) As a consequence of (d), hardening is connected to the thermal component of stress (flow stress can be expressed as $\tau = \tau_{th} + \tau_{ather}$). The forest dislocation model should be temperature dependent.
(f) Slip lines at this stage gradually shorten; the probable explanation has to do with the increase in Lomer-Cottrell obstacles.

The stress resisting the expansion of a loop is:

$$\tau = \alpha G\,(nb)\left(\frac{NL}{n}\right)^{1/2} \tag{4.24}$$

N is the number of dislocation loops per unit volume, n is the number of dislocations in each pile-up, L the size of the loop and N/n is the number of pile-ups per unit volume. Clearly, $\left(\frac{NL}{n}\right)^{-1/2}$ is the average distance between the pile-ups.

As previously, (nb) is the vector of a super-dislocation, where $\alpha = \frac{1}{k\pi}$. The strain for a small change in N, moving a distance L, is:

$$\mathrm{d}\varepsilon = \mathrm{b}\mathrm{L}^2\mathrm{dN} \tag{4.25}$$

Following the above and assuming that (nb) is independent of strain, the strain-hardening rate is given as:

$$\frac{d\tau}{d\varepsilon} = const.G\,(nb)^{1/2} \tag{4.26}$$

Seeger estimated that a pile-up containing ~25 dislocations stops growing. Equation 4.26 is for linear hardening, under the assumption that (nb) does not depend on strain.

Equation 4.26 may be obtained as follows:

(i) differentiate Eq. 4.24 with respect to dN to obtain:

$$d\tau = dN\alpha G(nb)\left(\frac{L}{n}\right)^{1/2}\frac{1}{N^{1/2}} \tag{4.27}$$

(ii) the derivative of (4.27) with respect to dε and the use of Eq. 4.25 is:

$$\frac{d\tau}{d\varepsilon} = \frac{dN\alpha G}{bL^2 dN}(nb)\left(\frac{L}{n}\right)^{1/2}\frac{1}{N^{1/2}} \tag{4.28}$$

(iii) with proper manipulation, one can write:

$$\frac{d\tau}{d\varepsilon} = \frac{\alpha G}{L^2}\left(\frac{n^2 L}{n}\right)^{1/2}\frac{b^{1/2}}{b^{1/2}N^{1/2}} = \frac{\alpha G}{L^2}(nbL)^{1/2}\left(\frac{1}{bN}\right)^{1/2} \tag{4.29}$$

(iv) this can also be written as:

$$\frac{d\tau}{d\varepsilon} = const.G(nb)^{1/2} \tag{4.26}$$

and the constant is:

$$const = \alpha\left(\frac{L}{bN}\right)^{1/2}\frac{1}{L^2} \tag{4.30}$$

4.4.2.2 Friedel's Pile-Up Model

Another pile-up model was suggested by Friedel using the Lomer-Cottrell system. Friedel's basic assumptions are as follows:

(a) The glide distance is constant, but the number of dislocations in each pile-up is proportional to the stress. (This is contrary to Seeger's approach, who assumed that glide distance varies and the number of dislocations in a pile-up is constant).
(b) The Lomer-Cottrell locks are formed between dislocations in different slip systems.
(c) Dislocations pile up behind Lomer-Cottrell locks and, therefore, the sources are surrounded by loops and further glide is hindered.
(d) The distribution of dislocations in a pile-up was calculated by Eshelby, Frank & Nabarro.

The number of dislocation loops in the pile-up of a slip plane of length L, into which n dislocations can be pushed, is:

$$n = \frac{k\pi\tau_S L}{Gb} \tag{4.31}$$

The stress is:

$$\tau_S = \frac{Gbn}{k\pi L} \tag{4.32}$$

The strain is:

$$\varepsilon = \mathrm{nNL^2b} \tag{4.33}$$

Again, N is the number of acting sources per unit volume of pile-ups. Clearly, L^2 is the area of a loop. Eliminating n from Eqs. 4.32 and 4.33, one can write:

$$\frac{\tau_S k\pi L}{Gb} = \frac{\varepsilon}{bNL^2}$$

and for τ_S, one can write:

$$\tau_S = \frac{G\varepsilon}{k\pi L^3 N} \tag{4.34}$$

The derivative is:

$$\frac{d\tau}{d\varepsilon} = \theta_{II} = \frac{G}{k\pi N L^3} \cong \frac{G}{NL^3} \tag{4.35}$$

Using realistic values for N and L, Friedel obtains the rate of strain hardening during stage II, expressed as:

$$\theta_{II} = \frac{d\tau}{d\varepsilon} = \frac{G}{2} \quad \theta_{II} \approx \frac{G}{2} \tag{4.35a}$$

The constant, k, takes into account whether a dislocation is an edge or screw type.

4.4.2.3 Another Approach by Seeger to the Pile-Up Model

Figure 4.21 shows dislocation loops emitted from a source. In (a) the expanding loops move until they reach Lomer-Cottrell barriers and in (b) the same is shown on two different, neighboring planes.

As previously, N is the number of pile-ups per unit volume, n the number of dislocations in each pile-up. For strain, one can write:

$$\varepsilon = \mathrm{N\pi L^2\,(nb)} \tag{4.36}$$

where L^2 is the area swept out by the dislocation before becoming stuck. The assumption is that such pile-ups may be considered 'super-dislocations', separated one from another by a distance l:

$$l \sim \frac{1}{(2LN)^{1/2}} \tag{4.37}$$

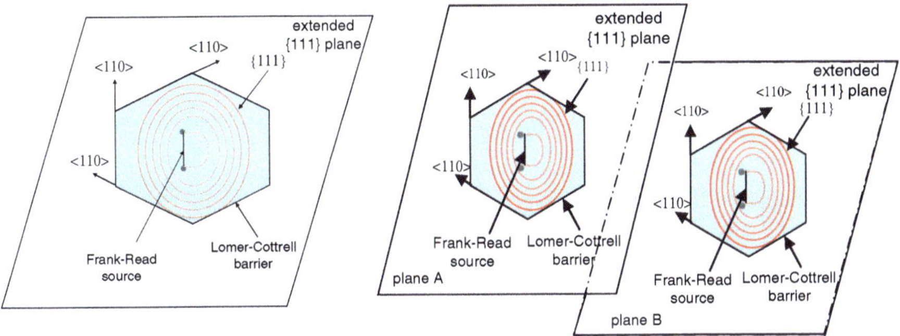

Fig. 4.21 Dislocation pile-ups emitted from the Frank-Read source dislocation loops, stuck at a Lomer-Cottrell-type obstacle: (**a**) a {111} type plane and (**b**) the same as (**a**), but the sources are operating in two planes and interacting with each other

Stress opposing movement may be expressed as:

$$\tau = \left(\frac{1}{2\pi}\right)\frac{G(nb)}{l} = \alpha\frac{G(nb)}{l} \tag{4.38}$$

After eliminating n from Eqs. 4.38 and 4.36, stress can be given as:

$$\tau = const.G\varepsilon \tag{4.39}$$

and the hardening rate becomes:

$$\frac{d\tau}{d\varepsilon} = \theta_{II} = const.G \tag{4.40}$$

Note that the slope of the stress-strain curve is constant and this relation is linear, as observed in experimental results. The constant may be expressed as:

$$const. = \left(\frac{1}{2\pi^2}\right)\left(\frac{1}{NL^2}\right)\left(\frac{1}{l}\right) \tag{4.41}$$

One can express n, in the manner of Eshelby, Frank and Nabarro, as:

$$n = \frac{k\tau L}{bG} \tag{4.42}$$

and L, the glide distance, as:

$$L = \frac{\Lambda}{(\varepsilon - \varepsilon_0)} \tag{4.43}$$

Mott assumes that the dislocation sources are on primary slip systems and are jogged by the dislocations operating on the secondary slip systems. The stress required to operate a jogged source is:

$$\tau = \alpha G b m_v \tag{4.45}$$

$\alpha \sim 0.2$, m_v is the number of 'vacancy-jogs' (i.e., vacancy generating jogs) per unit length. Assuming that the number of loops, dn, of mean radius, R, which are emitted on a primary slip plane during an increment of strain is:

$$d\varepsilon = b\pi R^2 dn \tag{4.46}$$

If the slip on the secondary plane is g times that on the primary plane and m is the number of jogs per unit length on a source or other stationary dislocations, an expression for dm may be written as:

$$dm = g\pi R^2 dn \tag{4.47}$$

Only a small fraction of these jogs are vacancy jogs:

$$dm_v = f dm \tag{4.48}$$

From Eqs. 4.45 and 4.47, one obtains:

$$d\tau = \alpha G b\,(fg)\,\pi R^2 dn \tag{4.49}$$

Equation 4.49 is obtained as follows:

i. Take the derivative of Eq. 4.45:

$$d\tau = \alpha G b\, dm_v \tag{4.50}$$

ii. Substitute Eq. 4.48 into Eq. 4.50 to obtain:

$$d\tau = \alpha G b f dm \tag{4.51}$$

iii. Substitute Eq. 4.47 into Eq. 4.51 to get:

$$d\tau = \alpha G b\,(fg)\,\pi R^2 dn \tag{4.51a}$$

iv. Then, substitute Eq. 4.46 into Eq. 4.49 (or Eq. 4.51a) giving:

$$d\tau = \alpha G f g\, d\varepsilon \tag{4.52}$$

or:

$$\frac{d\tau}{d\varepsilon} = \alpha Gfg \tag{4.53}$$

With $\alpha \sim 0.2$, $f = 1/20$ (the ratio of vacancy-jogs to interstitial-jogs on the sources) and $g = 0.5$ (the ratio of secondary to primary slip), one gets $(d\tau/d\varepsilon)/G \approx 1/200$, which was found to be in agreement with most metals.

The Mott model successfully explains the many experimental observations listed below:

(a) Hardening arises due to an increase in the density of jogs in primary slip systems.
(b) After unlocking the sources, slip occurs with the formation of slip lines in its wake.
(c) Jogs can be thermally activated which explains why the required applied stress is relatively small at high temperatures.
(d) Edge dislocations glide larger distances than those of screw type, since screw dislocations are associated with vacancy-jog formation.

The theory does not explain, among other things:

(a) the high stress necessary for slip in secondary slip systems (g is relatively high; see, for example, Eq. 4.53);
(b) why slip in secondary slip systems is strongly orientation-dependent, i.e., there is a large variation in θ_{II}. Practically, the change in θ_{II} is small with orientation.

4.4.2.6 Hirsch's Model

On the basis of TEM observations, dislocation tangles were observed that were not considered in the aforementioned theories. On the basis of Hirsch's approach, consideration is given to the following:

(a) The formation of dislocation pile-up groups.
(b) The eventual relaxation of these groups by slip on secondary slip systems and the formation of dislocation tangles.
(c) Tangles are the main obstacles for further slip.
(d) New sources start to operate when dislocations pile up beyond tangles.
(e) These new pile-ups also relax and form additional tangles.
(f) The number of obstacles (tangles) is proportional to the number of sources emitting groups of dislocation pile-ups.

The spacing l between dislocation tangles is inversely proportional to the density of the sources, i.e., their number per unit volume is:

$$N \sim \frac{1}{l^3} \tag{4.54}$$

Stress is proportional to the number of sources, given as:

$$\tau \propto N^3 \tag{4.55}$$

The change in strain is related to the change in the number of sources:

$$\mathrm{d}\varepsilon = \mathrm{L_1L_2nbdN} \tag{4.56}$$

n is the number of loops which have edge and screw components and the slip distance of these components are L_1 (the slip distance of the edge component) and L_2 (the slip distance of the screw component). From the geometry of the model one gets:

$$\mathrm{L_1L_2RN} \sim 1 \tag{4.57}$$

where R is the radius of the obstacle and is related to the fraction, f, of the pile-ups, having stress field:

$$R = f\frac{nbG}{2\pi\tau} \tag{4.58}$$

where nb is the Burgers vector of a super-dislocation of pile-up length. With the above equations, it can be shown that the slope (i.e., the strain-hardening rate) is:

$$\theta_{II} = \frac{d\tau}{d\varepsilon} = 3\frac{f}{2\pi}G \tag{4.59}$$

Though given in the literature, $\theta_{II} = \frac{f}{6\pi}G$ is probably erroneous. Equation 4.59 can be derived as follows:

1. Take the derivative of Eq. 4.55 to obtain:

$$dN = \frac{d\tau}{3N^2} \tag{4.55a}$$

2. Substitute for dN into Eq. 4.56 to obtain:

$$d\varepsilon = L_1L_2nb\frac{d\tau}{3N^2} \tag{4.56a}$$

3. Replace L_1L_2 from Eq. 4.57 in Eq. 4.56a to obtain:

$$d\varepsilon = \frac{nb}{RN}\frac{d\tau}{3N^2} \tag{4.56b}$$

4. Replace R from Eq. 4.58 in Eq. 4.56b and express N^3 in terms of Eq. 4.55. Thus, the expression for $d\varepsilon$ becomes:

$$d\varepsilon = \frac{2\pi\tau}{fGN}\frac{d\tau}{3N^2} = \frac{2\pi\tau}{fG}\frac{d\tau}{3N^3} = \frac{2\pi}{3fG}d\tau \tag{4.56c}$$

5. Finally, Eq. 4.59 can be obtained as:

$$\frac{d\tau}{d\varepsilon} = \theta_{II} = 3\frac{fG}{2\pi} \tag{4.59a}$$

The Hirsch's expression is:

$$\theta_{II}/G = k_1/3\pi \tag{4.60}$$

It can be shown that:

$$k_1 = \frac{9f}{2} = 4.5f \tag{4.61}$$

For the derivation of Eq. 4.60, the reader is referred to publications of Hirsch on work hardening.

4.4.3 *Stage III*

The main features of stage III are as follows:

(a) The strain-hardening rate is appreciably smaller than in stage II because cross slip sets in.
(b) The hardening rate decreases because cross slip enables the overcoming of obstacles.
(c) The stress required for stage III is strongly temperature-dependent since cross slip is a thermally activated process.
(d) For cross slip to occur in metals with low stacking-fault energy, higher stresses and strains are required.
(e) The slip lines in stage III are wide and consist of jogged segments and traces of cross slip.
(f) In the stage II jog model, it was indicated that vacancy jogs form strong obstacles hindering glide. However, when such jogs become partially mobile (especially at sufficiently high temperatures), the initiation of cross slip, under the appropriate conditions, may result in a decreased hardening rate.
(g) In the pile-up model, decreased hardening in stage III may be explained by the cross slipping of the screw dislocation component, thus by-passing the obstacles

(Lomer-Cottrell pile-ups). As such, only edge components of loops that cannot cross slip contribute to the hardening process.

Extensive research is still being done on the various stages of strain hardening in order to explain as yet unexplained experimental observations. The desire to find an acceptable, general, unifying theory, a common denominator, which might explain all the aspects of strain hardening, is a major challenge to those doing theoretical analyses of these experimental observations. Considering the many differences in the very fine details requires a broad overview of the strain hardening field and of the research that is still in progress. Discussing all or most of the research publications on this subject is out of the purview of this textbook.

4.5 Strain Hardening in Polycrystalline Materials

Because polycrystalline materials are composed of many grains, the slip system (i.e., slip direction and slip plane) varies from one grain (crystal) to another. Consequently, Schmidt's law does not apply directly to the entire aggregate and, therefore, the CRSS varies from grain to grain. These are major factors which determine the entire behavior of the material. Thus, on application of a stress, the crystal (grain) with the highest CRSS will yield, being the first to deform. Less favorably oriented crystals of the polycrystalline aggregate will yield at a later stage in the deformation process. Therefore, one cannot expect to have well-developed three-stage behavior in polycrystalline materials. Nevertheless, slip within each crystal occurs in the same fashion as in single crystals. The strain in each grain must either match or coincide with the boundaries in order to maintain the integrity of the material. All the factors mentioned above affect polycrystalline materials along with the additional influence of the grains or grain boundaries. Thus, temperature, purity, structure and size all exert their influences on the characteristics of the stress-strain relation. The concept of orientation is relevant only in aggregates, where preferred orientation or texture is present. The accepted term for the stress-strain relation is **'flow stress'** and, like the observation of slip in single crystals, the applied stress depends on the dislocation density (see, for example, Eq. 4.8). It increases with the increase of dislocation density. In comparison with single crystals, polycrystalline materials require higher stresses for deformation. Furthermore, as in single crystals, yielding in polycrystalline BCC materials is temperature-dependent, whereas FCC metals are less sensitive to temperature. Purity and specimen size also influence the stress-strain relation in a manner similar to that discussed in the case of single crystals.

In Chap. 1, the relevant features of a tensile test were discussed, including: yield stress, flow stress, UTS, fracture stress and ductility properties. Tensile properties have been measured over a period of many years in various metals. The behavior of polycrystalline materials under applied tensile stress differs distinctly from the behavior of single crystals of the same quality, when tested under the same conditions. Al is a typical example of a widely-studied polycrystalline FCC metal (see Reed, for example), characterized by the absence of stage I, which is a common feature of polycrystalline materials. According to Conrad, using the original proposal of Seeger, the flow stress, τ, consists of two components, one expressing temperature dependence and the second is an athermal contribution. This may be expressed as:

$$\tau = \tau * (T, \gamma) + \tau_G \tag{4.62}$$

This expression is commonly used in almost in every model of work hardening in single crystals also. The second term is the athermal contribution, when the temperature dependence of the shear modulus, G, is taken into account (i.e., corrected). Recall that τ (namely τ_G) is proportional to G and the dislocation density (see, for example, Eq. 4.8 above). The dependence of flow stress on temperature is often shown as consisting of three stages, shown schematically in Fig. 4.23. The ratio of flow stress is calculated according to the Cottrell-Stokes law. Cottrell and Stokes suggested that the deformation for flow-stress evaluation should be determined first at the high temperature and then, only after unloading, τ_0 at the lower temperature should be determined. This way, any possible change which may have occurred at the high temperature will not affect the characteristics at the lower temperature, where no further changes are expected to occur. Figure 4.24 is a collection of experimental data showing the dependence of flow stress on temperature. Three regions are shown in Figs. 4.23 and 4.24. Region II is almost temperature-independent or only insignificantly sensitive to it. In single crystals, stage II is considered to be almost insensitive to temperature (see, for instance, Sect. 4.2.1.3 and Figs. 4.6 and 4.7). It is assumed that this region is represented by the second term of Eq. 4.62. The third region is believed to be associated with cross slip, but diffusion processes may also be involved in lowering the flow stress. Thermal fluctuations assist in overcoming obstacles. Some researchers assume that these obstacles are jogs, because they are temperature-dependent and, as such, facilitate the constriction of jogs, allowing cross slip to occur, as mentioned above. Regarding region I in the above figures, it is assumed that the strong thermal term in Eq. 4.62 plays an important role in reducing flow stress with temperature. As discussed in Sect. 4.4, the obstacles to dislocation glide may be dislocation pile-ups, forest dislocations and jogs.

Related to the temperature effect, note that Cottrell and Stokes coined the term 'work softening' to describe the results of their observations. They observed that, when unloading a specimen from the deformation to stage II and then reloading it again, an appreciable drop in yield stress takes place. This is the reason that in Figs. 4.23 and 4.24 the flow stress is determined at the high temperature and, thus, a constant defect structure is maintained for the performance of low-temperature measurement. A thermally unstable dislocation structure is thus avoided. For more on work softening, the work of Cottrell-Stokes may be consulted.

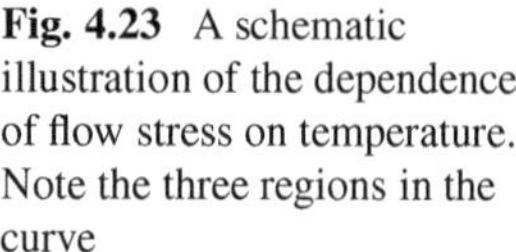

Fig. 4.23 A schematic illustration of the dependence of flow stress on temperature. Note the three regions in the curve

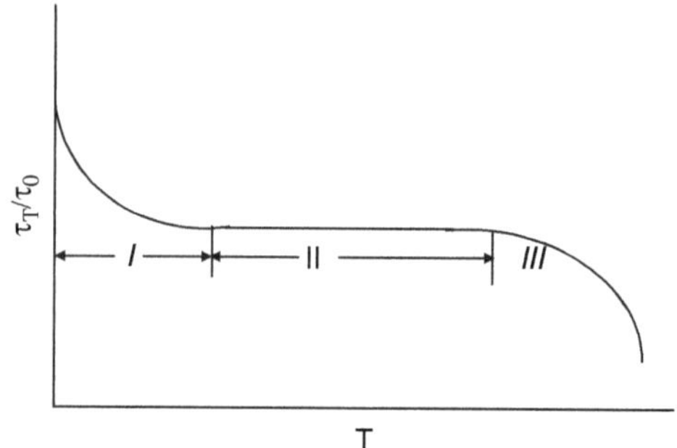

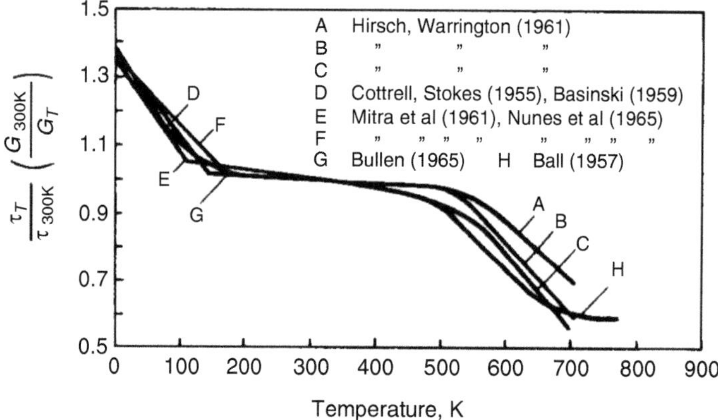

Fig. 4.24 Reversible temperature dependence of flow stress corrected for the temperature dependence of the shear modulus, where τ_T is the shear stress at temperature T, τ_{300K} is the shear stress at 300 K, $G_{300\,K}$ is the shear modulus at 300 K and G_T is the shear modulus at temperature T. Cottrell and Stokes, Hirsch and Warrington, Basinski, and Mitra et al. results contain single crystal measurements (From Read (1972). With permission of Elsevier Ltd.)

4.6 Solid Solution Strengthening

4.6.1 Introduction

Solid solutions can be formed by interstitial atoms or substitution atoms and, thus, one can speak about 'interstitial solid-solution strengthening' (hardening) or strengthening by substitution atoms. Recall that interstitial solutions are located in the voids formed between the matrix atoms and, for this section, it will be assumed that they are distributed randomly at the appropriate sites. The most familiar interstitial solid solution is C in iron, either in tetrahedral or octahedral voids (in γ iron). By geometrical necessity, these kinds of atoms must be small in their dimensions. It is usual to express interstitial solutions by a ratio of the radii between the largest void of the matrix atom to that of the matrix atom itself. The ratio in FCC and HCP is 0.41 and in BCC, it is 0.29. These ratios place a limit on the atoms that can be dissolved in these structures. H and B can occupy tetrahedral sites, whereas N, C and O usually settle in the largest octahedral voids in FCC structures.

The easiest way to form substitution solid solutions is by replacing the matrix atoms with others having the same radii. This may enable the replacement of a large percentage of the matrix atoms in order to form a desirable alloy. If the replacement atoms differ in size from the matrix atoms, then the known Hume-Rothery rules determine the degree of solubility, as based on experimental observations.

4.6.2 Strengthening by Interstitial Atoms

The general purpose of alloying is to impede dislocation motion and, thus, make materials stronger and harder. Sometimes interstitial atoms are in the matrix unintentionally. The formation of BCC metals is a very sensitive process, since it is rather difficult to avoid the incorporation of some interstitials during their production. As indicated above, the classic example is the alloying of iron with C atoms. C atoms have a strain field which reacts with the strain fields associated with dislocations, causing strengthening. The accepted mechanism is pinning the dislocations by forming an atmosphere (as discussed in Sect. 3.6.5 above and illustrated in Figs. 3.21 and 3.22). The unpinning of dislocations requires critical stress (previously discussed Chap. 3) as given in Eq. 3.30 and rewritten as:

$$\sigma_{crit} = \frac{0.65}{{r_{\max}}^2 b^2} \tag{3.30}$$

After their unpinning, dislocations can glide. For details on the Cottrell-Bilby mechanism, consult Chap. 3, Sect. 3.6.5. Note that interstitial atoms with tetragonal symmetries can react with the strain fields of both edge and screw dislocations. Dissolving C, N or an alloying element in αFe or γFe introduces lattice distortion with the consequent deviation of atoms from their equilibrium positions and elastic distortion sets in. As the result of elastic interaction, solid-solution strengthening is produced, requiring higher stress for yielding.

4.6.3 Strengthening by Substitution Atoms

Atoms usually diffuse to sites in the crystal where they can reduce the strain of dislocations. Contrary to interstitial atoms, solid-solution atoms are located under the edge dislocation to reduce the compressive strains. An atom of larger size than a matrix atom was shown in Fig. 3.21d and is reproduced here as Fig. 4.25. Either Cu or Al can serve as an example of strengthening by solute atoms in FCC metals.

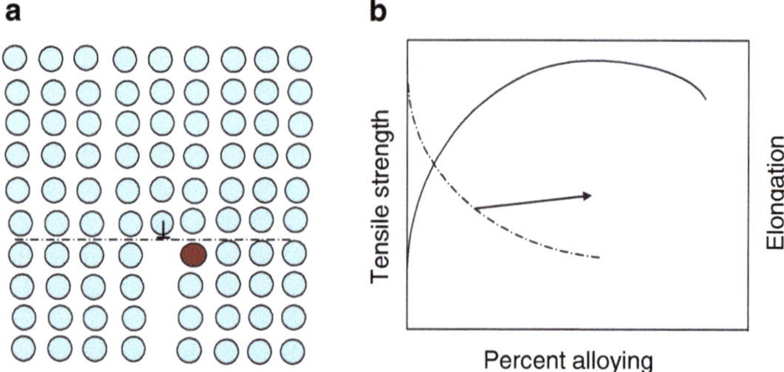

Fig. 4.25 (**a**) The location of a substitution solute atom under the dislocation and in its vicinity and (**b**) a typical tensile stress-strain relation. The strength increase is accompanied by a decrease in elongation

The relatively soft Cu, when alloyed with Sn (to form a bronze) or with Zn (to get a brass), becomes much stronger. Similarly, one can form various aluminum bronzes by alloying with Mn, Ni or Si. With increased strength by alloying, elongation decreases as expected. This is shown schematically in 4.25b.

Interstitial or substitution atoms produce stress fields around themselves, which interact with the stress fields of the dislocations and anchor them. Pure metals are generally softer than their alloys. The solutes modify the stress-strain relation by affecting the frictional resistance to dislocation motion by hindering them in their glide, unless higher stress is applied. Two main causes should be noted for the strengthening effect of solutes: (a) the size difference between parent and solute atoms and (b) their concentration. Solute distortion can be tetragonal or spherical. Distortion by interstitial atoms is non-spherical, whereas, by substitution solutes, distortion is spherically symmetrical. The significant strengthening by interstitial atoms is a consequence of their ability to interact with edge, screw and mixed dislocations, due to their non-spherical distortion (C in iron). Spherical distortion has no shear-stress component; therefore, substitution solute atoms do not interact with the shear-stress field characteristics of screw dislocations.

Of the many possible ways in which solutes may interact with dislocations, particular attention should be paid to their effect via stacking faults. The segregation of solutes in various areas, e.g., in stacking faults, is a result of their energy reduction. With the reduction of stacking-fault energy, the partials become even farther apart, causing a need for increased stress to continued deformation. Recall that cross slip is a function of the width of the stacking fault.

Many theories have been proposed by Fleischer, Friedel and others to relate flow (or shear) and yield stress to the concentration of the solute. They relate shear stress to concentration, c, by a parabolic function and to strain, ε, with an exponent of 1.5. More specifically, it is given as:

$$\Delta\tau = Gb\varepsilon^{3/2}c^{1/2} \tag{4.63}$$

Another relation was suggested by Arsenault et al.:

$$\tau = KG\varepsilon^{4/3}c^{1/2} \tag{4.64}$$

K is a constant equal to 0.0694. For details on this relation and on the effect of solute strengthening, Arsenault's paper may be consulted. Increasing applied stress to move a dislocation clearly increases the yield strength of the material, since plastic deformation sets in only after yielding. In both Eqs. 4.63 and 4.64, the shear modulus appears; this means that, in order to achieve significant strengthening by solid-solution strengthening, one should alloy with solutes having a higher shear modulus. In addition to the concentration of the solute and the shear modulus, its size is of considerable interest, since the strain developed is proportional to the size difference between the solute and the matrix atoms.

4.7 Grain Boundaries and Grain Size

Single crystals, to some extent, represent the ideal case for materials, since they enable studying the properties without the influence of grain boundaries. Although extensive research has been done both on single crystals and on grain boundaries, predictions of the mechanical behavior of polycrystalline materials are still incomplete. Grain boundaries play an important role in strengthening polycrystalline materials, thus their influence must be understood and taken into account. This is not a simple problem and so a major portion of the theories are based on experimental observations.

Grain boundaries introduce discontinuities in materials and, therefore, act as obstacles to dislocation motion by hindering the continuity of slip from grain to grain. For strengthening by grain boundaries, these obstacles are expected to be strong and able to sustain pile-ups generated and emitted by sources. If (a) the transfer of dislocations through boundaries to continue gliding on the same slip plane is improbable; (b) maintaining a whole aggregate together is a prerequisite and; (c) grain boundary sliding is to be eliminated (creep), then the response of a material is to generate new dislocations in other neighboring grains. In Sect. 3.10 (Fig. 3.29), slip is shown schematically in two neighboring grains. Dislocations nucleate at the grain boundary, which acts as a source for generating dislocations in the new glide plane. During the application of stress, each grain has to undergo deformation under the constraints imposed by the boundary and the limitations due to the changing orientation of the grains. The influence of grain boundaries manifests itself in several ways as indicated below:

(a) The boundaries induce higher initial hardening, since glide cannot proceed across a boundary from grain to grain without perturbation.
(b) Slip is complex and, in each grain, must involve multiple slip to allow each grain to adjustment appropriately to its individual slip, so that coherency is maintained with its neighbors.

(c) The constraints imposed by the boundaries cause slip to be irregular, unlike the deformation in single crystals.
(d) Grain size is a significant factor, because the number of boundaries is determined by grain size.
(e) In general, the yield stress is higher than in single crystals, because of the resistance of the boundaries and the complexity of slip in each grain.
(f) Usually, the strain hardening is larger (the material is stronger), because grain boundaries are obstacles to dislocation motion.
(g) The stress-strain relation is affected by the presence of boundaries and stage I is suppressed, since slip occurs by multiple slip from the beginning.

It is expected that the influence of grain boundaries will be higher in HCP crystals, because there is usually one slip plane, provided that only basal slip occurs and no twinning sets in at the higher stresses required for deformation. This also means that no consideration is given to the possibility that the higher stress required for deformation may induce slip in other systems, e.g. prismatic or pyramidal. In other structures, such as FCC or BCC, which have many slip systems, almost every grain is likely to be in appropriate orientation for slip under the influence of applied stress. Thus, deformation can proceed at little expense in neighboring grains, as well. Contrary to this, in HCP, only some grains are in the appropriate orientation for slip, as indicated earlier.

One of the theories describing strain hardening in polycrystalline materials considers the dislocation pile-ups formed at grain boundaries. Yielding occurs first in preferentially-oriented grains, where CRSS is reached. Pile-ups trigger yielding in neighboring grains. Their size is related to yielding by the Hall-Petch relation (discussed above in Sect. 3.10 and given in Eq. 3.42), reproduced here as:

$$\sigma_y = \sigma_0 + \frac{k_y}{\sqrt{d}} \tag{3.42}$$

σ_0 represents the resistance to dislocation glide and, often, it is termed 'friction stress'. A measure of the degree of pile-up behind an obstacle (i.e., the grain boundary) is k_y and d is the size of the grain. Another way of looking at k_y is by considering it to be the resistance of the grain boundaries to dislocation slip across the boundaries. As mentioned in Sect. 3.10, k_y can be determined from deformation experiments with varying grain sizes. The relation of the yield stress to the grain size (actually vs. $d^{-1/2}$) is linear, k_y is the Hall-Petch slope (see Fig. 3.30) and σ_0 is determined from the intercept of the line with the σ_y axis.

Details of this process may be envisaged as follows. On application of a tensile stress, the resolved shear stress acts on some sources in favorably-oriented grains (note that the boundaries themselves are sources of dislocation) and, on reaching the yield stress, dislocations pile up behind the boundaries, acting as obstacles. At a sufficiently large stress, other sources in unfavorably-oriented grains start to operate, emitting additional dislocations. Then, general yielding of the entire specimen will commence. Grain boundaries, in general, are considered as barriers limiting the free path of dislocations and, thereby, increasing strain hardening. Thus, the Hall-Petch model basically indicates the stress required to activate dislocation sources in neighboring grains. For a review of the hardening mechanism at grain boundaries, George's work may be consulted.

In Eq. 3.42, stress consists of two terms: the first is independent of grain size and the second is grain-size dependent. The first term at a certain strain, ε, is associated with flow stress in the interior of a grain, while $\frac{k_y}{\sqrt{d}}$ represents the contribution to the strength, as a consequence of the grain-boundary resistance to dislocation movement into another grain (Hansen). The contribution of the grain interior, $\sigma_0(\varepsilon)$, is related to the density of the dislocations accumulated at dislocation boundaries. This is detailed in Sect. 2.15 on dislocation pile-ups, especially Eqs. 2.59b, 2.59c and 2.60, applying them in terms of grain boundaries acting as obstacles to dislocation motion. By rewriting Eq. 2.59b, the relation between τ_e and τ_p is obtained:

$$n\tau_e = \tau_p \tag{2.59b}$$

τ_e represents an effective stress and τ_p is the stress assumed to exist at the head of the pile-up acting at the boundary. τ_p is n times larger than the effective stress and n is the number of dislocations in the pile-up. The number of dislocations given as Eq. 2.60, was calculated by Eshelby, Frank & Nabarro in terms of the effective stress as:

$$n = \frac{\pi L \tau_e k}{Gb} \tag{2.60}$$

where $k = 1$ for screw and $(1 - \nu)$ for edge dislocations, respectively. In the above, the distance L may be considered as a giant dislocation with Burgers vector nb in a pile-up of n dislocations (i.e., it is the length of the pile-up). L cannot be larger than the grain size and its value is usually taken to be d/2. The effective stress, in terms of the applied stress, τ_a, and all the contributions resisting dislocation motion, such as back stress and friction stress, τ_i, can be expressed (see Eq. 2.59c) as:

$$\tau_e = (\tau_a - \tau_i) \tag{4.65}$$

At the critical shear stress of the applied stress, yielding occurs ($=\tau_y$) and dislocations are nucleated at the head of the pile-up for slip in a neighboring grain (across the grain boundary). At this stage, Eq. 2.59b may be written as:

$$n\tau_e = \tau_p = \tau_c \tag{4.66}$$

Substituting Eqs. 2.60, 4.65 into Eq. 4.66 results in:

$$\tau_c = \frac{k\pi d \, \tau_e{}^2}{2Gb} \tag{4.67}$$

Expressing τ_e from Eq. 4.67, one obtains:

$$\tau_e = \left(\frac{2Gb\tau_c}{\pi kd}\right)^{1/2} = \left(\frac{2Gb\tau_c}{k\pi}\right)^{1/2} d^{-1/2} \tag{4.68}$$

From Eq. 4.65, when $\tau_a = \tau_y$ as indicated, and based on Eq. 4.68, one gets:

$$\tau_e = \tau_y - \tau_i = \left(\frac{2Gb\tau_c}{k\pi}\right)^{1/2} d^{-1.2} \tag{4.69}$$

or:

$$\tau_a = \tau_y = \left(\frac{2Gb\tau_c}{k\pi}\right)^{1/2} d^{-1.2} + \tau_i \tag{4.70}$$

This equation is similar to the Hall-Petch relation and, in order to make it equivalent in terms of the yield stress, σ_y, under tension, τ_y is multiplied by the Taylor factor, M, to obtain Eq. 3.42 as:

$$\sigma_y = \sigma_0 + \frac{k_y}{\sqrt{d}} \tag{3.42}$$

k_y is:

$$k_y = M\left(\frac{2Gb\tau_c}{k\pi}\right)^{1/2} \tag{4.71}$$

M is ~3.1 in FCC metals. For further such calculations, consult the work of Rollett et al.

The Hall-Petch model is widely used up to the present day and, with appropriate modification of Eq. 3.42, its applicability to various metals has been amply demonstrated. This flow stress model of deformed metals is applicable as long as the strength contributed by boundaries is introduced as a variable parameter and not as a constant (as in Eq. 3.42 of the Hall-Petch relation). Hansen has taken this into account. The Hall-Petch flow stress model is equivalent to Eq. 3.42 and is given as:

$$\sigma(\varepsilon) = \sigma_0(\varepsilon) + \frac{k(\varepsilon)}{\sqrt{d}} \tag{4.72}$$

However, experimental observations on several polycrystalline materials indicate that not at all conditions σ_0 and k are constants and they are strain dependent, as shown in Fig. 4.26.

However, the relation between stress and (grain size)$^{-1/2}$ at a certain strains remains constant, as required by the function. There are indications that this is not always the case, as shown in Fig. 4.27 for stainless steel, type 316 L (Kashyap and Tangri).

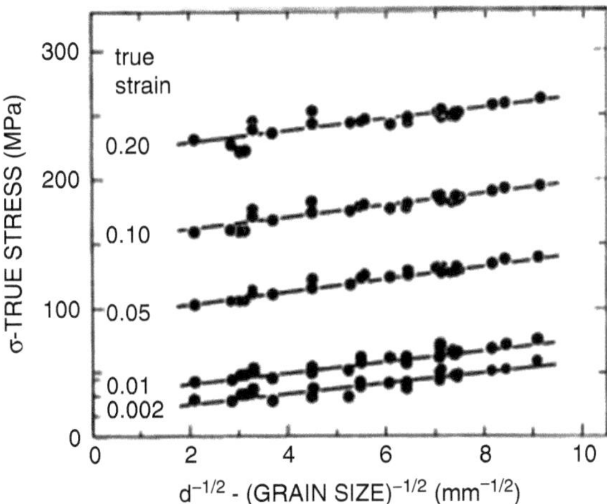

Fig. 4.26 The variation of the flow stress with (grain size)$^{-1/2}$ at various strains (Niel Hansen (2005). Copyright Wiley-VCH Verlag GmbH & Co. KGaA. Reproduced with permission of Wiley-VCH and the author)

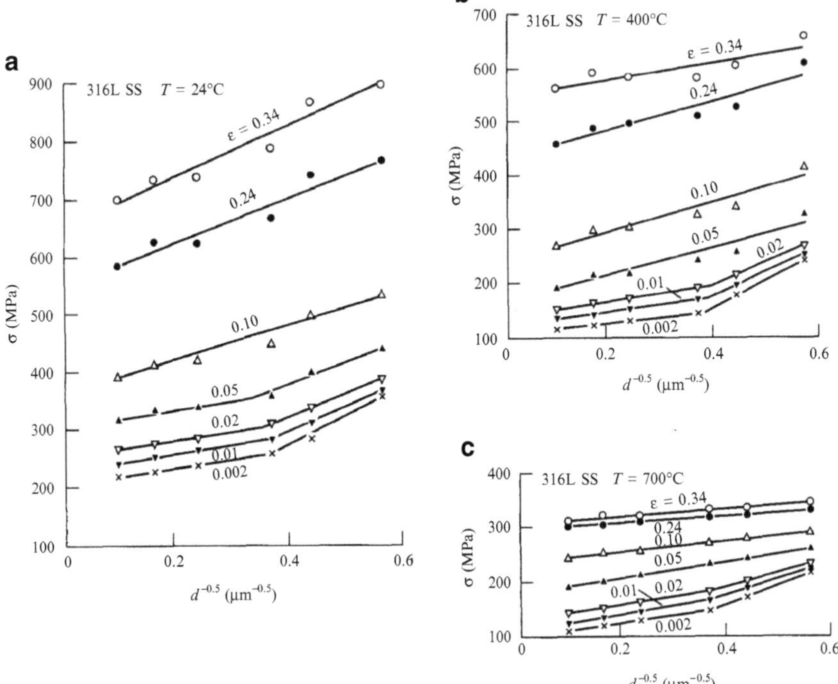

Fig. 4.27 Hall- Petch plots at the strains indicated at three temperatures. Note that below a certain strain level the Hall Petch relation is not represented by a single linear relation (Kashyap and Tangri (1995). With permission of the authors and Elsevier Ltd.)

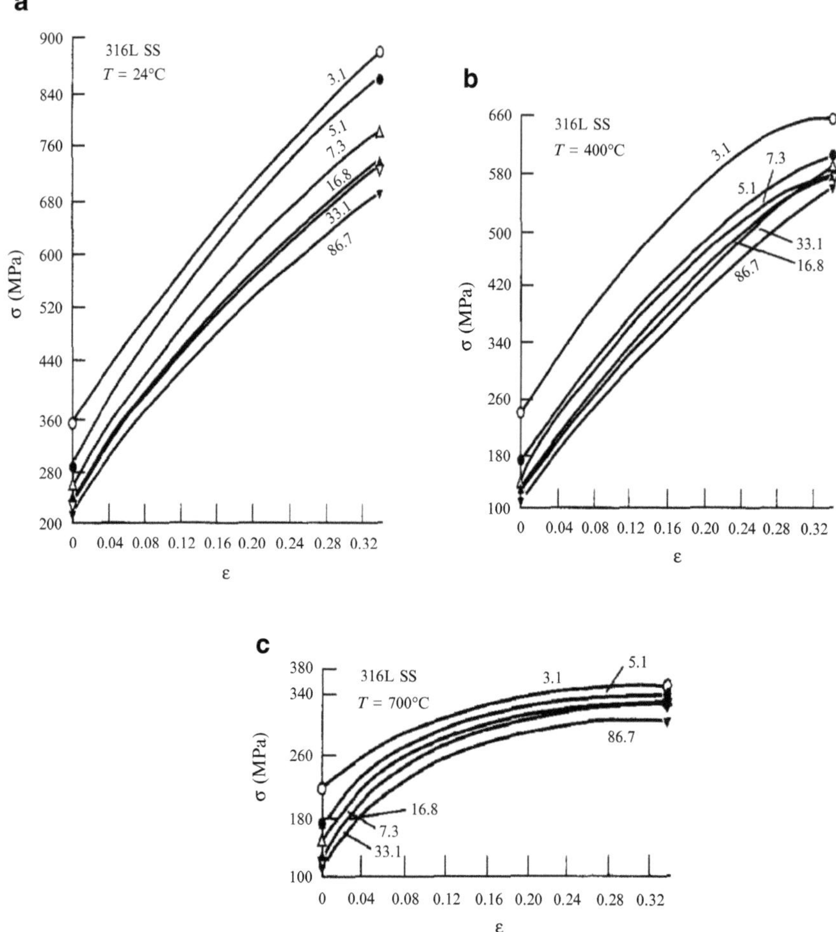

Fig. 4.28 Stress vs. strain curves at various grain sizes at the temperatures shown. The character of the $\sigma - \varepsilon$ relation depends on grain size and temperature (Kashyap and Tangri (1995). With permission of the authors and Elsevier Ltd.)

This figure shows a break in the Hall-Petch relation at three temperatures at small strains, which is attributed to the development of a substructure during deformation. The data from Fig. 4.27 are based on $\sigma - \varepsilon$ curves for three temperatures and various grains sizes, as seen in Fig. 4.28.

The flow stress is seen to be higher for smaller grain sizes. The difference in flow stress, between small- and large-grained specimens, generally increases with strain at room temperature and at 673 K. At the high temperature, there is a leveling off of the flow stress, but the effect of grain size remains similar, though less pronounced. The characteristics of the high temperature $\sigma - \varepsilon$ relation are thought to be associated with sub-grain boundary formation. It is also assumed that, at such high temperatures, recovery processes may set in and that the cross slip of screw dislocations may contribute significantly.

The main disagreement regarding applications of the Hall-Petch model is whether a strengthened layer in front of the boundary is responsible for the boundary acting as an obstacle to dislocation slip or whether the grain size and its orientation cause the main resistance. In other situations, the transfer of dislocations occurs indirectly, due to the activation of dislocation sources near boundaries under stress field of incoming dislocations. More on the grain-boundary contribution to hardening in polycrystalline materials and on the fine details of their characteristics may be found in the literature, for example, in Kumar et al., Caillard et al., Hansen, Qiao et al. and Bargmann et al.

In further consideration of dislocation pile-up models, Li considered forest dislocations, through which dislocations must move at sufficiently high stress when yielding sets in. Forest dislocations are generated by dislocation sources near grain boundaries. The suggested relation is similar to Eq. 3.42, but starts with a dislocation-density function which is inversely proportional, in many cases, to grain size. The two forms expressing Hall-Petch-like behavior, in terms of dislocation density, are:

$$\tau_c = \tau_i + \alpha G b \sqrt{\rho} \tag{4.73}$$

and:

$$\sigma_y = \sigma_0 + \alpha G b \rho^{1/2} \tag{4.73a}$$

In Eq. 4.73, α is a factor in the range 0.1–0.3 and the CRSS to move a dislocation through a distribution of dislocations varies with the dislocation density, ρ, as:

$$\rho_y = (ky/\alpha G b)^2 (1/d) \tag{4.73b}$$

The applicability of this Hall-Petch-like relation may be appreciated for the many uses of this function (by attributing the appropriate parameters for the model being used). Thus, for instance, an equation of Hall-Petch form is also often useful for predicting the ductile-brittle transition temperature:

$$T_B = T_0 - K_B d^{-1/2} \tag{4.73c}$$

Another feature of such equations is that they were also found to be applicable to the deformation of materials with very small grain size. However, there are indications (see, for example, Caillard and Couret) that, in some cases in which mobile dislocations encounter boundary structures composed of cell walls (see Figs. 2.9, 2.10, 2.11, 2.12), these Hall-Petch-like equations break down. Where heavily deformed materials are concerned (such as deformed Al), the relation is given by $\tau \propto \frac{1}{d}$, rather than $\tau \propto \frac{1}{d^{1/2}}$. This provides a parabolic, rather than a linear, relation. In general, heavily deformed materials contain both cell walls and high-angle boundaries that should induce an average variation, intermediate between $\tau \propto \frac{1}{d}$ and $\tau \propto \frac{1}{d^{1/2}}$. The difference between the Hall-Petch laws arises from the fact that the strength of cell walls, τ_B, depends on the pre-straining stress, instead of being constant. τ_B is the strength of the cell walls.

Due to the absence of stage I in polycrystalline materials undergoing deformation and the fact that hardening resembles stages II and III, one has to assume that strengthening by boundaries must involve complex slip, i.e., multiple slip as it previously occurs at the beginning of the stage II hardening of single crystals. In other words, the complexity of strain hardening and the incompatibility of grains at the boundaries necessitate the operation of several slip systems simultaneously in each grain. Cross slip is another expected feature of hardening in polycrystalline materials.

4.8 Second-Phase Hardening (Precipitates and/or Other Particles)

4.8.1 Introduction

The strengthening of a material is a consequence of dislocation interaction with an obstacle. The details of this strengthening mechanism may differ, depending on the characteristics of the obstacles, but the essence of the process is the same, namely resistance of the obstacle to dislocation motion. In the previous sections, the obstacles considered were grain boundaries and solid-solution hardening. In this section, there is discussion on second-phase obstacles, which include: precipitates (obtained by precipitation or age hardening); non-soluble particles, such as various oxides, nitrides, borides and carbides and; stable and non-soluble impurities. One may also add vacancies to this list, since they can be considered as particles of atomic dimensions without the characteristic mass. However, the effect of vacancies on hardening will not be considered separately, but rather in general terms, together with the precipitates or particles. The methods for forming and distributing particles, such as precipitation hardening, dispersion hardening or other well established metallurgical techniques, are not discussed here, since they are well-known and covered in books on physical metallurgy. However, it should be emphasized that second-phase particles (precipitates) are added to materials to toughen them and make them stronger by impeding dislocation motion, since pure metals often tend to be quite soft.

The main factors to be dealt with to some degree are the size, shape and distribution of these particles. In the following, it will be assumed that the shape of the particles is spherical. They may be soft or hard and, accordingly, dislocation lines may cut through them or may curve around each particle forming a loop (known as 'Orowan looping') during an obstacle-dislocation reaction. These particles are characterized in terms of size, expressed for spherical particles by their radius, r. These two concepts of reactions, among others, are discussed below.

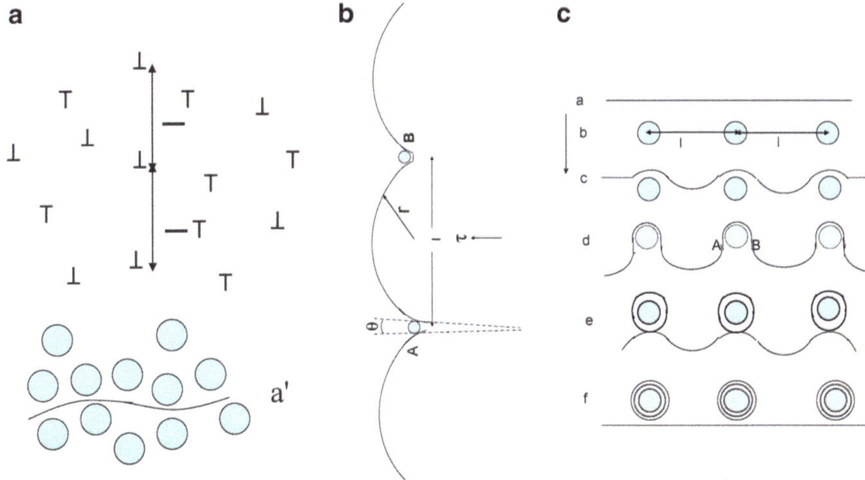

Fig. 4.29 A dislocation advances in the direction of the *arrow* in its slip plane, encountering obstacles in the form of second-phase particles (**c**). The distance between the particles is *l*. In this schematic, the various stages of the dislocation passing between widely-spaced precipitates (*particles*) is indicated: In the figure: (**a**) an array of dislocations with average spacing *l* between them; (**a′**) a schematic illustration of a dislocation curling around the stress field of precipitates; (**b**) dislocations curving under applied stress, τ, forming segments with radius, *r*; and (**c**) Orowan looping around particles

4.8.2 Orowan Loop Formation

The well-known schematic representation of a dislocation line curving round particles and forming Orowan loops is illustrated in Fig. 4.29c. Figure 4.30 shows a TEM micrograph of Orowan loops around alumina particles in a single crystal of Cu-30wt.% Zn deformed 6% in stage I at 77 K. According to the Orowan mechanism, the following stages are observed in Fig. 4.29c:

(a) The dislocation line encounters obstacles in the form of particles.
(b) The dislocation starts to curl around the particles, assumed to be at a distance l from each other.
(c) When the dislocation bulge becomes semi-circular (as shown in Fig. 2.40b), yielding starts, since the stress has increased beyond a critical value. The dislocation becomes unstable and tends to expand (see Sect. 2.14 above).
(d) At the next stage, the advancing dislocation must expand further, curving to form a loop around the particles. In Fig. 4.29c at stage d, the dislocations at points A and B have opposite signs and cancel each other out by forming a loop around the particles at the farthest expansion.

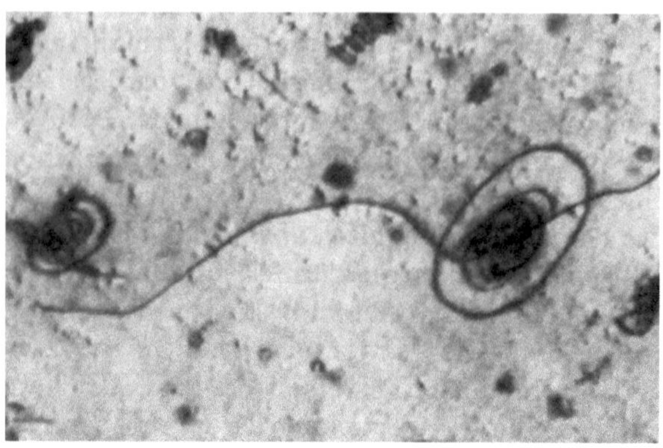

Fig. 4.30 Orowan loop around an alumina particle in a single crystal of Cu-30 wt.% Zn (Humphreys and Hirsch 1975. Reproduced with the permission of the publisher and professor Hirsch)

(e) Orowan loops are formed around the particles and the dislocation resumes movement during the deformation (Fig. 4.29c, stage e). Stage f shows that dislocations loops are re-formed around the same obstacles when they are encountered again. Thus, successive dislocation rings form around the particles and these dislocation rings repel each other, since they have the same sign, which contributes additional hardening to the obstacles.

The Orowan loops make the dislocation motion more difficult, resulting in more hardening of the material. Dislocation energy increases when a straight segment of the dislocation line is curved, due to an increase in its length, as illustrated in Fig. 2.32.

One can write some of the expressions as follows:

$$T = \alpha G b^2 \tag{2.27}$$

The force in the y direction on the dislocation line is balanced by the line tension, which tends to shorten the dislocation line (as shown in Fig. 2.32):

$$\tau b l = 2T \sin\frac{\theta}{2} \tag{2.28}$$

with the substitution for T from Eq. 2.27 and, when $\sin\frac{\theta}{2}$ is 1 (one), the stress is at its maximum or:

$$\tau_{\max} = \tau_0 = 2\frac{\alpha\, Gb}{l} \tag{4.74}$$

and l is the distance between the particles. Considering Sect. 2.8, Eq. 4.74 may also be expressed in terms of the radius of the curved dislocation as:

$$\tau = 2\frac{\alpha G b}{r} \tag{4.74a}$$

This expression shows that applied stress is inversely proportional to the radius of curvature of the dislocation and that the radius in equilibrium is:

$$r = 2\frac{\alpha G b}{\tau} \tag{4.74b}$$

It is clear from Eq. 4.74 that, if the distance between the particles increases, the applied stress required to force a dislocation past the obstacles decreases and vice versa. Since the size of particles is proportional to the distance between them, there is a limit to effective particle size as a strengthening medium. In order to optimize the strengthening of particles (precipitates in age-hardened material), their size and spacing must be controlled. Thus, in general, there is an optimum dispersion of particles that are not too big and not too far apart. If the second phase is dispersed to such an extent that the spacing between the particles is small, the dislocation is unable to bend, but when the distance between particles is sufficiently large, the dislocation can move between them, as shown in Fig. 4.29a′.

4.8.3 The Strength of Obstacles and Break-Away Stress

In Fig. 4.31, precipitates are shown with interacting dislocations. The central region reproduces Fig. 2.32, showing the dislocation curving when blocked in its movement by obstacles (particles). The acting force is balanced by the line tension, T, of the dislocation. The right hand side and the central region of this schematic illustrate the critical break-away stress angle, ϕ, at which the force, τbl, acting on the dislocation, overcomes the obstacle's opposing strength of the break-away . This process is depicted below:

Each obstacle resists the forward force, τbl, acting on the dislocation segment (shown in the central part of the figure). The resistance exerted by each particle at some point on the dislocation segment is given by the force, P, (in Chap. 2, force is denoted by F, i.e., P = F), which may be expressed in terms of the angle ϕ as

$$P = 2T\cos\frac{\phi}{2} = 2\alpha G b^2 \cos\frac{\phi}{2} \tag{4.75}$$

In Eq. 4.75, T was substituted from Eq. 2.27, namely $T = \alpha G b^2$, to obtain the right side term of Eq. 4.75. The angle, ϕ, is obtained between the dislocations, which are bent to a certain degree. At a critical value of the angle, ϕ_c, the dislocations break through the obstacles, as suggested by Brown and Ham. This occurs at the force which overcomes the resisting force, P_c, of the obstacles. The obstacle strength, τ_{obs}, is obtained by the condition:

$$\tau_{obs} bl = P_c \tag{4.76}$$

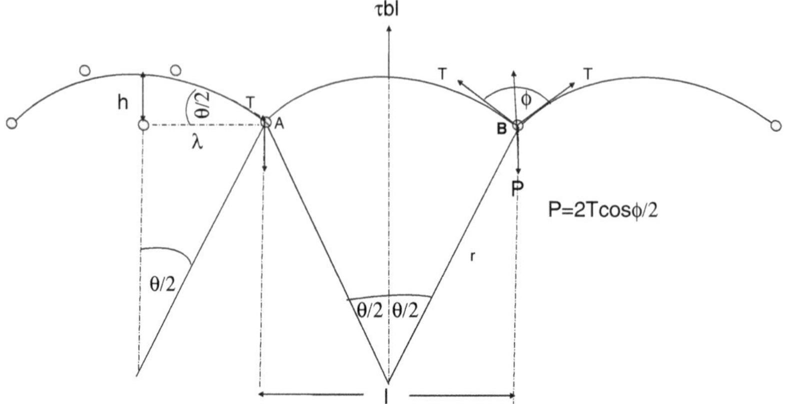

Fig. 4.31 The stress required to overcome obstacles depends on the actual spacing l (some average value) between them and the angle ϕ

$$\tau_{obs} = \frac{P_c}{bl} = \frac{2\alpha Gb}{l}\cos\frac{\phi_c}{2} \tag{4.77}$$

For strong obstacles, $\phi = 0$ and $\cos\phi_c/2 = 1$, giving:

$$\tau_{obs} = \frac{2\alpha Gb}{l} \approx \frac{Gb}{l} \tag{4.77a}$$

and for weak obstacles, $\phi \rightarrow 180°$.

Taking another approach, one may consider the left side of Fig. 4.31. It is possible to derive the following relations for the critical cutting stress by assuming that the particle was cut through by the dislocation, which has moved forward (curving) a distance, h. Then, one can write approximately for the area swept out:

$$\text{area swept out} \sim h\lambda \tag{4.78}$$

λ is the mean distance between the two particles along the line shown in Fig. 4.31. If l is the average inter-particle distance, and it is assumed that the area swept out, hλ, is about equal to an area $(l/2)^2$ defined by l (the average distance between the particles), then it is possible to write:

$$h\lambda \approx (l/2)^2 \tag{4.78a}$$

From Fig. 4.31

$$tg\frac{\theta}{2} = \frac{h}{\lambda} \tag{4.79}$$

For small angles, tg(θ/2) ~ sin(θ/2) ~ θ/2 ~ h/λ, and from Eq. 4.78a:

$$\sin(\theta/2) \sim \theta/2 \sim h/\lambda \sim \left(\frac{l/2}{\lambda}\right)^2 \tag{4.79a}$$

In Eq. 2.28, the line tension is related to the force and, for the critical force, P_c, to cut through a particle, we can rewrite:

$$P_c = 2T \sin\frac{\theta}{2} \tag{4.80}$$

from Eq. 4.79a:

$$P_c = 2T\left(\frac{l/2}{\lambda}\right)^2 \tag{4.80a}$$

and:

$$\lambda = \frac{l}{2}\left(\frac{2T}{P_c}\right)^{1/2} \tag{4.80b}$$

In Eq. 2.28, l must be replaced by the mean distance between the particles, λ, and then it is possible to rewrite the expression for the critical force balanced by the tension in the two particles as:

$$P_c = \tau_c b\lambda = 2T \sin\frac{\theta}{2} \tag{2.28}$$

or:

$$\tau_c = \left(\frac{P_c}{b\lambda}\right) \tag{2.28a}$$

and with Eq. 4.80b:

$$\tau_c = \left(\frac{P_c}{b}\right)\frac{2}{l}\left(\frac{P_c}{2T}\right)^{1/2} \tag{2.28b}$$

In Eq. 2.26, T ≈ Gb^2 and, with this value of T, Eq. 2.28b may be expressed as:

$$\tau_c = \frac{2^{1/2}P_c^{\,3/2}}{b^2 l G^{1/2}} = \frac{P_c^{\,3/2}}{b^2 l}\left(\frac{2}{G}\right)^{1/2} \tag{4.81}$$

This is one of the expressions for particle-cutting stress. This equation differs from the ones describing the occurrence of looping (Eqs. 4.74 and 4.74a). The dislocation segment breaks away from the obstacles when this force overcomes the maximum resisting force of the obstacles, P_c. The above may also be expressed in terms of $\cos(\phi_c/2)$ as:

$$\tau_c(\text{strong obstacle}) = \frac{Gb}{l}\cos\frac{\phi_c}{2} \tag{4.82}$$

$$\tau_c(\text{weak obstacle}) = \frac{Gb}{l}\left(\cos\frac{\phi_c}{2}\right)^{3/2} \tag{4.83}$$

Equation 4.83 can be obtained by substituting the value of Eq. 4.77 back into Eq. 4.81 as:

$$P_c = 2\alpha Gb^2 \cos\frac{\phi_c}{2} \cong 2Gb^2 \cos\frac{\phi_c}{2} \tag{4.71a}$$

and then Eq. 4.82 is clearly equivalent to Eq. 4.77 as:

$$\tau_{obs} = \tau_c == \frac{2\alpha Gb}{l}\cos\frac{\phi_c}{2} \cong \frac{Gb}{l}\cos\frac{\phi_c}{2} \tag{4.71b}$$

When the break-away angle, ϕ, is 0, the particle behaves like an impenetrable obstacle; when $\phi > 0$, the particle can be sheared by a dislocation with shearing force, P.

4.8.4 Cutting Through the Second Phase

In Fig. 4.32, a dislocation is shown cutting through a particle. As a consequence, new surfaces are formed in the deformed particle and the distance of displacement along the slip plane is one b vector. Cutting through particles by means of dislocations is an option when the particles are small (or soft), producing a deformation in the particle itself by gliding through it. Of the several interaction mechanisms involving dislocations and particles that cause particle cutting, here, consideration is given to the case in which a dislocation and a particle are relatively close to each other, in the vicinity of ~10b or less.

In earlier Sect. 2.10.1 (see also Eq. 2.42), the force per unit length for glide in the glide plane was given as:

$$F = \tau b \tag{2.38}$$

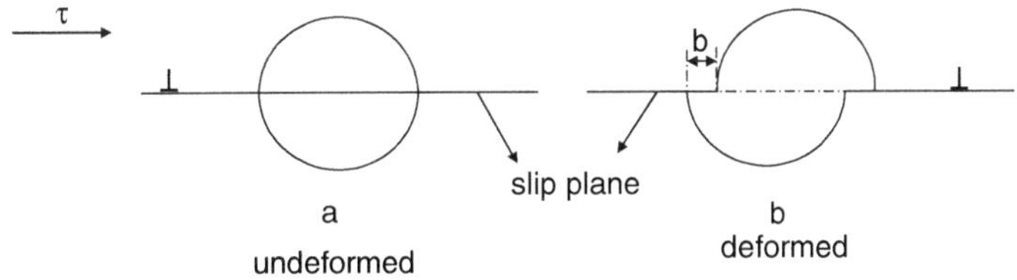

Fig. 4.32 Particle cutting by the slip of a moving dislocation

The force required in the presence of particles, compared to that in Eq. 2.38, is different and may be expressed as $\Delta\tau b$. If the radius of the particle is greater than b, i.e., $r > b$, it is possible to give the area of the cut particle as πr^2 and express $\Delta\tau b$ by:

$$\Delta\tau b = n\pi r^2\gamma \tag{4.83a}$$

n is the number of particles per area and γ is the surface energy per area of the particle. The number of particles, n, is given by:

$$n = \frac{3f}{2\pi r^2} \tag{4.84}$$

f is the volume fraction of particles. Substituting Eq. 4.84 in Eq. 4.83a, one obtains:

$$\Delta\tau = \frac{3f}{2b}\gamma \tag{4.85}$$

This expression gives the increment of hardening. Furthermore, Gleiter and Hornbogen provided the expression:

$$\Delta\tau = 0.28\frac{f^{1/3}\gamma^{3/2}}{G^{1/2}b^2}r^{1/2} \tag{4.86}$$

4.8.5 *The Mott-Nabarro Concept*

One cannot finish this section without mentioning the original concept of second-phase hardening in the pioneering ideas of Mott and Nabarro. Strengthening by solute atoms and particles is usually based on their theories. The exact nature of their theory is not discussed here. However, they suggested a misfit model to explain hardening in precipitation-hardened materials, expressing this as:

$$\Delta\tau = \mu\varepsilon' f = G\varepsilon' f \tag{4.87}$$

μ is equivalent to G in our notation. $\Delta\tau$ is the shear stress increase, due to the interaction between dislocations and particles. Critical shear stress increases linearly with f, the volume fraction of precipitate. ε is the misfit between the matrix and the precipitate (in their work, it is the misfit between the solute and the matrix atoms). In terms of the applied stress, σ, their relation may be given as:

$$\sigma = 2\mu\varepsilon c \tag{4.88}$$

where c is the solute concentration.

Note that critical shear stress increases linearly with the volume fraction of the precipitate or the solute concentration Eq. 4.88. The misfit gives rise to an internal stress field, which resists dislocation movement. This relation does not take particle size into account, thus, the distance between particles is ignored. At a later stage, Mott and Nabarro improved their model, assuming that dislocations are bowed out between precipitate pinning points. They have found that strengthening is nonlinear both in concentration, c, (i.e., fraction, f) and in misfit strain. The limiting radius of the curvature to which a dislocation can be bent by the stress is:

$$r = \frac{\alpha\mu b}{\sigma} \tag{4.89}$$

where $\alpha \approx 0.5$. When the distance between particles is of the same order as the radius of curvature, i.e., $l \sim r$, the dislocation can avoid the particles and take the form shown in Fig. 4.29a′. This means that Eq. 4.84 becomes:

$$l = \frac{\alpha\mu b}{\sigma} \tag{4.89a}$$

Dislocation loops can move independently of each other. Glide is realized by taking each loop over a potential hill, without any help from the other forces on the other loops. The effective internal stress is the same as the arithmetic average. An alloy in such a condition is hard. Equation 4.89a defines the critical distribution or the size of particles as l_c, which is obtained by introducing Eq. 4.88 into Eq. 4.89a as:

$$l_c = \frac{\alpha b}{2\varepsilon c} \tag{4.89b}$$

Note that thermal fluctuations help dislocations overcome these obstacles.

When $l \ll r$, a dislocation loop is unable to bend sufficiently in the regions having low internal stress. In such cases, these dislocations must overcome the stress fields either by scaling over the potential hills or, sometimes, by skirting them via the valleys. The internal stresses acting on the dislocations cancel out and only a small force is left to hinder dislocation motion; the required applied stress is low, because the specimens are soft. When $l \gg r$ (i.e., $\gg \frac{\alpha\mu b}{\sigma}$), which is equivalent to the dispersion of the particles in an over-aged specimen, the dislocation is believed to move between the widely-spaced particles using the mechanism suggested by

Orowan (discussed above and shown in Fig. 4.29c). Note that the yield stress, σ_0, is the stress needed to bend a dislocation line into a loop of radius ~l/2 as in:

$$\sigma_0 = \frac{2\alpha\mu b}{l} \tag{4.83c}$$

From the brief description in this section, it is clear that the valuable ideas developed by Mott and Nabarro in the field of metal physics serve as the basis for modern concepts of second-phase particle strengthening.

4.8.6 Summary of Second-Phase Strengthening

One well-known method for strengthening materials is by precipitation hardening. The above discussion on second-phase strengthening applies to precipitates resulting from precipitation hardening, as well. As a matter of fact, the original Mott-Nabarro model was developed for particles obtained by precipitation hardening. In precipitation hardening, a fine distribution of second-phase particles is required to get effective strengthening. These particles can be coherent with the matrix and dislocation can cut through such small particles. During this process, interactions occur between the strain fields of precipitates and dislocations, resulting in the strengthening of the material due to the hindrance of dislocation movement. When the particles become incoherent due to over-aging, the dislocations are forced to bend around the particles, producing Orowan loops. In tensile testing, the Orowan mechanism can be related to yielding as follows. A dislocation curves between two particles. One assumes that yielding has set in when the curved dislocation becomes semi-circular in shape. Following yielding, the dislocation leaves Orowan loops around the particles. The formation of these Orowan loops makes dislocation motion more difficult. This, in turn, produces much work-hardening. The CRSS for the Orowan mechanism is given as $\tau_o \approx Gb/l$ (see Eq. 4.74).

Summarizing the size effect in terms of the obstacles, one may state that, at small sizes, dislocations cut through particles at a lower stress than the Orowan bowing stress. For larger particles, higher cutting stresses are required. At large particle sizes, dislocations bend around particles more easily than they cut through them and no cutting is observed. Note that larger particles mean fewer particles per area (per unit volume), hence lower stresses for flow. The main concept is that the strength of ductile materials is governed by dislocation glide against obstacles. Therefore, strength can be designed by controlling the density and nature of the obstacles to dislocation motion. In the practical design of alloys, particles are introduced by precipitation hardening, dispersion hardening and other means to form materials with proper particle size and density to achieve high strength.

To properly learn of the effects of second-phase strengthening, it is best to study single crystals. In polycrystalline materials, strengthening by grain boundaries masks the effects of particles on strength.

References

L.V. Azàroff, *Elements of X-ray Crystallography, International Student Edition* (McGraw-Hill Book Company, New York, 1968)

L.M. Brown, R.K. Ham, in *Strengthening Mechanism in Crystals*, ed. by E. Kelly, R. B. Nicholson (Wiley, New York, 1971), p. 10

B.D. Cullity, *Elements of X-Ray Diffraction* (Addison-Wesley Publishing Company, Inc, Reading, 1956)

R.L. Fleischer, *The Strengthening of Metals* (Reinhold Publishing Co., New York, 1964), p. 93

J. Friedel, Dislocation interaction and internal strains, in *Internal Stresses and Fatigue in Metals* (Elsevier, Amsterdam, 1959), p. 220

J. Friedel, *Dislocations* (Pergamon Press, Oxford, 1964)

P.B. Hirsch, *Relation Between the Structure and Mechanical Properties of Metals* (H.M.S.O, London, 1963)

J.P. Hirth, J. Lothe, *Theory of Dislocations* (Krieger, Malabar, 1992)

R.W.K. Honeycombe, *The Plastic Deformation of Metals* (Edward Arnold, London, 1975)

F.J. Humphreys, P.B. Hirsch, Proc. Phys. R. Soc. **A318**, 73, 205 in P.B. Hirsch "Work Hardening" in *The Physics of Metals*, 2nd edn., ed. by P.B. Hirsch (Cambridge University Press, London, 1975), p. 207

N.F. Mott, F.R.N. Nabarro, in *Bristol Conference on Strength of Solids* (The Physical Society, London, 1948), p. 1

A. D. Rollett, H. Garmestani, G. Branco, Polycrystal plasticity-multiple slip, *Advanced Characterization and Microstructural Analysis* (Carnegie Mellon, Department of Materials Science and Engineering, Spring, 2005)

A. Seeger, Glide and work hardening in face-centered and hexagonal close-packed metals, in *Dislocations and Mechanical Properties of Crystals* (Wiley, New York, 1957), p. 243

Further References

K. Akhtar, A. Teghtsoonian, Acta Met. **19**, 655 (1971)

H. Alexander, P. Haasen, Acta Metall. **9**, 1001 (1961)

P.N.B. Anongba, J. Bonneville, J.L. Martin, Acta Metal. Matter. **41**, 2897 (1993)

C.J. Ball, Philos. Mag. **2**, 1011 (1957)

S. Bargmann, M. Ekh, B. Svendsen, K. Runesson, Tech. Mech. **30**, 316 (2010)

Z.S. Basisnski, Philos. Mag. **4**, 393 (1959)

J.L. Bassani, Texture Microstruct. **14–18**, 1097 (1991)

F.P. Bullen, C.B. Rogers, Philos. Mag. **11**, 191 (1965)

D. Caillard, A. Couret, Microsc. Res. Tech. **72**, 261 (2009)

H. Conrad, J. Metals **16**, 582 (1964)

A.H. Cottrell, R.J. Stokes, Proc. R. Soc. Lond. A **318**, 45 (1970)

A.M. Cuitino, Mater. Sci. Eng. **A216**, 104 (1996)

G. Das, T.E. Mitchell, Metall. Trans. **4**, 1405 (1973)

J. Diehl, Z. Metall. **47**, 331 (1956)

M. Ekh, M. Grymer, K. Runesson, T. Svedberg, Int. J. Numer. Method Eng. **72**, 197 (2007)

J.D. Eshelby, F.C. Frank, F.R.N. Nabarro, Philos. Mag. **42**, 351 (1951)

L. Evers, W. Brekelmanns, M. Geers, J. Mech. Phys. Solid **52**, 2379 (2004)

R.L. Fleischer, Acta Metall. **9**, 996 (1961)

R.L. Fleischer, Acta Metall. **10**, 835 (1962)

R.L. Fleischer, Acta Metall. **11**, 203 (1963)

C.P. Frick, B.G. Clark, S. Orso, A.S. Schneider, E. Arzt, Mater. Sci. Eng. A **489**, 319 (2008)

J. Friedel, Philos. Mag. **46**, 1169 (1955)
J. Friedel, Proc. R. Soc. **A242**, 147 (1957)
A. George, Rev. Phys. Appl. **23**, 479 (1988). H.O.K.K
H. Gleiter, E. Hornbogen, Mater. Sci. Eng. **2**, 285 (1967–1968)
F. Guiu, P.L. Pratt, Phys. Stat. Sol. **16**, 539 (1966)
P. Haasen, Czech. J. Phys. **B38**, 494 (1988)
N. Hansen, Scr. Mater. **51**, 801 (2004)
N. Hansen, Adv. Eng. Mater. **7**, 815 (2005)
P.B. Hirsch, Philos. Mag. **7**, 67 (1962)
P.B. Hirsch, Disc. Faraday Soc. **38**, 111 (1964)
P.B. Hirsch, F.J. Humphreys, Proc. R. Soc. **A242**, 147 (1957)
P.B. Hirsch, D.H. Warrington, Philos. Mag. **6**, 735 (1961)
B.P. Kashyap, K. Tangri, Acta Metal. Matter **43**, 3971 (1995)
A.S. Keh, Philos. Mag. **12**, 9 (1965)
H.O. Kirchner, Rev. Phys. Appl. **23**, 475 (1988)
U.F. Kocks, H. Mecking, Prog. Mater. Sci. **48**, 171 (2003)
R.C. Koo, Acta Metall. **11**, 1083 (1963)
D. Kuhlmann-Wilsdorf, N.R. Comins, Mater. Sci. Eng. **60**, 7 (1983)
R. Kumar, L. Nicola, E. Van der Giessen, Mater. Sci. Eng. A **527**, 7 (2009)
A. Lasalmonie, J.L. Strudel, J. Mater. Sci. **21**, 1837 (1986)
F. F. Lavrent'ev, Yu. A. Pokhil, V. I. Startsev, Translated from Problemy Prochnosti, **10**, 60 (1972)
J.C.M. Li, Trans. TMS-AIME **227**, 239 (1963)
M.A. Meyers, E. Ashworth, Philos. Mag. **A46**, 737 (1982)
T.E. Mitchell, Prog. Appl. Mater. Res. **6**, 119 (1964)
T.E. Mitchell, W.A. Spitzig, Acta Metall. **13**, 1169 (1965)
T.E. Mitchell, R.A. Foxall, P.B. Hirsch, Philos. Mag. **8**, 1895 (1963)
S.K. Mitra, J.E. Dorn, Trans. AIME **224**, 1062 (1962)
S.K. Mitra, P.W. Osborne, J.E. Dorn, Trans. AIME **221**, 1206 (1961)
N.F. Mott, Philos. Mag. **43**, 1151 (1952)
N.F. Mott, Trans. Metall. Soc. AIME **218**, 962 (1962)
N.F. Mott, F.R.N. Nabarro, Proc. Phys. Soc. **52**, 86 (1940)
N.F. Mott, F.R.N. Nabarro, *Bristol Conference on Strength of Solids* (The Physical Society, London, 1948), p. 1
B. Nicklas, H. Mecking, in ed. by P. Haasen et al., *Proc. 5th ICSMA,* Aachen, Germany (Pergamon Press, Oxford, 1979), p. 351
A.C. Nunes, A. Rosen, J.E. Dorn, Trans. ASM **58**, 38 (1965)
W. Püschl, Prog. Mater. Sci. **47**, 415 (2002)
X.G. Qiao, M.J. Starink, N. Gao, Mater. Sci. Eng. A **513–514**, 52 (2009)
R.P. Reed, Cryogenics **12**, 259 (1972)
W. Schröter, H. Siethoff, *Z. Metallk.***75**
C. Scwink, Rev. Phys. Appl. **25**, 395 (1988)
H. Siethoff, W. Schröter, Z. Metallk. **75**, 475 (1984)
R.J. Stokes, T.L. Johnston, J.C.M. Li, Trans. AIME **218**, 655 (1960)
H. Suzuki, S. Ikeda, S. Takeuchi, J. Phys. Soc. Jpn **11**, 382 (1956)
G.I. Taylor, Proc. R. Soc. **A145**(362), 388 (1934)
E. Votava, Phys. Status Sol. **5**, 421 (1964)
H. Wiedersich, J. Metals **16**, 425 (1964)
B. Wielke, Phys. Status Sol. (a) **83**, 241 (1976)
T.-Y. Wu, J.L. Bassani, C. Laird, Proc. R. Soc. Lond. A **435**, 1 (1991)

Chapter 5
Time Dependent Deformation – Creep

5.1 Introduction

In the previous chapters, the time element was not considered in the determination of the stability of materials exposed to a continuous force. However, this aspect is of considerable importance in designing structural components that are required to maintain dimensional stability over a long period of time while in service. Some structural materials are used at various temperatures; therefore, temperature plays a major role in creep deformation. Usually, these structures operate at temperatures higher than 0.5 T_m (T_m being the melting temperature) under loads which would not cause plastic deformation at room temperature (loads or stresses below the yield point). Therefore, strain depends not only on stress, but also on temperature and time. The strain rate of a material in creep is a function of σ, T and t as:

$$\dot{\varepsilon} = f(\sigma, t, T)$$

Instinct dictates that the melting temperature of a material is of significance and that some materials (high-melting) will be more appropriate than others (low-melting) for use when resistance to creep is an essential requirement. Generally, refractory metals, having high melting points are natural candidates for applications where resistance to creep is essential. The well-known problem, however, of the refractory metals is their poor resistance to oxidation and, therefore, coating is critical. Thus, creep depends on stress, time and temperature. Clearly, strain rate is a factor to be considered while creep deformation is in progress. In order to study creep as an important metallurgical phenomenon, without other additional effects that usually contribute to overall creep deformation, it is essential to investigate single crystals. This way grain-boundary sliding, which is a significant contributor

J. Pelleg, *Mechanical Properties of Materials*, Solid Mechanics and Its Applications 190, DOI 10.1007/978-94-007-4342-7_5,

to overall creep is eliminated. In polycrystalline material grain boundary sliding generally occurs with a considerable contribution to creep strain. In this chapter, the following topics will be discussed:

1. Creep in single crystals;
2. Creep in polycrystalline materials;
3. Mechanisms of creep;
4. Grain-boundary sliding;
5. Creep rupture;
6. Recovery (stress relaxation);
7. Prediction of life-time (parametric method);
8. Concepts of designing (selecting) creep resistant materials.

Deformation occurs by dislocation motion. The most essential requirement in developing creep resistance in materials is to slow down the dislocation movement. Engineers need to account for this aspect of deformation when designing systems for creep applications.

5.2 Creep in Single Crystals

A creep test is performed by applying a force to the test specimen while measuring its dimensional change, i.e., the strain, as a function of time with exposure to some temperature. The test can be performed either by constant stress or constant load. The temperature of the test depends on the material itself and on consideration of its homologous temperature. The 'homologous temperature' is defined as the ratio of the absolute temperature, T, to its melting point, namely:

$$\text{homologous temperature} = T/T_{mp} \tag{5.1}$$

Low-temperature creep at or below $0.5T_m$ is believed to be governed by non-diffusion controlled mechanisms, whereas high-temperature creep, above $0.5T_m$, is diffusion controlled.

The generally accepted method for recording the results of a creep test is by plotting strain vs. time, as shown schematically in Fig. 5.1a. The schematic curve is composed of three stages known as 'primary', 'secondary' and 'tertiary' creep and an instantaneous elongation upon applying the force. 'Primary creep' is also known as 'transient', or simply 'stage I creep'. In Fig. 5.1c, experimental transient creep is compared with a theoretical curve. For most metals and alloys, creep at low temperatures occurs in a single transient stage, in which the creep rate decreases continuously over time. This is seen schematically in Fig. 5.1b. The decrease in strain rate is associated with increases in the dislocation density or changes in the characteristics of the dislocation structure. Figure 5.1d shows the experimental creep-strain rate vs. time for several samples loaded at different rates at the given stress. Strain hardening occurs during transient creep, which is induced by pure

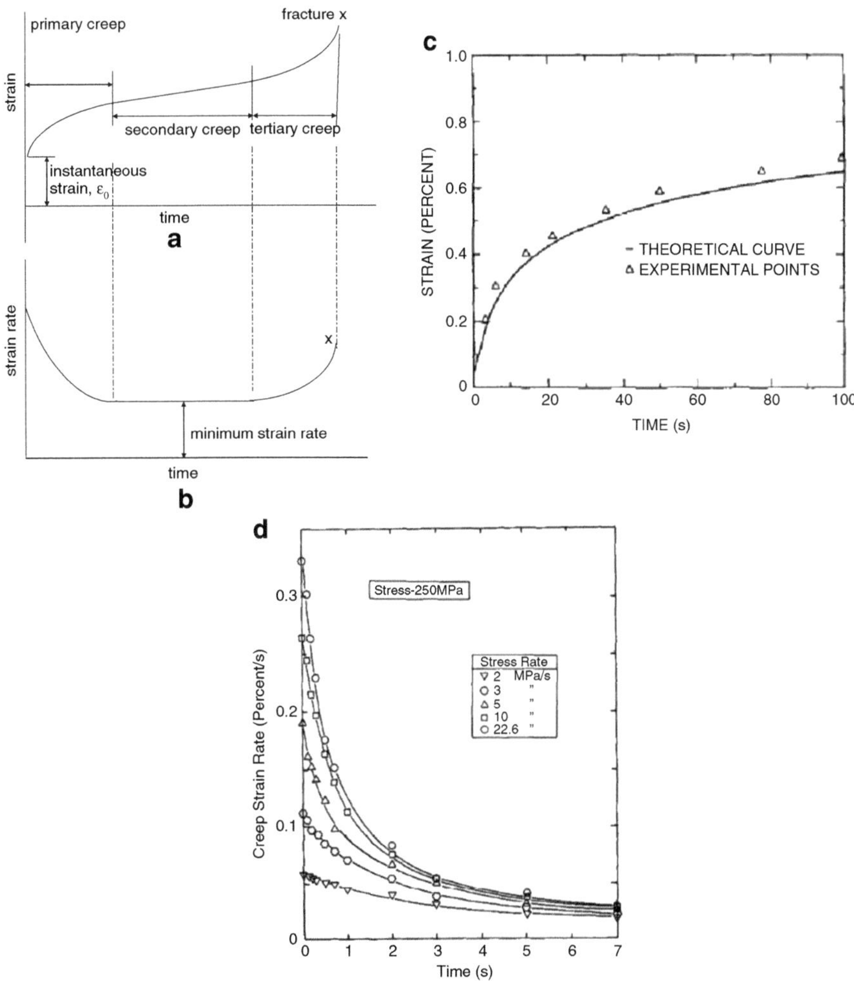

Fig. 5.1 (**a**) A schematic creep curve showing three stages of creep and an instantaneous elongation on application of load; (**b**) schematic strain rate plot vs. time; experimental data: (**c**) experimental transient creep compared with theoretical; (**d**) creep strain rate vs. time of 304 stainless steel (Alden 1987. With permission from Elsevier)

glide. Mobile dislocations, present at the start of creep, continue to move under the influence of an effective stress, which slowly declines as the mobile dislocations are trapped into the network. The total dislocation density, equal to the sum of the mobile and network densities, remains constant. It is clear that strain is a function of stress and increases with stress. Figure 5.2 compares the experimental data with the theoretical line

'Secondary creep' or 'stage II' is often referred to as the 'steady state' and also as 'linear creep'. During 'tertiary creep' or 'stage III', the creep rate begins

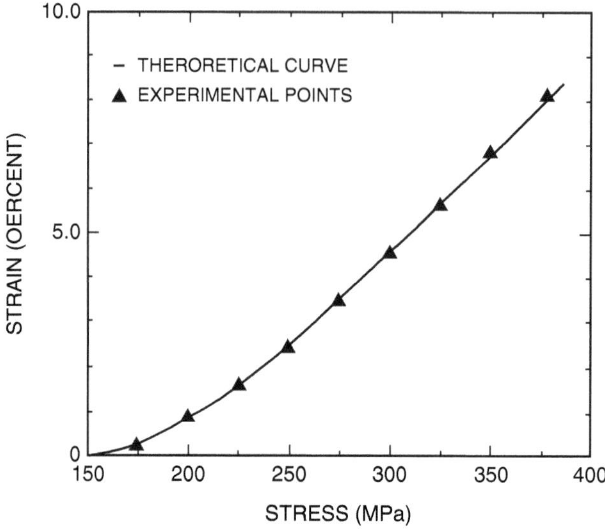

Fig. 5.2 Experimental strain vs. stress relation for 304 stainless steel. The curve was determined at a stress rate of 3.0 MPa/s. The theoretical curve is fitted to the experimental data. The curve was determined in a soft tensile machine at a stress rate of 3.0 MPa/s. The theoretical curve is fitted to the data (Alden 1987. With permission from Elsevier)

to accelerate as the cross-sectional area of the specimen decreases due to necking, which decreases the effective area of the specimen. If stage III is allowed to proceed, fracture will occur. The instantaneous strain, $\varepsilon_{0,}$ is obtained immediately on loading; it is not a creep deformation, since it is not dependent on time and is, by its nature, elastic. However, plastic strain also contributes. The strain rates characterizing these stages are as follows: in stage I, the strain rate, $\dot{\varepsilon}$, is decreasing, while in stage II, the strain rate is constant, due to the balance between the recovery processes and strain hardening. In stage III, the creep is accelerated and the strain rate continuously increases until fracture sets in. Figure 5.1b shows the variation of strain rate with time. The creep test is usually employed to determine the minimum creep rate in stage II. 'Logarithmic creep' also exists. Under certain conditions, the plastic extension of a sample subjected to a constant stress is (to a good approximation) proportional to the logarithm of the time (Nabarro).

Stress and temperature affect the form of the creep curve (as shown in Fig. 5.1a). Schematically, this effect is shown in Fig. 5.3a for low, medium and high temperatures. At medium temperature (A in the figure), the commonly observed creep curve is shown schematically, which is similar to the one seen in Fig. 5.1a. Schematic creep curves, at a constant temperature and with stress as the varying parameter, are shown in Fig. 5.3b. At the stress indicated by σ_3, the curve is similar to the ideal creep curve shown in Fig. 5.1a and has all three stages. The effect of stress and temperature on the shape of creep curves is similar. In Fig. 5.3b, the effect of stress at a constant temperature is shown schematically and similar curves may be obtained

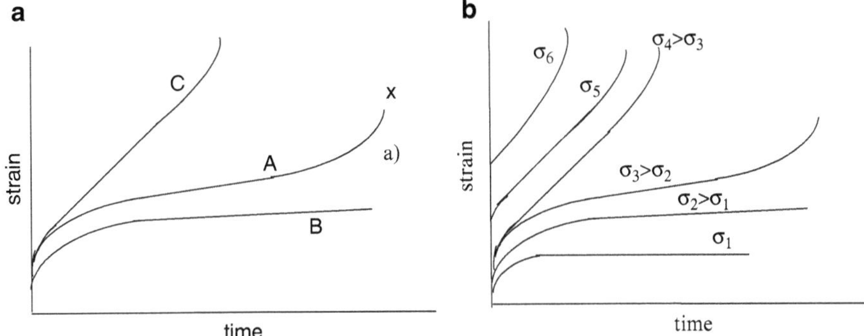

Fig. 5.3 (**a**) The shape of creep curves: (*A*) the common creep curve; (*B*) a creep curve at low temperature and stress and; (*C*) a high-temperature and high-stress curve. (**b**) Schematic creep curves at constant temperatures with stress as a parameter. Note that σ_3 represents the standard creep curve having all three stages

when the stress is kept constant and the temperature is the varying parameter. The transient-creep stage gradually diminishes with increased stress and, at a sufficiently high stress level, it disappears and the steady state dominates the shape of the strain-time relation.

The interest in single-crystal studies is not only a consequence of understanding creep phenomena without the influence of grain-boundary sliding, for example, but also for practical reasons, such as engineering applications. Thus, blades in the aircraft industry are one example of single-crystal use. Automotive components, space vehicles and nuclear power systems are additional beneficiaries of single-crystal materials. The common denominator of all these applications is the requirement that the structural materials be stable at the high temperatures commonly used in the above industries. At such high temperatures, it is critical to use materials that have excellent creep capabilities and can resist dimensional changes during lengthy applications. Superalloys are materials that have excellent creep capabilities and a high load-bearing capacity almost to their melting points. They retain their strength through long periods of exposure to high temperatures and, at the same time, have good low-temperature ductility. The superalloy families are based on nickel, cobalt and iron- based alloys. Despite the complexity of producing these alloys, which contain multiple elements as additives, they are available as single crystals. Examples of superalloys are: hastelloy, inconel, udimet, rene, etc. Elimination of oxidation is vital for good creep resistance, which can be achieved by various coating techniques. From experiments performed at a constant load in vacuum and air on 2.25Cr-1Mo Steel (Levi de Oliveira Bueno et al.), it is obvious that the maximal elimination of oxidation problems greatly improves creep resistance, as seen in Fig. 5.4.

In all these cases, the curves in the vacuum are displaced to higher stress levels than in air, with a clear indication that the drop in creep strength over time is much lower for the tests carried out in vacuum than in air. Creep life in the vacuum test is

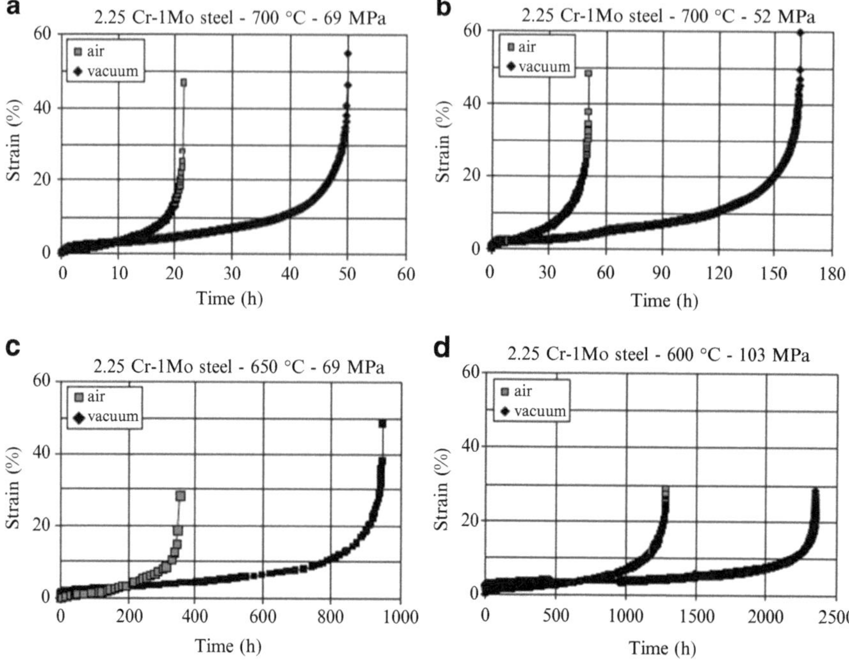

Fig. 5.4 Creep curves in the 600–700°C range of 2.25Cr-1Mo Steel at various stress levels (Levi de Oliveira Bueno et al. 2005. With permission of Materials Research and the authors)

about twice of that observed in the air test. The implication of these experiments for creep application is that coatings must be used to eliminate environmental effects. Superalloy products that are subjected to high working temperatures and corrosive atmospheres (e.g., the high-pressure turbine region of jet engines) should be coated with various kinds of protective coatings.

In Sect. 5.1, it was stated that dislocation motion should be slowed down as much as possible for good creep resistance. In polycrystalline materials, the grain boundaries act as barriers to dislocation motion; in single crystals, stable second-phase particles or precipitates act in a similar way. In nickel-based superalloys, the γ' phase [Ni_3(Al, Ti)] replaces the role of the grain boundaries, forming a good barrier to dislocation motion.

As expected, single crystals are anisotropic and the superalloys are no exception. Therefore, their orientations are significant. The creep behavior of single-crystal superalloys is highly anisotropic and orientation-dependent. From a design point of view, it is important to use an orientation utilizing maximum strength for creep resistance. There are experimental investigations of Ni based superalloy single crystals indicating that creep properties are best along the <001> orientation, compared to <011> and <111> (Prasad et al.; Caron and Khan). However, other experiments show that this is not always the case (Lukáš et al.). Lukáš et al. have performed creep tests on superalloy single crystals with different orientations at 1,223 K, both under tension and under compression at constant loads and at constant stress. It was found that creep resistance strongly depends on crystal orientation. Creep under tension decreases in the sequence <111>, <001>, <011>, while creep under compression decreases in the sequence <001>, <111>, <011>. Constant stress tests under tension generally reveal higher creep resistance than constant load tests under tension. Under compression, the differences in the mode of testing produce only small and unsystematic differences in creep behavior.

The orientation anisotropy and the various reports on the best creep-resisting properties in various crystallographic directions of Ni-based superalloy single crystals basically depends on many factors, such as: the type of loading of the specimen, the test temperature, the creep stage, but, first of all, the precipitates of the test piece. Such alloys are two-phase materials and the precipitates, known as the γ' phase, play a crucial role in strengthening for creep resistance. In their work, Pollock and Argon considered the effects of γ' and of the three-dimensional dislocation network. In their study, creep deformation in (001)-oriented nickel-based superalloy single crystals were considered in an effort to assess the factors contributing to the overall creep resistance of superalloys with high volume-fractions of γ' phase. In principle, creep occurs (like the deformation processes discussed hitherto) by dislocation motion, which can be either glide or climb. Pollock and Argon have found that, in the temperature range of 1,073–1,173 K and in stress ranges of interest for creep, the γ' precipitates are dislocation-free and are un-deformable until the very late stages of tertiary creep, after some percentage of creep deformation has accumulated. Dislocations move through materials by bowing out through the γ phase (matrix) on $\{111\}$ planes. In the early stages of primary creep, all dislocations have the same Burgers vector, indicating their origin at a common source. Steady-state creep is characterized by a three-dimensional dislocation network, containing multiple Burgers vectors. The network fills the matrix material and diffusion-controlled static recovery is ineffective at causing rearrangements in the network at temperatures of 1,123 K or less. For this reason, a stress-dependent dynamic recovery mechanism appears to be responsible for maintaining the creep process. The principal cause of the high creep resistance of CMSX-3 crystals is the non-deformability of the γ' particles, forcing the γ phase to undergo complex flow patterns in narrow channels. Resistance to deformation is a consequence of the mean normal stresses, that develop during creep in the narrow channels of the γ phase between the rigid γ' particles. In Fig. 5.5, a dislocation structure is seen in the early stages of primary creep. In Fig. 5.5a, the dislocation distribution in the early stages of the primary creep in the γ matrix is seen, while, in 5.5b, the same is shown at a later stage of primary creep.

The increase in dislocation density is a result of the increased stress to induce strain at this level. Figure 5.6 shows the dislocation network in steady-state creep. Mainly matrix dislocations are seen located at the γ matrix and the γ' precipitate interface. In such Ni-based superalloys composed of two phases, the resistance to creep is associated with the precipitate; for continued creep, cutting of the second-phase particles must occur. This is the reason why such Ni-based superalloys are in

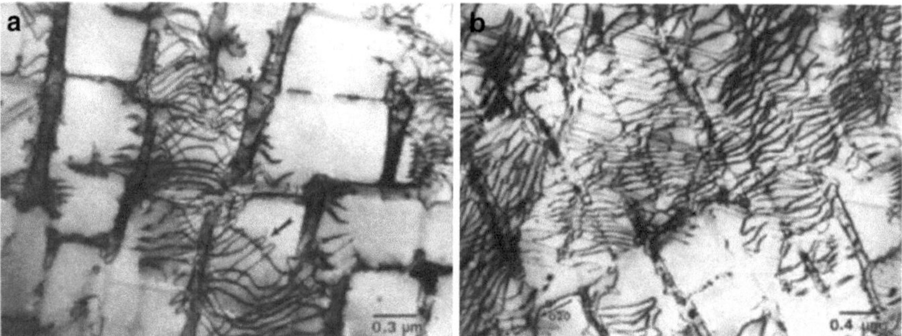

Fig. 5.5 Dislocation distribution in: (**a**) the early stages of primary creep. Some matrix regions are still free of dislocations, and spreading continues by bowing dislocations through the matrix, as shown in the area marked by the arrow; (**b**) typical dislocation distribution in the latter stages of primary creep. The density of dislocations has increased and few channels are left unpopulated (Pollock and Argon 1992. With permissions of the authors and Elsevier LTD)

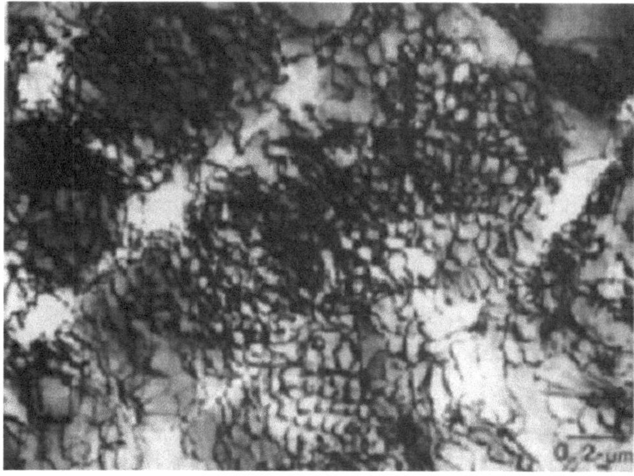

Fig. 5.6 Matrix dislocation networks during steady state creep. A very organized three-dimensional nodal network is present locally, and many of the dislocations are associated with the γ/γ' interface (Pollock and Argon 1992. With permissions of the authors and Elsevier LTD)

practical use for structural components and the construction of systems in which high-temperature performance is a prerequisite. In steady-state creep, multiple slip systems are involved, consequently forming a three-dimensional dislocation network resembling cell structure (see Fig. 2.9b). In the early stages of creep deformation, the interfaces of the γ' particles are sharply defined along the $\{001\}$ planes, whereas, in the steady state, these are not as sharply defined.

Diffusion-controlled static recovery is ineffective at causing rearrangements in the network at temperatures of 1,123 K or less. The principal cause of the high-creep

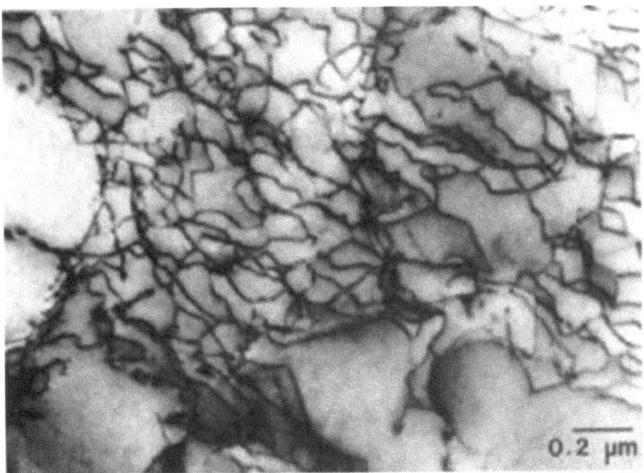

Fig. 5.7 Recovered dislocation networks in the matrix after 750 h of bulk static annealing at 850°C. The dislocations are largely associated with the γ/γ' interfaces (Pollock and Argon 1992. With permissions of the authors and Elsevier LTD)

resistance of CMSX-3 crystals is the non-deformability of the γ' particles, forcing the γ phase to undergo complex flow patterns in narrow channels. In Fig. 5.7, recovered dislocation networks are seen; however, the γ' phase does not deform before the late tertiary stages of creep, when the internal stresses become very large. This alloy tends to be relatively stable, even in the creep regime of 1,073–1,173 K and when the networks have persisted, even after extended periods of annealing.

A basic, accepted fact is that steady-state creep is a balance between work hardening and recovery processes. In Ni-based superalloys, such as CMSX-3 crystals, the three-dimensional dislocation networks in the γ phase channels are very resistant to recovery. Steady-state creep is characterized by a three-dimensional network and diffusion-controlled static recovery is ineffective at causing rearrangements in it at temperatures of 1,123 K or less. The resistance to recovery at the temperatures mentioned is the main reason for the application of Ni-based alloys at high temperatures.

In Fig. 5.1b, a minimum constant-creep rate, which is an important design parameter, is shown. The magnitude of the minimum-creep rate on the strain-time relation (see Fig. 5.1a) is associated with steady-state creep and is stress- and temperature-dependent. Two criteria are in common use: (a) the stress to produce a creep rate of 0.1×10^{-3}%/h (or 1% in 1×10^4 h) or (b) the stress to produce a creep rate of 0.1×10^{-4}%/h, namely 1% in 100×10^3 h (or approximately 11.5 years). Criterion (a) is used for turbine blades, while (b) is in common use for steam turbines.

Several empirical models have been suggested for creep. Andrade was the first to consider creep as early as 1914. He considered it to be the superposition of transient and viscous creep terms (discussed in the next section below). Since creep is a thermally-activated process, the minimum secondary-creep rate may be described by an Arrhenius equation. One of the forms (see McLean) is:

$$\frac{d\varepsilon}{dt} = \dot{\varepsilon} = A \exp - \left(\frac{Q_0 - \alpha\sigma}{kT} \right) \tag{5.2}$$

A and α are constants and Q_0 is the activation energy for creep at zero stress. A is also known as the 'frequency' or 'pre-exponential factor'. An additional expression for the creep rate, where the stress and the temperature terms are separated, is given as:

$$\dot{\varepsilon} = B\sigma^n \exp - \left(\frac{Q}{kT} \right) \tag{5.3}$$

In this expression, the stress affects the frequency factor, B.

Much experimental data indicates that creep rate, in the early stages, can be expressed by a function suggested by Cottrell as:

$$\frac{d\gamma}{dt} = \dot{\gamma} = At^{-n} \tag{5.4}$$

or

$$\left(\frac{d\varepsilon}{dt} = \dot{\varepsilon} = \mathrm{Bt}^{-\mathrm{n}} \right) \tag{5.4a}$$

A (or B) and n are constants with $0 \leq n \leq 1$.

It was claimed that the transient regime of creep curves can be fitted to a logarithmic law. This can be obtained from Eq. 5.4, when n = 1. In the extreme case, when n = 1, which is often observed experimentally, one obtains the logarithmic creep law as:

$$\gamma = \alpha \ln t \quad (t > 1) \tag{5.5}$$

One can see that Eq. 5.4 adequately describes the experimental creep data, since the creep rate in the primary stage (transient) decreases over time, as shown in the schematic illustration in Fig. 5.1b. Various values of n, in the range 0–1, may be observed experimentally, but, very frequently, the value of 2/3 is preferred. Thus:

$$\frac{d\gamma}{dt} = \dot{\gamma} = At^{-2/3} \tag{5.4b}$$

and integration gives the equation for strain as:

$$\gamma = \beta t^{1/3} \tag{5.6}$$

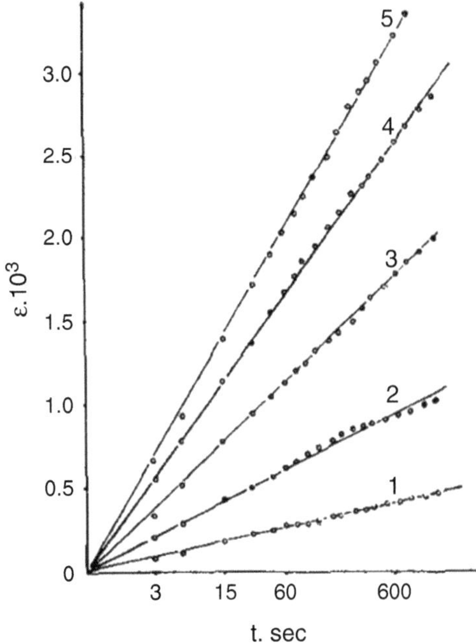

Fig. 5.8 Stress dependence of the logarithmic creep at 77 K and $\tau = (1)$ 180 gfmm^{-2}, (2) 310 gfmm^{-2} (3) 450 gfmm^{-2}, (4) 600 gfmm^{-2} and (5) 720 gfmm^{-2} (Stratsev et al. 1975. With kind permission from Springer Science)

Equation 5.6, representing transient creep, is often referred to as 'β-creep' or 'Andrade creep', since Andrade was the first to show that it applies to many materials. The creep behavior obeying Eq. 5.5 is often called 'α' or 'logarithmic creep'. Uchic et al., taking into account instantaneous strain, γ_0, also suggested a logarithmic relation for the low-temperature creep of $Ni_3(Al, Ta)$ as:

$$\gamma = \gamma_0 + \alpha \ln(\beta t + 1) \tag{5.7}$$

where α and β are constants. Figure 5.8 shows logarithmic creep curves at various stresses in Pb single crystals.

Figure 5.1a shows that stage II creep is linear; thus, the function describing this region must be linear. Much of the creep data is expressed by functions taking this linear contribution into account as:

$$\gamma = \gamma_0 + \beta t^{1/3} + \kappa t \tag{5.8}$$

where, clearly, the linear contribution is $\gamma_{II} = \kappa t$ (Cottrell and Aytekin). This relation well-describes many creep experiments. In experiments performed at high temperatures, in particular, usually transient and steady-state creep occur together; the combination of these stages is seen graphically in Fig. 5.9. The instantaneous strain, γ_0, obtained on loading (not creep strain) is not shown.

For high-temperature applications, superalloys, mainly two-phase materials, were of considerable interest and, in a large number of publications, this topic

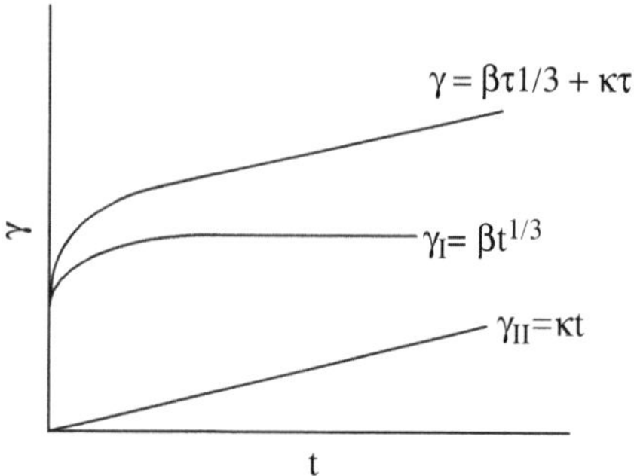

Fig. 5.9 A graphic presentation for obtaining Eq. 5.8 without γ_0 from the combination of transient ($\gamma_I = \beta t^{1/3}$) and steady-state ($\gamma_{II} = \kappa t$) creeps

was widely investigated. However, for understanding the creep phenomenon itself, various pure metals were studied in order to get an insight into the mechanism prevailing under various experimental conditions. Thus, all the structures, FCC, HCP and BCC materials, were objects of these studies. Even low-melting lead single crystals received attention in low-temperature creep studies. In recent years, it was revealed that an appreciable creep exists in metals at liquid-helium temperatures. This effect is independent of temperature and the curves obtained from extrapolation to 4.2 K give what is known as 'non-zero creep rates'. This peculiar low-temperature creep was observed in Cd (Glen), Zn (Natsik et al., Osetski et al.), and others. Pb has a melting point (T_m) of 600.6 K, and since creep is of interest at a level of ~0.5 T_m, namely at about 300 K, the illustrations below are in this range or below it. The experimental curves for Pb at a constant stress with temperature as a parameter and at a constant temperature with various stress parameters may be seen in Figs. 5.10 and 5.11 at lower temperatures.

These experimental curves show the effect of temperature on the shape of the creep curves when the stress is constant (Fig. 5.10a) and also illustrate the case when the temperature is kept constant as the stress changes (Fig. 5.10b). The stress dependence of transient low-temperature creep was studied throughout the temperature range in which logarithmic creep was observed, both as a function of creep strain-time and creep strain-lnt. In Fig. 5.11a, the effect of temperature on the creep curves at much lower temperatures is shown, while the stress is kept constant. At low temperatures, logarithmic curves express the state of affairs and, thus, Fig. 5.11b shows strain in terms of ε vs. lnt relation at constant stress for several low temperatures.

In tertiary creep, the strain and strain rate increase until fracture occurs. Entering into this stage occurs when there is a reduction in cross-sectional area due to necking or due to internal void formation. Void formation effectively causes a reduction in area. Thus, tertiary creep is important in engineering, because it is often associated with the formation of structural instability, as indicated by voids, crack formation and necking, leading to failure by fracture. The onset of tertiary creep occurs at the end of steady-state creep. It is easier to locate the onset of tertiary creep from the $\dot{\varepsilon}$-t relation than from ε-t, as seen from Fig. 5.1b, since the location of the deviation from the minimum creep rate is well defined. It is clear that the minimum creep rate parameter must limit allowable stress in practice to prevent the onset of tertiary creep.

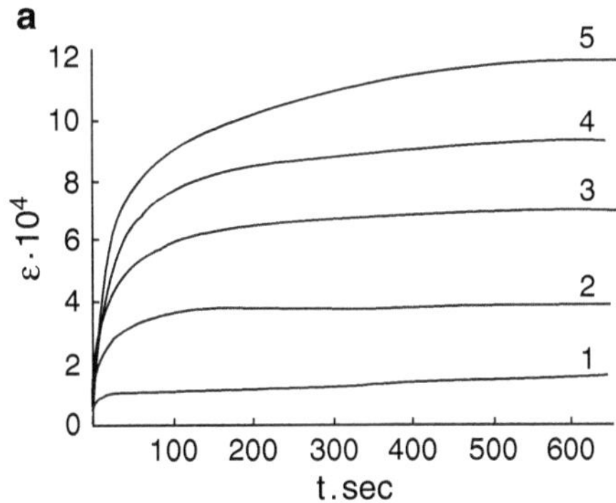

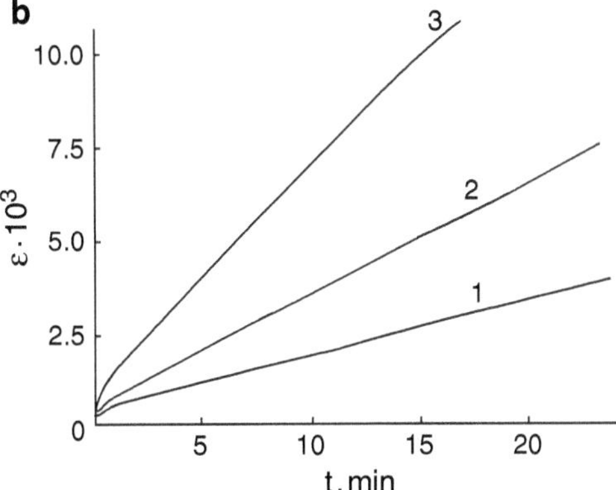

Fig. 5.10 Creep curves in the range 180–260 K: (**a**) The effect of temperature. The curves were obtained under constant stress, $\tau = 260$ gfmm^{-2} and the following temperatures (*1*) 255 K, (*2*) 240 K, (*3*) 225 K, (*4*) 210 K and (*5*) 190 K. (**b**) Creep curves at 296 K and the stresses of τ (*1*) 250 gfmm^{-2}, (*2*) 280 gfmm^{-2} and (*3*) 340 gfmm^{-2} (Stratsev et al. 1975. With kind permission from Springer Science)

Several investigators have shown that the start of tertiary creep and the rupture life are related for various alloys by the relation (e.g., Garofalo et al.):

$$t_2 = At_r^{\alpha} \tag{5.9}$$

where t_r is the rupture life, t_2 is the starting time of the tertiary creep, A and α are constants, often of ~1. This relation is one of the many expressions for creep, in general, and for tertiary creep, in particular, and is widely used for various materials under consideration for high-temperature applications. Thus, tertiary-creep behavior was investigated in various alloyed steels, such as Cr-Mo steels (Klueh). Other expressions are common in creep studies, such as power, exponential and logarithmic functions. For example, these three functions are shown as:

$$\varepsilon_{III} = \dot{\varepsilon}_{\min} t + At^{g} \tag{5.10}$$

$$\varepsilon_{III} = \theta_3 (\exp[\theta_4 t] - 1) \tag{5.11}$$

$$\varepsilon_{III} = -(\ln[1 - C\dot{\varepsilon}_{\min} t])/C \tag{5.12}$$

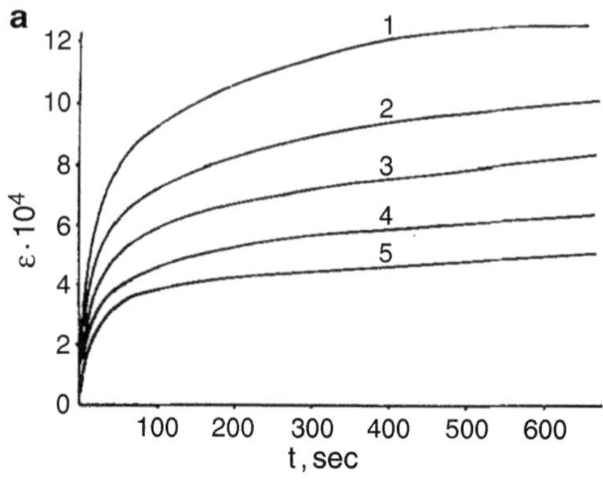

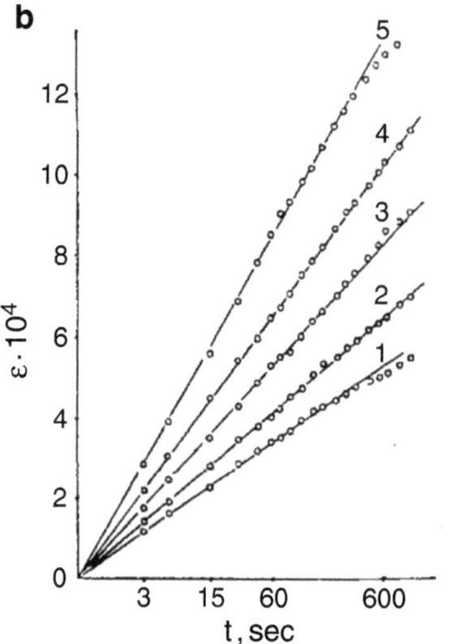

Fig. 5.11 Creep curves represented as (**a**) ε-t and (**b**) ε-lnt. The deformation at constant stress: τ = 260 gfmm^{-2} at temperatures (*1*) 20 K, (*2*) 77 K, (*3*) 106 K, (*4*) 130 K and (*5*) 184 K (Stratsev et al. 1975. With kind permission from Springer Science)

In these expressions, $\dot{\varepsilon}_{min}$ is the minimum-creep rate, A, g, θ_3, θ_4 and C are parameters. These expressions are for tertiary creep without the primary stage. Creep curves at higher applied stress, which have a pronounced tertiary stage, may be successfully described by all three equations. Dobeš has indicated that the calculated value of g of ~7–10 is higher than the one proposed by Graham and Walles as $g = 3$.

5.3 Creep in Polycrystalline Materials

Principally, all the factors discussed above apply to polycrystalline materials with the additional contribution of grain boundaries. These boundaries have a dual effect on creep. On one hand, their presence strengthens the material, but, on the other hand, it accelerates creep via grain-boundary sliding and grain-boundary migration (discussed in a later section).

The first unified creep relation was suggested by Andrade, who may be considered the father of such deformation. From the many relations given in the literature, all basically associated with the concept of the empirical relation given by Andrade, one gets:

$$\varepsilon = \varepsilon_0(1 + \beta t^{1/3})\exp(\kappa t) \tag{5.13}$$

When $\kappa = 0$, β creep is obtained, so called because of the constant β:

$$\varepsilon = \varepsilon_0(1 + \beta t^{1/3}) \tag{5.14}$$

This represents transient creep, since the creep rate is a decreasing function of time. By differentiating Eq. 5.14 with respect to time, one obtains (5.15):

$$\frac{d\varepsilon}{dt} = \dot{\varepsilon} = \frac{1}{3}\varepsilon_0 \beta t^{-2/3} \tag{5.15}$$

When $\beta = 0$ in Eq. 5.13:

$$\varepsilon = \varepsilon_0 \exp(\kappa t) \tag{5.16}$$

This is κ creep, describing the stationary stage. Clearly, the strain rate (as obtained from Eq. 5.16) gives a linear relation, as seen in Eq. 5.17 and, thus, describes steady-state creep:

$$\dot{\varepsilon} = K\varepsilon_0 \exp(\kappa t) = \kappa\varepsilon \tag{5.17}$$

Andrade postulated that β creep is related to dislocation glide within the grain, while κ flow is related to slip along grain boundaries. Ascribing κ flow to grain-boundary sliding is known to be in error. Equation 5.16 may also be expressed as:

$$\ln \varepsilon = \ln \varepsilon_0 + \kappa t \tag{5.16a}$$

A schematic plot of this function at different temperatures is seen in Fig. 5.12. From the intercept of such curves, ε_0, and from the slopes κ for various temperatures can be evaluated:

The term 'viscous creep' is often used for creep at high temperatures and low stresses. Two mechanisms were proposed to describe such creep in polycrystalline materials. The one known as 'Nabarro-Herring creep' conceives of a stress-directed, diffusional migration of vacancies, while the other, originally suggested by Mott and subsequently elaborated by Weertman, is based on a dislocation-climb model. Extensive experimental evidence also exists to support the dislocation-climb model. Viscous creep in Al, near its melting point, was discussed in a publication by

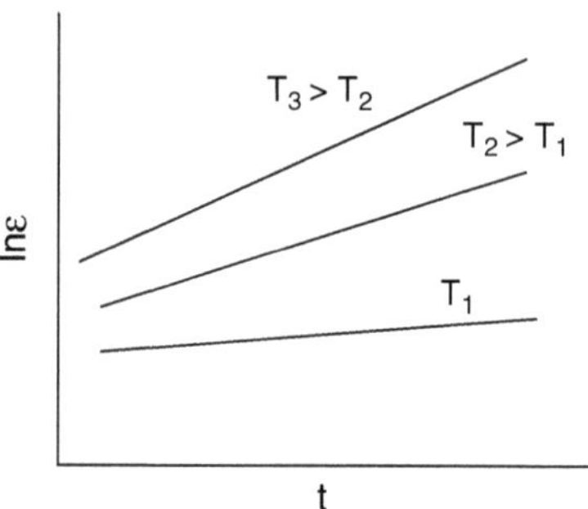

Fig. 5.12 Creep strain vs. t for a few temperatures

Harper and Dorn. It was suggested by the dislocation-climb model that the activation energy for the creep in many metals at high temperatures is equal to their activation energy for self-diffusion. According to both models, the activation energy for high-temperature low-stress creep should be equal to that of self-diffusion. The Nabarro-Herring model of creep due to the stress-directed diffusion of vacancies is given as:

$$Z = \dot{\varepsilon} \exp\left(\frac{\Delta H}{RT}\right) = A\sigma \tag{5.18}$$

where Z = Zener Hollomon parameter. A is a constant and ΔH is the activation energy. As seen in Eq. 5.18, Z is related to the strain rate as affected by temperature and it is constant at constant strain. The above relation was given by Sherby as obtained from constant stress creep tests at 0.5 T_m and higher as:

$$\varepsilon = f(\theta) \tag{5.18a}$$

with:

$$\theta = \int_0^t \exp\left(-\frac{\Delta H_c}{RT}\right) dt \tag{5.18b}$$

Equation 5.18 is obtained by differentiating Eq. 5.18b with respect to time at constant stress. Sherby et al. performed experiments at constant stress and stress-differential creep tests, in which the stress was abruptly changed.

An empirical relation for the creep laws of dislocation-climb models was given as:

$$Z = \dot{\varepsilon}_s \exp\left(\frac{\Delta H}{RT}\right) = f(\sigma) \tag{5.18c}$$

where $\dot{\varepsilon}_s$ is the secondary-creep rate and f (σ) is a function of stress and structural changes. By plotting Z as a function of stress, it is revealed that Z increases linearly up to a certain stress, in agreement with the Nabarro-Herring theory. Above this stress value (13 lb/in^2) Z increases linearly with the fourth power of the stress in agreement with Weertman's model. They have concluded that in the case of Al, creep occurs by a dislocation-climb mechanism at all stresses at high temperatures. Furthermore, Sherby et al. indicate, in the results of their experiments on pure Al, that the data obtained seriously question the validity of the theories for creep which postulate stress- or strain-dependent activation energies for creep.

One universal creep function can hardly fit all the experimental data; thus, many functions were suggested to provide an empirical description of creep phenomena. As seen above, functions for logarithmic, transient and linear creep were suggested in various forms and combinations. One cannot disregard an additional empirical function suggested by Wyatt, based on his experiments on polycrystalline copper, presented as a combination of logarithmic, power and linear functions, given as:

$$\varepsilon = a \log t + bt^n + ct \tag{5.19}$$

a, b, c are constants and $n \cong 1/3$. Although this function is empirical and was not obtained on the basis of some physical model, it is flexible in describing creep phenomena for practical use in materials in which creep resistance is required.

Much has been said above on Ni-based superalloy single crystals and the role of γ' in the two-phase structures (matrix is γ). The cost of producing single crystals, compared to polycrystalline materials, is high. Consequently, efforts have been made to produce polycrystalline materials, approaching single crystals in their performance. Directionally-solidified material approaches the performance of single crystals at much lower cost. The basic idea of directionally-solidified alloys is to avoid weakening due to transverse-grain boundaries. Directionally-solidified alloys contain columnar grains with boundaries parallel to the growth direction. The creep life of such alloys, in comparison with conventional polycrystalline materials, is greatly extended. At high temperatures, where turbine blades are used, the grain boundaries are relatively weaker than the matrix. The concept of equicohesive temperature [henceforth: ECT] relates the strength of the grain boundary to that of the matrix. At ECT, these are equal. Below ECT, small grain-sized materials are stronger than the matrix and strengthen the material, due to the high density of the grain boundaries. Above ECT, large-grain materials are stronger, due to the lower tendency toward grain-boundary sliding. Extrapolation of the size to a value in which only one grain is present, i.e. a single crystal is obtained. This is superior for high temperature use, e.g., Ni-base superalloys. Yuan et al. have used directionally-solidified Ni-based superalloys (DZ17G) and found that all of the creep curves exhibit negligible primary creep, relatively little steady-state creep and a dominant tertiary-creep stage. This appears in Fig. 5.13.

Because of the high activation energy for creep (540 kJ/mol), they have assumed that creep is controlled by the Orowan dislocation bowing process.

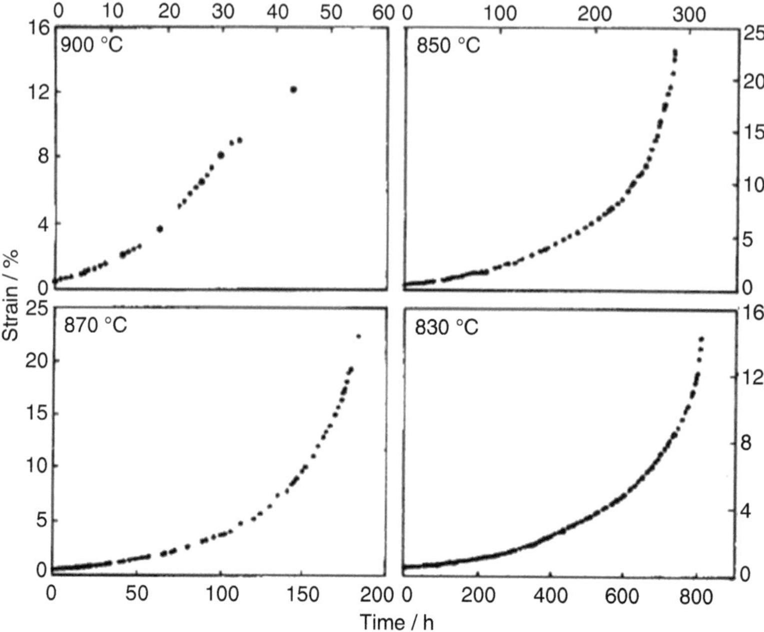

Fig. 5.13 Creep curves of a Ni based superalloy (DZ17G) at ~830–900°C under constant load of 380 MPa (Yuan et al. 1998. With permission of the publisher and Dr. J. Guo for the authors)

The test specimens broke by intergranular fracture as a consequence of creep-cracks propagation along the transverse boundaries, perpendicular to the applied stress axis.

Of the many functions, most of which were empirical, Yuan et al. analyzed the results of primary and secondary creep at high temperatures using Garofalo and Li's expressions; Yuan et al. concluded that the relation previously described by Garofalo and Li fit their results better. The Garofalo and Li's respective equations are:

$$\varepsilon = \varepsilon_0 + \beta\left[1 - \exp(k't)\right] + \dot{\varepsilon}_s t \tag{5.20}$$

and:

$$\varepsilon = \varepsilon_0 + \ln\{1 + \eta\,[1 - \exp(kt)]\}\,\frac{\dot{\varepsilon}_s}{k} + \dot{\varepsilon}_s t \tag{5.21}$$

where ε_0 is the instantaneous strain, β is the primary strain, k' is a constant related to the nucleation of primary creep, $\dot{\varepsilon}_S$ is the steady-state creep rate and k is the dislocation multiplication rate constant. $\eta = \frac{\varepsilon_i - \dot{\varepsilon}_s}{\dot{\varepsilon}_s}$ and ε_i is the initial creep rate. The derivatives of Eqs. 5.20 and 5.21 given respectively are:

$$\ln(\varepsilon - \dot{\varepsilon}_s) = -k't + \text{constant} \tag{5.20a}$$

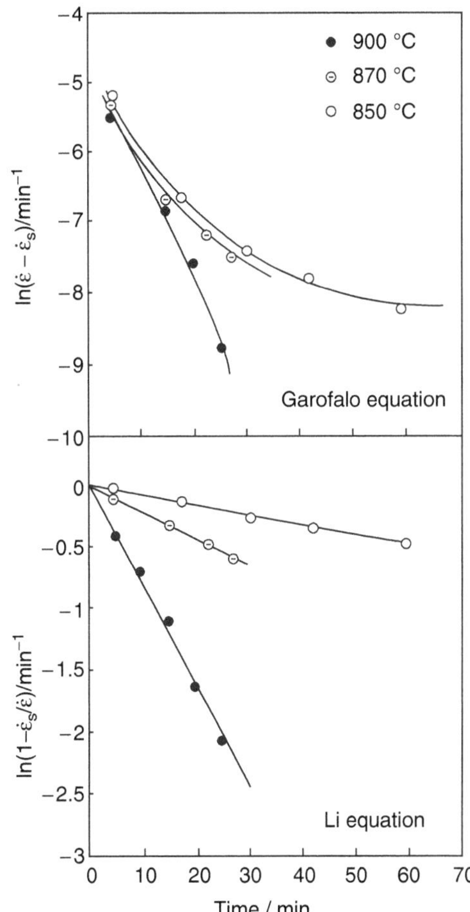

Fig. 5.14 Strain rate as a function of time according to Garofalo and Li. The graphs at all temperatures are linear on the basis of Li's relation, but not according to Garofalo (Yuan et al. 1998. With permission of the publishers and Dr. J. Guo for the authors)

and:

$$\ln\left(1 - \frac{\dot{\varepsilon}_s}{\varepsilon}\right) = -kt + \text{constant} \tag{5.21a}$$

$\ln = (\varepsilon - \dot{\varepsilon}_s)$ and $\ln\left(1 - \frac{\dot{\varepsilon}_s}{\varepsilon}\right)$ vs. t should be linear. Although Garofalo's expression was widely used for primary and secondary creep, Li's relation, which is based on dislocation kinetics, provides a more satisfactory description of the results for DZ17G Ni-based superalloy, as shown in Fig. 5.14.

Using an equation, such as Eq. 5.3, for the secondary-creep rate, namely:

$$\dot{\varepsilon}_s = A\sigma^n \exp-\left(\frac{Q_c}{RT}\right) \tag{5.3a}$$

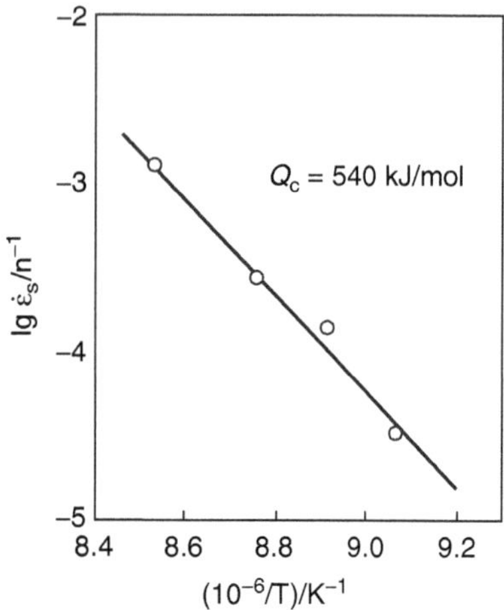

Fig. 5.15 Dependence of steady state creep on temperature according to expression 5.3a (Yuan et al. 1998. With permission of the publishers and Dr. J. Guo for the authors)

an apparent activation energy for creep, Q_c, was derived from Fig. 5.15. In Eq. 5.3a (following a creep-dislocation model), A is a structure-sensitive constant, σ is the applied stress and n is the stress exponent. The activation energy for creep, which was supposed to be in the range of Ni self-diffusion, is higher than the expected [in N_3Al (300 kJ/mole) or in austenite (~265–285)]. This suggests that creep, in the case of DZ17G Ni-based superalloy, is not a diffusion-controlled process. Again, Yuan et al. had concluded that creep is controlled by an Orowan bowing process.

Tertiary creep eventually leads to fracture, but only after the extended creep life of a directionally-solidified alloy. Thus, the tendency to replace single crystals with directionally-solidified polycrystalline materials was successful in reducing the production costs of creep-resistant alloys even for high-temperature applications, as confirmed by various experiments (e.g., Yuan et al. extensively considered above).

Creep-resistant steels are used in power plants, steam turbines, weldments, etc. The steels considered for creep-resistant applications are ferritic, austenitic and martensitic, all alloyed with various elements which form carbides, nitrides, etc. Creep-resistant steels must be reliable over very long periods of time, at high temperatures and, very often, in severe environments. Their microstructures must be very stable, regardless of the type of application at the high temperature. It is important to estimate the long-term creep strength in order to ensure the safe operation of the materials intended for use. Steels containing 9% Cr have been of interest for such applications, mainly for power-plants. The oxidation resistance in the steam of 9% chromium steels is not sufficiently high for their use as thin-walled components at temperatures of 600°C and above. There is potential for the development of steels with a higher chromium content (above 11%), to produce an improvement in their steam-oxidation resistance, while maintaining the creep resistance level of the 9% chromium steels. Recent advances in creep-resistant steels for power plant applications were discussed by Ennis and Czyrska-Filemonowicz, concentrating on thermal treatment and the resulting microstructure. The microstructure depends both on the temperature and on the duration of creep deformation. Figure 5.16 presents

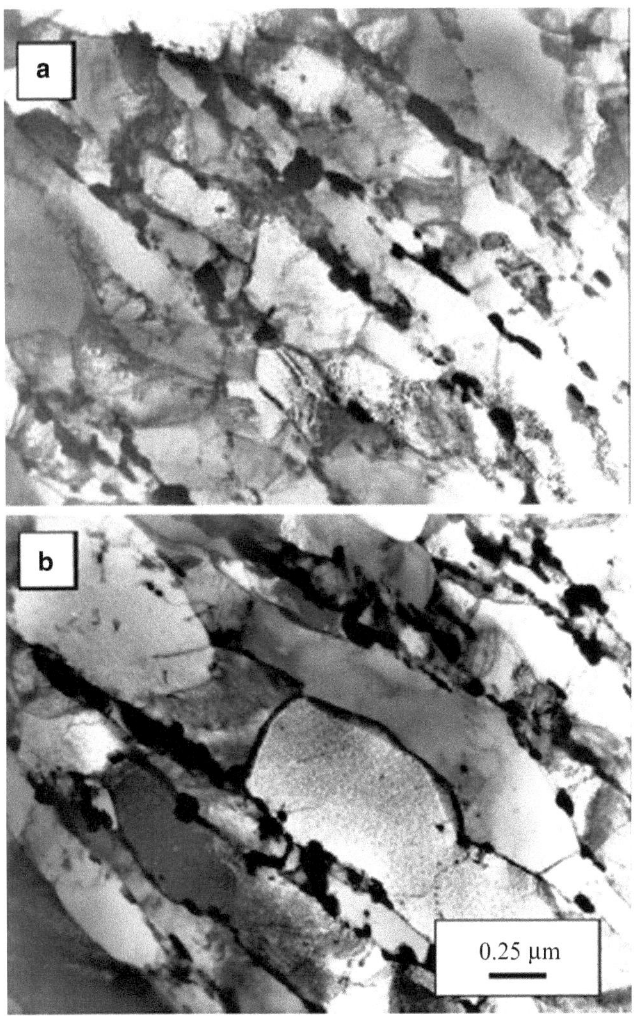

Fig. 5.16 TEM micrograph of P92 (9. 02% Cr) steel after creep deformation at: (**a**) 600°C for 33,000 h and (**b**) 650°C for 27,500 h (Ennis and Czyrska-Filemonowicz 2003. With kind permission from Springer Science and the authors)

a TEM micrograph after creep deformation at 600°C and 650°C up to 32,909 h and 27,433 h, respectively. Well-developed subgrains of low dislocation density in the interiors are characteristic features of long-term exposed specimens. Similar effects were observed for the P91 (8.10% Cr) and E911 (9.16% Cr) steels. In Fig. 5.17, dislocation interaction with precipitates can be seen after creep deformation. The precipitates (carbides, nitrides and carbonitrides) are distributed intragranularly and act as obstacles to moving dislocations, thus contributing to increased creep strength. The dislocation densities within the subgrains decrease with increasing

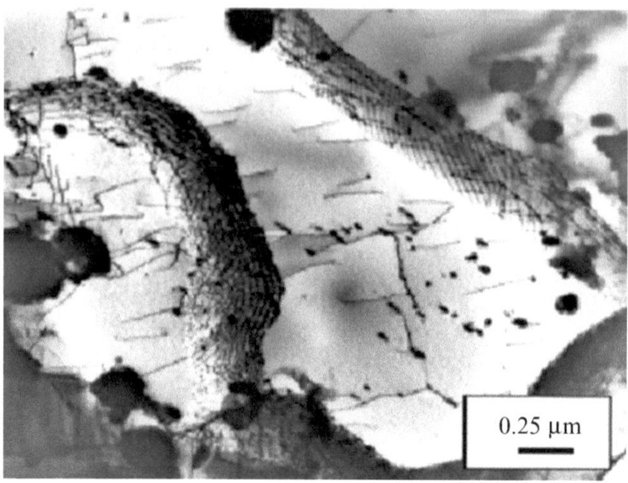

Fig. 5.17 TEM micrograph of P92 (9. 02% Cr) steel after creep deformation at 650°C for 6,468 h. Dislocations interaction with fine carbide precipitates are indicated (Ennis and Czyrska-Filemonowicz 2003. With kind permission from Springer Science and the authors)

creep exposure. In the extensive literature, it is reported that creep-resistant steels can also be improved by the addition of various carbides, such as VC, TiC, NbC, Mo_2C or $Cr_{23}C_6$.

The aim of Ennis and Czyrska-Filemonowicz was to combine, in one and the same steel, the strength of the 9% Cr steels and the steam-oxidation resistance of the 12% Cr steels for the temperature range up to 625°C. These steels, under development, contain increased amounts of V and N and are given an appropriate heat treatment to produce finely dispersed nitrides. Austenite-stabilizing elements, allowing the production of fully martensitic structures, are necessary in order to obtain the optimum microstructure for high-creep rupture strength and improved steam-oxidation resistance.

Secondary creep-rate curves are seen in Fig. 5.18. The secondary-creep rate is derived by taking the slope of the straight line portion of each curve. Figure 5.18 shows the creep strength plotted as iso-stress curves. Under intergranular creep conditions, it is often observed that the steady-state or minimum-creep rate, $\dot{\varepsilon}_s$, and the time, t_f, to fracture for many metals and alloys may be simply related by means of an equation of the form:

$$\dot{\varepsilon}_s^m t_f = C_{GM} \tag{5.22}$$

C_{MG} is called the 'Monkman-Grant constant' and lies in the range 0.05–0.5. Differentiating Eq. 5.22, the logarithm of the strain rate is obtained as:

$$\ln \dot{\varepsilon}_s = -\left(\frac{1}{m}\right) \ln t_f + \frac{1}{m} \ln C_{GM} \tag{5.23}$$

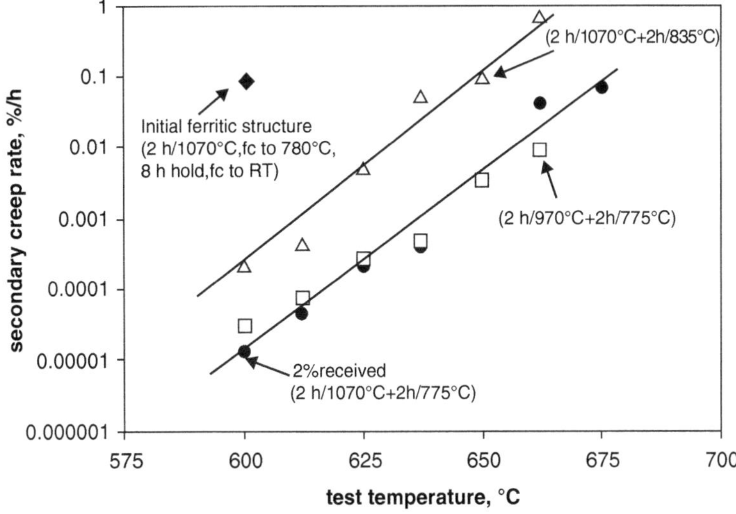

Fig. 5.18 Iso-stress plots at 120 MPa of secondary creep rate for P92 (9.02% Cr) in as-received condition and after different heat treatments: 2 h = 970°C + 2 h = 775°C; 2 h = 1,070°C + 2 h = 775°C; 2 h = 1,070°C + 2 h = 835°C; 2 h = 1,070°C; furnace cool to 780°C; 8 h = 780°C, furnace cool to RT (ferritic structure, no martensite) (Ennis and Czyrska-Filemonowicz 2003. With kind permission from Springer Science and the authors)

From a plot of $\ln \dot{\varepsilon}_s$ against $\ln t_f$, which should be linear, one can evaluate the constants m and C. The Monkman-Grant plots for P91, P92 and E911 are shown in Fig. 5.19 and indicate that a rupture life of 1,00,000 h at 600°C corresponds to a secondary-creep rate of 1.5×10^{-5}%/h.

It was indicated earlier that creep is a deformation phenomenon in which continued plastic strain operates over long periods of time at loads below the yield point. Creep failure may take years to occur. Applications of materials are typically limited by creep, since, for power-plants, they are designed to operate at high temperatures without creep failure for periods up to ~30 years. Meeting such requirements requires that these steels have a stable microstructure, containing carbides, nitrides, etc., able to resist dislocation motion, though changes are inevitable over a long service time. Sufficient solid-solution strengthening is also required to ensure long-term creep resistance and sufficient recovery processes. Given the large number of interacting variables, other requirements, such as weldability, corrosion and oxidation resistance, are also essential. It is, nevertheless, difficult to quantify the design process, given the large number of factors to be considered. The prediction of long-term strength, by means of any extrapolation procedure, is difficult owing to several factors, among them, of great concern, is the scatter in the experimental data. Further microstructural changes are inevitable and are likely to occur during long-term testing, done specifically to determine applicability and use in actual structures.

There is a large variety of heat-resistant steels. The engineering design of power-plants is based on the ability to support a stress of 100 MPa for 10^5 h at the service

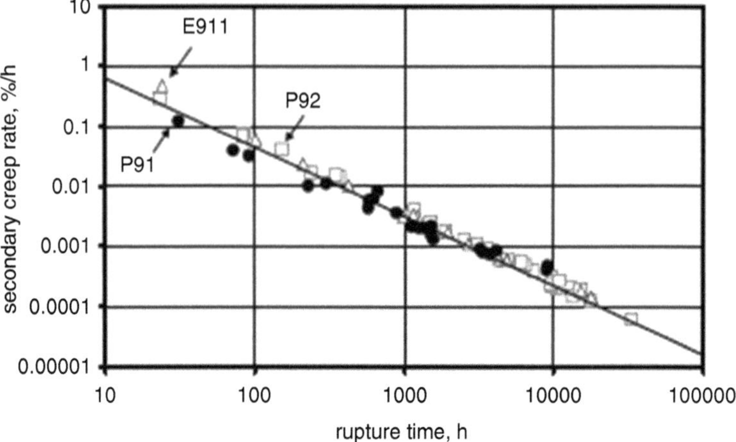

Fig. 5.19 A plot of the Monkman-Grant relation for a few Cr containing creep resistant steels for power plants applications (Ennis and Czyrska-Filemonowicz 2003. With kind permission from Springer Science and the authors)

temperature. Many attempts have been made to develop a quantitative design of heat-resistant steels and welding alloys that might serve as good and reliable alloys for creep applications. There are many types of creep-resistant steels currently being used in the manufacture of different components for power-plants, to serve in temperatures up to 650°C. Roughly, the steels of interest are: (a) ferritic (with pearlitic structure) for use up to ~480–500°C; (b) bainitic-martensitic, in the range of 560–600°C; (c) martensitic, 650°C and; (d) Austenitic, ~750°C. Nowadays, the most widely-used steel of these types is the martensitic creep-resistant steel, due to its higher strength at higher service temperatures.

5.4 Mechanisms of Creep

There are several basic mechanisms that can contribute to creep in materials. These are listed below:

1. dislocation slip;
2. climb, which can lead to subgrain formation;
3. grain-boundary sliding; and
4. diffusion flow by vacancies.

A short summary of the above contributions to creep follows:

1. Glide by slip strengthens materials as they deform. In primary creep, stress is constant, while strain increases to a certain extent (see Fig. 5.1a), but strain rate decreases (Fig. 5.1b).

2. During steady-state creep, the increased strain energy stored in the material, due to deformation, together with the high temperature, provide the driving force for the recovery process. There is, therefore, a balance between the processes of work hardening and recovery. Recovery involves a reduction in dislocation density and the rearrangement of dislocations into lower energy arrays, such as sub-grain boundaries. For this to occur, dislocations must climb, as well as slip and this, in turn, requires atomic movement or self-diffusion within the lattice. Hence, it is often said that the activation energies for self-diffusion and for creep are almost the same.

Vacancies must be located at a site where climb is supposed to occur, to enable climb by means of a vacancy-atom exchange. As the temperature increases, atoms gain thermal energy and the equilibrium concentrations of these vacancies in the metals increase exponentially. The number of vacancies, n, is given (see, for example, Damaks and Dienes) as:

$$n = N \exp\left(-\frac{E_F}{kT}\right) \tag{5.24}$$

N is the number of lattice sites and E_F is the energy of vacancy formation. The activation energy for the jump rate, J, is given by the sum of the energy of vacancy formation and the vacancy's energy for motion, E_M, $(Q = E_F + E_M)$:

$$J = J_0 \exp -\frac{Q}{kT} \tag{5.25}$$

J_0 represents the respective entropies. The diffusion coefficient, D, may be given as:

$$D = D_0 \exp -\frac{Q}{kT} \tag{5.26}$$

D_0, the pre-exponential factor, is equivalent to J_0 and Q is the overall activation energy for self-diffusion. The rate of steady-state creep increases with temperature, as does the essential number of vacancies for effective vacancy-atom exchange for climb (see earlier Sect. 2.6.2).

3. Grain-boundary sliding is considered in detail in Sect. 5.5 below.
4. Diffusion flow by vacancies must be considered, since the mechanism of creep depends on both temperature and stress. The various methods detailed below involve some sort of diffusion occurring with vacancy-atom exchange. This can occur either by lattice diffusion or grain-boundary diffusion, or both may be involved. Bulk diffusion assisted creep occurs during the processes listed in (a–d) below, where the kinetics of atom-vacancy exchange is performed by lattice diffusion. Afterwards, creep involving grain-boundary diffusion will be considered (e).

   (a) Nabarro-Herring creep;
   (b) climb, in which the strain is actually obtained by climb;

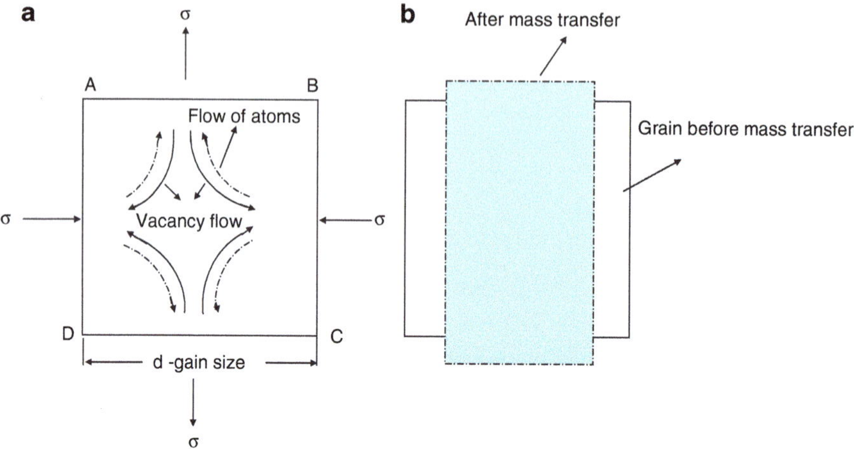

Fig. 5.20 The Nabarro-Herring concept of creep: (**a**) a schematic of vacancy and mass flow; (**b**) the elongated grain in the tensile-axis direction after mass flow

(c) climb-assisted glide, in which climb is a mechanism allowing dislocations to bypass obstacles;
(d) thermally-activated glide via cross slip;
(e) Coble creep, involving grain-boundary diffusion.

Before entering into a detailed discussion of the above list, and based on what has been said thus far on the subject, one can summarize briefly: (a) creep in metals, namely time-dependent plastic deformation, may occur during mechanical stresses well below the yield stress and (b) in general, two major creep mechanisms characterize the time-dependent plastic-deformation process: dislocation creep and diffusion creep. Now, a detailed discussion of paragraphs (a–d) follows.

5.4.1 Nabarro-Herring Creep

One type of creep, in which creep is diffusion-controlled, is Nabarro-Herring creep. In this type of creep, atoms diffuse through a lattice, causing grains to elongate along the stress axis. Mass transport (i.e., the diffusion of atoms) takes place in regions ranging from lower to higher tensile stress. A common illustration may be seen in Fig. 5.20. This schematic figure illustrates the flow of vacancies and atomic movements as induced by tensile stress, σ. During creep-deformation, vacancy-atom exchanges take place to and from the grain boundaries. One would expect that, during creep under tension, atoms would tend to diffuse from the sides of the specimen in the direction shown in Fig. 5.20 (a counter-flow of vacancies), causing sides to lengthen. Assume that local equilibrium of the vacancy concentration exists at the boundaries of the crystal when no stresses are acting on it. Also note that grain boundaries serve as vacancy sources or sinks. In this mechanism lattice diffusion occurs within the grain and the creep rate (strain rate) is assumed to be proportional to the vacancy flux. It will be shown below that the strain rate is inversely proportional to the square of the grain size, i.e., $\dot{\varepsilon} \propto \frac{1}{d^2}$ (Rieth et al., Garofalo and Ruoff). In Eq. 5.24, the number of vacancies was given. Vacancy concentration at equilibrium is given as:

$$\frac{\mathrm{n}}{\mathrm{N}} = \mathrm{C_V}^0 = \exp\left(-\frac{\mathrm{E_F}}{\mathrm{kT}}\right) \tag{5.27}$$

The energy to create a vacancy under acting stress is given by:

$$\mathrm{E_F} + \sigma \mathrm{V} \tag{5.28}$$

where V is the atomic volume (here, it is the volume of a vacancy) and E_F is defined by Eq. 5.27. There is a small concentration difference in vacancies between the faces of AB and BC shown in the above figure, where tensile and compressive stresses are acting, respectively. Denoting the vacancy concentrations at the respective faces as C_V^+ and C_V^- and their difference ΔC, one can write for each of them, by means of Eqs. 5.27 and 5.28, respectively:

$$\mathrm{C_V}^+ = \exp - \left(\frac{\mathrm{E_F} - \sigma V}{\mathrm{kT}}\right) = C_V{}^0 \exp\left(\frac{\sigma V}{kT}\right) \tag{5.29}$$

$$C_V{}^- = \exp - \left(\frac{E_F + \sigma V}{kT}\right) = C_V{}^0 \exp\left(-\frac{\sigma V}{kT}\right) \tag{5.30}$$

$$\Delta C = C_V{}^+ - C_V{}^- = \frac{\alpha}{V}\left\{\exp\left(-\frac{E_F}{kT}\right)\left[\exp\left(\frac{\sigma V}{kT}\right) - \exp\left(-\frac{\sigma V}{kT}\right)\right]\right\} \tag{5.31}$$

Recalling that sinhx = 1/2 [exp(x) – exp(–x)], Eq. 5.31 can be rewritten as:

$$\Delta C = \frac{2\alpha}{V}\exp\left(-\frac{E_F}{kT}\right)\sinh\left(\frac{\sigma V}{kT}\right) = \frac{2\alpha}{V} C_V{}^0 \sinh\left(\frac{\sigma V}{kT}\right) \tag{5.32}$$

where $C_V{}^0 = \exp(-E_F/kT)$ and E_F is the energy of vacancy formation in the absence of stress.

As indicated, there is a flow of atoms from the tensile to the compressed faces, and an opposite flow of vacancies. When a concentration gradient exists, diffusion flux will occur. This flux of vacancies may be expressed as:

$$\mathrm{J} = -\mathrm{D_V}\Delta \mathrm{C} = -\frac{\alpha D_V(\Delta C)}{d} \tag{5.33}$$

D_V is the diffusion coefficient of the vacancies and α is a geometrical factor. The corresponding transport of matter occurs in the opposite direction and produces a creep strain under the applied stress. In a unit time, Jd^2, atoms in the crystal leave the faces under compression and are added to the faces under tension. (Recall that J is the number of atoms in a unit time per unit area; thus, multiplying this value by the square of the grain size, d, one gets the number of atoms per unit time). Consequently, the grain lengthens in the tensile-axis direction and gets thinner in the transverse direction. The change in grain size may be written as:

$$\Delta d = \frac{(Jd^2)\,V}{d^2} = JV \tag{5.34}$$

where V is the atomic volume (often given as Ω). Since the strain rate is given as:

$$\dot{\varepsilon} = \frac{\Delta d}{d} = \frac{JV}{d} \tag{5.35}$$

An expression for the strain rate, given by Eq. 5.36, is obtained by substituting the value of Δd from Eq. 5.34 into Eq. 5.35, followed by inserting J from Eq. 5.33 into Eq. 5.35 to get:

$$\dot{\varepsilon} = \frac{\alpha D_V \Delta C}{d}\frac{V}{d} = \frac{\alpha D_V \Delta C\, V}{d^2} \tag{5.36}$$

With Eq. 5.32 substituted into Eq. 5.36, it is possible to write:

$$\dot{\varepsilon} = \frac{2\beta}{V}\frac{D_V V}{d^2}\exp\left(-\frac{E_F}{kT}\right)\sinh\left(\frac{\sigma V}{kT}\right) \tag{5.37}$$

For small values of stress, and since the nominator is always smaller than the denominator, the quotient is small and sinh(σV/kT) = σV/kT. Substituting this value into Eq. 5.37, one obtains:

$$\dot{\varepsilon} = \frac{2\beta D_V}{d^2} C_V{}^0 \frac{\sigma V}{kT} \tag{5.38}$$

D_V is the diffusion coefficient of vacancies. $D_V C_V{}^0$ is, D_S, the self-diffusion coefficient. Thus, Eq. 5.38 may also be expressed as:

$$\dot{\varepsilon} = \frac{2\beta D_S}{d^2}\frac{\sigma V}{kT} \tag{5.39}$$

More exact calculations, in terms of shear strain (i.e., γ = 2b/d) and macroscopic shear stress, τ (i.e., σ = βτ and β is close to unity and recalling that the shear stress at 45° is given by $\tau = \sigma\sqrt{2}$) gives:

$$\dot{\gamma}_s = \frac{32\alpha\beta D_s \tau V}{\pi d^2}\frac{1}{kT} \tag{5.40}$$

This relation defines a simple, ideal, viscous solid. One sees that increasing grain size reduces creep rate. Creep-rate change is proportional to d^{-2}. Nabarro-Herring creep is a low-stress and high-temperature process.

A somewhat alternate method for showing that $\dot{\varepsilon} \propto \frac{1}{d^2}$ is as follows. Based on Eqs. 5.27, 5.28, 5.29, and 5.30, the difference in concentration may be expressed as:

$$\Delta C = C_V{}^{+} - C_V{}^{-} = \frac{\alpha}{V}\left\{\exp\left(-\frac{E_F}{kT}\right)\left[\exp\left(\frac{\sigma V}{kT}\right) - \exp\left(-\frac{\sigma V}{kT}\right)\right]\right\} \quad (5.31)$$

The flux of the vacancies, going from the tensile to the compressive regions, is:

$$J_V = -D_V\frac{\Delta C}{\Delta x} \quad (5.41)$$

where Δx is the distance in the x direction, so that $\Delta C/\Delta x$ is a gradient. One has to bear in mind that the atomic, J, is in the opposite direction to the vacancy flux, J_V, and, therefore, $D\Delta C = -D_V\Delta C_V$. In our case, the diffusion distance is l.

Stress is not constant along the grain faces, therefore, the diffusion paths are shorter near the corners. Due to stress relaxation, one may assume that $\sigma = \beta\sigma_S$ at distance d/4 from the boundaries (when σ_S is the macroscopic shear stress and β is nearly unity). The length of the diffusion path through this point is $l = \pi/2(d/4)$. The atomic flux across the area of a single atom is given by:

$$J = \alpha D_V\frac{\Delta C}{l} = \alpha D_V\frac{8\Delta C}{\pi d} \quad (5.42)$$

The previous expression is the result of substituting for the value of $l = \pi/2(d/4)$. D_V is the diffusivity of vacancies. One may rewrite Eqs. 5.34 and 5.35 as:

$$\Delta d = \frac{(Jd^2)\,V}{d^2} = JV \quad (5.34)$$

$$\dot{\varepsilon} = \frac{\Delta d}{d} = \frac{JV}{d} \quad (5.35)$$

substituting from Eq. 5.42 for J gives:

$$\dot{\varepsilon} = \alpha D_V\frac{8\Delta C}{\pi d}\frac{V}{d} \quad (5.43)$$

and from Eq. 5.32:

$$\Delta C = \frac{2\alpha}{V}\exp\left(-\frac{E_F}{kT}\right)\sinh\left(\frac{\sigma V}{kT}\right) = \frac{2\alpha}{V}C_V{}^{0}\sinh\left(\frac{\sigma V}{kT}\right) \quad (5.32)$$

When the argument in the hyperbolic function is small, as mentioned earlier, it is equal to the argument; thus, for the strain rate one can write:

$$\dot{\varepsilon} = 16\alpha \frac{D_V C_V{}^0}{\pi d^2} \frac{\sigma V}{kT} = \frac{16\alpha D_S \sigma V}{\pi d^2 kT} \tag{5.44}$$

D_S is the self-diffusion coefficient and is equal to $C_V{}^0 D_V$. Again, the strain rate is proportional to d^{-2}.

One sees in Eq. 5.40 and Eq. 5.44 that the strain rate is linearly proportional to the stress and inversely proportional to the grain size. In Eq. 5.40, the expression is given in terms of shear strain and macroscopic shear stress. The above expressions explain why large-grained materials are preferential for creep applications at high temperatures.

5.4.2 Dislocation Creep and Climb

Bulk diffusion-assisted creep occurs in the processes listed above, namely in (b) climb; (c) climb-assisted glide and; (d) thermally-activated glide via cross slip. All these are obviously associated with dislocation motion.

High stress, below yield stress, causes creep by conservative dislocation motion, namely by dislocation glide within its slip plane. This readily occurs at high temperatures above 0.3 T_M for pure metals and at about 0.4 T_M for alloys, where the dependence on strain rate becomes quite strong. A formulation used for such creep is:

$$\dot{\gamma} \sim \left(\frac{\sigma_S}{G}\right)^n \tag{5.45}$$

where n has a value of 3–10 in high-temperature regimes. Since n is in the exponent, this creep is referred to as 'power law creep'. At high temperatures, obstacle-blocked dislocations can climb, in addition to glide. If gliding dislocations are blocked by some obstacle, climbing may release them to move on until they meet another obstacle, where the same process is repeated. Climb is performed by the diffusion of vacancies through the lattice or along the dislocation core, diffusing into or out of the dislocation core. By climbing, dislocations change their slip planes, enabling them to bypass their obstacles. Dislocation glide is responsible for most strain, while the average dislocation density is determined by the climb step in the deformation process. This mechanism is known as 'climb-controlled creep'.

5.4.3 Climb-Controlled Creep

At relatively high stresses beyond the elastic or the shear moduli, creep is controlled by dislocation-glide movement and by glide in adjacent planes following climb.

Real materials contain various internal obstacles (such as dislocations) or external ones (introduced intentionally, such as solutes and particles, or unintentionally by the fabrication process) which block dislocation glide in their respective slip planes. The dislocation motion is also hindered by the crystal structure itself, namely by crystal resistance, an internal obstacle. At high temperatures, obstacle-blocked dislocations can be released by dislocation climb. Creep arises as a consequence of climb, when further deformation by glide is enabled by means of vacancy-atom exchange. Creep rate is a function of several factors, usually given as:

$$\sum \dot{\varepsilon} = f(\sigma, T, S, GS, P)$$

S is the structure, GS is the grain size and P represents material properties, such as lattice parameter, atomic volume, etc. Vacancies increase with increasing temperature and are likely to diffuse to dislocations, thus, decreasing the overall free energy of the system. By the diffusion of vacancies to locations in which dislocations are blocked by obstacles, climb becomes possible, letting dislocations bypass those obstacles. Climb allows further glide in an adjacent slip plane to occur and, by such deformation, creep strain arises.

A steady-state-based model for edge-dislocation climb (Rieth et al., Garofalo) has been suggested by Weertman. He assumed that strain hardening occurs whenever dislocations are hindered in their motion by some obstacle and pile up behind it. The dislocations beyond the barrier, such as a Lomer-Cottrell lock, might escape by climbing. However, climb beyond Lomer-Cottrell barriers would lead to the generation of new dislocation loops and to a steady-state creep rate (which is applicable to FCC and BCC structures, but not to HCP). In an additional communication, Weertman further suggested that edge dislocations of opposite signs, gliding on parallel slip planes, would interact and pile-up when a critical distance of 2r between them is not exceeded. In such a case, as in the prior case, dislocations could escape from the piled-up array by climb. Dislocation pile-ups lead to work hardening, whereas climb is a recovery process. A steady state is reached when the hardening and recovery rates are equal. The creep rate will, therefore, be controlled by the rate at which dislocations can climb. This climb mechanism requires the creation of vacancies or their destruction at the obstacle-blocked dislocations (in this case, at the pile-up) in order to maintain the equilibrium concentration required to satisfy the climb rate. At the tip of a pile-up dislocation, a non-vanishing, hydrostatic stress $\pm\sigma_i$ may develop, exerting a force on the dislocation in a normal direction to the slip plane and causing a positive (up) or negative (down) climb. Vacancies will be absorbed where the stress is compressive and they will be created where the stress is tensile. A change in vacancy concentration develops in the vicinity of the dislocation line and a vacancy flux is established between the segments of the dislocations, acting as sources or segments of sinks.

The vacancy concentration, C_e, in equilibrium with the leading dislocation in the pile-up, is given by:

$$C_e = C_o \exp\left(\frac{\pm 2L\sigma_S{}^2 b^2}{GkT}\right) \tag{5.46}$$

2L is the length of the dislocation pile-up and C_0 is the equilibrium concentration of the vacancies in a dislocation-free crystal. The vacancy concentration at a distance, r, from each pile-up is assumed to be equal to C_0. The rate of climb, $\dot{X}$, is given (Garofalo) as:

$$\dot{X} = \frac{2C_0 D_V \sigma_S{}^2 L b^4}{GkT} \tag{5.47}$$

D_V is the vacancy-diffusion coefficient and $2Lb^2\sigma_S{}^2/GkT < 1$.

When self-diffusion occurs due to the vacancy mechanism, C_0D_V may be replaced by:

$$C_0 D_V = D_S = \frac{\nu}{b} \exp\left(\frac{\Delta S}{R}\right) \exp\left(-\frac{\Delta H}{RT}\right) \tag{5.48}$$

and $\dot{X}$ is given by:

$$\dot{X} = \frac{2\sigma_S{}^2 L b^3}{GkT} \nu \exp\left(\frac{\Delta S}{R}\right) \exp\left(-\frac{\Delta H}{RT}\right) \tag{5.49}$$

ΔH is the activation energy for self-diffusion, ν is a frequency factor and S is an entropy term. Equation 5.49 is obtained under the assumption that vacancies are easily destroyed or created and that an equilibrium concentration exists between pile-ups of dislocations. However, the diffusion of flux vacancies may be different in specific climb processes.

In an additional model by Weertman, the rate of dislocation climb is also given by Eq. 5.47 or Eq. 5.49 and the steady-state creep-rate model, in this case, becomes:

$$\dot{\gamma} = NAb\frac{\dot{X}}{2r} \tag{5.50}$$

N is the density of the dislocations participating in the climb process (or the density of the sources), A is the area swept out by a loop in a pile-up and 2r is the separation between the pile-ups. The stress necessary to force two groups of dislocation loops to pass each other on parallel slip planes must be greater than $\frac{Gb}{4\pi\sigma_S}$ (in terms of shear stress it is $\frac{Gb}{4\pi\tau}$). When this relation is satisfied, an estimate for r can be made:

$$r = \frac{Gb}{4\pi\sigma_S} \tag{5.51}$$

The probability, p, of blocking the dislocation loops generated from one source by means of loops emanating from three other sources is given by:

$$p = \frac{8\pi NL^2 r}{3} = \frac{2NL^2 Gb}{3\sigma_S} \tag{5.52}$$

Using Eqs. 5.47, 5.50, 5.51, 5.52 and setting p = 1 and A = 4πl², the creep rate at low stresses becomes:

$$\dot{\gamma}_S = \frac{C\pi^2 {\sigma_S}^{4.5} D_S}{\sqrt{bN} G^{3.5} kT} \tag{5.53}$$

C is a numerical constant of the order of 0.25 and D_S is the coefficient of self-diffusion.

Equation 5.53 has been substantiated experimentally for pure metals to a greater extent than other theoretical relations. Exceptions to the exponent 4.5 were obtained, but this value is very close to the observed experimental values.

5.4.4 *Glide via Cross-Slip*

Edge dislocations climb when their motion is hindered. The non-conservative motion of screw dislocations is by cross slip, since they cannot climb. The ease of cross slip is stacking-fault dependent (see Chap. 2). Materials with high stacking-fault energy cross slip readily, but not so when the stacking-fault energy is low. For a screw-oriented dislocation, the Burgers vector is parallel to the dislocation line (see Chap. 2) and, therefore, it can move in any plane in which it lies (in isotropic materials). In real crystals (which are in most cases anisotropic), screw dislocations may favor certain planes with the lowest energy. Cross slip can occur without diffusion, but thermal activation helps cross slip movement from the original to other slip planes. Climb and cross slip are recovery processes. Recall that steady-state creep is a deformation process, balanced by work hardening and dynamic recovery. The temperature dependence of creep is:

$$\dot{\varepsilon} \sim \exp - \left(\frac{Q_c}{kT}\right) \tag{5.54}$$

One of the known equations for steady-state creep, indicating stress dependence (Friedel) is:

$$\dot{\varepsilon}_s = A\sigma^n \exp - \left(\frac{Q_c}{kT}\right) \tag{5.55}$$

Here, Q_c is the activation energy for creep, n is the stress exponent. A similar expression may be given for climb-controlled creep:

$$\dot{\varepsilon}_s = A\sigma^n \exp - \left(\frac{Q_c}{kT}\right). \tag{5.56}$$

But in this case, Q_c is independent of applied stress (Friedel). At lower temperatures, the cross slips of screw dislocations is the process by which obstacles in the slip plane may be bypassed.

Since the study of cross slip is more informative in single crystals, many experiments have been performed on single crystals with various structures. Thus, in order to investigate the glide system in FCC metals, Al single crystals were deformed by compression parallel to [00I] at temperatures between 225°C and 365°C at strain rates between 9×10^{-6} and 9×10^{-4}/s (Le Hazif and Poirer). (Note that Al has high stacking-fault energy and readily cross slips). The stress-strain curves exhibit three stages, which have been correlated with observations of slip lines and dislocation structure. The unique observation was that, after a small percent of deformation cross slips of $\frac{a}{2}\ \langle 1\bar{1}0\rangle$, screw dislocations from $\{111\}$ to $\{110\}$ occurred, that might be responsible for the $\{110\}\ \langle 1\bar{1}0\rangle$ slip. The deformation at the first stage occurs, as expected in FCC metals on the $\{111\}$ planes, but after a small deformation, slip on $\{110\}$ sets in once the stress reaches a critical value, σ_{110}. This stress is thermally activated and decreases with temperature increase. Why dislocations cross slip on $\{110\}$ planes, rather than on $\{111\}$ planes (as is usually the case), is not clear, though several explanations have been given. An activation energy for the creep rate, $\dot{\varepsilon}$, of 28 kcal/mol, determined at a constant stress of σ_{110}, is close to the reported cross slip in Al. It is likely that these observations are compatible with the mechanism of cross slip by screw dislocations from $\{111\}$ to $\{110\}$ planes and that a stacking fault stable at high temperatures stabilizes slip in $\{110\}$. The possibility of a stacking fault in $\{110\}$ is explained on geometrical grounds and the dislocation proposed is expressed as:

$$\frac{a}{2}[110] = \frac{a}{12}[110] + \frac{a}{3}[110] + \frac{a}{12}[110]$$

Stacking fault energy, which determines the separation of the partial dislocations, improves creep resistance if it is low. Contrary to the high stacking-fault energy observed in Al (in which cross slip or climb occurs readily), in low-energy stacking-fault materials with large separation, cross slip by creep or climb is suppressed. This was observed by Suzuki et al. in their work on Mg-Y alloys with added Zn. The addition of small amounts of Zn has a beneficial effect on creep resistance, because it widens the separation between the partials by decreasing the stacking-fault energy. The average separation of partials in this alloy is given as:

$$d_S = \frac{Gb_1b_2}{8\pi\gamma}\left(\frac{2-\nu}{1-\nu}\right)\left(1 - \frac{2\nu\cos(2\alpha)}{2-\nu}\right) \tag{5.57}$$

where d_S is the separation width between the partials, γ is the stacking-fault energy, ν is the Poisson ratio and α is the angle between the total Burgers vector and the dislocation line. A large stacking-fault energy drop has been calculated, compared with pure Mg. Magnesium alloys are being used to an increasing extent in applications where the components are subjected to elevated temperatures. Consequently, research is being focused on the development of alloys able to withstand high stresses at temperatures up to 300°C, depending on the application. Thus, for example, in other Mg alloys improved creep properties are produced by the addition of rare earth alloys (Mordike). At low temperatures, a climb mechanism for edge dislocations exists, whereas, at higher temperatures, the cross-slip mechanism of screw dislocations is believed to operate. Opinions about the cross-slip mechanism are not unanimous, but a majority of the researchers support it. Whether the acting mechanism is climb or cross slip, it is most likely that the beneficial effect of alloying stems from the fact that they both widen the separation between the partials.

Much work was has been done on Ni-based alloys for high-temperature use. For example, the steady-state creep of Ni_3Al alloys containing 10 at% Fe below 680°C creep was thought (Nicholls and Rawlings) to occur by cross slip from octahedral $\{111\}$ planes onto cube $\{100\}$ planes. Above this temperature, up to 930°C, the creep-rate controlling mechanism is considered to be the removal/production of antiphase domain boundaries during climb. The activation energy of 3.27 eV atom^{-1} was derived for this process. Schoeck and Seeger expressed the creep rate for cross slip as:

$$\dot{\varepsilon} = C \exp - \left[\frac{\Delta H_0 - c \ln (\sigma/\sigma_c)}{kT} \right] \tag{5.58}$$

ΔH_0 is the energy for cross slip, σ_c is the critical resolved shear stress, σ is the applied stress and C and c are constants. A model for creep controlled by cross slip from $\{111\}$ to $\{100\}$ in the temperature range of 53–680°C over the stress range 360–600 MN m^{-2} was found to be in good agreement with the experimental results. The energy to form a restriction between the partials, namely to recombine the Shockley partials, was evaluated on the basis of Dorn's expression (Dorn). (See also Hemker et al. for the creep mechanism at intermediate temperatures in Ni_3Al).

In a recent work on high-temperature low-stress creep, it was observed in Ni-based superalloy single crystals (Zhang et al.), that during the primary-creep stage, the large lattice misfit in TMS-138 superalloys causes a large misfit stress. The superposition of misfit stresses and externally applied stress drives loops of matrix dislocations to move by cross slip through the matrix channels. In the alloy TMS-75(+Ru), the lattice misfit value is small and the force driving the dislocations to overcome the Orowan resistance is inadequate, so the dislocations move by climbing around the γ' cuboids. Figure 5.21 shows the morphology of the dislocations in the alloys TMS-75(+Ru) and TMS-138 after creep for 2 h. Some dislocations were propagating and spreading out in these two specimens. In Fig. 5.21a, from the dotted dislocation lines in TMS-75(+Ru), it was determined that many dislocations

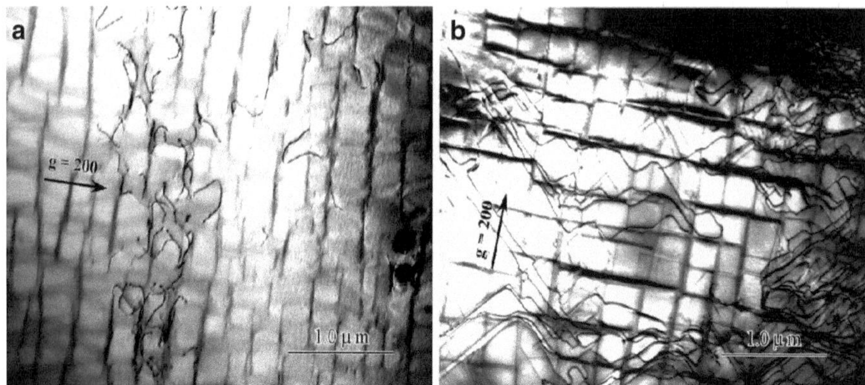

Fig. 5.21 Microstructure during creep; t = 2 h. (**a**) TMS-75(+Ru). Dislocations are moving by a process of gliding and climbing. (**b**) TMS-138. Dislocations are moving by cross-slip (Zhang et al. 2005. With permission of the authors and Elsevier)

were inclined in the TEM thin film. However, most dislocation lines were curved. Stereomicroscopy analysis confirmed that these dislocations were gliding in $\{111\}$ planes in the γ matrix or climbing around the cuboidal γ' precipitates along the γ/γ' interfaces. The Burgers vector was determined to be $b = 1/2[10\bar{1}]$. At this stage, most of the area of this specimen is free of dislocations. Creep dislocations are spreading into the previously dislocation-free areas from some dislocation sources. In Fig. 5.21b, the morphology of the dislocations in the TMS-138 specimen after creep for 2 h is seen. Their morphology is considerably different from those in TMS-75(+Ru). These dislocations have considerable length in the TEM specimen, indicating that the dislocation lines are parallel to the (001) plane. Stereo imaging indicates that all of these dislocations are in the γ channels, which are normal to the [001] tensile axis. The expanding dislocation loops assume very long, narrow shapes with zigzagged characteristics.

Most of these dislocations have a Burgers vector $b = 1/2[\dot{1}01]$. The dislocation characteristics indicate that these dislocations are moving by cross slip in the horizontal γ channels. In Fig. 5.22, the dislocation morphology is shown when the creep has been carried out for 5 h. It was found that the density of the dislocations in the γ channels continued to increase and most parts of these two specimens were full of dislocations. In Fig. 5.22a, b, the dislocation configurations in TMS-75(+Ru) are shown from two orientations, [001] and [011]. These dislocations below are still in the course of climbing and gliding; the configurations characterize the climbing of the dislocations along the surface of the γ' cuboids. However, the dislocations in Fig. 5.22c have a remarkably different morphology than those shown in Fig. 5.22a. Cross gliding in the horizontal γ channels is the main characteristic of dislocation motion in TMS-138. The deposited slip dislocations on the surfaces of the γ' cuboids are oriented in the [110] and $[1\dot{1}0]$ directions. In contrast, in some areas,

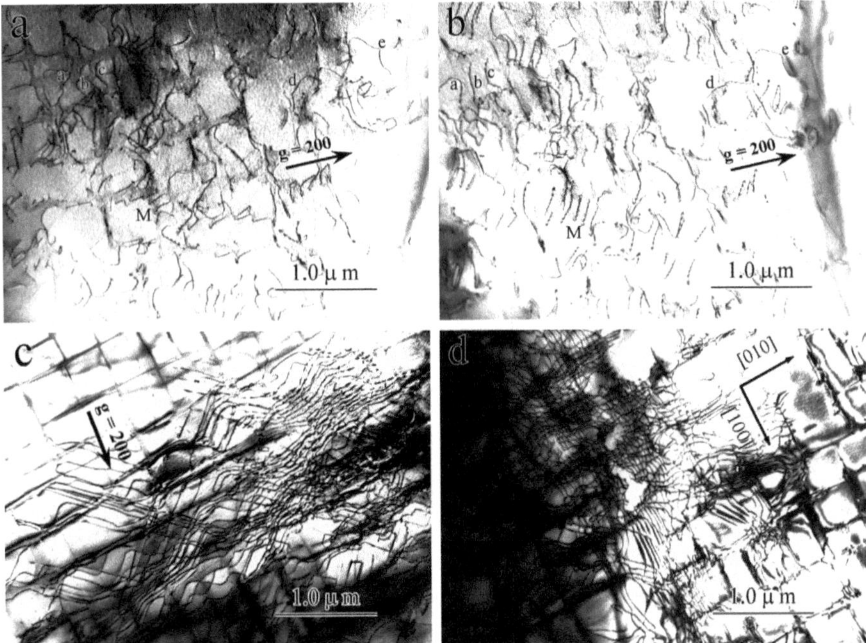

Fig. 5.22 Morphology of dislocations; creep t = 5 h. (**a, b**) TMS-75(+Ru). The observation was made from the [001] orientation and the [011] orientation, respectively. These dislocations are climbing around the γ' cuboids in the γ matrix. (**c, d**) TMS-138. The beam directions are along [001]. The cross slip of dislocations in the horizontal γ channel is dominant (**c**). In some areas, dislocation networks appear (**d**) (Zhang et al. 2005. With permission of the authors and Elsevier)

γ/γ' interfacial dislocation networks have formed (as shown in Fig. 5.22d), which means that these dislocations have completed the reorientation from the deposition direction <100> to the mismatch direction <100> in the (001) γ/γ' interfacial plane. In summary of their work in an alloy with a large, negative lattice misfit, these dislocations are able to move smoothly by cross slip in the horizontal γ channels. But in alloys with small lattice misfits, the dislocations move by a combination of climbing and gliding processes and the resultant γ/γ' interfacial dislocation network is incomplete. A good explanation of creep curves is obtained from these different microstructures.

In general, creep at temperatures below 0.5 T_M is not thought to occur by the diffusion-controlled mechanism.

Seldom does a lone creep mechanism operate at any given time. Creep mechanisms can operate simultaneously (in parallel) or independently. For two mechanisms, one can write:

$$\dot{\varepsilon} = \sum_i \dot{\varepsilon}_i \tag{5.59}$$

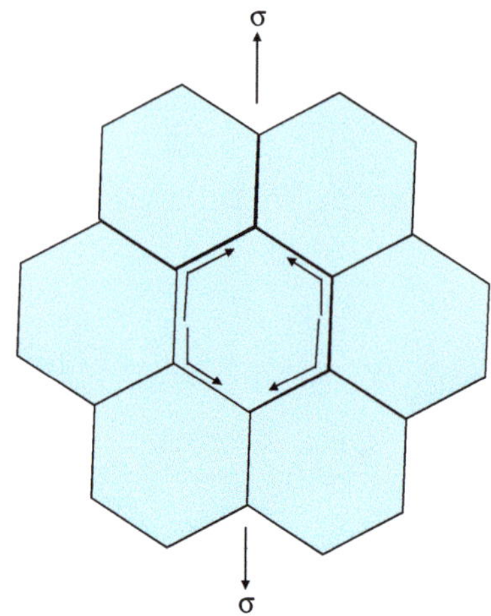

Fig. 5.23 Seven grains are shown in a two-dimensional hexagonal array before creep deformation. Following diffusion, the grains elongate in one direction and decrease perpendicularly to the tensile axis. A void formation develops between the grains, but grain-boundary sliding, which may accompany this process, removes these voids

or:

$$\frac{1}{\dot{\varepsilon}} = \sum_i \frac{1}{\dot{\varepsilon}_i} \tag{5.60}$$

In the case of parallel creep mechanisms, the fastest mechanism will dominate the overall creep, whereas when they operate in sequence, the slowest process controls the creep deformation.

5.4.5 *Coble Creep*

Coble creep is also a diffusion creep, but involves grain-boundary diffusion. The diffusion of atoms along grain boundaries produces a change in dimensions, due to the flow of the material. Of the two kinds of self-diffusions in polycrystalline materials, the one occurring at low temperatures is grain-boundary dominated, whereas lattice diffusion occurs at high temperatures. Figure 5.23 is an illustration of ideal grain structure, showing the flow of atoms along the boundaries under the influence of a tensile stress. In a polycrystalline matrix, the grain shape is not as indicated in Fig. 5.23 (for an ideal structure), but varies in orientation, making it difficult to analyze.

Coble, in his original paper, used a spherical grain (apparently following the Nabarro-Herring approach for lattice-controlled diffusion creep). In Coble's analysis of creep, a spherical grain was used once again. Based on experimental results for Al_2O_3, where it was observed that the Al ion-diffusion coefficient is larger by orders than that of oxygen ions and, since the creep rate in lattice-controlled diffusion is limited by the least mobile species, it was expected that the O^{2-} species would determine the rate of creep. Coble suggested that grain-boundary diffusion, rather than lattice diffusion, might control creep deformation. He proposed that Al diffuses in the lattice and O^{2-} in the grain boundaries, where its diffusion coefficient is enhanced in comparison with the values in the lattice. It was assumed that the spherical grain maintains constant volume and, thus, the areas of the vacancies at the source and the sink must be equal (grain boundaries may act as source or sink for vacancies). The average gradient of the spherical grain, having a radius R, is given as $\frac{\Delta C}{(R\pi/2)}$. The problem is to evaluate the concentration gradient at the 60° boundary, which, for equal areas of rotational symmetry, lies at 60° below the pole of a hemisphere.

For steady-state creep, where Fick's law applies, the flux at the 60° boundary is:

$$J_{vac\ \sec^{-1}} = D_V N \left[\frac{\Delta C}{(R\pi/2)} \right] (W) 2\pi R \sin 60 \tag{5.61}$$

Here, D_V is the diffusion coefficient of the vacancies in the boundary, N is a proportionality constant relating the average vacancy gradient, $\Delta C/(R\pi/2)$, and the maximum gradient, $1/R(dc/d\theta)_{\theta=60}$, W is the effective boundary width and $(2\pi R\sin 60)$ is the length of the zone in which the diffusion flux is at a maximum. Thus, the cross-sectional area for diffusion is $2\pi RW\sin 60$. After a detailed and lengthy evaluation of the relevant parameters, Coble arrives at the final expression for creep rate, given as:

$$\dot{e} = \frac{148\sigma(D_b W) a_0{}^3}{(GS)^3 kT} \tag{5.62}$$

where $a_0{}^3$ $(=\Omega)$ is the atomic volume of a vacancy. Other expressions are given for Coble's creep, the difference being in the coefficient representing the assumptions in each case. Thus, a common expression, by writing the coefficient as a constant, is:

$$\frac{d\varepsilon_{gb}}{dt} = \dot{\varepsilon} = A \frac{\sigma \Omega D_{gb} \delta}{l^3 kT} \tag{5.63}$$

The subscripts refer to the grain boundaries, Ω is the atomic volume (of a vacancy), δ is the grain-boundary width and l is the grain size. (In Nabarro-Herring, the grain size was denoted by d). D_S of expressions (5.40) or (5.39) is replaced in Coble's equation as $D_{gb}\delta$. The factor 1/l represents the density of the cross-section of the grain boundaries per unit area, hence δ/l is the cross-sectional area of the grain boundaries per unit area. In a realistic structure, A depends on the grain structure and the way the average grain size is determined. Creep by grain-boundary diffusion has a stronger dependence on grain size than on lattice diffusion. In terms of shear strain and shear stress (Rieth et al.) the expression is:

$$\dot{\gamma} = 42 D_S \frac{\pi \delta \tau \Omega}{d^3 kT} \tag{5.64}$$

Here, d is equivalent to l and D_S to D_{gb}. When creep deformation is influenced by both lattice- and grain-boundary diffusion, an expression may be derived as follows. Equation 5.44 may be written with the same designations used in Eq. 5.63 as:

$$\dot{\varepsilon} = 16\alpha \frac{D_S \sigma \Omega}{\pi l^2 k t} \tag{5.44}$$

By adding Eqs. 5.44 and 5.63, one gets:

$$2\dot{\varepsilon} = \frac{16\alpha D_S \sigma \Omega}{\pi l^2 k T}\left(1 + \frac{A\pi}{16\alpha}\frac{D_{gb}\delta}{l D_S}\right) \tag{5.65}$$

Designating $16\alpha/2\pi = B$ and $A\pi/2 \times 16\alpha = C$ gives:

$$\dot{\varepsilon} = \frac{B D_S \sigma \Omega}{l^2 k T}\left(1 + \frac{C D_{gb}\delta}{D_S l}\right) \tag{5.66}$$

An expression for creep may be given in terms of shear-strain rate and shear stress, when both lattice- and grain-boundary diffusion are involved in the deformation. For most polycrystalline materials, diffusion in grain boundaries is more rapid than in the lattice.

To summarize this section, it may be stated that in Coble creep the atoms diffuse along the grain boundaries and elongate the grains along the stress axis. This causes Coble creep to have stronger grain-size dependence than in Nabarro-Herring creep. Since the grain boundary is the controlling diffusion mechanism in Coble creep, the process occurs at lower temperatures than the Nabarro-Herring creep. Coble creep is still temperature dependent and, as the temperature increases, so does the grain-boundary diffusion. It also exhibits a linear dependence on stress, as does the Nabarro-Herring creep.

5.5 Grain-Boundary Sliding

Grains and their sizes are very important variables characterizing the microstructure of polycrystalline materials. Grain-boundary movement plays a significant role in the characteristic behavior of materials for creep application. Basically, grain-boundary sliding [henceforth: GBS] is a process in which grains slide past each other along their common boundary. It has also been observed that sliding may occur in a zone immediately adjacent to the grain boundary (Wadsworth et al.). As seen in earlier chapters, the role of grain size in the work-hardening mechanism, (strengthening is crucial), as given by the Hall-Petch relation:

$$\sigma_y = \sigma_0 + \frac{k_y}{\sqrt{d}} \tag{3.42}$$

In primary creep, the required stress increases due to work hardening, which also acts in steady-state creep, but is balanced by various recovery processes. Decreasing grain size should indicate a stronger material, since higher stress is crucial for continued deformation. Thus, one may expect materials with small grain sizes to show better creep resistance, while increasing grain size should cause an increased secondary-creep rate. This is attributed to the decrease in boundary barriers with increasing grain size (less strengthening media exists, because there are less grain-boundary obstacles). However, this is true as long as no undesirable processes occur at the grain boundaries. For example, large-grained materials with a small number of grain boundaries are low sources of vacancies and, therefore, dislocation climb will be reduced compared to small-grained materials.

The major structural changes occur at the start of tertiary creep. Damage is initiated by the formation of cavities, either wedge-shaped or rounded. The wedge-shaped cavities are seen mainly at grain boundaries and their coalescence is the unmistakable sign for creep rupture to occur. It is believed that GBS is a prerequisite for the nucleation of voids and cavities and it occurs when a sufficiently high stress concentration develops to create new surfaces. Cavitations increase with increasing strain at high temperatures. The stresses causing GBS are the shear stresses acting on the boundaries. Whether void formation is associated with or a consequence of GBS has not yet been completely determined, since the experiments found in the literature seem to support both concepts.

GBS has been the subject of numerous investigations, in light of the importance of grain boundaries for many aspects of material applications. Understanding the physics of the complex behavior of grain boundaries is of great interest in regard to topics, such as: growth, crystallization, recovery deformation, just to mention a few. A general review of the properties of grain boundaries may be found, for example, in the work of Valiev et al. Here, GBS is of interest in order to gain better practical and theoretical understanding; experiments were performed on various bi-crystals and analyzed theoretically. Due to the complex nature of GBS even in bi-crystals, HCP bi-crystals, such as Zn (Kaibyshev et al.) or Cd (Fukutomi et al.) were chosen for study. HCP crystals possess only one preferential-slip plane (the basal plane). A state of deformation provides the opportunity to distinguish between the effects of the interactions of deformation processes in the boundaries and in the grain interior. Thus, for example, cutting specimens in different angles results in obtaining bi-crystals with the same boundary, but in different orientations to the loading directions. The different ratios of the stresses in the basal plane and in the grain boundary during mechanical tests might be studied.

The sliding may be with or without concurrent intragranular slip. In the case of proper or pure GBS (without intragranular slip), the shear stress applied is along the grain-boundary plane and sliding arises from the grain shift directly along the grain boundary, when grains are displaced with respect to one another. In other cases, local stresses, caused by intragranular slip, play important roles. Studies at low stresses were found to be closest to the conditions of pure GBS. Some conclusions listed

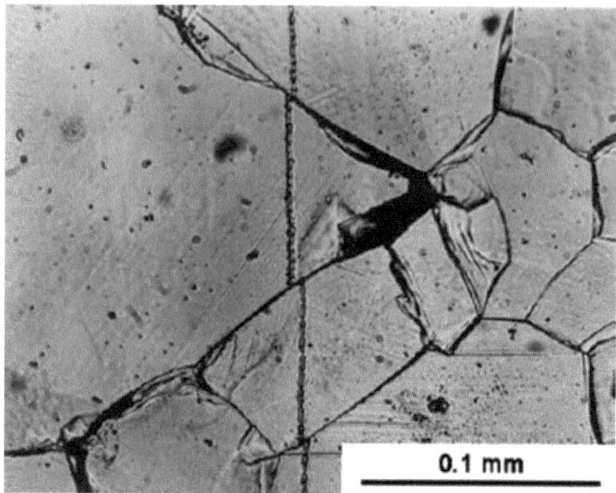

Fig. 5.24 Grain boundary sliding revealed by the boundary offsets in a transverse marker line in a Mg-078% Al alloy tested under creep conditions at 473 K under a stress of 17.2 MPa (From Bell and Langdon 1967. Reproduced from Langdon 2006. With kind permission from Springer Science and the author)

below were drawn from experiments done with HCP bi-crystals in which pure GBS occurred:

(a) There is non-uniform sliding along the boundary. No dislocation emission from the boundaries was observed. Therefore, the non-uniformity of grain-boundary displacement may only be a result of dislocation motion and accumulation in the grain boundaries.
(b) The rate of GBS is proportional to the shear stress at the boundary.
(c) The crystallographic grain-boundary parameters influence the rate of GBS. Under the same deformation conditions as special boundaries, the rate of GBS can be by orders lower than in random ones.
(d) The activation energy of pure GBS is close to that of grain-boundary diffusion.
(e) High stresses and curved grain-boundary surfaces during pure GBS are accompanied by accommodational, intragranular slip, which changes the process.

During pure GBS, there is no interaction between grain boundaries and lattice dislocations. In most experiments, however, GBS is accompanied by intragranular dislocation slip. It is clear that in polycrystalline materials the role of grain boundaries in influencing deformation behavior is of great significance. Deformation is not only influenced by many external factors, but also by other lattice defects.

Evidence of GBS is the displacement of scratch lines during creep testing. Figure 5.24 shows scratch lines displaced across a grain boundary; transverse markings are inscribed perpendicular to the tensile axis. Clear offsets can be seen in the transverse marker line in an Mg-0.8%Mg alloy strained to 2.49% at a temperature of 473 K and at an applied stress of 17.2 MPa. The tensile axis in this experiment is horizontal. An alternate method to evaluate GBS is by using interferometry. An example of the offsets of the same alloy revealed by interferometry is seen in Fig. 5.25.

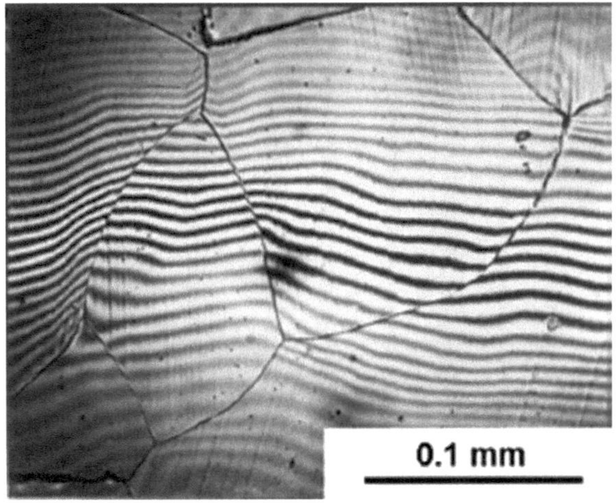

Fig. 5.25 Offset revealed by interferometry in a Mg-078% Al alloy pulled to an elongation of 1.5% at 473 K under a stress of 27.6 MPa (From Langdon 1993. Reproduced from Langdon 2006. With kind permission from Springer Science and the author)

A quantitative estimate of the contribution of GBS to overall strain, ξ, used by Tam and Tam following Langdon's proposal, is:

$$\xi = \frac{\varepsilon_{GBS}}{\varepsilon_t} \tag{5.67}$$

ε_t, the total strain at high temperatures, is expressed as:

$$\varepsilon_t = \varepsilon_g + \varepsilon_{GBS} + \varepsilon_{dc} \tag{5.68}$$

ε_g is the strain in the grain, due to processes taking place within the grain, ε_{GBS} is the strain due to GBS and ε_{dc} is the strain due to diffusion creep. In practice, experiments are often performed where there is negligible contribution of diffusion creep and, thus, Eq. 5.68 reduces to:

$$\varepsilon_t = \varepsilon_g + \varepsilon_{GBS} \tag{5.68a}$$

Damage leading to failure, in the form of stress rupture, is initiated by void and crack formation. The tertiary creep per se is a sign that some sort of structural damage has occurred. Round or wedge-shaped voids, known as 'r-type cavities' and 'w-type cavities', are seen, at first, along grain boundaries and, when they coalesce, creep fracture occurs. As indicated above, the mechanism of void formation is associated with GBS and occurs due to shear stresses acting along the boundaries. A commonly used illustration of a w-type crack initiation by GBS, its formation and growth (first presented by Chang and Grant and found in almost every publication), is shown in Fig. 5.26.

A number of w-crack configurations have been observed experimentally, formed at triple points. In this illustration, crack nucleation at grain boundaries is formed

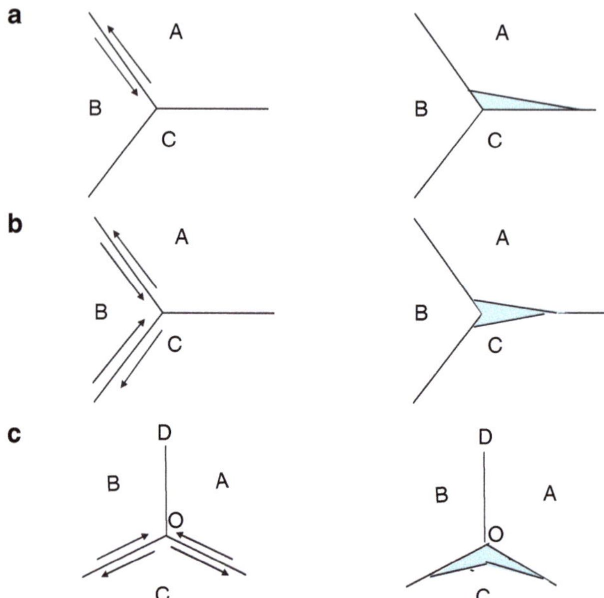

Fig. 5.26 Schematic representation of a w-type crack formation initiated by GBS (From Chang and Grant 1956)

when shear stress acts along the boundaries. The tensile axis in configuration (c) of the Figure is shown by DO. Wedge-type crack formation (shown in (c) or (b)) at triple points was initially suggested by Zener as early as 1948. According to Zener, at sufficiently high temperatures, grain boundaries behave in a viscous manner and when near triple points under an applied tensile stress, wedge-type cracks develop due to the high stress concentration. Another configuration for the initiation of intergranular cracks (somewhat more complex) is shown in Fig. 5.27. The propagation of a typical w-type intercrystalline crack in Al-20%Zn alloy is shown in Fig. 5.28.

GBS may be considered as a deformation mechanism above 0.5 T_m. The strain rate is important to the type of failure by GBS. It has been shown that in commercially pure Ni at 700°C, r-type intergranular cavities form at a strain rate of 10^{-9} s^{-1}, which, with increased strain rate, transforms to w-type and then to transgranular fracture with increasing strain rate to a level of 0.1 s^{-1} (Gandhi and Raj).

Alloying additions can decrease the tendency for w-type-cavity formation. Both type cavities are a result of GBS (Raj). GBS may result in grain-boundary (intergranular) cracking when the grain interior is stronger than the boundaries. GBS can be reduced by adding intergranular particles or by serrated grain boundaries. These provide obstacles to GBS, apparently due to an increase in friction between boundaries. Cavities have been seen to form at grain and phase boundaries

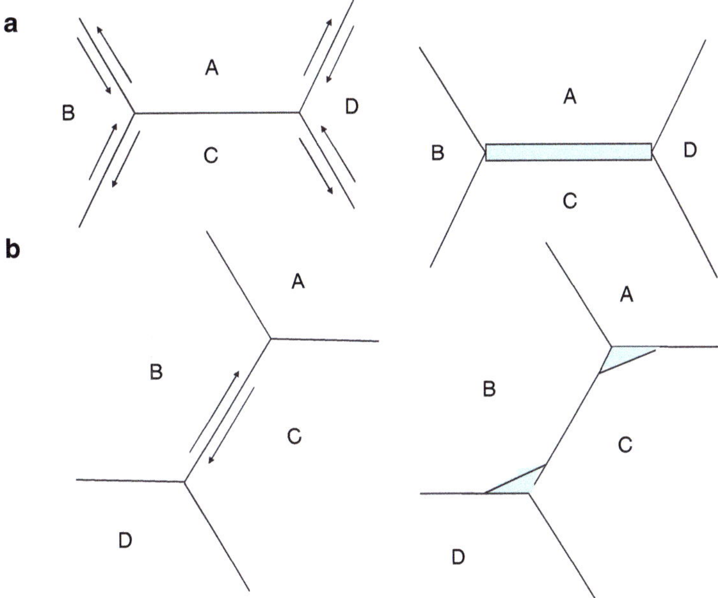

Fig. 5.27 Schematic views showing a more complex intergranular crack initiation by GBS (From Chang and Grant 1956)

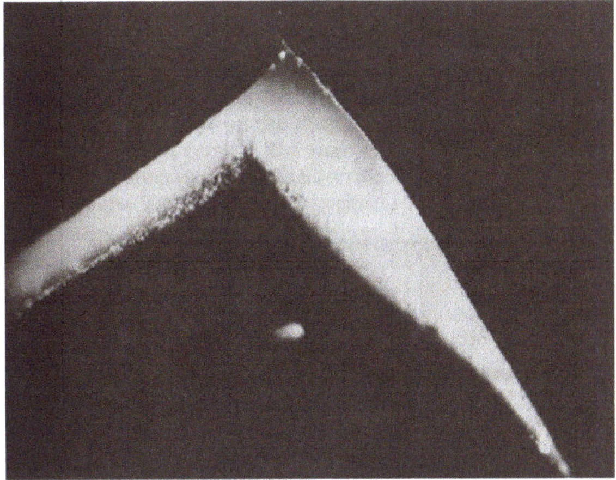

Fig. 5.28 Intercrystalline crack in an Al-20% Zn alloy deformed by creep at 500°F and 2,000 psi (From Chang and Grant 1956)

preferentially at interfaces or triple points. The process of cavitations is associated with GBS and cavity nucleation probably occurs at points of stress concentration in the sliding boundaries or interfaces.

Creep failure takes place by the nucleation, growth and coalescence of creep cavities on the boundaries predominantly perpendicularly-oriented to the applied stress. An increase in the number of cavitated boundaries over creep exposure time

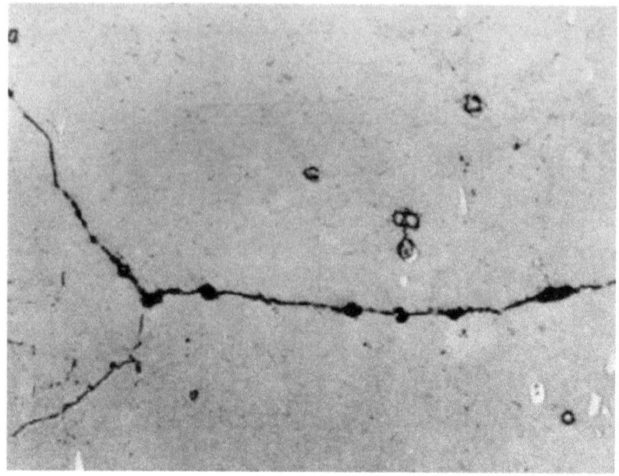

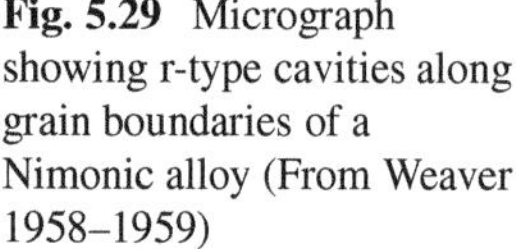

Fig. 5.29 Micrograph showing r-type cavities along grain boundaries of a Nimonic alloy (From Weaver 1958–1959)

supports the mechanism of continuous cavity nucleation and growth. Some believe, on the basis of experimental observations, that there are probably pre-existing cavities, voids or pores, previously introduced by the forming operations and they are actually responsible for creep cavitations in engineering alloys during long-term service at low stresses and elevated temperatures. Many experiments show that GBS is a necessary condition for cavity nucleation. GBS is a key factor, not only in the growth of the pre-existing voids, but also in nucleating voids for cavity formation.

Round-type cavities were illustrated by Weaver and shown in Fig. 5.29. Here, the r-type cavities are along a grain boundary in a Nimonic alloy tested at 750°C. Usually, these cavities develop in boundaries normal to the applied load. Such r-type cavities have been observed in FCC, HCP and BCC metals or alloys. Either type of cavities developed during GBS leads to creep failure.

In many polycrystalline metals at elevated temperatures, GBS contributes significantly to the total strain. GBS can be markedly reduced by introducing alloying elements, which form precipitates (such as nitrides, carbides, borides, etc.) at the grain boundaries. This mechanism of strengthening is used in high-temperature alloys to provide safe use when creep deformation is a problem.

Another method for improving creep resistance in materials is by the evolution of serrated grain boundaries. Serrated grain boundaries are effective in improving the creep-strength properties of heat-resistant alloys. For example, it was demonstrated in several commercial Ni-based alloys that a material with irregular serrated grain boundaries has improved resistance to creep crack growth compared with one having smooth grain boundaries. This is explained as a consequence of the difficulty in GBS and the increase in the path of grain-boundary diffusion (Beddoes and Wallace).

The strengthening mechanisms of serrated grain boundaries are principally the result of: (1) the inhibition of grain-boundary sliding; (2) the retardation of grain-boundary crack initiation, caused by the decrease of stress concentration at grain-boundary triple points, due to the decrease of sliding grain-boundary length

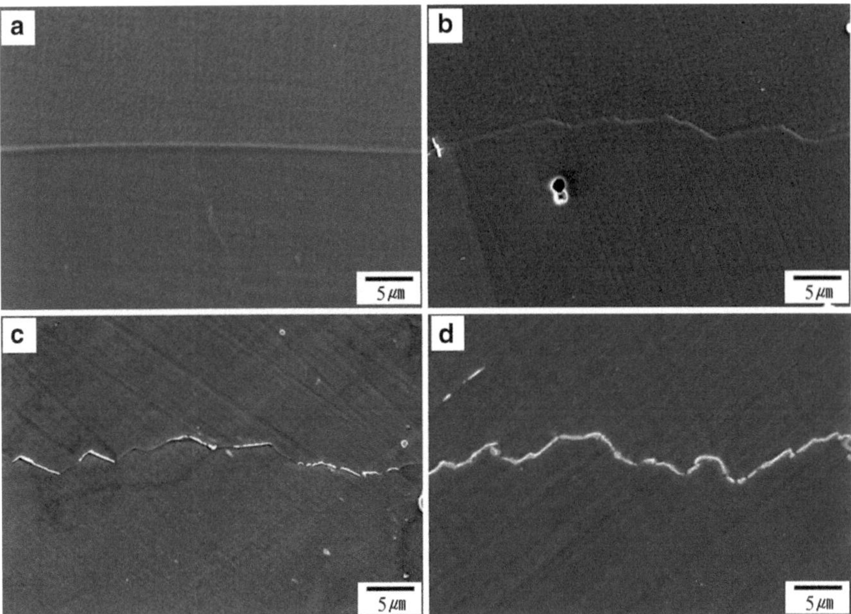

Fig. 5.30 The sequential development of grain boundary serration and subsequent carbide precipitation: (**a**) after solution-treatment (1,150°C/5 min/WQ), the grain boundary is flat (**b**) solution-treatment followed by slow cooling to 1,000°C at 10°C/min, (**c**) solution treatment followed by slow cooling to 800°C at 10°C/min, and (**d**) solution-treatment followed by slow-cooling to 800°C at 10°C/min and aged for 8 h (Hong et al. 2010. With permission of Material Science Forum and the authors)

and; (3) dynamic recovery at serrated boundaries. In a recent publication (Hong et al.), experimental information concerning the beneficial effect of serrated grain boundaries for reducing GBS was evaluated in a Nimonic 263 Ni-based alloy and the role of γ' precipitate was explored (Fig. 5.30). Nimonic 263 is a wrought Ni-based superalloy used in combustion chambers of gas turbines and potentially applicable to outlet headers and steam lines of advanced coal-fired power plants.

The precipitate, γ', was with a volume fraction of $\sim$10% in the γ matrix. Grain-boundary serration occurs prior to the formation of $M_{23}C_6$ and without interaction with γ' particles and is accompanied by the modification of grain-boundary carbide characteristics. This occurs when a specimen is slow-cooled from the solution treatment temperature. The high-resolution observation of the lattice image of the serrated grain boundary suggests that the grain boundaries tend to serrate with specific segments approaching one {111} low-index plane at a boundary, in order to have lower interfacial free energy. This finding implies that serration may be related to the local movement of grain boundaries, in order to have lower interfacial free energy.

Figure 5.30 shows non-serrated and serrated grain boundaries obtained by two thermal treatments of the specimens. The serrated boundary was obtained by a

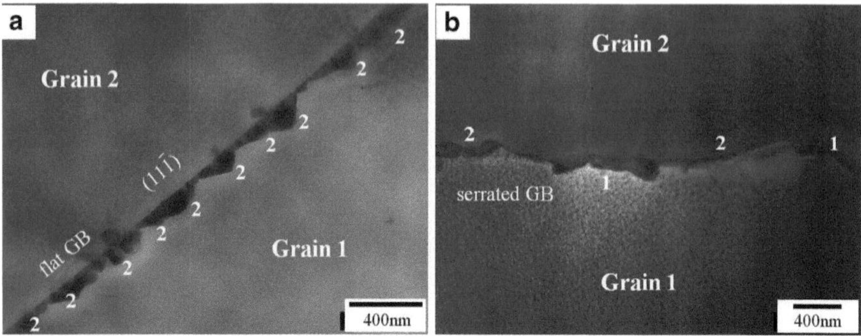

Fig. 5.31 Comparisons of the grain boundary morphology and that of the carbide characteristics after formation of the serration: (**a**) fine granular carbides on flat grain boundary in the standard heat-treated specimen and (**b**) planar carbides on serrated grain boundaries in the special heat-treated specimen. Note that each digit in (**a**) and (**b**) indicates the grain sharing of the coherency with each carbide formed at the grain boundaries (Hong et al. 2010. With permission of Material Science Forum and the author)

special heat treatment; after serration, carbides (identified as $M_{23}C_6$) precipitated and grew along the serrated grain boundary, as shown in (b) through (d). The grain-boundary serration led to a change in the $M_{23}C_6$ carbide characteristics. The carbide morphology changed from granular to planar, its density was lowered and the coherency pattern of two neighboring grains was altered, becoming a zigzag boundary. Figure 5.31 illustrates these changes, comparing the specimens receiving the standard treatment with those receiving the special heat treatments to produce the serrated grain boundaries. Improvement of creep resistance was observed by the introduction of grain-boundary serration without deterioration of the basic mechanical properties. This improvement in creep resistance by serration is associated with a lower rate of cavitation and crack propagation, through the modification of carbide characteristics, as well as grain-boundary configuration.

Figure 5.32 shows the improvement in the serrated grain boundary and that the creep life observed in the serrated sample is about 40% longer than in the non-serrated sample. There is a distinct difference in the fractured surfaces, shown in Fig. 5.33. The zigzag array of cavity formation in the serrated sample (d) makes it more difficult for the cavities to interlink and form an intergranular path for crack propagation than in the non-serrated sample. A lower rate of crack propagation along grain boundaries is expected in the serrated sample and, as a consequence, it can be inferred that the serrated sample is highly resistant to damage by cavity formation. The grain-boundary serration in wrought nickel-based superalloys has significant potential for improving both creep resistance and in-service life-time.

We can summarize this section by noting that GBS may account for 10–65% of the total creep strain, depending on the alloy and the conditions of its use in service (temperature, load, etc.). The contribution to the creep strain increases with increasing temperature, stress and reduced grain size. Above about 0.6 Tm, the grain-boundary region is thought to have lower shear strength than the grains

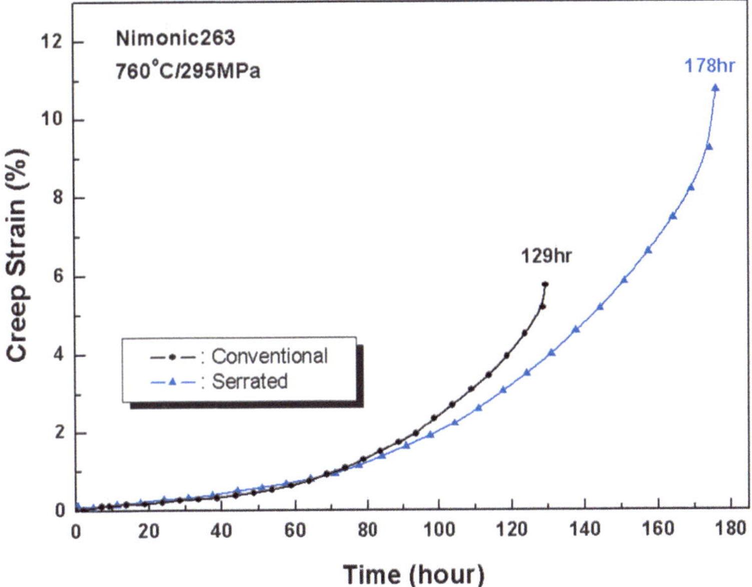

Fig. 5.32 Comparison of creep resistance between the two samples crept at 760°C/295 MPa (Hong et al. 2010. With permission of Material Science Forum and the authors)

themselves, probably due to the looser atomic packing at the grain boundaries. GBS can be reduced by introducing precipitates or grain-boundary serrations, which resist GBS and significantly reduce cavity formation of the types indicated above, which are a major factor in creep failure.

5.6 Creep Rupture

Stress rupture tests are used to determine the elapsed time until failure. Generally, higher stresses than in conventional creep tests are used for creep-rupture testing and these tests are carried out until the specimen fractures. The objectives of the respective tests are to determine the minimum creep rate at stage II creep, on one hand, and to evaluate the time at which failure sets in, on the other hand. Such information is essential so that the proper materials will be selected to eliminate failure in service and to evaluate the time-period of safe use during high-temperature applications of machine elements or structures. Pressure vessels, boilers, gas-turbine engines and other structural components are some of the systems that operate at high temperatures and experience creep. Like the creep test, stress-rupture testing involves the same testing equipment (for example, a tensile specimen) and is performed under a constant load (or stress) at a constant temperature. Not surprisingly, creep failures may appear ductile or brittle. Cavities,

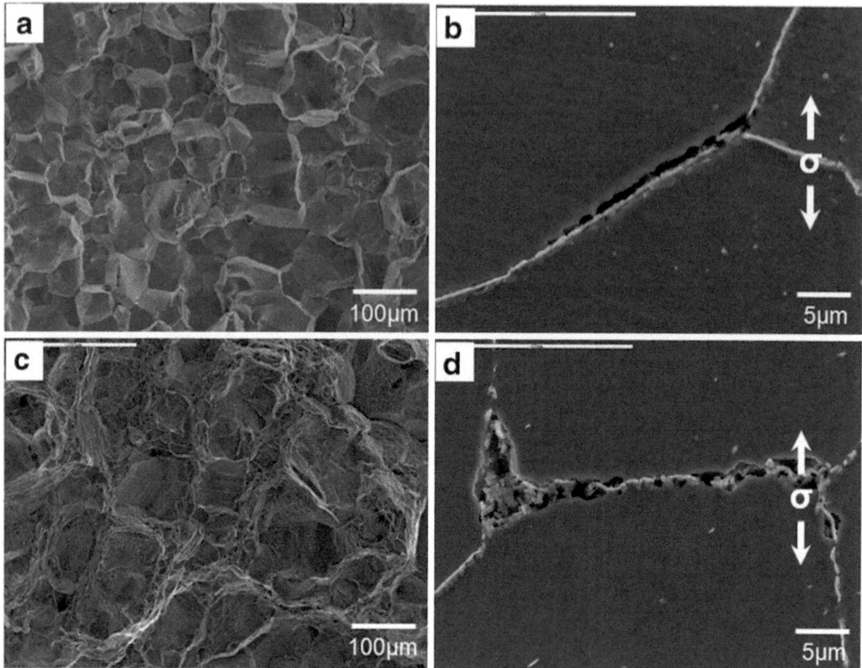

Fig. 5.33 Comparisons of the fractured surfaces and cavitation behaviors along the grain boundaries between two samples crept at 760°C/295 MPa: (**a**) and (**b**) for the unserrated sample, (**c**) and (**d**) for the serrated sample (Hong et al. 2010. With permission of Material Science Forum and the authors)

which are believed to be responsible for cracking by cavity coalescence, can be either r-type or w-type, either transgranular or intragranular. Figure 5.34 is a schematic illustration (based on Fig. 5.1), showing where creep damage starts. The location of creep damage coincides with the place where tertiary creep sets in and represents the minimum creep rate (also seen in Fig. 5.1b). This is one of the concepts regarding the time at which cavitation, either as microcavities or voids, develops. However, there are experimental indications (density measurements) that intergranular cavities can be observed before tertiary creep and are well-developed at the end of second-stage creep.

Experimental data indicate that, of the great variety of creep curves described by various laws, their shape is close to linear relation when the data representation is on a log strain vs. log time basis. Time is expressed in hours. This can be seen in Fig. 5.35 showing castings of Ni-based superalloy gas-turbine engine components.

Here, the best results are obtained for single-crystal components. Indeed, the usual practice in the industry is to fabricate single-crystal parts for such applications. This is especially true in the high-pressure turbine region of the engine, where actual metal surface temperatures of the turbine blades can exceed ~2,000 F (~940°C), which is ~0.75T_m, making creep and stress-rupture properties very important

Fig. 5.34 A schematic creep curve (see Fig. 5.1); ε_f and t_f are the strain and time to creep failure

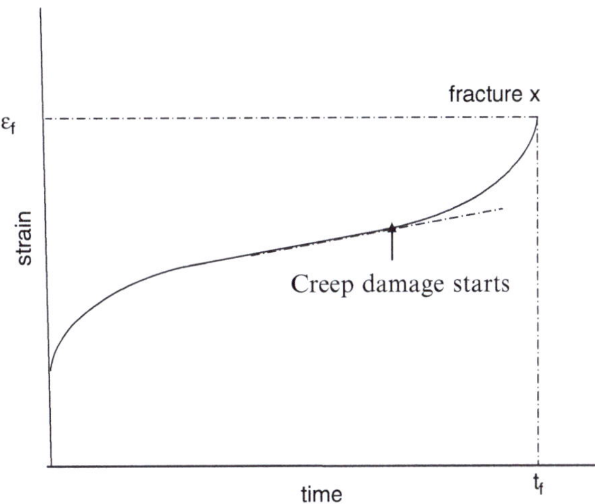

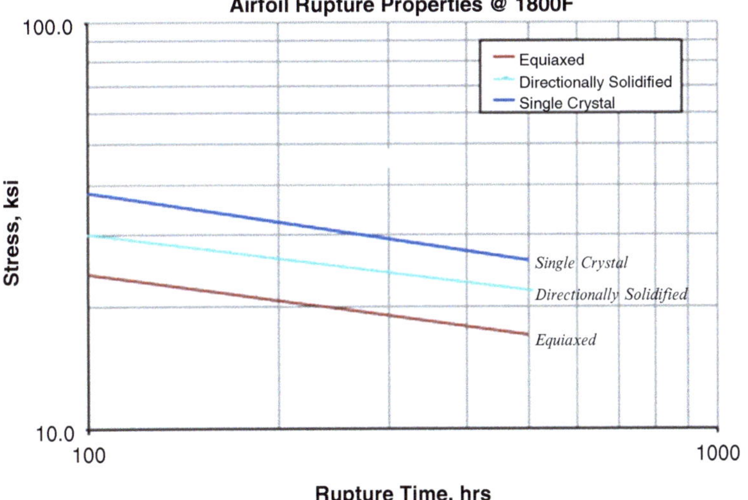

Fig. 5.35 Improved stress rupture life for the castings indicated (Pridemore 2008. With kind permission of Springer)

variables in the overall life of the blade in use. However, use of directionally-solidified alloys is the next best alternative for high-temperature use (as mentioned in Sect. 5.3) and is the present trend, mainly due to cost considerations.

Creep and creep-rupture data can be presented as curves in different ways. One example is given in Fig. 5.35 in terms of stress vs. rupture time. Very often, instead of using the rupture time, the time until reaching a steady-state or minimum creep is preferred, because a much shorter period of time involved in the tests to collect creep data. Stress-rupture tests are used to determine the failure time, as mentioned above.

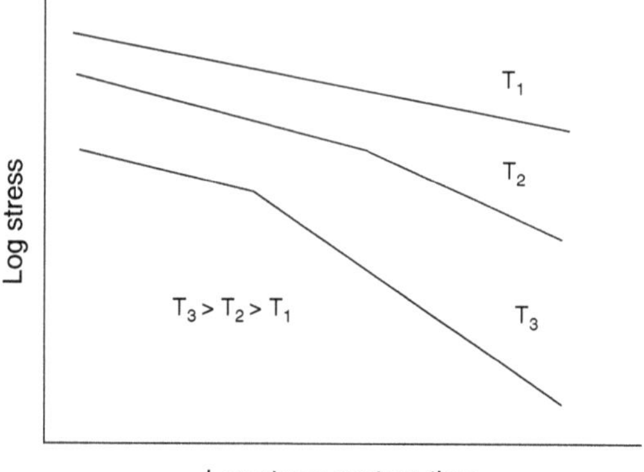

Fig. 5.36 A schematic illustration of logσ vs. logt_r curves at three temperatures. At the higher temperatures, a change in slope occurs, indicating possible structural changes

The data are plotted as log-log curves. A straight line is usually obtained at each temperature if no structural changes occur. This information can then be used to extrapolate time-to-failure for longer times. However, with increased temperature, structural changes are likely to occur, changing the creep resistance to rupture. It is important to be aware of such changes in material behavior. If structural changes do occur, then it is not safe to extrapolate the data of such curves to longer times. A schematic illustration (shown in Fig. 5.36) indicates a change in slope due to structural changes. Experimental plots of stress vs. rupture-time in logarithmic form, as logσ vs. logt_r, are shown (in Fig. 5.37) for high Cr ferritic steels, indicating change in slope above 600°C as a consequence of changes, probably structural; therefore, unreliable time-to-rupture values will result at higher temperatures. Only the extrapolated times up to 10^4 h at 600°C are reliable for creep life-time prediction in this alloy. The test specimens are Cr steels with increasing Cr content, between 9% (S9) and 10.5% (S10) and up to 12% Cr (D12).

The creep data in Fig. 5.37 provide one of the many methods for obtaining creep-rupture time, t_r, by relating it to the activation energy of creep, Q_c:

$$t_r = t_0 \sigma^{-n} \exp\left(\frac{Q_c}{RT}\right) \tag{5.69}$$

where t_0 is a material constant and n is the stress exponent. As mentioned earlier, long-term creep-rupture time is predicted by extrapolation from short-term creep data. This is reasonably safe when no structural changes occur. In the Cr ferritic steels (Fig. 5.37), premature creep failure occurs due to some recovery.

The schematic curves, illustrated in Fig. 5.38 at various temperatures, may be used to evaluate the predetermined design life of a material to be used against creep failure. Note the following results from these tests: (a) no change in slope occurred, thus, allowing extrapolation from the test time to longer times; (b) the schematic illustration shows the test lines for three temperatures. At the higher temperature, the stress of the creep test is lower and it changes over time; and (c) the extrapolation of the test lines to some desired life-time in service shows the stress

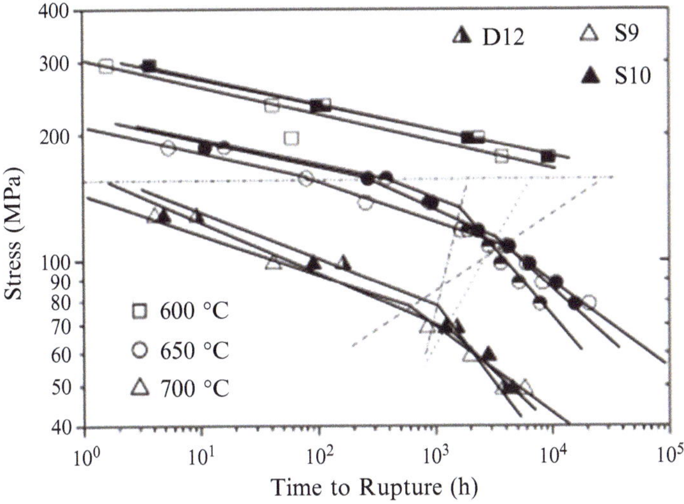

Fig. 5.37 Plot of creep stress versus creep rupture life for Gr. 122 type steels. The onset of premature creep failure is indicated by dashed, dotted and dash-dotted lines for S9, S10 and D12 steels, respectively (Armaki et al. 2010. With kind permission of Elsevier and R. Chen for the authors)

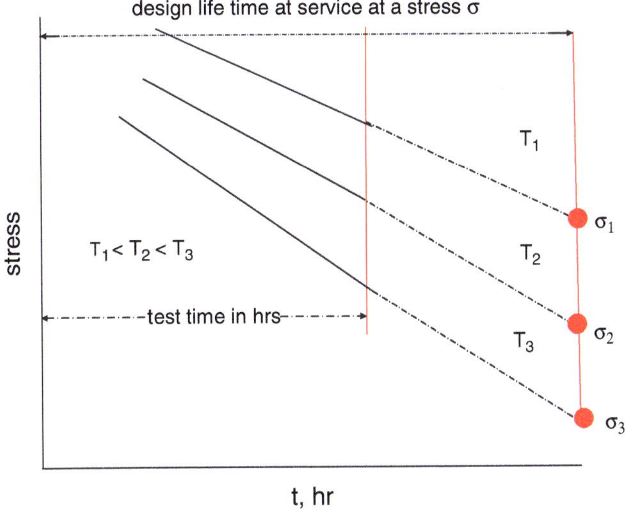

Fig. 5.38 A schematic extrapolation of creep-test data, from the time indicated to the desired, service life-time; extrapolations at three temperatures for the desired, applied stresses

that a material can sustain at the temperatures indicated (the point of intersection). Such creep tests help to predict the actual creep life of a material without failure throughout its designed life-time. These tests are performed with the same creep strain, ε, at each temperature and the temperature level of the test is set higher than the expected service temperature. Another way to obtain the same goal is by using a family of similar curves (as in Fig. 5.38), but instead of a family of curves at different temperatures, the family has different strains. The ordinate is presented as a log strain axis.

Often in creep testing a minimal 1% (of the expected life-time) criterion is used. However, a 10% criterion is preferable in order to obtain a meaningful prediction of the usable life-time for creep applications. For critical applications, such as for turbine components, an ~25% criterion is recommended and, indeed, preferred so as to avoid failure by creep rupture.

Considering Eq. 5.69 and graphs based on such equations, one may immediately recall the well-known Norton-Bailey relation, originally suggested for second-stage creep (or minimum creep rate) and used by many to predict the creep life-time for engineering materials intended for high-temperature applications. A brief summary of the essence of the Norton-Bailey approach follows. The steady-state or minimum creep rate $(d\varepsilon/dt)_{min}$ relates to applied stress and temperature by an empirical Arrhenius power law, given as:

$$\left(\frac{d\varepsilon}{dt}\right)_{\min} = A\sigma^n \exp\left(-\frac{Q}{RT}\right) \tag{5.70}$$

Multilinear regression may then be performed to determine the constants A, n and Q for each material of interest. Such an analysis implies that the same creep mechanism is acting as the dominant, rate-controlling mechanism at all temperatures and stresses.

The constant thermal creep behavior, ε, is a function of:

$$\varepsilon_c = f(\sigma, T \text{ and } t) \tag{5.71}$$

It is usually assumed that this function may be separated into:

$$\varepsilon_c = f_1(\sigma) f_2(t) f_3(T) \tag{5.71a}$$

Norton used what is called the ‘power law’ to describe stress dependence over time under constant stress (but many other expressions are also available). This power law function is given as:

$$f_1(\sigma) = A\sigma^n \tag{5.72}$$

The parameters A and n are material constants. Bailey suggested a time function as:

$$f_2(t) = Dt^m \tag{5.73}$$

(usually $1/3 \leq m \leq 1/2$) and:

$$f_3(T) = C \exp\left(-\frac{\Delta H}{RT}\right) \tag{5.74}$$

ΔH is activation energy in an Arrhenius-type relation. By combining these relations Eq. 5.75 is obtained (which is about the same as Eq. 5.70):

$$\varepsilon_c = B\sigma^n t^m \exp\left(-\frac{\Delta H}{RT}\right) \tag{5.75}$$

At isothermal conditions, Eq. 5.75 is given as:

$$\varepsilon_c = B\sigma^n t^m \tag{5.76}$$

which is Norton-Bailey's law. Equation 5.76 was developed for constant stress and is useful in describing all creep stages. Differentiating Eq. 5.76 for $n \gg 1$ and $m \leq 1$ yields the creep rate for varying stresses. Here, the strain rate is expressed as a function of stress and time (known as the 'time-hardening rule'). Another form (known as the 'strain hardening rule') may be derived by taking t from Eq. 5.76, namely:

$$\frac{d\varepsilon}{dt} = \dot{\varepsilon} = Bm\sigma^n t^{(m-1)} \tag{5.77}$$

$$t = \left(\frac{\varepsilon_c}{B\sigma^n}\right) \tag{5.78}$$

and inserting it into Eq. 5.77, resulting in:

$$\dot{\varepsilon} = Bm\sigma^n \left(\frac{\varepsilon}{A\sigma^n}\right)^{\frac{(m-1)}{m}} \tag{5.77a}$$

which may also be written as:

$$\dot{\varepsilon} = B^{\frac{1}{m}} m(\sigma)^{\frac{n}{m}} (\varepsilon)^{\frac{(m-1)}{m}} \tag{5.79}$$

The parameters B, n and m depend on material and temperature. They can be determined by means of a uniaxial test.

Norton's power law and an Arrhenius-type equation are the most common expressions describing the stress and temperature dependence of the steady-state creep rate. Of the many relations suggested to describe creep-rupture life, the steady-state strain-rate relation has been extended for the calculation of rupture strength and is used to predict life-time. To improve the results obtained by using the Norton-Bailey rule for life-time prediction, various researchers suggested modifications of the original approach. Fehér et al., for example, used the Graham-Walles creep equation, composed of three Norton-Bailey-style strain terms for each creep stage added until reaching overall creep strain. This expression is:

$$\varepsilon_c = K_1\sigma^{n1}t^{m1} + K_2\sigma^{n2}t^{m2} + K_3\sigma^{n3}t^{m3} \tag{5.80}$$

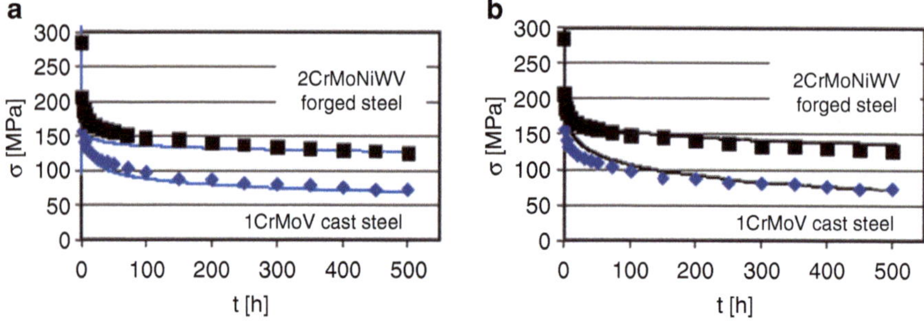

Fig. 5.39 Residual stress curves computed using Norton-Bailey (**a**) and Graham-Walles (**b**) creep models with experimental values (Fehér et al. 2009. With the kind permission of Elsevier)

Figure 5.39 compares the creep results for 10% CrMoV steel of the improved Graham-Walles and Norton-Bailey approaches.

Another example may be mentioned – that of Murali et al., who studied the creep-rupture properties of structural materials used in sodium-cooled fast reactors. Austenitic stainless steel was the structural material used in the high-temperature components, due to its good high-temperature mechanical properties, compatibility with the sodium coolant and adequate weldability. Murali did creep-rupture-life prediction based on the Norton-Bailey approach for creep-dissipation energy analysis.

To summarize this section, note that: (a) stress-rupture tests are used to determine the time necessary to produce failure, thus this testing is always done until failure occurs; (b) the data is plotted as log-log (Figs. 5.35, 5.36, 5.37, and 5.38); (c) a straight line or the best fit curve is usually obtained at each temperature of interest; (d) this information may then be used to extrapolate time-to-failure for longer durations. A typical set of stress rupture curves appear above; (e) changes in the slope of a stress-rupture line are due to structural changes in the material. These are significant changes in material behavior, because they may result in large errors when extrapolating the data.

5.7 Recovery (Relaxation)

Recovery is associated with the reduction of stress. Usually, during such processes, the number of dislocations decreases due to their annihilation or rearrangement into low-energy configurations. Recovery is one of the three stages of annealing in cold-worked materials. These stages are commonly known as 'recovery', 'recrystallization' and 'grain growth'. In usual annealing processes following cold work, the term 'static recovery' is often used in reference to the first structural changes taking place in the material. Softening processes occur during recovery. However, during high-temperature deformations, such as creep, test specimens or structural materials are simultaneously exposed to two effects, thermal and deformation. Recovery that does not follow annealing after deformation, but rather occurs simultaneously with both effects acting, is often referred to as 'dynamic recovery'. Softening processes, similar to static recovery, may occur during hot working. Major physical processes that lead to recovery are dislocation climb and dislocation cross slip, the former being thermally activated, whereas cross slip is a form of dynamic recovery. Cross slip can occur at temperatures lower than those required for climb. The principal difference between these softening processes is that during dynamic recovery stored energy continues to be introduced by deformation into the material, although in decreased amounts, due to the recovery process. At a certain creep strain, the rates of work hardening and recovery come into balance, as these two competing mechanisms reach dynamic equilibrium. When this occurs, a steady state is obtained. Structural changes occur by the formation and growth of subgrains. Within the subgrains, the processes of work hardening and recovery lead to the continuous formation of low-angle boundaries and to a constant density of unbound or free dislocations within the subgrains.

Heavily deformed metals contain a very large number of dislocations, predominantly as tangles or forests and the dislocation distribution after deformation is largely random. Metals with moderate or high stacking fault [henceforth: SF] energy, aluminum for example, tend to form a cellular structure, in which the cell walls consist of dislocation tangles. The interiors of the cells, however, have a reduced dislocation density (see Figs. 2.7, 2.8, 2.9, 2.10, 2.11, and 2.12). The concept, supported by experimental evidence, is that a deformed structure is often a 3-dimensional (3-D) cellular structure with walls consisting of dislocation tangles. As recovery goes on, these cell walls undergo transition to a subgrain structure. This occurs by means of the gradual elimination of extraneous dislocations and the rearrangement of the remaining dislocations into low-angle grain boundaries. When the temperature is increased to about one third of the absolute melting point, dislocations become mobile and are able to glide, cross-slip and climb. Climb, cross slip and dislocation annihilation cause softening both in static and in dynamic recovery. During cross slip, screw dislocations often encounter dislocations having an opposite sign, causing them to annihilate each other. A perfect local region is formed by reduced dislocation density and lower strength.

Dynamic recovery by cross slip is influenced by SF energy and, if it is high, the process occurs more readily. Dislocation climb occurs at higher temperatures than cross slip and is governed by three rates: of vacancy formation, of jog formation and of the vacancies' arrival at jogs. The annihilation of opposite-signed dislocations occurs not only at encounters in their common slip plane, but also in other slip planes following cross slip or climb. Annihilation decreases the strain-hardening rate. In polycrystalline materials, grain-boundary relaxation can also contribute to the softening of the material.

The Bailey-Orowan expression is one of the ways to express steady-state creep rate, $\dot{\varepsilon}_s$, (i.e., the balance between recovery and hardening) at elevated temperatures for pure metals and some alloys and is given as:

$$\dot{\varepsilon} = \frac{r}{h} \tag{5.81}$$

It has been shown that the recovery and hardening rates are given by:

$$r = -\frac{\partial \sigma}{\partial t} \tag{5.82}$$

$$\mathrm{h} = \frac{\partial c}{\partial \varepsilon} \tag{5.83}$$

and then the creep rate (from Eqs. 5.81, 5.82, and 5.83) is:

$$\dot{\varepsilon} = -\frac{\partial \sigma / \partial t}{\partial \sigma / \partial \varepsilon} \tag{5.84}$$

The following shows how Eq. 5.84 may be obtained. In steady-state creep, the assumption is that the properties of a material remain constant as a consequent of the balance between hardening and thermal softening. The hardening coefficient, $\mathrm{h} = \frac{\partial \sigma}{\partial \varepsilon}$, and thermal softening, $r = -\frac{\partial \sigma}{\partial t}$, are as shown in Eqs. 5.82 and 5.83. When they balance, $\frac{\partial \sigma}{\partial \varepsilon} = -\frac{\partial \sigma}{\partial t}$. It is possible to write the differential of σ as:

$$d\sigma = \frac{\partial \sigma}{\partial \varepsilon} d\varepsilon + \frac{\partial \sigma}{\partial t} dt = 0 \tag{a}$$

from which:

$$\frac{\partial \sigma}{\partial \varepsilon} \frac{d\varepsilon}{dt} = -\frac{\partial \sigma}{\partial t} \frac{dt}{dt} \tag{b}$$

and rearranging gives Eq. 5.84:

$$\frac{d\varepsilon}{dt} = \dot{\varepsilon} = -\frac{\partial \sigma}{\partial t} / \frac{\partial \sigma}{\partial \varepsilon} = \frac{r}{h} \tag{5.84}$$

The coefficient of hardening is measured from the slope of a stress-strain curve in the absence of softening, while thermal softening is measured from the drop in yield stress over time when a work-hardened metal is annealed.

Stress-relaxation curves may be used to determine r. Such curves are shown for pure polycrystalline magnesium in Fig. 5.40. The values of the strain-hardening coefficient, h, are obtained from the Bailey-Orowan relation. The stress and temperature dependencies of steady-state creep were determined for 200°C in the 20-50 MPa range and for 150–250°C at 35 MPa. Dynamic recovery is given as logarithmic plots. The recovery process is the rate-controlling step in their high-temperature plastic deformation. These results, for pure polycrystalline magnesium, support the proposed dislocation-network model (see below) for the recovery creep of single-phase materials.

Several steady-state models have been suggested. Weertman suggested that the steady-state creep rate is controlled by climb, which represents the escape rate of

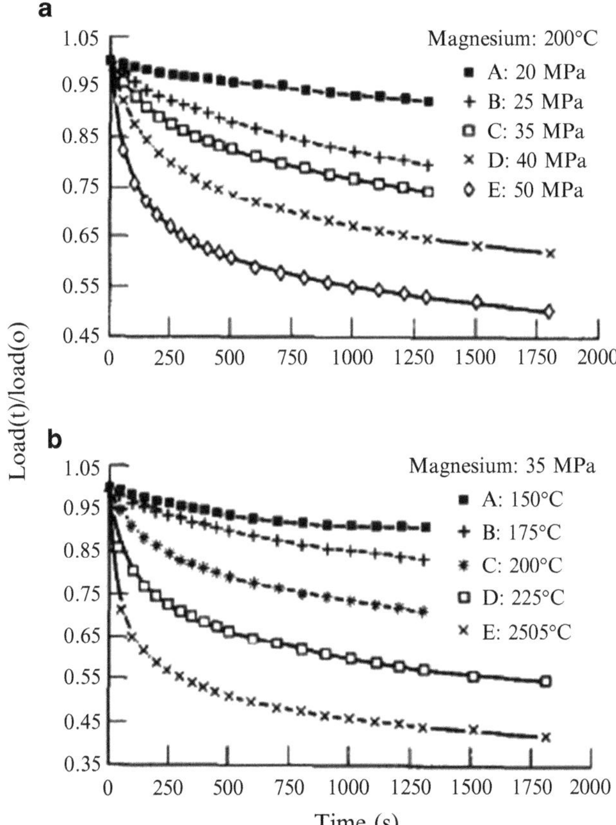

Fig. 5.40 Stress relaxation from load vs. time plots; (**a**) at 200°C in the stress range of $\sigma = 20 - 50$ MPa, (**b**) at $\sigma = 35$ MPa in the range of T = 150–250°C. Load is normalized to the initial load of steady-state creep (Shi and Northwood 1994. With kind permission of Elsevier)

the leading dislocations in a pile-up. He also suggested a viscous-glide model, applicable to solid solutions. On the basis of Friedel's assumption, Mitra and McLean suggested a network-recovery model. A group of researchers considered a jog-dragging model of screw dislocations, as originally discussed by Mott. Nabarro suggested a diffusion model, in which the climb of the edge dislocations determines the creep rate. Lindroos and Miekk-oja suggested a sub-boundary recovery model, based on a climb model along sub-boundaries.

The McLean model will be discussed as an example of steady-state creep involving a dynamic balance between hardening and softening. McLean assumed that, during creep, dislocations form a three-dimensional network. The steady-state creep may be considered as consecutive hardening and recovery processes of the dislocation network. Strength (hardening) is provided by the dislocation junctions when they are linked together. As a result of thermal fluctuations, some the weakest links of these three-dimensional network junctions break free and are capable of moving a certain distance before being held up again by the network and, thus, producing a strain increment (creep). Some strain hardening also occurs, since, on the average, the links of the network shorten and, therefore, are less likely to break free. Recovery consists of a steady increase in the average link length and, at some point, the links become weak again and are then likely to break free. This process repeats itself. Note that dislocation formation during creep exists (the three-dimensional network) and during the primary stage dislocation density increases, i.e., strain hardening occurs though recovery, at this time, is not yet significant. However, creep rate decreases with further increase of dislocation density (see Fig. 1b). Recovery becomes important when the deformation enters the steady-state creep, at which point a balance is reached between recovery and hardening, as indicated above by McLean's approach.

In general, the recovery theory (whichever model is chosen) is based on the concept that metals harden with strain and soften over time upon heating. During steady-state creep, a balance exists between these two phenomena. This relation holds for pure metals and single-phase alloys.

5.8 The Prediction of Life-Time (Parametric Method)

There are parametric methods to determine the creep life-time of materials. Such methods are based on evaluating the -rupture behavior. In essence, the results of short-duration, high-temperature tests are correlated with the performance of long-term tests at lower temperatures. The most popular parametric methods are: (a) Larson-Miller; (b) Manson-Haferd; (c) Orr-Sherby-Dorn and, (d) Monkman-Grant.

5.8.1 The Larson-Miller Approach

Of the above mentioned methods, the Larson-Miller approach is applicable to a variety of materials and is most commonly used because of its simplicity. This relation is:

$$T\,(C + \log t) = P \tag{5.85}$$

T is given in degrees Rankin (i.e., F + 460), t in hours and the constant $C \approx 20$. The value of C seems to be applicable to many cases and materials, but deviation from this value has been observed and its value can be in the range of 15–30, depending on the material. Selecting the proper C value, which may be determined for a material of interest, can narrow the scatter by reducing the scatter problem. Equation 5.85 is a stress-dependent, temperature-compensated rupture-life function. The Larson-Miller relation may be obtained as follows. Suppose that creep rate is adequately described by an Arrhenius-type equation, since creep is a thermally-activated process and the minimum secondary-creep rate can be described by an Arrhenius equation (see Eqs. 5.2 or 5.3). Thus:

$$\dot{\varepsilon} = A \exp - \left(\frac{Q}{kT} \right) \tag{5.86}$$

$$\ln \dot{\varepsilon} = \ln A - \frac{Q}{kT} \tag{5.86a}$$

by rearranging Eq. 5.86a as:

$$\ln A - \ln \dot{\varepsilon} = \frac{Q}{kT} \tag{5.86b}$$

Assuming that the creep strain to rupture, ε_r, is a constant over the temperature range of interest, and if the strain is predominantly in the steady-state creep regime, then the average creep rate for the specimen life-time until rupture, namely, t_r, is given by:

$$\dot{\varepsilon} = \frac{\varepsilon_r}{t_r} \tag{5.87}$$

and Eq 5.86b may be written as:

$$\ln A - \ln \frac{\varepsilon_r}{t_r} = \ln A - \ln \varepsilon_r + \ln t_r = \frac{Q}{kT} \tag{5.88}$$

Write $\ln A - \ln \varepsilon_r = C$ and rearrange Eq. 5.88 to get:

$$T(C + \ln t_r) = \frac{Q}{k} = P \tag{5.89}$$

$P = T(C + \ln t_r)$ is the Larson-Miller parameter [henceforth: LMP] and is a $f(\sigma)$. Under the assumption that activation energy is independent of applied stress, this equation may be used to relate the difference in rupture life to differences in temperature for a given stress. In many cases, C is indeed ~20, which is obtained from the intercept with the $\log t_r$ axis of a $\log t_r$ vs. 1/T plot. The slope of such a plot, namely Q/k (=P), is a function of stress as seen in the schematic figure below. For a graphic presentation, Eq. 5.89 may be written as:

$$\log t_r = 0.434 \frac{Q}{kT} - \log C \tag{5.89a}$$

As shown in Fig. 5.41, the intercept on $\log t_r$ is –C.

The parametric method is commonly used to predict creep-rupture strength at longer times (often up to 10^5 h or even longer) based on tests performed for much

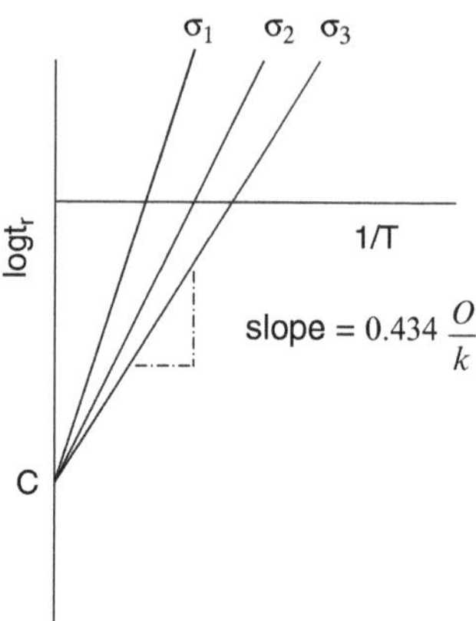

Fig. 5.41 A schematic plot according to Eq. 5.89a for various stresses

shorter times. The relationship between stress and the LMP is used to predict a probable time for the onset of creep-rupture failure. If one knows the temperature and stress at which a material is operating, the predicted time to creep-rupture failure for that set of conditions can be determined from a plot of LMP versus stress. The procedure using the LMP method requires a family of stress-rupture curves representing different test temperatures of a given material, which are then re-plotted on a revised temperature-compensated time axis, i.e., the LMP. The family of curves chosen is superimposed on a single master curve. Figure 5.42 shows creep-rupture curves for the AX52 die-cast alloy (Mg-Al-Ca alloy) tested at four temperatures, where the plot of stress vs. time- to-rupture is indicated. In Fig. 5.43, the data for this alloy are plotted on one single curve, as stress vs. the LMP. All the data shown in Fig. 5.43 fall on one single line, independent of the creep-testing conditions, i.e., temperature, applied stress and the rupture-life range, spanning a period of a few minutes to 10,000 h. This indicates that the value of the LMP is uniquely described by the logarithm of the applied stress for the alloy, irrespective of the creep-testing temperature, when the Larson-Miller constant is set at 20. Figure 5.44 shows the correlation between applied stress and the LMP for the AX52 die-cast alloy, with both parameters expressed on a linear scale.

A linear relation for long-term creep was also obtained for pure titanium with an HCP crystal structure (Maruyama et al.). The following equation was derived from the applied stress and the LMP value in a low-stress regime:

$$\sigma\,(\mathrm{MPa}) = 461 - 0.037\,\mathrm{LMP} \tag{5.90}$$

This may be used to predict the long-term creep-rupture life for an alloy considered under given conditions of temperature and applied stress.

The stress vs. LMP plot is effective in predicting the creep-rupture life for heat-resistant Mg-Al-Ca alloys, irrespective of the method used for their production. The Larson-Miller method applies to other materials as well; however, one has to remember that the constant value of 20 (used in the present example) may be different in other systems.

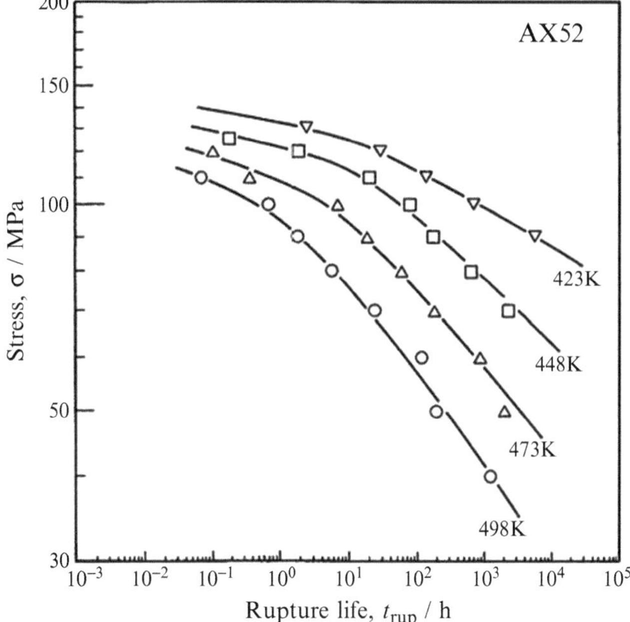

Fig. 5.42 Plots of stress against creep rupture lives for the AX52 (Mg-Al-Ca alloy) die-cast alloy at temperatures between 423 and 498 (Terada and Sato 2010. With kind permission of Elsevier and Professor Terada)

5.8.2 The Manson-Haferd Approach

Another parametric method for predicting the creep-rupture properties of materials is the Manson-Haferd approach, claimed to be very successful. The Manson-Haferd parameter [henceforth: PMH] is based on the observation that constant stress plots of logt vs. T are more linear than logt vs. 1/T curves. It is, therefore, often referred to as a 'linear parameter method'. Manson and Haferd determined that isostress lines converge at a point on a logt vs. T curve; their parameter is given by:

$$\mathrm{PMH} = f(\sigma) = (T - Tm) / (\log(t_r) - \log(t_m)) \tag{5.91}$$

where t_m and T_m correspond to the coordinates where the logt vs. T curves converge at a certain point which is constant for a given material. This is shown schematically in Fig. 5.45, in which t_a and T_a correspond to t_m and T_m, respectively, as in Eq. 5.91.

An example is shown below of the use of this parametric method when testing a 2.25Cr-1Mo steel in the 500–700°C range (dos Reis Sobrinho and de Oliveira Bueno). Figure 5.46 is a plot showing log stress vs. log t at various temperatures. As shown in Figs. 5.46 and 5.47, the hot tensile data were also incorporated and converted (dos Reis Sobrinho and de Oliveira Bueno) into equivalent creep data by

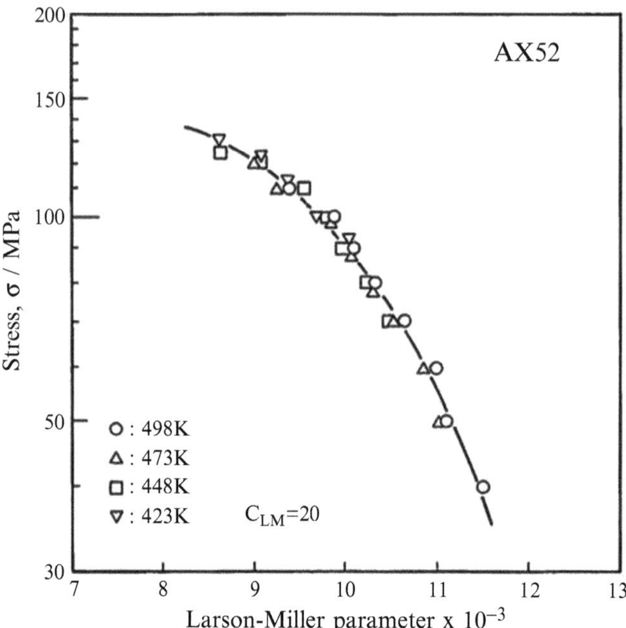

Fig. 5.43 Correlation between applied stress and the LMP for the Mg-Al-Ca die-cast alloy AX52, in which the Larson Miller constant is chosen as 20 (Terada and Sato 2010. With kind permission of Elsevier and Professor Terada)

using the following conversion rules: (a) the strain rate of a tensile test is equivalent to the minimum strain rate in a creep test; (b) the ultimate stress in a tensile test is equivalent to the applied stress in a creep test; and (c) the time of the occurrence of the ultimate stress (the onset of necking) is equivalent to the rupture time in a creep test. Figure 5.47 shows a plot of the log time-to-rupture (t_r) vs. the PMH. These constants were obtained from the convergence point of a $\log t_r$ vs. T plot (Fig. 2b in the work of dos Reis Sobrinho and de Oliveira Bueno) with a linear regression fit in each of the 15 isostress levels used for the creep tests and the points corresponding to the hot tensile data obtained in their work. $R^2 = 0.9927$ was obtained for the line representing the Manson-Haferd approach by means of a linear regression fit. Figure 5.47 shows the curve obtained from the analysis of the isostress lines resulting from the $\log t_r$ vs. T plots (Fig. 2b of their work). Figure 5.48 is a similar plot, but using minimum creep-rate data. The results are very similar to those obtained by the analysis of the rupture-time data (in Fig. 5.47).

Again, the best result was given by the Manson-Haferd method ($R^2 = 0.9915$). In the illustrations, CSR refers to the ‘constant strain rate’ tests. The authors compared five techniques for the parametric prediction of creep-rupture properties and material life-times and claim that the best results are obtained by using Manson-Haferd analysis. They also state that the hot tensile results show very good compatibility with the creep results and may be used as a very helpful complement in the determination of parametric curves for creep data.

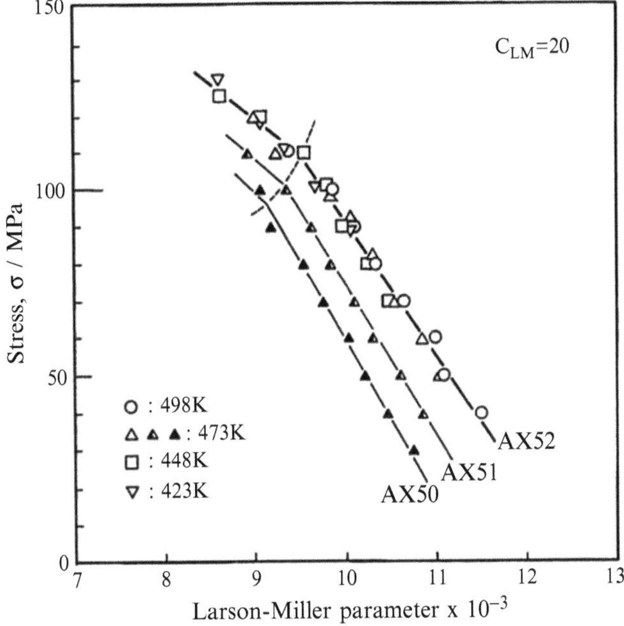

Fig. 5.44 A correlation between applied stress and the Larson-Miller parameter for the Mg-Al-Ca die-cast alloys AX50, AX51, and AX52, in which the Larson-Miller constant was chosen as 20 (Terada and Sato 2010. With kind permission of Elsevier and Professor Terada)

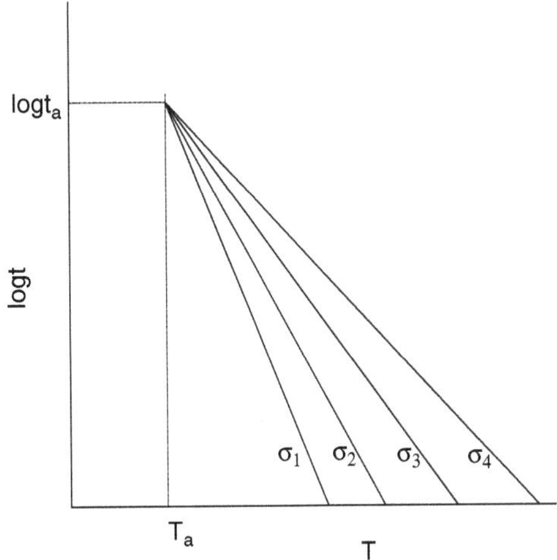

Fig. 5.45 A schematic representation of the Manson-Haferd method for the evaluation of the constants, defined by the intersection point, where the plots of the given stresses converge

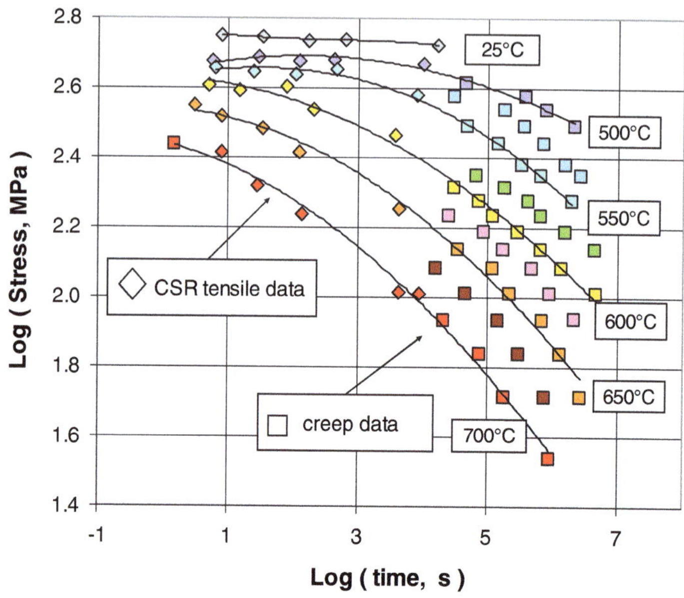

Fig. 5.46 The variation of stress with rupture time in creep tests is plotted together with the ultimate tensile stress and the time for its occurrence in the tensile tests (dos Reis Sobrinho and de Oliveira Bueno 2005. With kind permission of the Revista Matéria and the authors)

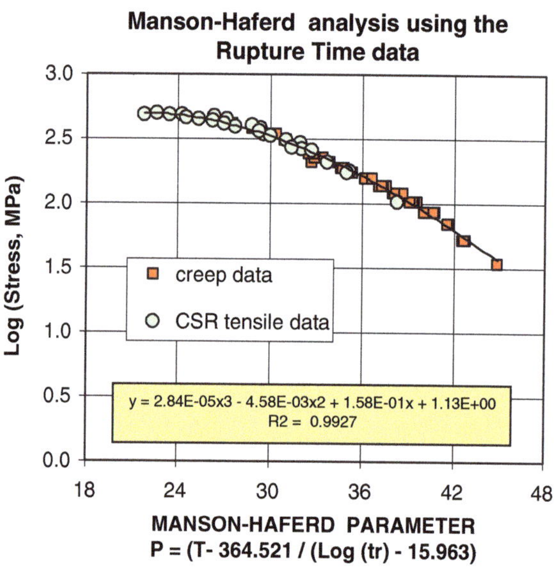

Fig. 5.47 Creep by rupture time data and CSR tensile data based on the $\log t_r$ versus T diagrams (dos Reis Sobrinho and de Oliveira Bueno 2005. With kind permission of the Revista Matéria and the authors)

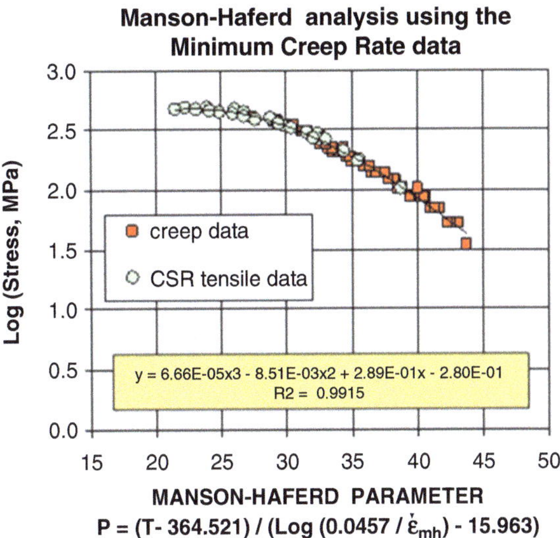

Fig. 5.48 Curve for minimum creep rate and CSR tensile data based on the $\log t_r$ versus T diagrams (dos Reis Sobrinho and de Oliveira Bueno 2005. With kind permission of the Revista Matéria and the authors)

5.8.3 The Orr-Sherby-Dorn (OSD) Approach

This approach for predicting the creep life-time of materials is expressed as:

$$P_{SD} = t \exp - \left(\frac{Q}{RT}\right) \tag{5.92}$$

where P_{SD} is the Orr-Sherby-Dorn [henceforth: OSD] parameter. By expressing this equation in a logarithmic form one gets:

$$\ln P_{SD} = \ln t - \frac{Q}{RT} = \log P_{SD} = \log t - 0.434\frac{Q}{RT} \tag{5.92a}$$

Rearranging Eq. 5.92a, one gets:

$$\log t = \log P_{SD} + 0.434Q/RT \tag{5.92b}$$

with Q being the activation energy for creep. In the above relations, t is either the time-to-rupture or the time to reach a given creep strain. Over the years, this expression has been applied to a variety of metals with great success, especially for evaluating life-times before stress rupture occurs. Like the Larson-Miller approach, the OSD method assumes that log t is a linear function of 1/T. Yet, rather than assuming a common intersection point of the log t vs. 1/T lines (see Fig. 5.41), no intersection occurs and parallel isostress lines are obtained, each representing a single stress value (as seen in Fig. 5.49). This is based on Dorn's assumption that

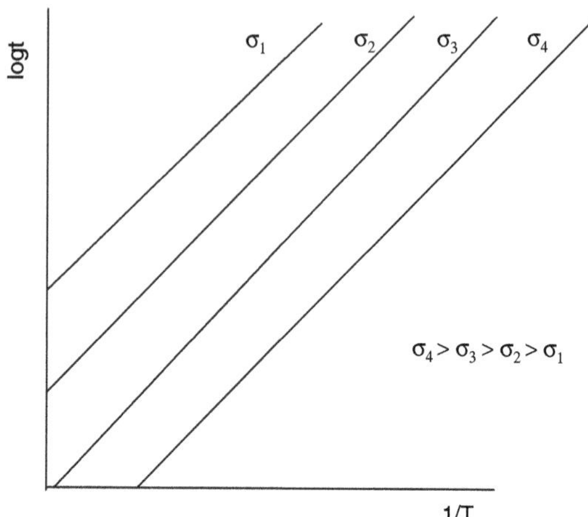

Fig. 5.49 A schematic presentation of OSD creep method for obtaining the parameter

only one creep process is expected to operate at high temperatures, which may be expressed by a temperature-compensated time parameter given as:

$$\theta = \text{texp}(-\Delta H/RT) \tag{5.93}$$

This is basically Eq. 5.92, the commonly used OSD parameter, ($\theta = P_{SD}$ and $\Delta H = Q$). Q (or ΔH) is the activation energy for creep, assumed to be about the value of self-diffusion.

This parametric method, for creep life-time prediction for industrial applications, has been successfully used especially with pure metals (the original work was done on Al). Note that the OSD technique is based on Norton's empirical power law for obtaining the dependence of σ on the life-time, θ, expressed as:

$$\log \sigma = A - \frac{1}{n} \log \theta \tag{5.94}$$

where θ is given (as seen above) by:

$$\theta = t_r \exp\left(-\frac{\Delta H}{RT}\right) \tag{5.93}$$

The OSD parameter is stress dependent, and a master curve for this relation has to be created. In order to determine an empirical model for this relation, creep data are used to find the best fit between stress and the OSD parameter. For instance, a Ni-based superalloy, often used at high temperatures, in which creep conditions prevail, may be considered. Figure 5.50 (following OSD parametric creep-life prediction) presents a temperature-compensated, creep-rupture life-time diagram for two alloys of Ni-based (Heilmaier and Reppich). The master curves shown in this Figure are based on the experimental results (shown in Fig. 5.51) for the MA 6000 alloy at

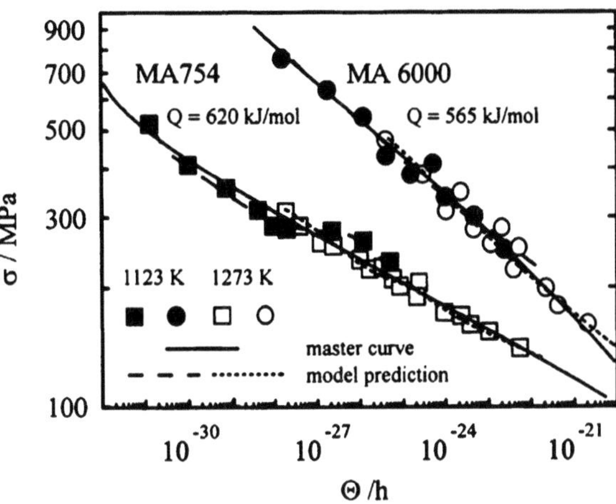

Fig. 5.50 Temperature-normalized, creep-rupture life-time according to Sherby Dorn for two Ni based alloys of MA754 and MA 6000. Note that the master curves almost coincide with the model predictions (Heilmaier and Reppich 1996. With kind permission of Elsevier)

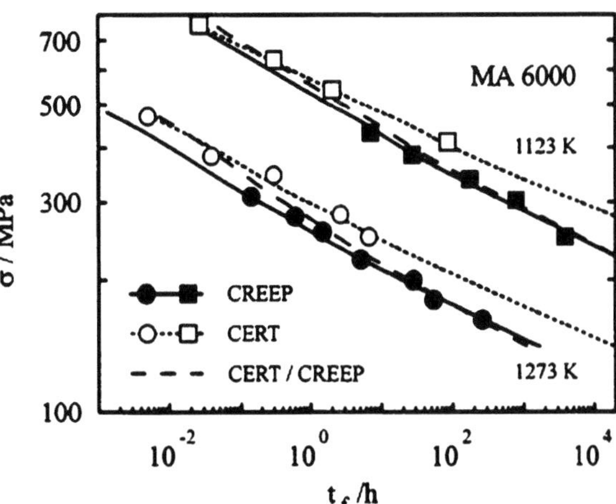

Fig. 5.51 Creep rupture-lifetime diagram of MA 600 M (Heilmaier and Reppich 1996. With kind permission of Elsevier)

two temperatures, where stress is plotted vs. rupture life-time, t_f. The term CERT in the illustration means 'constant extension variable tests', which were also used, in addition to creep tests. The OSD parameter, P_{SD}, is emphatically stress-dependent and the master curve for this relation must still be constructed. Note that one may use an empirical model for this relation in order to find the best fit between stress and P_{SD}. Statistical creep models have been created for stainless-steel materials. In order to correlate the results of short-duration, elevated-temperature tests with long-term service performance at more moderate temperatures, different creep-prediction models, among them the OSD model, were considered (He).

Comparisons between the various creep-prediction models have been carried out for a range of stresses and temperatures. The best relation between P_{SD} and stress is a polynomial model (where $R^2 = 0.9993$), given as:

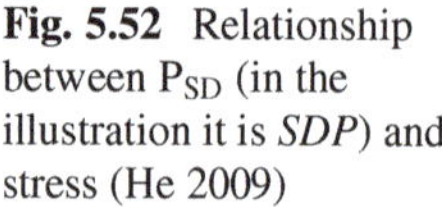

Fig. 5.52 Relationship between P_{SD} (in the illustration it is *SDP*) and stress (He 2009)

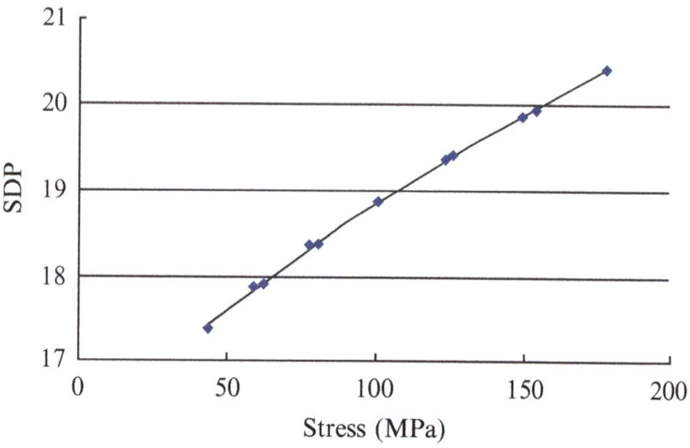

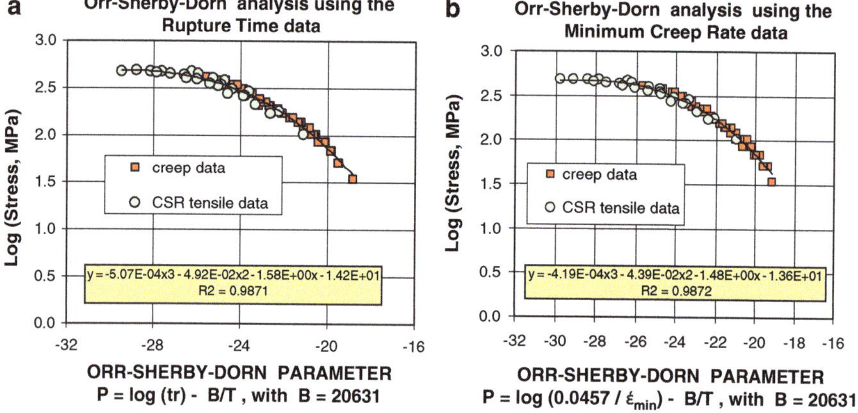

Fig. 5.53 Parameterisation curves for creep and CSR tensile data: (**a**) analysis using creep rupture time data; (**b**) analysis on the basis of minimum creep data (dos Reis Sobrinho and de Oliveira Bueno 2005. With kind permission of the Revista Matéria and the authors)

$$P_{SD} = -0.00005\sigma^2 + 0.0325\sigma + 16.056 \tag{5.95}$$

The expression ($P_{SD} = \theta$), from Eq. 5.95, may be used after some rearrangement and after taking a logarithm to obtain t_r in Eqs. 5.94 and 5.93. Figure 5.52 shows a relation between P_{SD} and stress based on the statistical evaluation of the best fit for obtaining P_{SD} for an AISI 316 steel.

Creep data parameterization, done according to the OSD method for 2.25Cr-1Mo steel from 500–700°C, was also performed by dos Reis Sobrinho and de Oliveira Bueno on the basis of rupture-time and minimum creep-rate data. Figure 5.53 shows stress as a function of the OSD parameter. In the above illustration and also in earlier Figs. 5.46, 5.47 and 5.48, the significant fact is that the hot tensile results show very good compatibility with the creep results in all the parameterization methods investigated, further confirming the consistency in the criteria adopted for converting tensile results into creep results.

The method for obtaining the temperature-compensated time of P_{SD} is done in the following way. First of all, a high-temperature stress rupture is produced at a specific stress level for a short time. Then, at the same stress level, long-term rupture behavior is predicted by means of linear extrapolation (at a temperature lower than that of the shorter-duration test). This linear extrapolation is based on the assumption that the activation energy determined by the short-term data does not change for long-term creep. In this type of accelerated testing, it is assumed that the failure mechanism does not change and, hence, is not a function of either temperature or time. In addition, an assumption is made that Q (ΔH) is both stress and temperature independent. This is particularly useful, since it is often necessary to extrapolate to real service lives (which tend to be from years to several decades) from laboratory creep-test data with limited durations (from days to years). Time-temperature parameters are useful for this purpose, in particular, for the evaluation of new alloys in regard to creep resistance applications.

5.8.4 The Monkman-Grant Approach

The Monkman-Grant relation between minimum creep rate, $\dot{\varepsilon}_{mcr}$, and time-to-fracture, t_f, is given as:

$$\dot{\varepsilon}_{mcr}^{m} t_f = C \tag{5.96}$$

$$\log t_f = -m \log \dot{\varepsilon}_{mcr} + \log C \tag{5.96a}$$

where m and C are material constants. This relation holds for Nimonic PE 16 alloys at all environments tested, as indicated below in Fig. 5.54. The calculated values of m, using least-squares analysis, are 0.92, 0.89 and 0.93 for CO_2, air and vacuum environments, respectively. This relation is useful for industrial applications when one knows the constants m and C of a material, since the above expression evaluates the fracture time on the basis of the minimum creep-rate data. There is no need for long-term creep testing of high-temperature materials, since generally, the time required to reach steady-state creep is much shorter than the time-to-fracture. This is particularly important regarding newly developed materials for high-temperature applications. Figure 5.54 is a logarithmic plot of Eq. 5.96 applied to different environments. No significant effect of environment was found on the failure time for this alloy.

The Monkman-Grant relation was experimentally applied to various alloys at various temperatures and stresses and an example of its use to predict the creep life-time of a Ni-based superalloy is found in the work of Koul et al. For this alloy, Fig. 5.55 summarizes the evaluation based on the Monkman-Grant relation. According to Koul et al., the scatter seems to be quite large, so they conclude that their proposed prediction method is better able to evaluate the actual service time

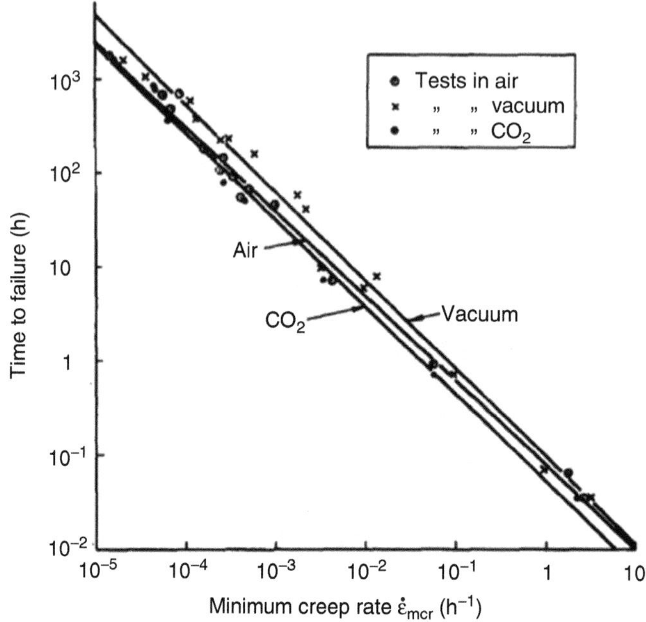

Fig. 5.54 Monkman-Grant plot for Nimonic PE 16 at 700 C (Lobb 1979. With kind permission of Elsevier)

than the Monkman-Grant method. Nevertheless, the relation expressed in Eq. 5.96a is satisfied by a correlation factor of 0.98. They expected to reduce the scatter by normalizing Eq. 5.96a by introducing ε_r, as suggested by Dobes and Milicka, namely

$$\frac{t_r}{\varepsilon_r}\dot{\varepsilon}^{m^*} = C^* \tag{5.97}$$

with m^* and C^* being constants. This correction did not improve the scatter seen in Fig. 5.55. A different approach for predicting creep life-time was suggested in the work of Dobes and Milicka, dealing with the service induced degeneration effects in a turbine blade, which is out of the scope of this section.

Another example of the use of the Monkman-Grant parametric method is for an as-quenched Austenitic Fe-30Mn-9Al-1 C alloy.

The creep-deformation behavior and rupture property of this alloy has been studied at 923, 948, and 973 K under applied stresses ranging from 50 to 350 MPa. Mn and C are added to this alloy to form the Austenitic structure, essential for good mechanical properties at both elevated and cryogenic temperatures. Furthermore, these alloys exhibit superior oxidation resistance, due to the high concentration of aluminum and, thus, may hold some promise for applications at elevated

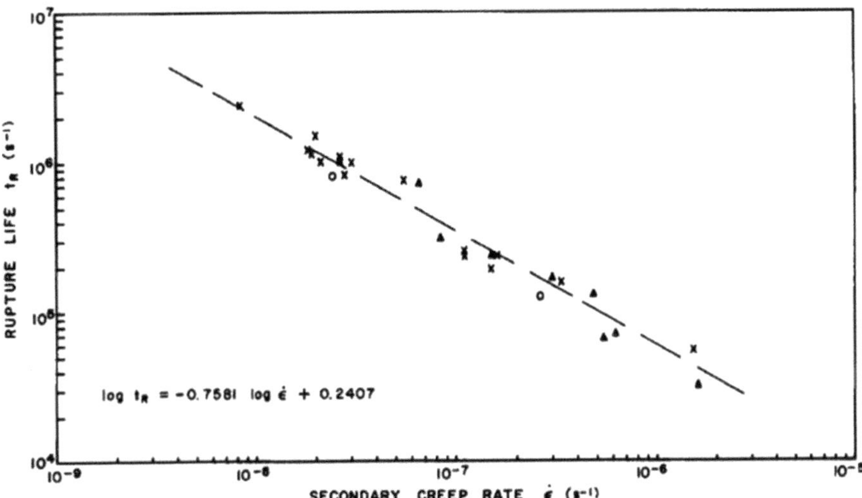

Fig. 5.55 Creep rupture life vs. creep rate plot for a new IN-738LC turbine blade by the Monkman-Grant relation. The plot includes service-exposed creep data also (▲-13,000 h, o-20,000 h) (Koul et al. 1984. With kind permission of Elsevier)

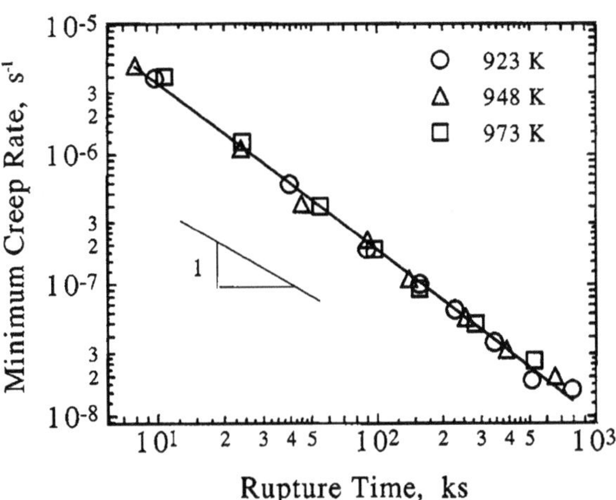

Fig. 5.56 A plot of the Monkman-Grant relationship for an Austenitic Fe-30Mn-9Al-1 C alloy (Zhu and Tjong 1998. With kind permission of Springer)

temperatures, as creep-resistant alloys. Figure 5.56 shows the minimum creep rate vs. creep-rupture time. The plot is linear, according to:

$$t_r = \frac{C_{MG}}{\dot{\varepsilon}_{\min}^{\alpha}} \tag{5.98}$$

which is the Monkman-Grant relation seen earlier ($t_r = t_f$, $\dot{\varepsilon}_{\min} = \dot{\varepsilon}_{mcr}$, and $C_{MG} = C$).

The Fe-30Mn-9Al-1C alloy exhibits a reasonably high resistance to intergranular creep cavitations. Consequently, creep cavitations were observed only in specimens that have crept to final failure. Creep cavitations tend to occur at grain boundaries and appears to be primarily round in specimens fractured in low- and intermediate-stress regimes, while wedge-type cracking at the triple junctions of grain boundaries is more obvious in specimens fractured in a power-law breakdown regime. TEM observations of deformed specimens reveal that low-density dislocations, coarse dislocation networks and profuse slip bands develop under low stress, power law and power-law breakdown regimes, respectively. On a Monkman-Grant plot, all the data points merge into a single line (as seen in Fig. 5.56). However, the slope of the line (α) is determined as 0.78, which is considerably lower than unity. Zhu and Tjong are skeptical about the use of the Monkman-Grant plot for this alloy, despite the fact that all the data points line up. Nevertheless, this alloy exhibits a reasonably high resistance to intergranular-creep damage.

Various parameters have been suggested for the extrapolation of time-to-rupture with varying success; certain ones have been used to predict the in service life-time of a component operating at high temperatures. Of these methods, the four most popular ones were discussed in this section. The reader may turn to the professional literature in order to find the most appropriate method for some specific application (perhaps in another approach not discussed here).

5.9 Concepts of Designing (Selecting) Creep-Resistant Materials

General designs for universal creep-resistant purposes, based on theoretical considerations, do not exist, despite the vast experimental results and the great understanding of the creep behavior of materials under various conditions. Primary design problems are complex, due to the many material parameters, the various stages of creep and the varying service conditions. Today, the decision to select appropriate materials for use or to design and develop new materials rely heavily on dependable experimental observations and their interpretations according to physical principles. The following may be stated regarding creep, as a high-temperature deformation–it is associated with: (a) the presence of dislocations in materials and dislocation motion under stress; (b) temperature; (c) stress; (d) time; and (e) structure. Points (b–d) may be considered as environmental conditions, but, to these environmental factors, one should add the ambient atmosphere prevailing during the intended service. These factors will now be discussed briefly.

(a) The role of dislocations. Dislocation slip is hindered by obstacles, such as: grain boundaries (in polycrystalline materials), precipitates or impurity particles, stress fields around solute atoms in solid solutions, strain fields of other dislocations and pile-ups. Various interpretations of dislocation motion under stress are considered by various researchers in the field, based on the stages of creep and the ways that they interpret the contribution of dislocations to creep strain. Thus, dislocations can (i) glide, leading to slip in their slip planes; (ii) climb, leading to subgrain formation; (iii) cross slip; or (iv) slide at grain boundaries, considered by many as being induced by slip to satisfy strain compatibility at those grain boundaries (see, for example, Matsunaga et al.). Actions (ii) and (iii) are ways of overcoming the resistance of obstacles, to allow further strain to occur.

(b) The temperature. The creep resistance of a material depends on its melting point, T_m, and, at a given temperature, the higher the T_m of a material, the longer its life-time will be. This is simply related to the rate of self-diffusion, which is slower in high-T_m materials. A specimen may be exposed to levels above or below $0.5T_m$, respectively. Low-temperature creep, at or below $0.5T_m$, is believed to be governed by non-diffusion-controlled mechanisms, whereas high-temperature creep (above $0.5\ T_m$) is diffusion-controlled. At and above $0.3T_m$, creep becomes a significant factor in materials, to be considered when designing creep-resistant materials. Creep resistance is an important material property in high-temperature design, but it is difficult to suggest a function for calculating this property. Therefore, parametric methods are used to achieve the desired long-term creep resistance of specific materials. Since climb is one of the mechanisms for overcoming obstacle resistance and produces creep strain in high-T_m materials, climb will not readily occur below $0.5T_m$.

(c) Stress. The role of stress is obvious, since it is responsible for creep deformation. However, one can talk about low-stress creep and high-stress creep. The effect of stress at constant temperatures is shown schematically in Fig. 5.3b. Accordingly, when increasing stress from a low level to higher values, the creep curves pass through all the stages, from transient to tertiary creep. At a specific stress, all three stages of creep can be obtained.

(d) Time. Creep is a time-dependent deformation. Unlike brittle fracture, creep deformation does not occur suddenly upon the application of stress. Instead, strain develops as a result of the application of stress for a long time. At a given stress and temperature, the resulting creep strain depends on the duration of time. The other parameters (σ, T) determine the rate of creep. Given the aforementioned parameters (σ, T and t), one may state that the rate of deformation is a material property.

(e) Structure. Generally, microstructural changes may occur in materials as a consequence of thermal effects in the absence of stress. Such changes are augmented when stress operates at high temperatures (creep conditions). Microstructural changes that may occur in an alloy under the combined long-term effects of acting stress and temperature are of prime concern when designing creep-resistant alloys. Designers must be familiar with the structural properties and the possible changes that may occur in the wake of long-term exposure to the combined effects of stress and temperature.

Intentional structural changes are often induced, in order to improve the duration of material life-times. Solid solutions have a dual effect, namely they strengthen materials (solid-solution strengthening) and the stress field around solute atoms acts as an obstacle blocking dislocation motion. Furthermore, precipitates obtained in alloys by precipitation hardening also have a dual effect (as mentioned above) with regard to solid solutions. The size of these precipitates is significant (a) for preventing dislocation bowing between the precipitates and (b) because if they are very small, they may dissolve into the matrix, being unstable at the service temperature. Dispersion-hardened precipitates, such as oxides, nitrides, etc., are stable at high temperatures, generally insoluble in the matrix and are also stable at high temperatures; they are more effective material strengtheners for creep resistance.

Additional structural aspects to be considered are: (i) the crystallinity of materials; (ii) orientation, in the case of single crystals; (iii) the SF energy; and (iv) grain size. Concerning crystallinity: (i) The material may be single crystal or polycrystalline. Single crystals, although not as strong as polycrystalline materials due to the absence of boundaries, have been widely used for high-temperature applications (turbine blades, etc.), but cost considerations are an important factor in industry. Polycrystalline materials obtained by directional solidification are almost as good for creep resistance as single crystals. There is a tendency to replace single crystals, wherever possible, by directionally-solidified material. (ii) From a design point of view, it is desirable to use an orientation in which single-crystalline material has the best strength. In Ni-based superalloys (see Sect. 5.2), single crystals with <001> orientation are preferred. MacKay and Maier also report that Ni-based superalloy crystals, having an orientation within $\sim 25°$ of [001], exhibit significantly longer life- times when their orientation is closer to the [001]-[011] boundary than [001]-[$\bar{1}$11]. Thus, orientation showing the best strength properties of some structure is preferential for creep resistance applications. (iii) High SF-energy materials, such as Al, are inferior to those with low SF-energy for creep-resistant applications. The restriction of partials is performed more readily in high-energy SF materials, enabling climb and cross slip (see Chap. 2, Sects. 2.15.2 and 2.15.3) and avoiding creep-resisting obstacles. Note that it is advantageous to use solutes for solution strengthening, which can also simultaneously reduce the SF energy. (iv) Grain size has a dual effect. On one hand, small-grain materials are stronger than large-grain materials, but, on the other hand, larger grains improve creep resistance. This is a consequence of vacancy formation and flow. In fine-grained materials with a large number of grain boundaries, climb may be rapid, because the grains are sources (not only sinks) of vacancies. For climb to occur readily, vacancies must be available. In large-grained materials, with many fewer boundaries, the number of vacancies (that might induce climb) is considerably lower and, therefore, climb is slower with the consequent lower creep rate. Thus, the effect of grain size on creep rests on the diffusion rate of the vacancies in polycrystalline materials. A compromise must be made between the strengthening effect of small grains and the reduction of the number of vacancies required for climb by using coarser grains.

Summing up the accepted rules for producing creep-resistant material, the approach should be such that: (i) high-melting materials are preferential; (ii) if one-phase alloys are required, the matrix should be strengthened by effective solutes; (iii) low SF energy should be chosen (sometimes a solute which reduces the SF energy may also be used for solution strengthening); (iv) dispersed stable, non-soluble particles may be added to achieve a specific creep application; and (v) stable structures should be chosen for certain high-temperature applications, thus, short-duration tests can be extrapolated for long-term use in service.

References

ASM, *Creep and Recovery* (American Society of Metals, Metals Park, 1957)
A.C. Damask, G.J. Dienes, *Point Defects in Metals* (Gordon and Breach, New York, 1971)
J.E. Dorn, in *Energetics in Metallurgical Phenomens*, ed. by W.M. Mueller, vol. 1 (Gordon and Breach, London, 1964)
R.W. Evans, B. Wilshire, *Creep of Metals and Alloys* (The Institute of Metals, London, 1985)
J. Friedel, *Dislocations* (Addison-Wesley Publishing Company, Inc, Reading, 1964)
F. Garofalo, *Fundamentals of Creep and Creep-Rupture in Metals* (The Macmillan Company, New York, 1965)
D. McLean, *Mechanical Properties of Metals* (Wiley, New York, 1962)
D. McLean, K.F. Hale, *Structural Processes in Creep*. Special Report No. 70 (Iron and Steel Institute, London, 1961)
F.R.N. Nabarro, *Report on a conference on strength of solids* (Physical Society of London, London, 1948), p. 75
F.R.N. Nabarro, H.L. De Villers, *The Physics of Creep* (Taylor & Francis, London, 1995)
F.H. Norton, *The Creep of Steels at High Temperatures* (McGraw-Hill, New York, 1929)
A.L. Ruoff, *Materials Science* (Prentice-Hall, Inc, Englewood Cliffs, 1973)
C. Zener, *Elasticity and Anelasticity* (University of Chicago Press, Chicago, 1948)

Further References

T.H. Alden, Acta Metall. 35, 11, 2621 (1987)
H.G. Armaki, R. Chen, K. Maruyama, M. Igarashi, Mater. Sci. Eng. A 527, 6581 (2010)
A.N.da C. Andrade, Proc. R. Soc. Lond. **A 84**, 1 (1910)
A.N.da C. Andrade, Proc. R. Soc. Lond. **A 904**, 329 (1914)
A.N.da C. Andrade, in *Creep and Recovery*, ed. by ASM (American Society of Metals, Metals Park, 1957), p. 176
A.N.da C. Andrade, B. Chalmers, Proc. R. Soc. Lond. **A 138**, 348 (1932)
R.W. Bailey, J. Inst. Met. **35**, 27 (1926)
J. Beddoes, W. Wallace, Metallography **13**, 85 (1980)
R.L. Bell, T.G. Langdon, J. Mater. Sci. **2**, 313 (1967)
W. Blum, Phys. Stat. Sol. (b) **45**, 561 (1971)
P. Caron, T. Khan, Mater. Sci. Eng. **61**, 173 (1983)
H.C. Chang, N.J. Grant, Trans. AIME **206**, 544 (1956)
R.L. Coble, J. Appl. Phys. **34**, 1679 (1963)
A.H. Cottrell, J. Mech. Phys. Solid. **1**, 58 (1952)

F. Dobeš, J. Mater. Sci. **33**, 2457 (1998)
F. Dobeš, K. Milicka, Met. Sci. **10**, 382 (1976)
J.E. Dorn, J. Mech. Phys. Solid. **8**, 85 (1954)
J.F. dos Reis Sobrinho, L. de Oliveira Bueno, Rev. Matér. **10**, 463 (2005)
B.F. Dyson, M. Mclean, Acta Met. **31**(17), 27 (1983)
P.J. Ennis, A. Czyrska-Filemonowicz, Sādhānā **28**, 709 (2003)
A. Fehér, S. Linn, M. Schwienheer, A. Scholz, C. Berger, Mater. Sci. Eng. **A 510–511**, 29 (2009)
H. Fukutomi, H. Takatobi, R. Horiuchi, Jpn. Inst. Met. **46**, 755 (1982)
C. Gandhi, R. Raj, Met. Trans. **A12**, 515 (1981)
F. Garofalo, R.W. Whitemore, W.F. Domis, F. von Gemmingen, Trans. AIME **221**, 310 (1961)
I.M. Glen, Philos. Mag. **1**, 400 (1956)
A. Graham, K.F.A. Walles, J. Iron Steel Inst. **179**, 105 (1955)
J. Harper, J.E. Dorn, Acta Met. **5**, 654 (1957)
X. He, IEEE **978**, 1049 (2009)
M. Heikmaier, B. Reppich, Met. Mater. Trans. A **27A**, 3861 (1996)
K.J. Hemker, M.J. Mills, W.D. Nix, Acta Met. Mater. **39**, 1901 (1991)
C. Herring, J. Appl. Phys. **21**, 437 (1950)
H.U. Hong, B.G. Choi, H.W. Jeong, I.S. Kim, Y.S. Yoo, C.Y. Jo, ed. by T. Chandra, M. Ionescu, N. Wanderka, W. Reimers, Mater. Sci. Forum THERMEC, **638**, 2245 (2010)
O.A. Kaibyshev, R.Z. Valiev, V.G. Khairullin, Fiz. Metallov i Metallovedenie **56**, 577 (1983)
R.D. Kissinger, D.J. Deye, D.L. Anton, A.D. Cetel, M.V. Nathal, T.M. Pollock, D.A. Woodford (eds.), *Superalloys* (The Minerals, Metals and Materials Society, Warrendale, 1996). 283, 290 pp
R.L. Klueh, J. Nucl. Mater. **79**, 363 (1979)
A.K. Koul, R. Castillo, K. Willett, Mater. Sci. Eng. **66**, 213 (1984)
T.G. Langdon, Mater. Sci. Eng., A166, 67 (1993)
T.G. Langdon, J. Mater. Sci. 41, 597 (2006)
F.R. Larson, J. Miller, Trans. ASME **74**, 765 (1952)
R. Le Hazif, J.P. Poirer, Acta Met. **23**, 865 (1975)
G.R. Leverant, H.B. Kear, J.M. Oblak, Met. Trans. **4**(335), 362 (1973)
J.C. Li, Acta Met. **21**, 1195 (1963)
V.K. Lindroos, H.M. Kiekk-oja, Philos. Mag. **17**, 119 (1968)
R.C. Lobb, Mater. Sci. Eng. **38**, 249 (1979)
P. Lukáš, J. Čadek, V. Šustek, L. Kunz, Mater. Sci. Eng. **A208**, 149 (1996)
R.A. MacKay, R.D. Maier, Met. Trans. A **13A**, 1747 (1982)
K. Maruyama, K. Sawada, J. Koike, ISIJ Int. **41**, 641 (2001)
T. Matsunaga, T. Kameyama, E. Sato, IOP Conf. Ser. Mater. Sci. Eng. **3**, 012014 (2009). IOP publishing
D. McLean, Rep. Prog. Phys. **29**, 1 (1966)
S.K. Mitra, D. McLean, Proc. R. Soc. **285A**, 288 (1966)
E.C. Monkman, N.J. Grant, Proc. Am. Soc. Test. Mater. **56**, 593 (1956)
B.L. Mordike, Mater. Sci. Eng. **A324**, 103 (2002)
N.F. Mott, Philos. Mag. **43**, 1151 (1952)
N.F. Mott, Proc. R. Soc. Lond. **A 220**, 1 (1953)
V. Murali, M.D. Mathew, K.B.S. Rao, V. Ganesan, S. Ravi, S. Baskar, N. Angaraj, Trans. Indian Inst. Met. **63**, 635 (2010)
F.R.N. Nabarro, Philos. Mag. **16**, 231 (1967)
F.R.N. Nabarro, Mater. Sci. Eng. **A309**, 227 (2001)
V.D. Natsik, A.I. Osetski, V.P. Soldatov, V.I. Stratsev, Phys. Status Sol. (b) **34**, 99 (1972)
Levi de Oliveira Bueno, Vitor Luiz Sordi, Luiz Marino, Mater. Res. **8**, 401 (2005)
E. Orowan, J. West Scot. Iron Steel Inst. **54**, 45 (1947)
R.L. Orr, O.D. Sherby, J.E. Dorn, Trans. ASM **46**, 113 (1954)
A.I. Osetski, V.P. Soldatov, V.I. Stratsev, V.D. Natsik, Phys. Status Sol. (b) **22**, 739 (1974)
W.D. Pridemore, J Fail. Anal. Prevent, **8**, 281 (2008)

T.M. Pollock, A.S. Argon, Acta Metall. Mater. **40**, 1 (1992)
S.C. Prasad, I.J. Rao, K.R. Rajagopal, Acta Mater. **53**, 669 (2005)
R. Raj, Met. Trans. **A12**, 1089 (1981)
W. Raman, Met. Trans. **A17**, 1100 (1986)
M. Rieth, A. Falkenstein, P. Graf, S. Heger, U. Jantsch, M. Klimiankou, E. Materna-Morris and H. Zimmermann, Inst. For Materialforschung Program Kernfusion, Forschungszentrum, Karlsruhe, in der Helmholtz-Gemeinschaft Wissenschaftliche Berichte, FZKA 7065, November 2004
V. Sass, U. Glatzel, M. Feller-Kniepmeier, Creep anisotropy in the monocrystalline nickel-base superalloy CMSX-4, in *Superalloys* (The Minerals, Metals and Materials Society, Warrendale, 1996)
G. Schoeck, A. Seeger, *Report on the Conference on Defects in a Crystalline Solid* (Physical Society of London, London, 1955), p. 340
O.D. Sherby, J.E. Dorn, Creep correlations in alpha solid solutions of aluminum. Trans. AIME **4**, 959 (1952)
O.D. Sherby, J.L. Lyton, Trans. AIME **206**, 928 (1956)
O.D. Sherby, R.L. Orr, J.E. Dorn, Trans. AIME **200**, 71 (1954)
O.D. Sherby, J.L. Lyton, J.E. Dorn, Acta Met. **219** (1957)
I. Shi, D.O. Northwood, Acta Met. Mater. **42**, 871 (1994)
V.I. Stratsev, A.I. Osetski, V.P. Soldatov, E.E. Badyan, Pham Duyen, A.M. Gorki, J. Mater. Sci. **10**, 387 (1975)
M. Suzuki, T. Kimura, J. Koike, K. Maruyama, Mat. Sci. Eng. **A387**, 706 (2004)
J.C. Tan, M.J. Tan, Mater. Sci. Eng. **A339**, 81 (2003)
Y. Terada, T. Sato, J. Alloys Compd. **504**, 261 (2010)
M.D. Uchic, D.C. Chrzan, W.D. Nix, Intermetallics **9**, 963 (2001)
R.Z. Valiev, V.Yu. Gertsman, O.A. Kaibyshev, Phys. Status Sol. (a) **97**, 11 (1986)
J. Wadsworth, O.A. Ruano, O.D. Sherby, Met. Mater. Trans. **33A**, 219 (2002)
C.W. Weaver, J. Inst. Met. **87**, 126 (1958–1959)
J. Weertman, J. Appl. Phys. **26**, 1213 (1955)
J. Weertman, J. Appl. Phys. **28**(196), 1185 (1957a)
J. Weertman, J. Appl. Phys. **28**(196), 362 (1957b)
J. Weertman, Trans. ASM **61**, 680 (1968)
O.H. Wyatt, Proc. Phys. Soc. **B66**, 495 (1953)
C. Yuan, J. Guo, S. Wang, J. Mater. Sci. Technol. **14**, 219 (1998)
J.X. Zhang, J.C. Wang, H. Harada, Y. Koizumi, Acta Mater. **53**, 4623 (2005)
S.M. Zhu, S.J. Tjong, Met. Mater. Trans. A **29A**, 299 (1998)

Chapter 6
Cyclic Stress – Fatigue

6.1 Introduction

The most common failure that occurs in materials, such as metals, is caused by fatigue. The simplest way of looking at fatigue is by considering a specimen which is being repeatedly stressed under tension and compression. Not only tensile stresses that are repeatedly applied can cause fatigue failure, but any force which is acting in a reverse direction may ultimately result in such a failure. Loading a test specimen repeatedly by applying a force acting axially, torsionally or flexurally can induce fatigue failure. The danger in fatigue failure is that it may occur without any warning at stress levels considerably below the yield stress. Over the years, much experience has been accumulated by exploring the possible reasons for fatigue failure and tests have been suggested to evaluate the propensity for the failure of machine elements, exposed to pulsating or vibrational stresses. Nevertheless, the problem in predicting fatigue failure is due to the wide statistical spread of the results; often a deviation of ~50% from the average value is observed. In contrast, other mechanical tests, such as yield stress, for example, do not deviate from an average value by more than ~2–3%. Thus, many test specimens must be used in fatigue experiments in order to reach a meaningful average value, below which the probability for fatigue fracture is quite low. Later on in this chapter, a common method for evaluating the endurance of specimens is presented by way of repeated stress tests involving reverse loading versus the number of cycles. The results of such tests are collected into S-N curves (stress vs. cycles). Not only alternating stress can cause failure. Materials, such as glass, when exposed to long-term stress, fracture without any plastic deformation; the term for such behavior is 'static fatigue'. The term 'thermal fatigue' is assigned to the failure of materials caused by repeated changes in stress due to the rise and fall of thermal gradients for various reasons involving the restriction of thermal expansion or contraction. In the following sections, various aspects related to fatigue are discussed.

J. Pelleg, *Mechanical Properties of Materials*, Solid Mechanics and Its Applications 190, DOI 10.1007/978-94-007-4342-7_6,

6.2 The Endurance Limit; S-N Curves

Various wave forms of cyclic stresses may be applied to a specimen for testing its suitability to act for a specific purpose in which good fatigue resistance is required. Figure 6.1 is an illustration of stress cycles to which machine elements in industrial applications may be exposed. The first three illustrations in Fig. 6.1 represent sinusoidal-type cycles and the last one is a repeated cycle, possibly representing a specimen exposed to repeated stress. Airplanes and outer-space vehicles are not exposed to such regular stresses. Usually, the stresses acting, e.g., on the wings of airplanes, are irregular by nature.

Various types of machines are available on the market for applying any of the three usual types loading (mentioned in Sect. 6.1), namely, tension-compression, torsion and flexure. The most common of these tests is when tension-compression-type loading is applied, since it is easier to perform such investigations. The results of such tests are generally recorded as stress versus the number of cycles applied. The term for the resulting graphs, constructed from the recorded data of these tests, is 'S-N curves'. Figure 6.2 shows schematic S-N curves.

Metals and alloys showing a definite fatigue limit, such as the ferrous metals or titanium, are characterized by their ability to endure a large number of stress-cycles at the stress of the horizontal line and below it. The stress at this line is known as the 'endurance limit'. In most of the non-ferrous metals, no definite limit exists and the number of cycles that they can endure without failure increases monotonically with

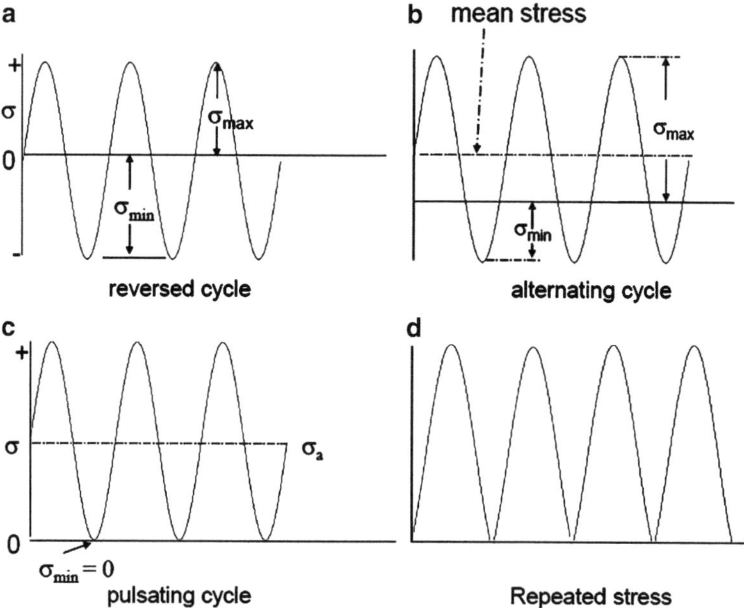

Fig. 6.1 Various forms of cyclic stresses: (**a**), (**b**) and (**c**) are sinusoidally varying cycles and (**d**) represents a repeating cycle

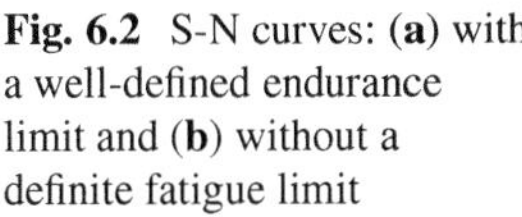

Fig. 6.2 S-N curves: (**a**) with a well-defined endurance limit and (**b**) without a definite fatigue limit

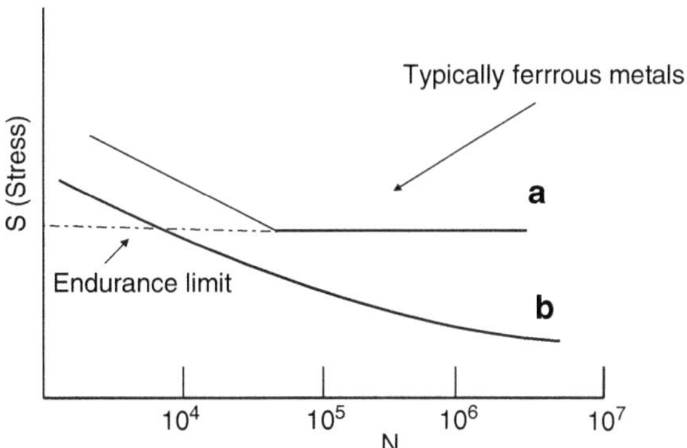

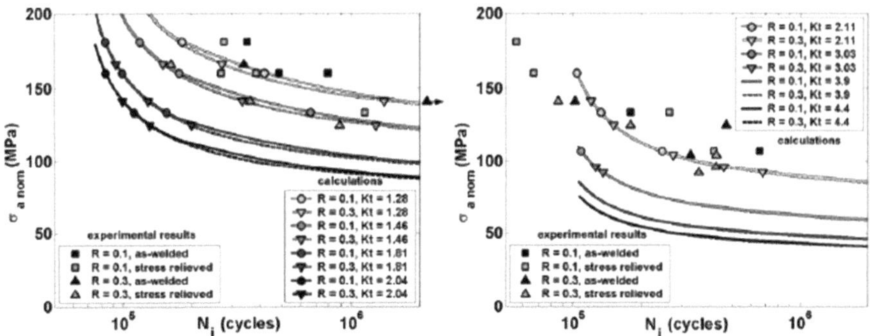

Fig. 6.3 Comparison of experimental and calculated (by finite elements) S-N curves in welded structure of fine grained steel (S355NL) on both sides "front side" and "rear side". R in the experimental and calculated results shown in the *inset* relates to the radii of the specimens. Kt is the concentration factor (Thevenet et al. 2008 (By courtesy of Dr. D. Thevenet)

decreased applied stress. In Fig. 6.3, experimental S-N curves are shown for fine-grained steel (S355NL) welded joints of as-welded and stress-relieved specimens. The legend in the insets indicates the radii and the stress-concentration factors, Kt, of the specimens tested. All of these fatigue tests were conducted under constant amplitude loading in each configuration. The weld geometry is characterized by a weld-toe radius and a weld-toe angle.

Based on Fig. 6.1, one may define several concepts, useful for fatigue studies, as follows: The safety factor allowed (σ_{al}) and applied (σ_{ap}) stress may be given as:

$$FS = \frac{\sigma_{al}}{\sigma_{ap}} \tag{6.1}$$

the mean stress as:

$$\sigma_{mean} = \frac{\sigma_{\max} + \sigma_{\min}}{2} \tag{6.2}$$

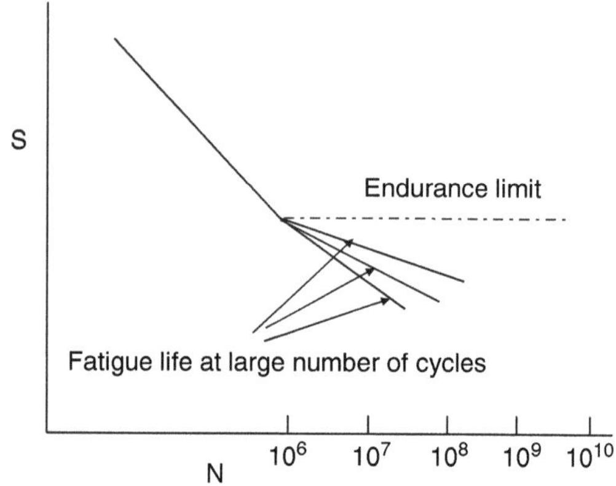

Fig. 6.4 The concept of a finite-endurance limit for materials, such as steel, where it is assumed that a life-time is infinite, is not necessarily true for high-cycle (gigacycle) testing, as seen in the illustration (After Marines et al.)

the ratio of load or stress by:

$$R = \sigma_{min}/\sigma_{max} \tag{6.3}$$

and the stress range by:

$$\sigma_r = \sigma_{max} - \sigma_{min} = \Delta\sigma \tag{6.4}$$

If $\sigma_{max} = -\sigma_{min}$, then the stress cycle is reversed. In such cases, the mean stress is zero.

The stress amplitude, σ_a, is one half the stress range and is given as:

$$\sigma_a = \frac{\Delta\sigma}{2} = \frac{\sigma_{max} - \sigma_{min}}{2} \tag{6.4a}$$

In the above illustrations, the number of cycles does not exceed 10^7. In conventional fatigue tests, seldom more than 10^7 cycles are applied to check fatigue lives. However, in some industries, the required design life-time of many components often exceeds 10^8 cycles (see, for example, Marines et al.). Thus, in aircraft (gas-turbine disks, 10^{10} cycles), automobiles (car engines, 10^8 cycles) and trains (high-speed trains, 10^9 cycles) fatigue life beyond the conventional 10^7 cycle tests are crucial. Time constraints usually prevent the performance of such extended tests. Here, the possibility of the accelerated testing of specimens is considered by way of high-frequency, cyclic loading (see, for example, Marines et al., B. Pyttel et al.). Their results seem to indicate that materials with definite endurance limits, such as ferrous metals, do not necessarily show infinite-fatigue life when tested at cycles above 10^7 and, as Pytell et al. state: "The fatigue limit cannot be a general material property. The term fatigue limit should not be used and it is better to substitute it by fatigue strength at a definite number of cycles" (as indicated in Fig. 6.4).

An example of fatigue testing by means of a large number of cycles may be seen in the case of cast iron (consult Fig. 6.5). In this case, no definite endurance limit

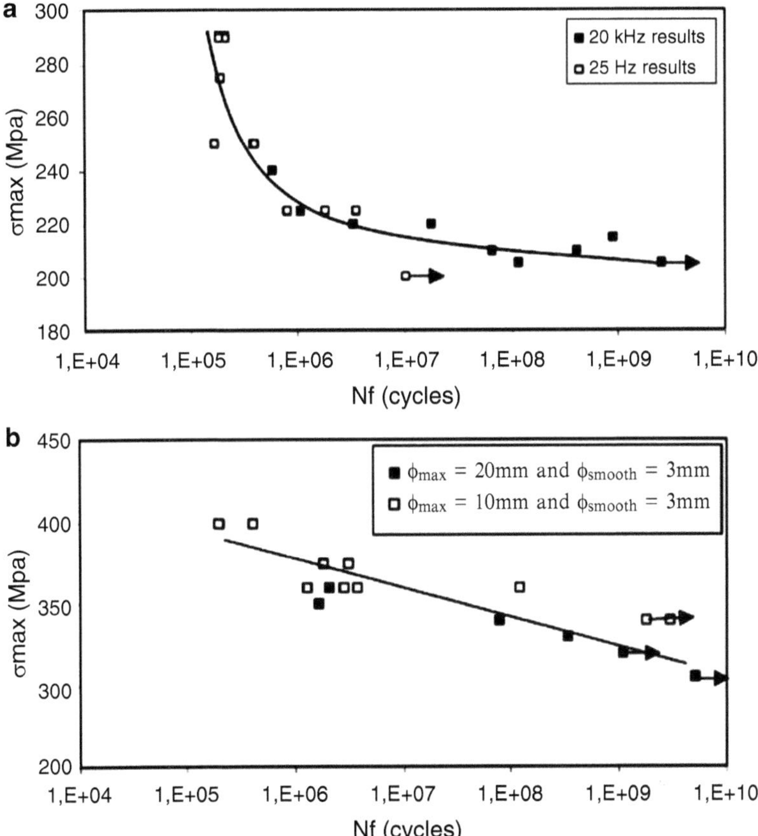

Fig. 6.5 Cast iron loaded in a wide range of cycles beyond 10^6 cycles: (**a**) shows the frequency effect, (**b**) shows the size effect. Note in (**a**) no definite endurance limit exists. Nf is the cycles to failure (Marines et al. 2003a. With kind permission of Elsevier)

exists and the stress-to-failure decreases with the increased number of cycles. It may be stated (Pyttel) that the use of fatigue strength at a definite number of cycles is more appropriate than the endurance limit, which is horizontal to the axis of the number of cycles.

The total fatigue life-time of an Al alloy is shown in Fig. 6.6 in two modes. Note that the scatter of experimental data is large, so one must be careful when interpreting the fatigue data. The statistical approach to fatigue tests is a consequence of the spread in experimental results. Mode A is related to the initiation of failure by the presencc of inclusions, which occurs at high stress with a low number of cycles below 10^6. Mode B refers cases in which the initiation of failure by cracks occurs at slip bands at lower σ_{max} and when the fatigue life is beyond 10^6 cycles. Persistent slip bands are believed to be the main mechanism causing fatigue failure in this case. Figure 6.7 illustrates mode B failure in slip bands. In light of

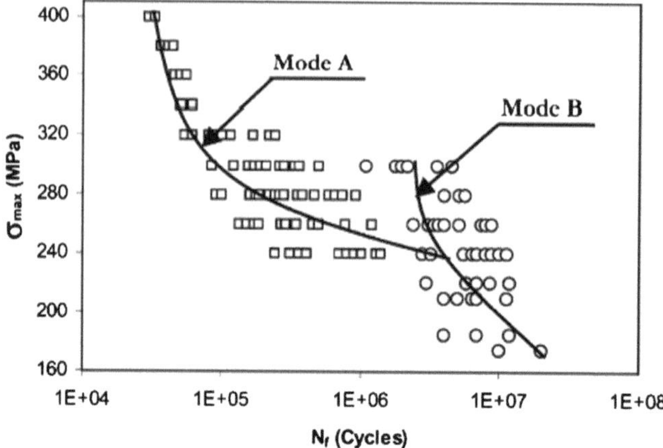

Fig. 6.6 S-N curve of a 2024/T3 Al alloy presented in two modes (Marines et al. 2003a. With kind permission of Elsevier)

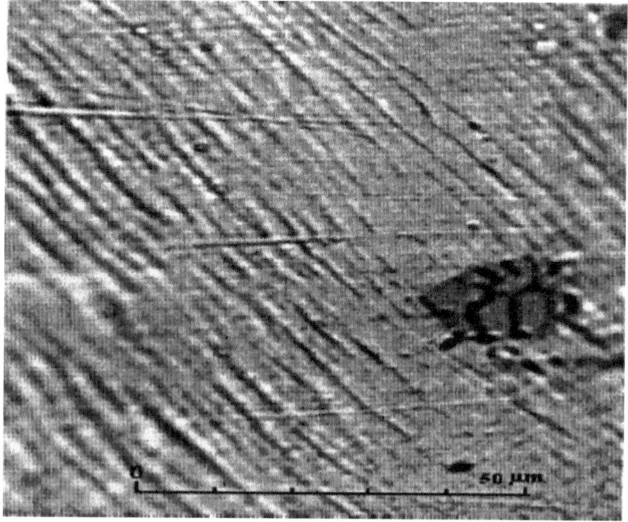

Fig. 6.7 Illustrates the slip bands in the Al alloy of Fig. 6.6. The inclusions broke. $\sigma_{max} = 260$ MPa (Marines et al. 2003a. With kind permission of Elsevier)

the manivaried experimental results, it is clear that a different approach to fatigue testing is necessary, one that can be performed by modern equipment and is specially designed for such purposes.

Determining the material life-time by establishing its endurance limit may be accomplished by the accelerated testing of specimens by means of high-frequency cyclic loading. Piezoelectric-fatigue systems suggest one approach for the

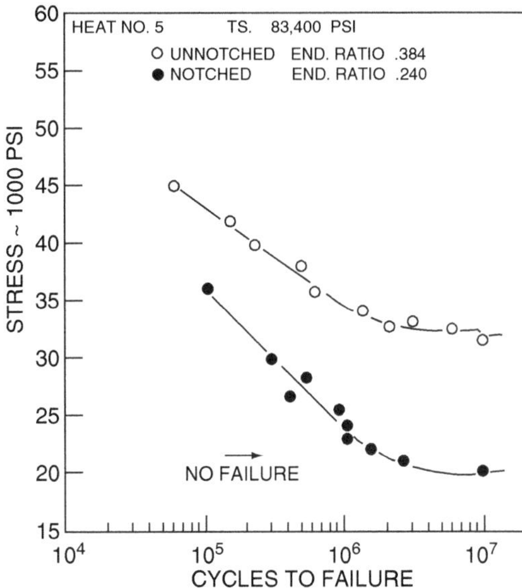

Fig. 6.8 S-N notched and unnotched specimens (Moore) are compared in the illustration. The endurance limit is indicated (Vishnevsky and Wallace 1967. With the kind permission of the Steel Founders Association)

performance of such tests. Another testing method may be by means of ultrasonic techniques, as used by Marines et al. to perform ultrasonic fatigue tests on bearing steel AISI-SAE 52100 at frequencies of 20 and 30 kHz. Xue et al. investigated cast iron in the very high cycle regime of 10^{10} also using ultrasonic technique. Today, there are also high-speed servohydraulic fatigue-testing machines, enable 10^{10}-cycle fatigue tests to be completed within 10 days or 3 months (Furuya et al.). As mentioned above, some industries require the design of machine components for life-times in the gigacycle range. Therefore, it is absolutely necessary to determine safe fatigue strength for these parts experimentally.

6.2.1 The Endurance Limit in Ferrous Metals

In Fig. 6.2, one of the S-N curves shows a sharply-defined knee, characterizing ferrous metals and a few others, such as Ti alloys. The endurance limit, or as it is often called, the 'fatigue limit', is typically parallel to the x-axis, which, as indicated in the above graph, represents the number of cycles-to-failure at a given stress. In theory, a material characterized by the endurance limit can perform at this stress level indefinitely (when the proper safety factors are taken into account) and can withstand in service a large number of cyclic stresses (as often stated as an infinite number). Figure 6.8 shows experimental S-N curves for normalized and tempered Ni-Cr-Mo (8630) cast steel. Notched and unnotched specimens (Moore) are also compared.

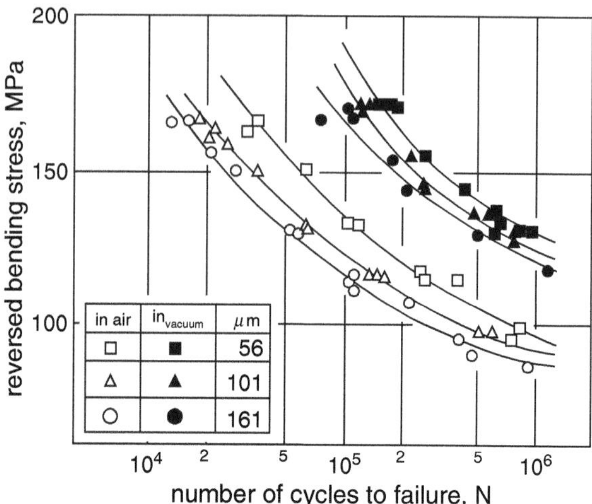

Fig. 6.9 S-N curves of Cu at different grain sizes in air and vacuum (Wang et al. 1994. With kind permission of Elsevier)

However, all these tests were performed at cycles not exceeding 10^6 or, at best, in the range of 10^6–10^7. Recent techniques (mentioned above) indicate that when the number of cycles is large, up to the gigacycle range, the concept of endurance limit does not apply and no well-defined knee exists in ferrous metals.

One must realize that the allowable stress level decreases with increased cycles, so talk of a well-defined fatigue limit is meaningless (see Fig. 6.4). The design of machine elements for operation over a very large number of cycles must not take for granted that any machine part will endure a certain stress for an unlimited number of cycles.

Fatigue tests are very sensitive to test conditions: the type of loading, the environment, the dimensions, the type of specimen and surface conditions (following machining). In Fig. 6.9, the effect of a notch is illustrated in comparison with unnotched specimens and one realizes that the presence of a notch greatly reduces both the stress level and the endurance limit.

6.2.2 *The Endurance Limit in Non-ferrous Metals*

No sharply defined knee exists in most of the non-ferrous metals. Thus, no true endurance limit exists. Usually, structures to be used as machine parts are tested by the application of a stress-level value corresponding to an arbitrary fatigue strength, often given at the number of cycles of $N = 5 \times 10^8$. The S-N relation for non-ferrous metals (shown schematically in Fig. 6.2) has been verified experimentally in a large number of cases. Of the many cases reported in the literature, Cu provides a good example; the experimental results for Cu are presented in Fig. 6.9. In this illustration, various grain size specimens were tested both in air and in vacuum. As can be seen, the stress values are about an order of magnitude higher for the same

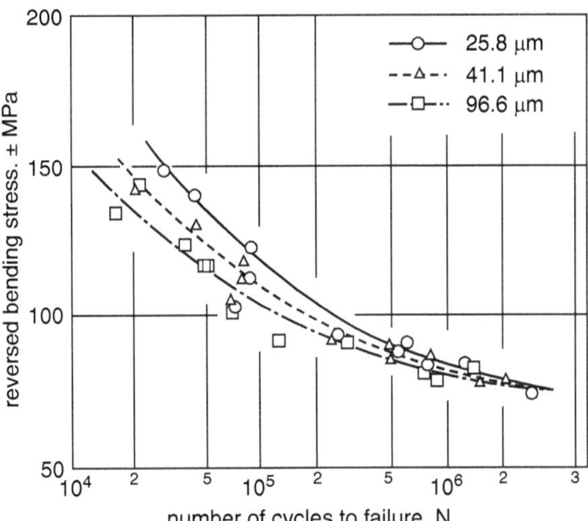

Fig. 6.10 The effect of grain size on the S-N curves of OFHC Cu (Wang et al. 1994. With kind permission of Elsevier)

number of cycles in vacuum for all the grain sizes. The better fatigue resistance in vacuum manifests itself in lower crack density and crack length in the dominant cracks responsible for failure. Furthermore, the fatigue life is one order of magnitude higher in vacuum. The S-N curves in an oxygen-free, high-conductivity copper [henceforth: OFHC], as affected by grain size, may be seen in Fig. 6.10. The loads applied to the specimens in Figs. 6.9 and 6.10 are reversed-bending stress.

The purpose of fatigue tests is to predict the resistance of materials to fatigue failure, namely to determine their life-times. Various methods are used to evaluate fatigue resistance, the performance of materials under cyclic loads and the fatigue life for practical design purposes; many ways are used to express the fatigue property of a material. Thus, any of the definitions in Eqs. 6.1, 6.2, 6.3, and 6.4 were used for this purpose, as are the empirical relations suggested by researchers, in order to reduce the data scatter and, thus, to improve the experimental results (briefly discussed in following chapters).

6.3 The Stress Cycles

Cyclic loads may be applied in various forms, as mentioned earlier. The load applied may be tensional, torsional or flexural. Any of these can induce fatigue failure. Rather than focusing on these various loading possibilities, this section will deal with the range of the cycles. A machine element may be exposed either to a large number of cycles or, often, to a small number of low cycles, inducing failure. The purpose of a designer is to obtain a long life-time in service; therefore, testing must be aimed at obtaining information on the propensity of a material to endure a large number of cycles without failure.

6.3.1 Low-Cycle Fatigue Tests

As the title implies, specimens may be tested during a relatively small number of cycles by repeated stress or strain until failure sets in. The number of cycles, determined arbitrarily, is in the range of 10^4–10^5 cycles, which may be considered the upper limit for low-cycle tests. Experience has taught an important lesson that structural materials, particularly steel components for ship construction, fail due to low-cycle fatigue, which imposes a serious structural problem. Performing tests at high cycles incurs time- and cost-related problems; therefore, low-cycle tests are often performed at high stresses. Low-cycle fatigue is strain-controlled and frequently the number of cycles does not exceed 10^3. The purpose is to evaluate fatigue life more precisely and to identify when fatigue cracks form.

The Ramberg-Osgood relation is commonly used to represent cyclic, as well as monotonic, stress-strain curves. For monotonically-loaded specimens, when elastic and plastic strains occur, the tensile behavior of metallic materials is given by means of the Ramberg-Osgood relation, as:

$$\varepsilon = \varepsilon_e + \varepsilon_p = \frac{\sigma}{E} + \left(\frac{\sigma}{K'}\right)^{\frac{1}{n'}} \tag{6.5}$$

In this relation, ε is the sum of the elastic and plastic strains, ε_e and ε_p, i.e., the total amplitude of deformation. The first term on the right side is the value of ε_e in terms of stress and the elastic modulus. The second term, related to plastic strain, may be obtained from the Zener-Hollomon telation, given as

$$\sigma = K{\varepsilon_p}^{n} \tag{6.6}$$

or along the line of the notation in Eq. 6.5:

$$\sigma = K'{\varepsilon_p}^{n'} \tag{6.6a}$$

By expressing ε_p from Eq. 6.6a, one gets:

$$\varepsilon_p = \left(\frac{\sigma}{K'}\right)^{\frac{1}{n'}} \tag{6.6b}$$

and by substituting for ε_p, on the left side of Eq. 6.5, one obtains Eq. 6.5. E is the modulus of elasticity, K' is the cyclic-strength coefficient and n' is the cyclic strain-hardening exponent.

For cyclic loading, cyclic stress, in terms of cyclic strain, is expressed by explicit relations using the cyclic-strength coefficient. The cyclic σ (at typically less than 10^3 cycles) is associated with plasticity in metals; therefore, a strain-based parameter should be used for fatigue-life prediction in metals and alloys. Thus, testing is conducted with constant strain amplitudes. In the case of fully reversed cycling in the area of low-cycle fatigue, the Manson-Coffin approach has been widely adopted in cases of crack initiation. This Manson-Coffin relation may be used to evaluate the damage, in terms of the total strain amplitude, as:

$$\varepsilon_a = \frac{\sigma'_f}{E}\left(2N_f\right)^b + \varepsilon'_f(2N_f)^c \tag{6.7}$$

ε_a is an alternating strain. This relation expresses the fatigue resistance of a material and is known as the 'Basquin-Manson-Coffin equation'. It may be obtained as follows. The relation between the amplitude of total deformation and the number of cycles-to-failure of a test specimen (see, for example, Junek et al.) is:

$$\sigma_a = \sigma'_f\left(2N_f\right)^b \tag{6.8}$$

The plastic-strain amplitude, in terms of the number of cycles, is:

$$\varepsilon_p = \varepsilon'_f\left(2N_f\right)^c \tag{6.9}$$

Rewriting Eq. 6.5, where ε and ε_a are the same, using Eq. 6.8 for the stress and Eq. 6.9 for the plastic strain, one obtains:

$$\varepsilon = \varepsilon_a = \varepsilon_e + \varepsilon_p = \frac{\sigma}{E} + \varepsilon_p = \frac{\sigma'_f\left(2N_f\right)^b}{E} + \varepsilon'_f\left(2N_f\right)^c \tag{6.7}$$

which is Eq. 6.7.

Metallic materials have b values between −0.05 and −0.12 and, in many cases, σ'_f is about the true-fracture stress (sometimes given as $1.75\sigma_{UTS}$). The fatigue-ductility exponent, c, usually varies between −0.5 and −0.7. N_f is the cycles-to-failure. Note that for σ in Eq. 6.5, σ_a from Eq. 6.8 is used. The strain amplitude, as a function of cycles-to-failure, is shown in Fig. 6.11 for an Al alloy (AA7175-T1), according to Eq. 6.7. The test results of this figure relate to zero mean strain and zero mean stress. The zero mean strain during cyclic loading indicates that the maximum strain is equal in absolute value to the minimum strain observed during cyclic loading. These data allow for the determination of fatigue strength, σ'_f, and ductility coefficients, ε'_f, and their exponents (as indicated in Eq. 6.7). A cyclic stress-strain curve may be determined from a stable hysteresis loop, allowing for the determination of the cyclic strain etc. up as indicated. Strain-hardening exponent, n′, and the strength coefficient, K′. The Ramberg-Osgood relation is used to evaluate tension-test information, such as the strength coefficient, the strain-hardening exponent, etc. Furthermore, the tension tests are conducted in order to determine: the modulus of elasticity E, yield strength, tensile strength, etc. Usually, experimental data are used to evaluate the number of cycles to fatigue failure. The tests mentioned above are conducted at zero mean stress and strain. Empirical relations were developed to evaluate fatigue life behavior at mean-strain values other

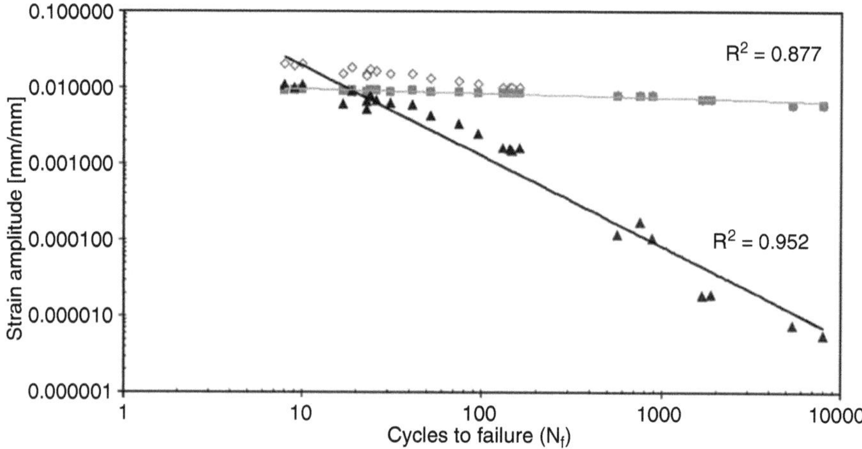

Fig. 6.11 Strain amplitude versus number of cycles to failure at zero mean strain and stress (R = −1) (Salerno et al. 2007. With kind permission of Elsevier)

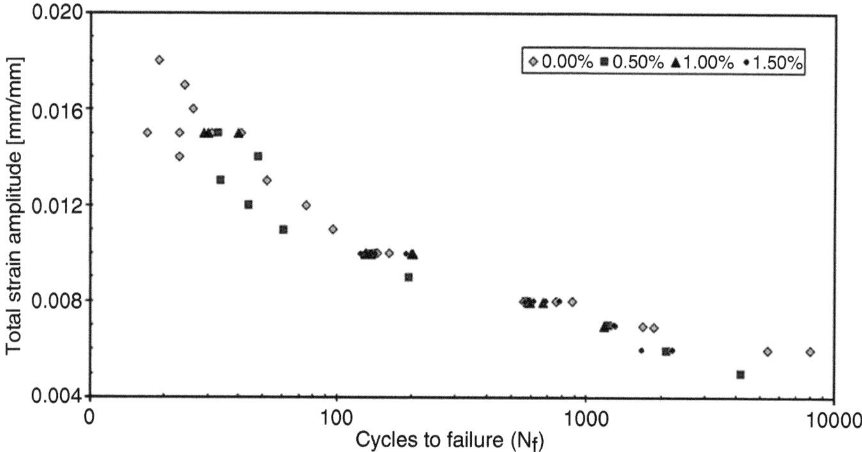

Fig. 6.12 Strain amplitude versus cycles to failure at different mean strains (Salerno et al. 2007. With kind permission of Elsevier)

than zero. Experimental data under different mean strains, for an Al alloy (AA7175-T1), is shown in Fig. 6.12, where the total strain versus cycles-to-failure is indicated.

Such data are useful for the application of various empirical parameters. One of the empirical relations developed to evaluate fatigue-life behavior at mean-strain values other than zero is known as the 'Smith, Watson and Topper' parameter [henceforth: SWT]. The SWT parameter is based on the assumption that the product

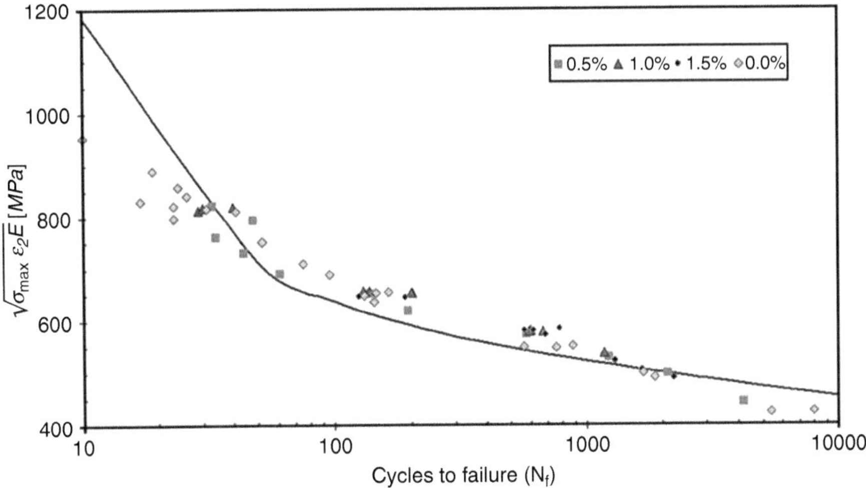

Fig. 6.13 SWT parameter versus cycles to failure in fatigue tests results ($R^2 = 0.846$) (Salerno et al. 2007. With kind permission of Elsevier)

of σ_{max} and ε_a is a constant, i.e., $(\sigma_{max}\varepsilon_a E)^{1/2} =$ constant for a given number of cycles-to-failure. Different combinations of strain amplitude and mean stress may be used, as seen in the relation below (Smith):

$$\sigma_{max}\varepsilon_a E = {\sigma'_f}^2 (2N_f)^{2b} + \sigma'_f \varepsilon'_f E (2N_f)^{b+c} \tag{6.10}$$

The SWT parameter is very successful in predicting the effect of mean stress in low- and high-cycle fatigue for many materials. Figure 6.13 shows the SWT parameter versus the number of cycles-to-failure in an AA 7175-T1 aluminum alloy. Many other parameters have been suggested for the evaluation of material fatigue life-times.

One may also mention the Walker and the Morrow parameters. The Walker relationship for Al is expressed as:

$$\varepsilon_a = \frac{\sigma'_{fw}}{E}\left[2N_f\left(\frac{1-R}{2}\right)^{\frac{1-\gamma}{b_w}}\right]^{b_w} + \varepsilon'_{fw}\left[2N_f\left(\frac{1-R}{2}\right)^{\frac{1-\gamma}{b_w}}\right]^{c_w} \tag{6.11}$$

with:

$$R = \frac{\sigma_{min}}{\sigma_{max}} \tag{6.12}$$

$$N^* = N_f\left(\frac{1-R}{2}\right)^{\frac{1-\gamma}{b_w}} \tag{6.13}$$

As seen in Eq. 6.11, the Walker equation presents N_f, found during an applied strain amplitude, ε_a, at any R ratio. Strain amplitude may be related to the number of cycles obtained under zero-mean stress, according to Eq. 6.13. The Morrow relation may be expressed as:

$$\varepsilon_a = \frac{\sigma'_f}{E}\left[2N_f\left(1-\frac{\sigma_m}{\sigma'_f}\right)^{\frac{1}{b}}\right]^b + \varepsilon'_f\left[2N_f\left(1-\frac{\sigma_m}{\sigma'_f}\right)^{\frac{1}{b}}\right]^c \tag{6.14}$$

Equation 6.14 may also be expressed as:

$$\varepsilon_a = \frac{\sigma'_f}{E}\left(1-\frac{\sigma_m}{\sigma'_f}\right)(2N_f)^b + \varepsilon'_f\left(1-\frac{\sigma_m}{\sigma'_f}\right)^{\frac{c}{b}}(2N_f)^c \tag{6.15}$$

In Eq. 6.16, the cycles-to-failure at zero-mean stress is defined by:

$$N^* = N_f\left(1-\frac{\sigma_m}{\sigma'_f}\right)^{\frac{1}{b}} \tag{6.16}$$

When expressing Eq. 6.15 in terms of Eq. 6.16, the similarity of Morrow's relation to that of Manson-Coffin in Eq. 6.7 is seen with N^* replacing N_f:

$$\varepsilon_a = \frac{\sigma'_f}{E}(2N^*)^b + \varepsilon'_f(2N^*)^c \tag{6.17}$$

The above relations for predicting the life-times of various materials have been used and their advantages and disadvantages have been compared. Comparisons were made for steel and Al alloys between the SWT approach, the Walker model and the Morrow equation. It was concluded that the SWT approach is a good choice for general use. SWT is quite accurate for aluminum alloys and is acceptable for steels, although it is not quite as good as Morrow's equation (see, for example, Dowling). It has also been stated that the Walker mean-stress equation, with adjustable constant γ, provides superior results where γ is known or can be estimated. These empirical methods for evaluating approximate fatigue life-times are quite convenient to use. However, designers must also be aware of the disadvantages imposed by each type of alloy and choose a method accordingly, which is expected to best describe the particular experimental results. Empirical relations, that are able to predict the total number of cycles to fatigue failure under nonzero mean-strain cyclic loadings, should be analyzed on the basis of laboratory tests of materials of interest. Low-cycle fatigue tests are of particular importance when interest lies in new materials for fatigue applications, for which no earlier experimental results are available.

6.3.2 High-Cycle Fatigue Tests

Today, various machines are available for high-cycle fatigue testing. A load-controlled servo-hydraulic test rig is one such machine, commonly used in these tests, with frequencies of around 20–50 Hz. Resonant, magnetic machines are also in use as are rotating, bending machines. In such fatigue tests, a constant bending stress is applied to a round specimen, combined with the rotation of the sample around the bending-stress axis until its failure. The cycle range of these tests is 10^3–10^8. High-cycle fatigue strength may be described by stress-based parameters. High-cycle tests are usually performed until failure. The common procedure during these tests is to start testing specimens at high stress until their failure, which usually sets in after a relatively small number of cycles. Then, in tests done to the next specimens, the stress level is successively decreased, to the point where the test samples do not fail. When a few specimens do not fail under applied stress throughout all the cycles of the test, the data are noted and it is said that 'runout' has occurred. In most cases, this runout refers to at least 10^7 cycles. Thus, runout represents the highest stress of non-failure at the specified cycles. In ferrous alloys, like steel or Ti alloys, the point at which runout occurs is termed the 'endurance limit' (as mentioned earlier), which is not observed in most non-ferrous alloys. Therefore, these tests are usually terminated at $\sim 10^8$ cycles, considered as being appropriate for safe use in practice.

6.3.3 Very High Cycle Tests

As mentioned above (in Sect. 6.2), fatigue tests often exceed 10^8 cycles for justifiable reasons. An example of such a test appears in Fig. 6.5. Specifically, many structural components used in industrial applications are exposed to high frequencies, stresses acting at very high cycles and often at low amplitudes. Fatigue resistance under such conditions is not predictable unless preliminary evaluation has been performed. To this end, special tests are performed under those conditions that provide adequate information and tools, so that the designers of various components can avoid failure during service in high-cycle regimes. One method for studying fatigue life at very high-cycle regimes is by ultrasonic fatigue testing. Figure 6.14 illustrates an S-N curve in the high-cycle range, where the experimental data at 20 kHz were collected for a 10^9–10^{10}-cycle regime. As expected, fatigue life increased as the applied maximum stress was decreased. Similar to Fig. 6.5 for cast iron, no well-defined fatigue limit exists for this iron-based alloy. Again, the concept of infinite fatigue life, a claim based on the horizontal knee of an S-N curve in ferrous alloys, is not always found to be correct. Therefore, it is essential to perform tests in the gigacycle range in order to assess in service performance. These kinds of tests were first performed by French (e.g., Bathias et al., Marines et al., Pyttel et al.) and Japanese (Nishijima, Murakami et al., Sakai et al.) schools

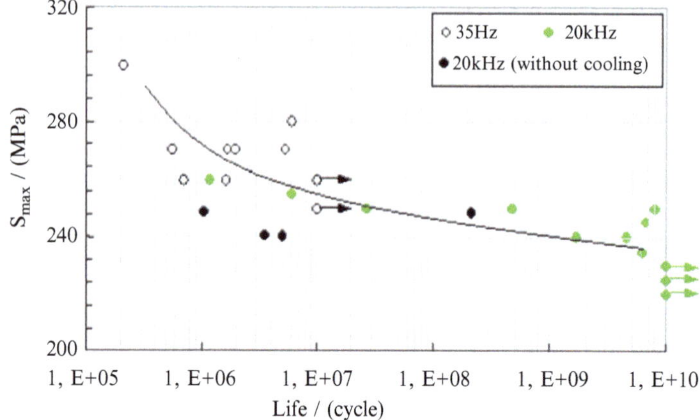

Fig. 6.14 S-N curve of a nodular cast iron at 35 Hz and 20 kHz with R = −1 in air at room temperature (Xue et al. 2006. By courtesy of Bayraktar)

that studied fatigue in the gigacycle range for various metals. From a design point of view, the determination of the endurance limit of ferrous and other alloys (where the endurance limit was observed at cycles not exceeding 10^7) is insufficient to ensure fatigue resistance in those systems when exposed to very high cycles.

6.4 Fatigue Life

Many factors, in addition to stress, strain, amplitude, etc., influence the fatigue life of materials. Here, the effects of (a) stress-based and (b) strain-based evaluations of fatigue life will be considered for low- and high-cycle cases. In the past, but also at present, the stress-based approach is quite common.

6.4.1 The Stress-Based Approach

The stress-based evaluation of fatigue life is clearly related to the S-N curves observed in the various metals. The applied-stress range or rather the stress amplitude, σ_a, of S-N curves is the significant factor determining life-time.

Equation 6.4a states that $\sigma_a = \frac{\Delta\sigma}{2}$. Thus, stress amplitude may be expressed as:

$$\sigma_a = \frac{\Delta\sigma}{2} = \frac{\Delta(E\varepsilon_e)}{2} = \frac{E\Delta\varepsilon_e}{2} = \sigma'_f(2N_f)^b \tag{6.18}$$

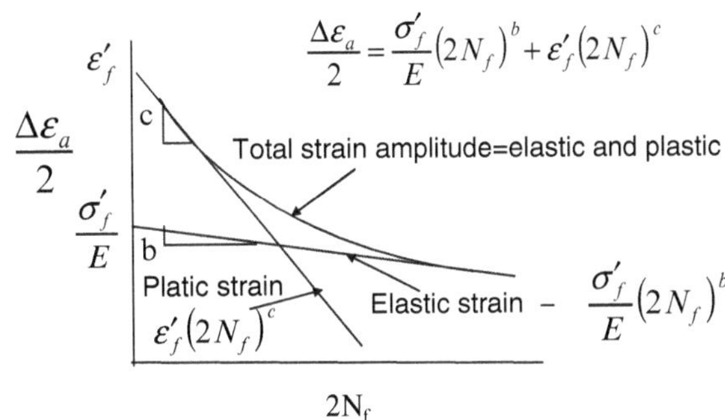

Fig. 6.15 A schematic strain-life curve, typical of many materials. The elastic- and plastic-strain amplitudes are included

Equation 6.18 is a consequence of Eq. 6.8, where stress amplitude is related to the number of cycles-to-failure. σ'_f is the fatigue strength and b is the Basquin exponent. $2N_f$ represents load reversal to failure. One immediately recognizes that the last term of Eq. 6.18 is basically the elastic-strain component of the Manson-Coffin relation, given in Eq. 6.7, where it is divided by E. The stress-based approach for evaluating fatigue life is generally used in high-cycle fatigue tests. In such tests, the cycles are above 10^4 and the applied stress is primarily within the elastic range of the material. Thus, the application is for long fatigue lives.

6.4.2 *Strain-Based Life-Times*

Strain-life fatigue analysis is more appropriate in low-cycle applications, where the applied strains have a significant plastic component, due to high load levels. The total amplitude of strain may be given as the sum of the elastic and plastic strains, according to Eq. 6.5, i.e., the Ramberg-Osgood relation:

$$\Delta\varepsilon = \Delta\varepsilon_e + \Delta\varepsilon_p \tag{6.19}$$

The Manson-Coffin relation, given above in Eq. 6.7, is now rewritten as:

$$\Delta\varepsilon = \Delta\varepsilon_e + \Delta\varepsilon_p = \frac{\sigma}{E} + \Delta\varepsilon_p = \frac{\sigma'_f(2N_f)^b}{E} + \varepsilon'_f(2N_f)^c \tag{6.7a}$$

or rather:

$$\frac{\Delta\varepsilon}{2} = \frac{\Delta\varepsilon_e}{2} + \frac{\Delta\varepsilon_p}{2} = \frac{\sigma}{E} + \frac{\Delta\varepsilon_p}{2} = \frac{\sigma'_f(2N_f)^b}{E} + \varepsilon'_f(2N_f)^c \tag{6.7b}$$

Figure 6.15. is a schematic illustration of Eq. 6.7b, representing a strain-life curve. The total strain-amplitude curve is the sum of the elastic- and plastic-strain amplitudes.

In Eq. 6.7b on the right side, the two terms represent the elastic- and plastic-strain amplitudes, each of which should be a separate linear function of the total strain amplitude. This is, indeed, shown in the schematic (Fig. 6.15) for the elastic- and plastic-strain amplitudes. The plot of σ_a versus the number of cycles of the elastic portion should be linear on a logarithmic scale, to enable the determination of σ'_f and b; whereas, a log-log plot of $\Delta\varepsilon/2$ versus $2N_f$ provides ε'_f and the exponent c. Thus, the four material constants σ'_f, ε'_f, b and c can be evaluated. More specifically, the elastic component may be written according to Eq. 6.18 as:

$$\frac{\Delta\varepsilon_e}{2} == \frac{\sigma'_f}{E}(2N_f)^b \tag{6.18a}$$

$$\log\frac{\Delta\varepsilon_e}{2} = \log\frac{\sigma'_f}{E} + b\log(2N_f) \tag{6.18b}$$

At $(2N_f) = 1$ (i.e. $10^0 = 1$), the intercept on a log-log plot is $\frac{\sigma'_f}{E}$ and the slope of log $\Delta\varepsilon_e/2$ versus log $(2N_f)$ is b, as indicated on the graph in Fig. 6.15. One can also say that the slope is the first derivative of Eq. 6.18b, namely:

$$\frac{d\left(\frac{\Delta\varepsilon_e}{2}\right)}{d\left(2N_f\right)} = b \tag{6.18c}$$

Note from Eq. 6.18 that the stress amplitude may be also written as $E(\Delta\varepsilon/2)$. Similarly, the plastic-strain amplitude, namely the second term of (6.7b), may be expressed by a log-log plot of $\Delta\varepsilon/2$ versus $(2N_f)^c$, resulting in ε'_f as the intercept at $(2N_f) = 1$ and c as the slope. The plastic component, namely the Manson-Coffin relation of Eq. 6.9, rewritten as presented in Eq. 6.7b, is given as:

$$\frac{\Delta\varepsilon_p}{2} = \varepsilon'_f(2N_f)^c \tag{6.19a}$$

$$\log\frac{\Delta\varepsilon_p}{2} = \log\varepsilon'_f + c\log(2N_f) \tag{6.19b}$$

$$\frac{d\left(\frac{\Delta\varepsilon_p}{2}\right)}{d\left(2N_f\right)} = c \tag{6.19c}$$

Fatigue life is expressed in terms of these four constants. As mentioned above, σ'_f is the fatigue strength and b is the Basquin exponent. Note that the total strain-life curve approaches the plastic strain-life curve in the low-cycle region. Experimental strain-life curves in the form of log-log plots of the total strain amplitude versus $2N_f$ (the load reversal to failure) are illustrated for AlMgSi aluminum alloys 6082 and 6060, respectively, in Fig. 6.16 and for invar in Fig. 6.17.

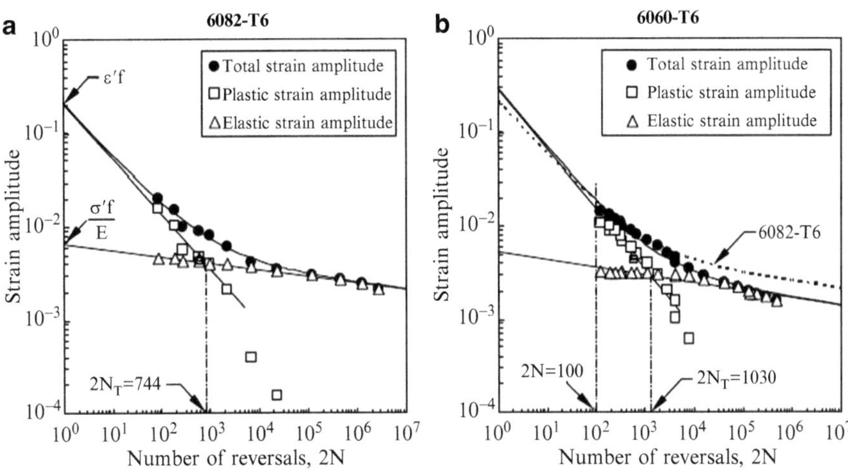

Fig. 6.16 Strain-fatigue life relationship; life as a function of elastic, plastic and total strain amplitudes (**a**) 6082-T6 alloy, (**b**) 6060-T6 alloy (Borrego et al. 2004. With kind permission of Elsevier and Professor Borrego for the authors)

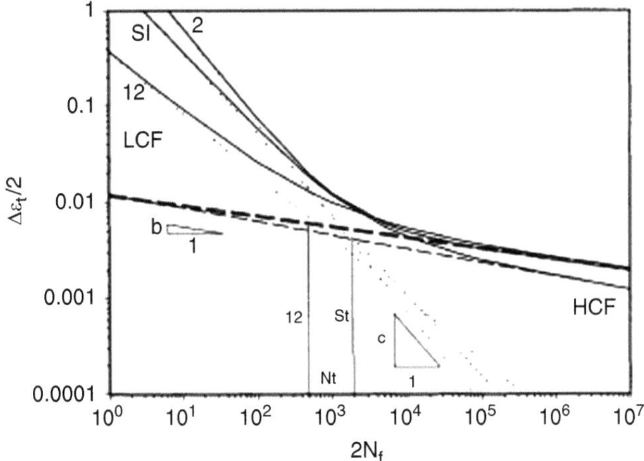

Fig. 6.17 The total strain amplitude versus life curve obtained from the superposition of the elastic and plastic strain amplitudes for an invar alloy. Asymptotic lines are shown by *dotted* and *dashed lines* for plastic and elastic strain components, respectively (Vinogradov et al. 2003. With kind permission of Elsevier and professor Vinogradov)

Invar is a ferrous metal containing nickel in various proportions. In Fig. 6.17, the invar is a Fe-36%Ni alloy having various fabrication treatments. In Figs. 6.16 and 6.17, the straight, asymptotic lines correspond to the Manson-Coffin relation and Basquin's law, respectively. One may define a 'transition life', N_t as the number of reversals-to-failure at which the elastic- and plastic-strain amplitudes are equal, namely:

$$2N_t = \left(\frac{\varepsilon'_f}{\sigma'_f}\right)^{\frac{1}{b-c}} \tag{6.19d}$$

This relation can be obtained from the two straight lines representing the elastic- and plastic-strain amplitudes, respectively. At the intersection of the two asymptotic lines, the life values are equal. In terms of the transition life, N_t, the two relevant equations from the right side of Eq. 6.7b are:

$$\frac{\sigma'_f}{E}(2N_t)^b = \varepsilon'_f(2N_t)^c \tag{6.20}$$

and

$$\frac{\varepsilon'_f E}{\sigma'_f} = \frac{(2N_t)^b}{(2N_t)^c} = (2N_t)^{b-c} \tag{6.21}$$

$$\left(\frac{\varepsilon'_f E}{\sigma'_f}\right)^{\frac{1}{b-c}} = 2N_t \tag{6.22}$$

Equation 6.22 is the same as Eq. 6.19d above. Note that $2N_t$ is at $\sim 10^3$ cycles.

Manson and Manson and Hirschberg proposed two methods for evaluating life-times in high- and low-cycle fatigue, known as 'four point correlation' and 'slope methods' are presented below

1. The four-point method. According to Eq. 6.7b (expressed schematically in Fig. 6.15), one can expect the total-life curve to approach the plastic strain-life line in the low-cycle region, while, in the high-cycle region, the elastic strain-life line is approached. The four-point correlation method is based on the fact (discussed above) that elastic and plastic lines which asymptotically approach the total strain-life time curve are linear. Thus, by selecting two points on each of the linear curves, four parameters may be evaluated. At $\sim 10^3$ cycles, the elastic and plastic strains are related, as seen in Figs. 6.16 and 6.17. Taking this point, or rather a point at 10^4 cycles as taken by Manson, and selecting another point on the elastic line at 1 cycle, the corresponding points on the y ordinate may be obtained. Recall that the value at 1 (i.e., 10^0 cycle) is $\frac{\sigma'_f}{E}$ and, according to Manson, the ordinate is 2.5 $\frac{\sigma_f}{E}$, where σ_f is the true-fracture stress, obtained by dividing the load at failure by the actual area during a tensile test. Another point chosen was at 10^5 cycles, when the ordinate was 0.9 σ_U and σ_U is the ultimate tensile stress (σ_{UTS}) of the material being tested. Two points were also selected on the plastic line, at 10 and 10^4 cycles, and their ordinates were expressed in terms of the logarithmic ductility, D, and the reduction of area, RA, and is

given as $D = \ln(1\text{-}RA)^{-1}$. Note that at 10^4 cycles, the elastic and plastic strains are related to each other and the total strain at this point is about 1% for all materials (Manson and Hirschberg). For further details on evaluating ordinates one is referred to the original research published by Ong or Cui. In any event, the relation developed by Manson and Coffin is given for Eq. 6.7b as

$$\frac{\Delta\varepsilon}{2} = A\left(2N_f\right)^b + B\left(2N_f\right)^c \tag{6.7c}$$

where:

$$b = \frac{\log\left[\frac{2.5[1+\ln(1/(1-RA))]}{0.9}\right]}{\log\left[1/\left(4x10^5\right)\right]}$$

$$A = \frac{1}{2}x10^{b\log 2+\log\left[\frac{2.5\sigma_U(1+\ln(1/(1-RA)))}{E}\right]}$$

The strain on the elastic line, $\Delta\varepsilon_e{}^{@}$, located at 10^4 cycles and used to evaluate the exponent c of Eq. 6.7c, is given as:

$$\Delta\varepsilon_e{}^{@} = 10^{b\log\left(4x10^4\right)+\log\left[\frac{2.5\sigma_U[1+\ln(1/(1-RA))]}{E}\right]}$$

c is expressed as:

$$c = \frac{1}{3}\log\left[\frac{0.0132 - \Delta\varepsilon_e{}^{@}}{1.91}\right] - \frac{1}{3}\log\left[\frac{1}{4}\left(\ln\left(\frac{1}{1-RA}\right)\right)^{3/4}\right]$$

and B is given as:

$$B = \frac{1}{2}x10^{c\log\frac{1}{20}+\log\left[\frac{1}{4}(\ln(1/(1-RA)))^{\frac{3}{4}}\right]}$$

2. The slope method. Manson's assumption is that the slopes of the elastic and plastic lines for all materials are the same. As indicated above (in Sect. 6.3.1) with respect to Eq. 6.7, for metallic material, b, is –0.05 to –0.12 and the fatigue-ductility exponent is between –0.5 and –0.7. For b and c, Manson has used –0.12 and –0.6, respectively, as seen in Eq. 6.23 below. He also showed that the intercept of the elastic line at life cycle $N_f = 1$ is $[\ln(1\text{-}RA)^{-1}]^{0.6}$ and the intercept of the plastic line at $N_f = 1$ is $3.5(\sigma_{UTS}/E)$. Thus:

$$\Delta\varepsilon = 3.5\frac{\sigma_U}{E}N_f{}^{-0.12} + \left[\ln\left(\frac{1}{1-RA}\right)\right]^{0.6}N_f{}^{-0.6} \tag{6.23}$$

or rewriting Eq. 6.23 in terms of the total strain amplitude:

$$\frac{\Delta\varepsilon}{2} = 1.9018\frac{\sigma_U}{E}\left(2N_f\right)^{-0.12} + 0.7579\left[\ln\left(\frac{1}{1 - RA}\right)\right]^{0.6} 2N_f^{-0.6} \quad (6.24)$$

Other methods have been proposed for the evaluation of fatigue life-times, for instance, Socie et al. and other approaches, as indicated in the work of Troshchenko and Khamaza. These methods are outside the scope of this section and the reader should consult the original works.

Summarizing this section, low-cycle fatigue applications are those with less than $\sim 10^4$ loading cycles during the component life. High-cycle fatigue is associated with component lives greater than $\sim 10^5$cycles. The transition life between low- and high-cycle fatigue depends on the material being considered and is usually between 10^4 and 10^5cycles. In general, the stress-life approach should not be used to estimate fatigue lives below 10^4 cycles.

6.5 Work Hardening (Softening)

The work hardening of materials occurs in a manner similar to monotonic testing during deformation. Under cyclic stress, in addition to work hardening, the softening of materials is also observed, called 'work softening'. It has been and is still customary to compare dynamic and static deformation results. Furthermore, there have also been attempts to present the fatigue behavior of materials in terms of ratios of dynamic-to-static test data, mainly the ratio of the fatigue limit to the tensile strength. Miller states that attempting such a correlation is understandable, since the ability to predict fatigue behavior from a single monotonic test is highly desirable. However, Miller claims that correlating static and dynamic test results is dangerous and unwise and that static data should not be used by designers to assess fatigue behavior in materials, since cyclic softening, cyclic hardening or both may occur to some extent, depending on the strain range, strain rate, temperature, material composition, etc. Figure 6.18a illustrates the rapid and substantial cyclic-hardening behavior of an aluminum alloy, En 32B, which occurs during the first few cycles of life. A steady state is soon established which predominates for a considerable life span. Figure 6.18b illustrates the cyclic softening of a steel specimen, En 25, showing the effect of the strain range on the degree of softening for various twist ranges during strain cycling. Figure 6.18 is based on torsion tests. Figure 6.19 illustrates the differences between cyclic and static behavior for the indicated materials. Analysis of Fig. 6.19a shows that maximum softening occurs at that specific plastic-strain range where there is a maximum torque-range differential between the monotonic and cyclic characteristics of the material. Moreover, Fig. 6.19a shows that, at high-strain ranges, cyclic hardening occurs.

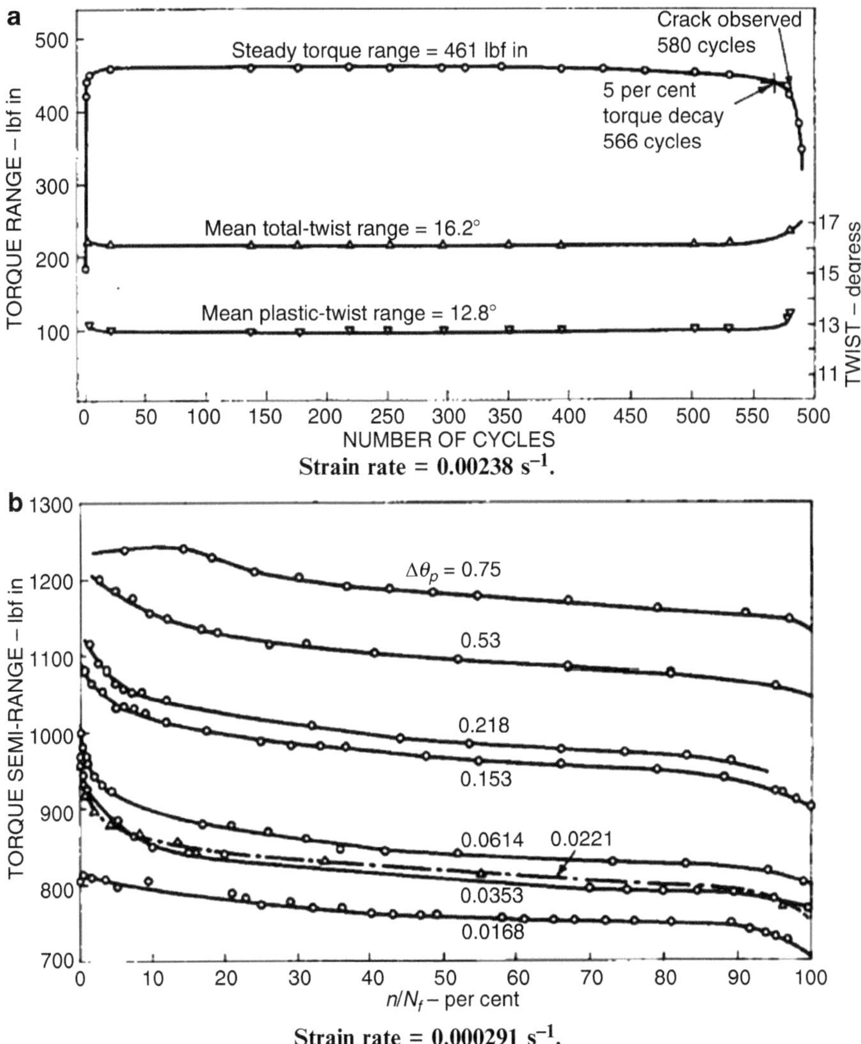

Fig. 6.18 (**a**) Hardening of an En 32B specimen; (**b**) softening at various twist ranges (After Miller 1970)

Miller indicated that these curves underline the inability of static tests to predict the behavior of materials under cyclic conditions so that designers are well-advised to use cyclic characteristics instead. Nonetheless, a controversy remains, whether monotonic deformation, such as tensile or other modes, should be used to evaluate fatigue performance.

The following effects can be noted:

(a) The effect of strain amplitude. Results of cyclic deformation indicate that cyclic hardening and softening depend on the strain amplitude. It was observed

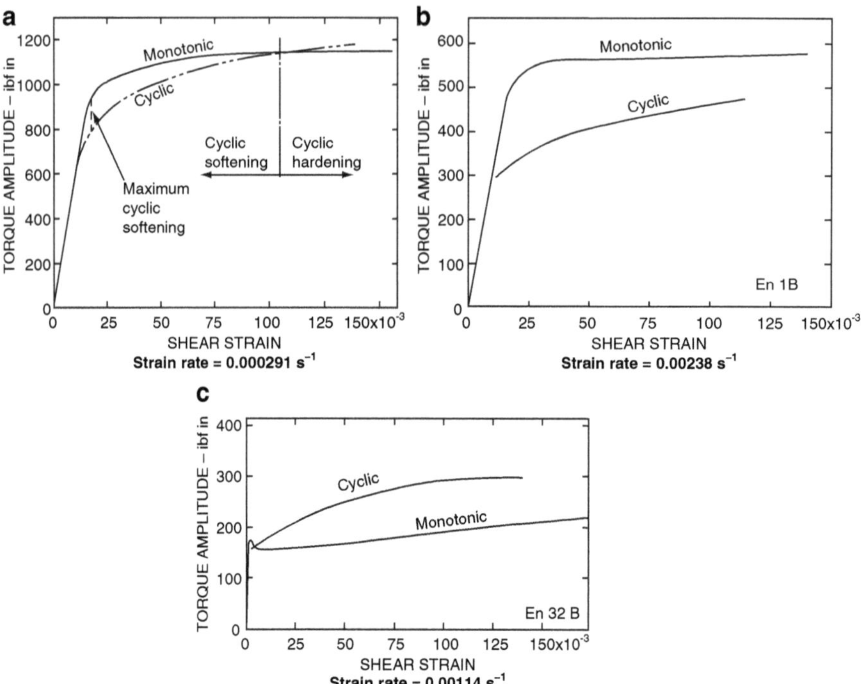

Fig. 6.19 Monotonic characteristics are also displayed: (**a**) cyclic softening and hardening of En 25 at the same strain rate; (**b**) cyclic softening of an En IB cold-drawn mild steel; and (**c**) cyclic and monotonic torque vs. strain for an En 32B alloy (After Miller 1970)

in AlMgSi aluminum alloys with a T6 heat treatment that, at low-strain amplitudes ($\Delta\varepsilon/2 < 0.8\%$), cyclic softening and, at high-strain amplitudes ($\Delta\varepsilon/2 > 0.9\%$), cyclic work hardening, occur. For intermediate-strain amplitudes ($0.8\% \leq \Delta\varepsilon/2 \leq 0.9\%$), the initial-hardening stage is followed by long-term cyclic softening. Figure 6.20 illustrates the effect of cyclic hardening-softening for various strain amplitudes in two alloys. The difference in these alloys is in their monotonic properties, alloy 6082 having higher tensile and yield stress (lower elongation) than that of 6060, as a result of different compositions. It is clear that the properties of the uniaxial tensile test (unidirectional deformation) influence the cyclic behavior. The initial increase in hardening rate, represented by the increased rate of stress amplitude during the strain cycles, increases with the strain range in both alloys. Furthermore, the softening rate, as indicated by the decreasing rate of stress amplitude, increases with decreasing strain range. The short cyclic-hardening stage observed during the initial stages of the fatigue test is followed by a long cyclic-softening stage.

(b) The effect of grain size. The cyclic response of TiC particles in reinforced and unreinforced, naturally-aged aluminum materials exhibits three regions, as shown in Fig. 6.21: a region of cyclic hardening, a saturation region and a region

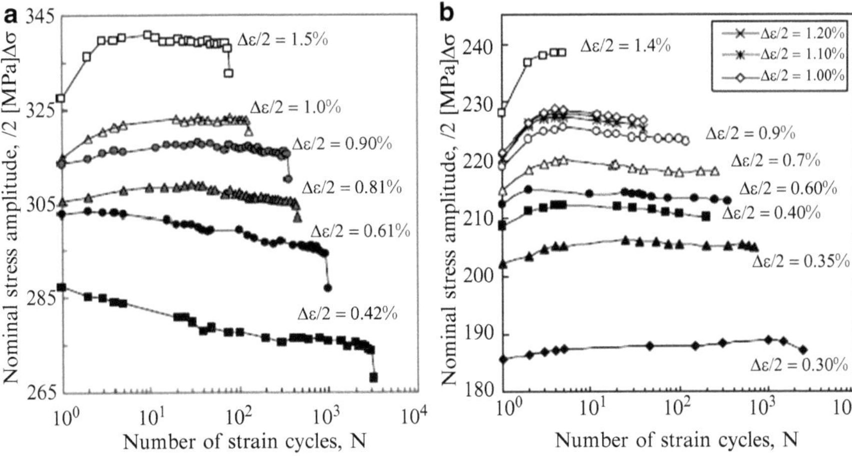

Fig. 6.20 Variation of the stress amplitude for strain cycling: (**a**) alloy 6082-T6; (**b**) 6060-T6 (Borrego et al. 2004. With kind permission of Elsevier and Professor Borrego for the authors)

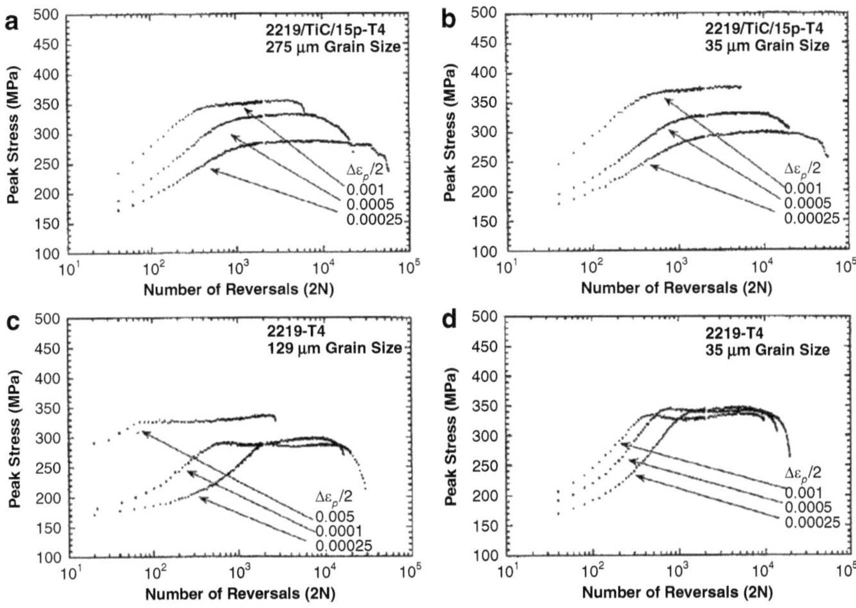

Fig. 6.21 Cyclic response with various grain sizes (Vyletel et al. 1995. With kind permission of Professor van Aken for the authors)

of cyclic softening. Grain-size effects are indicated in these graphs. Decreasing grain size was found to modestly increase the peak stress of the composite, while significantly increasing the peak stress for the unreinforced material. The saturation-stress level of the unreinforced, naturally-aged material was found

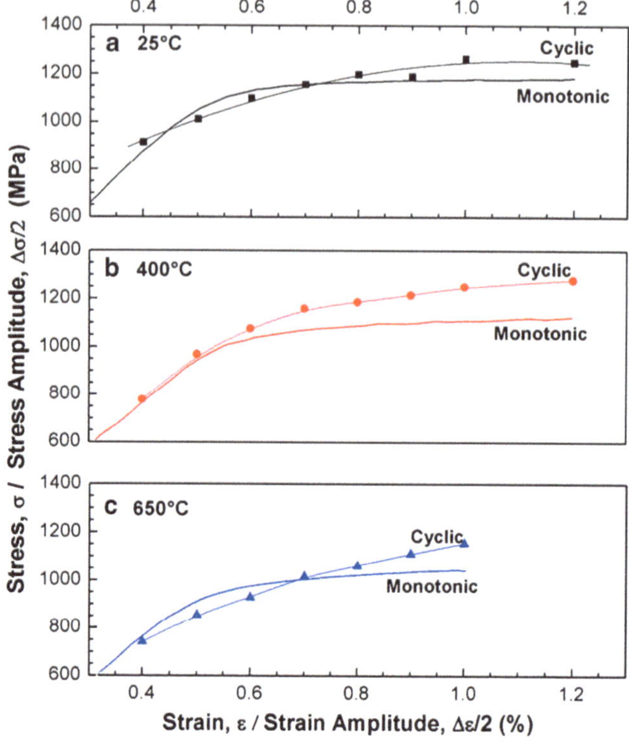

Fig. 6.22 Low cycle fatigue behavior of 720Li (Gopinath et al. 2009. With kind permission of Elsevier and Dr. Gopinath for the authors)

to be virtually independent of the plastic-strain amplitudes investigated (see Fig. 6.21d). In both small- and large-grained composites, the saturation stress increased as the plastic strain was increased. The naturally-aged materials were found to be cyclically unstable, showing significant hardening and, sometimes, cyclic softening. Unreinforced, naturally-aged materials show evidence of a plateau in the cyclic stress-strain curves, attributed to the localization of strain in the persistent slip bands.

(c) The effect of temperature. Precipitation-strengthened superalloys, such as Ni-based ones, have excellent monotonic strength up to high fractions of their melting point. However, the use of such alloys in turbine applications is limited by fatigue considerations. Attempts have been made to improve their fatigue properties by controlling the composition of the microalloying elements, in particular, the content of B and Zr. This Ni-based superalloy was designated as 720Li.

The cyclic-stress response (Fig. 6.22) is influenced by $\Delta\varepsilon/2$ and temperature. No appreciable hardening or softening occurs at $\Delta\varepsilon/2 \leq 6\%$. However, at 25°C and 650°C at $\Delta\varepsilon/2 \geq 0.6\%$, this alloy exhibited initial hardening, followed by softening (Fig. 6.23). Softening following hardening was not observed at 400°C

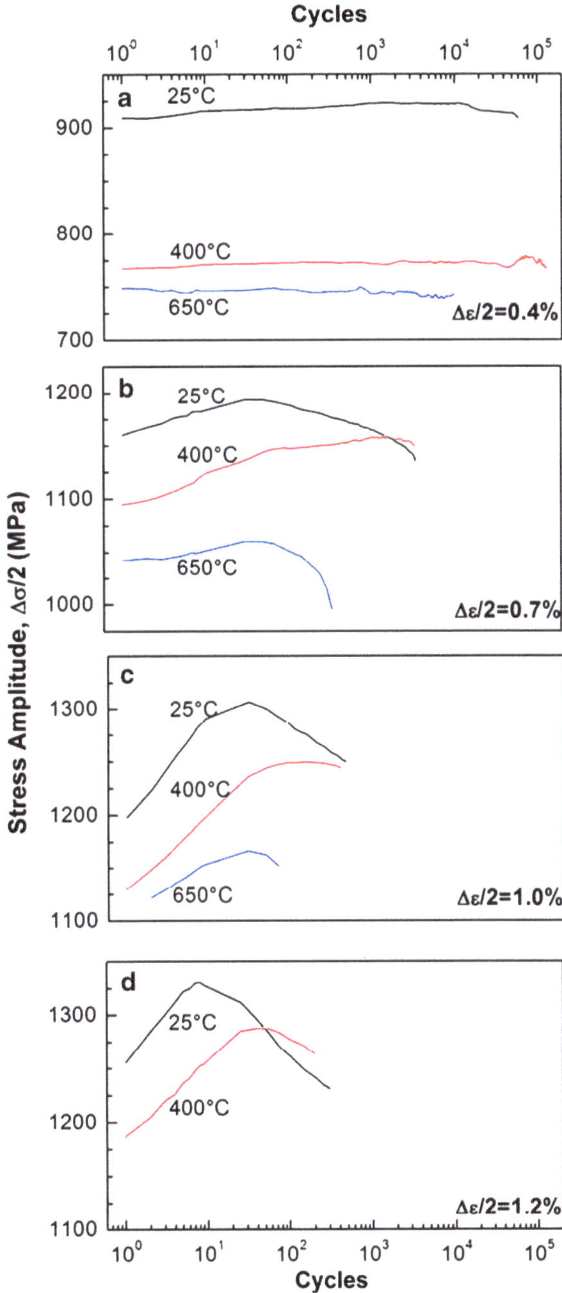

Fig. 6.23 Strain and strain amplitude $\Delta\varepsilon/2$ (%) comparing monotonic and cyclic stress-strain at the temperatures indicated (Gopinath et al. 2009. With kind permission of Elsevier and Dr. Gopinath for the authors)

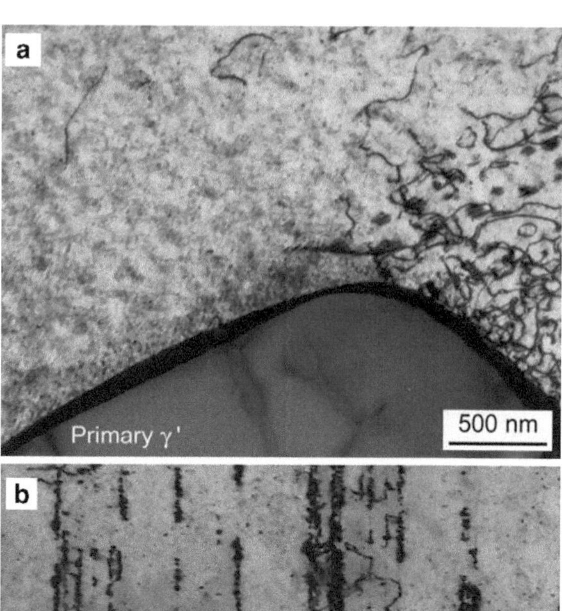

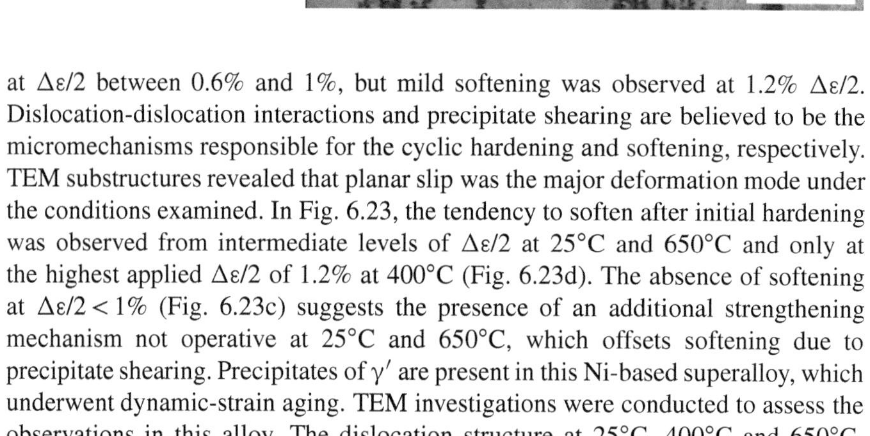

Fig. 6.24 TEM micrographs of deformation substructure in specimens fatigued at 25°C. (**a**) $\Delta\varepsilon/2 = 0.4\%$ ($\Delta\varepsilon_p/2 = 0.006\%$), $N_f = 60{,}047$ cycles; general bright-field image of low dislocation activity limited to grain boundaries and primary γ' boundaries. (**b**) $\Delta\varepsilon/2 = 1.0\%$ ($\Delta\varepsilon_p/2 = 0.4\%$), $N_f = 467$ cycles; bright-field image showing intense slip bands along (111) planes (Gopinath et al. 2009. With kind permission of Elsevier and Dr. Gopinath for the authors)

at $\Delta\varepsilon/2$ between 0.6% and 1%, but mild softening was observed at 1.2% $\Delta\varepsilon/2$. Dislocation-dislocation interactions and precipitate shearing are believed to be the micromechanisms responsible for the cyclic hardening and softening, respectively. TEM substructures revealed that planar slip was the major deformation mode under the conditions examined. In Fig. 6.23, the tendency to soften after initial hardening was observed from intermediate levels of $\Delta\varepsilon/2$ at 25°C and 650°C and only at the highest applied $\Delta\varepsilon/2$ of 1.2% at 400°C (Fig. 6.23d). The absence of softening at $\Delta\varepsilon/2 < 1\%$ (Fig. 6.23c) suggests the presence of an additional strengthening mechanism not operative at 25°C and 650°C, which offsets softening due to precipitate shearing. Precipitates of γ' are present in this Ni-based superalloy, which underwent dynamic-strain aging. TEM investigations were conducted to assess the observations in this alloy. The dislocation structure at 25°C, 400°C and 650°C, under the fully-reversed ($R = -1$) constant-strain amplitude ($\Delta\varepsilon/2$) cycling is seen in Figs. 6.24, 6.25, and 6.26.

(a) At all three temperatures studied, at low $\Delta\varepsilon/2$ (60.6%), the cyclic-stress response was stable from the initial cycles. This is linked to the low level

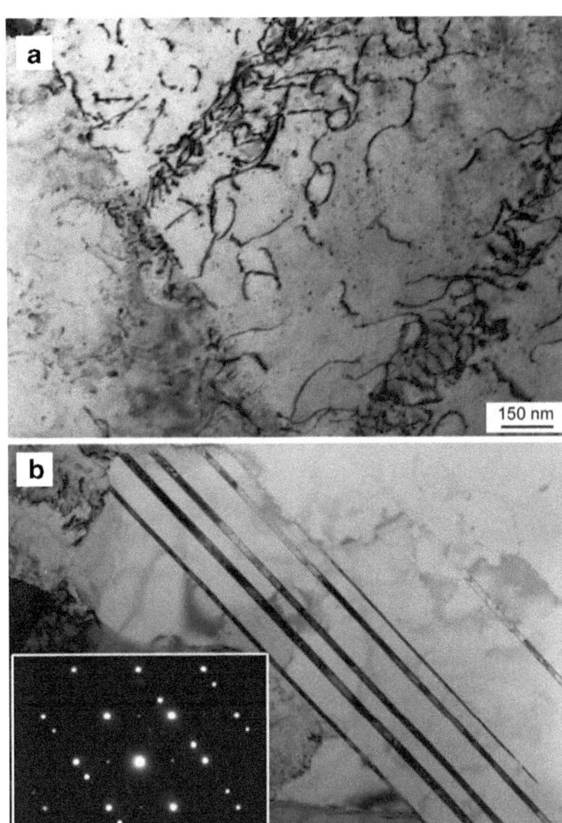

Fig. 6.25 General bright-field TEM micrographs of deformation structure in specimens fatigued at 650°C, $\Delta\varepsilon/2 = 0.4\%$ ($\Delta\varepsilon_p/2 = 0.01\%$), $N_f = 11{,}132$ cycles: (**a**) slip bands and low dislocation activity in the inter-band regions; (**b**) multiple twins within a grain. SAD pattern in the *inset* taken in (110) zone axis orientation clearly shows twin diffraction spots (Gopinath et al. 2009. With kind permission of Elsevier and Dr. Gopinath for the authors)

of dislocation activity that prevailed. This is seen in Fig. 6.24 for 25°C. In specimens subjected to $\Delta\varepsilon/2 = 0.4\%$ (mid-life $\Delta\varepsilon_p/2 = 0.006\%$) at 25°C, the grain interiors were relatively free of dislocations. Minimal dislocation activity was observed at grain boundaries which were decorated with primary γ' precipitates. Isolated slip bands with relatively low dislocation density were also occasionally observed. In specimens subjected to $\Delta\varepsilon/2 = 1.0\%$ (mid-life $\Delta\varepsilon_p/2 = 0.4\%$) at 25°C, most grains showed the presence of many intense slip bands along one family of {111} planes. High-angle tilting experiments suggested that dislocation activity between the slip bands was minimal (Fig. 6.24b).

Activation of secondary slip along other sets of {111} planes was also noticed only in a few grains. Deformation is, thus, heterogeneous and mainly confined to slip bands, but numerous bands, spread throughout the grain, were noted. Figure 6.25 represents the dislocation structure at 650°C. The dislocation structure at $\Delta\varepsilon/2 = 0.4\%$ (mid-life $\Delta\varepsilon_p/2 = 0.01\%$) at 650°C revealed the presence of a few slip bands and homogeneously distributed dislocations of low density in the interband regions (Fig. 6.25a). In addition, fine twins (50–100 nm

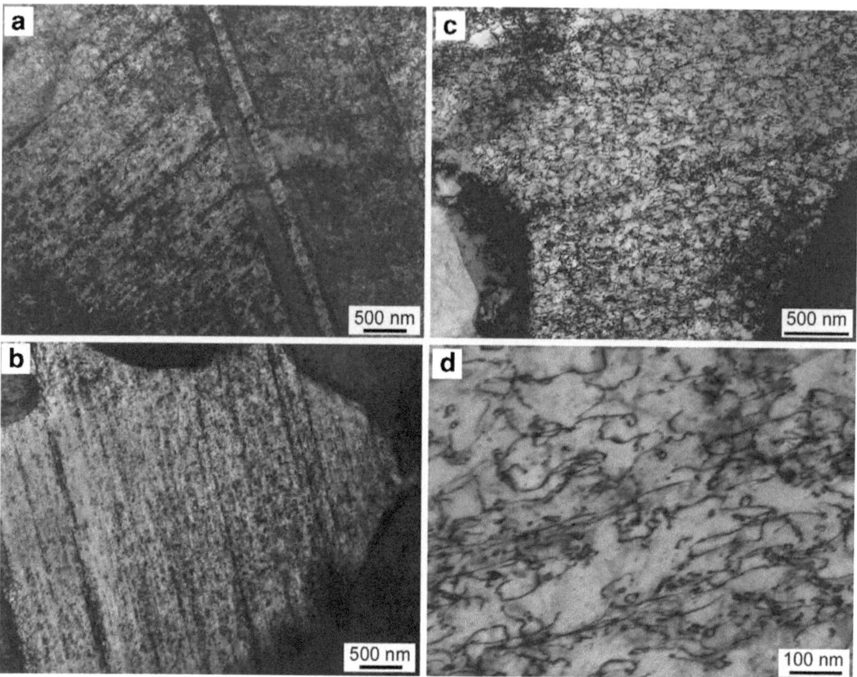

Fig. 6.26 General bright field TEM of deformation substructure in specimens fatigued at 650°C and $\Delta\varepsilon/2 = 1.0\%$ ($\Delta\varepsilon_p/2 = 0.38\%$), $N_f = 196$ cycles: (**a**) intense slip bands and homogeneous dislocation activity in the inter-band regions; (**b**) intense slip bands spread through out the grain; (**c**) isolated slip band and homogeneous dislocation distribution; (**d**) homogeneous dislocation activity, same grain as (**c**), at higher magnification (Gopinath et al. 2009. With kind permission of Elsevier and Dr. Gopinath for the authors)

thick) were observed in numerous grains, many of them with multiple, parallel twins (Fig. 6.25b). The high density of fine twins in fatigued specimens suggests that they are deformation induced. Figure 6.26 shows the structure of specimens subjected to $\Delta\varepsilon/2 = 1.0\%$ (mid-life $\Delta\varepsilon_p/2 = 0.38\%$) at 650°C. Intense slip bands (Fig. 6.26a–d), as well as significant dislocation activity between the bands, are seen.

(b) Furthermore, at 25°C and 650°C, at intermediate and higher $\Delta\varepsilon/2$ ($\geq 0.7\%$), this alloy initially exhibited cyclic hardening, followed by softening until fracture. Interactions of dislocations with other dislocations and precipitates are considered to account for the initial hardening and the shearing of precipitates is considered key to subsequent softening.

The monotonic stress of Fig. 6.22 was evaluated according to Eq. 6.6, which is rewritten below, while the cyclic stress was expressed by Eq. 6.6a, modified for half a life-time as Eq. 6.25:

$$\sigma = K\varepsilon_p{}^n \tag{6.6}$$

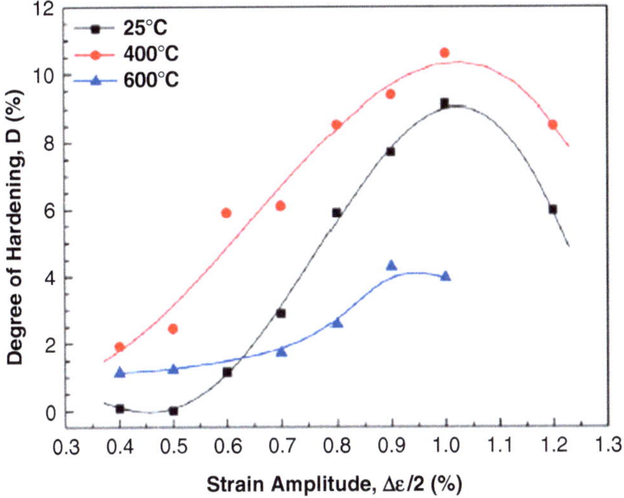

Fig. 6.27 Degree of hardening as a function of strain amplitude in fatigue tests (Gopinath et al. 2009. With kind permission of Elsevier and Dr. Gopinath for the authors)

$$\sigma = K' \varepsilon_p{}^{n'} \tag{6.6a}$$

$$\frac{\Delta\sigma}{2} = K'\left(\frac{\Delta\varepsilon_p}{2}\right)^{n'} \tag{6.25}$$

Thus, σ is the true stress and $\Delta\sigma/2$ and $\Delta\varepsilon_p/2$ are stress amplitude and plastic-strain amplitude at half-life. The increased work hardening in fatigue is associated with the values of n′, which are larger than the n in monotonic hardening. Even at 650°C, $n' = 0.118$ compared to $n = 0.037$.

The degree of hardening was given by:

$$D\,(\%) = \left[\frac{(\Delta\sigma/2)_{max} - (\Delta\sigma/2)_1}{(\Delta\sigma/2)_1}\right] \tag{6.26}$$

$(\Delta\sigma/2)_{max}$ is the maximum stress amplitude exhibited during cycling and $(\Delta\sigma/2)_1$ is the stress amplitude from the first cycle. In many publications, H is used for hardening instead of D. Thus, for example:

$$H = \frac{\sigma_a^{sat} - \sigma_a^1}{\sigma_a^1} \tag{6.26a}$$

Recall that, according to Eq. 6.18, $\sigma_a = \Delta\sigma/2$ and, thus, the hardening relations are the same. $\sigma_a{}^{sat}$ and $\sigma_a{}^1$ are the stress amplitude at the half-life time and in the first cycle, respectively. Figure 6.27 is a plot of the hardening, D, versus $\Delta\varepsilon/2$. D

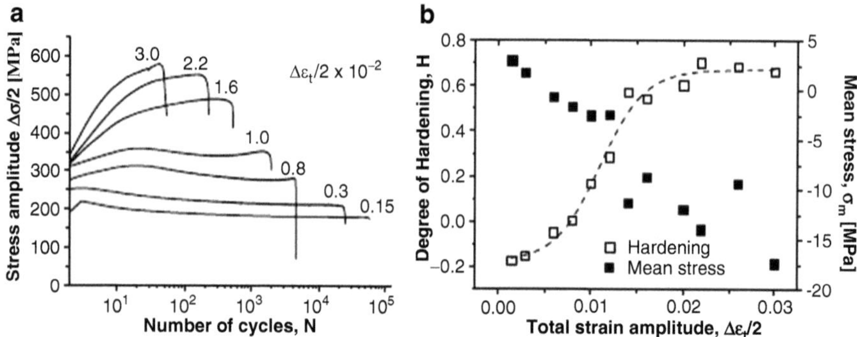

Fig. 6.28 (**a**) Cyclic deformation curves at different total strain amplitude; (**b**) degree of cyclic hardening H and the mean stress as a function of total strain amplitude (Glage et al. 2009. Free access article and with the permission of Dr Glage)

depends on both T and $\Delta\varepsilon/2$. When $\Delta\varepsilon/2 < 0.6\%$, D is low and it starts to increase when $\Delta\varepsilon/2 \geq 0.6\%$. It reaches a maximum at $\Delta\varepsilon/2 \cong 1\%$ for all temperatures. At 650°C, D is comparatively lower than at 25°C and 400°C and the drop in D after maxima could not be observed, since the tests were not continued to 1.2%. The trend in the variation in D was similar at the three Ts; the magnitude of D at 400°C was clearly the highest among the three Ts studied.

The effect of total-strain amplitude on the degree of hardening in a high-alloyed metastable, austenitic cast TRIP-steel can be seen more specifically in Fig. 6.28b, where the variation in the mean stress is also plotted. A small reduction of the mean stress, σ_m, taken at half the number of cycles-to-failure, may be observed by increasing total-strain amplitude. This behavior can probably be attributed to the larger specific volume of the α'-martensite, compared to the austenite. Moreover, in 6.28a, cyclic deformation curves (i.e., the cyclic stress at different total-strain amplitudes versus the number of cycles) is also shown. Thus, as in the previous case, also in this high-alloyed austenitic steel, cyclic-deformation behavior reveals combinations of cyclic hardening, saturation and cyclic softening, depending on the applied, cyclic total-strain amplitude.

The resulting cyclic stress-strain curve may be divided into three different regimes of cyclic plastic deformation caused by differing microstructures. Low-strain amplitudes result predominantly in planar dislocation structures. Conversely, at high-strain amplitudes cell/wall structures dominate. Furthermore, with increasing strain amplitude, the amount of lamellas inside the deformation bands increases, resulting in cyclic hardening.

The dislocation structures related to Fig. 6.28 were obtained by high-resolution ECC-images (i.e., electron channeling contrast imaging) of the developed dislocation microstructure and were taken from specimens cycled at low and medium total-strain amplitudes. The micrographs in Fig. 6.29 display a remarkable difference in the developed dislocation structures. Whereas Fig. 6.29a illustrates a

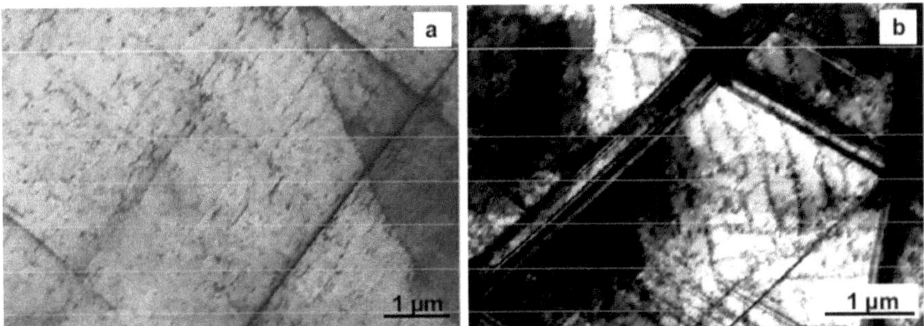

Fig. 6.29 ECC-images of the dislocation structure: (**a**) planar dislocation structure at low total strain amplitude ($\Delta\varepsilon/2 = 6 \times 10^{-3}$); (**b**) deformation bands with lamellar structure and typical wavy dislocation glide structure with walls/cells between the deformation bands at medium total strain amplitude ($\Delta\varepsilon/2 = 1.2 \times 10^{-2}$). Loading axis is *horizontal* (Glage et al. 2009. Free access article and with the permission of Dr Glage)

well-developed planar-dislocation structure at low-strain amplitudes, Fig. 6.29b shows a pronounced cell/wall structure of dislocation-dense regions and dislocation-free channels, which is typical for wavy-glide materials. In fact, both the density and the thickness of the deformation bands increase with increasing total-strain amplitude. Additionally, Fig. 6.29b demonstrates very well that the deformation bands consist of a large number of very fine lamellas. The thickness of these lamellas, measured perpendicularly to their length, varies in a range <50 nm. Typical values are about 20 nm.

6.6 Hysteresis

The directionality of work (strain) hardening associated with the Bauschinger effect was considered in Sect. 3.7 and illustrated graphically in Fig. 3.23.

This figure is reproduced below in Fig. 6.30a and also in Fig. 6.30b somewhat modified and relabeled. As may be seen in Fig. 6.30b, the process of loading a specimen at the plastic region and unloading it at point B, slightly above the elastic region along line AB, and then reversing the loading along BCD to point D, causes yield stress (point C) to be smaller than σ_y (the original specimen load) following the load reversal. In Fig. 6.30, σ_{max} is the stress at the start of reverse loading and σ_f is the failure or fracture stress in the unidirectional loading. The process of loading-unloading does not have to occur only under tension/compression and any forward and backward deformation (e.g., shear deformation) will cause a Bauschinger effect. If the loading process (shown in Figure 6.30b) is continued from $-\sigma_{max}$ to $+\sigma_{max}$, then a closed loop will result known as a hysteresis loop. Such a loop is shown in Fig. 6.31 for a magnesium alloy. The hysteresis loop defines a single fatigue cycle. The area within the loop is the energy per unit volume dissipated during a cycle.

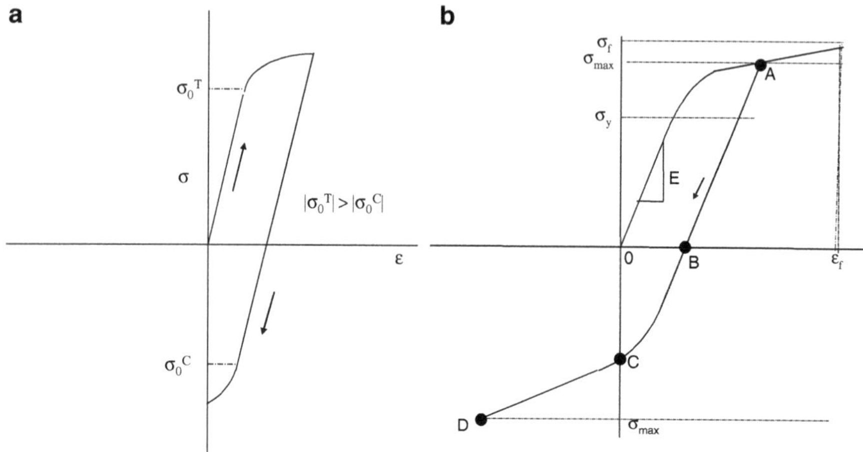

Fig. 6.30 The Bauschinger effect: (**a**) a specimen under tension is pulled beyond its yield point into the plastic region and then reloaded under compression. The yield point under compression, $\sigma_0{}^C$, is lower than the yield point under tension, $\sigma_0{}^T$; (**b**) is illustration (**a**) redrawn and relabeled

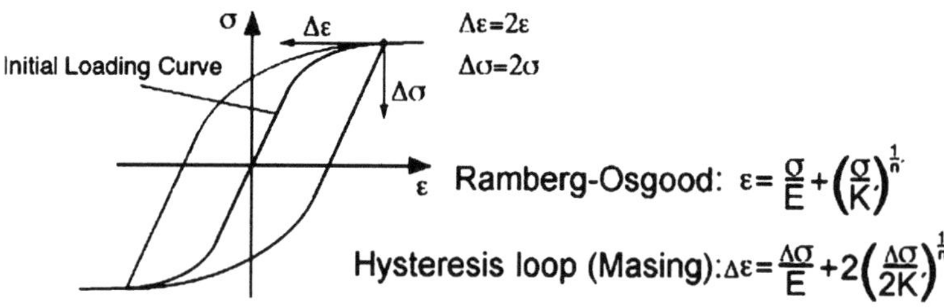

Fig. 6.31 A hysteresis loop with the Masing calculation of the curve (Zenner and Renner 2002. With kind permission of Elsevier)

When a material is subjected to cyclic loading, its stress-strain response may change with the number of applied cycles. If the maximum stress increases with the number of cycles the material is said to 'cyclically harden'. If maximum stress decreases over the number of cycles, the material is said to 'cyclically soften'. If the maximum-stress level does not change, the material is said to be 'cyclically stable'. Transient material behavior tends to stabilize after a relatively small number of cycles, typically less than 10% of the total life. The topics of cyclic hardening, softening and saturation were discussed in the previous section. Figure 6.32 shows actual stress-strain hysteresis curves of annealed and cold-worked copper and Cu-7.5% Al. These hysteresis loops show typical changes with increasing cycles derived from cyclic stress-strain experiments for both annealed and cold-worked Cu-7.5% Al. In the case of copper, the hysteresis loops become symmetrical under tension and compression after only a few cycles. However, the loops for Cu-7.5% Al are asymmetrical under tension and compression with the flow stress during

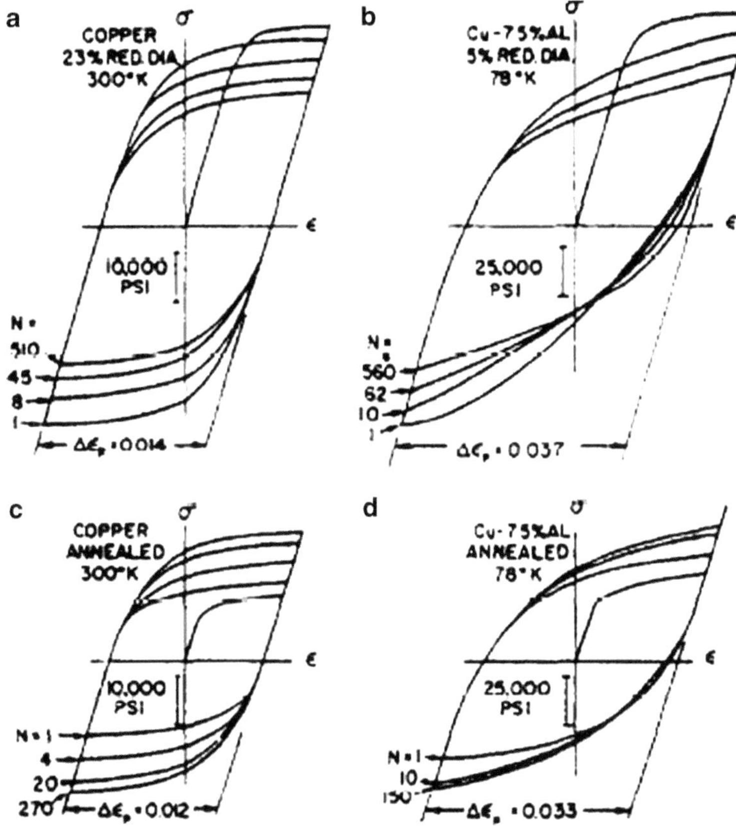

Fig. 6.32 Actual cyclic stress-strain response of copper and Cu-7.5% Al with different fabrication histories (Feltner and Laird 1967. With kind permission of Elsevier)

compression always being lower than the flow stress during tension, when these tests are started under tension; the reverse is true when the tests are started under compression. This effect was more prominent at 78 K in both the annealed and cold-worked conditions. The dislocation structure of the cyclically-stressed specimens (Fig. 6.32) is indicated in Figs. 6.33 and 6.34. The annealed copper, before cycling, contains only a few stray dislocations, which are either grown in or introduced by handling. The dislocation structure, after cycling, shows a cell structure in the annealed and cold-worked copper having an interior relatively free of dislocations. The size of the cells observed, depends, to a significant degree, on both the temperature and the amplitude of the testing. Cell size increases with temperature and the amplitude decreases (see Feltner and Laird). Unlike the structures seen in copper, cell structure was never observed in Cu-7.5% Al. Instead, the dislocations are dissociated and are generally arranged in bands or planar arrays. A typical

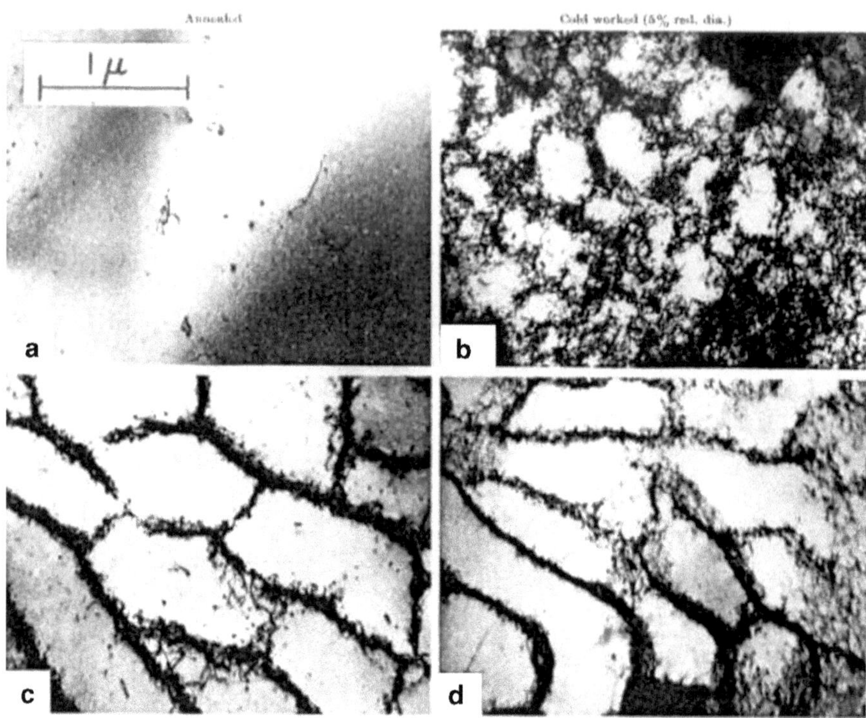

Fig. 6.33 Cu specimens before and after cycling. (**a**) Before cycling only a few stray dislocations present; (**b**) cold worked Cu before cycling shows a rough cell structure and the interior of the cells are relatively free of dislocations; (**c**) and (**d**) annealed and cold worked specimens after cycling show well developed cell structure with their interior almost completely free of dislocations (Feltner and Laird 1967. With kind permission of Elsevier)

example of the structure before and after cycling ($\Delta\varepsilon_p = 0.05$ at 78 K) in both the annealed and cold-worked conditions is seen in Fig. 6.34.

Before testing, annealed specimens have a few grown-in or accidentally-introduced dislocations, as seen in Fig. 6.34a, while the cold-worked specimens show bands containing a high density of dislocations with a lower density of faults and perfect dislocations between the bands. In a later section, the effect of faults on cyclic-deformation behavior and the results are discussed. After cycling, these cold-worked structures are similar in kind, consisting of bands of dislocations with a lower density of faults and dislocations between the bands. The dislocations are not always confined to slip planes. The number of bands in the cold-worked material after cycling is greater than the number before cycling, while the interband dislocation density is lower after cycling than before. The number of bands in the cycled, cold-worked material is less than the number in the cycled, annealed material.

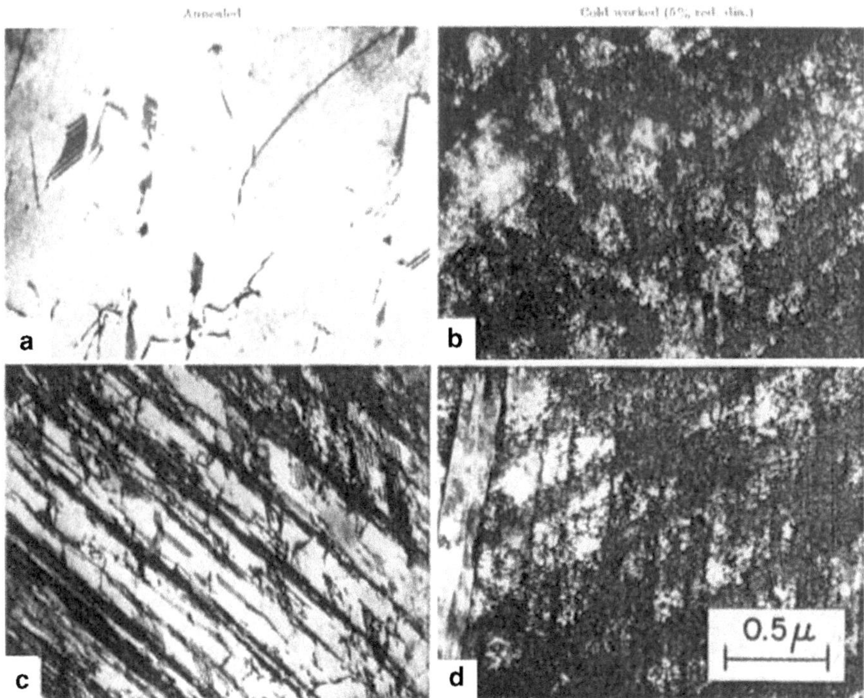

Fig. 6.34 Cu-7.5% Al specimens before and after cycling. (**a**) before cycling only a few stray dislocations are present; (**b**) cold worked Cu-7.5% Al specimens before cycling show bands containing a high density of dislocations with a lower density of faults and perfect dislocations between the bands; (**c**) and (**d**) annealed and cold worked specimens after cycling all the structures are similar in kind consisting bands of dislocations with a the latter not always confined to slip planes (Feltner and Laird 1967. With kind permission of Elsevier)

For detailed structural analysis of annealed and cold-worked Cu and Cu-7.5%Al, the reader is advised to consult the original work of Feltner and Laird. Regarding cell formation, one may note the accepted concept that the cyclic softening of cold-worked materials (or those having an initial high-dislocation density) occurs when the annihilation rate of dislocations is greater than their generation rate. This causes a net decrease in dislocation density. Softening also occurs when the rearrangement into cells takes place, resulting in an increased mean free path of dislocations. A threshold strain is required for softening to take place.

The hysteresis loops may be stable or not. Masing has pointed out that, in the tension-compression testing of metallic materials, either branch of the stress-strain hysteresis loop is geometrically similar to the monotonic stress-strain curve by a scale factor of two. More specifically, in his original work entitled: "Self Stretching and Hardening for Brass", Masing indicated that, if the force-displacement curve for a cyclically-stabilized system at initial loading is described by:

$$f(r, x) = 0 \tag{6.27}$$

where r is the restoring force corresponding to the displacement x of the system, then the unloading and reloading branches of the steady-state hysteretic response of the system are geometrically similar to the initial-loading curve, except for a twofold magnification, and are described by:

$$f\left(\frac{r - r_0}{2}, \frac{x - x_0}{2}\right) = 0 \tag{6.28}$$

In Eq. 6.28, (x_0, r_0) is the load-reversal point for that particular loading branch. The above is usually referred to as 'Masing's hypothesis' for steady-state, cyclic-hysteretic response. A schematic diagram illustrating Masing's hypothesis is shown in Fig. 6.31. It should be mentioned, however, that Masing's hypothesis is just an approximation of the cyclically-stabilized stress-strain response of some metals. Many experiments on fatigue deformation were done to verify Masing's hypothesis on the stability of hysteresis loops or deviation from it. Thus, for example, Zenner and Renner investigated the cyclic material behavior of magnesium die-castings and extrusions in an attempt to verify Masing behavior. In order to do so, stress-strain hysteresis was calculated from the cyclic stress-strain curves determined by the strain-controlled, incremental step test [henceforth: IST] and by the test series for the four magnesium alloys under investigation. Masing behavior occurs if the $\sigma - \varepsilon$ variation during load reversal is described by a curve which corresponds to the doubled initial cyclic loading curve (cyclic $\sigma - \varepsilon$ curve) with respect to stress and strain (Fig. 6.31). A new $\Delta\sigma - \Delta\varepsilon$ coordinate system is, thereby, defined for each point of reversal. In this study, it was observed that the magnesium extrusions show a totally different behavior under tension than under compression. Experimental hysteresis differs from calculated hysteresis for the cyclic stress-strain curve. Masing behavior cannot be observed due to the strong anisotropic behavior under tension and in the direction of compression. In the case of the magnesium die-castings, an approximate Masing behavior was observed.

We can, thus, summarize that a material is said to exhibit Masing behavior when the tips of the hysteresis loop, recorded at different total-strain amplitudes, coincide at peak compressive stress. The ascending branches follow a common envelope. Figure 6.35 shows hysteresis loops at various conditions for AISI 300 LN stainless steel. Only specimens of the 30% previously cold-worked material presented ideal Masing behavior. The material in the other conditions (indicated in Fig. 6.35) did not show an ideal Masing behavior. The differences in dislocation structures at different strain amplitudes are believed to be the reason for this observation.

The observation of cell formation and its observed relation to SFs was considered by Plumtree et al. for a range of ferrous and non-ferrous metals and related to Masing behavior. In Figs. 6.36 and 6.37, steady-state hysteresis loops from constant-strain amplitude tests translated to a common point of maximum, compressive stress are shown. Figure 6.36, of heat treated AISI steel (containing ~30% martensite), represents a group of dispersion-hardened materials, including among others the precipitation-hardened aluminum alloys 2024-T6 and Al-4% Cu with small interparticle spacing. These alloys contain closely-spaced ($\leq$0.5 μm) particles. In the case

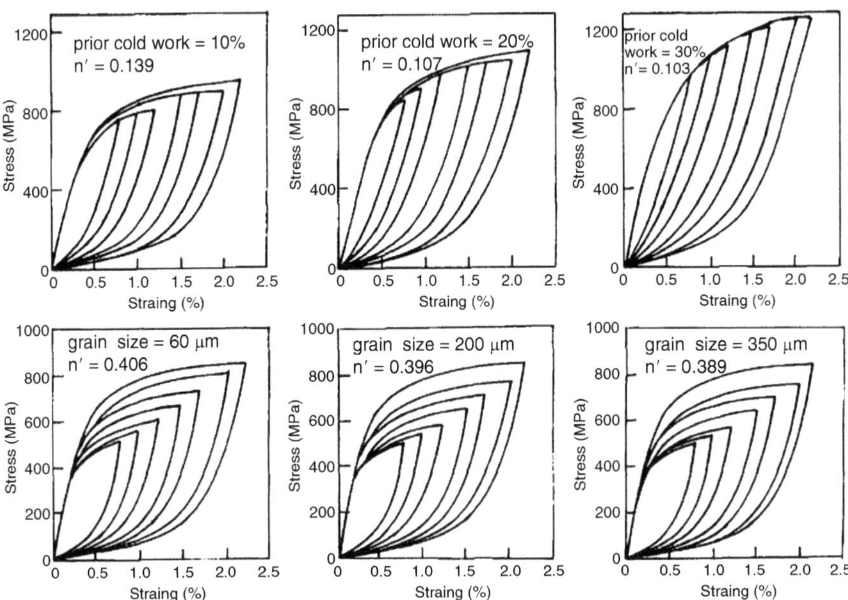

Fig. 6.35 Superimposed hysteresis loops with matched lower tips for materials processed by different conditions as indicated (Ganesh et al. 1996. With kind permission of Elsevier)

of the low-SF energy austenitic AISI 304 stainless steel, the substructure consists of planar arrays of dislocations at room temperature. All these metals exhibit Masing behavior. The hysteresis-loop branches fall on the doubled cyclic stress-strain curve for all strain levels. The cyclic stress-strain curve magnified by two accurately describes the loop shape.

Materials represented by the ferritic-pearlitic AISI 1018 HR (hot-rolled) steel in Fig. 6.37 exhibited Masing behavior only at low-strain levels. This group also contains, AA 1175 aluminum, CDA 102 copper and the Al–4% copper alloy with large interparticle spacing. Above a critical strain range, the saturated hysteresis loops for each of these metals deviate from a doubled cyclic stress-strain curve. At certain strain ranges, non-Masing behavior is found. Below these strain ranges, the substructure contains patches of dislocation dipoles, whose density governs the friction stress. Figure 6.37 shows that at strain ranges greater than the threshold, the loop branches of the saturated-hysteresis loop rise above the doubled cyclic stress-strain curve only to join it at the tensile loop tips. At these cyclic-strain ranges, the substructure consists of dislocation cells. By translating the loops in Fig. 6.37 along the elastic slope from the origin (as shown in Fig. 6.38), it is possible to accurately match all the upper branches. The lower tips coincide with each other only below the threshold strain range. However, all these tips lay along the linear elastic line. The linear part of the hysteresis curve is not constant and increases with the size of the loop. This is related to dislocation density and cell size. With the onset of saturation,

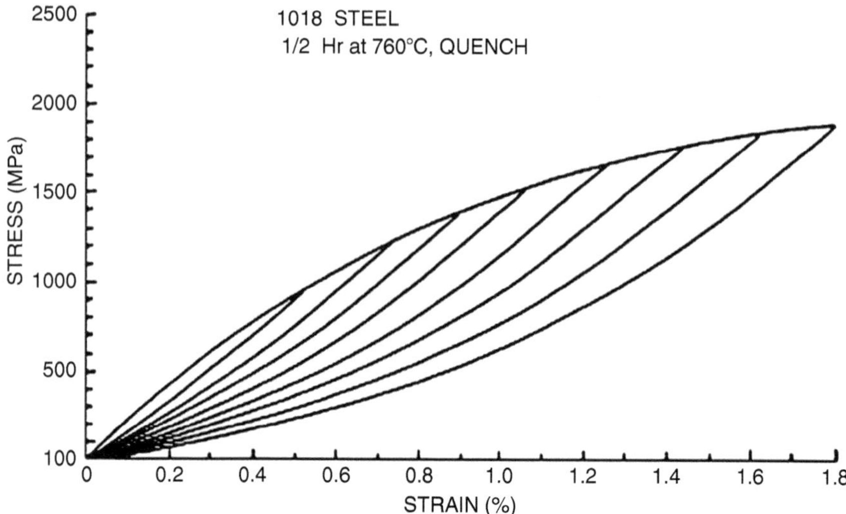

Fig. 6.36 Stable stress-strain hysteresis loops giving a common cyclic stress strain curve. AISI 1018 steel water quenched (Plumtree and Abdel-Raouf 2001. With kind permission of Elsevier and Professor Plumtree)

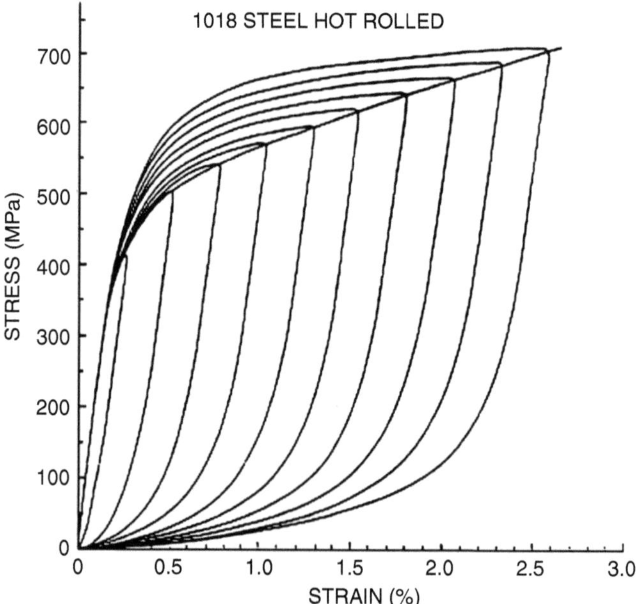

Fig. 6.37 Stable stress-strain hysteresis loops of AISI 1018 steel and the cyclic stress-strain curve is the locus of *upper* branch tips (Plumtree and Abdel-Raouf 2001. With kind permission of Elsevier and Professor Plumtree)

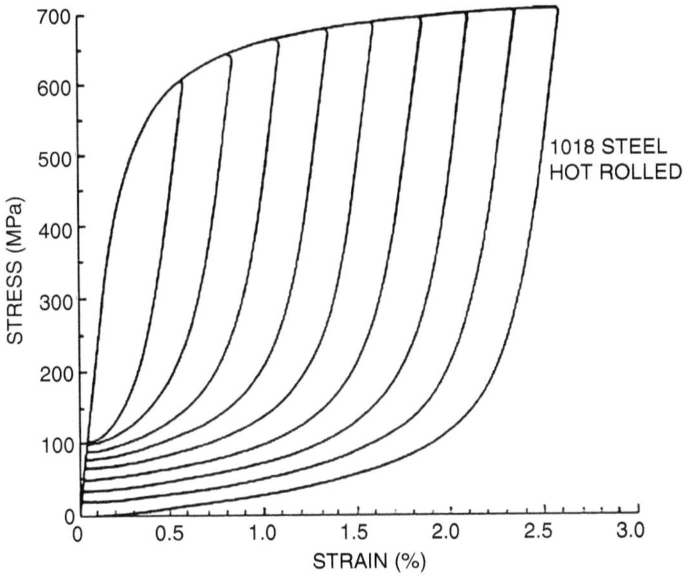

Fig. 6.38 The loops of Fig. 6.37 moved on to a common cyclic stress-strain curve (Plumtree and Abdel-Raouf 2001. With kind permission of Elsevier and Professor Plumtree)

an elementary-dislocation cell structure with thick, uncondensed walls is observed, as seen in Fig. 6.39a. On further cycling, these cell walls become narrow and well-defined, as seen in Fig. 6.39b, indicating that a stable low-energy substructure has developed.

After full life, this structure showed no significant change in low temperature annealing. Although the effect of cycling caused the dislocations to be rearranged into a lower-energy cellular configuration (i.e., a dynamic recovery process), the size remained relatively constant and was related to the saturation stress corresponding to the plastic-strain amplitude. A relation describing the dependence of normalized-saturation stress, σ_s, on dislocation cell size, d, was developed and presented in Eq. 6.29 as:

$$\frac{\sigma_s - \sigma_0}{E} = A\frac{b}{d} \tag{6.29}$$

In Eq. 6.29, A is a material constant and σ_0 is the friction stress. The line plotted by Eq. 6.29 should be linear (the plot is against b/d), passing through the origin with the slope, A. This relation is plotted in Fig. 6.40 for a group of alloys.

The formation of a cellular structure is an efficient means for storing high-dislocation density and cell size indicates the average distance of dislocation glide. Upon loading, high stress must be applied so that the dislocation sources in the cell walls will operate. Once they are operational, however, there will be little resistance to dislocation motion in the comparatively dislocation-free body of the cell, until

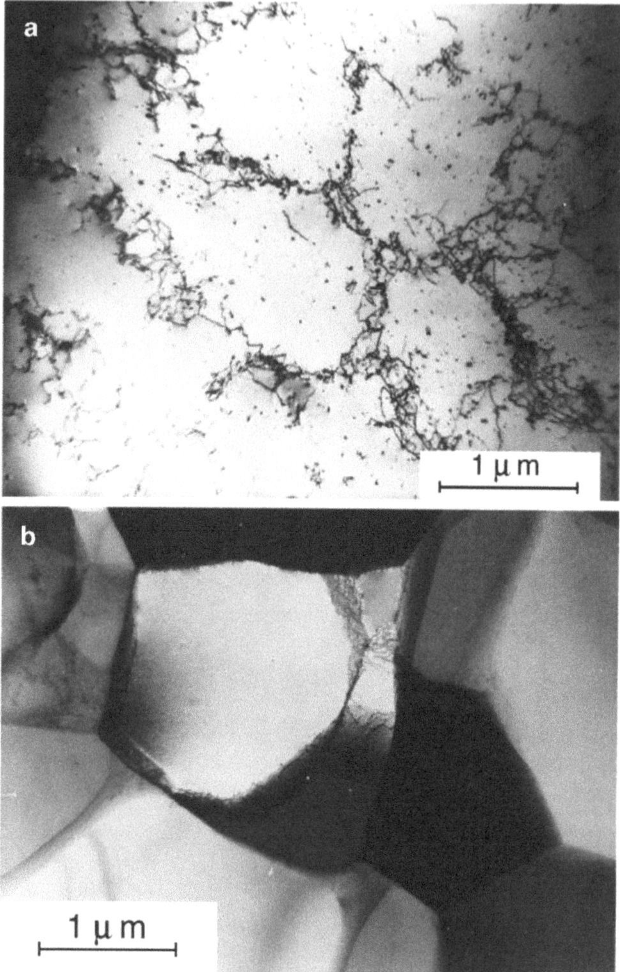

Fig. 6.39 Dislocation cells in aluminum cycled at $\Delta\varepsilon = 0.010$: (**a**) after 30 cycles; onset of saturation, (**b**) at failure after 11,300 cycles (Plumtree and Abdel-Raouf 2001. With kind permission of Elsevier and Professor Plumtree)

adjacent walls are encountered, where interactions and annihilations occur, lowering the internal energy and increasing the stability of the substructure. This process will repeat itself with each strain reversal. The ease of dislocation movement and cross-slip play a significant role in the cyclic stress-strain response. In high-SF energy metals, in which cyclic plastic straining gives rise to dislocation cells, non-Masing behavior is exhibited. Above a threshold-strain level, which coincides with that of dislocation-cell formation, the flow stress of a stable hysteresis loop exceeds that of the cyclic stress-strain curve. The characteristic increase of cyclic flow stress with the increasing strain range and subsequent decrease in cell size provide a

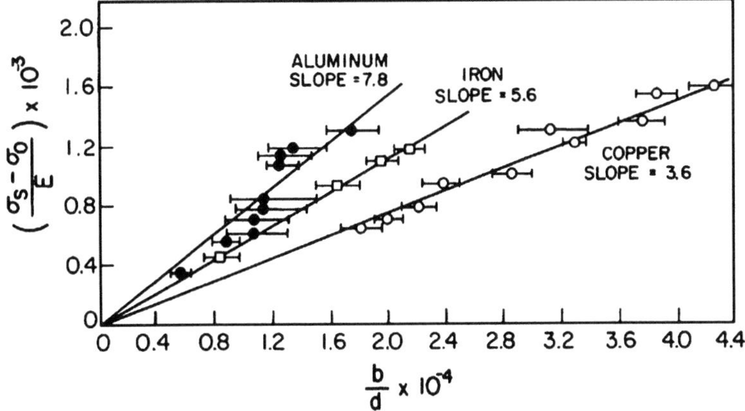

Fig. 6.40 Normalized saturation stress σ_s cell size for aluminum, iron and copper (Plumtree and Abdel-Raouf 2001. With kind permission of Elsevier and Professor Plumtree)

basis for describing non-Masing behavior. At strain ranges below the threshold, the hysteresis-loop branches lay along the doubled cyclic stress-strain curve, resulting in Masing behavior. It is anticipated that Masing behavior will be observed in low-SF energy metals, where dislocation motion is planar and cell formation is difficult.

In precipitation-hardened metals, the spacing between the particles determines whether Masing or non-Masing behavior is anticipated. In addition to the effect of the SF energy mentioned above, non-Masing behavior is likely to be observed when particle spacing is wide enough to allow deformation control by the matrix, whereas, when deformation is controlled by closely-spaced, non-sharable particles, Masing behavior is likely to be observed in alloys containing particles as a second phase.

To summarize this section, the effects of dislocation configurations in the structure of material indicate the influence of factors, such as: cell structure, dislocation density within cell boundaries and SF energy. These factors determine whether a Masing or non-Masing behavior of hysteresis loops prevails.

6.7 The Mean Stress

The mean stress was defined in Eq. 6.2 as:

$$\sigma_{mean} = \sigma_m = \frac{\sigma_{\max} + \sigma_{\min}}{2} \tag{6.2}$$

A large number of experiments on fatigue were performed with completely reversed cycles (indicated in Fig. 6.1a), where the mean stress is zero ($\sigma_{\max} = \sigma_{\min}$). Often in components exposed to fatigue, a pattern (such as is shown in Fig. 6.1b) is observed, resulting from the superposition of a static preload during the reversed

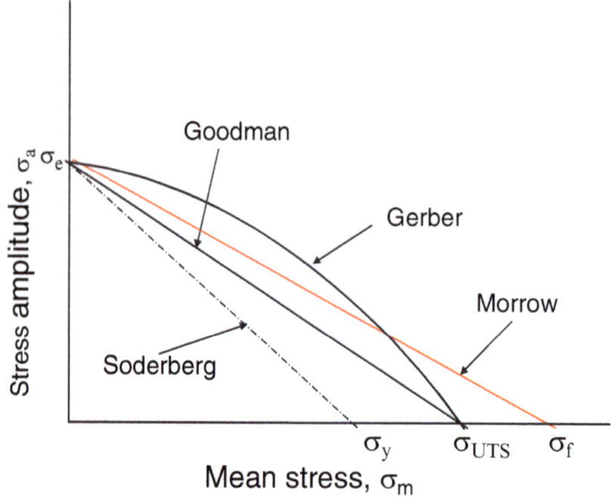

Fig. 6.41 A schematic diagram relating stress amplitude to the mean stress by several approaches

cycle (see Fig. 6.1a; also note that the names of the cycle-patterns shown in Fig. 6.1 may have different nomenclature in the literature). It is often stated differently, namely that the mean stress, σ_m ($=\sigma_{mean}$), represents a steady-state stress, while the alternating stress is a variable stress. This stress cycle is asymmetrical, since the sum of σ_{max} and $\sigma_{min} \neq 0$. Clearly, machine parts in service exposed to cyclic stresses may experience particular conditions, in which $\sigma_{min} \geq 0$ or $\sigma_{max} \leq 0$.

To obtain an S-N curve, a very large number of tests is essential. In determining fatigue-stress levels using standard testing equipment, the test specimens are subject to alternating and reversed stress levels, as indicated in Fig. 6.1. Cyclic stress varies from σ_a, tensile, to σ_a, compressive. The mean stress $= 0$.

To overcome the difficulty posed by the large number of experiments, methods have been suggested, one of them by Goodman and known as the 'Goodman diagram', in which the alternate stress is plotted against the mean stress. Another, similar approach was suggested by Gerber. Their research on the effect of mean stress on fatigue was performed as early as the nineteenth century, with a focus on the fatigue limit. The emphasis in their research was on tensile-mean stress with respect to fatigue life. Now, various combinations of stresses (such as mean stress or stress amplitude) are plotted against N_f or σ_a is plotted against mean stress for various values of N_f. An accepted schematic drawing of the Goodman relation is shown in Fig. 6.41 below. In this diagram, other approaches are also indicated, as suggested by Gerber and Soderberg. The relations expressed by the lines in Fig. 6.41 are almost the same for Goodman, Gerber and Soderberg. In addition, Morrow's relation is also indicated where, instead the UTS, σ_{UTS}, or the yield stress, σ_y, the fracture stress, σ_f, is included:

$$\sigma_a = \sigma_e \left[1 - \left(\frac{\sigma_m}{\sigma_{UTS}}\right)\right] \tag{6.30}$$

$$\sigma_a = \sigma_e \left[1 - \left(\frac{\sigma_m}{\sigma_{UTS}}\right)^2\right] \tag{6.31}$$

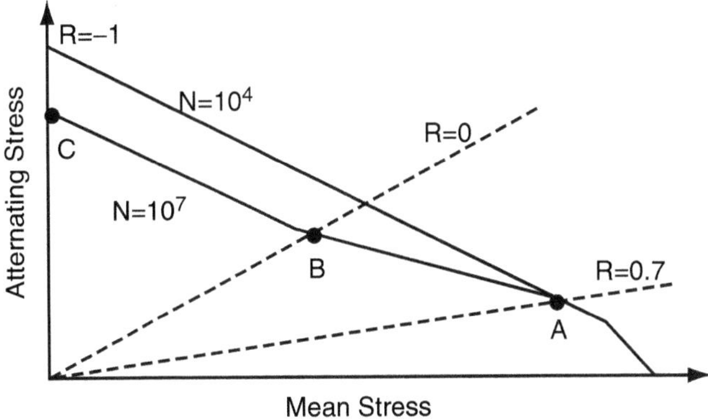

Fig. 6.42 A Goodman diagram (Nicholas and Zuiker 1996). With kind permission of Springer)

$$\sigma_a = \sigma_e \left[1 - \left(\frac{\sigma_m}{\sigma_y} \right) \right] \tag{6.32}$$

$$\sigma_a = \sigma_e \left[1 - \left(\frac{\sigma_m}{\sigma_f} \right) \right] \tag{6.33}$$

Here, σ_e is the endurance limit for fully-reversed cyclic loading.

As mentioned above, a Goodman diagram is a plot of alternating stress versus mean stress. It represents the locus of stress states corresponding to a runout test or a given number of cycles to failure, e.g., 10^7 or even above it, in the range of 10^8–10^9. Equation 6.30 expresses the alternating stress as a function of the mean stress. Although, in Goodman's original work, the plot was a straight line connecting $\sigma_{UTS}/3$ on the y axis with σ_{UTS} on the x-axis, here it is expressed differently, i.e. σ_a is a function of σ_{UTS}. This diagram is sometimes referred to as the 'modified Goodman diagram', but more often it is called simply the 'Goodman diagram'. The input data for constructing a Goodman diagram may be an S-N curve.

Ti fans, compressor blades and disk lugs sometimes experience high-cycle fatigue failure for various reasons, among them fretting. A Goodman diagram, with appropriate data, may even serve as a reliable design tool in cases suffering high-cycle fatigue. Such a Goodman diagram is presented in Fig. 6.42 (Nicholas and Zuiker), based on an S-N curve at different stress ratios. It is practical to choose a region in which the S-N curve is flat or nearly flat, when no knee exists the number of cycles selected must be such that they exceed the number of cycles encountered in real service. In this figure, the locus of points with 10^7 cycles-to-failure is illustrated. Recall that $\sigma_a = \sigma_{max} - \sigma_{min}$, usually defined as half the vibratory range or ½($\sigma_{max} - \sigma_{min}$). Points A, B and C of Fig. 6.43 are shown on the Goodman diagram corresponding to stress ratios of 0.7, 0 and −1, respectively. Lines corresponding to these ratios are also shown with the vertical axis being the R = −1 line for fully reversed cycles (Fig. 6.42).

Fig. 6.43 Typical S-N curves which were used to construct the Goodman diagram of Fig. 6.42 (Nicholas and Zuiker 1996. With kind permission of Springer)

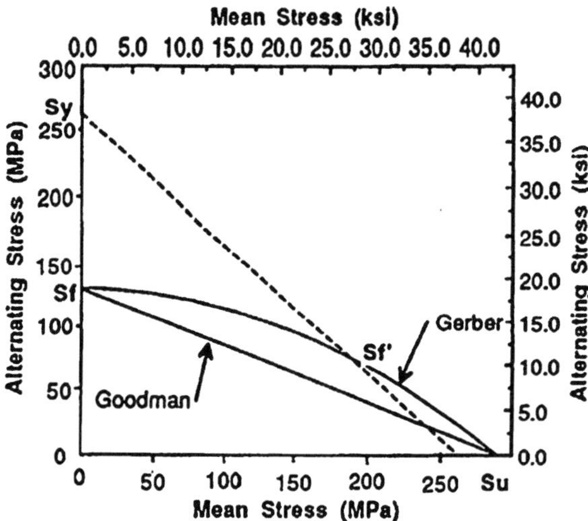

Fig. 6.44 Comparison of Gerber and Goodman mean stress corrections at 200,000 cycles (Reproduced from Yahr 1993)

It is important to bear in mind that the Goodman diagram represents ideal test conditions. Ideal test conditions preclude corrosion, shock loading, nicks or scrapes in the wire, etc. Therefore, to get reliable information, all these aspects must be taken into account, including the material properties. Because of the statistical nature of fatigue, a Goodman diagram should represent average data, where a 2σ or 3σ confidence limit is applied.

Experimental studies of aluminum show that the mean-stress effect is better described by the Gerber equation (Yahr). The modified Goodman diagram is generally used, because it is simpler to apply than the Gerber equation and is more conservative. However, for materials that have a high-yield strength to tensile-strength ratio, such as 6061-T6 aluminum or high-strength steel bolts, the modified Goodman approach tends to become overly conservative.

Figure 6.44, based both on the Gerber equation and on the linear Goodman equation, compares these two approaches. The alternating stress, σ_a is shown on the ordinate and the mean stress is represented on the abscissa. The straight, solid line drawn from the fatigue strength for a fully reversed loading for 200,000 cycles (129 MPa) to the tensile strength (290 MPa) is a graphic representation of the Goodman prediction of the effect of mean stress. The Gerber prediction is indicated by the curved line drawn from the same strength values. The dashed line drawn between the cyclic-yield stress (262 MPa) on both axes represents the boundary of sustained cyclic conditions.

Yielding will occur when the combination of alternating stress and mean stress is above the dashed line. Therefore, the intersection of the dashed line with the Goodman or Gerber lines gives the maximum effect of mean stress, since yielding would reduce any higher mean stress.

For the 200,000-cycle case (Fig. 6.44), the Goodman approach predicts a maximum reduction in fatigue strength to 22 MPa, while the Gerber approach gives the maximum reduction as 75 MPa. These values can be read off the y axis of this diagram. Here, the choice of the value 262 for the yield stress was based on studies showing that 6061-T6 cyclically hardens a small amount at room temperature to 262 MPa. This determines the maximum effect of mean stress, which is lower than would have been obtained if the monotonic-yield strength of 241 MPa had been used, rather than that of the cyclic. This also corresponds to lower values of allowable alternating stress than would have been obtained if the monotonic yield strength of 241 MPa had been used. For example, the maximum mean-stress reduction in the 200,000-cycle fatigue strength would have been to 93 MPa, instead of to 75 MPa.

6.8 Underloading (UL), Overloading (OL), Coaxing and Cumulative Damage

Very often a machine component working under cyclic conditions may experience overstressing or understressing under common loading conditions. Wings of airplanes are typical cases where common loading may be influenced by the absence of load constancy, due to overloading [henceforth: OL] or underloading [henceforth: UL]. Should designers be concerned when such phenomena occur? The aim of this section is to consider such possibilities, whether accidental or uncontrollable. Understanding the consequences on fatigue life and crack propagation is of considerable interest. Compressive or tensile stress may be applied, but the importance of compressive stresses on fatigue cracking is emphasized by many researchers.

6.8.1 Underloading (UL)

Cyclic loading may be performed under various conditions. Figure 6.1 illustrated some cycles and Eqs. 6.2, 6.3, and 6.4 defined certain important parameters that might be relevant to loading conditions. Of the many reports on loading conditions in this section, some principal, experimental results will be discussed.

Designers wish to increase the fatigue resistance of machine components exposed to cyclic (or dynamic) stresses. The end result of such exposure may be failure, believed to begin with a crack. Hence, the effect of various loading sequences and modes are usually related to crack initiation and growth, which are inherent steps in the progress to failure. OLs are known to retard crack growth, while ULs accelerate crack growth relative to the background rate. It has been stated that fatigue ULs are known to have a detrimental effect on both fatigue-crack initiation and growth from notches. These effects are thought to be due to the production of a tensile, plastic zone at the crack tip and the flattening of the crack tip surface roughness leading to a reduction in crack-closure effects (Geary). The loading of test specimens may be done by single or periodic ULs. In both cases, when applied to aluminum alloys, crack-growth acceleration is observed. In aluminum testing, similar behavior is also observed (Stephens et al.). Apparently, the reduced contribution of the crack-closure mechanism is involved in the accelerated crack growth. A single UL gives only a temporary increase in the crack-growth rate; however, it has been found that periodically repeated ULs result in noticeable crack-growth acceleration. Repeated ULs keep the crack-growth mechanism active. Thus, if in fatigue there may be frequent ULs, such as occur in airplane wings, then crack-growth acceleration will continue promoting premature failure. Furthermore, when UL occurs during an otherwise constant-amplitude loading, the crack growth following the UL will be greater than that occurring during constant-amplitude loading. The period of crack-growth acceleration is short and the normal, constant amplitude growth rate is quickly restored, unless, as indicated, the machine element (or the test specimen) is exposed to repeated UL. Note that if UL immediately follows OL, the degree of crack retardation due to OL is reduced, but not eliminated. An UL applied prior to an OL, on the other hand, has little effect on the degree of crack retardation.

Tests done on aluminum 7010 and 8090 alloys (Zitounis and Irving) have indicated that fatigue-crack growth rates under loading sequences containing tensile UL show crack-acceleration effects, compared with growth rates calculated from the linear summation of constant-amplitude growth rates. In aluminum 7010, the acceleration of cracking was in the range of 30%, whereas in Al 8090 alloys it was up to a factor of 12 (i.e. 1,200%). The number of reversed cycles determines crack acceleration. The experiments on these alloys were performed at large cycles of $R = 0$ and small cycles of $R = 0.9$ to create load spectra similar to those observed in helicopter lift frames and rotor heads. These parts contain large numbers of small amplitude cycles, $R = 0.8$ and 0.9, interspersed with much larger amplitude cycles of R approaching zero.

Reports over the last decades state that the compression part of the loading, under tensile-compression tests (negative R ratios), may have a significant effect on the results of crack initiation and propagation. Examples of components with this type of loading include those in helicopters and gas turbines. Topper studied crack-opening stresses under spectra consisting of original constant-amplitude small cycles of a range of R ratios, followed by tension-compression ULs. He found that the crack-opening stress level was reduced immediately after the application of the UL and then gradually increased with subsequent constant-amplitude cycling, reaching a steady-state level after a large number of cycles. This observed crack-opening behavior had a significant acceleration effect when the R ratio was low. At high R ratios (0.8), changes in -opening stress had no effect on the crack-growth rates, because the crack-opening stress was always less than the minimum stress of the loading cycle.

Several mechanisms have been suggested for crack-growth rate, among them: fatigue-crack closure, strain hardening ahead of the crack tip (caused by the ULs) and the influence of mean stress, etc. Topper and Yu suggested that the compressive ULs caused the squeezed crack tip to have decreased crack-closure stress, thus increasing the effective stress-intensity factor range of the baseline constant-amplitude cycles at a low R ratio. The stress-intensity factor is discussed below (in Sect. 6.8.2).

To conclude this section, note that the loading conditions of a particular material greatly influence the outcomes when applying ULs to specimens of that material. Experimental-loading conditions are meant to simulate the loads to which components may be exposed during service. Various loading-UL sequences have been studied, but the overall consequence of UL is that crack growth is accelerated to different degrees, depending on several factors commonly encountered during the test pattern.

6.8.2 *Overloading (OL)*

It is a well known fact that load-cycle interactions can have a very significant effect on fatigue-crack growth. This has practical import for the potential prediction of fatigue life. There is a vast literature indicating that tensile OLs, when applied over constant-amplitude loading, can cause the retardation or arrest of crack growth and that compressive OLs may sometimes affect the rate of subsequent crack propagation. Qualitatively, the fatigue-crack propagation rate is retarded after the application of one or several tensile-peak OLs. These delaying effects, caused by OLs, are of particular interest in design, since they occur frequently and represent the most common operating conditions. A large amount of research on the delaying effects of OLs has been conducted and several explanations have been proposed. The mechanisms most often discussed in regard to this delaying phenomenon are: (a) residual stresses in the vicinity of a crack tip; (b) crack-tip closure; and (c) crack-tip blunting; other relevant approaches are: (d) irregular crack front and (e) strain hardening at the crack tip (see, for example, Shin and Hsu).

Many machine components and structures for engineering applications, such as airplane wings, engine crankshafts, etc. often experience both tensile and compressive OLs in service. Therefore, the effects of both kinds of OL are of much interest. Topper and Yu investigated the effects of tensile and compressive OLs on the threshold-intensity level and crack-closure behavior of several alloys, i.e., Al alloy 2024-T351 and SAE 1045, SAE 1010 and G40 21 steels. Accordingly, several

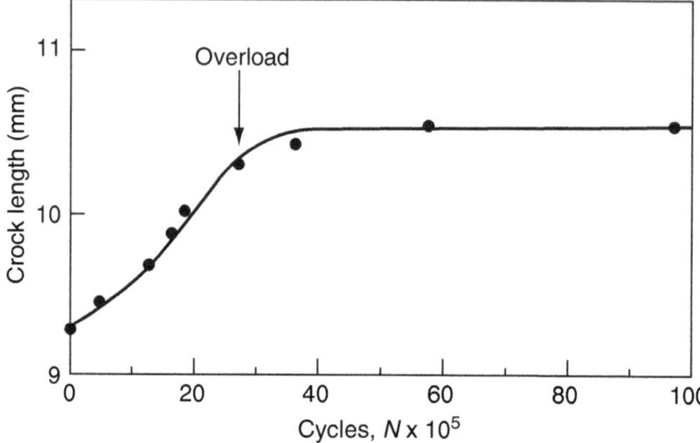

Fig. 6.45 Crack length vs. number of cycles for tensile overload test (Topper and Yu 1985. With kind permission of Elsevier and Professor Topper)

tensile OLs significantly decreased the crack-propagation rate while increasing the threshold-stress intensity. An initially decreased and then increased opening stress was mostly responsible for the delayed retardation and crack arrest. However, intermittent, compressive OLs significantly accelerated crack propagation and decreased the threshold-stress intensity, which was a function of the frequency OL. It was often stated that crack-initiation life is always decreased by tensile and compressive OLs; however, it has also been reported that a single tensile OL can cause a delay in fatigue-crack propagation and even crack arrest. Crack-growth delay may be increased by an increased level of positive OL. Experimental data suggest that retardation due to OL is a consequence of residual compressive stress, which develops in the crack-tip region and is responsible for crack-closure effects. Furthermore, intermittent compressive OLs significantly reduce threshold-stress intensity and increase the crack-propagation rate. Large, compressive OLs decrease crack-opening stress to below zero and a significant number of cycles are required to initiate a crack-opening stress to a stable level, where it can reopen. Local strain at the crack tip increases with the tensile-load level and decreases with compressive loading. In Fig. 6.45, the effect of crack length, as a function of the number of cycles, is seen and the influence of OL on crack propagation is indicated. It is clear that the crack length increased significantly before applying the OL, but afterwards there was only a slight, temporary increase in crack length; then the crack was unaffected, remaining at an almost stable level (starting at a $\sim 3.94 \times 10^6$ cycles).

Figure 6.46 shows the crack closure before, during and immediately after the OL. All the curves were recorded during OL. During OL and immediately after it, the crack-opening stress decreased, but then, as the fatigue cycles increased, the crack-opening stress level gradually increased. As can be seen in the inset, the cycling load was 45 MPa, the same as before applying the OL.

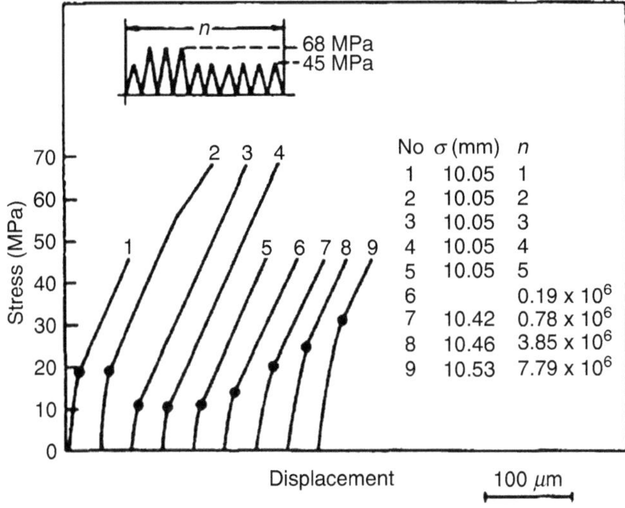

Fig. 6.46 Stress vs. displacement for tensile load (Topper and Yu 1985. With kind permission of Elsevier and Professor Topper)

The effective stress ratio, before applying OL, was calculated according the relation given below, giving 0.58, decreasing as the fatigue cycles increased to 0.31. The stress ratio is given as:

$$U = \frac{S_{max} - S_{op}}{S_{max} - S_{min}} \tag{6.34}$$

S is the stress and the subscript, op, stands for the crack-opening stress. Another factor to consider is the stress-intensity factor, ΔK, which is given as:

$$\Delta K_{th} = \Delta K_{in} + \Delta K_{op} \tag{6.35}$$

$$\Delta K_{in} = K_{max} - K_{op} \text{ and } \Delta K_{op} = K_{op} - K_{min} \tag{6.35a}$$

or:

$$\Delta K_{th} = \Delta K_{max} - \Delta K_{min} \tag{6.35b}$$

Here, ΔK_{th} is the threshold-intensity factor, ΔK_{in} is the intrinsic or basic stress intensity required for the extension of a crack opening and ΔK_{op} is an additional crack-opening stress intensity required to resist crack closure and maintain a fully opened crack. The increased opening-stress intensity caused by tensile OL is an important factor in crack-growth retardation.

The aforementioned effect of compressive stress is illustrated in Fig. 6.47.

Curve 1 in Fig. 6.47 represents the crack-closure behavior before the application of compressive OLs. The crack-opening stress was above zero. While applying the

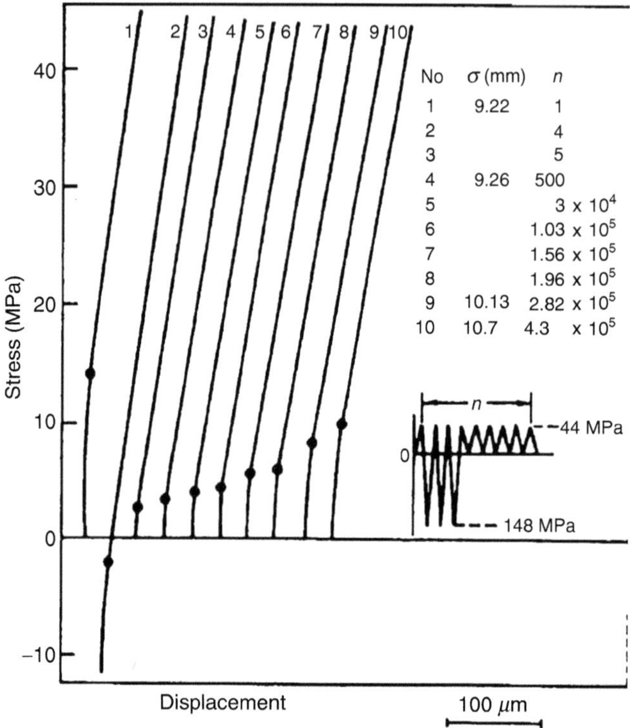

Fig. 6.47 Stress vs. displacement curves for overload compressive tests (Topper and Yu 1985. With kind permission of Elsevier and Professor Topper)

OL, the crack-opening stress decreased to below zero, as seen in curve 2. During the subsequent zero-to-tension cycling (curves 3–10) the crack-opening stress gradually increased. The crack-propagation rate gradually decreased from an initially high level to the stable condition.

An important parameter for OL crack retardation is the OL ratio, defined as:

$$\mathrm{OLR} = \frac{\mathrm{K_{OL}} - K_{\min}}{\mathrm{K_{max}} - K_{\min}} \tag{6.36}$$

K_{max} and K_{min} are the maximum and minimum applied-stress intensities of the constant baseline loading and K_{OL} is the maximum applied-stress intensity during OL. It was observed that OL retardation in AISI 304 stainless steel, under different ΔKs, OL ratios and ratios, R, have an effect on the influence of OL. It was also found that the degree of crack retardation increases with the OL ratio, deceases with R and that the relation with the baseline, ΔK, is complex (Shin and Hsu). The major mechanism causing crack retardation by OL is plasticity-induced crack closure. Crack-tip blunting and residual-compressive stress ahead of the crack tip are significant at high R, when crack closure is suppressed.

Nonetheless, other experimental results (Zheng) for low-carbon, low-alloyed steels indicate that OLs decrease the fatigue life of smooth (unnotched) specimens; OLs decrease the endurance limit and shorten fatigue life. This means that the OL effect contributes negative values by reducing the number of cycles that a component can withstand at some lower stress.

Thus, one can conclude that the effect of OL on fatigue resistance, as it is manifested by crack initiation and propagation is complex and depends on many parameters, such as: the type of OL (tensile or compressive), the interaction between loading amplitudes prior to and after applying OL, the material properties and dimensions, etc. The sometimes contradictory results mentioned in the literature merely reflect the complexity of the subject, as revealed by the various tests applied to the study of fatigue.

6.8.3 Coaxing

The fatigue resistance of some metals showing a fatigue limit (endurance limit) may be improved by a process known as 'coaxing'. Improvement by coaxing is achieved by gradually increasing the fluctuating stress, starting from a level below the fatigue limit. Thus, the process of improvement consists of understressing a metal, followed by gradually increasing the amplitude of the alternating stress in small increments. One of the first reports on coaxing was published by Sinclair, who found that the fatigue resistance of ingot iron, SAE 1045 and 2340 steels is considerably increased, while 75S-T6 aluminum alloy, annealed 70–30 brass and strain-aged ingot iron were not improved. It is quite clear that those alloys that showed considerable improvement after coaxing (in Sinclair's experiments) have a well-defined endurance limit (e.g., ferrous alloys), whereas non-ferrous alloys, such as those of aluminum and copper, do not show the flat endurance limit of the S-N curves. Murakami et al. confirmed Sinclair's observations that the 70–30 brass does not show an improvement due to coaxing. There are reports that, even in composite materials, a very strong coaxing effect exists. Berg, for example, reported a fivefold improvement, compared to a specimen that had not received coaxing treatment.

Sinclair linked the increase in fatigue resistance to strain aging. Of all the materials he tested, only those metals that usually show strain aging responded to coaxing. Fatigue damage seemed to be related to slip. Structural studies by TEM indicate that the initiation of damage by crack formation is preceded by slip-line formation, an indication that some plastic deformation has occurred. Such studies have also revealed that cracks form in localized zones, which have suffered plastic deformation; the cracks grow within and parallel to the slip lines. Any mechanism that can hinder slip formation or continuation will also increase fatigue resistance by strengthening such local regions. In an attempt to explain the role of understressing, it has been suggested that the elastic limit of these soft spots (local weak regions) is increased. Similarly, the resistance to slip is also increased somehow in areas where fatigue damage begins. Sinclair reasoned that one should reject the concept of work hardening and suggested that strain aging offers a better explanation of this phenomenon. Common ferrous metals, such as mild steel, are known to strain age at room temperature after some small plastic deformation, e.g., produced by tension. The elastic limit increases after strain aging, compared to the same material, but without strain aging. Such an increase will also increase the fatigue strength. 70–30 brass is not amenable to strain aging. In essence, the coaxing effect is due to time-dependent, localized strengthening by strain aging, which occurs during cyclic stressing at small stress amplitudes. At this stage, slip occurs only locally at a few weak spots (regions favorably-oriented for slip). The elastic strength of these deformed areas gradually increases by strain aging, due to subsequent cycles of stress. The result is that these regions are more resistant to slip than their surroundings. When the load is increased (coaxing), the yield stress is also increased and slip occurs in new regions, which are, in turn, also strengthened by strain aging. This process gradually improves fatigue resistance by the continued strengthening of the weakest areas, until a stage is reached in which no more strengthening can occur due to strain aging and slip is initiated, leading to the beginning of a fatigue crack. When the degree of aging increases during rest periods (i.e., when the stress increment is decreased in value), when more time is provided for continued strain aging, localized elastic strength increases, resulting in greater improvement.

The strain-aging concept is not universally accepted, but is still widely used (see, for example, Murakami et al. or Miller and O'Donnell). However, based on some experiments, it is claimed that coaxing does not strengthen fatigue resistance (Lerch et al.). The research by Lerch, et al. deals with high-cycle fatigue-strength step tests of gamma TiAl. They also claim that the S-N curve is relatively flat and that at least two alloys exhibit a horizontal S-N curve. Such a horizontal S-N curve characterizes alloys with well- defined endurance limits (such as ferrous alloys). Increased resistance is of a statistical nature; the statistical nature of fatigue and the importance of defects in reducing fatigue life cannot be ignored. It is possible that the higher fatigue life of coaxed samples is simply due to the sampling of the 'cleaner' regions of a specimen. The results for gamma TiAl indicate that the higher fatigue strength observed when more steps are used is simply due to natural material scatter, rather than to coaxing.

Approaches other than strain aging relate the strengthening effect either to work hardening or strain-induced martensite formation that strengthen local areas by slip (Nakajima). Figure 6.48 is an indication of coaxing in a 304-type stainless steel, in which strain-induced martensitic transformation has occurred during fatigue.

For additional discussion on coaxing, the vast literature on the subject may be consulted.

6.8.4 Cumulative Damage

Closely related to the effects of stress levels (discussed above), the cumulative damage caused by the application of various stress levels to machine elements has

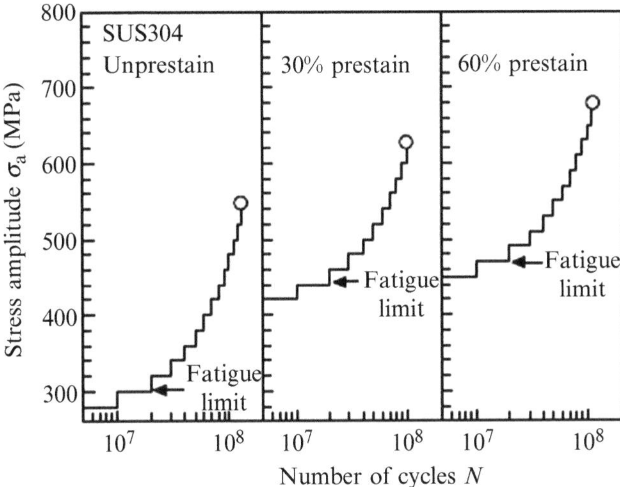

Fig. 6.48 Shows the stress-incremental test results. The failure stresses of all specimens increase compared with the fatigue limits, indicating that the coaxing effect occurs remarkably regardless of the prestrain level (Nakajima et al. 2009. With kind permission of Elsevier and Professor Nakajima for the authors)

been assessed by various approaches. The Miner approach is one such attempt to calculate, or rather to approximate, the portion of life consumed and to ascertain what is to be expected during continued cyclic-stress application. 'Miner's rule' assumes that a total life can be approximately evaluated by adding the percentages of life consumed by each overstress. This rule is also aptly named the 'Palmgren-Miner linear damage hypothesis', since Palmgren proposed it before Miner. It is widely used to estimate cumulative damage, probably due to the simplicity of this damage model, expressed as:

$$\frac{n_1}{N_1} + \frac{n_2}{N_2} + \frac{n_3}{N_3} + \cdots = \sum_{i=1}^{k} \frac{n_i}{N_i} = C \tag{6.37}$$

In the above relation, k indicates the different stress levels in stress spectrum $1 \le i \le k$. The average number of cycles to failure at the i-th stress is n_i (S_i) and N_i (S_i) is the number of cycles to failure. Thus, n_i (S_i) is the number of contributed cycles. C is experimentally found to be between 0.7 and 2.2. Usually for design purposes, C is assumed to be 1 and then Eq. 6.37 may be written as:

$$\sum_{i=1}^{k} \frac{n_i}{N_i} = 1 \tag{6.37a}$$

Thus, failure, according to this rule, will occur when:

$$\sum_{i=1}^{k} \frac{n_i}{N_i} \geq 1 \tag{6.37b}$$

In essence, Miner's rule indicates what portion of life is consumed by stress reversal at each magnitude then forming a linear combination for their overall effect.

Though, in general, Miner's rule is not accurate, it is a useful approximation under many circumstances, despite the many limitations of the model. This method is accepted by engineers in order to arrive at preliminary figures, with appropriate safety factors, for evaluating the degree of damage. One of the important limitations, not taken into account by the model, is the order in which the cyclic stress is applied. Thus (as was stated in Sect. 6.8.1), applying a low-cyclic stress (UL) followed by a higher stress causes more damage than predicted by the linear combination of Miner's rule, which does not consider the effects of OL or high stress on the assumed results according to Eq. 6.37. However, high stress followed by low stress may cause less damage, due to the possible presence of compressive residual stress. Although it was indicated that the loading sequence was not taken into account in Miner's rule, many tests have been performed, mostly in a two-step load history, in order to verify the validity of this rule. These two-step load tests are known as 'high-low experiments' or 'low-high experiments', depending on the order of stress application. In high-low load tests $S_1 > S_2$, whereas in low-high experiments, $S_1 < S_2$. In Eqs. 6.37, 6.37a, and 6.37b, the ratio, often given as the damage, is:

$$D_i = \frac{n_i}{N_i} \tag{6.37c}$$

To overcome the shortcomings of Miner's linear rule, many nonlinear damage approaches have been proposed. One of them is the Marco-Starkey method, which expresses the ratio of the number of cycles to the cycles of failure by a damage parameter, p, given as:

$$D_i = \left(\frac{n_i}{N_i}\right)^p \tag{6.38}$$

The value of this parameter is considered to be greater than zero and less than or equal to 1.0, with the value of the nonlinear parameter increasing with stress level. At the value of $p = 1$, this formula is equivalent to that of Miner. The nonlinear method describes observed material behavior relatively well, but again a large number of tests are required to evaluate the constant.

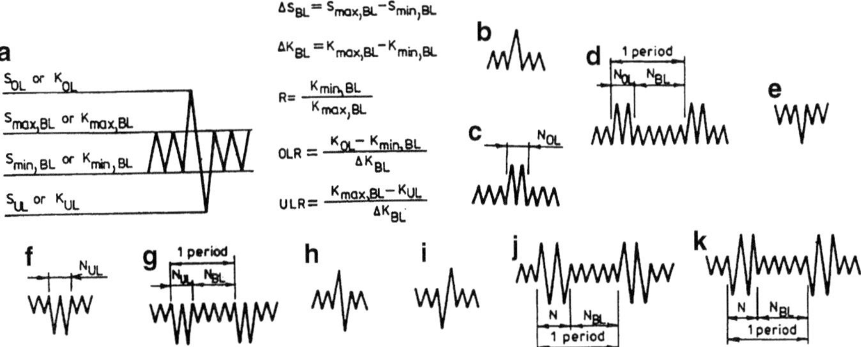

Fig. 6.49 Typically used variable amplitude load sequences assembled with overloads (*OL*) and underloads (*UL*): (**a**) associated load parameters; (**b**) single OL; (**c**) OL block; (**d**) periodic OL blocks; (**e**) single UL; (**f**) UL block; (**g**) periodic UL blocks; (**h**) single OL–UL; (**i**) single UL–OL; (**j**) periodic OL–UL blocks; (**k**) periodic UL–OL blocks (After Skorupa 1998. With kind permission of John Wiley and Sons)

6.8.5 Variable-Amplitude Loading (Intermittent Loading)

In many practical applications, complex loading conditions occur and, in service, no fixed loading is encountered. Therefore, fatigue tests are performed to understand the effects of non-constant amplitude loading on fatigue resistance. The most commonly known condition of non-uniform loading, in practice, occurs to the wings of airplanes during flight, when repeated, non-uniform vibrations may be catastrophic, if not properly understood during the design stage. For instance, at the National Aeronautical Establishment of the National Research Council (Ottawa Canada), complete airplanes are exposed to cyclic-fatigue tests, simulating actual stresses incurred during flight, to address the problem of potential, variable-amplitude loading during service. Such tests are performed under various conditions and cyclic patterns in order to understand the effects of non-uniform amplitude on fatigue life. The technique involves the combination of various patterns of OL and UL test specimens. Such combinations and their interactions were studied by Skorupa, providing several possible models (reproduced schematically in Fig. 6.49). In the illustration, K is the stress-intensity factor, ΔK represents the stress-intensity factor range, ΔK_{th} is the fatigue threshold, ΔS is the stress range and R is the ratio. The subscript, BL, refers to the baseline. All the relations shown in Fig. 6.49 have been discussed above (in Sect. 6.8.2). Another common, two-level block loading is seen in Fig. 6.50.

In Fig. 6.51, simple load histories, containing combinations of OL and UL cycles, is shown. Most available test results suggest that an UL applied immediately after an OL reduces the post-OL retardation more significantly than an UL which immediately precedes an OL. Usually, faster, transient fatigue-crack growth rate

Fig. 6.50 Two-level block loading. (**a**) high–low; (**b**) low–high; (**c**) step in both maximum and minimum load (Skorupa 1998. With kind permission of John Wiley and Sons)

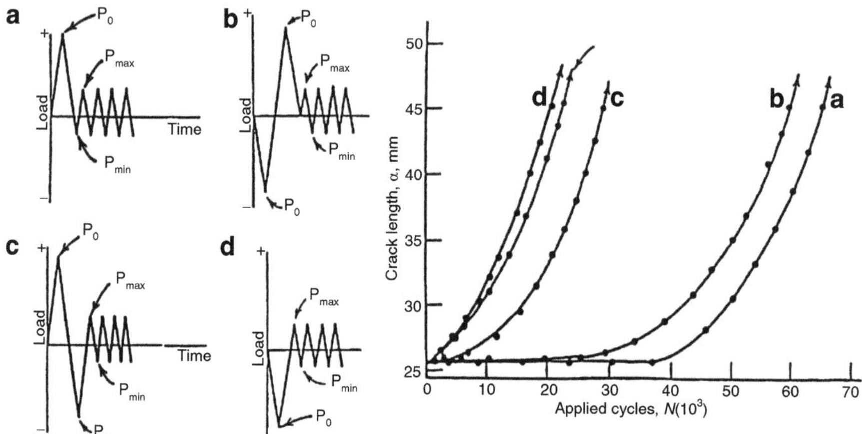

Fig. 6.51 Effect of UL position on post-OL retardation in 2024-T3 Al alloy (Skorupa 1998. With kind permission of John Wiley and Sons)

[henceforth: FCGR] and, therefore, shorter crack-propagation lives are observed, e.g., in Al alloys, under the load sequences indicated in Fig. 6.49 for (h) and (j) than for those under (i) and (k), respectively.

In Fig. 6.51, the crack length versus the applied cycles is indicated. Note that, for a given crack length, the alloy can endure much larger cycles when an OL precedes loading marked by (a), with the next best case being (b), where an OL precedes an UL, followed by a baseline loading sequence.

The effect of the loading sequence under various conditions was investigated in structural steel. The test results are summarized in Fig. 6.52. It can be seen that some of the loading patterns of a single block of OL–UL produce less retardation during subsequent baseline cycles than a block OL-block UL sequence with the equivalent number of major cycles. In both types of pattern sequence, crack retardation increases with increasing block size, but the retardation effect is markedly reduced, compared with the block OL case. In contrast to the above mentioned results, more retardation was produced by a single OL-UL, [Fig. 6.49h] than by a single UL-OL, [Fig. 6.49i]. In Fig. 6.52, (a) the effect of UL position and magnitude is indicated. If, instead of a single UL, a block of 10 ULs followed the OL, faster crack growth would occur, as indicated in Fig. 6.52b. Also note (Fig. 6.52) that for the sequences

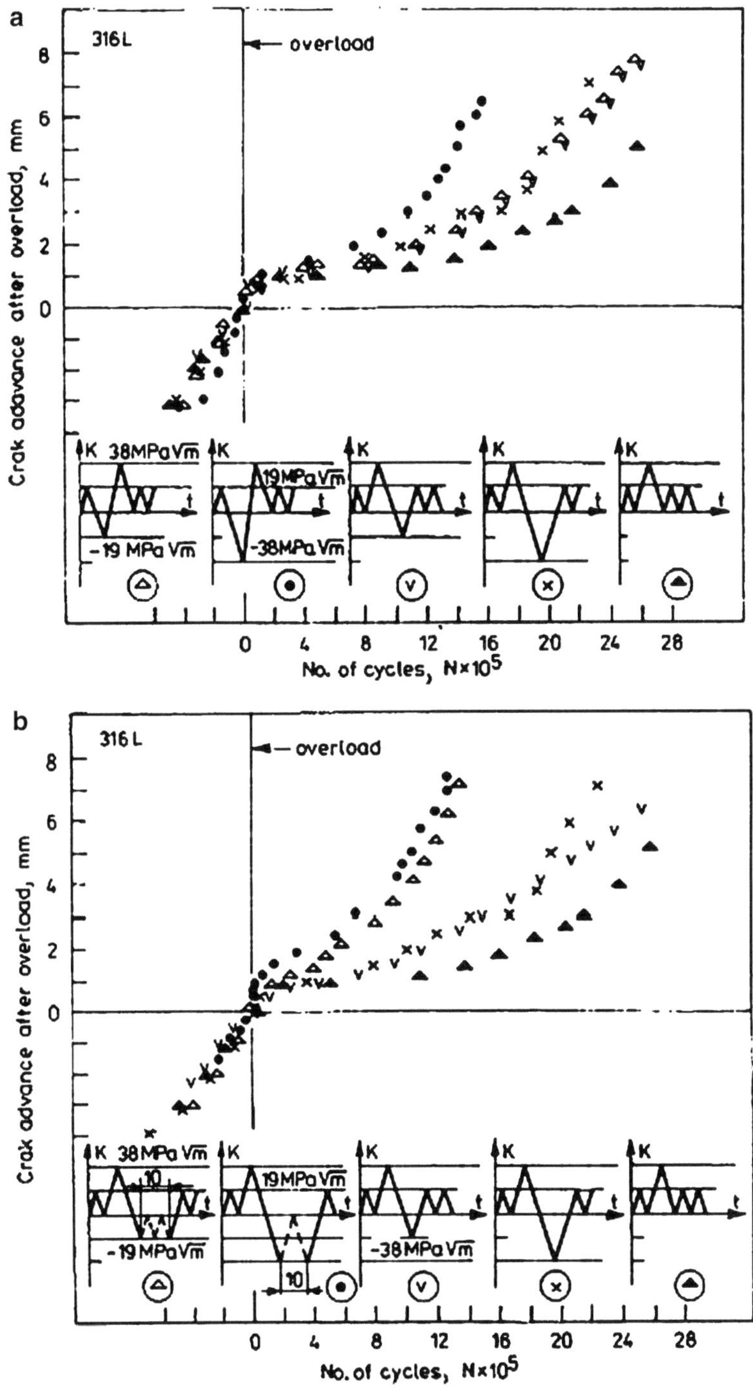

Fig. 6.52 Reduction of post-OL retardation by application of an UL in stainless steel: (**a**) effect of UL position and magnitude; (**b**) effect of number and magnitude of UL cycles following an OL (Skorupa 1998. With kind permission of John Wiley and Sons)

in which a single UL or multiple ULs were applied after the OL, the effect of the UL magnitude was small. On the contrary, with an UL preceding the OL, significantly less retardation was noted.

In summary, fatigue-crack growth under variable-amplitude loading is usually accompanied by load interaction, resulting in a FCGR for a given load cycle which may differ from the growth rate observed for that same cycle in constant-amplitude tests. The retardation or acceleration of crack growth depends on many factors beside the pattern sequences indicated. Among these highly influential factors are: loading amplitude, test specimen size and geometry, environment and material properties, such as microstructure and dislocation reactions.

6.9 Structural Observations in Fatigued Specimens

6.9.1 Progression Markings (Beach Marks) and Striations

Since Forsyth's earliest work, is has been known that slip bands associated both with intrusions and extrusions occur on metal surfaces, when they are subjected to cyclic loading. This was an important observation, since slip-band intrusions form concentrations, which may become the locations in which cracks develop (as discussed later in the chapter on fracture). The process of reverse slip, which causes slip-band ridges and grooves on fatigued-material surfaces, is thought to be the mechanism by which extrusions occur. Additional structural features observed in fatigued machine parts are striations and fatigue-progression markings, otherwise known as 'beach marks'. A typical beach mark may be seen in Fig. 6.53, obtained from the fractured crankshaft of an aircraft engine operating under fatigue conditions.

In this specific case, it seems that under cyclic loading, these fatigue cracks were initiated at stress-concentration points, leading to the fracture of the shaft. However, focus will now be turned to the structural aspects, as observed in fatigued specimens and in fatigue-progression markings. In Fig. 6.53, the point indicated by the arrow is a high stress-concentration site. Many structural inhomogeneities can act as stress raisers, such as: a groove, a lubrication hole, the bottom of a thread, a shaft fillet, etc. Often, at a location where a small preexisting or accidental inclusion is found, a stress concentration will develop, influencing the fatigue resistance of a material. Pronounced progression markings (beach marks) on the fractured surface of an Al alloy are shown in the scanning electron micrograph [henceforth: SEM] photo in Fig. 6.54. These markings originate from an inclusion.

Striations and other structural characteristics are crack-front markings and they are believed to be associated with the cycle applied during fatigue. Thus, it is thought that each striation (observed by examining fractured specimen surfaces after fatigue) was produced by one stress cycle. High-resolution TEM and SEM studies show that striations often have a sawtoothed profile and that one side of the sawtooth has

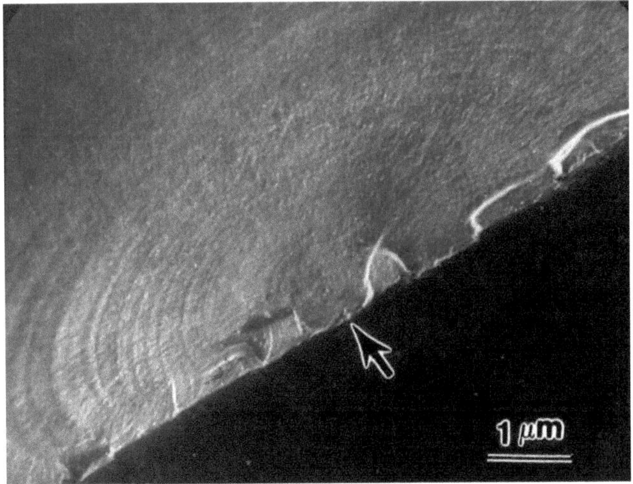

Fig. 6.53 SEM fractograph of a fractured crankshaft. Fatigue beach marks and the crack origin region are seen. The origin of the crack is indicated by the *arrow* (Bhaumik et al. 2002. With kind permission of Elsevier and Dr. Bhaumik for the authors)

Fig. 6.54 SEM of fracture surface of an Al alloy showing pronounced progression markings. The markings radiate from a surface breaking inclusion marked by the *arrow* (Lynch 2007. With kind permission of Elsevier)

a more rumpled appearance, exhibiting more slip traces than the other side. The appearances and profiles of striations can vary widely, depending on the material, ΔK, and other variables, and may even be different in adjacent regions of the same fracture surface, as indicated in Fig. 6.55.

At low ΔK, striation spacings are often about the same (in the range 80–300 nm depending on the material), regardless of the ΔK value and may be up to several orders of magnitude greater than macroscopic crack-growth rates per cycle at near-ΔK

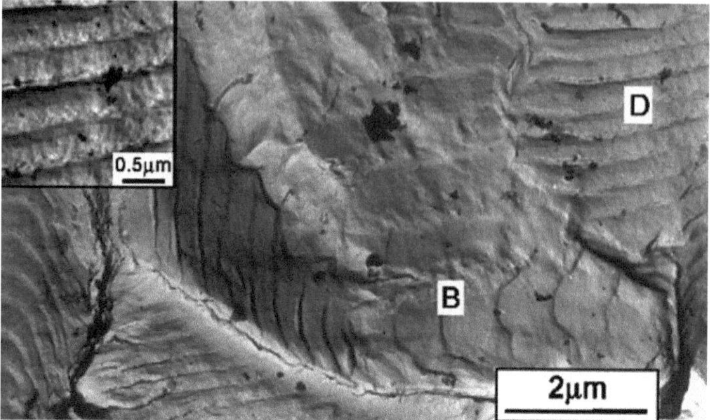

Fig. 6.55 TEM of direct-carbon replica of constant amplitude fatigue fracture surface of an age hardening high purity Al-Zn-Mg alloy showing striations. Ductile striations at *D* and brittle striations at *B* with widely differing spacings and orientations. Fine scale roughness superimposed on striations is seen in the *inset* (Lynch 2007. With kind permission of Elsevier)

threshold values (Eq. 6.35b) for a wide range of materials and environments. The characteristics of striations at low ΔK appear to be similar to those produced at intermediate ΔK and it is widely assumed that each striation is produced by one cycle of stress (as occurs at intermediate ΔK), as indicated above. However, it is thought that, as ΔK decreases, an increasing number of stress cycles are required before a crack-growth/blunting/re-sharpening event occurs, because microstructural changes or damage must develop ahead of the cracks before crack growth can occur. Observed changes include the formation of dislocation-cell structures and the resolution of precipitates, but other changes, such as the formation of vacancy clusters/nano-voids ahead of the cracks may also be involved. Alternative scenarios for striation formation at low ΔK have been proposed and further research is clearly required to gain understanding of the processes occurring in this ΔK regime.

6.9.2 The Dislocation Structure in Fatigue

The following TEM micrographs compare dislocation structures of Al alloys obtained by regular tension tests and fatigue. The modifying effects of Mg additions on dislocation configuration in both deformation modes are also shown. The variation of dislocation density for the Al-Mg alloys under strain is shown in Fig. 6.56 and compared with that of pure aluminum.

Some micrographs are shown in Figs. 6.53, 6.54, 6.55, 6.56, 6.57, 6.58, 6.59, 6.60, 6.61, and 6.62. Pure aluminum pulled by tension to a 10% extension is shown in Fig. 6.56. This level of deformation corresponds to the formation of a sharply-defined cell structure. It was claimed that such a cell structure already exists after

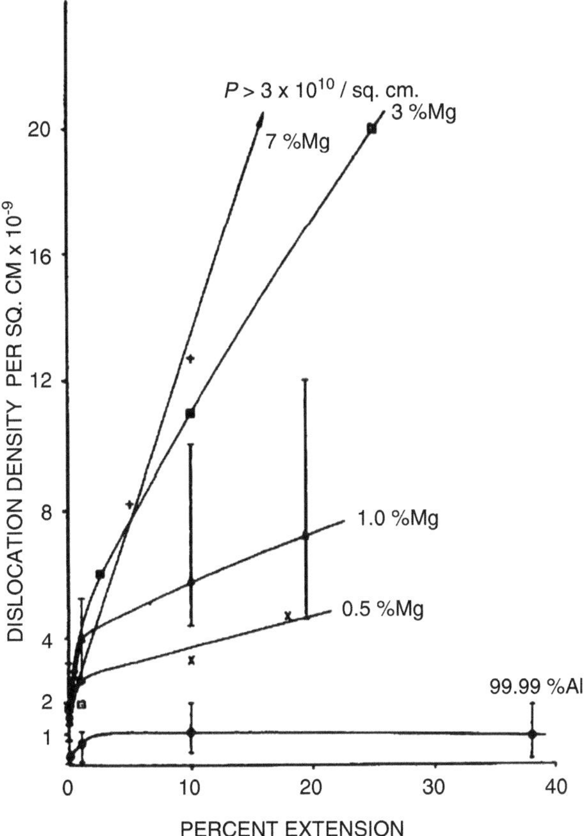

Fig. 6.56 Variation of dislocation density with strain for aluminum and aluminum-magnesium alloys (Waldron 1965. With kind permission of Elsevier)

about 2.5% strain in annealed aluminum deformed by tension. In Chap. 2, Figs. 2.8, 2.9, and 2.10 also showed a cellular structure in pure aluminum deformed to 5% and 15% by rolling followed by annealing. These results are compared with those of unstrained, pure aluminum (Fig. 2.7). As may be seen in Fig. 6.56, the addition of magnesium, even at a 0.5% level, strongly increases the dislocation density. The substructure is more pronounced, although the extension level is the same in the specimens presented in Figs. 6.57 and 6.58.

Well-defined cell structure was obtained by fatiguing an Al-0.5% Mg alloy, as shown in Fig. 6.59. This alloy exemplifies the condition after 10^5 cycles at a relatively high load level. A rough substructure develops from the dislocation clusters. The cell walls are often roughly aligned, parallel to the trace made by a [111] plane in the surface of the foil (see Fig. 6.60). The complex structure of Fig. 6.60 is probably associated with multiple slip, since traces of a second striation

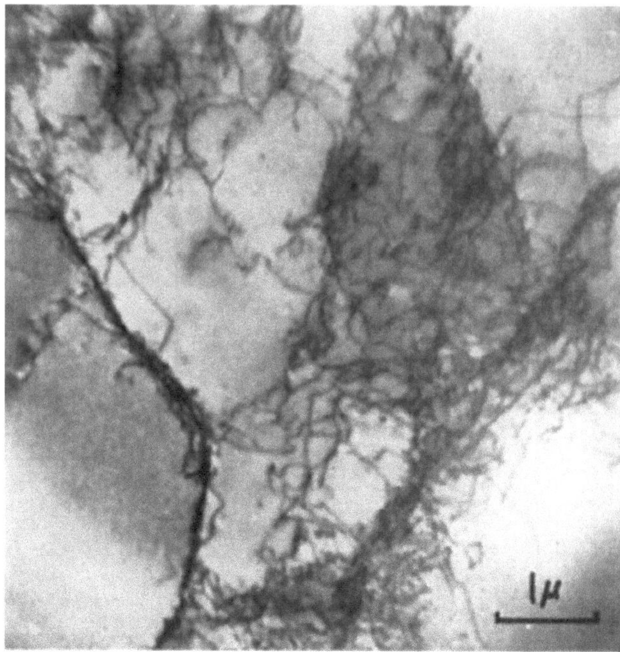

Fig. 6.57 Sub-grains in aluminum pulled to 10% extension (Waldron 1965. With kind permission of Elsevier)

system appear on the fatigued surface of this disc under optical scrutiny prior to the final piercing polish for the TEM examination. In an aluminum-0.5% magnesium specimen that failed after only 2×10^3 cycles at 3.0 tsi, the aligned walls form well-defined, elongated subgrains, some of which extend over tens of microns (see Fig. 6.61). The equiaxed subgrains, also present in this Figure, are typical of many seen in other areas examined. Furthermore, in an Al-3% Mg specimen that had not failed after 1.65×10^7 cycles at 3.0 tsi (well below the fatigue limit of 4.5 tsi, at about the macroscopic yield stress of tensile specimens), the fatigued structure consists of many uniformly-distributed, elongated loops and jogs.

Foils from materials, stressed at the fatigue limit for 8.73×10^6 cycles, contain higher dislocation densities and there is a tendency for clusters to form. It is possible to sum up the main points observed in Al-Mg alloys pulled by uniaxial tension and those that were fatigued as follows. The effect of Mg additions on pure Al is to increase the dislocation configurations, such as jogs and loops, following deformation (as indicated in Fig. 6.56). It is also evident that the dislocation density of tension-deformed alloys increases with Mg additions. It was observed that the formation of a cell structure is increasingly inhibited by magnesium additions. In an Al-7% Mg alloy pulled to failure, no cells are formed and the dislocation density is $>3 \times 10^{10}/cm^2$. In fatigued specimens, the distribution of jogs and loops is uniform. For the cell-structure formation the jogs and loops cluster and by the

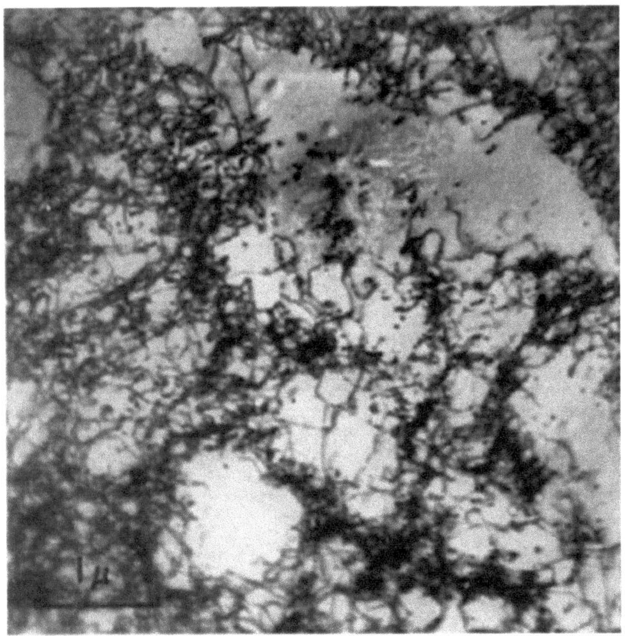

Fig. 6.58 Rough sub-structure in Al-0.5%Mg pulled to 10% extension (Waldron 1965. With kind permission of Elsevier)

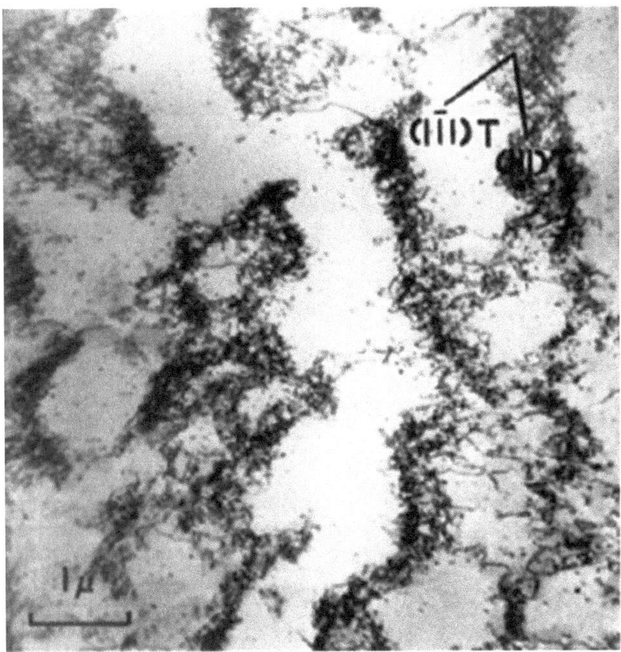

Fig. 6.59 Sub-boundary formation in Al-0.5% Mg alloy fatiguing after 10^5 cycles at 2.22 tsi. (1 tsi ~13.8 MPa on USA scale) (Waldron 1965. With kind permission of Elsevier)

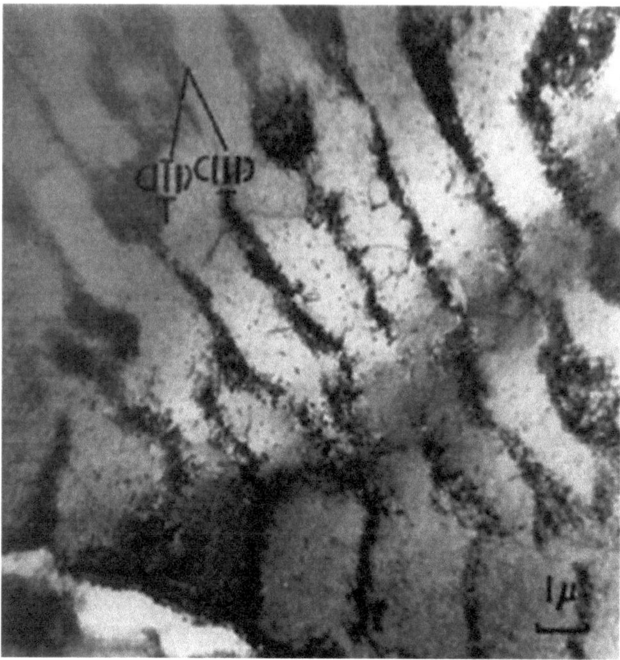

Fig. 6.60 Complex sub-boundary structure in Al-0.5% Mg alloy after 10^5 cycles at 2.22 tsi (Waldron 1965. With kind permission of Elsevier)

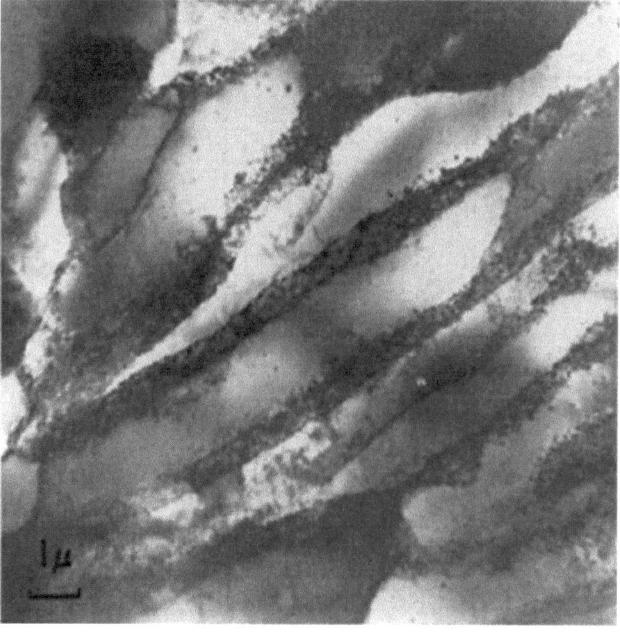

Fig. 6.61 Elongated and equi-axed sub-grains in Al-0.5% Mg alloy failed after 2×10^5 cycles at 3.0 tsi (Waldron 1965. With kind permission of Elsevier)

linking of these clusters a sharply defined cell structure develops. The consequence of increasing Mg additions is to raise the fatigue stress at which a given dislocation structure is formed. Moreover, Mg additions interfere with cross-slip processes, believed to control dislocation distribution in both tensile and fatigue deformation. Dislocation configurations may be the same in specimens deformed by tensile and fatigue deformation and their roles in promoting/hindering plastic flow may be the same and occur by means of the same mechanism. It is believed that the effect of the solute in tensile and fatigue deformation is to interfere with cross slip. The addition of Mg increases the fatigue resistance of the aluminum (also in specimens deformed by tension).

Cellular structure seems to be a general feature of fatigue deformation. Below (in Fig. 6.62a) the cellular structure of austenitic steel is shown. Fragmented and cellular structures have formed in the failure zone (Fig. 6.62). It is evident that the cellular substructure formed during the initial test cycles becomes fragmented later on.

Aluminum exemplifies the behavior of other FCC and BCC metals. However, other reports, based on TEM observations, indicate that cyclic deformation is accompanied by higher dislocation densities (Bolshakov et al.). Thus, contradictory information exists on the density of dislocations in fatigued specimens. The opinion of the author of this book is that this discrepancy reflects the experimental conditions, rather than a basic difference.

Forsyth was among the first to study localized cyclic, plastic straining and he observed extrusions and intrusions on the surfaces of several fatigued materials. These extrusions and intrusions, observed in most single-crystal and polycrystalline materials, are an indication of the surface relief evolving in fatigued test specimens. Wood's concept of persistent slip bands, observed on polished surfaces where microcracks are expected to be initiated, is shown schematically in Fig. 6.63. When pre-existing stress concentrations (originating from machining operations: grinding, milling etc.) are present, they may very well generate localized plastic deformation when a stress is applied to induce slip-band formation. Slip bands were observed experimentally in many materials. The optical micrograph in Fig. 6.64 illustrates persistent slip bands in an Al 7.5%–Zn 2.5% magnesium alloy. The test specimens were deformed by torsion and the persistent slip bands and associated transcrystalline fatigue cracks were revealed and marked A and B, respectively. TEM has revealed that precipitate resolution and substructure formation occurred in these bands, which were up to 500Å wide, as seen in Fig. 6.64, and where A marks the precipitate resolution and B indicates the substructure. Both transcrystalline and intercrystalline deformation were observed on specimens heated to 200°C for 20 min and then immediately fatigued at that temperature. Re-electropolishing such test pieces revealed precipitation along the persistent slip bands, as shown in Fig. 6.65.

Associated fatigue cracks (B) may also be seen; (C) is a grain boundary. Grain-boundary fatigue cracks were also observed on specimens fatigued at this temperature. The optical micrograph (in Fig. 6.66) revealed these features following fatigue deformation by torsion. Furthermore, in this micrograph, the persistent slip

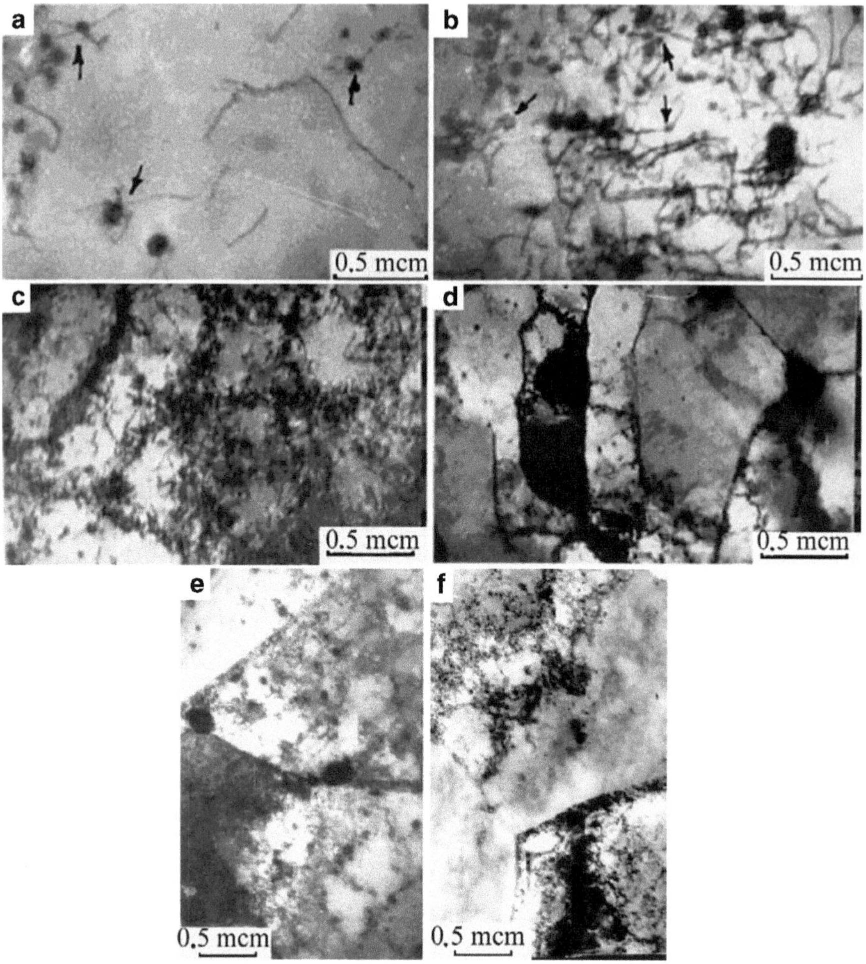

Fig. 6.62 Dislocation substructure types in steel 08H18N10T; (**a** and **b**) in initial state, (**c** and **d**) after low-cycle tests, (**e** and **f**) after low-cycle tests, electro stimulation and subsequent low-cycle tests before failure. Note the cell structure after fatigue (**c** and **d**) (Sosnin et al. 2005. With kind permission of Elsevier and Dr. Gromova for the authors)

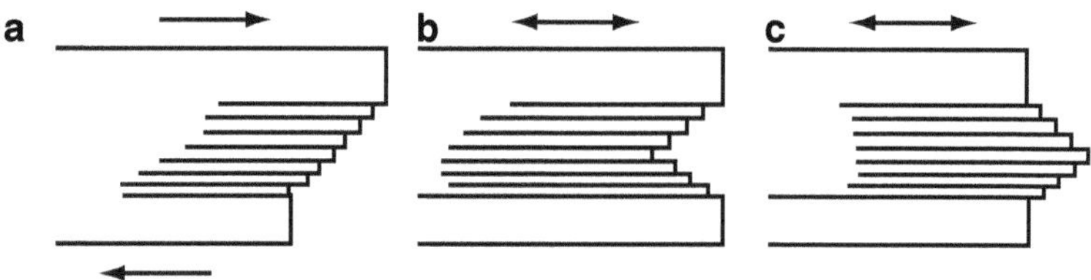

Fig. 6.63 Wood's concept of micro-deformation leading to formation of fatigue crack: (**a**) static deformation, (**b**) fatigue deformation leading to surface notch known as intrusion, (**c**) fatigue deformation leading to slip-band extrusion (After Novovic et al. 2004. With kind permission of Elsevier)

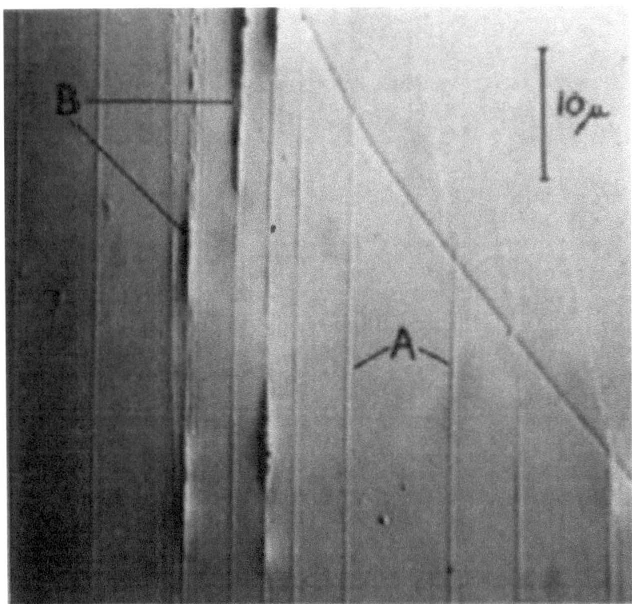

Fig. 6.64 Al 7.5%–Zn 2.5% magnesium alloy solution treated at 450°C, boiling water quenched and aged 1 h at 150°C. Optical micrograph showing persistent slip bands (*A*) and fatigue cracks (*B*) on test piece re-electropolishod after room temperature torsional fatigue. X 1,200 (Stubbington and Forsyth 1966. With kind permission of Elsevier)

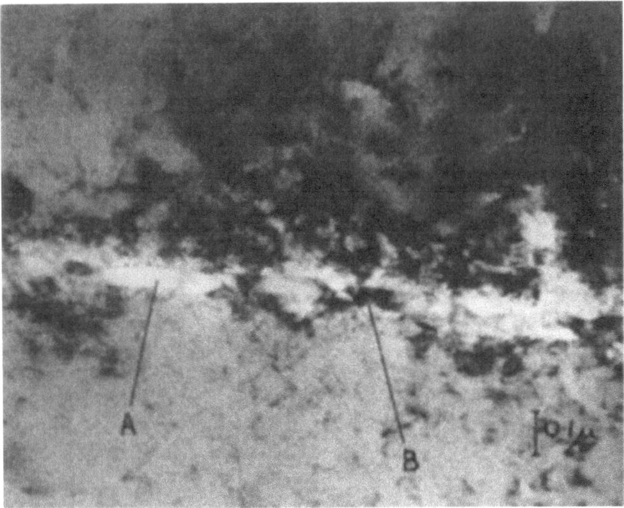

Fig. 6.65 TEM micrograph of Al 7.5%–Zn 2.5% magnesium alloy, solution treated at 450°C, boiling water quenched and aged 2 days at 150°C. Precipitate resolution and substructure in a persistent slip band is marked by (*A*) and (*B*) respectively. Obtained by room temperature torsional fatigue. X 52,000 (Stubbington and Forsyth 1966. With kind permission of Elsevier)

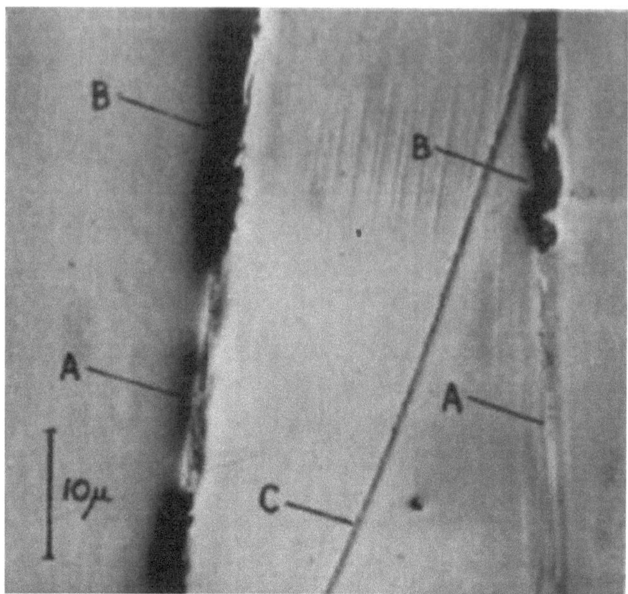

Fig. 6.66 Al 7.5%–Zn 2.5% magnesium alloy; solution treated at 450°C, cold water quenched and aged 16 h at 150°C. Heated to 200°C and immediately fatigued in torsion at that temperature. Optical micrograph showing persistent slip bands (*A*), fatigue cracks (*B*) and grain boundary (*C*) on subsequently re-electropolished test piece surface. X 1,200 (Stubbington and Forsyth 1966. With kind permission of Elsevier)

bands on the specimens fatigued at and above 150°C are revealed as ridges on the subsequently electropolished test piece surfaces. Persistent slip-band grooves are also observed on electropolished specimens immediately after fatigue at room temperature. Vacancies are retained in the persistent slip bands. Groove formation is basically equivalent to notch formation, which is a stress raiser on an atomic scale and may well be the initiating site of failure.

It was suggested that vacancies with different concentrations in the bands are involved in interaction with arrays of dislocations; persistent-slip bands were revealed as ridges, where vacancies are thought to be annihilated, and as grooves, where the vacancies are retained in the slip bands.

Further studies using AFM were performed by Polák et al. in polycrystalline austenitic and ferritic stainless steels on persistent-slip markings emerging from persistent-slip bands formed by extrusions and intrusions as surface reliefs. Surface-slip markings in austenitic steel are already detectable at 20 cycles, when the stress amplitude reaches its maximum and the loop-shape parameter approaches its minimum. Since these markings become persistent, they are aptly called 'persistent slip markings'. Figure 6.67 shows that well-developed markings were detected at 50 cycles and became more pronounced during fatigue life

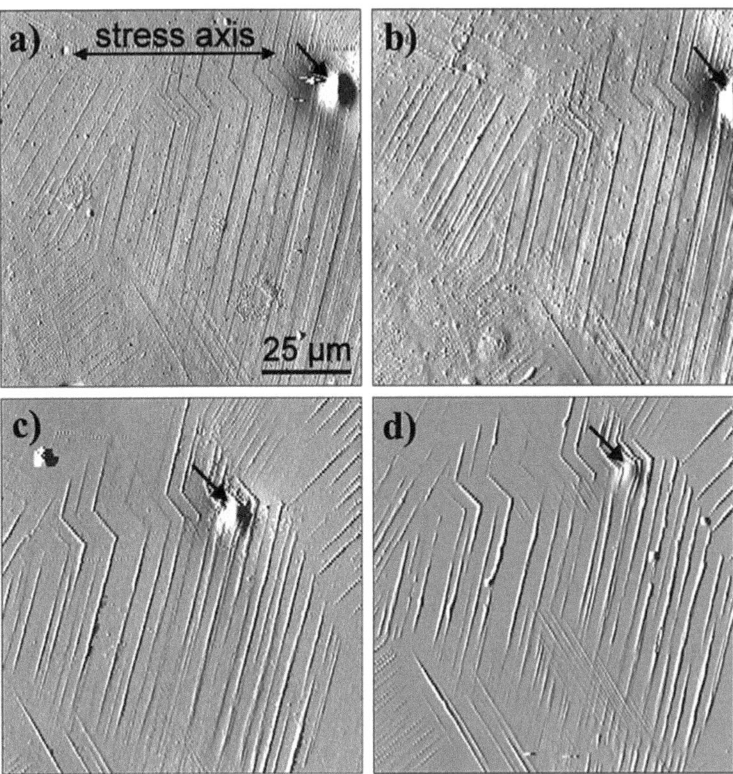

Fig. 6.67 Evolution of the persistent slip markings in austenitic stainless steel observed by AFM. Cycling with $\varepsilon_{ap} = 2 \times 10^{-3}$; (**a**) N = 50 cycles, (**b**) N = 500 cycles, (**c**) N = 5,000 cycles, (**d**) N = 20,000 cycles (Polák et al. 2003. With kind permission of Elsevier and Professor Polák)

The evolution of these surface markings in ferritic steel is shown in Fig. 6.68. In comparison with austenitic steel, their appearance is slightly delayed and the first persistent-slip markings were detected at 50 cycles. AFM can quantitatively detect the height of an extrusion, measuring it during the fatigue lives of austenitic and ferritic steels. Figure 6.69a shows the height of an extrusion measured in three different sections as a function of the number of cycles. Since this extrusion is band-like, its height varies only slightly along the persistent-slip markings and the measurements from all three sections give fairly similar results. The initial extrusion growth rate is higher than the latter. For most of the fatigue life, approximately linear growth of the extrusion height with the number of loading cycles is observed. The average growth rate is evaluated at 2.5×10^{-11} m/cycle. The situation in ferritic steel is more complicated, since the height of an extrusion fluctuates along a persistent-slip mark. Therefore, the mean-extrusion height has been evaluated as a mean-integral value along each of the persistent-slip markings denoted as B, C, D, E, F and G.

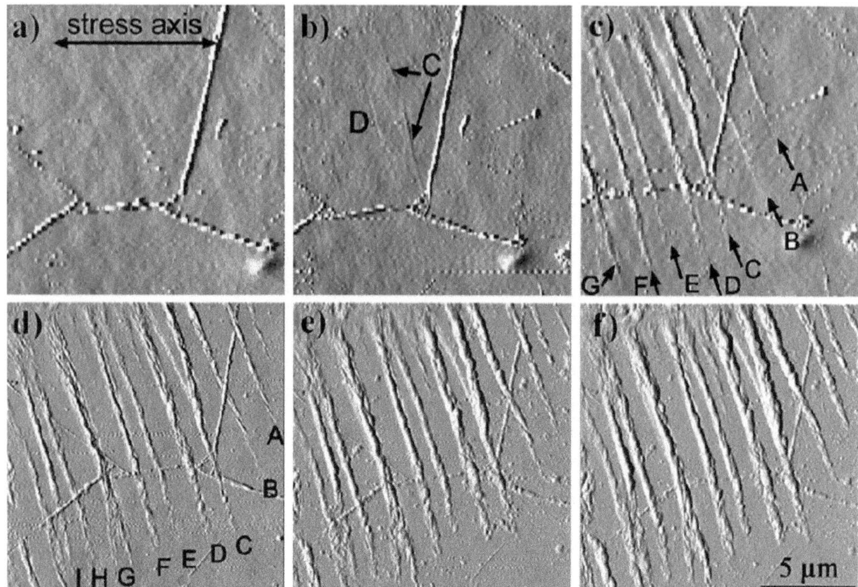

Fig. 6.68 Evolution of the persistent slip markings in ferritic stainless steel in cycling observed by AFM with $\varepsilon_{ap} = 2 \times 10^{-3}$; (**a**) N = 20 cycles, (**b**) N = 50 cycles, (**c**) N = 500 cycles, (**d**) N = 1,000 cycles, (**e**) N = 3,000 cycles, (**f**) N = 9,000 cycles (Polák et al. 2003. With kind permission of Elsevier and Professor Polák)

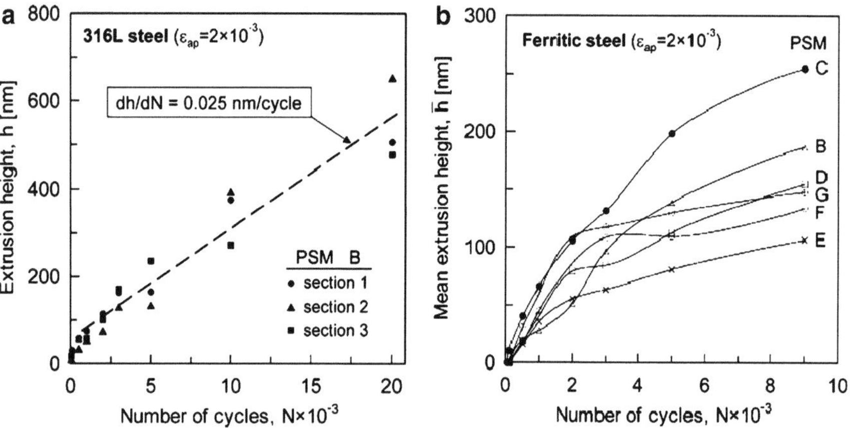

Fig. 6.69 Extrusion height and the mean extrusion height vs. number of loading cycles in cycling with constant plastic strain amplitude $\varepsilon_{ap} = 2 \times 10^{-3}$; (**a**) in austenitic steel (**b**) in ferritic steel. *PSM* stands for persistent slip markings and *B* is one location on it. The same is with the PSM markings *B*, *C*, etc. in 6.64b (Polák et al. 2003. With kind permission of Elsevier and Professor Polák)

The mean-extrusion height of these persistent-slip marks is plotted in Fig. 6.69b vs. the number of loading cycles. The initial growth of an extrusion is high also in ferritic steel and later decreases as in the austenitic steel. If the growth of extrusions in ferritic steel during most of its fatigue life could be approximated by a linear dependence, then the average growth rates would be in the interval (6-12) 10^{-12} m/cycle.

It was previously stated and suggested by many experimental investigations that vacancies or point defects are involved in extrusion formation. The experimental extrusion growth in poly-crystals supports the crack-nucleation model. This model is based on the redistribution of the matter between the persistent-slip bands and the matrix, due to point-defect migration and annihilation and predicts a continuous growth of extrusions and the nucleation of cracks from the deep, narrow, adjacent intrusions.

6.10 The Notch Effect

Study of notched specimens has practical significance, because fatigue in engineering components starts at the surface, at those places where stress raisers are present in the form of 'micro-notches' (i.e., all the traces of machining, the artificial defects); in the actual applications, such components are never perfectly smooth. Corners, welding toes, fillets, holes, etc. all act as stress raisers. Such geometries cause stress concentrations in the component and reduce the fatigue strength and life of a structure. As seen in Sect. 6.2.1 (Fig. 6.8), the experimental S-N curves for notched and un-notched specimens of normalized and tempered Ni-Cr-Mo (8630) cast steel were compared. The stress level of the notched specimens in the S-N curves was substantially lower than that of the smooth specimens with reduced endurance limits. One can state, based on numerous experimental observations, that notches are one of the main factors that determine the fatigue strength of structures. This notch-effect is expressed in terms of a notch factor, K_f, which plays an important role in the estimation of fatigue life and the strength of some structures exposed to fatigue deformation. A common definition of the notch factor relates the strength of smooth to notched specimens and is given as:

$$K_f = \frac{\text{fatigue strength of smooth specimen}}{\text{fatige strength of notched specimen}} \tag{6.39}$$

Based on the definition of K_f, many expressions for the notch factor have been developed over the years, differing on the basis of their various assumptions regarding: notch geometry (e.g., sharpness), notch location (edge or centered), crack length and location, and material properties, such as tensile stress, etc. Traditional fatigue analysis of notched specimens is done using empirical approaches, though various models have been developed to consider the notch effect in regard to the strength and fatigue life of structures.

Of the many approaches, classic fatigue theory regarding the notch effect will be discussed primarily, but additional concepts will be mentioned. The classic approach emphasizes the S-N relation-based approach. Another concept is based on crack-growth analysis (e.g., Ciavarella and Liu), which is essentially a fracture mechanics-based approach, which considers the fatigue process to actually be a crack-growth process.

Peterson proposed the formula below for the concentration factors of Eq. 6.39:

$$q = \frac{K_f - 1}{K_t - 1} = \frac{1}{1 + \frac{C_P}{\rho}} \tag{6.40}$$

where K_t is the elastic-stress concentration factor, ρ is the notch-tip radius and C_P is a material parameter related to the ultimate-tensile strength. An empirical relation for C_p was given by Peterson for steels in which the tensile stress is $\sigma_R > 550$ MPa, as:

$$C_p = 1000(270/\sigma_R)^{1.8} \tag{6.41}$$

When there is no effect of the notch on the fatigue strength, $K_f = 1$, according to Eq. 6.39 and the notch-sensitivity factor in Eq. 6.40 is $q = 0$. If, however, $K_t = K_f$, then $q = 1$. The observed values of q are between these two extremes.

Another similar formula was proposal by Neuber as:

$$q = \frac{K_f - 1}{K_t - 1} = \frac{1}{1 + \sqrt{\frac{C_N}{\rho}}} \tag{6.42}$$

C_N is also a material constant. An empirical constant for evaluating C_N was suggested by Kuhn and Hardrath as:

$$\log \frac{C_N}{1000} = -\frac{\sigma_R - 134}{586} \tag{6.43}$$

This relation is valid for steels having $\sigma_R < 1{,}520$ MPa. In both relations, C_P and C_N are expressed in μm, while σ_R is expressed in MPa.

Furthermore, another approach was proposed by Heywood as:

$$\frac{K_f}{K_t} = \frac{1}{1 + 2\sqrt{\frac{C_H}{\rho}}} \tag{6.44}$$

In Eq. 6.44, C_H takes into consideration not only the material strength, but also the geometry of the specimen by:

$$C_H = \left(\frac{C}{C_R}\right)^2 \tag{6.45}$$

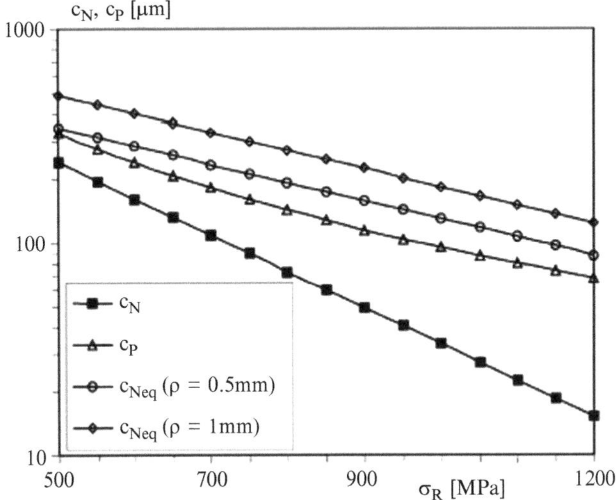

Fig. 6.70 Comparison between the constants. Equivalent Neuberg's constant is also shown for two tip radii, i.e. 0.5 and 1 mm, respectively (Ciavarella and Meneghetti 2004. With kind permission of Elsevier and Professor Ciavarella for the authors)

where C = 5, 4 or 3 for cylindrical specimens containing a transverse hole, a shoulder or a groove, respectively. Clearly these geometries are related to notch geometry and its effect on fatigue. For making the material constants C_N and C_P more comparable, an equivalent Neuber constant was suggested in the form of $C_{Neq} = \sqrt{C_N \rho}$. Figure 6.70 compares these constants for the range $\sigma_R = (500–1{,}000)$ MPa. Both constants are reasonably close in the range, $\sigma_R = (500–1{,}000)$ MPa, in particular for ρ equals 0.5 mm.

There are many other models with similar formulas to those mentioned above. These models are discussed by Yao in a review paper on the fatigue-notch factor, K_f. In Fig. 6.70, the Peterson and Neuber constants are plotted against tensile stress, σ_R.

The use of the above approaches is restricted to special cases, since the constants (C_P, C_N and C_H) are related to the ultimate-tensile stress of the materials without including fracture-mechanics constants, such as the fatigue threshold. There are many modern formulations of K_f that have been proposed, because Peterson's formula is not adequate for large K_f and the Heywood formula does not behave well for short notch-cracks. A better criterion can, therefore, be devised by using another concept known as the 'Atzori-Lazzarin [henceforth: AL] approach', given as:

$$K_f = \begin{pmatrix} \dfrac{1+a}{a_0}, \text{ for } a < a^* \\ K_t \text{ for } a > a^* \end{pmatrix} \tag{6.46}$$

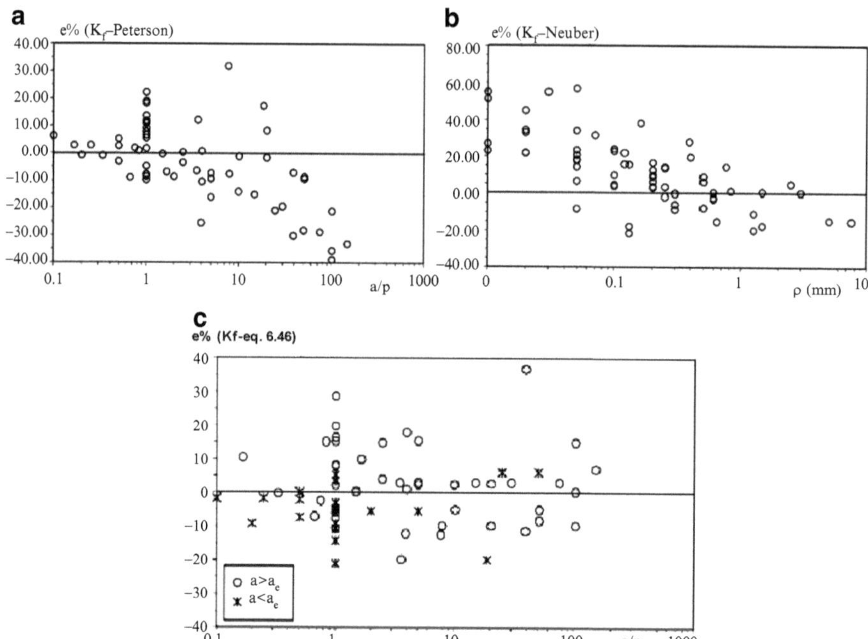

Fig. 6.71 Per cent error in *K*f prediction by using (**a**) Peterson formula (6.41) based on net section stresses, (**b**) by using Neuber formula (6.42), (**c**) *K*f prediction by using Atzori and Lazzarin criterion (6.46) based on gross section stresses (Ciavarella and Meneghetti 2004. With kind permission of Elsevier and Professor Ciavarella for the authors)

where $a^* = K_t^2 > a_c = a_0(K_t - 1)$. a_c is the characteristic notch or crack size and a_0 is a material constant used in fracture mechanics studies. As can be seen in the next illustration, the error is reduced when Eq. 6.46 is used.

The percentage of error is expressed by:

$$e\% = \frac{K_{f,predicted} - K_{f,experimental}}{K_{f,experimental}} \times 100 \qquad (6.47)$$

From Fig. 6.71, it is seen that Eq. 6.46 improves the fatigue-strength predictions with respect to the classic Peterson and Neuber approaches. Peterson's approach to high values of notch sharpness, e.g., $a/\rho > 10$, considerably underestimates K_f. Neuber's approach shows better agreement with the experimental results for $\rho > 0.5$ mm, but considerably overestimates K_f for $\rho < 0.5$ mm. Figure 6.71 also shows that the AL criterion gives a mean error of only 0.5% and a slightly higher standard deviation of 11.1%.

The exemplary data for steel, according to the classic S-N approach to notches is given in Fig. 6.72. A severe notch (i.e., with a 0.015 in. radius) at its bottom (see Fig. 6.72) will result in the steel having a much lower endurance limit.

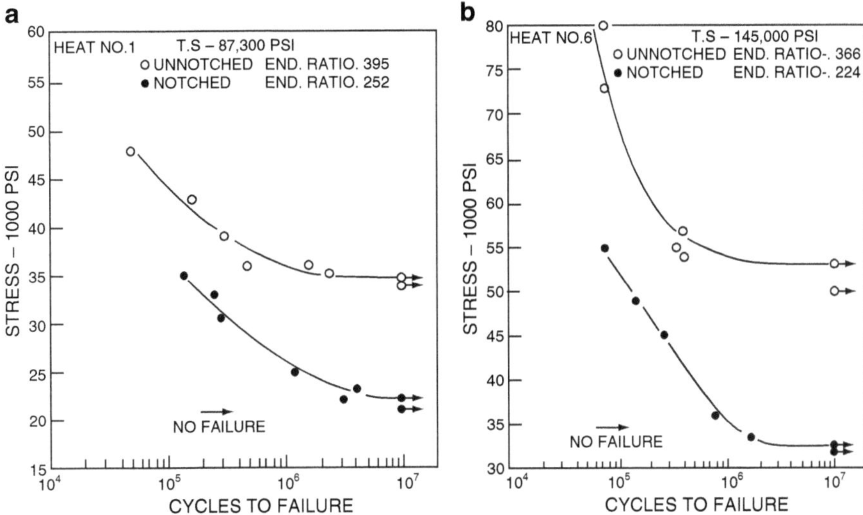

Fig. 6.72 Comparison of S-N curves of notched and smooth R. R. More specimens of Ni-Cr-Mo (8680) cast steel. (**a**) Normalized and tempered; (**b**) quenched and tempered. Notch of 0.015 in. radius (Vishnevsky and Wallace 1967. With the kind permission of the Steel Founders Association)

This reduction in the endurance limit can be ~36%, due to the presence of that notch. When the notch is made more severe, by reducing the radius at the bottom to 0.001 in., the endurance limit becomes even more reduced, in the range of 42–53%. A notch has a greater effect on quenched and tempered steel than on normalized and tempered steel. Observations indicate that the lower-strength cast steels (normalized and tempered) are less notch-sensitive (i.e., factor q) than the higher-strength cast steels (quenched and tempered). Nearly all structures contain stress raisers of one type or another. Smooth, highly-polished bars have the highest endurance limits. In order to experimentally evaluate notch-effect tests, various notch shapes and sizes were investigated at various locations, either edges of flat specimens, holes in flat plates, V-type notches in cylindrical specimens, etc. Note that all the components used for real service contain stress raisers, due to machining; thus, notch influence is studied by intentionally introducing notches in many shapes and sizes.

6.11 Failure Resulting from Cyclic Deformation (Fracture by Fatigue)

In almost all the sections discussed in this chapter, crack initiation and propagation were considered as essential steps leading to failure. Again, fatigue cracks usually start from the surface of a component, where fatigue damage begins as shear cracks on crystallographic slip planes. Surfaces show slip planes as intrusions and extrusions. This is stage I of crack growth. After a transient period, stage II crack growth takes place in a direction normal to the applied stress. Finally, the crack becomes unstable and fracture occurs.

A separate portion of Chap. 7 (devoted to fracture) will consider this aspect of fatigue. Fracture by fatigue is a major part of all failures encountered in service, where components of structural systems are operating.

6.12 The Effects of Some Materials and Process Variables

Fatigue properties are frequently correlated with tensile properties. Several parallels may be drawn between the effects of certain metallurgical variables on fatigue properties and the effects of these same variables on tensile properties. Thus, for example, most of the metallurgical processes applied to improve static properties hold true for cyclic deformation, as well. To cite one method, for instance, the common metallurgical technique for improving static-deformation features is solid-solution alloying. The effect of solute additions on the fatigue properties, e.g., of iron and aluminum, parallels nearly exactly their effect on the tensile properties. However, the greater structural sensitivity of those fatigue properties, compared with the tensile properties, demands added consideration of other material variables, those apparently causing the statistical data scatter observed in fatigue experiments (and in the service of machine components). It must be emphasized that the change in the stress-strain characteristics, when the direction of deformation is reversed, occurs in both single-crystal and polycrystalline materials, although this subject has not been given separate consideration. The greatest changes in material behavior due to cyclic action occur during the early life of a specimen; indeed, the first cycle may indicate the extent to which cyclic softening or hardening may eventually occur. The following is a brief discussion of some material properties, external and internal.

6.12.1 Surface Effects on Fatigue

Many investigations have been conducted in the past on the effects of the surface roughness on fatigue strength. Since fatigue cracks generally initiate on the specimen surface, the surface roughness of a specimen has a great effect on its fatigue strength. Surface roughness can originate from machining; to reduce or eliminate its effect as a stress raiser, surface-finishing effects of various kinds are introduced into the manufacturing process. Surface roughness, such as scratches of various sizes, act as micronotches with similar effects on fatigue-strength properties as notches (discussed earlier). There are certain metallurgical factors that must be considered in order to ensure the best fatigue performance of a particular metal or alloy, but these are common to the static and dynamic properties of materials. Material properties are of great importance and advanced manufacturing practices influence fatigue performance by affecting the intrinsic-fatigue strength of material near the surface, by introducing or removing residual stress in the surface layers and by introducing or removing irregularities on the surface that act as stress raisers. Important material properties to be considered are: grain size, grain boundaries, temperature, size, etc. Some metallurgical effects common to static- and dynamic-strength properties are listed as: alloying, adding second phases (dispersion of a second phase), work hardening and heat treatment. Alloying modifies the chemical composition. Second-phase particles dispersed in a metal, beside their known effect on strength, may change crack propagation in a fatigued test specimen, due to the strain in their vicinity and due to their influence on the nature of the bond. In addition, work hardening and heat treatment, both of which improve tensile properties, also increase fatigue strength. Among the effects of manufacturing practices on fatigue, one must consider: machining, drilling, grinding, surface compression and coating (plating). Machining cuts, scratches or other residual marks promote fatigue failure. Holes introduced by drilling clearly reduce fatigue strength merely by their presence. Grinding must be such that surface irregularities, introduced by machining, are removed to produce a smooth surface. The most effective surface-smoothing operation is electropolishing. Introducing compressive-residual stress on the specimen surface increases fatigue life. Among the well-known techniques are shot peening (not always beneficial) and the coating of hard layers to improve strength by introducing compressive-residual stress. However, soft coating impairs the fatigue properties. Nitriding or carburizing has the effect of a hard coating, but decarburizing softens the surface, thus its influence is similar to a soft coating. All these, from the material properties through the manufacturing process, result in modifying surface properties by changing the characteristics of the stress raisers. More detailed discussion will follow.

6.12.2 The Residual Stresses

As mentioned in previous sections, cracks always form on highly-stressed, free surfaces. The sign and level of residual stress on a surface layer determine the ease with which the crack-initiation process occurs. Residual-compressive stress should increase while tensile stress decreases the fatigue limit of fatigued materials. Manufacturing processes applied when forming machine components for engineering use usually generate residual stresses unintentionally. However, some processes are applied to introduce residual stresses intentionally. These stresses induce local-plastic deformation and often small cracks can form and grow if the object is exposed to cyclic stresses. It was observed that, in regions of compressive-residual stresses, fatigue-crack growth is retarded, while tensile-residual stresses promote crack opening and propagation. Eber studied crack closure and stated that this phenomenon is a direct consequence of the permanent tensile-plastic deformation following a propagating crack. Such plastic deformation results in a compressive

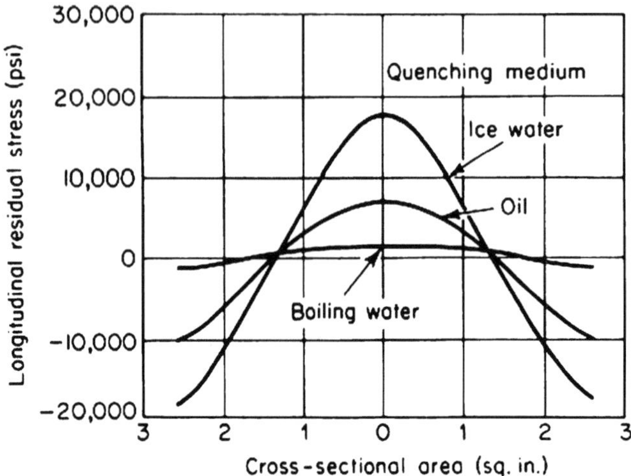

Fig. 6.73 The effect of quenching medium (quench severity) on the magnitude of the residual stress and its variation along the cross-sectional area (From Baldwin 1949)

residual-stress field. Surface-residual stresses introduced, for example, by shot peening, play a dominant role in retarding microcracks formed by fatigue, thus, determining total fatigue life. The presence of residual stresses is known to affect both the initiation and growth stages of fatigue cracks by means of the effective-mean stress. Tensile-residual stresses add to the applied-stress levels and lead to fracture at lower loads than would occur in their absence. Non-uniform deformation during working induces material flow and is a major source of residual stress. Differential-cooling rates in machine elements in adjacent regions are other sources of residual-stress fields and may possibly cause crack formation of which designers should beware. The temperature-induced strain which arises during cooling or heating is proportional to the thermal coefficient of expansion (contraction) of a material. These thermal stresses are dependent on the size of the material or a section of it, since the larger the section, the larger the stress. An illustration of the influence of section size in various quenching media may be seen in the following diagram (Fig. 6.73) reproduced from Baldwin's residual stresses in metals.

6.12.3 Introduction to Residual Stresses

A few methods are considered below for the modification of the surface state of a material. Residual stresses can be either compressive or tensile. Methods introducing residual-compressive strength are considered first, because (as mentioned in earlier sections) they can improve the fatigue life of materials. This is followed by a short discussion on the detrimental effects of tensile-residual stresses.

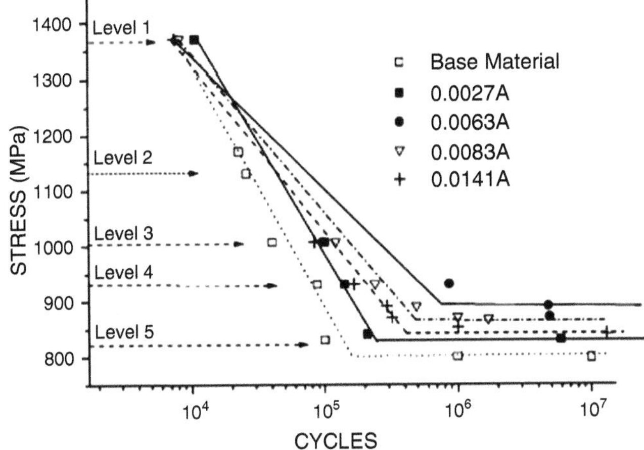

Fig. 6.74 S-N curves of a AISI 4340 steel shot peened are compared with the base metal (Torres and Voorwald 2002. With kind permission of Elsevier)

6.12.3.1 Shot Peening

The effect of shot peening in an AISI 4340 steel is indicated in Fig. 6.74 at various conditions. Here, the base metal is compared with a shot-peened steel surface.

This steel surface was treated with four intensities of shot peening in the range of 8–45 psi (0.0027A-8 psi; 0.0063A-13 psi; 0.0083A-18 psi and 0.0141A-45 psi). The specimens were tested in rotating, bending fatigue tests ($R = -1$) at a frequency of 50 Hz. Figure 6.74 shows that the shot peening improved the stress level and the endurance limit of the specimen compared to the base material. This improvement is a result of the compressive-residual stress induced to the surface. Crack formation is pushed below the surface, as can be seen in Fig. 6.75a, b. Crack initiation is marked by arrows. Since the compressive-residual stress field is bigger in the initial layers, it is expected that the crack source will be some distance from the surface, where the applied-tension stress can balance the compressive-stress effect created by the shot peening. Figure 6.76 illustrates a case in which the crack originated from the surface. All the specimens with shot peening in low-cycle fatigue or without shot peening had cracks that originated from the surface.

This may be explained by the fact that high, applied-tension stress always surpasses residual-compressive stress in low-cycle conditions. The experimental data regarding the effect of shot peening on fatigue-crack formation are summarized in Fig. 6.77. In this illustration, crack-source formation at and under the surface is shown as a function of the number of cycles to failure. The increase in shot-peening intensity resulted in increased residual-compressive stress, but the surface-residual stress was nearly independent of the peening conditions. Shot-peening treatment pushes the crack source beneath the surface in most of the medium- and high-cycle cases. The compressive-residual stress field improves the fatigue life of the AISI 4340 steel.

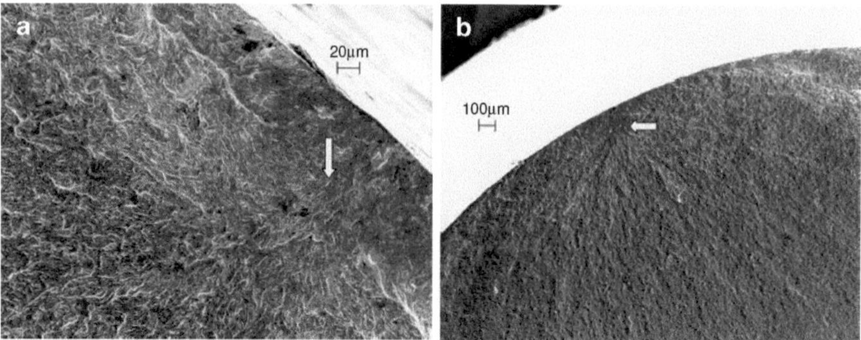

Fig. 6.75 Fatigue fracture surface: (**a**) of specimen 7D, (**b**) of specime 9D. The crack sources are beneath the surface (Torres and Voorwald 2002. With kind permission of Elsevier)

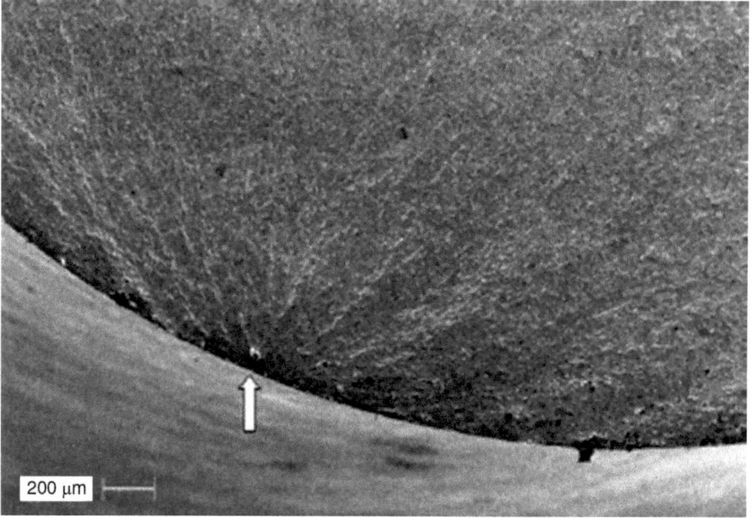

Fig. 6.76 Fatigue fracture surface of specimen 9B. The crack source is at the surface as indicated by the *arrow* (Torres and Voorwald 2002. With kind permission of Elsevier)

Introducing compressive-residual stress into materials other than steel has a similar effect on fatigue-life improvement. An example of the effect of shot peening on non-ferrous materials is found in the improvement of the fatigue life of wrought magnesium AZ 80, as reported by Zhang and Lindemann. Fatigue strength increased from 100 to 160 MPa under optimum conditions and an improvement of about 60% resulted. Figure 6.78a compares the microhardness-depth profile, while Fig. 6.78b shows the variation in stress over the number of cycles to failure.

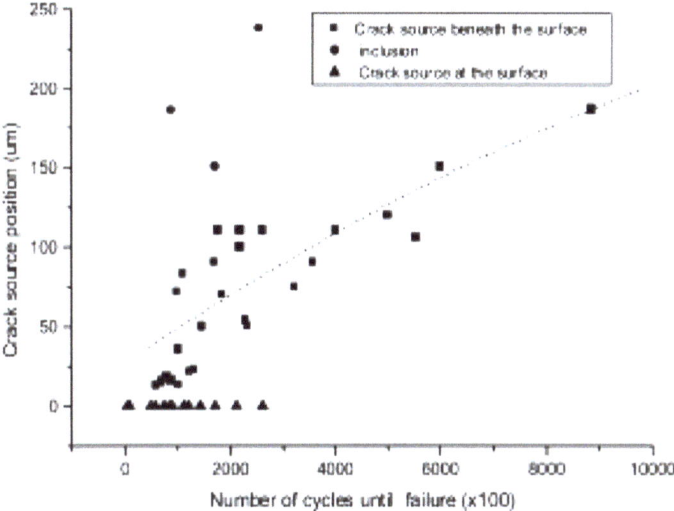

Fig. 6.77 Crack source position against the number of cycles to failure (Torres and Voorwald 2002. With kind permission of Elsevier)

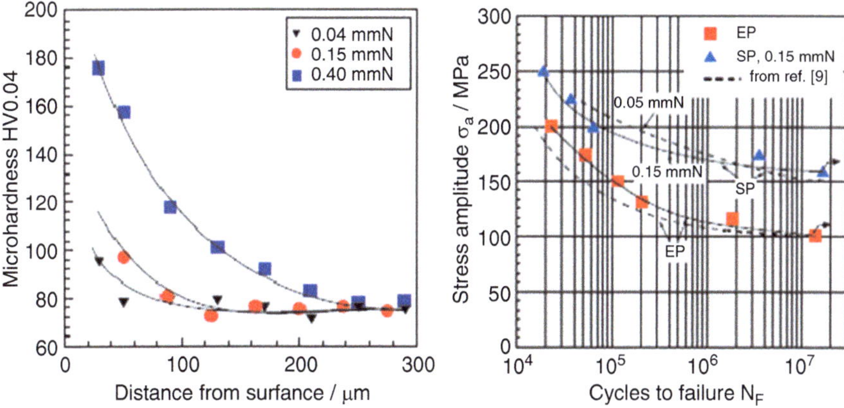

Fig. 6.78 Shot peening of AZ80 magnesium alloy: (**a**) Microhardness depth profile with after shot peening different Almen intensities indicated. (**b**) S-N curve (i.e. stress life) after optimum shot peening conditions (An Almen strip is a thin strip of SAE 1070 steel used to quantify the intensity of a shot peening process) (Zhang and Lindemann 2005. With kind permission of Elsevier)

Laser shot peening has been developed as a surface treatment of materials. It has been reported (see, for example, Hammersley et al.) that the introduction of surface-residual stress is comparable to conventional shot peening, but reaches a much greater depth. Laser shot peening is compared with conventional shot peening in Fig. 6.79 for inconel 718.

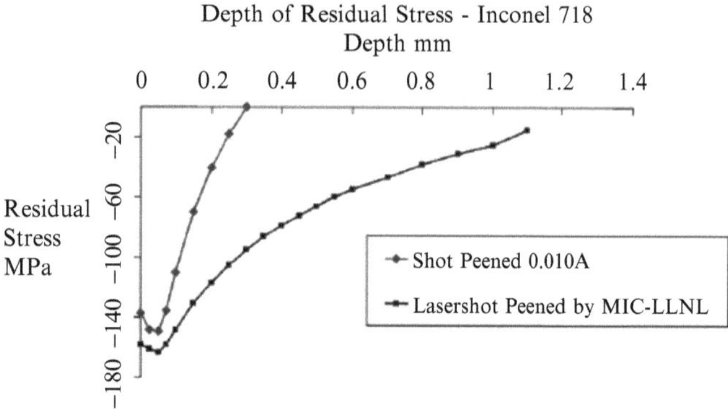

Fig. 6.79 Comparison of the he residual stress profile depth with the conventionally shot peened inconel 718 (Hammersley et al. 2000. With kind permission of Elsevier)

Similar tests of laser shot-peened fan blades of Ti-6Al-4V have been performed; their fatigue strength was reported to double the strength, compared with blades which had been conventionally shot peened.

Shot peening, like any other surface technique applied to improve fatigue life, is basically the introduction of compressive-residual stresses to surfaces.

6.12.3.2 Case Hardening

Case hardening refers to the hardening of metal surfaces, usually of low-carbon steels. This technique involves the introduction of an element into a surface by means of diffusion, forming a thin, hard layer. In the case of steel, when the hardening element is carbon, the term ‘carburizing’ is used. The hardening element can be nitrogen or boron. Examples of carburizing and nitriding to improve fatigue life are considered below.

Carburizing

Carburizing is of great importance for the development of fatigue resistance and good wear in steels. The relative depth of the case-hardened material is an important parameter in determining fatigue performance. The effect of relative case depth on fatigue performance is shown in Fig. 6.80, where stress is plotted versus the number of cycles for various case depths. The lowest stress level is for zero-depth, hardened steel, i.e., never carburized. Effective case depth, t (mm), is given by:

$$t = 0.41[\text{Carburizing Time, (hours)}]^{1/2} \tag{6.48}$$

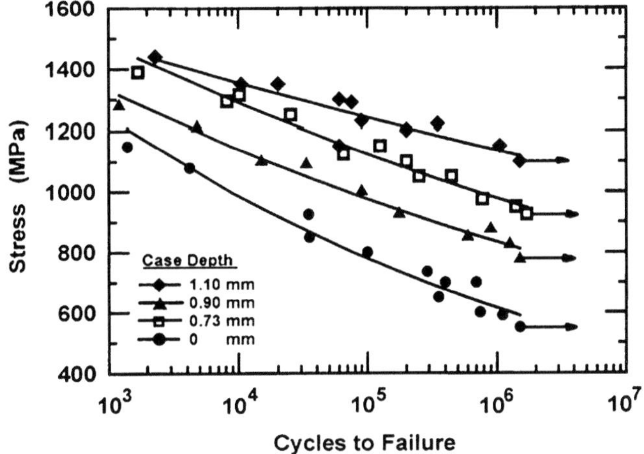

Fig. 6.80 Fatigue curves of heat treated and carburized AISI 8620 steel (Genel and Demirkol 1999. With kind permission of Elsevier)

An attempt was made to construct empirical relations between the parameters defining relative-case depth and fatigue strength of the steel, based on failure at 10^6 cycles and it was concluded that a power model might be used to represent the relation between the case profile and the fatigue performance of carburized specimens as:

$$\sigma_f = 590 + 36400(t/D)^{1.92} \tag{6.49}$$

The power function was found to be the best relation, reflecting the dependency between relative-case depth and the fatigue performance of the steel. As indicated above, carburizing, as other surface treatments, induces surface-residual compressive stress.

Nitriding

Gas nitriding is a thermochemical treatment commonly used to enhance wear, fatigue and corrosion properties of mechanical components, such as: gears, crankshafts, extrusion and forging dies, valves and springs, to name a few. Gas nitriding is generally performed between 500°C and 580°C in an ammonia-containing atmosphere. Nitriding hardens parts without or with minimal distortion, as compared to carburizing. As in carburizing or shot peening, increased surface hardness and compressive-residual stresses are generated. Surface hardening and compressive-residual stresses usually increase the fatigue limits of steels. Nitriding is beneficial during high-cycle fatigue and may result in a transition from surface failure to internal failure. In Fig. 6.81, the residual stress at three levels is shown.

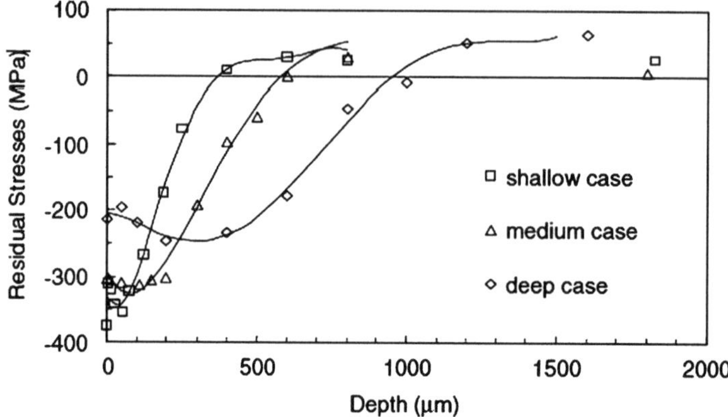

Fig. 6.81 Residual stress profile for three case depths in AISI 4140 steel (Limodin et al. 2003. With kind permission of John Wiley and Sons and Dr. Y. Verreman for the authors)

S-N data are plotted in Fig. 6.82. All the tests were conducted at 10 Hz frequency, except at 700 MPa, where it was necessary to reduce the frequency to 1 Hz in order to avoid specimen heating and premature fracture. Nitriding improves high-cycle fatigue life, whatever the case depth. In Fig. 6.82, a comparison is made between the base metal (un-nitrided) and the case-hardened surface at three levels. A shallow, nitrided case delays crack initiation and growth at the surface, but the improvement of surface fatigue life is small. Shallow-case specimens fail from the surface, except at the lowest stress levels. In medium- and deep-case specimens, the improvement of surface-fatigue life is large enough for failure to always occur from an internal crack. Medium-case nitriding is recommended for engineering applications, because deep-case nitriding is time consuming and more expensive.

6.12.3.3 Tensile Residual Stresses

The detrimental nature of tensile residual stresses decrease fatigue life considerably. Soft coatings, plating of layers or cladding various steels might be of interest in reducing environmental effects, but their side effect of reducing fatigue life can not be acceptable in parts operating under fatigue conditions. Of the factors that can modify surface conditions by introducing tensile residual stresses decarburization will be considered since it can readily occur in heat treatments of steels.

Decarburizing

Numerous reports are available in the literature on the effect of decarburization on the cyclic strength of steels. The detrimental effect of decarburization on

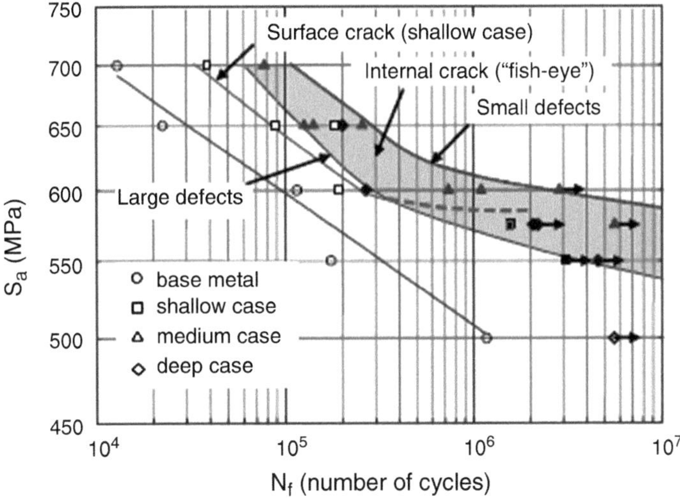

Fig. 6.82 S-N data under fully reversed loading (*open symbol*: surface failure; *full symbol*: 'fish-eye' crack) (Limodin et al. 2003. With kind permission of John Wiley and Sons and Dr. Y. Verreman for the authors)

otherwise strengthened surfaces, e.g., by shot peening, is to be noted if full advantage is to be expected from a strengthening treatment. For instance, study Fig. 6.83 to see the detrimental effect of decarburization on the fatigue limit. Here, 40CrNi2Si2MoVA steel (Chemical composition Fe-0.39C-1.82Ni-0.91Cr-1.61Si-0.69Mn-0.42Mo-0.07V-0.06Cu (wt.%)) specimens are shown after various surface treatments. Tensile-residual stresses are introduced by chromium plating and decarburization. Superimposed compressive-residual stress improves the decarburized surface, but its S-N curve still remains below the shot-peened curve. Thus, the ill effects of decarburization are remedied to a certain degree. This remedy brings the fatigue limit from a low level of 674 to 920 MPa.

In summary of this section, some techniques for surface treatment have been discussed. Techniques that introduce residual-compressive stresses increase fatigue life by pushing crack initiation to below the surface. Residual-tensile stresses, however, are detrimental and should be avoided in parts intended to operate under fatigue conditions.

6.13 Miscellaneous Variables

Various variables, such as: grain size, temperature, dimension and material type respond to environmental effects and are considered in the following.

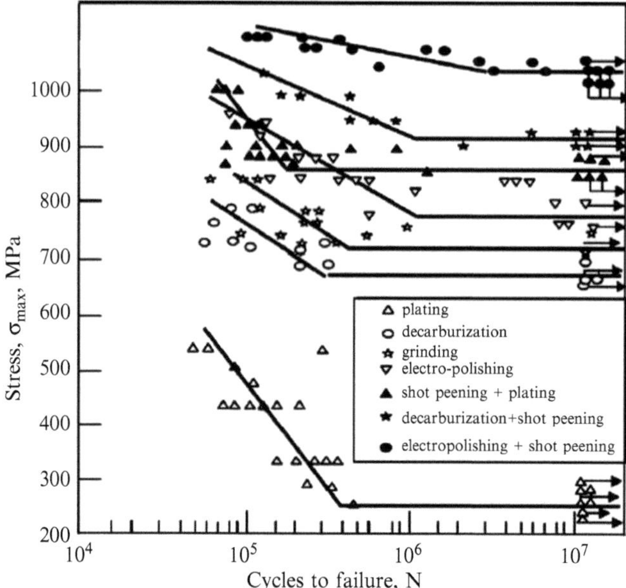

Fig. 6.83 S-N curves of 40CrNi2Si2MoVA steel with different final surface conditions (Gao et al. 2007. With kind permission of Elsevier)

6.13.1 Grain Size

Experimental results indicate that the resistance of metals and alloys to fatigue-crack initiation and propagation is influenced significantly by grain size. It is widely recognized that, when all the other structural factors are kept approximately fixed, an increase in grain size will generally result in a reduction in the fatigue-endurance limit. This observation parallels observations made in static deformation, namely that strength increases with decreasing grain size. Various grain sizes used to evaluate their effects on fatigue life are shown in Fig. 6.84. Their sizes are exemplified in Fig. 6.85, where the dependence of the fatigue behavior of ultrafine-grained AISI 304 stainless steel is given.

Clearly, decreasing grain size has a beneficial effect on the fatigue behavior of AISI 304 Austenitic stainless steel. Grain size has an effect on other materials, as well. For an example, the grain-size dependence of the fatigue properties of wrought Mg-Al-Zn alloys may noted. Its average grain size, ranging from 18 mm to 101 mm, was obtained by strain-annealing technique. Figure 6.86 illustrates the effect of grain size on the S-N curves in pure magnesium, AZ31 and AZ61 alloys. Fatigue tests at room temperature were performed under the constant-stress amplitude over a life range of 10^4 to 10^7 cycles. It was found that the grain-size dependence of the fatigue strength for pure magnesium and AZ31 alloys spans a life range of 10^4 to 10^7 cycles.

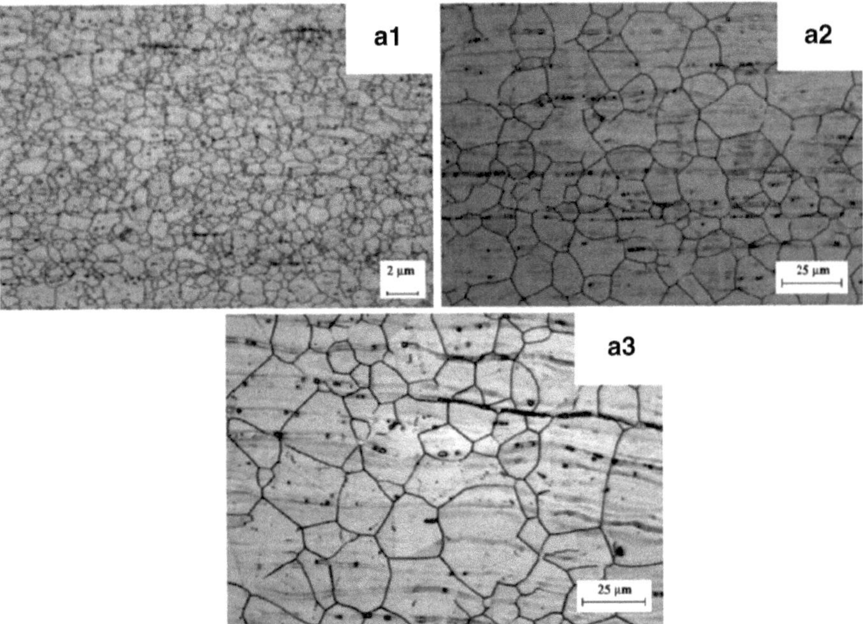

Fig. 6.84 Microstructures of the high-nitrogen austenitic steel samples annealed at 850°C for 30 s (*A1*), 1,100°C for 3 min (*A2*) and 1,100°C for 10 min (*A3*) (Schino and Kenny 2003. With kind permission of Elsevier)

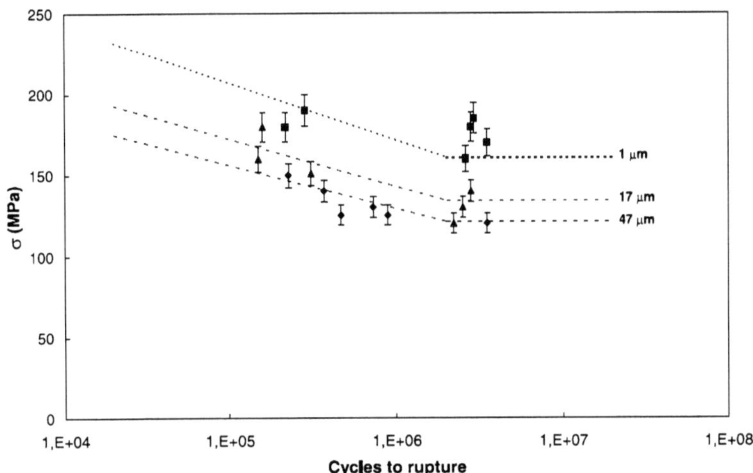

Fig. 6.85 Fatigue resistance of AISI 304 steel as a function of grain size (Schino and Kenny 2003. With kind permission of Elsevier)

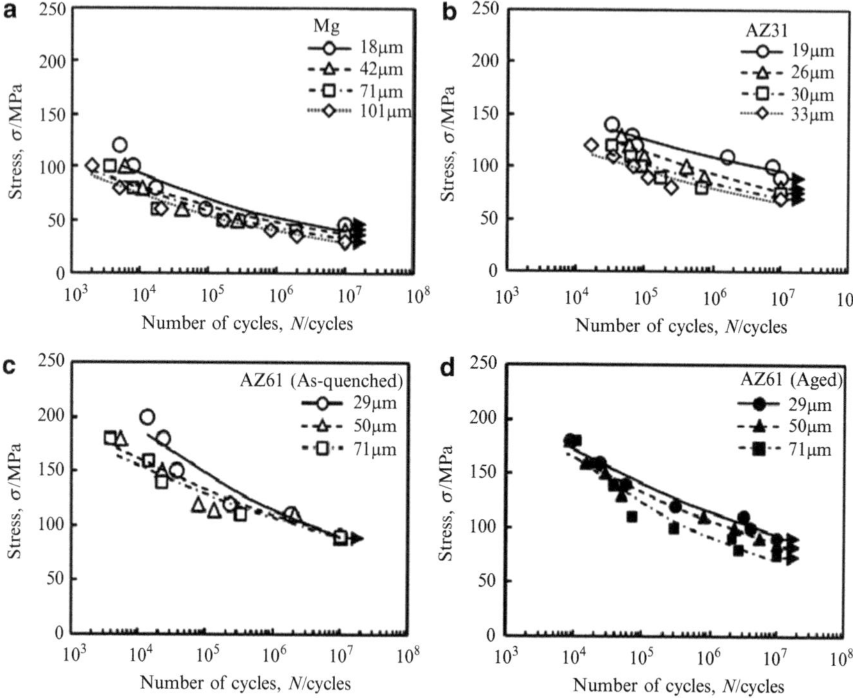

Fig. 6.86 The effect of variation in grain size on the S-N curves: (**a**) pure Mg, (**b**) alloy AZ31, (**c**) As quenched AZ61, (**d**) aged AZ61 (Ono and Nishimura 2009. With kind permission of Nagato Ono and Yoshito Nishimura)

The improvement of fatigue properties by means of grain refining is in good agreement with other publications on this subject. Fatigue-crack propagation and fracture were also examined in each specimen. The results of the above research suggests that the slip bands and the twin boundaries, as well as the grain boundaries, have significant effects on fatigue fracture, since they prefer to appear in the coarse grains, where local-stress concentration exists. This is due to the operation of a group of piled-up dislocations against the grain and twin boundaries under cyclic loading. However, the effect of grain size on the fatigue properties for as-quenched AZ61 alloys decreases with an increase of the number of cycles and then disappears beyond $\sim 10^5$ cycles, as shown in Fig. 6.86c. The grain-size independence at the region of long life is supposed to be associated with early precipitation during fatigue tests. Again, the resemblance between static and fatigue deformations has been pointed out. The grain-size influence on fatigue behavior may be expressed in terms of the Hall-Petch expression, relating grain diameter, $d^{-1/2}$, to the stress. In the present case, grain diameter is related to cyclic stress, as seen in Fig. 6.87 for all the Mg alloys tested. As observed in static deformation, this relation is linear, indicating an improvement in the strength properties with decreasing grain size (excluding the as-quenched AZ61 alloy).

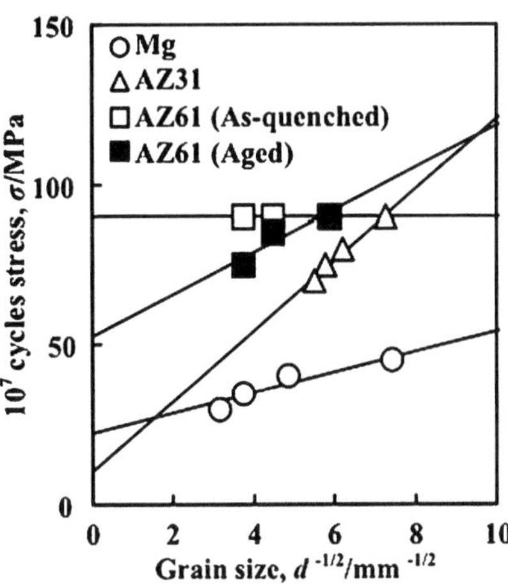

Fig. 6.87 Grain-size dependence on fatigue strength at 10^7 cycles for specimen of each Mg and its alloys indicates a Hall-Petch relation (Ono and Nishimura 2009. With kind permission of Nagato Ono and Yoshito Nishimura)

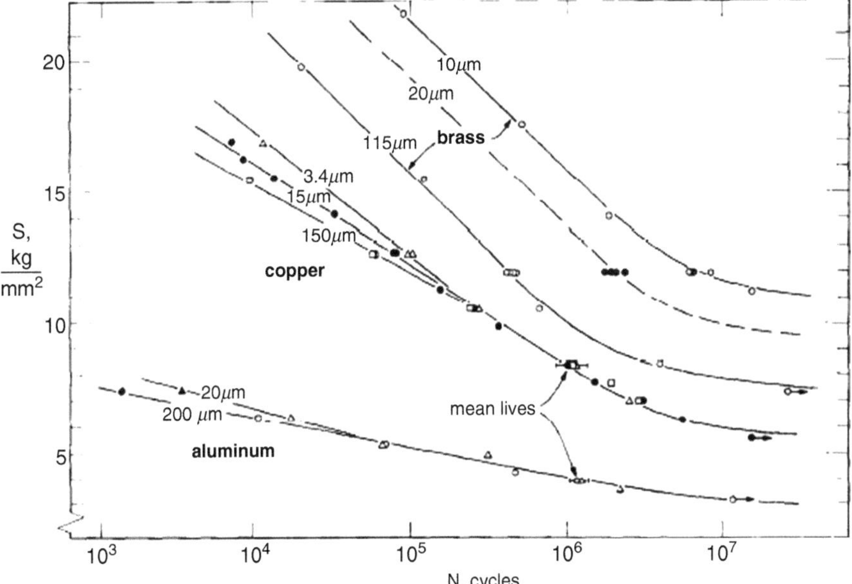

Fig. 6.88 S-N curves for copper, brass and aluminum. For copper and aluminum the curves converge at about 10^5 cycles (Thompson and Backofen 1971. With kind permission of Elsevier)

Thompson and Backofen investigated grain-size effect on fatigue. Their results are reproduced in Fig. 6.88 for several metals, such as alpha brass, copper and aluminum. This test was performed at constant-stress amplitude over a life-time of 10^4–10^7 cycles. The grain-size dependence at constant-stress amplitude during the long lives of aluminum and copper does not agree with that of brass. Thompson and Backofen claim that this difference is due to cross slip. In alpha brass, cross slip is difficult and decreasing grain size acts to increase fatigue life. However, in

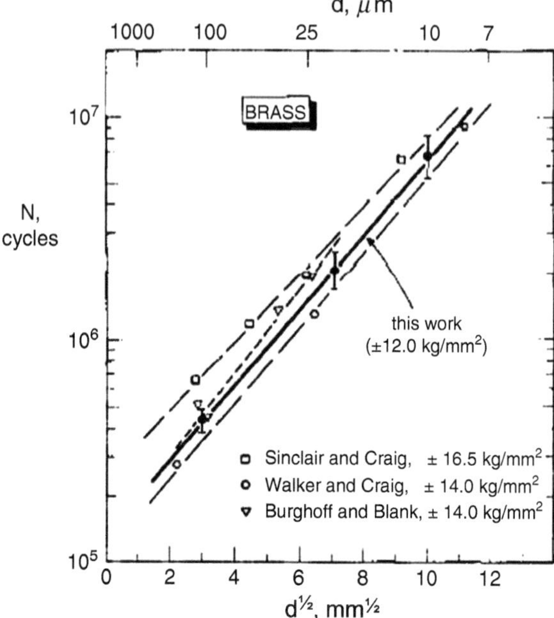

Fig. 6.89 Fatigue life for brass as a function of grain size is shown (Thompson and Backofen 1971. With kind permission of Elsevier)

copper and aluminum, in which cross slip is easy, material life is insensitive to grain size. It is argued that the cellular substructure in fatigued copper and aluminum has the effect of masking the grain-size effect. In brass, however, the absence of such a structure causes grain boundaries to affect cracking rate. Thus, the cracking rate in brass becomes grain-size sensitive. In Fig. 6.89, the number of cycles, as a function of $d^{-1/2}$, is plotted. A Hall-Petch type relation is obeyed in fatigue deformation as well and the linear relation may be expressed by:

$$\sigma = \sigma o + kd^{-1/2} \tag{6.50}$$

σ can mean either the endurance limit or fatigue strength; σ_0 and k are constants.

Therefore, grain refining is a recommended method for improving the fatigue behavior of materials. Nonetheless, one should bear in mind that grain size is not realized in all cases and in all materials, because other factors may mask its influence.

6.13.2 *The Effect of Temperature*

Fatigue strength is generally increased by any heat treatment that increases tensile strength. However, in this section, the focus is not on the thermal aspects of heat treatment and how they modify material properties, but rather the effect of

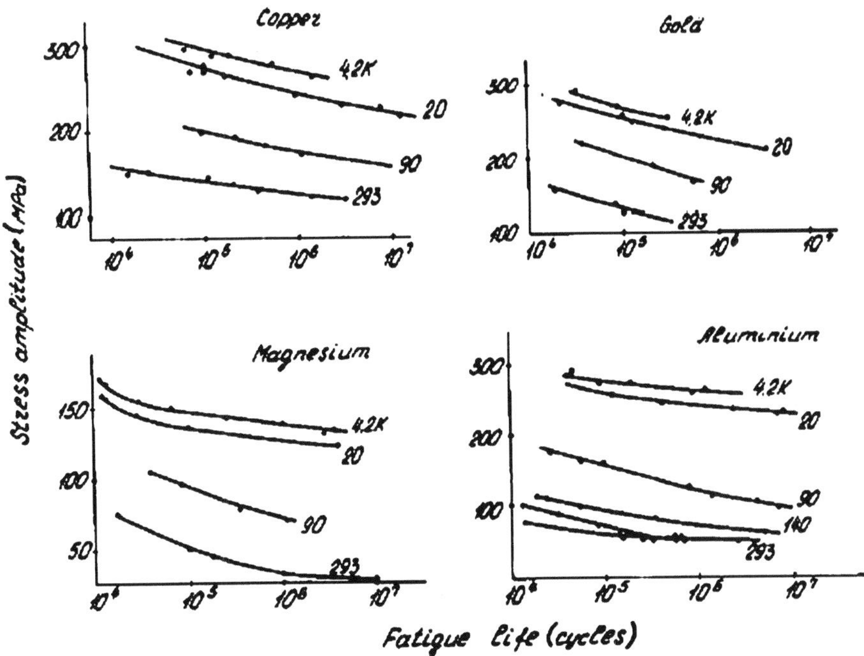

Fig. 6.90 Fatigue life as a function of stress amplitude for copper, gold, magnesium and aluminum below 293 K at the temperatures indicated (Verkin et al. 1983. With kind permission of Elsevier)

temperature directly on fatigue behavior. In general, as the temperature is lowered, the S-N curves move to higher stress levels; however, the effect of temperature depends on its level:

(a) **The effect at low temperatures**
Figure 6.90 shows the effects of temperature, below and at room temperature, on the S-N curves of several metals. It can be seen that the stress level of the metals tested at a specific cycle is increased with decreasing temperature. An exception is at 4.2 K for copper and gold, where, although the stress level is the highest, the cycles are only in the 10^5–10^6 range. It was reported that dislocation structures in FCC metals at low temperatures are very sensitive to SF energy and applied stress (but less to temperature). This was supported by the fact that, in copper, the dislocation-cell size is independent of temperature. These results indicate that a drop in the temperature of cyclic strain affects the substructure of FCC metals and that this variation has the same characteristics as in the effect of SF energy. This suggests that cross slip becomes more difficult at lower temperatures.

(b) **The effect at high temperature**
Increasing test temperature above room temperature generally decreases fatigue resistance. An exception is found in the fatigue observations of steel (considered below). An example of the effect of temperature on fatigue appears in E319-T7

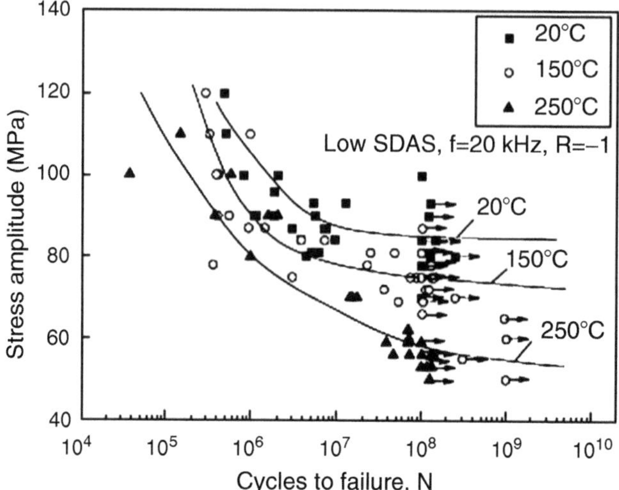

Fig. 6.91 Effect of temperature on fatigue resistance of an E319-T7 cast aluminum alloy (SDAS in the figure means secondary dendritic arm spacing) (Zhu et al. 2006. With kind permission of Elsevier)

cast aluminum alloy tested at ultrasonic frequencies, as illustrated in Fig. 6.91. The endurance limit at 10^8 cycles was decreased to 10 MPa from 20 MPa at 150°C and to 17 MPa from 150 MPa at 250°C. The dependence of the endurance limit, when microstructure is held constant, may be related to the temperature dependence of yield strength or tensile strength. One sees that there is considerable spread in the S-N data at each test temperature. This variation is likely to originate from the complex microstructure of the alloy.

The effects of temperature on steel are complex; Fig. 6.92 illustrates this for smooth and notched steel St.45 in the temperature range of 20–600°C. The fatigue limit of medium-carbon steel at 200–400°C is usually 18–30% higher than at room temperature and decreases at higher temperatures. This unusual observation is likely to be related to strain aging (or dynamic-strain aging), since the temperature is almost the same as the strain-aging temperature in steel.

During a strain aging-related increase in S-N curve level, the pinning of dislocations should occur more readily than at lower temperatures. The C atoms are sufficiently mobile, by diffusion to dislocation sites, to lock the dislocations and prevent their motion.

6.13.3 Specimen Size

Experimental investigations have proven that tensile and fatigue strength decrease with increasing specimen size. Griffith, in his studies on glass filaments, introduced

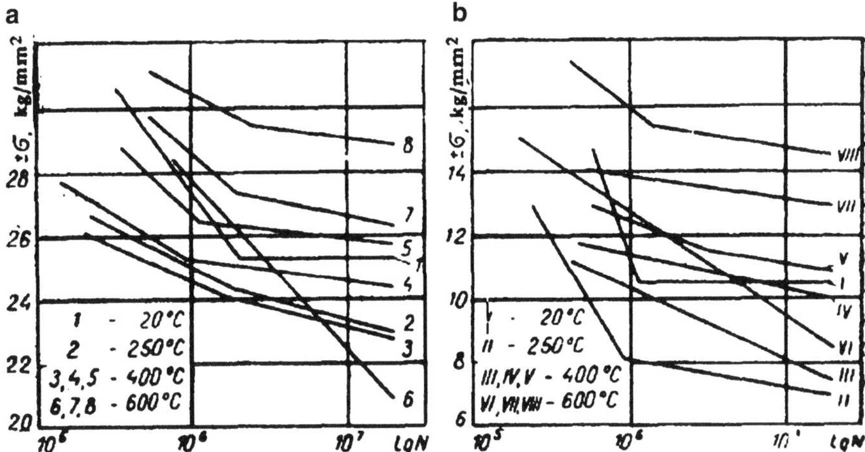

Fig. 6.92 Effect of temperature on the endurance limit of steel St. 45 specimens at the temperatures indicated. (**a**) Smooth specimens, (**b**) notched specimens (Nikolin and Karpenko 1967. With kind permission of Springer)

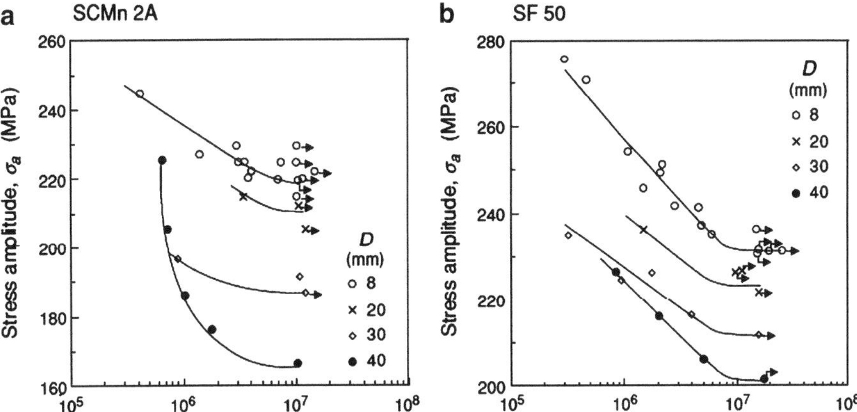

Fig. 6.93 S-N curves for two steels for different values of specimen diameter (Carpinteri et al. 2002. With the kind permission of John Wiley and Sons)

the concept that the size of a microcrack is proportional to the diameter of the filament. Thus, the volume of a specimen is related to crack initiation; the larger the specimen, the more likely it is to fail (i.e., the probability of failure increases with size). In Fig. 6.93, the effects of size on the S-N curves of two steels are shown. In both these steels, fatigue strength decreases by increasing specimen size. The fatigue limit decreases when D is increased from 8 to 40 mm, 24% for the SCMn 2A steel and ~13% for the SF 50 steel. This indicates that the fatigue strength for a given material is not a constant parameter, but depends on size when all other factors

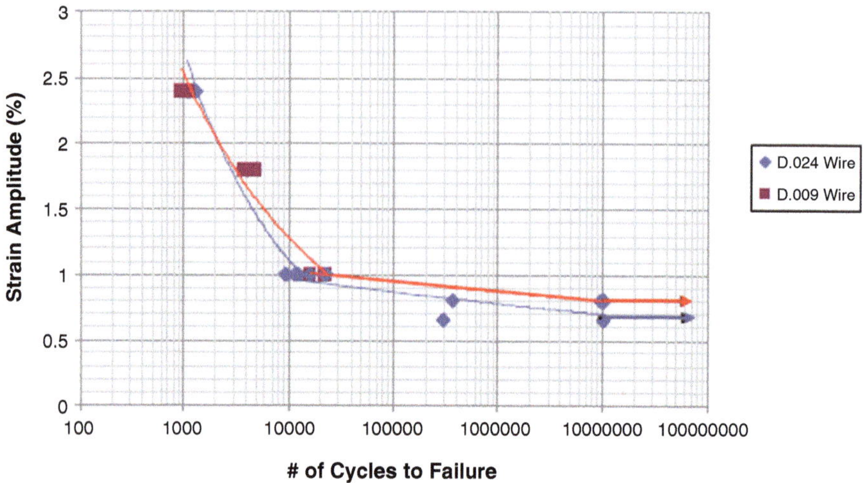

Fig. 6.94 S-N curve in terms of strain amplitude to failure for two diameters of heat treated TiN wires (Norwich and Fasching 2009. With kind permission of Springer)

remain unchanged. This size effect is of considerable importance in large structures, since it may lead to sudden fatigue failure if it is not taken into account during the design phase, to ensure safe operation under fatigue conditions.

The size effect is general, not specific to certain materials. In the following examples (shown in Figs. 6.94 and 6.95), the size effect is shown for TiN alloys. In Fig. 6.94, S-N curves are illustrated in terms of strain amplitude for two wire diameters. In Fig. 6.95, fatigue strength is indicated for several diameters. The tendency is the same as shown for the two steel specimens in Fig. 6.93.

6.13.4 *The Environment*

Fatigue life (i.e., crack growth) is influenced by the environment in which a machine component is operating, often in humid ambient conditions or perhaps in water with various NaCl content (e.g., ships). Therefore, tests have also been performed in various environments, in order to obtain valuable information on fatigue resistance under such conditions. Experiments have indicated that the fatigue resistance in vacuum conditions is the best when compared with those performed in other atmospheres. An example of crack growth in ambient and vacuum conditions for DIN Ck 45 steel (composition: 0.41C, 0.76Mn, 0.08Ni, 0.01Ti, 0.19Cu, 0.23Si) at various ratios is illustrated in Fig. 6.96. The Paris equation is used to evaluate this relation, expressed as:

$$\frac{da}{dN} = C\,(\Delta K)^m \tag{6.51}$$

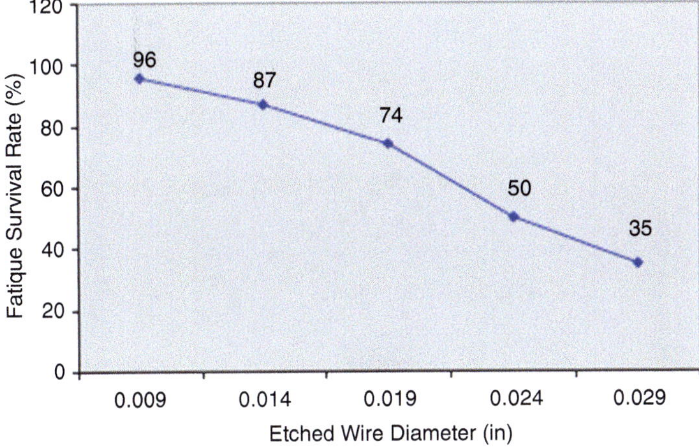

Fig. 6.95 Fatigue survival rate vs. etched TiN wire diameter beyond 80 k cycles at 0.8% strain (Norwich and Fasching 2009. With kind permission of Springer)

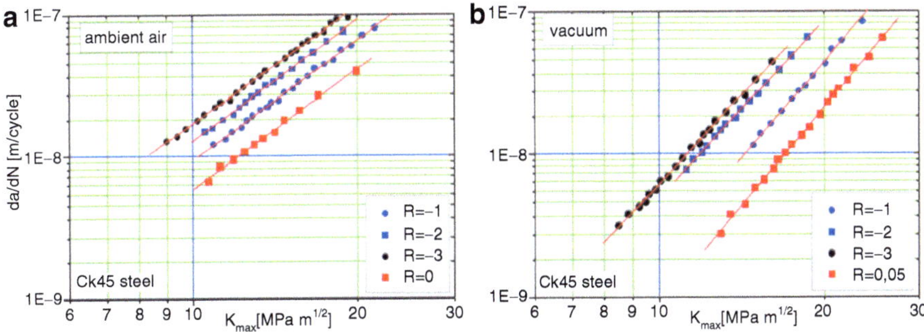

Fig. 6.96 Crack growth in (**a**) ambient at $R \leq 0$ and (**b**) vacuum at $R < 0$ (de Freitas et al. 2006. With kind permission of Professor M. de Freitas)

ΔK is the stress intensity factor (see Sect. 6.82 relations (6.34), (6.35), (6.35a)), da/dN is the crack-propagation rate and m is an appropriate exponent:

$$\Delta K = K_{max} - K_{min} \text{ for } R > 0 \tag{6.35c}$$

and:

$$\Delta K = K_{max} \text{ for } R \leq 0 \tag{6.35d}$$

Since the plot of Fig. 6.96 is for $R < 0$, the abscissa indicates K_{max}. It is clear that the life-time is better in vacuum, because the crack-growth rate (which controls the life-time) is significantly better than in ambient conditions. In another example, the detrimental effect of water vapor is shown to be even greater than in an

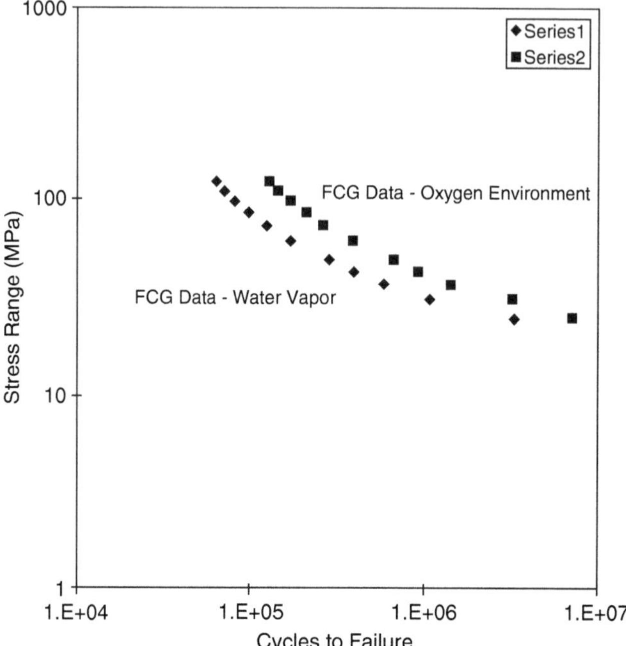

Fig. 6.97 Predicted stress-fatigue life for aluminum alloy 5456 H116 in the environment of water vapor and oxygen (Menzemer and Srivatsan 1999. With kind permission of Elsevier)

oxygen environment. In Fig. 6.97, a comparison is made of the S-N curves in the environmental fatigue tests with those of an aluminum alloy 5456. The fatigue-crack growth rates were found to be nearly three times greater in the intermediate-growth region in the laboratory environments having air and water vapor than in the oxygen environment. Crack-growth rate is expressed by Eq. 6.51.

6.14 Thermal Fatigue

Regardless of the origin of stress when cycling is applied, fatigue damage may result. Dangerous stress cycling is associated with thermal effects. Premature failure, resulting from cyclic stresses due to temperature changes, seems to be one way of looking at the problem of thermal fatigue, which may easily occur in high-temperature structural components. This problem is of great concern in many fields, in which structural components must operate at high temperatures, but it is of special importance for aeronautic gas-turbine blades and vanes, where internal cooling of hollow airfoils, to permit higher gas temperatures, exists. At high service temperatures, thermal gradients result and the propensity for thermal-fatigue cracking increases.

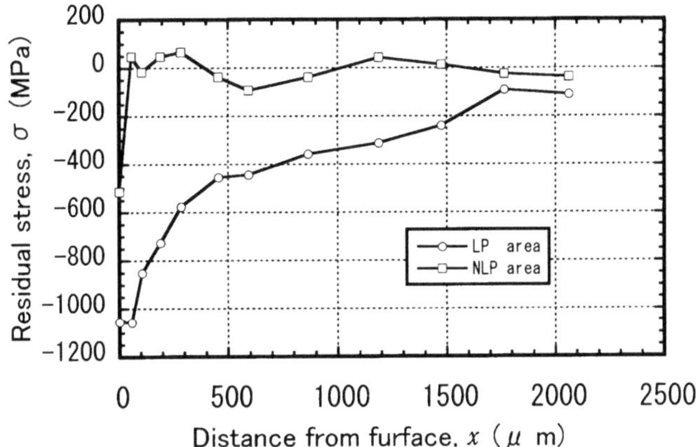

Fig. 6.98 Distribution of residual stress (Yatsushiro et al. 2003)

Thermal fatigue causes considerable softening of the surface layers of materials. Tool steels are no exception to this softening. Various hardening techniques are used to inhibit thermal-fatigue damage. The application of compressive stress and surface hardening are effective methods for preventing thermal fatigue. It is likely that their inhibiting effects are associated with both delayed- crack nucleation and crack growth, due to coatings that provide high, hot hardness and, at the same time, high residual-compressive stresses. High hardness reduces wear, possibly increasing the life-time, an important factor in materials, such as tool steels. A nitrided layer, for example, between TiN and tool steel is beneficial to low-cycle fatigue resistance. Among the methods for surface modification, shot peening is another technique used to improve the fatigue properties of mechanical parts. Usually, since the depth of residual stress is only 10 mm in quenched steel (e.g., hot-work dies), shot peening with nitriding are effective processes for application to hot-work dies. Modern shot-peening techniques, such as laser shot peening, may achieve more depth than mechanical shot peening and, thus, provide residual-compressive stress. Increased fatigue strength can be obtained. An example of improvement attained by the laser shot peening of hot-worked die steel is shown below (Yatsushiro) in Fig. 6.98. Here, LP indicates laser shot peening and NLP stands for no laser shot peening.

Figure 6.98 shows the residual-stress distribution from the surface to the inside of the specimen. In the LP area, the residual stress on the surface is over −1,000 MPa, which gradually decreases as the distance from the surface increases and achieves 0 MPa at 1,800 μm. In the NLP area, due to machining, the residual stress on the surface is about −500 MPa. Residual stress was not detected beyond 50 mm, which was reached by electropolishing. Figure 6.99 shows the changes in residual stress during the thermal-fatigue test. In the LP area, about −1,100 MPa high-compressive stress existed before the test. During the thermal-fatigue test, compressive stress decreases drastically to −600 MPa at 10^1 cycles. After these cycles, the residual stress decreases slightly as the test cycles increase. Finally, the residual stress

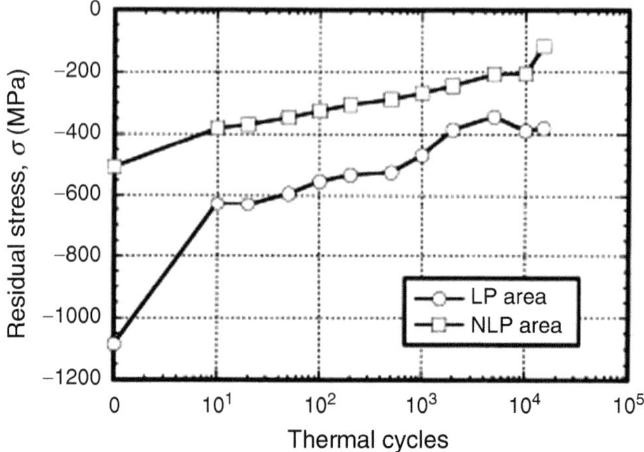

Fig. 6.99 Changes of residual stress during the thermal fatigue test (Yatsushiro et al. 2003)

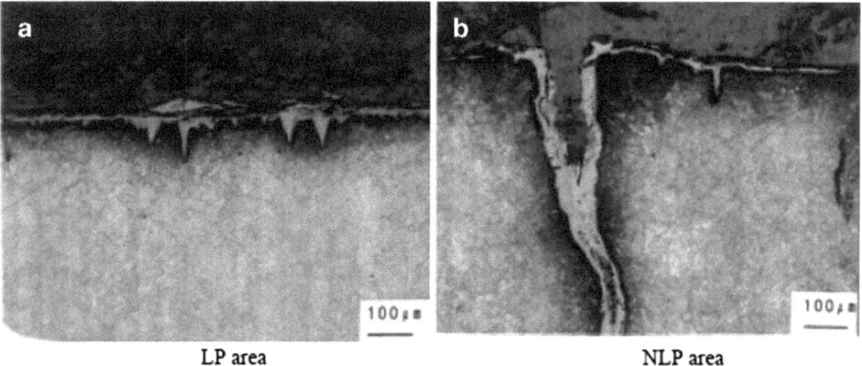

Fig. 6.100 Microscopic observation of sectional area of specimen after the thermal fatigue test (Yatsushiro et al. 2003)

converged at about −400 MPa. Nonetheless, in the NLP area, approximately −500 MPa residual stress existed before the test. The residual stress decreased slightly as the test cycles increased. After the 101 cycles, though the same tendency appears in each area, the residual stresses in the LP area were ~200 MPa higher than in the NLP area at every one of the test cycles.

Figure 6.100a, b show microphotographs of sectional areas for each specimen. In the LP area (a), several cracks of 100 μm length, containing oxide, were observed; a large number of fine cracks of 10–20 μm length were also observed. In the NLP area (b), cracks too wide for microscopic observation were also observed.

Figure 6.101 shows the curves of the residual-stress distributions for each area after thermal fatigue. The residual-stress result for the specimen before the test

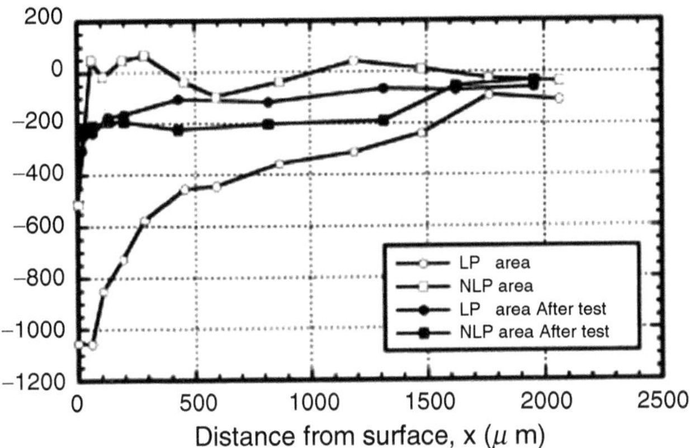

Fig. 6.101 Distribution of residual stress for the specimens after the thermal fatigue test (Yatsushiro et al. 2003)

is shown, as well. The residual stress in the LP area on the surface was about −200 MPa; it decreased slightly as the distance from the surface increased, similar to the tendency in the NLP area. Though, in the LP area, the residual stress became constant at ~400 mm below the surface, in the NLP area, the residual stress was approximately −200 MPa between the surface and ~1,300 mm. High compressive-residual stress, over −1,000 MPa, was observed on the surface of the LP area. During the thermal fatigue test, residual stresses in the LP area were ~200 MPa higher than in the NLP area for all the test cycles. The increase in crack resistance due to LP treatment was confirmed by microphotograph observation and crack measurement. Furthermore, the maximum crack length decreased to 170 μm, compared to the 1,800 μm crack length in the NLP area. The existence of residual stress in each area corresponds to the depth of the crack length in each area.

Another example of thermal fatigue is given below for 4Cr2NiMoV steel, which is used as a hot-forging die for shaping red-hot metal. This die must bear large compressive stress, local-tensile stress and additional bending stress, because of the severe impacts of the punch. The working-surface temperature may reach 300–400°C and even 600–800°C at some parts as the die contacts the red-hot metal (1,050–1,150°C). After each forging, these dies are cooled by sprayed water, such that the die surfaces undergo frequent shocks. Therefore, the failure of hot-forging dies is mainly due to wear and thermal cracks.

Thermal fatigue is a thermal stress that produces thermal strain, which, in turn, produces thermal-strain fatigue. It is customary to express thermal stress, σ, (Li Guobin et al.) as:

$$\sigma = y\alpha E \frac{\Delta T}{(1-2\gamma)} \tag{6.52}$$

where α is the linear-expansion coefficient, E is the modulus of elasticity, ΔT the difference in temperature, γ the Poisson's ratio and y the constraint coefficient.

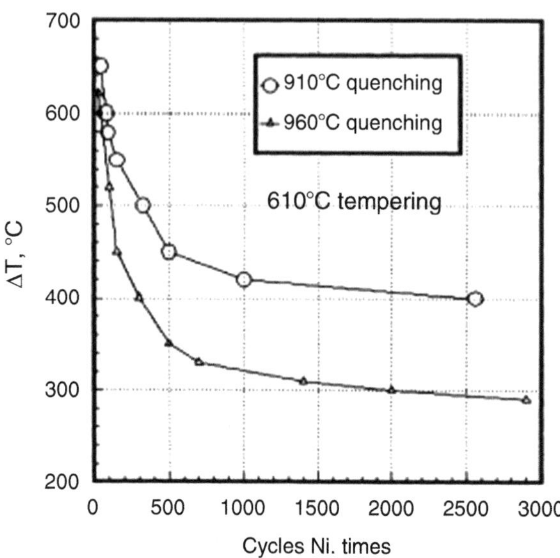

Fig. 6.102 The TFCN life for the 4Cr2NiMoV steel (Guobin et al. 2000. With kind permission of Elsevier)

Thermal strain, Δε, is given by:

$$\Delta\varepsilon = k_{\varepsilon} y \alpha \Delta T \tag{6.53}$$

where k_e is the stress-concentration factor. The thermal-fatigue crack nucleation [henceforth: TFCN] is expressed as:

$$N_i = k(\Delta DT - \Delta T_0)^{-2} \tag{6.54}$$

k being a constant dependent on material properties and notch geometry. ΔT is the range of surface-temperature change in the specimen and ΔT_0 is a critical temperature. When $\Delta T < \Delta T_0$, the TFCN life extends to infinity. Hot-forging dies are not damaged by thermal fatigue as long as $\Delta T < \Delta T_0$. Obviously, the greater the ΔT_0, the longer the TFCN life of the hot-forging dies. Figure 6.102 shows the TFCN life. One can see that the lower ΔT, the more cycles the specimen can endure.

Experiments show that if the temperature difference, ΔT, of the thermal-cycle tests is large, only one crack nucleates around the notch. In such a case, that crack propagates to form a main crack which is easy to observe and trace and the TFCN life is short. However, when the ΔT is not large, many tiny cracks nucleate at the notch root, which distribute along the strain sliding field, as shown in Fig. 6.103. The curves of thermal-fatigue crack propagation for notched specimens of 4Cr2NiMoV are plotted as crack length, a, vs. cycles, N, as shown in Fig. 6.104. The inset indicates the quenching and tempering treatments. It is clear that the resistance to TFCN of a specimen after 910°C quenching and 580°C tempering is greater than for the other quenching and tempering conditions. This is evident from the main-crack

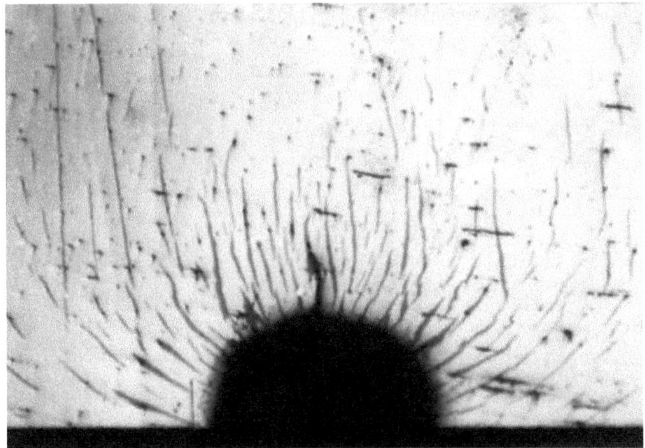

Fig. 6.103 Cracks nucleating along the strain sliding field (Guobin et al. 2000. With kind permission of Elsevier)

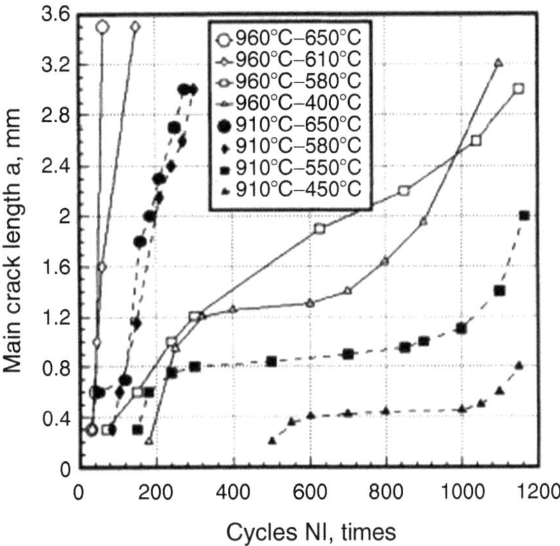

Fig. 6.104 Shows the propagation of a thermal fatigue crack for 4Cr2NiMoV steel (Guobin et al. 2000. With kind permission of Elsevier)

length. Figure 6.103 shows cracks nucleating along a strain field. Thermal-fatigue cracks propagate very rapidly when the thermal-cycle temperature is higher than the tempering temperature and vice verse.

The detrimental effects of thermal fatigue are indicated for two different temperatures (Fig. 6.104). Introducing residual-compressive stress by any of the known methods, such as nitriding or laser shot peening, reduces damage due to thermal fatigue.

6.15 Design for Fatigue

The basic characteristics of fatigue and some of the parameters affecting it have been discussed in this chapter. The resistance to fatigue failure, required to overcome crack initiation and propagation are clear. These aspects of fatigue should be translated to the design phase for machine components that will be operated under fatigue conditions. It is virtually impossible to provide design prescriptions for each specific case, due to the divergent applications of structural constructions, the different environments in which they operate and their various sizes (to mention only some of the factors that make general design guidelines difficult). Nevertheless, here are a few of basic principles that may apply to each case, as follows:

(a) In ferrous metals and Ti alloys, the stress applied should be kept below the endurance limit, at which the material is expected to have an infinite life-time. This design concept is also known as the 'infinite life-time concept'. In cases of non-ferrous alloys, the use of an empirically evaluated life-time at a specific stress is sound. As mentioned earlier, a stress level of $\sim 10^7$–10^8 cycles is a reasonable design criterion.
(b) Safe life design is a conservative approach. Parts are designed to operate for certain life-times, after which they are replaced by new parts. Thus, as per the designers' instructions, the empirically determined life-time is only used with the proper safety factor. Clearly, in either (a) or (b), the random appearance of an unexpected stress, due to some irregular-stress cycle cannot always be foreseen.
(c) Damage-related design requires periodic, nondestructive inspection of crack formation and growth. By using one of the nondestructive-testing methods, it is possible to perform such inspections to see if existing cracks are nearing the critical-crack size. There are mathematical methods for making reasonably-accurate predictions of crack growth between inspections and for evaluating the time left for further safe use of that part following its inspection.

Despite these crucial design rules, a fatigue problem exists, since failure is usually sudden and often occurs at a stress level much lower than the ultimate stress level. Frequent attempts have been made to relate static deformation data to fatigue, but experiments have indicated little direct relation between the fatigue limit and yield strength, ductility and other static-deformation properties. However, some connection between the fatigue limit and the tensile strength, σ_{UTS}, has been established for un-notched, polished specimens tested by the rotating-beam method. The commonly used relation (see, for example, Chaprin et al.) for the fatigue limit of steel, S_{FL}, is: approximated as one half of the tensile strength

$$S_{FL} = 0.5\sigma_{UTS} \tag{6.52}$$

Another empirical relation is:

$$S_{FL} = 140 + 0.25\sigma_{UTS} \tag{6.53}$$

In addition, empirical hardness and fatigue life relations do exist. Empirical relations for other metals have also been suggested. Attempts have been made to relate fatigue properties to static parameters, based on the similarity between monotonic and fatigue mechanisms, which control cyclic straining and plastic flow. Various degrees of stress concentration are usual features of fabricated-structural parts. Stress concentration acts at certain locations for fatigue-crack initiation. Residual stresses are common in machine parts (e.g., welded structures), but only compressive stresses do not contribute to crack initiation or propagation. Residual-tensile stress or tensile components of applied loads are to be avoided, in so far as possible while designing for fatigue, because of its damaging nature. It has been observed that, all other things being equal, increasing specimen dimensions results in decreased fatigue. One explanation is that smaller-sized specimens have fewer microcracks than larger ones. Fatigue performance designers generally correct for the dimensions of the various parts.

To summarize the requirements for fatigue design, the following is a list of some of the many factors that influence fatigue life and must be taken into account during the design process:

(a) The material. Usually materials with high melting points are preferred, since, in general, physical and mechanical properties are related to the melting point via the cohesive properties. Materials should be free of inclusion porosity and other voids that interrupt material continuity.
(b) The conditions of material processing. The cyclic properties, like the static ones, are dependent on the processing of the machine elements, whether they have been quenched, annealed, normalized or tempered.
(c) The surface conditions of a part. Parts should be scratch-free and polished specimens perform better than unpolished ones.
(d) The geometry of machine elements. Machine element geometry is an essential design parameter; length, width, thickness and diameter produce the size effects of materials. Furthermore, the radius (size and sharpness) and transition radius from location to location are of utmost importance, since they act as stress raisers. Designers must account for the effects of the characteristics (type) of the radii.
(e) The effect of the environment. Environmental effects are important design considerations. Corrosive environments are detrimental. Some structures (e.g., ships) are often located in the corrosive environment of salty water; appropriate design steps should be taken to select the most resistant materials for parts exposed to seawater. Corrosive environments may accelerate the growth of fatigue cracks and, therefore, reduce overall fatigue performance. Cathodic protection reduces the impact of this detrimental effect to some extent.
(f) The effect of temperature. Temperature is a significant factor to be considered. Combined creep and fatigue deformation may act at high temperatures.

To conclude this chapter, it may be stated that fatigue damage is one of most frequent causes of the breakdown of structural elements in service. All the previous sections explained the failure of machine parts due to fatigue during service and the means for improving their resistance by realizing the various acting factors. Every structural material experiences the problem of fatigue life and any differences that exist are dictated by overall material properties. Clearly, it is beyond the possibility of one textbook to consider each of the many materials used for machine parts and all the particular applications of these parts. Therefore, one or two examples of certain materials sufficed to serve as representatives of the wide spectrum of applications for structural use exposed to fatigue deformation.

References

W.M. Baldwin Jr., *Residual Stresses in Metals*, Proc. ASTM 49, 1949. p. 1

N.E. Dowling, *Mechanical Behavior of Materials*, Second edn. (Prentice Hall, Upper Saddle River, 1999), p. 650

N.E. Dowling, *Mean Stress Effects in Stress-Life and Strain-Life Fatigue*. Society of Automotive Engineers, Inc., 2004, F2004/51

J. Goodman, *Mechanics Applied to Engineering* (Longmans Green, London, 1899)

R.W. Hetzberg, *Deformation and Mechanics of Engineering Materials* (Wiley, New York, 1976), pp. 415–462. and 465-520

R.E. Heywood, *Designing Against Fatigue* (Chapman & Hall, London, 1962). Quoted by Ciavarella

P. Kuhn, H.F. Hardrath, An Engineering Method for Estimating Notch-Size Effect in Fatigue Tests on Steel. (NASA Tech Note 2805, 1952). Quoted by Ciavarella

B.J. Lazan, A.A. Blatherwick, Fatigue Properties of Aluminum Alloys at Various Direct-Stress Ratios, Part II Extruded Alloys. (WADC Technical Report, 52-307, December 1952). Approved for public release

S.S. Manson, Fatigue a Complex Subject-Some Simple Approximations. Exp. Mech. SESA **5**, 193 (1965)

S.S. Manson, M.H. Hirschberg, *Fatigue: An Interdisciplinary Approach* (Syracuse University Press, Syracuse, 1964), p. 133

G. Masing, Eigenspannungen und Verfestigung bei Messing, in *Proceedings of the 2nd International Congress of Applied Mechanics, Zurich, 1926*. Quoted by H. Zenner, F. Renner, in Int. J. Fatigue **24**, 1255 (2002)

J.D. Morrow, Cyclic Plastic Strain Energy and the Fatigue of Metals, in *Internal Friction, Damping and Cyclic Plasticity*. ASTM STP, 378 (American Society for Testing and Materials, Philadelphia, 1965)

H. Neuber, *Theory of Notch Stresses* (Springer, Vienna, 1958). Quoted by Ciavarella

N. Ono, Y. Nishimura, in *Proceedings of the 12th International Conference on Fracture* (Ottawa, CD ROM, 2009), pp. 1–10

R.E. Peterson, Notch Sensitivity, in *Metal Fatigue*, ed. by G. Sines, J.L. Waisman (MacGraw-Hill, New York, 1959), pp. 293–306. Quoted by Ciavarella

D.F. Socie, M.R. Mitchell, E.M. Caulfield, Fundamentals of Modern Fatigue Analysis. (Fracture Control Program, Report No. 26) (University of Illinois, Chicago, 1977)

R.I. Stephens, *Metal Fatigue in Engineering*, 2nd edn. (Wiley-Interscience Publication, New York, 2001)

R.I. Stephens, D.K. Chen, B.W. Horn, Fatigue Crack Growth with Negative Stress Ratio Following Single Overloads in 2024-T3 and 7075-T6 Aluminium-alloys, in *Fatigue Crack Growth Under Spectrum Loads*. ASTM STP 595 (ASTM, Philadelphia, 1976), pp. 27–40

S. Suresh, *Fatigue of Materials* (Cambridge University Press, Cambridge, 2001)

K. Walker, The Effect of Stress Ratio During Crack Propagation and Fatigue for 2024-T3 and 7075-T6 Aluminum, in *Effects of Environment and Complex Load History on Fatigue Life*. ASTM STP, 462 (American Society for Testing and Materials, Philadelphia, 1970), p. 1
H.Q. Xue, E. Bayraktar, C. Bathias, J. Achiev. Mater. Manufac. Eng. **18**, 251 (2006)
G.T. Yahr, Fatigue Design Curves For 6061-T6 Aluminum, Oak Ridge National Laboratory, U.S. Department of Energy under Contract No. DE-AC05-84OR21400 (1993)

Further References

B. Atzori, P. Lazzarin, R. Tovo, Österreichische Ingenieur-und Architekten- Zeitschrift **137**, 556 (1992)
O.H. Basquin, *Proc. ASTM,***10**, 625 (1910) quoted from W. Cui, J. Mar. Sci. Technol. **7**, 43 (2002)
C. Bathias, L. Drouillac, P. Le Francois, Int. J. Fatigue **23**, S143 (2001)
C.A. Berg, M. Salama, Fibre Sci. Technol. **6**, 125 (1973)
S.K. Bhaumik, R. Rangaraju, M.A. Venkataswamy, T.A. Bhaskaran, M.A. Parameswara, Eng. Fail. Anal. **9**, 255 (2002)
V.I. Bol'shakov, V.S. Zoteev, L.G. Orlov and M.A, Tylkin, Translated from Metallovedenie i Termicheskaya Orabotka Metallov (2), 45 (1974)
L.P. Borrego, L.M. Abreu, J.M. Costa, J.M. Ferreira, Eng. Fail. Anal. **11**, 715 (2004)
D. Brandolisio, G. Poelman, G. De Corte, J. Symynck, M. Juwet, F. De Bal, Rotating Bending Machine for High Cycle Fatigue Testing, March 26, 2009
F.P. Brennan, Int. J. Fatigue **16**, 351 (1994)
M.W. Brown, D.K. Suker, C.H. Wang, Fatigue Fract. Eng. Mater. Struct. **19**, 323 (1996)
D.A. Carpinteri, A. Spagnoli, S. Vantadori, Fatigue Fract. Eng. Mater. Struct. **25**, 619 (2002)
M.D. Chaprin, H. Miyata, T. Tagawa, T. Miyata, M. Fujioka, Mater. Sci. Eng. **A 381**, 331 (2004)
M. Ciavarella, G. Meneghetti, Int. J. Fatigue **26**, 289 (2004)
W. Cui, J. Mar. Sci. Technol. **7**, 43 (2002)
A.A. Dabayeh, T.H. Topper, Int. J. Fatigue **17**, 261 (1995)
N.E. Dowling, Fatigue Fract. Eng. Mater. Struct. **32**, 1004 (2009)
W. Elber, ASTM STP **559**, 45 (1974)
C.E. Feltner, C. Laird, Acta Met. **15**, 1621 (1967)
P.J.E. Forsyth, Nature **171**, 172 (1953)
M. de Freitas, F. Romeiro, M. da Fonte, Anales de Mecanica de la Fractura **2**, 641 (2006)
Y. Furuya, S. Matsuoka, T. Abe, K. Yamaguchi, Scr. Mater. **46**, 157 (2002)
S. Ganesh, S. Raman, K.A. Padmanabhan, Int. J. Fatigue **18**, 71 (1996)
Yu-kui Gao, Xiang-bin Li, Qing-xiang Yang, M. Yao, Mater. Lett. **61**, 466 (2007)
W. Geary, Int. J. Fatigue **14**, 377 (1992)
K. Genel, M. Demirkol, Int. J. Fatigue **21**, 207 (1999)
W.Z. Gerber, Z. Bayer Archit. Ing. Ver. **6**, 101 (1874)
A. Glage, A. Weidner, T. Richter, P. Trubitz, H. Biermann, European Symposium on Martensitic Transformations, ESOMAT 2009, 05007 (2009)
K. Gopinath, A.K. Gogia, S.V. Kamat, R. Balamuralikrishnan, U. Ramamurty, Acta Mater. **57**, 3450 (2009)
A.A. Griffith, Philos. Trans. R. Soc. Lond. **A221**, 153 (1921)
Li Guobin, Wu Jianjun, J. Yanfei, Li Guiyun, J. Mater. Process. Technol. **100**, 63 (2000)
G. Hammersley, L.A. Hackel, F. Harris, Opt. Lasers Eng. **34**, 327 (2000)
X. Huang, T. Moan, W. Cui, Int. J. Fatigue **30**, 2 (2008)
J.W. Jones, H. Mayer, J.V. Lasecki, J.E. Allison, Int. J. Fatigue **28**, 1566 (2006)
L. Junek, J. Bystriansky, L. Vlcek, B. Strnadel, Trans., SMiRT 19, Toronto, August 2007, Paper # G05/5, p. 1

K. Kanazawa, S. Nishijima, J. Soc. Mater. Sci. **46**, 1396 (1997)
J.M. Larsen, B.D. Worth, C.G. Annis Jr., F.K. Haake, Int. J. Fract. **80**, 237 (1996)
P. Lazzarin, R. Tovo, G. Meneghetti, Int. J. Fatigue **19**, 647 (1997)
B.A. Lerch, S.L. Draper, J.M. Pereira, Met. Mater. Trans. A **33A**, 3871 (2002)
N. Limodin, Y. Verreman, T.N. Tarfa, Fatigue Fract. Eng. Mater. Struct. **26**, 811 (2003)
Y. Liu, S. Mahadevan, Eng. Fract. Mech. **76**, 2317 (2009)
S.P. Lynch, Mater. Sci. Eng. A **468**, 74 (2007)
S.M. Marco, W.L. Starkey, Trans. ASME **76**, 627 (1954)
I. Marines, X. Bin, C. Bathias, Int. J. Fatigue **25**, 1101 (2003a)
I. Marines, G. Dominguez, G. Baudry, J.-F. Vittori, S. Rathery, J.-P. Doucet, C. Bathias, Int. J. Fatigue **25**, 1037 (2003b)
C. Menzemer, T.S. Srivatsan, Mater. Sci. Eng. **A271**, 188 (1999)
K.J. Miller, J. Strain Anal. **5**, 185 (1970)
M.A. Miner, J. Appl. Mech., Trans. ASME, **12**, A159 (1945)
Y. Murakami, Y. Tazunoki, T. Endo, Metall. Trans. A **15A**, 2029 (1984)
Y. Murakami, T. Namoto, T. Ueda, Fatigue Fract. Eng. Mater. Struct. **22**, 581 (1999)
M. Nakajima, M. Akita, Y. Uematsu, K. Tokaji, Proc. Eng. **2**, 323 (2009)
T. Nicholas, J.R. Zuiker, Int. J. Fracture. **80**, 219 (1996)
E.S. Nikolin, G.V. Karpenko, Mater. Sci. **3**, 487 (1967) [Fiziko-Khimicheskaya Mekhanika Materialov, 3, 667 (1967)]
D.W. Norwich, A. Fasching, J. Mater. Eng. Perform. **18**, 558 (2009)
D. Novovic, R.C. Dewes, D.K. Aspinwall, W. Voice, P. Bowen, Int. J. Mach. Manufac. **44**, 125 (2004)
J.H. Ong, Int. J. Fatigue **15**, 213 (1993)
A. Palmgren, ZVDI **68**, 339 (1924)
A. Plumtree, H.A. Abdel-Raouf, Int. J. Fatigue **23**, 799 (2001)
J. Polák, J. Man, K. Obrtlík, Int. J. Fatigue **25**, 1027 (2003)
J. Polák, J. Man, T. Vystavěl, M. Petrenec, Mater. Sci. Eng. A **517**, 204 (2009)
B. Pyttel, D. Schwerdt, C. Berger, Int. J. Fatigue **33**, 49 (2011)
T. Sakai, B. Lian, M. Takeda, K. Shiozawa, N. Oguma, Y. Ochi, M. Nakajima, T. Nakamura, Int. J. Fatigue **32**, 497 (2010)
G. Salerno, R. Magnabosco, C. de Moura Neto, Int. J. Fatigue **29**, 829 (2007)
C.S. Shin, S.H. Hsu, Int. J. Fatigue **15**, 181 (1993)
F.S. Silva, Int. J. Fatigue **29**, 1757 (2007)
G.M. Sinclair, ASTM Proc. **52**, 743 (1952)
M. Skorupa, Fatigue Fract. Eng. Mater. Struct. **21**, 987 (1998)
K.N. Smith, P. Watson, T.H. Topper, A stress–strain function for the fatigue of metals. J. Mater. JMLSA **57**, 67 (1970)
O.V. Sosnin, A.V. Gromova, Yu.F. Ivanov, S.V. Konovalov, V.E. Gromov, E.V. Kozlov, Int. J. Fatigue **27**, 1186 (2005)
C.A. Stubbington, P.J.E. Forsyth, Acta Met. **14**, 5 (1966)
D. Thevenet, N. Lautrou, J.Y. Cognard, PAMM Proc. Appl. Math. Mech. **8**, 10243 (2008)
A.W. Thompson, W.A. Backofen, Acta Met. **19**, 597 (1971)
T.H. Topper, M.T. Tu, Int. J. Fatigue **7**, 159 (1985)
M.A.S. Torres, H.J.C. Voorwald, Int. J. Fatigue **24**, 877 (2002)
V.T. Troshchenko, L.A. Khamaza, Strength Mater. **42**, 647 (2010)
B.I. Verkin, N.M. Grinberg, V.A. Serdyuk, L.F. Yakovenko, Mater. Sci. Eng. **58**, 145 (1983)
A. Vinogradov, S. Hashimoto, V.I. Kopylov, Mater. Sci. Eng. **A355**, 277 (2003)
C. Vishnevsky, J.F. Wallace, Fatigue of Cast Steels Part I – A study of the notch effect and of the specimen design and loading on the fatigue properties of cast steel, Steel Foundry Research Foundation, Ohio, April, 1967. Published and Distributed by Steel Founders' Society of America Westview Towers, 21010 Center Ridge Road Rocky River, Ohio 44 116
G.M. Vyletel, J.E. Allison, D.C. van Aken, Met. Mater. Trans. A **26A**, 3143 (1995)
D. Wagner, N. Ranc, C. Bathias, P.C. Paris, Fatigue Fract. Eng. Mater. Struct. **33**, 11 (2009)

G.W.J. Waldron, Acta Met. **13**, 897 (1965)
J.F. Wallace, A.M. Said, R&D Center Laboratory, Technical Report 13100, Improvement in the Fatigue Behavior of Tank Track Pins, U.S. Army Tank-Automotive Command Research and Development Center, Warre ,Michigan 48090, Aug 1985
Z. Wang, T. Nian, D. Ryding, T.M. Kuzay, Nuclear Instr. Method. Phys. Res. A **347**, 651 (1994)
T. Wehner, A. Fatemi, Int. J. Fatigue **13**, 241 (1991)
W. Yao, K. Xia, Y. Gu, Int. J. Fatigue **17**, 245 (1995)
K. Yatsushiro, M. Sano, K, Yamanashi, M. Kuramoto, From JCPDS – International Centre for Diffraction Data 2003, Adv. X-ray Anal. **46**, 92 (2003)
H. Zenner, F. Renner, Int. J. Fatigue **24**, 1255 (2002)
P. Zhang, J. Lindemann, Scr. Mater. **52**, 485 (2005)
X. Zhu, A. Shyam, J.W. Jones, H. Mayer, J.V. Lasecki, J.E. Allison, Int. J. Fatigue **28**, 1566 (2006)
V. Zitounis, P.E. Irving, Int. J. Fatigue **29**, 108 (2007)

Chapter 7
Fracture

7.1 Introduction

Atomic cohesion is the bond between atoms, holding them together to form an aggregate that does not disintegrate under the normal conditions characteristic of that specific material. Hence, a short look at the essentials of cohesion aids in understanding fracture, which occurs when a force of a certain magnitude is applied against the atomic bonding of the atoms to cause the disintegration of a material. Those forces that hold the groups of atoms or molecules of a substance together are called 'bonds'. The formation of bonds between atoms is mainly due to their tendency to attain minimum potential energy, thus reaching a stable state. In solid material, it is usually assumed that two types of forces act between the atoms: (a) an attractive force, which keeps the atoms together, forcing them to form solids and (b) a repulsive force, which comes into play when a solid is compressed. Figure 7.1a shows the concept of cohesion, based on the relation below, graphically:

In a material, such as NaCl, the overall cohesive energy may be expressed as:

$$F(r) = \frac{A}{r^M} - \frac{B}{r^N} \tag{7.1}$$

The first term is the attractive force, while the second term is the repulsive force. With decreasing distance between the atoms near the equilibrium position, the repulsive force (the second term in the relation) increases more rapidly than the first one and the exponent, N, must be greater than M. At the equilibrium distance, $r = r_0$ F(r) is zero, as seen in Fig. 7.1. One can express Eq. 7.1 in terms of r_0 as:

$$\frac{A}{{r_0}^M} = \frac{B}{{r_0}^N} \tag{7.2}$$

The equilibrium distance is:

$$r_0 = \left(\frac{B}{A}\right)^{\frac{1}{N-M}} \tag{7.3}$$

J. Pelleg, *Mechanical Properties of Materials*, Solid Mechanics and Its Applications 190, DOI 10.1007/978-94-007-4342-7_7,

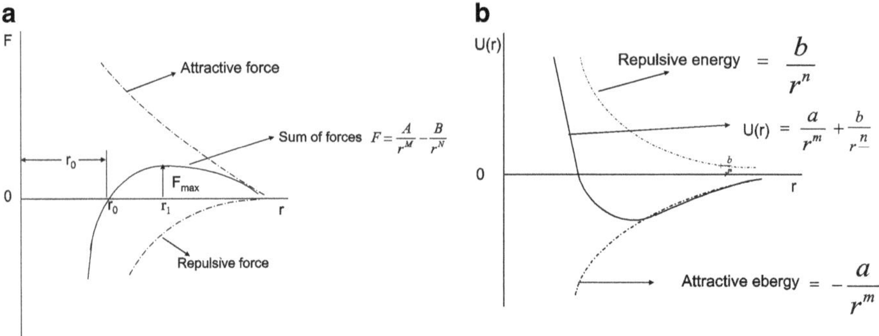

Fig. 7.1 The cohesion in a material, such as NaCl; (**a**) a plot of force against distance and (**b**) the energy variation with distance. The *solid curves* are the sum of the attraction and the repulsion between two atoms

Usually, the force is the negative derivative of the energy. Thus, integration of the force gives the cohesive energy as:

$$U(r) = \int F(r) = \int \left(\frac{A}{r^M} - \frac{B}{R^N} \right) dr \tag{7.4}$$

$$U(r) = -\frac{A}{r^{M-1}} \frac{1}{M-1} + \frac{B}{r^{N-1}} \frac{1}{N-1} + C \tag{7.5}$$

Equation 7.5 may be written as:

$$U(r) = -\frac{a}{r^m} + \frac{b}{r^n} + C \tag{7.6}$$

when U(r) = 0 and r goes to infinity, C = 0 and:

$$U(r) = -\frac{a}{r^m} + \frac{b}{r^n} \tag{7.7}$$

This relation is plotted in Fig. 7.1b, which gives the potential energy vs. the distance. Similarly to the above, at $r = r_0$, the potential energy is a minimum and one can write:

$$\left(\frac{dU}{dr} \right)_{r=r_0} = 0 = \frac{am}{{r_0}^{m+1}} - \frac{bn}{{r_0}^{n+1}}$$

$$r_0 = \left[\left(\frac{b}{a} \right) \frac{n}{m} \right]^{\frac{1}{n-m}} \tag{7.8}$$

The cohesion above is for ionic solids; covalent and metallic bonds are different. In covalent bonds, electrons between atoms are shared, whereas in metallic solids, atoms of the same (or different) elements donate their valence electrons to form an electron gas throughout the space occupied by the atoms. Giving up their electrons to a common pool, known as 'electron cloud' or 'electron gas', these atoms actually become positive (similar to positive ions). They are held together by forces similar to those of ionic bonds, but acting between ions and electrons. The electrostatic interaction between the positive ions and the electron gas holds metal together. Unlike other crystals, metals may be deformed without fracture, because the electron gas permits atoms to slide past one another, acting as a lubricant.

For further study of cohesive forces, one is referred to the literature. The main focus of this short introduction on cohesive energy and forces is on the shape of Fig. 7.1a, which is half a sinusoidal function (used to determine the theoretical cohesive strength of a crystal in Chap. 2, Sect. 2.2 via Eqs. 2.1, 2.2, 2.3, 2.4, 2.5, 2.6, and 2.7). Here, the critical amount of shear stress required to move adjacent atomic planes past one another has been calculated, i.e., the energy per unit area involved in shearing two atomic layers from their equilibrium configuration. Calculations based on Fig. 7.1a can be most useful when fracturing a material.

7.2 Fracture Types

Regardless of the kind of fracture (static, cyclic or creep), most types of failure by fracture are either brittle or ductile. Figure 7.2 illustrates schematically some types of fracture observed in metals, such as steel or aluminum, and in very ductile and soft specimens (e.g., Pb, Au.). 'Fracture' means the separation of a body into two or more parts due to stress acting on it at temperatures below the melting point. It was previously mentioned in Chap. 6 that failure by fracture consists of two steps: (a) crack formation and (b) crack propagation leading to complete fracture. The ability of a material to undergo complete or partial plastic deformation barely encompasses all the types of fractures that exist. In the next section, detailed consideration will be given to certain kinds of fractures. The first type of fracture to be analyzed is brittle fracture.

Some experimental micrographs of fracture are shown in Fig. 7.3. Figure 7.3a shows the cup-and-cone ductile fractures of spheroidized 1080 steel, that failed by ductile fracture (see Fig. 7.2b). Figure 7.3b is an experimental micrograph of shear fracture, showing a macrophotograph of a 7075-T4 sample, that failed by shear fracture (see Fig. 7.2c). Figure 7.3c is the macrophotograph of a 7075-T4 aluminum alloy, an actual representation of the schematic Fig. 7.2d, indicating complete ductility, fracturing by chisel-point fracture. For most practical purposes, metals yield when the effective stress reaches a critical value, that of the initial-yield stress, σ_y. Fracture is different when voids nucleate and grow or when cleavage cracks nucleate and propagate. The SEM micrographs of ductile fracture (shown in Fig. 7.3) clearly indicate this process for various types, namely for cup-and-cone

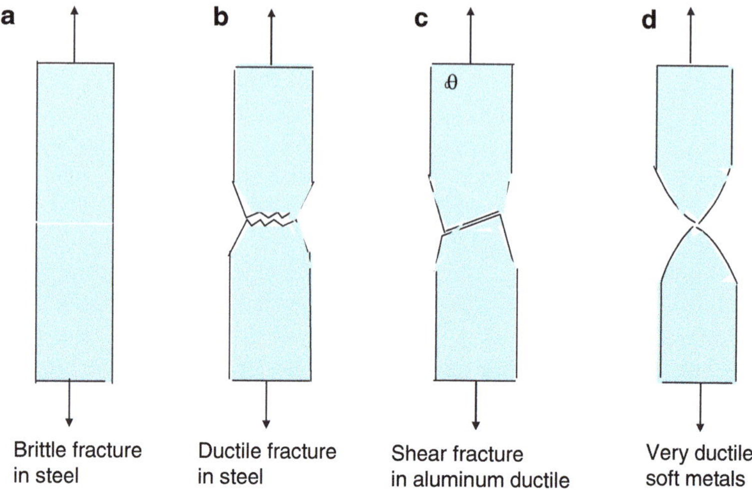

Fig. 7.2 A schematic illustration of some types of fractures observed in metals, such as steel or aluminum: (**a**) brittle fracture; (**b**) ductile fracture; (**c**) shear fracture; and (**d**) complete ductile fracture, also known as 'chisel point fracture'

and shear-type fractures. As the stress state is changed, the dominant fracture mode may switch from one mechanism to another.

Figure 7.3 illustrated various ductile fractures corresponding to the schematic drawings in Figs. 7.2b–d. Instructive examples of a brittle fracture and a ductile cup and cone may be seen in Fig. 7.4. Figure 7.4a is an ideal tensile fracture of the classic cup-and-cone-type. In comparison, in the brittle-tensile fracture shown in Fig. 7.4b, no necking has occurred and the percentage of elongation and % RA values are nearly zero. In this type of tensile fracture, the yield and tensile strengths are essentially identical.

Before considering brittle fracture in detail, some characteristic points are worth mentioning:

1. There is no gross, permanent deformation of the material.
2. The surface of the brittle fracture tends to be perpendicular to the principal tensile stress, although other components of stress may be factors.
3. Brittle fracture often occurs by cleavage fracture; the surface is characterized by flat facets with sizes normally of the grain size (in poly crystals).
4. A Riverside pattern consists of steps between parallel cleavages on parallel planes, that usually converge in the direction of the local crack propagation.
5. Characteristic crack advance markings frequently indicate where the fracture originated.
6. The path that a crack follows depends on the material's structure. In polycrystalline metals, transgranular and intergranular cleavages are important. Cleavage shows up clearly in micrographs such as by SEM.

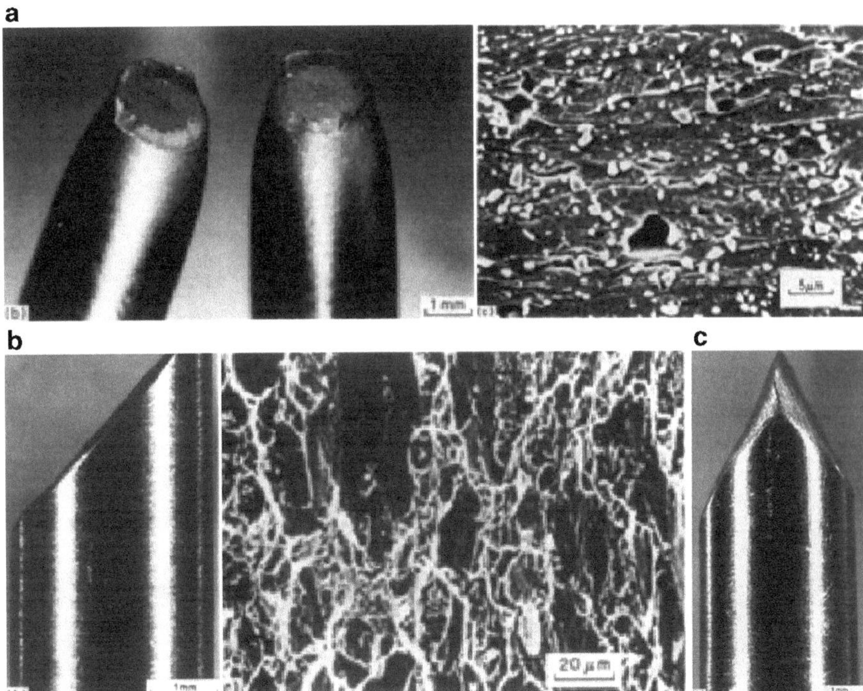

Fig. 7.3 Various ductile fractures obtained experimentally are shown: (**a**) cup and cone fracture in spheroidized 1080 steel and a SEM micrographs of the fractured surface showing damage development; (**b**) shear fracture of 7075-T4 aluminum alloy and SEM micrograph showing the void mechanism leading to shear fracture; (**c**) complete ductility down to a chisel point fracture (Teirlinck et al. 1988. With kind permission of Elsevier)

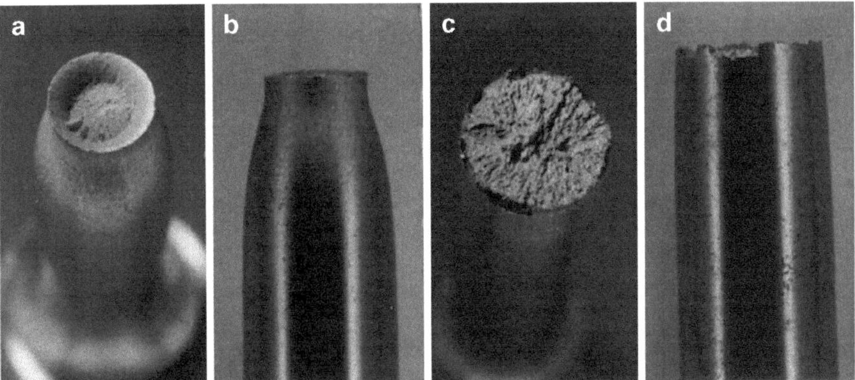

Fig. 7.4 Macroscopic appearance of ductile (**a**) and brittle (**b**) tensile fractures (Vander Voort 1987. With kind permission of ASM International. All rights reserved. www.asminternational.org)

7.3 Brittle Fracture

Structural elements fail with little or no plastic deformation, often without warning. Since brittle fracture can lead to catastrophic results, it has been studied more intensively. Begin by assuming that a solid is perfect and that it breaks by separating along atomic planes. Differentiating energy vs. distance results either in a force vs. distance relation (as indicated in Fig. 7.1a) or in a stress vs. distance relation, because force is related to stress by the area (Sect. 1.2, Eq. 1.4). In Fig. 7.1a, r_0 represents the equilibrium separation between two atoms. To move an atom even a small distance by deformation requires a force (stress) which increases over the distance which the atom is being forced to move. A maximum force is required to bring the atom to distance r_1.

Unlike slip, produced by dislocation motion (see Sect. 2.2), where a periodic motion, assumed to be sinusoidal, was applied, here, the periodic motion is not considered. In this case, the atom's continued movement beyond r_1 occurs with great ease, since the force beyond F_{max} is decreasing. Recall that the area under the curve represents the energy required to produce a fracture over a unit area. To clarify, Fig. 7.1a is redrawn below in terms of stress vs. distance (Fig. 7.5a), assuming that the curve is halved into two equal areas (Fig. 7.5b).

One may now express the above statements in the following way. The sum of the elastic areas (half-curves) equals the energy:

$$\text{energy} = 2 \times (\Delta \times \mathrm{r}\sigma_{\mathrm{max}})/2 = \Delta \times r \times \sigma_{\mathrm{max}} \tag{7.9}$$

This relation is a result of the assumption that each half area may be approximated as a triangle. Thus, with two triangle-like areas Eq. 7.9 is obtained. Now, $\Delta r = r_1 - r_0$, may be expressed in terms of strain, as given below:

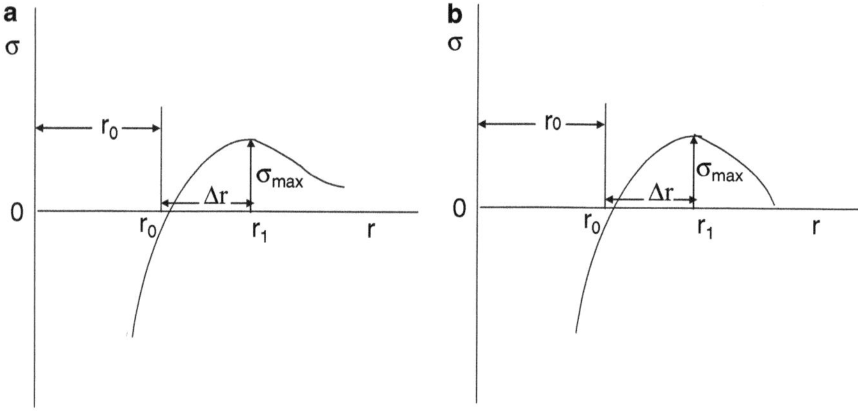

Fig. 7.5 The cohesive force (in terms of stress) as a function of the distance between atoms, representing (in this case) the stress needed to move an atom away from its neighbor: (**a**) the actual plot of stress vs. distance and (**b**) presenting as two equal halves

$$\varepsilon = \frac{\Delta r}{r_0} \tag{7.10}$$

Substituting for Δr in Eq. 7.9, Eq. 7.11 is obtained:

$$\text{Energy} = \varepsilon r_0 \sigma_{max} \tag{7.11}$$

Strain can be expressed in terms of Young's modulus, E($\sigma = \varepsilon E$). Overcoming the cohesive theoretical strength of a material by fracturing creates two new surfaces and the surface energy of each is γ. The energy of fracturing is balanced by both energies of the two surfaces formed. Thus, Eq. 7.11 may be written as:

$$2\gamma = \frac{r_0}{E}{\sigma_{max}}^2 \tag{7.12}$$

or:

$$\sigma_{max} = \sqrt{\frac{2\gamma E}{r_0}} \tag{7.13}$$

Equation 7.13 provides a value (see 'theoretical strength' in Sect. 2.2) which is ~20–100 times higher than the strength actually observed. It was suggested that the calculated fracture stress needed to create two new surfaces in reality does not apply except in the case of flawless, perfect brittle materials, such as whiskers (i.e., silica fibers), because various flaws, such as cracks, are usually present in engineering materials and are responsible for the lower-than- theoretical fracture strength. Giffith was among the first to explain the discrepancy between the theoretical and actual strengths of materials (but this will be discussed later on).

7.4 Theories of Brittle Fracture

Figure 7.6 is a schematic illustration of the cleavage plane. When a material is not amorphous, the fracture obtained by tension tests usually acquires crystallographic low-index planes, where the density of the atoms is high. These planes are usually referred to as 'cleavage planes' and the tensile test acts normally to such planes; similarly to slip it occurs on a plane in which a critical value is achieved. This law was formulated by Sohncke in the nineteenth century and states that brittle fracture occurs in single crystals when the resolved-normal stress on the cleavage plane reaches a critical value, σ_c.

Thus (as given in Sect. 3.2 for slip), in fracture, this relation may be formulated as:

$$\sigma_c = \frac{P\cos\phi}{A/\cos\phi} = \frac{P}{A}\cos^2\phi \tag{7.14}$$

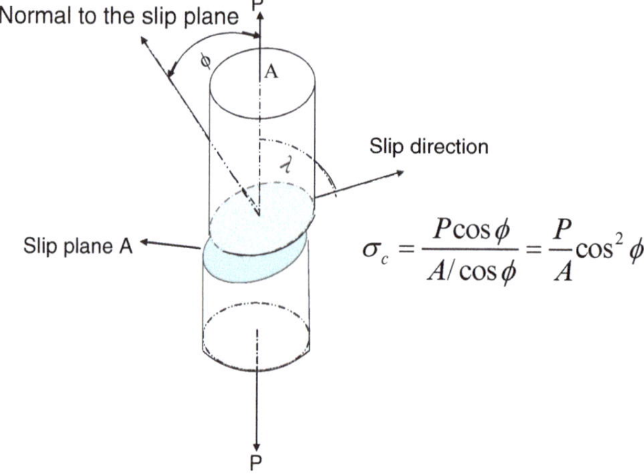

Fig. 7.6 A cleavage plane in single crystals related to normal stress, which is Pcosϕ

Equation 7.14 was obtained as follows. The normal stress, which is the tensile stress acting on the cleavage plane (in the Fig. 7.6 it is the slip plane), is given in terms of the load, P, as Pcosϕ. The area on which normal stress is acting is A/cosϕ. Therefore, the load per area gives the (critical) stress as shown in Eq. 7.14. The difference between fracture strength and cohesive strength is a consequence of the inherent flaws in a material (mentioned above), which lower fracture strength, as stated by Griffith.

This section will start with Griffith's classic concept of 'brittle fracture'.

7.4.1 Griffith's Theory on Fracture

The existence of cracks explains why actual strength is lower than theoretical strength. Crack formation requires energy to produce two new surfaces. Thus, Griffith's approach to brittle failure is often called an 'energy balance theory'. Observe the schematic illustration in Fig. 7.7 characterizing a crack. Here, the crack is shown at the center, though it may be at the sides of the specimen (half-cracks).

In his model, Griffith assumed that glass (a perfect, brittle material) contains small, flat (slot-like) cracks, which act as stress raisers. Together with Inglis' theory of a pre-existing crack and its growth, Griffith was able to show that theoretical, cohesive strength is reached locally at the crack tip. Crack growth is associated with strain-energy release during growth. (Griffith's relation is derived below).

Inglis' description of the crack (Fig. 7.7 is a plate) assumes that the applied stress, σ_a, is magnified at the end of the ellipse along the x axis (the major axis). The expression of this ellipse in the configuration shown in Fig. 7.7 is:

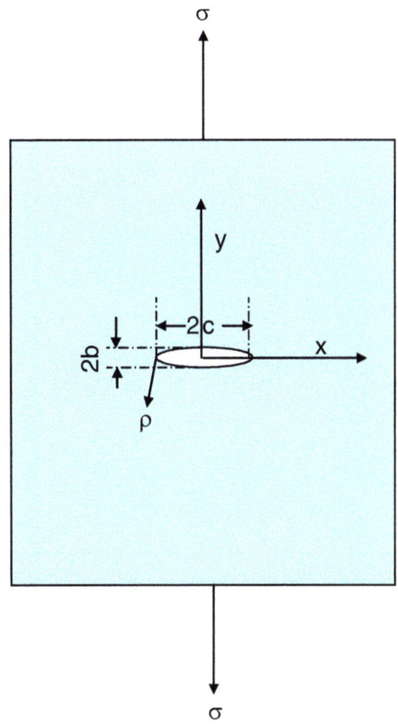

Fig. 7.7 The Griffith crack model for fracture. The flat, elliptical profile in the center of the *plate* represents a crack

$$\frac{x^2}{c^2} + \frac{y^2}{b^2} = 1 \tag{7.15}$$

The radius of curvature, ρ, is given (see, for example, Shumate, Jr.) as:

$$\rho = \frac{b^2}{c} \tag{7.16}$$

The stress concentration at the edge of the crack (the end of the major axis), where c is half the major axis and b = half the minor axis (i.e., half the dimensions of the crack), is given as:

$$\sigma_{\max}(c,0) = \sigma_a\left(1 + \frac{2c}{b}\right) = \sigma_a\left(1 + 2\left(\frac{c}{\rho}\right)^{1/2}\right) \tag{7.17}$$

The last term in Eq. 7.17 is a result of substituting for b from Eq. 7.16. When $b \ll c$ (e.g., a slot), Eq. 7.17 reduces to:

$$\frac{\sigma_{\max}(c,0)}{\sigma_a} \approx \frac{2c}{b} = 2\left(\frac{c}{\rho}\right)^{1/2} \tag{7.18}$$

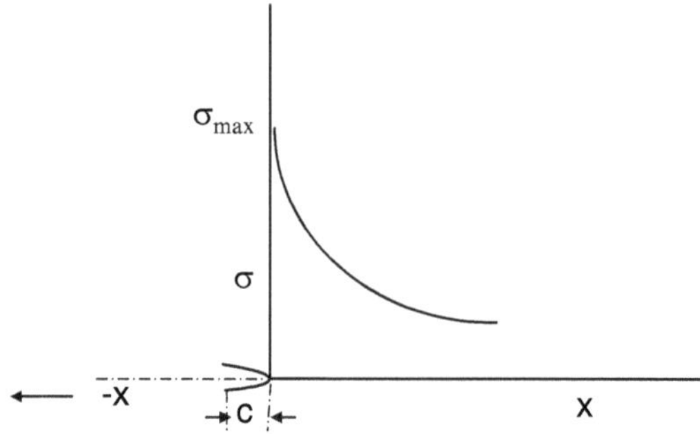

Fig. 7.8 Stress distribution away from the tip of a crack. Equation 7.18 relates σ_{max} to the crack by $\sigma_{max} = \sigma_a \sqrt{2\frac{c}{\rho}}$

For circular holes b = c, thus, from Eq. 1.17 (the central relation) one gets:

$$\frac{\sigma \max (c, 0)}{\sigma_a} \approx 3 \tag{7.19}$$

The stress distribution along x is shown in Fig. 7.8.

For a static-crack system, the total energy is the sum of three terms:

$$\mathrm{U} = (-\mathrm{W_L} + \mathrm{U_E}) + \mathrm{U_S} \tag{7.20}$$

U_E is the elastic energy, U_S is the surface energy and W_L is the mechanical energy of the body or, alternatively, the amount of work done by the applied loads. A decrease in strain energy results from the formation of a crack. Thermodynamic equilibrium is reached when:

$$\mathrm{dU/dc} = 0 \tag{7.21}$$

The mechanical energy of a body under constant, applied force during crack formation is:

$$\mathrm{W_L} = 2\mathrm{U_E} \quad \text{(at constant load)} \tag{7.22}$$

The Inglis' solution for stress and strain fields around a sharp crack is:

$$U_E = \pi c^2 \frac{{\sigma_L}^2}{E} \quad \text{(the plane stress for a thin plate)} \tag{7.23}$$

For a thick plate, the above should be modified to:

$$U_E = \pi \left(1 - \nu^2\right) c^2 \frac{{\sigma_L}^2}{E} \quad \text{(the plane stress for a thick plate)} \tag{7.24}$$

The surface energy of the crack in the plate is:

$$\mathrm{U_S} = 4\mathrm{c}\gamma \tag{7.25}$$

γ is the surface energy per unit area of the crack. In Eq. 7.25, the relation applies to a unit width of the crack. The total energy for the plane stress case is the sum of Eqs. 7.23 and 7.25, namely:

$$U = -\pi c^2 \frac{{\sigma_L}^2}{E} + 4c\gamma \tag{7.26}$$

The minus sign is a consequence of the fact that the growth of a crack releases strain energy. The tensile stress acts normally to the surface of the crack. Following Griffith's concept by using the equilibrium condition in Eq. 7.21 for Eq. 7.26, an expression for the constant load and plane stress conditions is obtained as:

$$\sigma_L = \sqrt{\frac{2\gamma E}{\pi c}} \tag{7.27}$$

For constant load and plane strain conditions one obtains:

$$\sigma_L = \sqrt{\frac{2\gamma E}{(1-\nu^2)\,\pi c}} \tag{7.28}$$

The change in stress with distance in the vicinity of the crack tip from Fig. 7.7 is illustrated schematically in Fig. 7.8.

Despite the great impact that Griffith's theory had on the understanding of fracture in truly brittle materials, such as glass, some problems remain unresolved. Why do large cracks tend to propagate more easily than small cracks? Is this because dimensional change also modifies the radius of curvature? Where from do cracks originate? Griffith's theory (discussed above in regard to perfect, brittle materials, such as glass) is entirely satisfactory for amorphous materials, but cannot, in principle, be extended to metals, because of the different nature of plastic deformation in metals.

7.4.2 Orowan's Fracture Theory

Orowan considered Griffith's theory. If no plastic deformation occurs, the radius of curvature at the tip of a crack must be equal to the atomic radius, "a", representing the sharpest crack. By using the stress-concentration factor at the tip of the elliptical crack, Orowan arrived at a similar relation to Griffith's. The stress-concentration factor has been defined (in an earlier chapter) as K_t, given as:

$$K_t = \frac{\sigma_{max}}{\sigma_a} \tag{7.29}$$

From Eq. 7.17:

$$\frac{\sigma_{\max}(c,0)}{\sigma_a} = \left(1 + \frac{2c}{b}\right) \tag{7.17}$$

and for b ≪ c (e.g., a slot), Eq. 7.17 reduces to Eq. 7.18 or:

$$K_t = \frac{\sigma_{\max}(c,0)}{\sigma_a} \approx \frac{2c}{b} \tag{7.18}$$

This relation is a consequence of b ≪ c and, thus, 1 can be neglected. When the radius of curvature ρ is of atomic dimensions (i.e., "a"), Eq. 7.16 may be written as Eq. 7.16a:

$$a = \frac{b^2}{c} \tag{7.16a}$$

Equation 7.18, in terms of "a", after substituting for b becomes:

$$K_t = \frac{2c}{\sqrt{ac}} \tag{7.30}$$

If it is assumed that the atomic radius is "a", then the fracture stress at the crack tip becomes equal to the theoretical stress (as calculated in Chap. 2, Sect. 2.2, Eq. 2.7) and is approximately:

$$\sigma_T = \frac{E}{10} - \frac{E}{20}$$

Then, with the value of ~E/10 (i.e., the theoretical stress) replacing $\sigma_{\max}$ (c,0) in Eq. 7.18 and by using K_t from Eq. 7.30, in Eq. 7.29 one obtains for σ_a:

$$\sigma_a = \left(\frac{E}{20}\right)\sqrt{\frac{a}{c}} \tag{7.31}$$

One of the questions that Eq. 7.31 poses is whether a crack of the size obtained by calculations from Eq. 7.31 really exists in glass. (For σ_T, an experimental strength of ~0.01–0.1 is often observed and with stress-concentration factors on the order of 10–100 crack sizes, 100–10,000 Å are calculated). Whether such cracks are present in freshly-formed glass remains an unanswered question in the theories of both Griffith and Orowan. Since glass is easily damaged, mechanically or during handling, it is likely that, in most commercial glasses, surface defects result from the above. An indication of such a possibility is the well-known fact that hydrofluoric acid (HF) glass etching results in higher strength; such etching removes surface flaws and, therefore, increased strength is observed. Etching away the surface of the glass is known as 'Joffé's effect'.

A definite size effect exists in brittle materials, such as glass, and their strength depends on the volume of the material. This volume-dependence is simply explained by the fact that the probability of finding proper-sized cracks increases with volume.

7.4.3 Brittle Fracture in Crystalline Materials

Numerous investigations, especially during recent years, have shown that even brittle fracture is preceded by some plastic deformation. Results confirm that fracture starts by some plastic deformation which produces microcracks. In line with this notion is the fact that data exist indicating that fracture can only occur after stress exceeds the yield point. Orowan showed that when plastic deformation is concentrated in a region of a material whose thickness is small, compared to the length of a crack, the work of plastic deformation can be treated as a contributing factor to the surface energy of the crack. The relation suggested by Orowan is given by:

$$\sigma_L = \sqrt{\frac{2E\left(\gamma + w_p\right)}{\pi c}} \tag{7.32}$$

In Eq. 7.32, w_p is the plastic work required to spread the crack, i.e. to increase its surface, which should also be taken into account. Rewriting Eq. 7.26 accordingly, gives:

$$U = -\pi c^2 \frac{\sigma_L{}^2}{E} + 4c(\gamma + w_p) \tag{7.26a}$$

Thermodynamic equilibrium is reached when:

$$\mathrm{dU/dc} = 0 \tag{7.21}$$

Clearly, instability occurs when dU/dc = 0 and the crack will always grow. Thus, Eq. 7.26a may be written as:

$$\frac{2\pi c}{E}\sigma_L{}^2 = 4(\gamma + w_p) \tag{7.33}$$

$$\sigma_L = \sqrt{\frac{2E\left(\gamma + w_p\right)}{\pi c}} \tag{7.34}$$

Orowan estimated the plastic work to be about 2,000 times larger than the surface energy, γ, which may, therefore, be neglected in comparison with w_p. Thus,

Fig. 7.9 A schematic illustration of Zener's idea on the nucleation of a crack with atomic dimensions at dislocation sites. Crack growth is illustrated by the coalescence of two or three dislocations

Orowan's relation incorporates the plastic-deformation energy in the layer near the fracture surface and neglects the surface energy, γ, of the crack, given as:

$$\sigma_L = \sqrt{\frac{2Ew_p}{\pi c}} \approx \sqrt{\frac{Ew_p}{c}} \tag{7.35}$$

It can be seen from Eqs. 7.27 and 7.35 that the stress that causes crack growth is proportional to $c^{-1/2}$.

7.4.4 *The Dislocation Theory of Brittle Fracture*

The next logical step in the discussion of fracture (following Orowan's hypothesis that plastic deformation is, to various degrees, a part of the fracturing process) is to consider the role of dislocations in this process. Zener was among the first to suggest the idea that fracturing is a consequence of the plastic deformation necessary for crack formation. His schematic illustration appears in Fig. 7.9, where a crack can be nucleated at the site of an edge dislocation. In Fig. 7.9b, c, the coalescence of two or three dislocations is illustrated, as they increase the size of the crack. The concept of crack origin at dislocation sites has been adopted and modified by various researchers. In essence, Zener has suggested that cracks nucleate at pile-ups of dislocations, where sufficient stress for the nucleation of cracks builds up at the obstacles. Of the many representations of the pile-up concept, the approach taken by Stroh and Cottrell is presented below.

Stroh presented a dislocation model for spontaneous microcrack formation, calculated the elastic energy associated with wedge deformation. The nucleation of a wedge crack, caused by the pile-up of dislocations on a slip plane was also described. In Fig. 7.10, one sees a two-dimensional crack dislocation of giant Burgers vector, nb, with length, c, extending to a barrier.

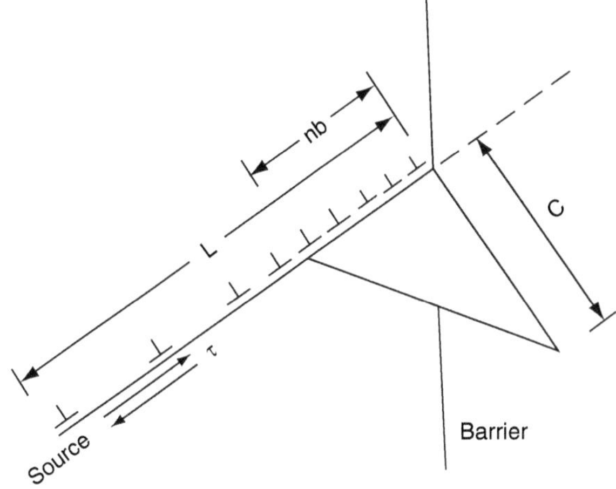

Fig. 7.10 Nucleation of a wedge crack due to piled-up dislocations on a slip plane (Stroh's model) (Sarfarazi and Ghosh 1987. With kind permission of Elsevier)

Stroh's expression for the elastic energy associated with wedge deformation is:

$$W_e = \frac{Gn^2b^2}{4\pi(1-\nu)}\ln\frac{4R}{c} \tag{7.36}$$

where G is the shear modulus (modulus of rigidity), nb is a giant Burgers vector and n is the number of dislocations comprising the giant vector. R is the bounding radius in the stress field. As previously, the surface energy term, $2\gamma_s c$, may be added to obtain the total energy of the system as:

$$W_S = \frac{Gn^2b^2}{4\pi(1-\nu)}\ln\frac{4R}{c} + 2\gamma_s c \tag{7.37}$$

Differentiating Eq. 7.37, the critical length, c_{min}, can be found:

$$\frac{\partial W_s}{\partial c} = 0 \tag{7.37a}$$

$$c_{\min} = G\frac{n^2b^2}{4\pi(1-\nu)}\frac{1}{2\gamma_s} \tag{7.38}$$

In polycrystalline solids, typical values of b, G, ν and γ_s are (Sarfarazi and Ghosh): $b = 2 \times 10^{-8}$ cm, $G = 10^{12}$ dynes/cm^2, $\nu = 1/3$ and $\gamma_s = 10^3$ dynes/cm, respectively, which gives for c_{min}:

$$c_{min} = 2.4 \times 10^{-8}$$

In his papers, Stroh also stressed a possible analogy to the microstructures of dislocation boundaries in the cases of screw and edge dislocation. A screw-dislocation

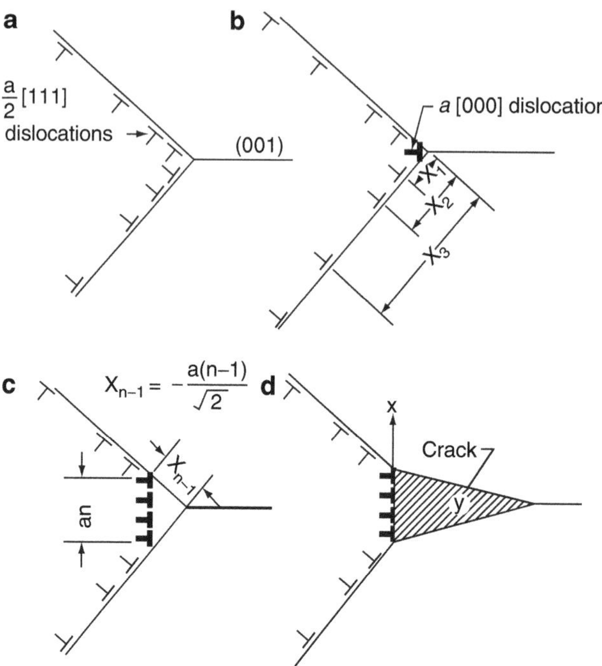

Fig. 7.11 A double pile-up model of crack initiation as proposed by Cottrell (Hahn and Rosenfeld 1966. With kind permission of Elsevier)

line may be described by a cylindrical hole with some surface-energy distribution, while an edge dislocation may form a tensile microcrack at its front.

Figure 7.11 shows Cottrell's concept of crack formation by dislocation pile-ups in a BCC metal. According to Cottrell's approach, a/2 <111> dislocations gliding on intersecting {101} planes (as shown in Fig. 7.11a, b) coalesce to form a <100> dislocations.

Furthermore, he suggested that a <100> dislocations act as the wedge to nucleate an unstable crack on a{100} planes, as shown in Fig. 7.11d. This type of wedge also appears in Fig. 7.9, just as Zener envisaged the role of dislocations. In Fig. 7.11, two types of crack formation by dislocations are indicated: the first shows coalesced a <100> dislocations wedging the crack; the second illustrates how other piled-up a/2 <111> dislocations confine coalesced dislocations. He calculates the breaking stress, σ_F, for fracture in a perfect crystal as:

$$\sigma_F = \frac{2\gamma}{na} \tag{7.39}$$

γ is the specific surface energy, a (=b) is the Burgers vector and n is the number of coalesced a{100} dislocations. According to Cottrell, since an a[001] edge dislocation has a large Burgers vector, but cannot decompose into glide dislocations without increasing its energy, it is likely to be incipiently cracked, the atomic bonds just beneath its half-plane being stretched beyond their critical strain, ε_m. The other dislocations in the slip bands (or twin bands) may use their stress fields to open up the incipient crack and enter it, in effect, forming a wedge in the (001) plane. Cottrell showed that the crack-nucleation stress required to form a microcrack may be quite small, especially when the intersecting dislocations are on low-energy slip planes. This reaction occurs on a cleavage plane in accordance with the following equation:

$$\frac{a}{2}[\bar{1}\bar{1}1] + \frac{a}{2}[111] \rightarrow a[001] \tag{7.40}$$

From all the experimental evidence, it may be concluded that plastic deformation precedes fracture. Cleavage-crack formation is associated with the nature of plastic flow and cracks form at the tips of slip bands (or twin bands).

7.5 Factors Causing Embrittlement

Various modes of brittle facture are widely discussed in the literature. As a consequence, the terminology for brittle fracture varies; the terms range from 'intergranular' and 'transgranular' to 'cleavage failure', all representing some sort of brittle fracture. Materials can be a priori ductile or brittle; however, it is often observed experimentally that a ductile material becomes brittle. In Chap. 1, Sect. 1.6, there was some discussion about the concepts of impact testing and the transient temperature in ductile-to-brittle failure. In this section, the focus is on other 'embrittling factors', factors causing a fracture to become brittle. A list of influencial embrittling factors and a brief discussion are given below.

7.5.1 Liquid Metal Embrittlement (LME).
7.5.2 Hydrogen Embrittlement (HE).
7.5.3 Aqueous Environment Embrittlement (AEE).
7.5.4 Temper Embrittlement.

7.5.1 Liquid Metal Embrittlement (LME)

The best-known form of embrittlement, liquid metal embrittlement [henceforth: LME] is usually induced by zinc in iron. However, LME has also been observed in many other systems, among them the embrittlement of Zn by mercury. Lynch has extensively considered various metals influenced by LME, including: aluminum alloys, nickel, iron-silicon, high-strength steels, magnesium and titanium alloys. Optical, SEM and TEM observations indicate that normally ductile metals are embrittled by LME. Intergranular and cleavage-like fractures are often associated with considerable, localized slip on planes intersecting crack tips and fractured

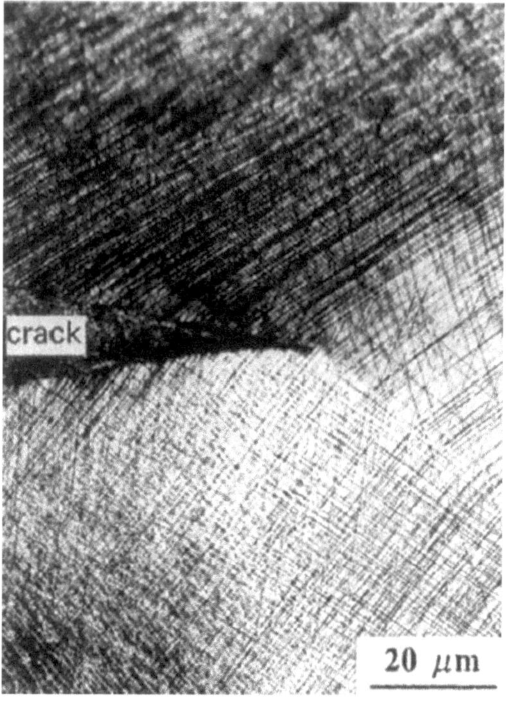

Fig. 7.12 Optical micrograph of aluminum single crystal (99.9% purity) cracked by liquid metal-alloy (a Bi-Pb-Sn-In-Cd eutectic at 60°C) showing slip around crack on the side surface of the specimen (Lynch 1989. With kind permission of Elsevier)

surfaces are sometimes dimpled. Exposure to liquid metals occurs when a generally ductile metal is coated with a low-melting material, such as tin, zinc or cadmium. (This occurs when a low-melting metal is in liquid form, such as during the galvanizing process of iron in a liquid-Zn tank.). A liquid metal may diffuse into a base metal or an alloy, probably along the grain boundaries in polycrystalline materials, making it brittle. Generally, cracking may start below the yield stress of the base material. Figures 7.12, 7.13, and 7.14 illustrate the effect of LME on otherwise ductile materials. Extensive slip, particularly on the planes are $\{112\}$ intersecting crack-fronts, was observed on specimen side surfaces and crack-tip opening angles are $\sim 8°$ after crack growth in hydrogen and liquid-metal environments (as shown in Fig. 7.14). Lynch did extensive metallographic and fractographic studies of environmentally-assisted cracking in: aluminum alloys, nickel, iron-silicon, high-strength steels, magnesium and titanium alloys, indicating that an adsorption-induced localized-slip process is responsible for cracking.

7.5.2 Hydrogen Embrittlement (HE)

Hydrogen is known to be one of the harshest embrittling agents in steel and its presence results in premature failures of structural components. Macroscopic

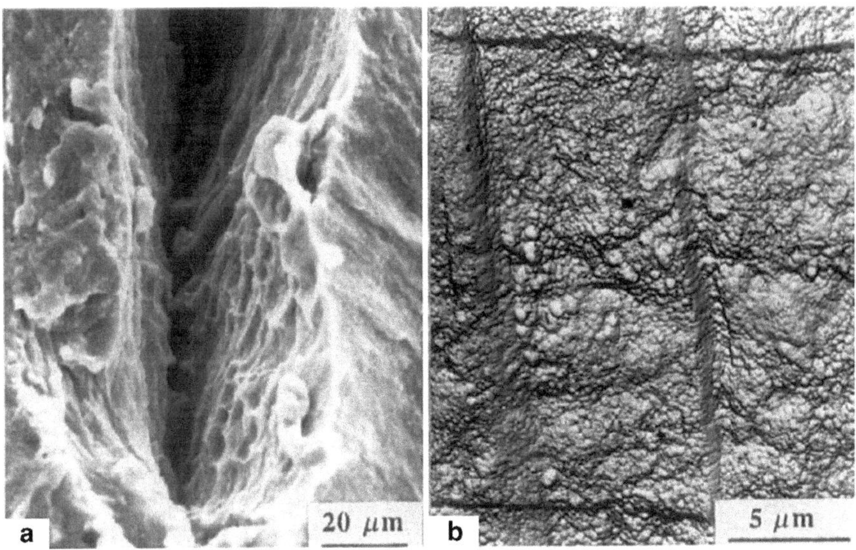

Fig. 7.13 Fracture surface of aluminum single crystal cracked in liquid Bi alloy at 60°. (**a**) SEM of secondary {100} crack normal to primary fracture surface showing dimples on the crack surface; (**b**) TEM of replica of primary fracture surface showing dimples and slip traces (Lynch 1989. With kind permission of Elsevier)

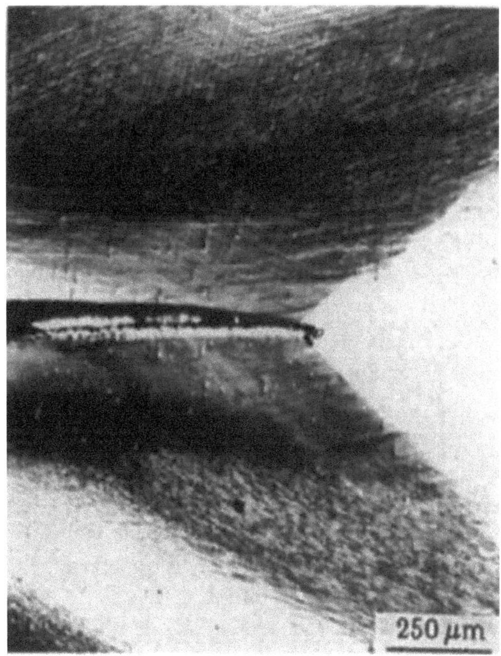

Fig. 7.14 Optical micrograph showing extensive slip around cleavage-like crack on the side surface of a F-Si single crystal after rapid crack growth (−1 mm/s) in liquid indium at ∼160°C (Lynch 1988. With kind permission of Elsevier)

observations of fractured surfaces indicate that brittle-intergranular fracture and cleavage facets exhibit tear ridges and patches of dimples. Fractured surfaces also show internal flakes or what is known as 'fisheyes'

Three major theories have been suggested to explain hydrogen embrittlement [henceforth: HE]. Despite the large amount of research that has been done on this subject, many aspects of this phenomenon still remain enigmas. Several theories have been proposed to explain HE (Quadrini). These theories may be grouped into three classes, as follows:

(a) The pressure theory. Embrittlement results from high pressure in the microvoids present in the material after the transformation of atomic hydrogen into molecular hydrogen.
(b) The decohesion theory (Troiano). Embrittlement is manifested as tensile atomic-bond breaking, occurring in the region with the highest stress triaxiality. This theory accepts that hydrogen diffuses under the influence of a stress gradient to regions having greater plastic deformation, thus reducing the interatomic cohesion forces.
(c) The surface-energy theory (Petch and Stables). This theory of embrittlement proposes that fracture stress is lowered by a reduction in the surface energy of a material at the internal surfaces of the microcrack.

Note the concept that hydrogen-assisted cracking is due to a hydrogen-induced microplasticity mechanism occurring ahead of the crack tip (Birnbaum). Another model attributed a material's loss of ductility to the temporary or permanent interaction of hydrogen with dislocations, inclusions, microcavities, point faults and grain joints (Pressouyre). With several theories on hand regarding HE, it is likely that actual conditions dictate the most prominent HE effect without being able to pin-point the exact mechanism in a particular case. One thing, however, is certain – all these theories share the common feature – that some critical concentration of hydrogen must be reached at potential crack sites in order for failure to initiate. This critical concentration is influenced by complex environmental and metallurgical interactions, such as the nature of the potential crack site, the mechanical resistance of the steel, the microstructure and the state of stress at the site.

The appearance of fracture surfaces produced by slow crack growth ($\sim 10^{-3}$ mm/s) in gaseous hydrogen at 25°C is shown in Fig. 7.15. Here, the fracture surfaces are macroscopically parallel to the {100} planes with crack growth in <110> directions. Figure 7.15, a TEM micrograph of replicas of fracture surfaces, reveals closely-spaced slip lines and small depressions between the tear ridges and steps. Extensive slip, particularly on the {112} planes intersecting the crack fronts, is observed on the specimen side surfaces and the crack-tip opening angles are $\sim 8°$, after crack growth in hydrogen, similar to a liquid-metal environment (Fig. 7.14). Crack growth in dry air, under equivalent testing conditions to those used in the hydrogen and liquid metal environment, produced ductile fractures.

Factors that greatly influence hydrogen embrittlement are strain-rate sensitivity, temperature dependence and the tendency for delayed fracture. Delayed fracture

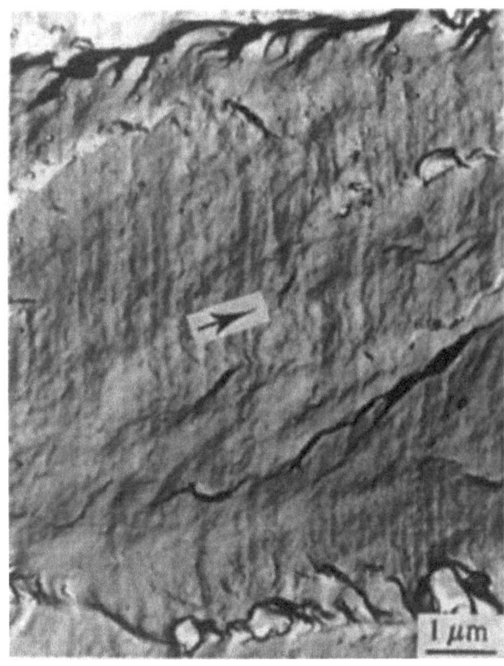

Fig. 7.15 TEM of fracture surface, produced in FeSi single crystal by slow crack growth ($\sim 10^{-3}$ mm/s) in hydrogen, showing closely spaced slip lines and small depressions (Lynch 1988. With kind permission of Elsevier)

is time-dependent and has been studied in glass, which is very susceptible to environmental effects.

The steels, that absorb large amounts of hydrogen, are prominent among the metals susceptible to delayed fracture. Therefore, a common method for studying hydrogen embrittlement is by investigating delayed fracture, also known as 'static fatigue'. A UNI 39NiCrMo3 steel was used for the delayed-failure test at a constant load. Since the hydrogen was introduced into the material by means of cathodic polarization in a sulphuric-acid solution at different pH values, the effects of the pH and current density on delayed fracture are of interest. The results obtained (Quadrini) show noticeable influences of the pH and current density on the threshold-load values and crack-incubation times. This was attributed to the effects of the pH and current density on the rate at which atomic hydrogen is adsorbed onto the material surface. In the next two figures (Fig. 7.16), delayed-fracture curves at various pH and current densities are illustrated. These curves show the same characteristics. Commonly-notched specimens are used for delayed-fracture studies. There is a region in the curves in which the time-to-fracture depends only slightly on the applied stress. The critical, minimum value, below which fracture will not occur (similar to the S-N curves), depends on the hydrogen content and, with its decrease, the static-fatigue limit increases. It may be seen from this curve that more hydrogen is adsorbed at low pH values. Since hydrogen was introduced to the specimen by cathodic polarization in a sulphuric acid solution, the pH is determined

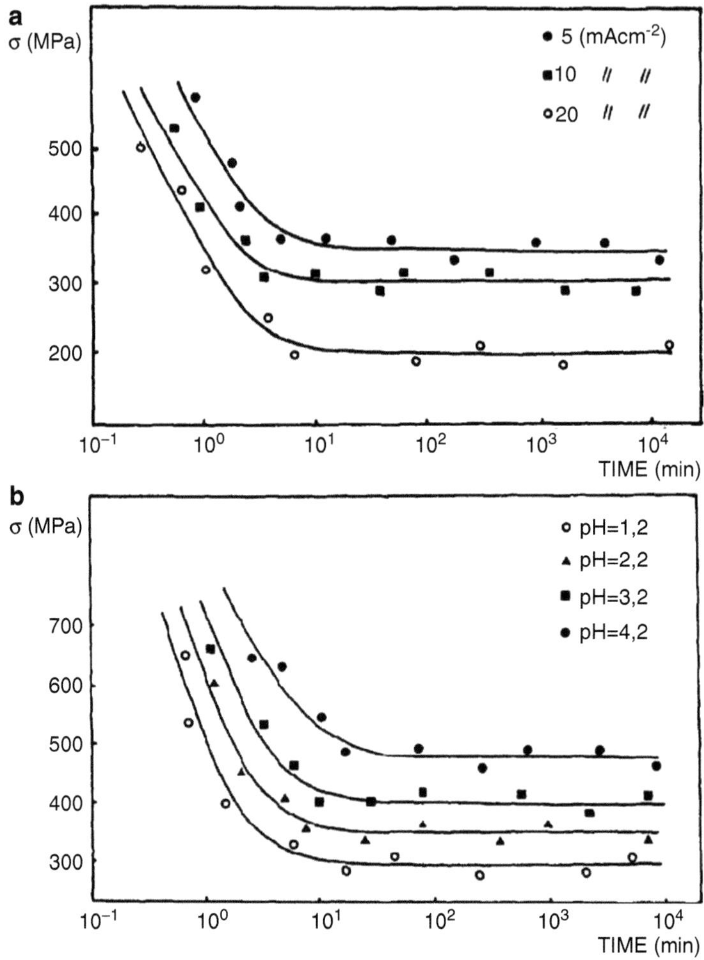

Fig. 7.16 Effect of the applied stress on time to failure, (**a**) for different current densities (**b**) for different pH values (Quadrini 1989. With kind permission of Elsevier)

by the current density, the value of which determines the pH and, hence, the amount of hydrogen absorbed on the specimen surface.

Any of the aforementioned theories can explain the observed HE effect responsible for the delayed-fracture curves, but the researchers offer diverse interpretations of how this mechanism works. Figure 7.17 is a SEM fractographic illustration of UNI 39NiCrMo3 steel, indicating a predominantly intergranular-type fracture, not modified by the test conditions. This is Austenitic steel, obtained after the mechanical finishing of the specimens, which were then Austenitized for 1 h at 870°C and oil-quenched.

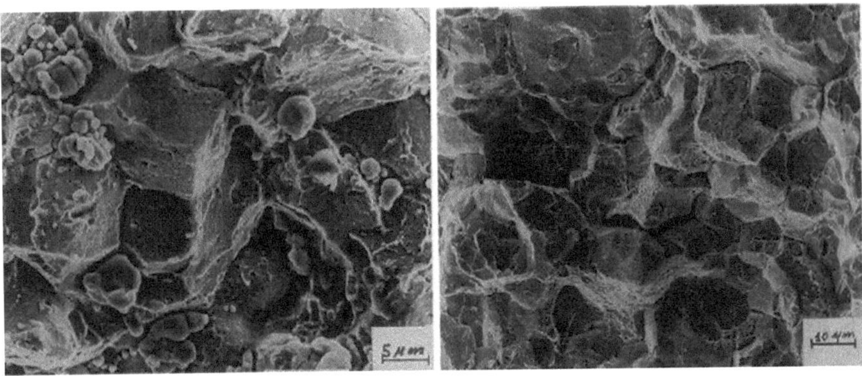

Fig. 7.17 SEM fractography with predominant intergranular fracture (Quadrini 1989. With kind permission of Elsevier)

7.5.3 Aqueous-Environment Embrittlement (AEE) or Stress-Corrosion Cracking

Basically, aqueous-environment embrittlement [henceforth: AEE] is an LME, where the liquid is aqueous. Though AEE might have been discussed above as part of the section on LME, a separate section is devoted to it, as an aspect of the broader subject of stress-corrosion cracking [henceforth: SCC]. Many materials in various industries are exposed to AEE, in particular, to some aspect of SCC. To name but a few, this topic is of critical interest in: the food industry, fermentation, waste disposal and, last, but not least, the ship-building industry. In this section, consideration is given to the combined influences of stress (e.g., tensile stress) and corrosive environments on machine elements. In AEE, there is no difference between directly-applied stress or residual stress acting on a part. One can imagine aircraft components or pipelines, in which residual stresses from fabrication may have remained. Almost every metal can be affected by AEE so that protection against sudden unexpected failure is of great importance. The AEE media can be acidic (such as chlorides) or caustic (e.g., ammonia, etc.). Usually, most of the surface remains unattacked, but with fine cracks penetrating into the material. In the microstructure, these cracks can have an intergranular or transgranular morphology. Since AEE happens unexpectedly and rapidly, leading to catastrophic failure or even fracture, it poses a problem of great concern.

An AEE is illustrated in Fig. 7.18 for two beta-brass alloys. Tensile tests were conducted at a strain rate of 0.02 min^{-1} in a polycrystalline material and the ductility of both alloys was reduced by 56%, compared with tests in air. Failure was transgranular with much secondary-transgranular cracking. The situation generally resembles SCC in other metals. The ternary alloy with Mn additions has a branch-like nature. The stress axis is along the horizontal and the cracks are emanating from a side surface. Crack propagation tends to be normal to the tensile stress.

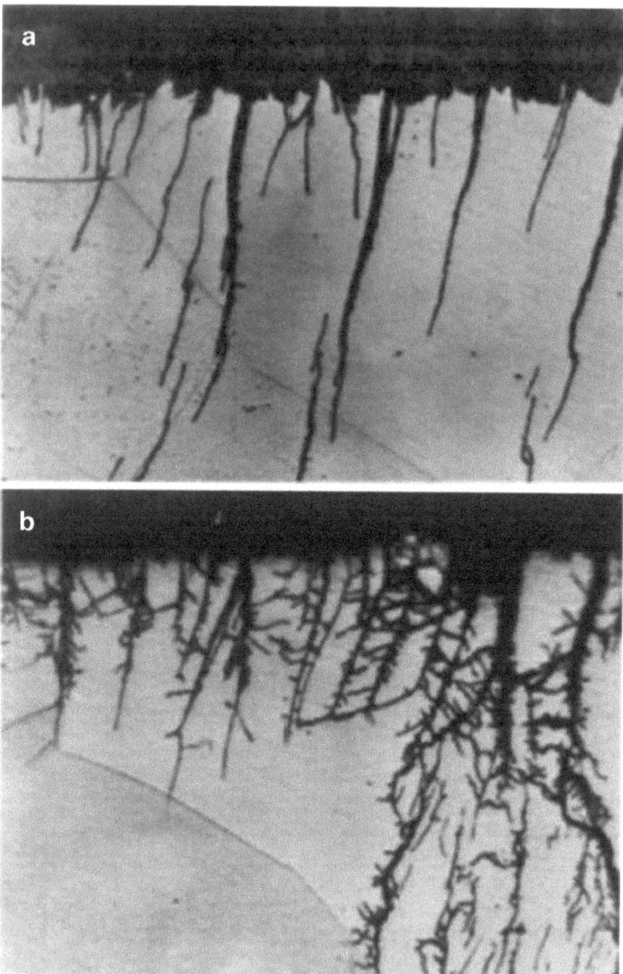

Fig. 7.18 Cracks (SCC) in 10M $NH_4OH + 0.12M\ Cu^+$ at 25°C, ×400 extending from side surface in two alloys; (**a**) Cu-48%Zn, (**b**) Cu-48%Zn-7%Mn (Shea and Stoloff 1973. With kind permission of Elsevier)

Note that the branching of the ternary alloy is a result of there being more effective sites for stress concentration as a consequence of restrictive cross slip. The absence of crack-branching in the binary alloy is an indication of easy cross slip. Mn is claimed to suppress cross slip. Figure 7.19 is another example illustrating cracks induced by water. The fracture plane in these metals is {100}, the facture direction <110> and the active slip planes {111} (Lynch). (Though LME and AEE can be broadly classified in the context of SCC, the scope of this textbook precludes a full discussion of all the aspects of SCC.)

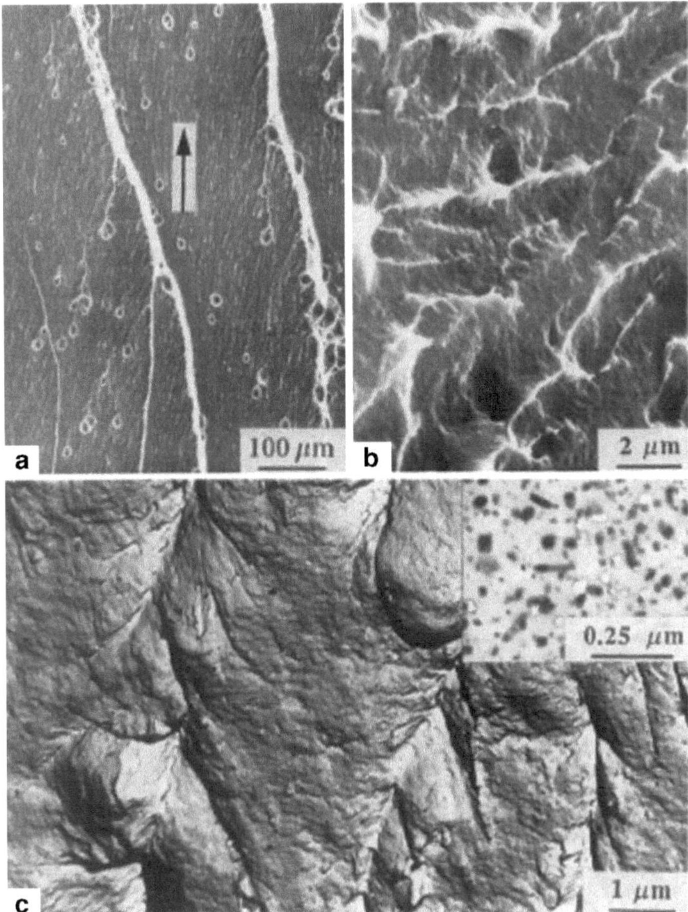

Fig. 7.19 Micrograph of overaged Al-6%Zn-3%Mg single crystal cracked in water at 20°C (velocity ~10 mm/s): (**a**) SEM at low magnification showing cleavage-like appearance; (**b**) SEM at high magnification showing small dimples; (**c**) TEM of replica (*inset*) showing smaller dimples than resolved by SEM (Lynch 1989. With kind permission of Elsevier)

One cannot finish this section without pointing out the very similar observations in LME, HE and AEE. Firstly, the metallographic, structural and fractographic similarities must be emphasized. They occur in a wide variety of materials and structures (FCC, HCP and BCC). These similarities suggest that the mechanism of embrittlement is likely to be the same. The existence of different mechanisms causing embrittlement in LME, HE or AEE is unlikely, considering that all their fractures have the same: crystallographic fracture planes, directions of crack growth, distributions of slip around the crack tips and characteristic appearance, consisting of dimples, tear ridges and slip-lines. One is tempted to conclude that a common mechanism is acting in all these cases.

Due to the very catastrophic effect that brittle failure causes, it is of utmost importance to control and eliminate embrittling phenomena. A few suggestions follow:

1. Select materials that are not susceptible to the environmental effects existing in their service environment.
2. Make sure that the cleaning materials used are not detrimental and that no residues remain on the cleaned parts. It is advisable to add inhibitors to the cleaning materials used during cleaning operations.
3. It is essential to reduce the stress concentration in manufactured parts by their sound and careful design.
4. Often, residual stresses can be reduced (if not completely eliminated) by heat treatment. Annealing is essential for this purpose.
5. Closed systems or vacuums are helpful for eliminating environmental effects.
6. Coatings may be effective means for protecting parts in service from harmful environmental effects.

7.5.4 Temper Embrittlement (TE)

Over the last century or so, the subject of embrittlement, including temper embrittlement [henceforth: TE] has been a major concern of various industries, due to the propensity for catastrophic failure; therefore, a vast literature has been accumulated on the subject. It is impossible to cover the entire range and all the fine details here; thus, a familiarizing discussion must suffice. TE occurs in some alloys that contain constituents, often referred to as 'tramp elements', e.g., phosphorus, arsenic, antimony, tin, etc. If any of these alloys are held (or heat-treated) for a period of time in a critical temperature range, then a tramp element may segregate into the grain boundaries, causing embrittlement. At that time, the concentration of the tramp element may increase within the grain boundaries, causing more embrittlement. Each alloy has its characteristic embrittlement temperature, but often the range of 350–570°C is quoted. In many steels, brittleness in the range of 230–370°C is known as 'blue brittleness'. (In the literature, sometimes a different range for blue brittleness is quoted). TE may occur whenever an alloy is held in this temperature range, a process known as 'tempering' (which should be avoided) or even when it is slowly cooled above this temperature range. Large parts are more susceptible to TE, because their interiors cool much more slowly than their exteriors. Fractures usually occur at grain boundaries, since embrittling impurities microsegregate there. Consequently, microsegregation has become an active field in the study of TE. Below, for example, is a discussion of phosphorus and antimony segregation-induced TE in steel.

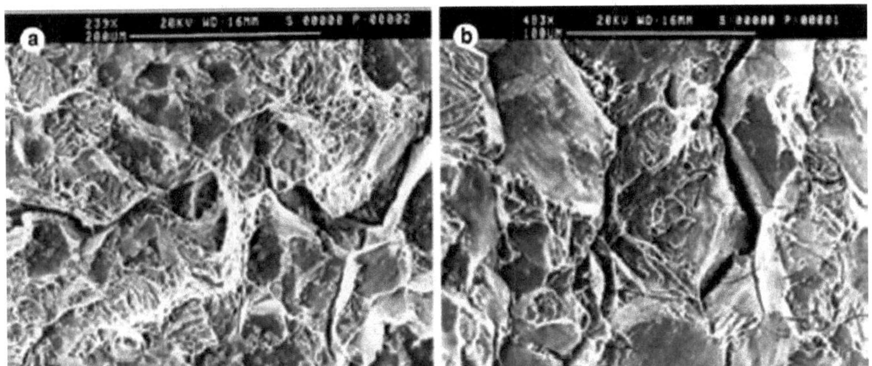

Fig. 7.20 Impact fracture SEM image of 12CrlMoV steel held at 540°C for 500 h: (**a**) impact at 10°C; (**b**) impact at −30°C (Qing-fen et al. 2005. With kind permission of Elsevier)

7.5.4.1 Phosphorus-Induced Embrittlement

The grain boundaries in materials produce a narrow zone of weakness, so when failure occurs in a material, it often occurs catastrophically by fracture along the grain boundaries. The results of experimental research indicate that phosphorus segregates into the grain boundaries in the austenitic phase (see, for example, Li et al., Ohtani and McMahon Jr.). The microsegregation of P into the grain boundaries can occur by equilibrium and non-equilibrium segregation. Whichever approach is accepted for grain boundary segregation, the fact is that the solubility of impurities is higher than in the matrix and is temperature-dependent. The enrichment of some impurities, such as P, at the grain boundaries, lowers the cohesion state across them, causing grain-boundary failure.

Figure 7.20 illustrates SEM images of fractured surfaces obtained by impact and the effect of P segregation to the grain boundaries. The reason for applying the temperature holding time at 500 h is because this is the time of maximum segregation of P at grain boundaries as shown in Fig. 7.21.

It has been found experimentally that reverse-temper embrittlement can occur, generally appearing after heating or slow cooling in the critical-temperature range (350–550°C) in low-alloy steels (see, for example, Li et al.); this is a metallurgical problem. Common indications of embrittlement are: the loss of toughness (i.e., a shift in the ductile-to-brittle transition to higher temperatures), fracture along grain boundaries and the segregation of impurities into grain boundaries. The presence of certain constituents in steel may augment temper embrittlement by P, due to a complex interaction at the grain boundaries. Mn is a well-known constituent of certain steels (e.g., 3.5 Ni-Cr-Mo-V steel, used as rotor steel) that causes a dramatic upward shift in ductile-to-brittle transition temperature, accompanied by severe-temper embrittlement. This phenomenon is attributed to the cooperative segregation of certain impurities and alloying elements into grain boundaries at temperatures below ~550°C (Smith).

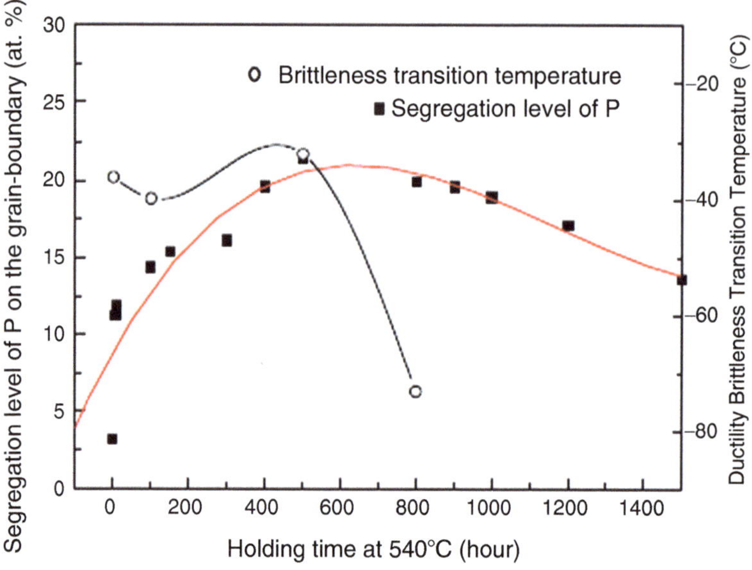

Fig. 7.21 Non-equilibrium grain boundary kinetics curves of P segregation and temper brittleness in steel 12Cr1MoV (Qing-fen et al. 2005. With kind permission of Elsevier)

7.5.4.2 TE by Antimony

Alloy steels loose their toughness progressively when heated or slowly cooled in the range ~350–550°C. TE was observed in a low carbon Ni-Cr steel doped with Sb (Fe-3.5% N-1.7% Cr and 0.07%Sb) or P, (Ohtani and McMahon, Jr.). This steel was quenched and tempered; the results indicate that primary austenite is not involved in the embrittling action by antimony (unlike the case of embrittlement by phosphorus segregation in the austenite boundaries). The microstructure shown in Fig. 7.22a, resulting from an iced-brine quench from 1,200°C (antimony-doped alloys) is a mixture of blocky ferrite and a lathe-type structure resembling upper bainite. This Ni-Cr steel alloy was tempered for 1 h at 625°C following the quench. The prior austenite boundaries are entirely obliterated (by transformation) and no signs remain after the high-temperature and quench treatments. TE is usually studied by impact testing to determine the ductile-to-brittle transition temperature. Figure 7.22b shows that when the specimen was quenched from 625°C (tempering temperature), the steel was in an unembrittled condition. The transition temperature was ~ –90°C and the fractures produced below this temperature were all cleavage fractures. The alloy became extremely embrittled by aging at 480°C. After a 1,000 h treatment at this temperature, the transition temperature was >300°C.

It was mentioned above that the grain boundaries of primary austenite are not the sites where antimony segregates. The dilemma -whether austenitic or ferritic grain boundaries are the sites where the embrittling constituent is located- existed for some time until it was settled. A metallographic approach has shown that antimony

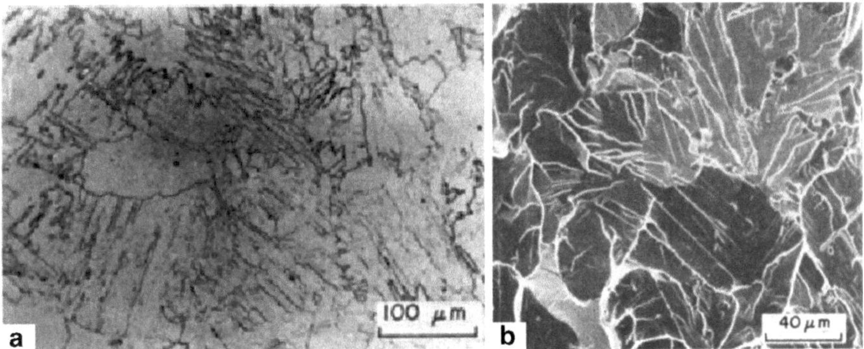

Fig. 7.22 Microstructure of Sb containing alloy after austenitization, quenching, and tempering for 1 hr at 626°C: (**a**) Blocky ferrite and a lath-type structure resembling bainite; (**b**) Fractograph of specimen (**a**), broken below the transition temperature, showing 100% cleavage (Ohtani and McMahon 1976. With kind permission of Elsevier)

is segregated in ferrite grain boundaries, as seen in Fig. 7.23a. This Figure shows a sample containing Sb, metallographically polished and then austenitized in vacuum and quenched. The prior austenitic grain boundaries are delineated by thermally-etched grooves. Figures 7.23b–d show the end of a crack, passed partly through the specimen (from the notch shown at the top of (a)), after it had been tempered at 625°C for 1 h and step-cooled (i.e., slow cooled). The fracture occurs along the prior austenitic boundaries, on and just beneath the free surface. However, further beneath the surface, the fracture path follows the rough boundaries of the transformed ferrite. The grain boundaries of the blocky ferrite and the upper bainite were formed during austenitic decomposition and the equiaxed ferrite formed by complete recrystallization below the eutectoid temperature. The almost classic aspect of the primary austenite grain-boundary location of TE, in the case of Sb, is actually not an essential feature of the problem. One can add, however, that prior austenite grain-boundary fracture can be induced in low-carbon steel (when temper-embrittled), if these boundaries are sufficiently well preserved during austenite decomposition, e.g., by very high quenching rates achieved near a free surface.

To summarize the essential aspects of TE:

1. In the 375–575°C range, most common low-alloy steels show an increase in their ductile-to-brittle transition temperatures, regardless of whether they are heated in this range or slowly cooled through it.
2. Certain tramp elements, such as P, Sb, As and Sn, induce embrittlement steels by segregating into grain boundaries, either of austenite (P) or ferri (Sb).
3. Certain constituents of steel may contribute to TE by their cooperative action with some of the well-known embrittling impurities. Manganese is susceptible to this at a certain concentration and temperature.
4. Steel with a high Cr content (15 wt.% or more) may be subject to embrittlement when held in the 400–500°C range for long enough time.

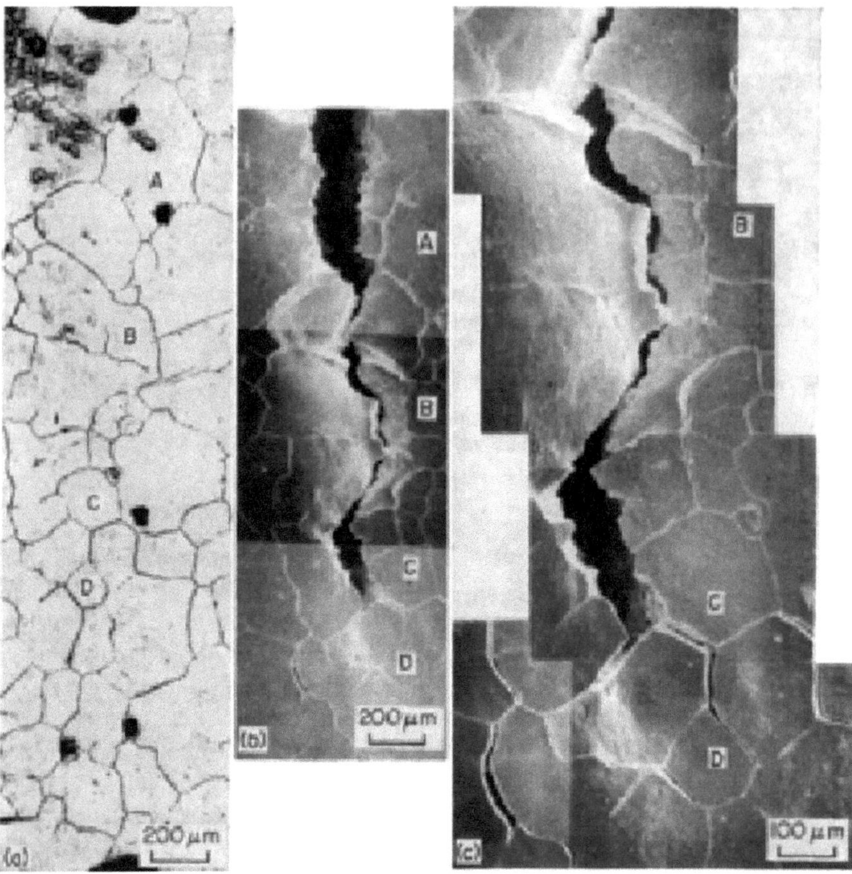

Fig. 7.23 (**a**) Prior austenite grain boundaries. Notches can be seen at *top* and *bottom*; (**b**) fractured sample of (**a**) tempered and step cooled before testing; note that on the fast-quenched surface the fracture occurred mainly along the prior austenite grain boundaries. (grains labeled *A*, *B* and *C* to locate fracture path.) (**c**) Portion of (**b**); note that fracture just beneath the fast-quenched surface does not follow prior austenite grain boundaries, but rather that of the ferrite (Ohtani and McMahon 1976. With kind permission of Elsevier)

To reduce TE:

(a) A plant engineer must make it sure to temper outside the critical temperature range.
(b) Alloys must be cooled rapidly through their embrittlement temperature range, to make certain that the tramp elements do not have sufficient time or energy to segregate into the grain boundaries.
(c) Embrittlement can be eliminated by a proper soak at a higher temperature.
(d) **TE is reversible!** If parts are suspected of being tempered or slow-cooled within the embrittlement range, the suspect parts can be re-tempered at a temperature outside the embrittling range. By re-tempering, TE can be drastically reduced.

7.6 Fracture Toughness

The stress-to-fracture in conventional tests, such as tensile tests, is not a safe guide for the prevention of fracture resulting from crack propagation in structures used in service since crack growth to catastrophic dimensions might often occur even below the tensile yield stress of the material. A critical-stress intensity factor, K_c, is a measure of fracture toughness. The fracture resistance of a material is known as its 'fracture toughness'. Fracture toughness generally depends on: temperature, environment, loading rate, the composition of the material and its microstructure, together with the geometric effects of the tip. This problem is of particular importance for welded joints, where the metallurgical and geometric effects are complex. The subject of fracture toughness is generally related to the amount of stress necessary to propagate a preexisting crack. Defects, in various forms, commonly exist in materials and may develop to failure by fracture. The I subscript is used to denote the crack-opening mode (see Fig. 7.24) in Eq. 7.41:

$$K_{Ic} = \sigma \sqrt{\pi a B} \tag{7.41}$$

This stress-intensity factor is a function of loading, crack size and structural geometry. σ is the applied stress, "a" is the crack length and B is a dimensionless factor, which depends on specimen geometry. Thus, fracture toughness is the resistance of a material to failure from fracture initiated by a preexisting crack. The occurrence of cracks in structural components indicates a certain threat to their reliable operation, because these cracks can grow during operation and reach critical sizes, leading to fracture. For a crack to grow, the stress at the crack tip must be greater than the strength of the material. The cohesive bond in the vicinity (front) of the crack is thus broken. However, this, by itself, is not a sufficient criterion for the growth of a crack to failure. An additional requirement is the release of energy that must occur when crack growth should be at the level needed to form two new crack surfaces. Both of these prerequisites are essential, i.e., if the force introduced to the system is of sufficient magnitude to break the bonds, but there is not enough energy to form new surfaces, the crack will not grow. On the other

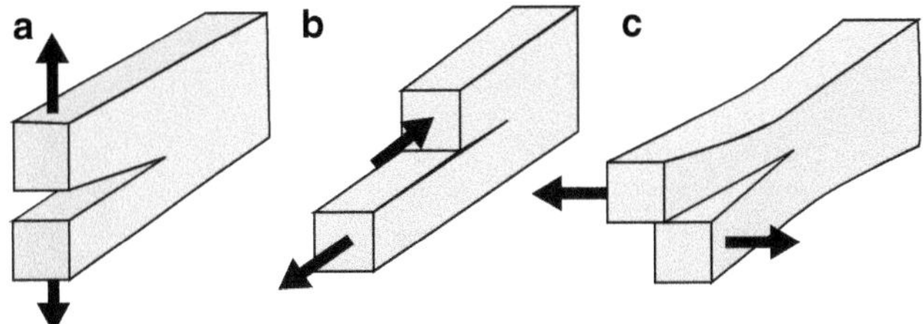

Fig. 7.24 Different failure modes: (**a**) Represents mode-I (tensile force), (**b**) represents mode-II (shear force) and (**c**) represents mode-III (torsional force) (Soderholm 2010. With kind permission of Elsevier)

hand, if there is not enough force introduced into the system to break the bonds, but there is enough energy to form the new surfaces, the crack will also not grow. The geometry of a crack, more precisely, its sharpness, is a decisive factor in establishing the magnitude of the stress-concentration factor, which develops at the crack tip. The stress-concentration factor at the tip of a sharp crack requires only sufficient energy to form new surfaces, since the magnitude of the stress at the tip is of the order of the theoretical strength of the material (its cohesive stress). Therefore, a relatively small load will provide the necessary conditions for crack propagation. When the crack tip is dull (i.e., blunt), it is relatively easy to provide the energy for the formation of two new surfaces, but the stress is insufficient at the crack tip, because the stress concentration is much lower compared to a sharp crack. In this case, higher loads are needed to provide the required stress for the growth of the crack.

Experimental evaluation of fracture toughness may be performed statically or dynamically; the most common method is dynamic testing. Four-point bending is applied to deform specimens, which are usually notched and fatigue pre-cracked. The effect of fatigue damage is removed by heat treatment. Bend-test specimens basically have the shape of impact specimens and the test is performed at various temperatures (Ritchie). There are several types of loading (Anderson) illustrated in Fig. 7.24. Below, A represents a situation in which a crack is being pulled apart by the application of a tensile load (mode I). Modes II and III represent shearing forces acting on the opening (crack). Mode II represents an in-plane shear, while mode III is related to a tearing-shear load.

Of these three modes of cracking, mode I (crack opening) is the most commonly discussed in the literature. Eftis and Leibowitz (following previous contributions on the effects of crack-tip stresses and after Westergaard's developments) show the opening-mode stresses in polar coordinates as:

$$\begin{aligned}
\sigma_y &= \frac{K_{Ic}}{\sqrt{2\pi r}} \cos\frac{\theta}{2}\left(1 + \sin\frac{\theta}{2}\sin\frac{3\theta}{2}\right) \\
\sigma_x &= \frac{K_{Ic}}{\sqrt{2\pi r}} \cos\frac{\theta}{2}\left(1 - \sin\frac{\theta}{2}\sin\frac{3\theta}{2}\right) \\
\tau_{xy} &= \frac{K_{Ic}}{\sqrt{2\pi r}} \left(\sin\frac{\theta}{2}\cos\frac{\theta}{2}\cos\frac{3\theta}{2}\right) \\
\tau_{yz} &= \tau_{xz} = 0 \\
\sigma_z &= n(\sigma_y + \sigma_x) \\
\sigma_z &= 0 \ \text{ for plane stress}
\end{aligned} \tag{7.42}$$

The stress distribution in the vicinity of a crack is shown in Fig. 7.25. Figure 7.7 is inset at the center of Fig. 7.25 and the origin of the coordinate system is located at the crack tip.

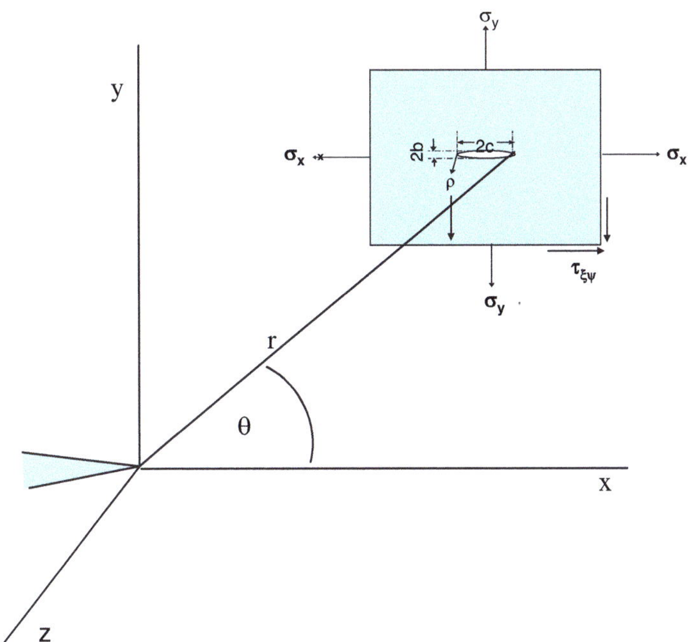

Fig. 7.25 Distribution of the stresses in the vicinity of the tip of a crack

Note that the stress approaches infinity as the crack tip is approached, since the denominator factor, $(2\pi r)^{-1/2}$, approaches infinity with r going to zero. The literature contains expressions for K (and α) for a large number of crack and loading geometries and both numerical and experimental procedures exist for determining the stress-intensity factor in actual, specific geometries. From Eq. 7.41, it is possible to express the critical stress as:

$$\sigma_f = \frac{K_{Ic}}{\sqrt{\pi a B}} = \frac{K_{Ic}}{\alpha\sqrt{\pi a}} \tag{7.41}$$

In the above relation, α $(=B^{1/2})$ is a geometrical factor equal to 1 for edge cracks and generally of the order of unity in other situations. In design analysis, it is assumed that a material can withstand a stress up to a critical value of the stress-intensity factor. Beyond this critical value of K_{Ic}, cracks propagate rapidly.

The dynamic testing method may be implemented by means of impact testing. This is a consequence of the fact that fracture behavior is usually strain-rate dependent. Various measures of toughness exist, including the simple and widely-used, qualitative Charpy impact test. Currently, instrumented impact tests are readily available and provide more quantitative results. Instrumented Charpy pendulum tests are widely used to determine the fracture properties of materials. A graph obtained by the dynamic testing of API X70 steel (used as high-Mn pipe-line steel) is shown in Fig. 7.26 and records the force-time relations.

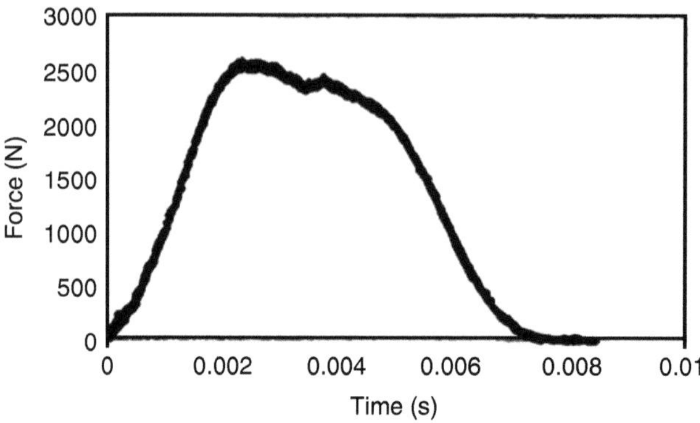

Fig. 7.26 Force time record from an impact test (initial velocity 0.6 m/s) of API X70 steel (used as pipe-line steel) (de Luna et al. 2000. With kind permission of Elsevier)

In such tests, standard V-notch Charpy-sized specimens, with an initial crack generated by fatigue, are used. The pre-cracking of specimens is carried out as usual in such tests. After fatigue pre-cracking, these specimens are side-grooved to promote straight-fronted, ductile-crack growth during the test. The energy absorbed, E_a, by the specimen is given by the area under the curve (de Luna, et al.). Thus, the integral is:

$$E_a = \int_0^{t_f} F(t) ds_c \tag{7.43}$$

t_f is the time at which the hammer velocity decays to zero. The displacement is s_c, at the load-application point of the specimen and ds_c is its variation. The J integral (a means of calculating the strain-energy release rate or work (energy)) per unit fracture surface area in a material is given by Eq. 7.44. It is independent of the path around a crack:

$$J = \frac{2E_a}{B(W - a_0)} \tag{7.44}$$

B and W are, respectively, the thickness and the width of the specimen and a_0 is the initial length of the crack (in Fig. 7.7 or in Griffith' theory, the length of the crack is noted as 2c). Figure 7.27 shows the conventional static and dynamic J-Δa curves (known as R-curves) derived from experimental results. In Fig. 7.28, the relationship between the J integral and the damage variable is shown for the dynamic tests at different velocities and also for the static tests. Relatively little difference between the curves is observed, indicating that low-blow tests (de Luna et al.) are appropriate for obtaining the dynamic-fracture toughness at intermediate-strain rates.

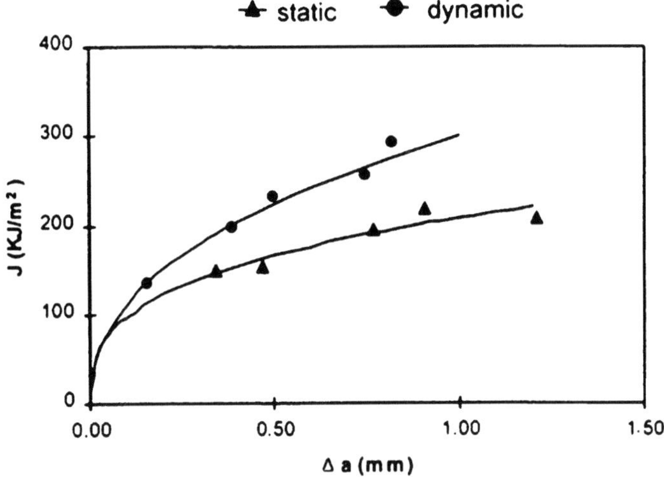

Fig. 7.27 Conventional static and dynamic R-curves (de Luna et al. 2000. With kind permission of Elsevier)

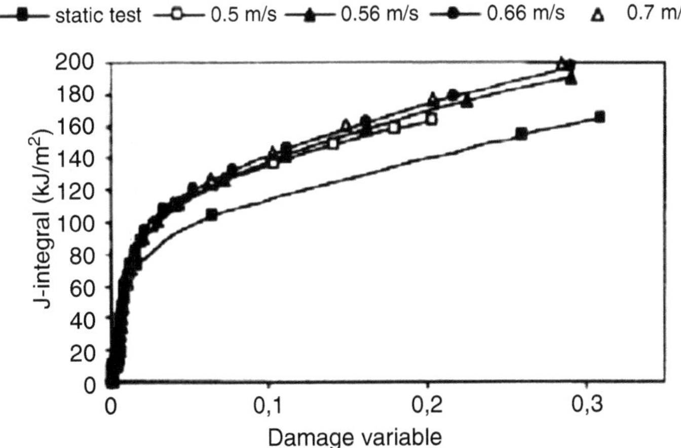

Fig. 7.28 J-D curves showing the relationship between the J-integral and the damage variable (de Luna et al. 2000. With kind permission of Elsevier)

Figure 7.28 reflects both static and impact three-point bending tests that were performed to determine the fracture toughness of the material at different strain rates. These test results show higher fracture toughness of the material under dynamic conditions than under static ones. Analyses of the blunting line were made for both the static and the dynamic conditions and the comparison between the experimental and the numerical analyses showed good agreement. Under dynamic conditions, the slope of the blunting line is higher than that of the static one. The damage parameter is given as:

$$D = \frac{R}{R_0} \tag{7.45}$$

where R represents the growth radius of a spherical microvoid at any time and R_0 is its initial radius (de Luna et al.), respectively.

There is a critical value of the fracture energy, J_{Ic}. The quantity, J_{Ic}, defines the point at which large-scale plastic yielding during crack propagation takes place under mode I loading. The J-integral is equal to the strain-energy release rate for a crack in a body subjected to monotonic loading. This is true under quasistatic conditions, both for linear elastic materials and for materials that experience small-scale yielding at the crack tip. K_{Ic}, on the other hand represents a critical-stress intensity factor.

Various measures of toughness exist, including the widely-used, but qualitative, Charpy impact test. Although it is possible to correlate Charpy energy with fracture toughness, a large degree of uncertainty is associated with such correlations, because they are empirical. It is preferable to determine fracture toughness in a rigorous fashion by one of these three methods: in terms of K (the stress-intensity factor); crack-tip opening displacement [henceforth: CTOD]; or the J integral. The most common specimen configuration is the single-edge notch bend [henceforth: SENB] specimen. A sharp-fatigue crack is inserted in a specimen, which is then loaded to failure. The crack-driving force is calculated for the failure condition, giving the fracture toughness.

In regard to the Charpy impact method for the evaluation of fracture toughness, note that the Charpy V-notch impact test has had considerable success in providing designers with some idea of the relative notch-toughness of various materials. The main advantage of this test is that the specimen used is relatively small. Transition temperature may be determined by means of a test (see Chap. 1), however, the inherent disadvantage is that the measurement taken is valid only specifically for the prevailing test conditions. The Charpy V-notch transition temperature may be altered by: a change in specimen size, a change in notch configuration or a change in the rate of loading. However, one must be aware of the disadvantages of using this method for practical-design purposes, as indicated below: (a) the Charpy V-notch impact test does not reproduce the triaxiality that occurs in thicknesses greater than the 10 mm-specimen size (used for the Charpy tests); (b) the notch is blunt by comparison with natural cracks occurring in materials; (c) the majority of brittle failures in service occur under static conditions and not under dynamic conditions (as in impact tests); and d) the material tested is usually taken from a test sample that is not always entirely representative of the material as a whole. Consequently, Charpy test results cannot be used to determine a safe working temperature for service applications, nor for safe design against brittle fracture. More sophisticated types of tests are required and linear elastic fracture mechanics [henceforth: LEFM]

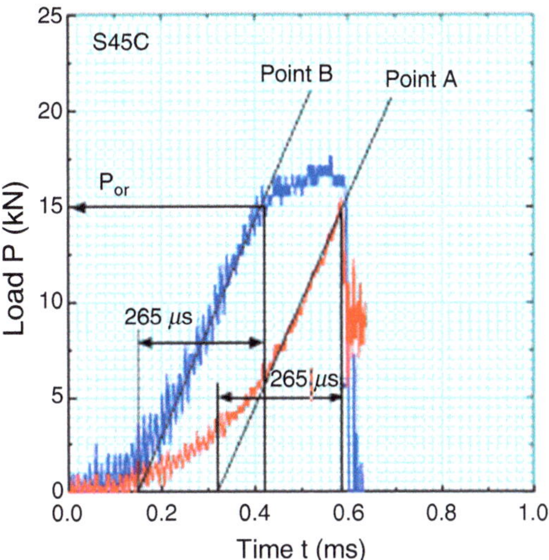

Fig. 7.29 Time histories of dynamic load (Wada et al. 2004. With the courtesy of Professor Kennedy for the authors)

have been found useful for predicting, in terms of a single parameter, the fracture stress of components containing sharp flaws.

When instrumented Charpy impact tests are used (as indicated in Figs. 7.26, 7.27, and 7.28), most of the disadvantages of the conventional Charpy tests are eliminated. Today, these instrumented tests often use modern machines that incorporate sophisticated, high-technology instrumentation, producing accurate results applicable for design purposes.

New techniques exist (Wada et al.), allowing the use of sharp-notched, round-bar specimens without fatigue pre-cracking for the determination of plane strain fracture-initiation toughness under dynamic loading. This technique evaluates fracture-initiation toughness for a wide spectrum of loading rates by combining the finite-element method and fracture experimentation. That is, the fracture-initiation time is measured by means of an electrical resistance strain gage during the fracture experiment and the stress distribution in front of the notch is analyzed by finite-element calculation. The effectiveness of this combined technique has been confirmed by comparison with the results available in the literature. This technique is performed by means of a weight-dropping-type dynamic tension test device, consisting of a tension jig, foundation and cylindrical tower. The cylindrical tower is a seamless pipe of 2,000 mm length with flanges welded at the ends and it is vertically fixed on the tensile jig. A dynamic tensile load is applied to the specimen installed in the tension jig as the weight that falls freely along the cylindrical pipe, colliding with the tension jig. The weight is freely-dropped from arbitrary height. The velocity of the colliding weight with the tension jig changes in proportion to the height of the fall. The weight is a carbon-steel cylinder with a diameter of 153 mm, length of 150 mm and mass of 21 kg. Figure 7.29 shows the time-history of the load at point A and point B. A phase difference in time is observed between the two. Therefore, the fracture time is determined from the signal of point A and is applied

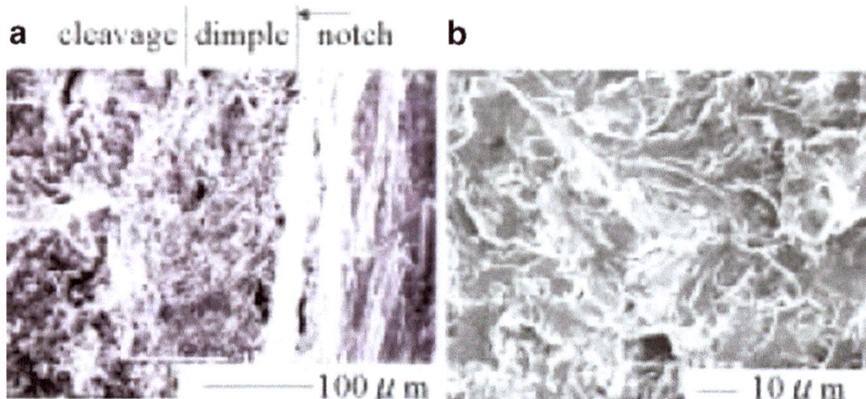

Fig. 7.30 SEM dynamic fractographs in the fracture surface of a dynamic test. (**a**) in the notch tip neighborhood; (**b**) in the center of the fracture surface (Wada et al. 2004. With the courtesy of Professor Kennedy for the authors)

to the signal of point B. Initially, the signal at point B is a straight line, but, after a period of time, the slope of the curve decreases. The load, at the point where there is an abrupt change in slope, is the 'critical load'. Most measurements are executed only by measuring point B. A SEM fractograph of a surface taken during a dynamic test is shown in Fig. 7.30. As for Fig. 7.30a, the dimple, which indicates ductile fracture, is about 80 μm from the notch tip. The fractograph in Fig. 7.30b shows a cleavage pattern, which indicates brittle fracture. Since this dimple-fracture area is small, compared with the remaining circular-ligament size, it is determined that brittle fracture is predominant in this case. Figure 7.30 shows an example of the time-history of the load measured in a quasistatic test. The fracture time and the critical load may be determined from the graph (Fig. 7.31).

Figure 7.32a shows a SEM fractograph of the notch-tip neighborhood and Fig. 7.32b shows a SEM fractograph of the center of the fracture surface. This SEM fractograph, similar to that obtained during the dynamic test (though excluding the tip vicinity), indicates brittle fracture. The dimpled area, in both cases, dynamic and static, is a result of heat generated during the machining of the notch. Again, this technique is simpler, since there is no need to introduce a pre-crack and the results are as good as the fracture-toughness measurements attained by more conventional techniques.

We can summarize this section as follows. Brittle failures cannot be predicted by simple engineering calculations, due to the catastrophic and unexpected nature of such structures. Efforts have been made to control this problem. From a macroscopic standpoint, brittle fractures are characterized by the following:

(a) Little or no plastic deformation precedes fractures.
(b) Fractures are generally flat and perpendicular to the surface of a component. In addition to the cleavage planes, river patterns are also observed.

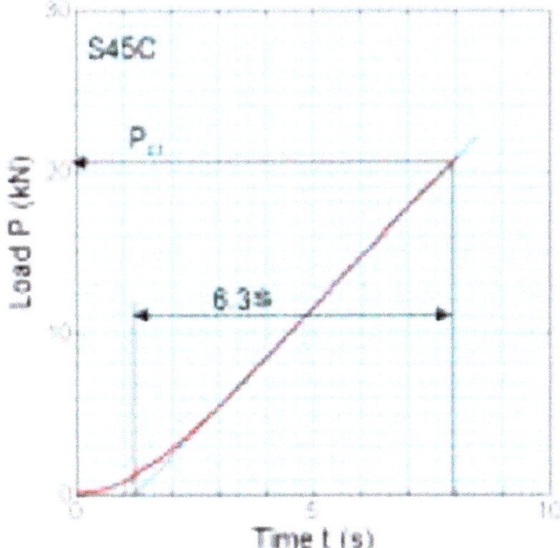

Fig. 7.31 Time history of a static load (Wada et al. 2004. With the courtesy of Professor Kennedy for the authors)

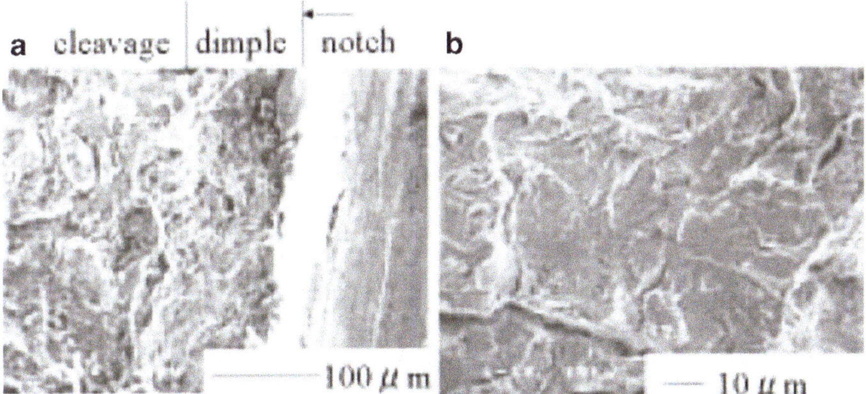

Fig. 7.32 SEM fractographs in the fracture surface of a static test (Wada et al. 2004. With the courtesy of Professor Kennedy for the authors)

(c) Fractures may appear granular or crystalline and are often highly reflective to light. Facets may also be observed, especially in coarse-grained steels, that can propagate rapidly.
(d) Herringbone or chevron patterns may be also be present.

A few structures of the kind mentioned above are illustrated below: Figure 7.33 shows a cleavage fracture in an Al-Li 8090 alloy (Al-2.43 Li-1.17 Cu- 0.74 Mg-0.12 Zr). Each cleavage part initiates in a large particle (1–5 μm) and propagates outside it, as seen in Fig. 7.33a. Figure 7.33b shows an initiated cleavage fracture in a large particle on a grain or subgrain boundary, propagating to one side. Squared etch pits on the cleavage-fracture surface indicate that it is a {100} plane. This specimen

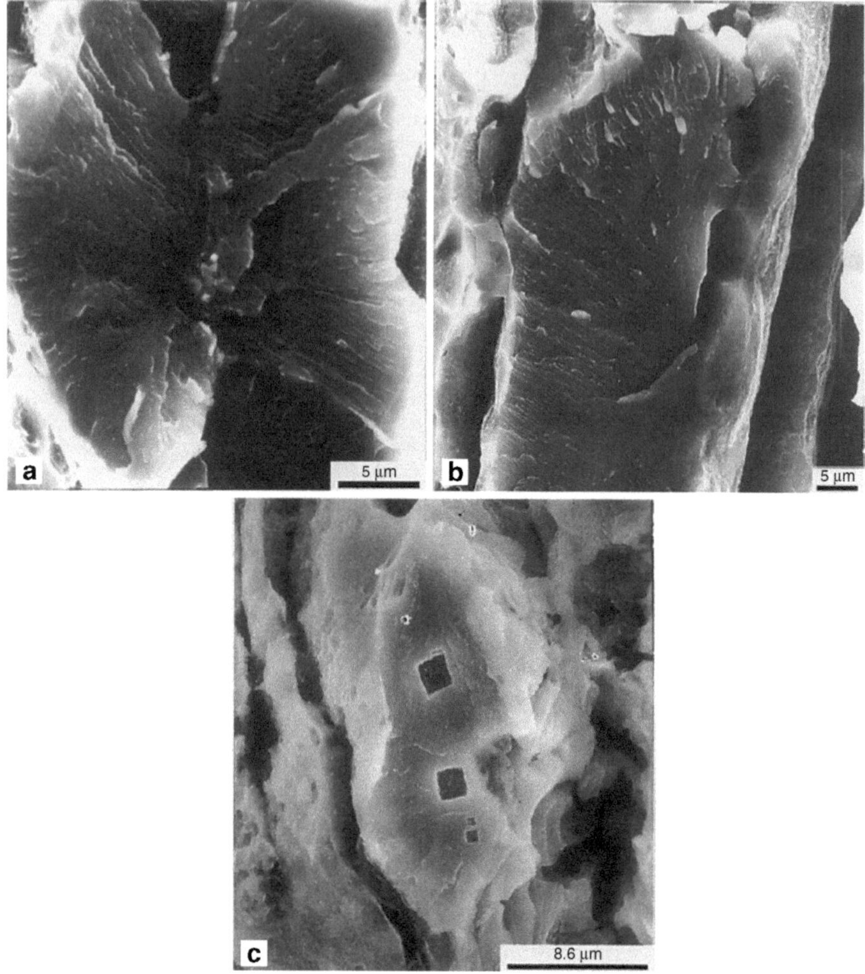

Fig. 7.33 SEM micrographs showing cleavage fracture in an Al-Li 8090 alloy: (**a**) crack initiates in the grain; (**b**) crack initiates on the grain; (**c**) etch on cleavage surface (Daming et al. 1996. With kind permission of Springer)

has undergone tensile tests at different strain rates. In this alloy, a crack initiated at inclusions (i.e., the alloy contains tramp elements); more specifically, the embrittling tramp element is identified as Na.

Another microstructure feature in brittle fracture is the so-called 'river pattern', shown in Fig. 7.34. In the SEM micrographs in Fig. 7.34, steps and river patterns are seen on the facet, also covered by fine pits. Since a crack in this alloy, experiencing low-temperature brittle fracture, propagates either through grain boundaries or transgrains, and since intergranular cracks form more easily than transgranular cracks, it is often the case that cracks prefer to propagate along grain boundaries.

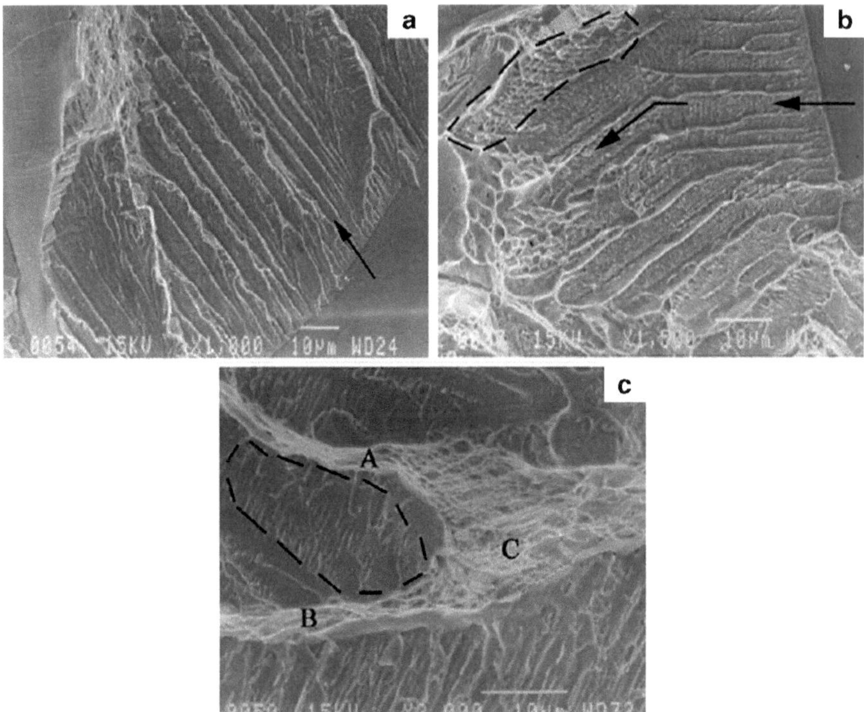

Fig. 7.34 SEM fractographs of transgranular fracture facets of high nitrogen austenitic steel fractured at 77 and 4 K: (**a**) tensile fractured at 4 K showing a facet with steps and river pattern; (**b**) steps and river pattern; (**b**) and (**c**) specimen fractured at 77 K (Liu et al. 2007. With kind permission of Springer)

When coming across another grain, the crack induces a transgranular crack. According to the river-pattern direction, one can identify the cracking direction of the transgranular fracture in the micro area, as indicated by the arrows. Careful observation indicates that steps and river patterns on the transgranular facet are clustered and basically straight, flat and parallel within each individual group, showing crystallographic characteristics. Figure 7.34a shows that one group of parallel steps traverses a grain from one side to the other. In Fig. 7.34b, transgranular fracture starts in a group of smaller, parallel steps and then the direction of the fracture turns to another group of larger parallel steps. Based on observations and morphological features, a mechanism has been proposed for the formation of the transgranular facet. When a stress is applied, a large amount of deformation structures are formed along {111} planes. Intersections of these deformation structures, on different {111} planes, cause stress concentration and originate microcracks, which then propagate and coalesce to form transgranular fractures. Furthermore, on cleavage surfaces structures may also appear known as 'herringbones' or 'chevrons'. Such features indicate that cleavage-fracture surfaces often run in two different directions, on

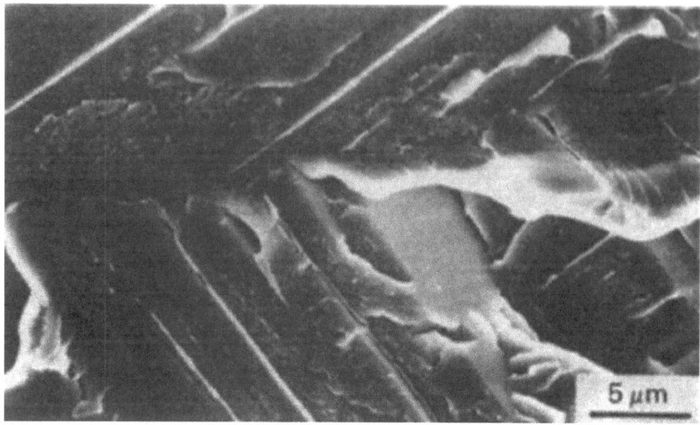

Fig. 7.35 SEM of fracture surface produced by rapid cleavage in Fe-2.6% Si single crystal in air at 25°C. Herringbone pattern (Lynch et al. 1986. With Kind permission of Elsevier)

either side of a central spine, producing a herringbone appearance. Herringbone patterns are commonly observed especially in environmentally-induced cleavage-like fractures of normally-ductile materials. A SEM fracture surface produced by rapid cleavage is illustrated in Fig. 7.35.

In all the cases in which herringbone structures were observed (Fe-Si single-crystals cracked in liquid lithium, Ni single-crystals cracked in hydrogen gas, Al-Zn-Mg single-crystals cracked in moist air), the fracture plane was approximately parallel to the {100} planes and crack growth occurred in mutually orthogonal <100> directions on either side of the spine.

Experimental observations suggest that hydrogen absorption is responsible for this embrittlement. Absorption weakens the interatomic bonds at the crack tips, thereby facilitating the nucleation of dislocations and promoting an alternate slip process.

7.7 Ductile Fracture

7.7.1 Introduction

Ductile fracture has been less extensively studied than brittle fracture, because, in this case, unexpected catastrophic failure does not pose a threat. In Chap. 1, the problem of necking (Sect. 1.2.7), before the onset of fracture and instability under tension (Sect. 1.2.8), was discussed. The central consideration was that necking occurs before fracture. In Fig. 7.2, three of the schematic illustrations show various forms of ductile necking preceding fracture. A few characteristics of ductile structure are listed below:

(a) There is permanent deformation at the tip of an advancing crack. Distinct patterns in microstructures, such as SEM images, characterize the mode of the fracture.
(b) As with brittle fractures, the surface of a fracture tends to be perpendicular to the principal-tensile stress, although other stress components may be involved (e.g., shear stresses).
(c) At a crack tip, shear stresses (microscopically resolved shear stresses) operate and are involved in crack expansion.
(d) Fracture surfaces are dull and fibrous.

For crack extension, a considerable amount of energy must be available.

As illustrated schematically in Fig. 7.2, the result of the necking process can appear in three forms, but in all successive necking event, cavities form in the necked regions. Cavities eventually coalesce and grow to form cracks. When these cracks have grown sufficiently large, they spread at an accelerated rate in a direction transverse to the applied stress. One of the ductile fractures is the cup-and-cone type as shown in Fig. 7.2b. At a certain stage, a crack spreads to the surface, inclined at a 45° angle to the tensile-stress axis (see Fig. 7.2c). When the material is very soft, the necking appears as shown in Fig. 7.2d.

One may state that, prior to fracture, the strain hardening of a material is less than the load-carrying capacity of its test specimen, due to the accelerated decrease in the cross-section. The localized necking or thinning down of the specimen occurs quickly until fracture sets in. Basically, the competition between strain hardening and decreased load-carrying capacity at some weak location determines the stability of a material during tensile deformation; when the load-carrying capacity is low, fracture occurs.

7.7.2 *The Process of Neck Formation*

The micrographs in Figs. 7.36 and 7.37 present the stages of neck formation after uniform strain in high-purity single crystals of 3% silicon iron pulled by tension. Fracture occurs either when the specimen draws down to a point, forming what is often termed a 'chisel edge' or by pronounced sliding along well-defined shear zones. A schematic representation of the plastic-rupture modes of crystals tested at 473 K (shown in Fig. 7.37) may be seen in Fig. 7.38. The tensile test of the crystals deformed at 293 and 473 K is shown in Fig. 7.39. The intersecting shear bands give rise to a peculiar specimen geometry, called 'cross-channel formation' (shown in Fig. 7.38b).

The single-crystal specimens represented in these figures were cut from a, large-grained, 0.4 mm-thick sheet. The plane of the sheet is parallel to $(1\bar{1}0)$ and the specimens are cut so that the tensile axis is parallel to [110]. For [110] crystals, two of the four (111) slip vectors, namely [111] and $[11\bar{1}]$, lay in the (110) plane of the sheet and the other two, $[\bar{1}11]$ and $[\bar{1}1\bar{1}]$, lay in the (110) plane, normal to the plane

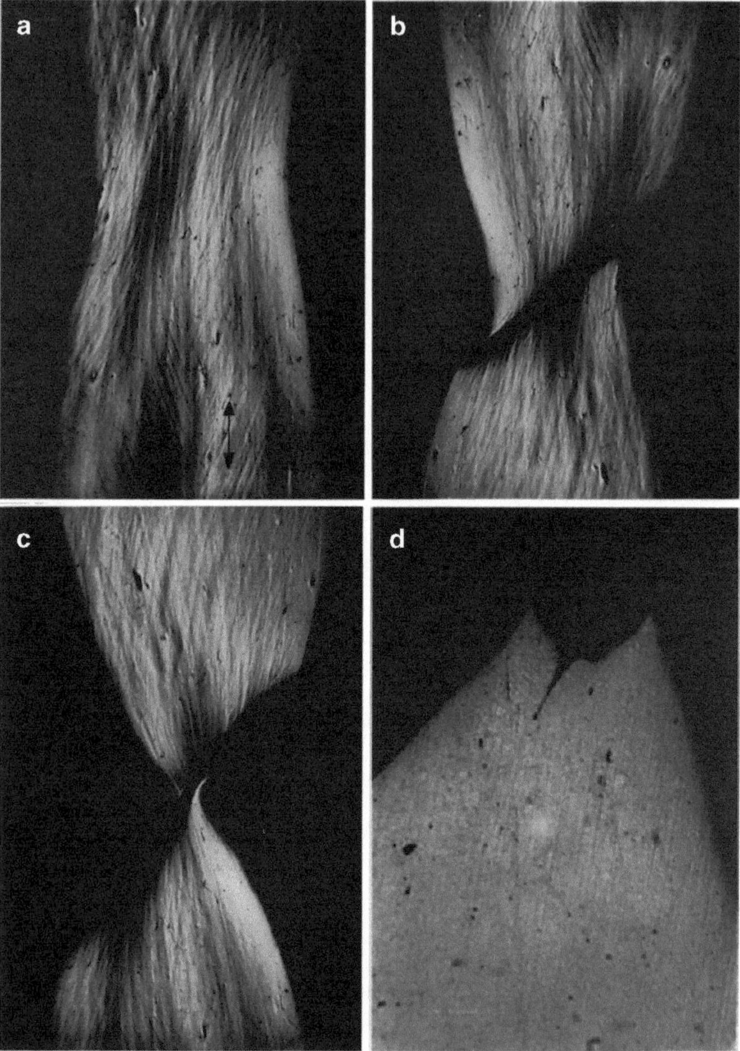

Fig. 7.36 Successive stages during necking at 293 K. *Arrow* indicates the direction of the [110] tensile axis. (**a**) Uniform neck formation, 27%, elongation, X60, (**b**) Localized massive sliding, 29.8% elongation, X60; (**c**) Further massive sliding immediately prior to rupturing at a chisel edge, 33% elongation, X60; (**d**) Polished and etched section of fractured crystal, X400 (Govila and Hull 1968. With kind permission of Elsevier)

of the sheet and the tensile axis. For uniaxial tension along [110], the resolved-shear stress on all the planes containing $[\bar{1}11]$ and $[\bar{1}1\bar{1}]$ is zero and the operative-slip vectors, [111] and$[11\bar{1}]$, lay on the $(1\bar{1}0)$ face of the crystal. During homogeneous, plastic deformation, all the plastic strain is confined to the (110) plane, so that the width of the crystal is reduced without any reduction in thickness. These conditions are equivalent to the plane strain.

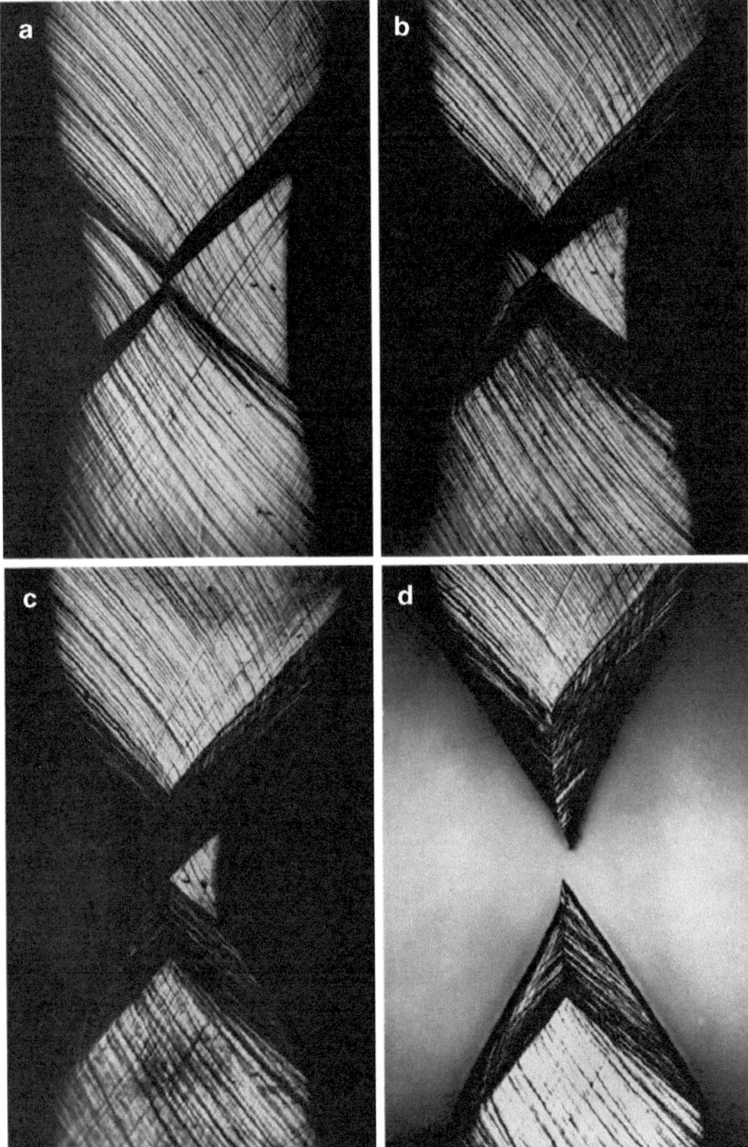

Fig. 7.37 Successive stages of necking at 473 K. (**a**) Uniform neck formation, 25% elongation; (**b**) Localized massive sliding, 28% elongation; (**c**) Localized massive sliding, 31% elongation; (**d**) Rupturing at a chisel edge, 32% elongationX400 (Govila and Hull 1968. With kind permission of Elsevier)

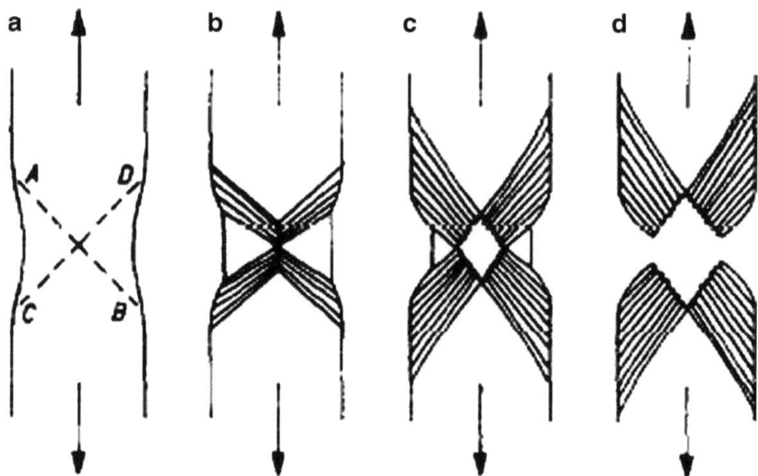

Fig. 7.38 Schematic representation of plastic rupture modes for crystals tested at 473 K: (**a**) Uniform neck formation due to shear on intersecting slip planes AB and CD; (**b**) Appearance of cross-channels in the neck; (**c**) Void nucleation at an inclusion, (**d**) Fracturing of the crystalX400 (Govila and Hull 1968. With kind permission of Elsevier)

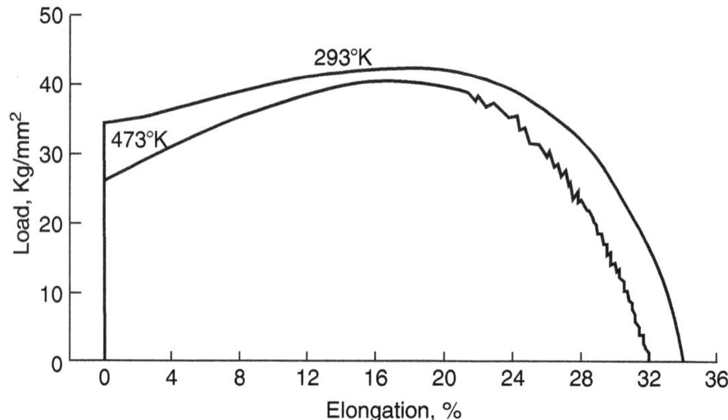

Fig. 7.39 Typical load-elongation curves for crystals deformed in tension in [110] direction at 293 and 473 KX400 (Govila and Hull 1968. With kind permission of Elsevier)

At both temperatures, 293 and 473 K, the crystals deform homogeneously to ~18% elongation after an initial yield propagation stage. Slip occurs on the planes having maximum-resolved shear stress, i.e., {112}. After the onset of plastic instability, two types of shear deformation are observed in the crystals tested at 293 K. The first type appears in Fig. 7.36. After a 27% elongation, a well-defined, uniform neck forms (Fig. 7.36a). Further plastic deformation occurs due to massive sliding of the material along one predominant shear plane (Fig. 7.36b) and, eventually, rupture occurs at a chisel edge (Fig. 7.36c). No cavities are observed

Fig. 7.40 Notch effect observed in a silicon iron crystal tested at 473% showing localized plastic flow at the notch, X 125X400 (Govila and Hull 1968. With kind permission of Elsevier)

on the surface of the crystals at any stage of necking. After fracture, polishing and etching of the crystal, it was observed that the fractured surface is associated with the formation of a cavity and a small longitudinal crack (Fig. 7.36d). The second shear mode at 293 K was also preceded by the formation of a uniform neck, followed by the development of a localized zone of deformation at the edge of the crystal on one side of the neck, producing a notch. This notch grew across the crystal and rupture occurred at the root of the notch. Similar observations were made in crystals tested at 473 K (see Fig. 7.40). The notch is accompanied by very well-defined shear bands (Fig. 7.40).

7.7.2.1 The Effects of Cavities

The mechanisms of plastic rupture at 473 K are summarized schematically in Fig. 7.38. After uniform neck formation by slip on planes parallel to AB and CD (Fig. 7.38a), further localization of strain occurs by what is known as ‘cross-channel formation’ in the neck, as seen in Fig. 7.38b. Subsequent behavior depends on the purity of the given material. In inclusion-free crystals, cross-channels widen continuously and fracture occurs at a chisel edge. In material containing inclusions, cavity formation occurs within this cross-channel region (Fig. 7.38c), leading to fracture (as shown earlier in Fig. 7.38d). During the early stages of necking in FCC metals, a cavity or void often develops in the central part of the neck, around

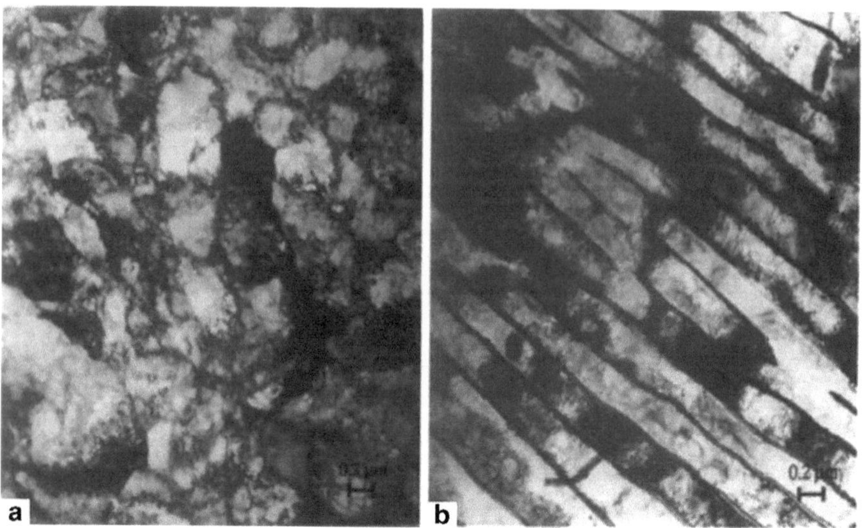

Fig. 7.41 TEM micrograph of dislocation cell structure in AISI 4340 steel: (**a**) in ferrite; (**b**) in pearlite (Cuddy and Bassim 1990. With kind permission of Elsevier)

inclusions or precipitate particles. Govila and Hull claimed that, contrary to the observations of FCC crystals, no cavity formation is observed either in slip bands or at slip-band intersections in materials not containing inclusions (Fig. 7.39).

The relation between the ductile-fracture process and the microstructural aspect of plastic deformation preceding fracture may be seen in the following illustrations. Fracture can originate in dislocation cells initiated by inhomogeneous plastic deformation (see, for example, Cuddy and Bassim; Gardner et al.). In Fig. 7.41, a TEM micrograph shows dislocation cells in AISI 4340 steel. Deformed substructures of ferrite and pearlite regions are seen. Figure 7.41a shows a well-defined dislocation-cell structure formed in proeutectoid, ferrite grains as a result of plastic deformation. Figure 7.41b relates to pearlite. Cementite (Fe_3C) platelets act as impervious barriers to dislocation movement and, thus, only an incipient-cell structure is observed. In some regions, however, a separation of the Fe_3C platelets is observed in conjunction with dense bands of dislocations (marked by arrows).

Figures 7.42 and 7.43 are SEM photomicrographs of a tensile surface. The fine distribution of voids (or dimples) shows that the 4340 steel failed by a fibrous mode of fracture. These voids exhibit a wide variation in size and shape. The elongated shape of some voids may indicate that localized-shear stress, in addition to tensile-shear stress, is also involved. Also note that many of the larger voids are actually composed of many smaller, shallow voids. There is no evidence for the presence of inclusions or particles in conjunction with these voids. In Figure 7.43a, the fracture-surface topography in regions containing pearlite is characterized by microcracks

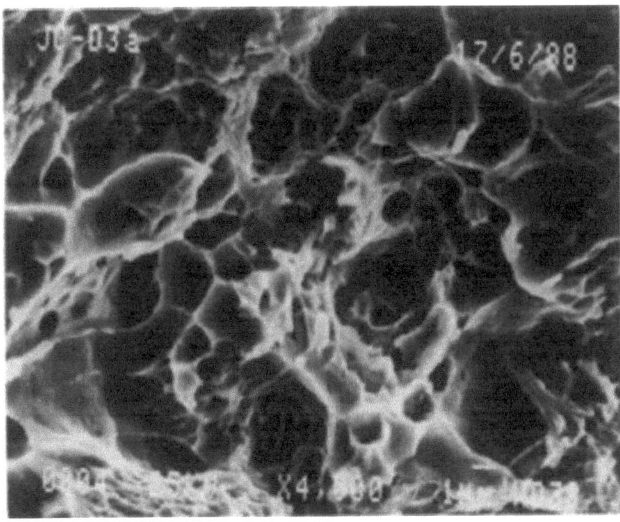

Fig. 7.42 SEM micrograph of fibrous structure in the ferrite (Cuddy and Bassim 1990. With kind permission of Elsevier)

running normal to the pearlite colonies. At higher magnifications (Figs. 7.43b, c), shallow microvoids, with sizes less than 0.5 μm, are visible in the interlamellar region between cementite platelets, appearing as bright, parallel ridges.

The topography of this fracture surface is different than that observed in ferrite grains (Fig. 7.42). Both in ferrite grains and in pearlite colonies, ductile fracture is controlled by void formation. The fracture topography is, however, different. In ferrite grains, plastic deformation leads to cell-substructure formation (Fig. 7.41). The final cell size at the onset of fracture closely coincides with the size distribution of the voids. The walls of well-developed dislocation cells are the probable nucleation sites of the voids in ferrite grains. Based on TEM micrographs of pearlite, the separation of cementite plates occurs prior to the fracture, in conjunction with the accumulation of dislocations into shear band in the adjacent ferrite (Fig. 7.41b). The model seen in Fig. 7.45 explains the process leading to the formation of voids and microcracks in pearlite as observed in the fracture surface. According to this model, the separation of the cementite plate occurs under the combined influence of applied-tensile stress and localized shear in the ferrite lamellae. Deformation becomes concentrated in a shear band, leading to void growth and coalescence, as depicted in Fig. 7.45.

The variation in cell size during plastic strain appears in Fig. 7.44. Cell size decreases with strain up to 0.4–0.5 μm, after which it remains constant, as predicted by the theory of work hardening.

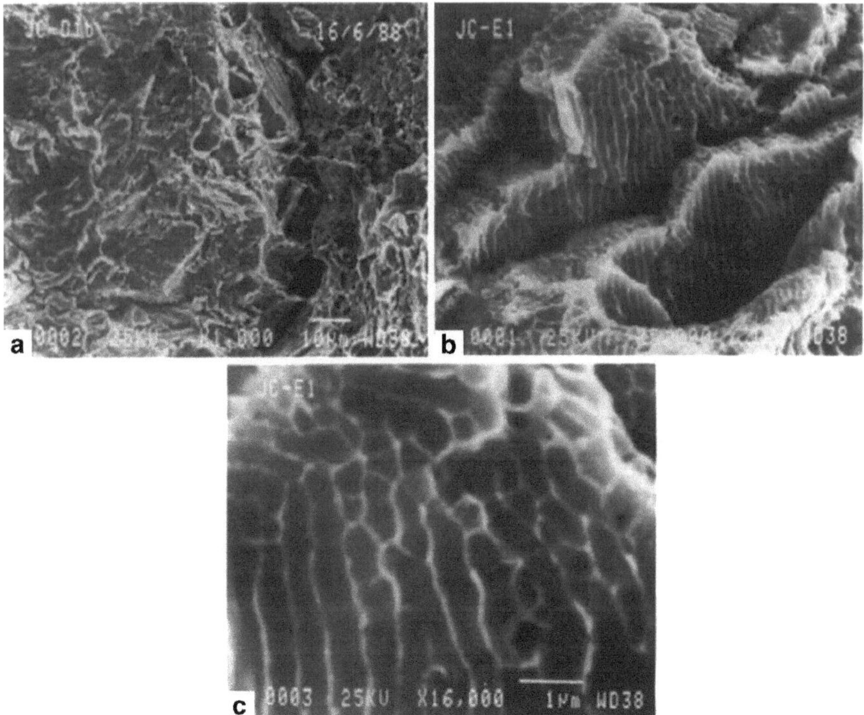

Fig. 7.43 Fracture surface of the pearlite: (**a**) evidence of cleavage and microcracks; (**b**) and (**c**) evidence of microvoids in interlamellar ferrite (Cuddy and Bassim 1990. With kind permission of Elsevier)

7.7.2.2 The Effects of Inclusions

Above, the role of dislocations (cell structure) in ductile-fracture formation was considered. However, the consequences of inclusions and their influence on failure in ductile fracture must be further, though briefly, elaborated here. The role of inclusions or second-phase particles in ductile fracture has been discussed by many researchers (see, for example, Broek). Electron-microscopic examination of fracture surfaces and micro-structures of 13 different aluminum alloys reveal that ductile rupture is initiated at small inclusions.

The average dimple spacing is equal to the average inclusion distance. Void initiation, followed by spontaneous void growth and coalescence in the vicinity of inclusions, is associated with ductile fracture. Dislocation loops, piling-up against particles, are related to void initiation and growth (Broek). In Fig. 7.46, large particles are seen, which are able to initiate voids at small strains. Final fracture takes place at strains on the order of 25%. Large inclusions are not essential to the fracture process, although they decrease the overall ductility of a material. However,

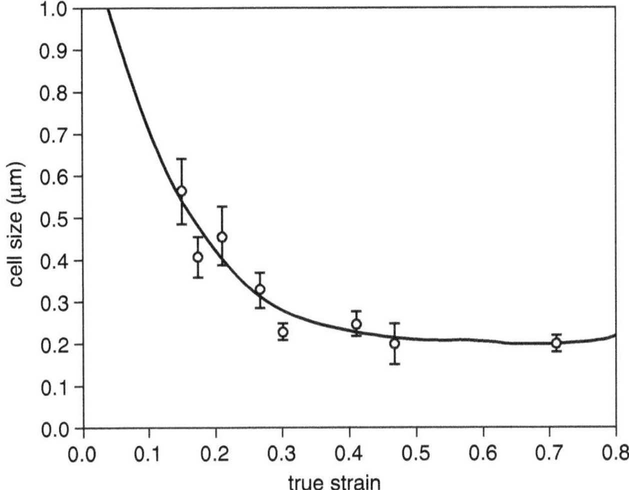

Fig. 7.44 Variation of dislocation cell size with plastic strain (Cuddy and Bassim 1990. With kind permission of Elsevier)

they can act as stress concentrations and, thus, may effect the location of ductile fracture.

In Fig. 7.47a, an electron fractograph shows a large, cleaved particle among small dimples. The large inclusions break at relatively low strains. Generally, the major part of a fracture surface consists of small dimples. The initiation and coalescence of these microvoids are critical events in ductile fracture. Smaller particles (visible via electron microscopy) appear to govern the fracture process. Depressions in dimples, representing small inclusions that initiate voids, are shown in Fig. 7.47b. Small particles have also been observed in TEM. If such particles initiate the small voids, then dimple size should be equal to the spacing of these small inclusions. This was investigated in 13 Al alloys.

It was found that the average spacing of these particles is, indeed, equal to the average dimple spacing. Thin foils and replicas of fracture surfaces of these alloys were examined by electron microscopy. Counts were made of about 1,000 dimples and 1,000 inclusions. Average dimple sizes and inclusion distances were determined by these counts. The average dimple size and average inclusion distance were defined as:

$$\bar{d} = \left(\frac{A}{N}\right)^{1/2} \quad \text{and} \quad \bar{s} = \left(\frac{A}{N}\right)^{1/2}$$

$\bar{d}$ and $\bar{s}$ are the average dimple size and inclusion distance, respectively. N is the count in an area A.

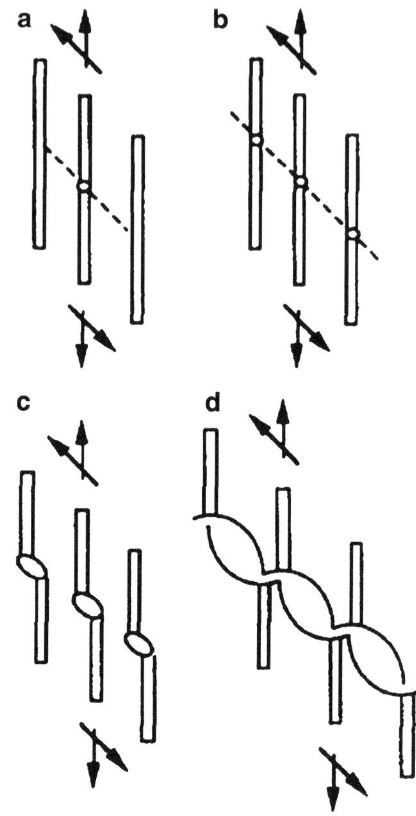

Fig. 7.45 Fracture mechanism in pearlite (after Miller and Smith): (**a**) cracking of cementite plate; (**b**) shear zone developing in ferrite causing cracking of adjacent planes; (**c**) and (**d**) void formation and coalescence (Cuddy and Bassim 1990. With kind permission of Elsevier)

From the straight line of the measured relation between the dimple size and the inclusion spacing, the expected one-to-one relation between them was confirmed (Broek). Thus, the justified conclusion was that small inclusions are responsible for the creation of voids. Void formation at particle interfaces was observed in only a few specimens. Figure 7.48, illustrates such voids on small particles in several Al alloys. However, the direct observation of voids on these particles was achieved only incidentally, contrary to reports of their general existence. Therefore, it must be assumed that void initiation in these materials does not take place until very late in the fracture process. The conditions for void growth must already be met before voids are formed; hence, immediate coalescence occurs. It appears that voids are usually formed by interface decohesion.

In order to experimentally observe in these Al alloys voids during the test, a large number of specimens must be considered by electron microscopy and by straining electrically so that cracking could be followed. Again, no voids are observed in the area of cracking. Some results of this test are shown in Fig. 7.49. In Fig. 7.49a, the crack approaches the boundary of an inclusion. The stress on the inclusion must be very high, yet no cavities are observed nor has any cracking of the particle occurred.

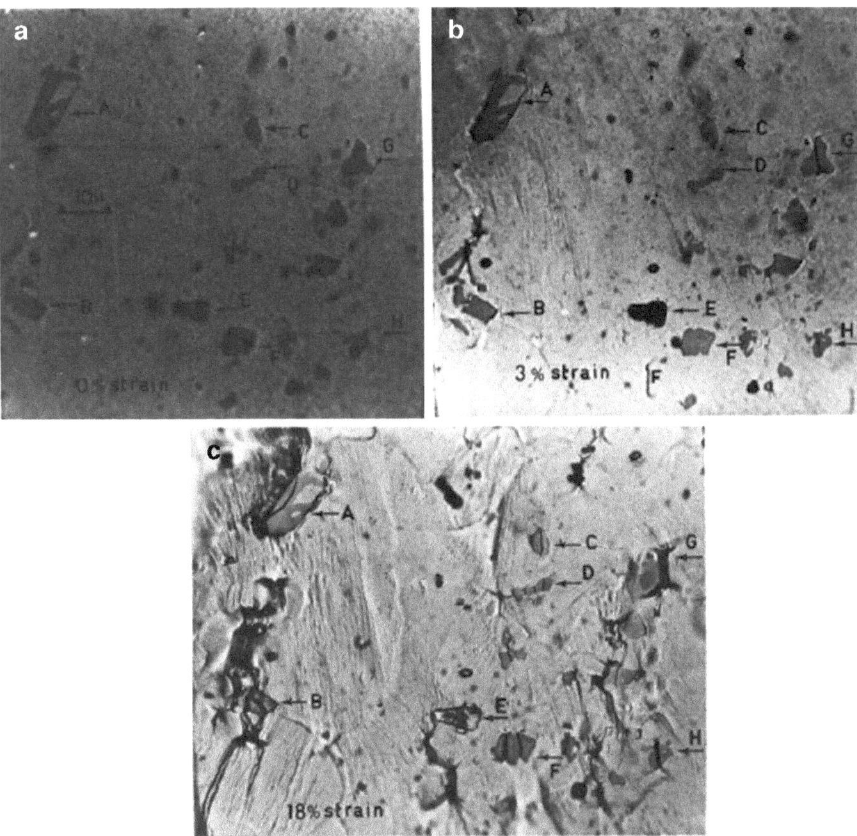

Fig. 7.46 Void initiation at large particles as observed under the optical microscope: (**a**) 0% strain; (**b**) 3% strain; (**c**) 18% strain (Broek 1973. With kind permission of Elsevier)

Actually, the crack has passed along the interface (Fig. 7.49d). Apparently, the cohesive forces between the matrix and particles are so extremely high that void initiation does not take place until very late in the fracture process. The scarcity of voids leads to the assumption that immediate coalescence must have taken place at the moment that a sufficient quantity of voids had formed. This also implies that voids may only be initiated at such high stresses and strains that the conditions for coalescence have already been fulfilled. Thus, a model of void initiation is required which can predict immediate and spontaneous void growth. A dislocation model has been suggested, based on the experimental observations of inclusions and voids (at whatever stages of straining the voids were formed in these Al alloys). Figure 7.50 presents the essential parts of this model. Void initiation and growth, due to dislocation loops piling-up against the particles, is a model compatible with the experimental observations. A quantitative evaluation of this model relates fracture stress and strain to the volume fraction of small particles.

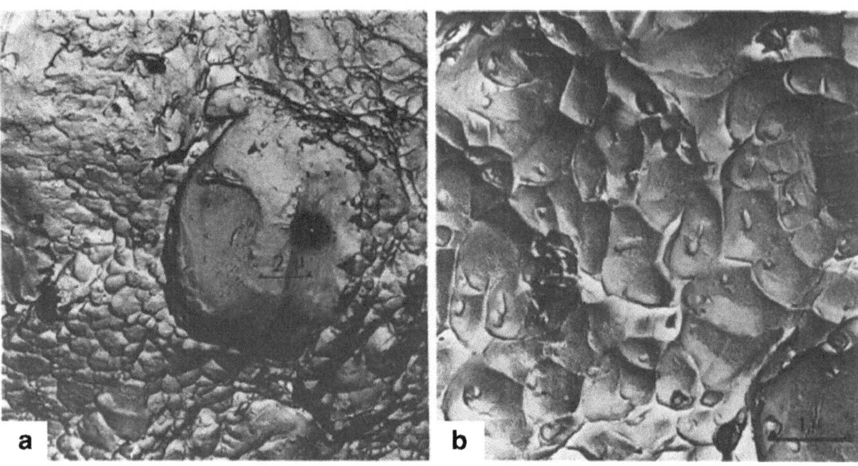

Fig. 7.47 Dimples in different Al alloys: (**a**) Cleaved inclusion among small dimples in 2024-T3 alloy (**b**) Inclusions visible in dimples in 7079 alloy (Al-Zn-Mg type) (Broek 1973. With kind permission of Elsevier)

During plastic deformation, dislocation pile-ups form at the particles. The pile-up loops are repelled by the particles through the action of their image forces. Nonetheless, the leading loop is pushed towards the particle by stresses initiated by the dislocations behind it and by their applied shear stress. When one or a couple of loops are pushed to the interface, decohesion of that interface will ultimately take place along AB (in Fig. 7.50c). If this occurs, a void is formed. As a result, the repulsive forces of the subsequent loops are drastically reduced and the greater part of the pile-up empties itself into the newly-formed void. The dislocation sources behind these loops, which had become inactive when constrained by the pile-up ahead, are able to resume their action. This process may lead to spontaneous void growth and coalescence, as soon as the voids have been initiated (Figs. 7.50c, d). Friedel assumes that the plastic relaxation of pile-up stresses begins when the stress exerted by the pile-up group on the secondary sources exceeds the elastic limit. He shows that pile-up groups in FCC crystals should be completely relaxed when they contain about five dislocations. This implies that many pile-ups must be formed on different slip planes, as in Fig. 7.50e. Ultimately, this situation will lead to interface decohesion, when one or more dislocations are pushed into the interface. The newly-formed voids can now grow as dislocations on other slip planes run into the void (Fig. 7.50f). Regular pile-ups have been observed by several investigators (but not in the work discussed here), although large clusters of dislocations around particles often occur (as shown in Fig. 7.49b).

Of the relations derived by Broek, the following expressions should be mentioned:

$$\sigma_f = \alpha(\nu)\varepsilon_f + \beta(\nu)\sigma_{YS} \tag{7.46}$$

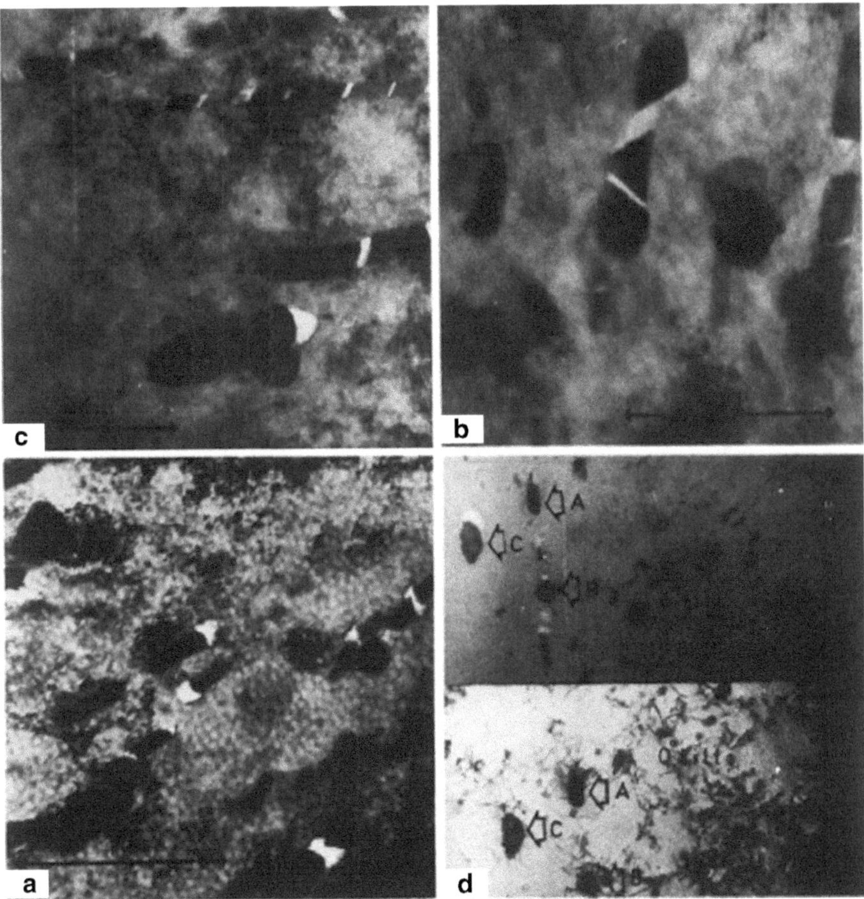

Fig. 7.48 Void formation at small particles in Al alloys: (**a**) 7079 alloy; (**b**) 7079 alloy; (**c**) 2024 alloy (**d**) 6061 alloy (Broek 1973. With kind permission of Elsevier)

where $\alpha(v)$ and $\beta(v)$ are functions of the volume and σ_f, ε_f and σ_{YS} are clearly the fracture stress, fracture stain and yield stress, respectively. The fracture toughness is expressed as:

$$K_{Ic} = CE(s)^{1/2} f(1/v) \tag{7.47}$$

Equation 7.47 predicts that fracture toughness depends on both the volume fraction and the distance, s, from the tip of the particles. C is a constant.

In summary of this section, note that void initiation and growth are due to dislocation loops piling-up against the particles; this model is compatible with the observations. A quantitative evaluation of this model relates fracture stress and strain to the volume fraction of small particles. Fracture toughness, however, also depends on the presence of large particles, since fracture of these particles causes strain concentrations.

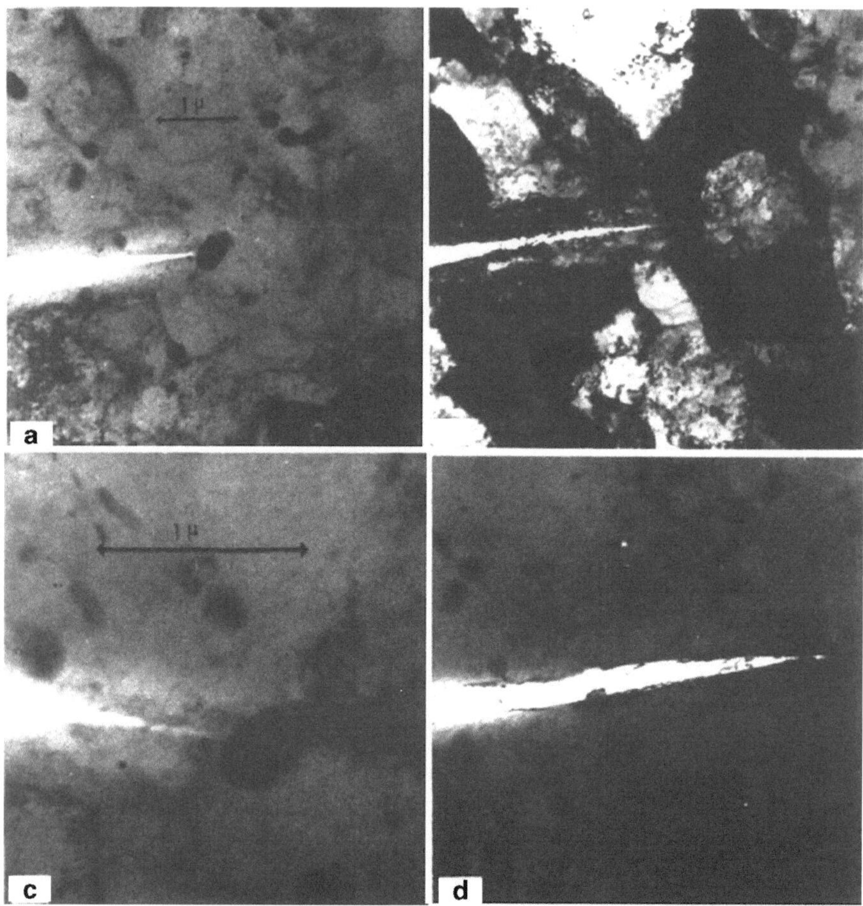

Fig. 7.49 Passage of crack along inclusion interface in thin foil specimen of 7075 alloy: Direction of straining from *top* to *bottom*. (**a**) Particle at *tip* of crack; (**b**) Same crack with dislocations fitted into contrast, showing formation of subgrain cells; (**c**) enlarged picture of (**a**), particle at *tip* of crack; (**d**) interface decohesion after passage of crack (Broek 1973. With kind permission of Elsevier)

7.8 Ductile-to-Brittle Transition (Transition Temperature)

7.8.1 Introduction

In Chap. 1, Sect. 1.6, transition temperature was illustrated in Figs. 1.41 and 1.42.

Those illustrations show the ductile-to-brittle transition as determined by the impact-transition temperature.

The concept of ductile-to-brittle transition is of significance, since temperature affects failure behavior in tested materials. It is of great practical importance to recognize this effect, especially since the forming operations of structural materials

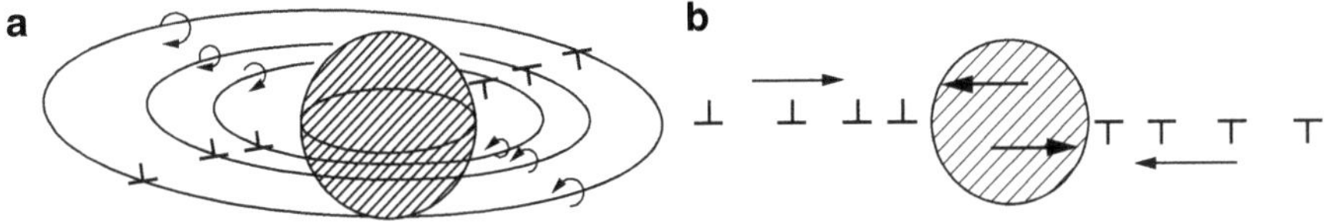

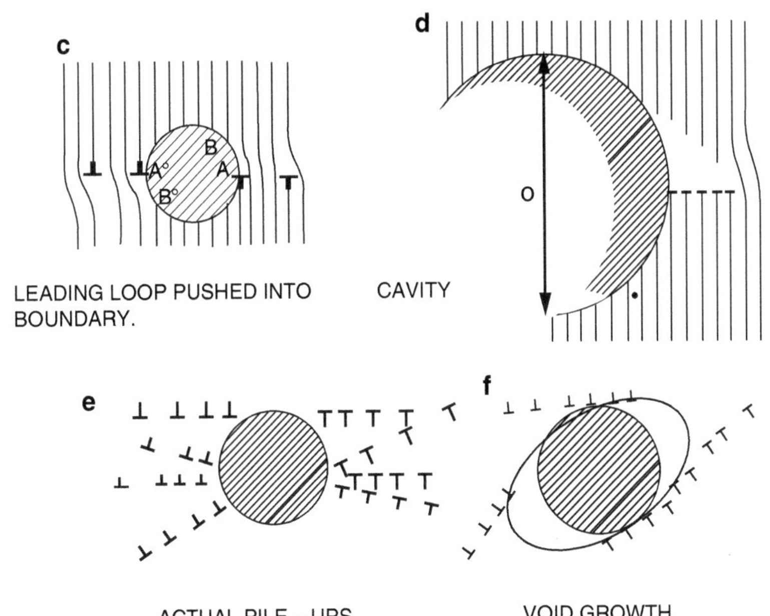

Fig. 7.50 Dislocation model for void initiation and growth and the role of inclusions (particles) (Broek 1973. With kind permission of Elsevier)

are usually performed at elevated temperatures in order to avoid fracture during processing and also to improve ductility (thus reducing the number of passes required to arrive at the desired dimensions, e.g., by rolling or drawing). Materials do not have to contain embrittling constituents if they are sufficiently pure, without particles or inclusions. Here, the concept of ductile-to-brittle transition will be considered in regard to executing safe manufacturing operations, while avoiding catastrophic, brittle failure.

7.8.2 The Features

In Chap. 1, the stress-strain relation was discussed. Contrary to room-temperature tests, the strength of metals, such as steel, increases when tested at low temperatures,

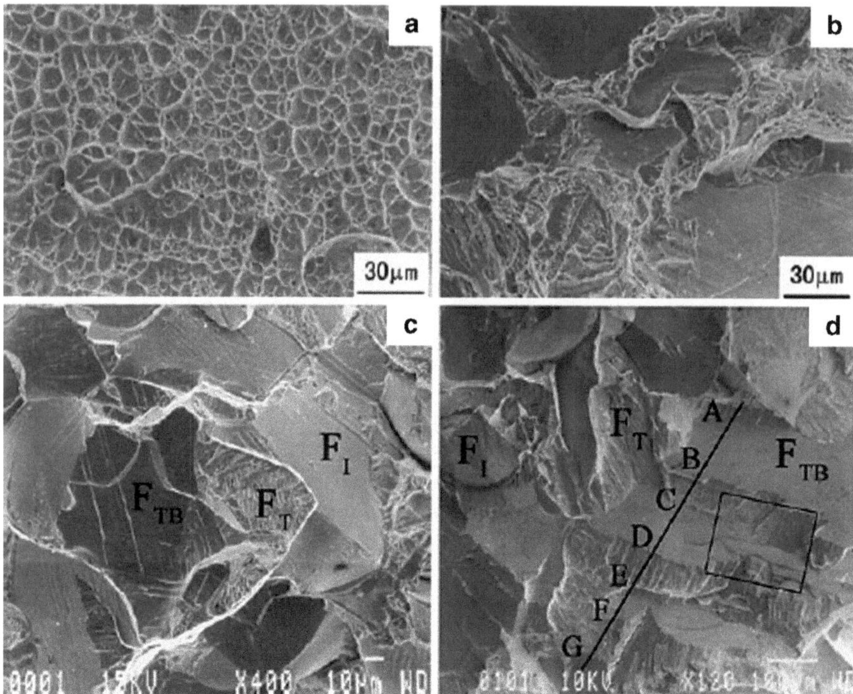

Fig. 7.51 SEM fractograph for specimens impact fractured (**a**) at 293 K; (**b**) at 193 K; (**c**) at 77 K (**d**) tensile fractured at 4 K. F_I transgranular fracture facet, F_T intergranular fracture facet and F_{TB} annealing twin boundary fracture facet. The specimen was sectioned along the ABCDEFG line for dual-surface observation (Liu et al. 2007. With kind permission of Springer)

while ductility and toughness decrease rapidly as the temperature is lowered. Such structures are good candidates for the study of ductile-to-brittle transition by the most common and effective method, namely by impact tests. Figure 7.51 shows the microstructure obtained by impact testing an FCC, polycrystalline, high-nitrogen austenitic steel. Note that the microstructure changes from a ductile appearance at room temperature (293 K) gradually to a brittle one (Fig. 7.51b, c), characterized by cleavage (b – 193 K and c – 77 K). In addition, tensile fracture at 4 K is also shown. Low-carbon steels having BCC crystal structures usually exhibit ductile-to-brittle transition with decreasing temperature, but this is also observed in FCC high-nitrogen steel. Accompanying the ductile-to-brittle transition, the appearance of the fracture surface, changes from dimple to cleavage rupture along specific crystallographic planes as indicated in Fig. 7.51. Here, the flat, smooth surfaces resemble the cleavage in brittle fracture. However, Liu et al. claim that these flat, cleavage-like surfaces represent three fracture modes marked as follows: an annealing twin-boundary fracture, F_{TB}; an intergranular fracture, F_I; and a transgranular fracture, F_T. Along the line ABCDEFG in Fig. 7.51d, the specimen was sectioned perpendicular to the macrofracture surface, so it would be possible to observe the dual-surface of both the fracture and the adjacent side by SEM. The

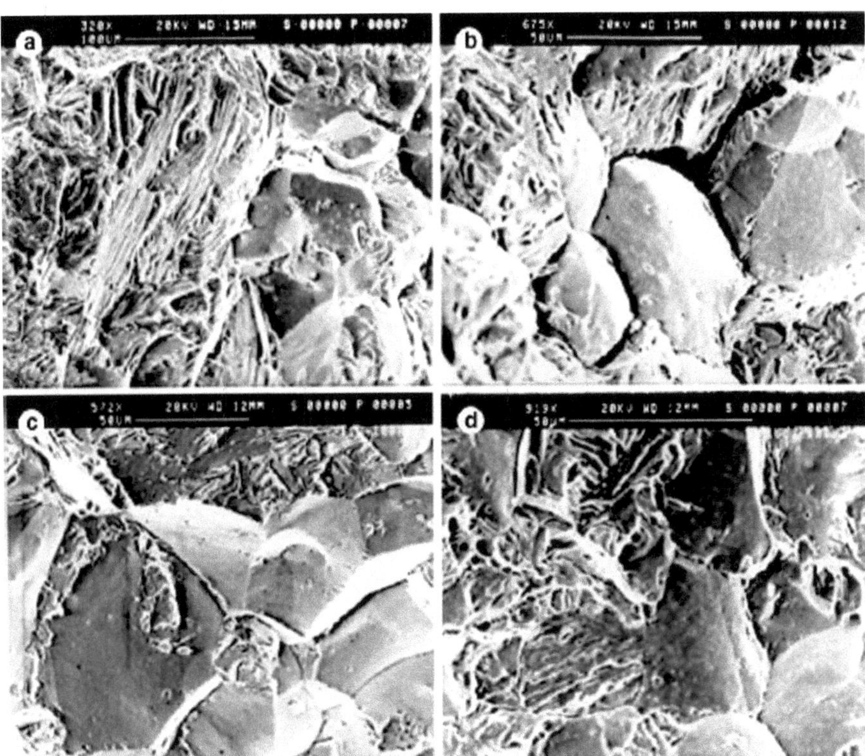

Fig. 7.52 Impact fracture SEM images of 12CrlMoV steel held at 540° for 500 h: (**a**) impact at 20°C, (**b**) impact at −30°C, (**c**) impact at −50°C and (**d**) impact at −60°C (Qing-fen et al. 2005. With kind permission of Elsevier)

results appear in Fig. 7.34 above, also indicating the presence of river patterns. Discussion of the cut specimen accompanies Fig. 7.34. The deformation structures on {111} planes show distinct crystallographic features characterized by brittle failure, in accordance with the analysis of transition temperature variation obtained by Charpy impact tests in the 77 K – ~293 K range using V-notch specimens. The ductile-to-brittle transition temperature is about 200 K, thus the dimple-like structure at 293 K (above the transition temperature) is understood.

Ductile-to-brittle transition produced by impact tests is a method for testing most embrittling phenomena. TE generally is not sensitive to tensile tests; therefore, TE is generally tested by means of impact tests. Figure 7.52 provides such an example in the case of 12Cr1MoV steel, where TE was detected due to phosphorus segregation. The ductile-to-brittle transition temperature is low, in the range of –36.18 to –73.18 K, depending on the isothermal holding time at 540°C, so these tests were performed at low temperatures to evaluate the transition temperature. Both cleavage and intergranular fracture are observed, although intergranular fracture dominates.

There is a discussion below on observations of isolated-cleavage regions [henceforth: ICRs] in fast-loading tests at the upper shelf of an impact test, above the

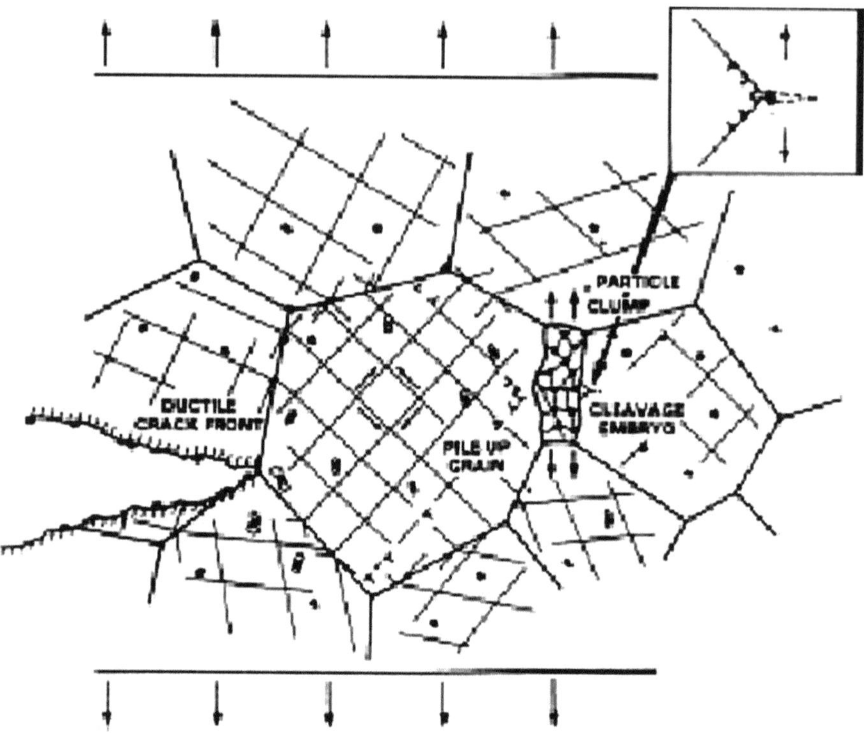

Fig. 7.53 Model for initiation of an isolated cleavage region (ICR) within a matrix of ductile fracturing (Armstrong 2010. With kind permission of Elsevier)

defined transition temperature. Fractographic observations show that the ICRs are formed at the ductile fracturing transition in nuclear vessel steels of various materials, when fractured by rapid load tests, such as Charpy V-notch, three-point bend tests or compact-tensile tests (Zhang et al.). These tests are performed at the upper-shelf energy absorption, where usually no cleavage fracture occurs. Nevertheless, a few isolated first-cleavage sites are observed surrounded by hole-joining fracture areas. It is expected that, at low temperatures in the transition range, failure by rapid-cleavage islands (or boundary regions), characterized by dimpled rupture, can also be observed. Furthermore, at intermediate temperatures, slow fracturing is followed by cleavage instability, while at higher temperatures, regions of slow dimpled rupture surround cleavage islands, as indicated above. First-cleavage initiation may occur at ICRs. Particle clumps may influence the advance of cracks at an ICR.

Armstrong suggested a dislocation pile-up model relating plasticity to both the static and the dynamic properties of a low-carbon steel, when the alloy is subjected to fracturing by means of a Charpy V-notch impact test. The elements of this model are illustrated in Fig. 7.53. This model is based on the experimental results of Zhang et al. and others. One main presumption in this model is that a particle clump, when loaded rapidly, can separate sufficiently abruptly to generate a tensile stress high enough to initiate cleavage in favorably-oriented, adjacent grains. These variations in stressed microstructures then produce corresponding irregularities in local strains and grain separations during ductile-crack growth, in turn, leading to rapid and relatively-brittle fracturing of the plastically-stiffer, carbide-particle clump, which is enhanced by dislocation pile-ups with greater effect in larger grains. These higher-strength particle clumps undergo relatively less uniform strain before becoming plastically unstable. Rapid load transfer is shifted to an adjacent grain of suitable crystallographic orientation, containing a potential nucleus for cleavage initiation and growth. Less favorable conditions in the surrounding microstructure environment halts the spread of the ICR. Lower test temperatures and/or enhancement of the strain rate enable more ICRs to form and others to spread further. ICRs are often initiated by a localized region of ductile hole-joining failures.

The occurrence of an ICR within a sea of ductile-fracturing elements is relatively rare, due to the combined conditional requirements indicated.

7.9 Fatigue Fracture

In Chap. 6, Sects. 6.3 (stress cycles) and 6.4 (fatigue life), the subject of fatigue was extensively considered. In Sect. 6.91, Figs. 6.53, 6.52, and 6.54 have shown beach marks and striations on fractured surfaces, which are associated with crack propagation. This section is completely devoted to failure as a consequence of cyclic deformation. Various factors, inclusions in particular, effect fatigue-crack initiation and fatigue growth leading to fracture.

7.9.1 Crack-Tip Blunting

The propagation of cracks eventually leads to failure. It is of great interest to extend the life-times of components operating under cyclic conditions. The process of crack growth is associated with crack-tip blunting. When a crack tip is sharp, only slow crack growth is involved in fatigue; however, when a crack tip is blunt, crack initiation is involved in addition to crack growth. Electron-microscopic studies of dislocation behavior at the crack tips of various metals have shown the existence of a dislocation-free zone [henceforth: DFZ], often located between the crack and the plastic zone. The growth or motion of a crack is accompanied by edge dislocations, generated at the crack tip primarily on planes that are inclined to the crack plane. The slip geometry of two inclined dislocation pile-ups, symmetric with respect to a crack, is occasionally observed; however, such a crack often emits dislocations on only one plane, which is inclined to the crack plane. Pile-ups of dislocations are occasionally observed close to the crack tip, which often emits dislocations. The dislocations emitted from a crack tip move through a DFZ and pile up in the plastic zone. The presence of a DFZ is attributed to difficulty in generating dislocations at the crack tip, as expressed in terms of the critical stress-intensity factor, K_g, required for dislocation generation. Chang's calculation shows how to evaluate the local stress-intensity factor at the crack tip in the presence of an inclined-dislocation pile-up of length e of the DFZ and length l of the plastic zone. The local stress-intensity factor of the components operating under cyclic conditions is a function of time. The fatigue-crack propagation per cycle has the order of magnitude of the crack blunting.

For cyclic loading, the load levels are lower than those used for crack propagation in monotonic loading to failure. Nevertheless, a similar expression is assumed for the use of the crack-propagation rate as:

$$\frac{da}{dN} = f(K_{\min}, K_{\max}) \tag{7.48}$$

In the case of cyclic loading, subcritical-crack growth can occur if the stress-intensity factor, K, is much lower than the fracture toughness, K_c. As usual, da/dN is the crack extension per load cycle, defined by K_{min} and K_{max}. Ks, the stress-intensity factor range, is commonly used instead, as was defined in Chap. 6 and given in a somewhat different form as:

$$\Delta K = \begin{cases} K_{\max} - K_{\min} & \textit{if } K_{\max} > 0 \\ K_{\max} & \textit{otherwise} \end{cases} \tag{7.49}$$

Recall that R was previously defined (Chap. 6) as $R = K_{min}/K_{max}$. In the case of cyclic loading, subcritical crack growth can occur if the stress intensity factor K is much lower than the fracture toughness K_c. For most engineering alloys, the total crack-resistance curve (i.e., a plot of the crack-growth rate, da/dN, vs. the stress-intensity factor range, $K = K_{max} - K_{min}$, on a logarithmic scale) exhibits a sigmoidal shape, in which three regimes may be distinguished. A schematic illustration of this sigmoid is shown in Fig. 7.54.

Note the resemblance of the curves to the sigmoidal variation of the fatigue-crack propagation rate (da/dN) with alternating-stress intensity, illustrated by Beevers et al. In regime A, the average crack growth per cycle is smaller than lattice spacing. Furthermore, a threshold-stress intensity-factor range exists (marked by the arrow), below which cracks remain dormant and above which the crack-growth rate increases with log ΔK. Regime B, known as the 'Paris regime', shows a linear variation of log da/dN with log ΔK. Regime C pertains to the range of high ΔK values, where the crack-growth rate increases rapidly, causing catastrophic failure.

Predicting fatigue-crack growth under time-dependent loading is usually done according to Paris' law, given as:

$$\frac{da}{dN} = C\Delta K^m \tag{7.50}$$

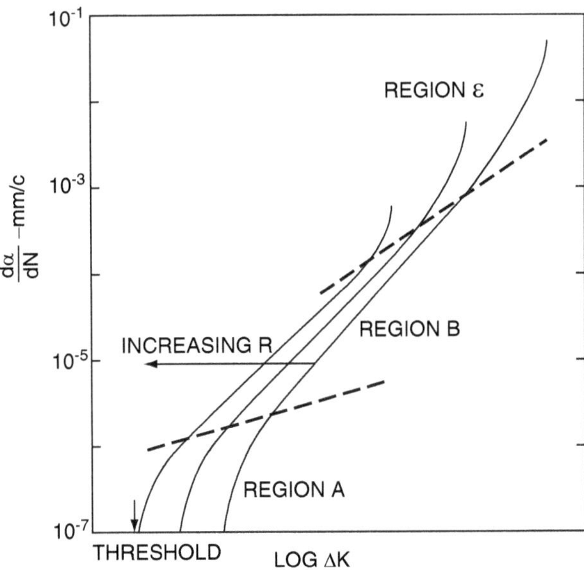

Fig. 7.54 Schematic shape of fatigue crack growth rate curves at three stress ratios. *Region A* low fatigue crack growth rates approaching threshold. *Region B* fatigue crack growth rates; linear log da/dN vs. log ΔK relationship. *Region C* high fatigue crack growth rates approaching the critical stress intensity (Vosikovsky 1979. With kind permission of Elsevier)

C and K are constants with m usually in the range 2–4. In essence, this is an empirical law, expressing the relation between the stress-intensity factor and the crack-growth rate, usually expressed in terms of crack-growth increment per load cycle. In Fig. 7.54, this relation is expressed by a straight line.

The crack-advance mechanism for crack growth has been the subject of considerable activity in the field of fatigue failure. Plastic blunting and re-sharpening are accepted to provide a mechanism for the process. In order to explain crack growth via striation formation, Laird performed direct observations of crack-tip geometry on some ductile metals corresponding to different stages in the stress cycle. The mechanism of crack propagation deduced from these observations is now called the 'plastic blunting process' or 'Laird's model' and is illustrated schematically in Fig. 7.55.

The initial zero load corresponds to a well-developed stage II crack with a fracture surface exhibiting striations (Fig. 7.55a). As the tensile load is applied, the metal yields plastically, due to high-stress concentration. This plastic deformation is highly concentrated in the slip zones along the planes of maximum shear stress (Fig. 7.55b). When the load is further increased, the slip zones at the tip broaden and the crack tip blunts to a semicircular configuration. The crack tip, thus, is effectively shifted (Fig. 7.55c). If the far-field stress is reversed, the crack tip re-sharpens by buckling and folding the newly-created surface into a double notch,

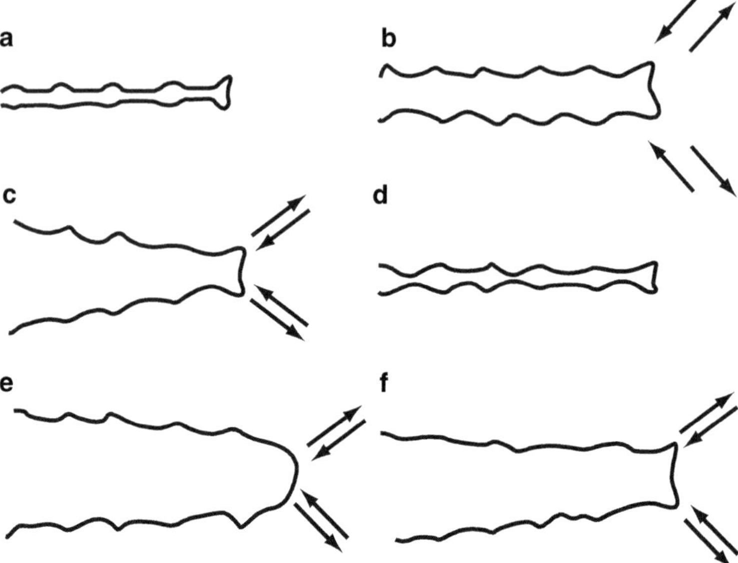

Fig. 7.55 Laird's scheme for stage II crack propagation via plastic blunting and re-sharpening (Levkovitch et al. 2005. With kind permission of Springer)

resulting in a striation formation (Figs. 7.55d, e). Since the closure of the crack during compression cannot fully negate the blunting and the attendant extension of the crack during the preceding tension load, net-crack growth occurs during a fatigue cycle, leading to the formation of a striation. Laird's model applies to a wide variety of ductile materials, including polymers, because of its general description of the plastic-blunting process. Figure 7.55 is an idealization of the plastic blunting and re-sharpening processes leading to stage II fatigue-crack growth in fully-reversed fatigue (after Laird). Note that the 45° arrows stand for the maximum shear stress indicated in Figs. 7.55b, c, d).

Once again, when the crack tip is sharp, only slow crack growth is involved in fatigue; when the crack tip is blunt, crack initiation is involved in addition to crack growth.

7.9.2 The Effect of Inclusion

Of the various particles present in materials, non-metallic inclusions are of special interest, due to their detrimental effects on fatigue resistance. Constant efforts have been and are still being made to improve processing technologies, so that they may successfully remove most, if not all, the inclusions. Improvement in steel production (steel is a very important structural material), with the consequent

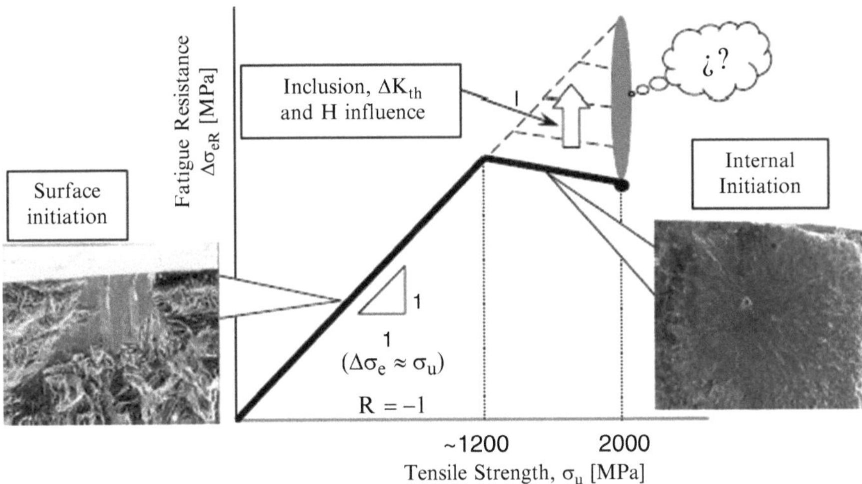

Fig. 7.56 Schematic relationship between fatigue resistance $\Delta\sigma_{eR}$ and ultimate tensile strength, σ_u for $R = -1$. Typical fractographic appearance of the initiation zones is also shown (Chapetti 2011. With kind permission of Elsevier)

reduction in the inclusion content, especially for HCF applications, has resulted in stronger components with increased fatigue life-times. Despite this improvement, the presence of inclusions in steels, for example, is of great concern, particularly in very HCF, where the number of cycles is in the range of $N > 10^7$.

Usually, fatigue resistance, σ_e, increases with tensile strength, σ_u (see, for example, Chapetti), as seen schematically in Fig. 7.56. In this Figure, the fractographic appearance of the crack-initiation zones is also shown. One can see that at a certain level of tensile stress, fatigue resistance decreases with further increase in tensile stress, as expected. At the lower strength, fatigue-crack initiation occurs at the surface, whereas internal-crack initiation occurs at a higher strength, corresponding to the lower fatigue resistance of the material. One can speak of two fatigue mechanisms in the two respective σ_e and σ_u regions. Fatigue resistance in the two competitive fatigue mechanisms may be seen experimentally in the form of S-N curves for steel (JIS SUJ2-QT steel), together with typical fracture-surface appearance, shown in Fig. 7.57.

The total fracture process, due to pure fatigue, may be divided into five well-defined steps (Chapetti), as indicated in Fig. 7.58, where the fatigue-crack initiation process is shown schematically from the surface. The illustrated steps are: (a) persistent slip-band formation and damage accumulation due to fatigue; (b) microcrack initiation; (c) microcrack propagation; (d) macrocrack formation with continued application of stress and; (e) propagation of the macrocrack and fracture. Figure 7.58 is a plot of the applied-stress range, $\Delta\sigma$ (or $\Delta\sigma_t$, i.e., the stress-threshold range for pure fatigue-crack propagation, which is a function of crack length) versus the crack length, "a". This diagram, taken originally from Kitagawa

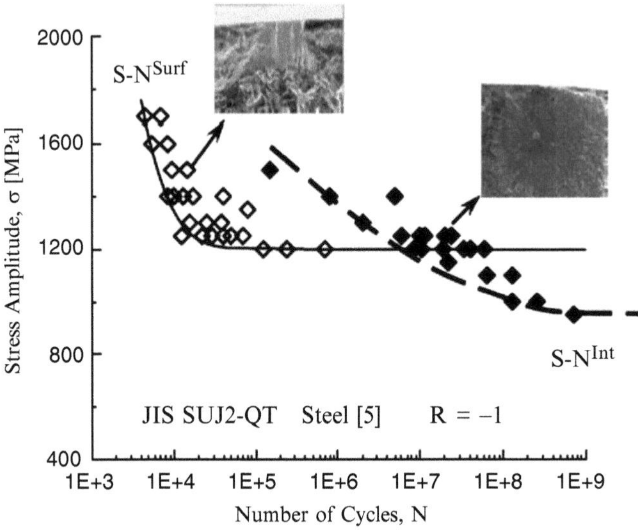

Fig. 7.57 Experimental results of the applied nominal stress as a function of the number of load cycles to failure N for the JIS SUJ2-QT steel (Chapetti 2011. With kind permission of Elsevier)

and Takahashi, may be used to deal with the fatigue-crack propagation threshold as a function of crack length. It is now recognized that the fatigue life-times of many structural components are determined by the rate of growth of small fatigue cracks (microcracks).

Analysis of the growth of these small cracks, by means of a traditional linear-elastic fracture mechanics approach, is not straightforward. The reasons for this complexity are as follows: (i) the transition from microcrack growth to macrocrack growth seems to be associated with the concept that, instead of stress, the intensity factor controls fatigue; (ii) crack growth is elastic-plastic in nature, rather than linear-elastic; and (iii) the crack-closure level increases from zero and increases in length in the small-crack region up to the level associated with a macroscopic-fatigue crack.

Among the significant factors that affect crack initiation by inclusions are: inclusion type, shape and size, residual stresses and H-content. Examples of types of inclusions are: A12O3, AlO3(CaO), MnS, TiN, etc. Often, rather than applying stress, hardness is used (see Vickers) in experimental-fatigue tests; a hardness test is simpler and cheaper to apply and good relations between stress and hardness exist in steels.

The size effect was first evaluated in pure-fatigue tests and the results were used to study the effect of inclusions on resistance (Chapetti). The threshold for pure-crack propagation for a load ratio R = –1 is expressed as:

$$\Delta K_{th} = 3.3x10^{-3}\,(H_V + 120)\left(\sqrt{area}\right)^{1/3} \tag{7.51}$$

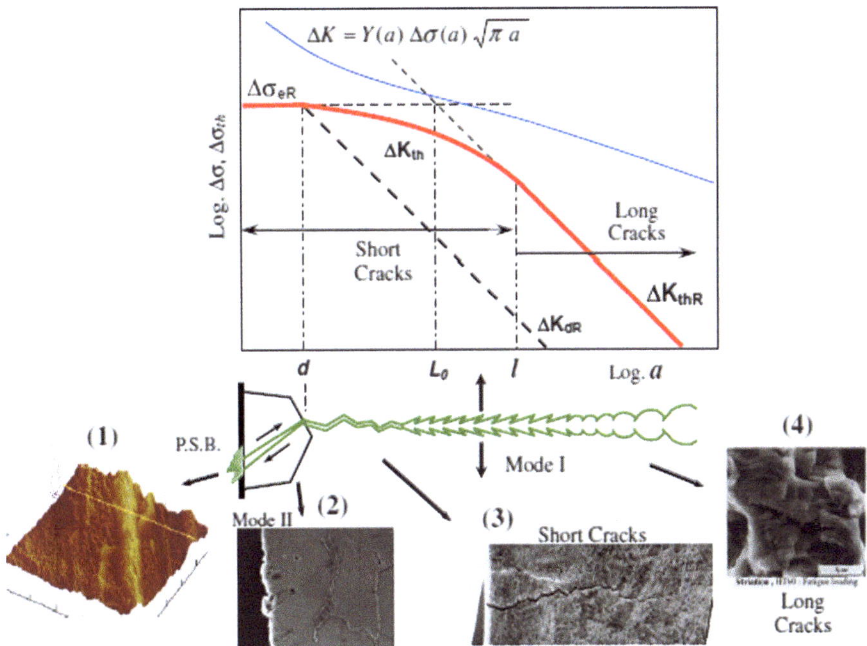

Fig. 7.58 Mechanical fatigue of materials without cracks or crack-like defects, based on the Kitagawa and Takahashi curve. Five stages are indicated: *1* initial cyclic damage in the form of cyclic hardening or softening, e.g., persistent slip bands, *2* microcrack initiation, *3* microcrack propagation to give rise to an initial engineering-sized flaw (0.5–1 mm), and *4* subsequent macroscopic propagation until final failure or instability (Chapetti 2011. With kind permission of Elsevier)

H_V is Vickers' hardness. In Eq. 7.51, according to Murakami, the threshold is proportional to the area of the defect (inclusion) as $\Delta K_{th} \propto (\sqrt{area})^{1/3}$. Thus, Eq. 7.51 is an expression in terms of Vickers' hardness. In Murakami's work, the region of $(area)^{1/2} \leq \sim 1{,}000$>. The relation between ΔK_{th} and $(area)^{1/2}$ is approximately linear, with a slope of 1/3 on a log-log scale and $\Delta K_{th} \propto (\sqrt{area})^{1/3}$. In Eq. 7.51, H_V is in kgf/mm^2 and area$^{1/2}$ is in μm, giving ΔK_{th} in MPam$^{1/2}$. Instead of the area, it is possible to use the size of the inclusions by their radii, R_i, assuming that the inclusion is spherical, giving:

$$\sqrt{area = \sqrt{\pi} a} \tag{7.52}$$

$$\sqrt{area}_{inclusion} = \sqrt{\pi} R_i \tag{7.52a}$$

Using Eqs. 7.51, 7.52, and 7.52a), one obtains the total fatigue-crack propagation threshold . (a function of crack length, a, where a is in μm).

The scatter of the σ–N data (see Fig. 7.57) is mainly due to the large scatter of inclusion size. The maximum inclusion contained in the specimen becomes the fracture origin. The fatigue crack propagation lives, N_P, of cracks initiated from the surface and from internal inclusions have been estimated and analyzed by using the previously proposed Chapetti model. The following expression was proposed to estimate the number of cycles necessary to generate the optical dark area [henceforth: ODA], i.e., the number of fatigue cycles that must be applied to obtain the critical fatigue-crack length, a_0^{pf}, from which the pure fatigue crack-propagation process occurs:

$$\Delta K_{th} = 4 \times 10^{-3}(H_V + 120)a^{1/3} \tag{7.53}$$

$$\frac{\sqrt{area}_{ODA}}{\sqrt{area}_{inclusion}} = \frac{a_0{}^{pf}}{R_i} = 0.25N^{0.125} \tag{7.54}$$

For a penny-shaped crack, ΔK can be written as:

$$\Delta K = \frac{2}{\pi}\Delta\sigma\sqrt{\pi a} \tag{7.55}$$

From Eqs. 7.53 and 7.54, the threshold stress and the crack length (in terms of hardness) are related as:

$$\Delta\sigma_{th} a^{-1/6} = 3.55\,(H_V + 120) \tag{7.56}$$

$$\sigma_{th} N^{1/48} = 4.473\frac{(H_V + 120)}{R_i{}^{1/6}} \tag{7.57}$$

In Eq. 7.57, it was assumed that a_o^{pf}, (i.e., the number of fatigue cycles that must be applied to obtain the critical fatigue-crack length) is almost similar to the total fatigue life, assuming that N_p ($N_i \gg N_p$), providing the expression for estimating the total life-time. The parameters are $\Delta\sigma_{th}$ in MPa, "a" in μm and H_V in kgf/mm^2. Clearly, the largest inclusions are the most detrimental to fatigue strength. Thus, R_i is given by the maximum radius of the inclusions contained in the material as R_i^{max}. In Fig. 7.59, the stress amplitude is a function of the number of cycles in JIS SUJ2-Q/T steel. The bold and dashed lines are calculated by Eq. 7.57.

In order to estimate the internal-fatigue resistance for very HCF, a simple model was first proposed for estimating the threshold nominal stress for obtaining fatigue fracture from cracks initiated at internal inclusions. This model took the threshold for pure fatigue-crack propagation and the inclusion size in the material in which the crack initiation takes place. Using Eq. 7.55 and with the applied ΔK equal to the pure-fatigue crack-propagation threshold, ΔKth, for a given crack length and

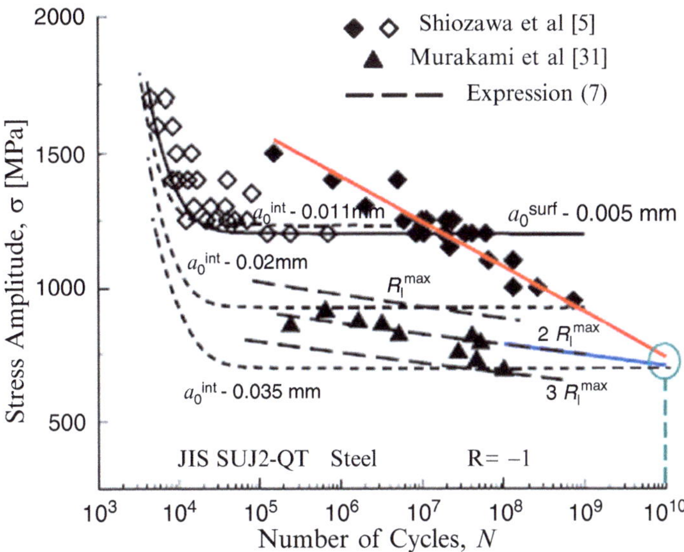

Fig. 7.59 Experimental fatigue life results for JIS SUJ2-Q/T steel from literature and estimations of the pure fatigue crack propagation life for surface (*bold line*) and internal (*dashed lines*) nucleated cracks are calculated by relation (7.57) (Chapetti 2011. With kind permission of Elsevier)

with $\Delta K = \Delta K_{th}$, $a = nR_i$ and $\Delta\sigma = 2\sigma_e^{int}$, the following expression for estimating fatigue resistance associated with internal-crack initiation is obtained:

$$\sigma_e^{int} = 444\frac{\Delta K_{th}}{\sqrt{nR_i}} \tag{7.58}$$

Figure 7.60 shows the fatigue limit vs. tensile strength. Equation 10 in Fig. 7.60 is parallel to Eq. 7.58 in the present notation and Eq. 9, given for R = −1, parallels Eq. 7.59 below:

$$\Delta K_{th-1} = -0.0038\sigma_u + 15.5 \quad R = -1 \tag{7.59}$$

In Eq. 7.58, n is a dimensionless factor, σ_e^{int} is internal (or subsurface) fatigue-resistance amplitude and ΔK is the applied stress-intensity factor range (i.e., the applied crack driving force).

Note the similar shapes in Figs. 7.56 and 7.60, respectively. Also note that the fatigue limit decreases with the initiation of crack formation at inclusions. Internal-fatigue resistance increases with decreasing the inclusion size, R_i (or the presence of hydrogen, known to induce HE).

Estimation becomes very important in the assessment of the fatigue resistance of components subjected to very HCF, in light of the fact that tests in the high-cycle range are both very expensive and time consuming. This is of particular interest in

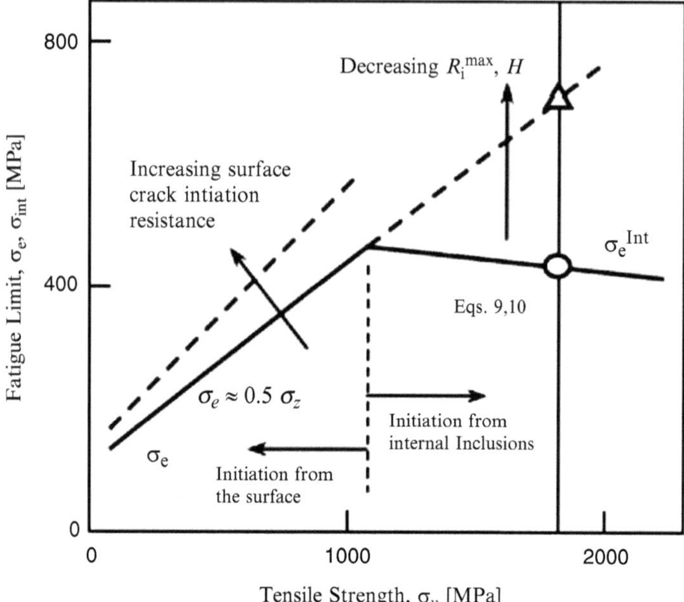

Fig. 7.60 Fatigue limit as a function of tensile strength and the influence of some important parameters (Chapetti 2011. With kind permission of Elsevier)

regard to structural materials, such as steels, that are expected to operate in this range in many modern applications. Since inclusions and various impurities cannot be completely eliminated, understanding their detrimental effects on fatigue resistance is of prime importance.

References

T.L. Anderson, *Fracture Mechanics: Fundamentals and Applications* (CRC Press, Boca Raton, 1991)

C.D. Beachem, Microscopic Fracture Processes, in *Fracture, an Advanced Treatise*, ed. by H. Liebowitz. Microscopic and Macroscopic Fundamentals, vol. 1 (Academic, New York, 1968)

A.H. Cottrell, *Dislocations and Plastic Flow in Crystals* (Oxford University Press, London, 1956)

D.K. Felbeck, A.G. Atkins, *Strength and Fracture of Engineering Solids*, 2nd edn. (Prentice-Hall, Englewood Cliffs, 1996)

A.A. Griffith, The phenomena of rupture, *Proceedings of the 1st International Congress, Applied Mechanics* (Delft, 1924), pp. 55–63

R.W. Hertzberg, *Deformation & Fracture Mechanics of Engineering Materials*, 4th edn. (Wiley, New York, 1996)

L.S. Li, G. Wang, *Introduction to Micromechanics and Nanomechanics* (World Scientific, Hackensack, 2008)

M.A. Meyers, K.K. Chawla, *Mechanical Behavior of Materials*, 2nd edn. (Cambridge University Press, Cambridge, 2009)
S. Suresh, *Fatigue of Materials* (Cambridge University Press, Cambridge, 1998)
A.S. Tetelman, *Fundamental Aspects of Stress Corrosion Cracking* (National Association of Corrosion Engineers, Houston, 1969), p. 446
H. Wada, N. Takimoto, I. Kondo, K. Murase, T.C. Kennedy, Estimation of Dynamic Fracture Toughness from Circumferentially Notched Round-Bar Specimens, Conference: 2004 SEM X International Congress and Exposition on Experimental and Applied Mechanics (2004)
C. Zener, The Macro-Mechanism of Fracture, in *Fracturing of Metals* (American Society of Metals, Metals Park, 1948), p. 3

Further References

R.W. Armstrong, Eng. Fract. Mech. **77**, 1348 (2010)
C.D. Beachem, Metal Trans. **3**, 437 (1972)
C.J. Beevers, R.J. Cooke, J.F. Knott, R.O. Ritchie, Metal Sci. **9**, 119 (1975)
H.K. Birnbaum, Hydrogen Related Failure Mechanisms in Metal, in *Environment-Sensitive Fracture of Engineering Materials*, ed. by Z.A. Foroulis (TMS-AIME, Warrendale, 1979), p. 326
D. Broek, Eng. Fract. Mech. **5**, 55 (1973)
M.D. Chapetti, Int. J. Fatigue **33**, 833 (2011)
A.H. Cottrell, Theory of Brittle Fracture in Steel and Similar Metals. Trans. Metal. Soc. AIME **212**, 192 (1958)
A.H. Cottrell, Proc. R. Soc. Lond. Ser. A Math. Phys. Sci. **276**(1) (1963)
J.K. Cuddy, M.N. Bassim, Mater. Sci. Eng. **A125**, 43 (1990)
J. Daming, W. Yinong, H. Bande, L. Tingquan, S. Fenglian, J. Mater. Sci. Lett. **15**, 1597 (1996)
S. de Luna, J. Fernández-Sáez, J.L. Párez-Castellanos, C. Navarro, Int. J. Press. Vessel. Pip. **77**, 691 (2000)
J. Eftis, H. Liebowitz, Int. J. Fract. Mech. **8**, 383 (1972)
R.N. Gardner, T.C. Pollock, H.G.F. Wilsdorf, Mater. Sci. Eng. **29**, 169 (1977)
R.K. Govila, D. Hull, Acta Metall. **16**, 45 (1968)
A.A. Griffith, Philos. Trans. R. Soc. Lond. A **221**, 163 (1920)
G.I. Hahn, A.R. Rosenfeld, Acta Metall. **14**, 1815 (1966)
E. Hayne Shumate Jr., The Radius of Curvature in the Prime Vertical. ITEA J. **30**, 159 (2009)
C.E. Inglis, Proc. Inst. Nav. Archit. **55**, 219 (1913)
C. Laird, G.C. Smith, Philos. Mag. **7**, 847 (1962)
V. Levkovitch, R. Sievert, B. Svendsen, Int. J. Fract. **136**, 207 (2005)
S. Liu, D. Liu, S. Liu, J. Mater. Sci. **42**, 7514 (2007)
S.P. Lynch, Scripta Metall. **1067** (1986)
S.P. Lynch, Acta Metall. **36**, 2639 (1988)
S.P. Lynch, Metallography **23**, 147 (1989)
L.E. Miller, G.C. Smith, J. Iron Steel Inst. **208**, 998 (1970)
Y. Murakami, S. Beretta, Extremes **2**, 123 (1999)
Y. Murakami, M. Endo, Fatigue **16**, 163 (1994)
H. Ohtani, C.J. McMahon Jr., Acta Metall. **23**, 377 (1976)
E. Orowan, Trans. Inst. Eng. Shipbuilders (Scotland) **89**, 165 (1946)
E. Orowan, Rep. Prog. Phys. **12**, 185 (1949)
P.C. Paris, M. Gomez, W.E. Anderson, Trend Eng. **13**, 1219 (1961)
G.M. Pressouyre, J. Dollet, B. Vieillard-Baron, Mémoires et Etudes Scientifiques de la Revue de Metallurgie **79**, 161 (1982)
L. Qing-fen, L. Li, L. Er-bao, L. Dong, C. Xiu-fang, Scr. Mater. **53**, 309 (2005)

E. Quadrini, Mater. Chem. Phys. **21**, 437 (1989)
R.O. Ritchie, J.F. Knott, J.R. Rice, J. Mech. Phys. Solids **21**, 395 (1973)
M. Sarfarazi, S.K. Ghosh, Eng. Fract. Mech. **27**, 257 (1987)
M.M. Shea, N.S. Stoloff, Mater. Sci. Eng. **12**, 245 (1973)
E. Smith, Int. J. Fract. Mech. **4**, 131 (1968)
J.F. Smith, J.H. Reynolds, H.N. Southworth, Acta Metall. **28**, 1555 (1980)
K.-J. Soderholm, Dent. Mater. **26**, e63 (2010)
L. Sohncke, Ann. Phys. Lpz. **137**, 177 (1869)
A.N. Stroh, Proc. Roy. Soc. Lond. **223A**, 548 (1955)
A.N. Stroh, Adv. Phys. **6**, 418 (1957)
D. Teirlinck, F. Zok, J.D. Embury, M.F. Ashby, Acta Metall. **36**, 1213 (1988)
A.R. Troiano, Trans. ASM **52**, 54 (1960)
G.F. Vander Voort, *Fractography*. ASM Handbook, vol. 12 (ASM International, Metals Park, 1987), p. 102
O. Vosikovsky, Eng. Fract. Mech. **11**, 595 (1979)
X.J. Zhang, R.W. Armstrong, G.R. Irwin, Metall. Trans. A **20A**, 2862 (1989)

Chapter 8
Mechanical Behavior in the Micron and Submicron/Nano Range

8.1 Introduction

Throughout this book, there has been frequent discussion about the effect of size on the mechanical properties of materials. Usually, strength properties increase with decreasing dimensions, while ductility decreases. Decreasing the dimensions of a material may decrease the size of the grains in polycrystalline materials. The size of single crystals depends on their growth conditions, but, also in this case, decreased size has the same influence on the mechanical properties. The expectation of improved mechanical characteristics, especially in the submicron/nanometer range, however, must be supported by experimental evidence. Experimental evidence has, indeed, indicated the outstanding mechanical properties of nanocrystalline (NC) materials that often show: superstrength, superhardness, improved specific strength and tribological performance (as attested in the literature). This pattern of reduced ductility with increased strength is also indicated in materials having small dimensions; however recently, some cases of substantial ductility were reported in superstrong NC materials undergoing 100% elongation or more without failure. These reported properties, the unique combination of high strength and good ductility, make such materials ideal for applications in a wide range of fields, such as the aviation, automotive and electronics industries, to name just a few. The aim of this chapter is to provide an overview of some of the mechanical properties discussed thus far regarding materials with small dimensions and to characterize their observed behavior.

8.2 Mechanical Behavior in the Small-Size Range

Unlike previous chapters, where experimental observations preceded theoretical explanations of the observed behavior, it was felt that, in this section, the order of presentation should be reversed, since, in all the chapters, especially Chap. 1,

J. Pelleg, *Mechanical Properties of Materials*, Solid Mechanics and Its Applications 190, DOI 10.1007/978-94-007-4342-7_8,

the testing methods for static, mechanical behavior were thoroughly discussed. Indeed, the initial focus here will be on the static properties and only later on will time-temperature (creep) and dynamic-cyclic (fatigue) properties be discussed. Thus, in the following, the reason for the size effect will be given, emphasizing the contribution of dislocations to the observed experimental data.

8.2.1 An Explanation of the Size Effect

Ever since the concept of dislocations was introduced (see Chap. 2), the physical and, in particular, mechanical behavior of materials has pivoted around their contribution to, or rather their responsibility for, a vast number of phenomena occurring in materials. Below, of a number suggestions explaining the size effect in terms of dislocations, the three most accepted explanation are considered:

8.2.1.1 Dislocation Starvation

When the size of a specimen is decreased, the surface-to-volume ratio increases. Because of the micron-nanometer/submicron dimensions of crystals during initial deformation, dislocations leave crystals before they can multiply by cross slip or other dislocation interactions. As a consequence, dislocation-density decreases during the deformation process, as the ability of these dislocations to multiply is reduced. Therefore, plastic deformation requires high stress levels for the repeated generation of dislocations (recall that in the absence of dislocations, the theoretical strength of the crystal is approached). Whiskers approach theoretical strength, due to the absence of dislocations. The question which arises is: How many glissile dislocations are present in micron-submicron/nanometer sized samples having the appropriate directions of the Burgers vectors for slip to occur readily? According to the argument that follows, only $\sim$1 single dislocation would be present in a volume of 10 μm^3. This estimation may be obtained if it assumed that the dislocation density is on the order of $\rho = 10^{10}$ m^{-2} (Hull and Bacon).

8.2.1.2 Dislocation Nucleation

Dislocation sources are essential for generating the dislocations required for plastic flow by slip. If the number of dislocation sources is limited, because of the small volume in a small-sized sample or, if the existing sources lay in unfavorable glide systems, then they must be activated to operate. This requires high flow stress. The critical shear stress required to cause a dislocation to bow out is given (Von Blanckenhagen) as:

$$\tau_{source} = \frac{\mu b}{2\pi} \frac{1}{s} \ln \frac{as}{b} \tag{8.1}$$

μ is the shear modulus, b the Burgers vector, s the source size and a is a numerical constant (see Chap. 2). It was found that the optimum source size, in terms of operating the source at a low-stress level, is between 1/3 and 1/4 of the limiting dimension, namely a pillar diameter or a half-beam thickness. Thus, there is an inverse relation of the flow stress to the pillar diameter (specimens usually have a pillar shape, hence this terminology is used in the tension-compression testing of small-sized specimens).

8.2.1.3 The Dislocation Pile-Up Model

Various obstacles can cause dislocations to pile up against them, thus forming pile-ups (Chap. 2). Due to the small number of dislocation sources available as specimen size decreases, dislocations become concentrated in a few glide systems. Consequently, dislocation pile-ups affect the dislocation sources due to strong back stresses. The stress necessary to operate a dislocation source is given as:

$$\sigma_A = \left(\frac{4\mu b\sigma_i}{\pi d} + {\sigma_s}^2\right)^{1/2} \tag{8.2}$$

where σ_A is applied stress, μ is the shear modulus, σ_i is the strength of the interface, d is the critical-length scale and σ_s is the source strength. The back stress from the dislocation pile-ups leads to an increased flow stress, scaling inversely with the square root of the critical-length scale.

8.3 The Static Properties

Advanced microfabrication techniques, developed over the past decade or so, enable the experimental testing of the mechanical properties of materials in the submicron range. Many studies (for example, Volkert, Greer, Uchic, Kim and others) have convincingly demonstrated that both the yield strength and the flow stress increase with decreasing sample size at micron and submicron (nano) scales. Of the above dislocation models, the theory of hardening by dislocation starvation was used to explain the observed size effects in cases of compression of FCC single-crystal nanopillars.

8.3.1 Single Crystals

Copper and Mo provide examples of stress-strain curves under compression and tension, as shown in Figs. 8.1 and 8.2. Note that the strain scales (x-axis) in

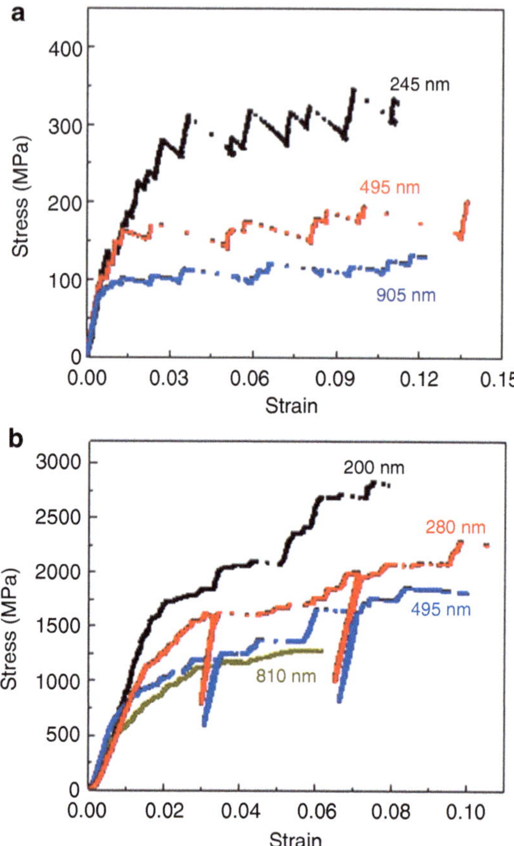

Fig. 8.1 Typical stress–strain curves in compression of (**a**) Au and (**b**) Mo with some intentional unloading and reloading segments (Kim and Greer 2009. With kind permission of Elsevier)

Fig. 8.1a, b are different. The tension samples with rectangular cross-sections were milled out of the well-annealed and electropolished (100) Au and Mo crystals, fabricated by using a focused-ion beam [henceforth: FIB]. Uniaxial tension and compression tests on single-crystalline Au and Mo were performed on (001) oriented nanopillars with effective diameters between 250 and 1 μm. By assuming volume conservation during plastic deformation, the true stress, σ, was evaluated from the known relation (see Chap. 1):

$$\sigma = \mathrm{PL}/\mathrm{A_0L_0} \tag{8.3}$$

Since the specimens tested were of rectangular cross-sections, an effective diameter was evaluated by:

$$d_{eff} = \sqrt{\left(\frac{4}{\pi}d_1 d_2\right)} \tag{8.4}$$

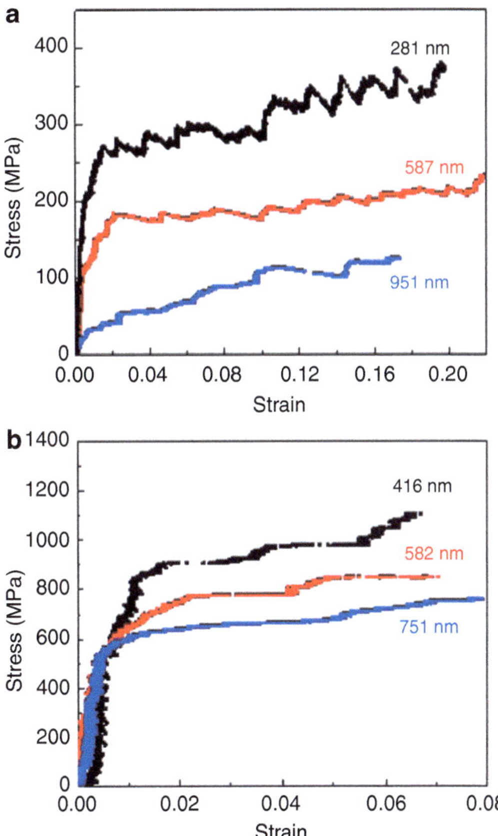

Fig. 8.2 Typical stress–strain curves in tension of (**a**) Au and (**b**) Mo. Note that the strain scales (x-axis) in (**a**) and (**b**) are different (Kim and Greer 2009. With kind permission of Elsevier)

d_1 and d_2 are the width and thickness of the rectangular cross-section. The tensile curves for Au show discrete strain bursts during plastic flow and size effects similar to those during compression.

In Fig. 8.3, SEM micrographs show the deformation results and the images of these small-sized specimens also resemble those previously seen in Figs. 7.2 and 7.3 for macroscopic test samples. Note the numerous slip lines in the middle of the gage length in Fig. 8.3a.

One can see that both the FCC Au and the BCC Mo work harden under compression, as it occurs in macrospecimens; however, in the Mo deformed by tension, there was very little work hardening. It is also clear that the size effect plays a role in the hardening process and that the degree of strain hardening is substantially greater in small-sized specimens. In regard to fabrication methods and various sophisticated tests of micron and submicron specimens, not discussed in this chapter, vast information is available in the literature. The flow stresses during the tension and compression of Au and Mo specimens are shown in Figs. 8.4 and 8.5, respectively.

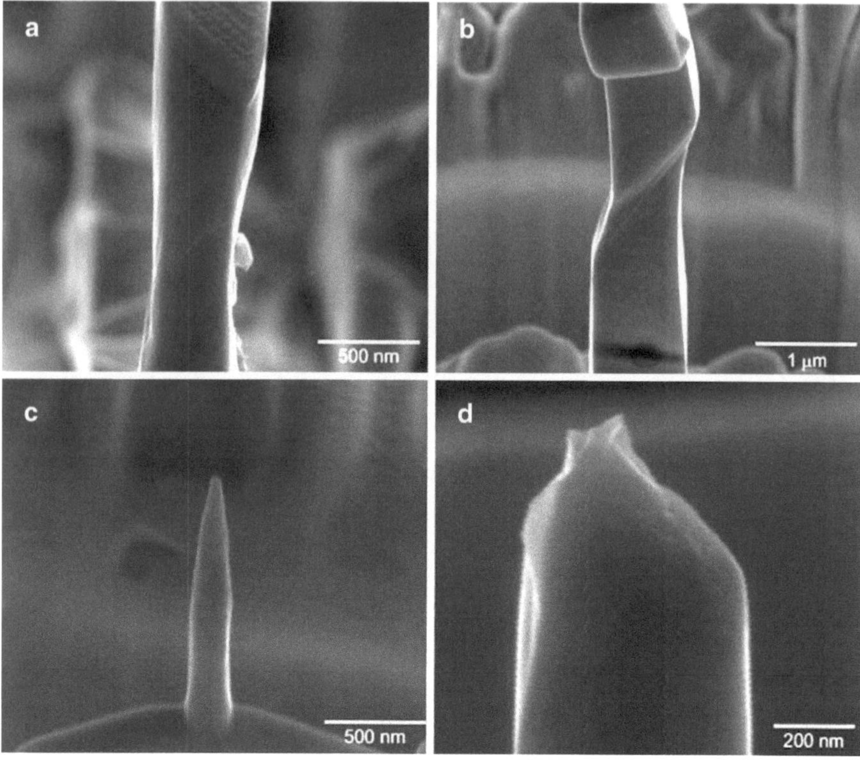

Fig. 8.3 SEM images showing (**a**) multiple slip lines in Au, (**b**) bending after severe deformation in Au, (**c**) Au sample drawn down to a point before rupture, (**d**) Mo fracture surface after quick necking and rupture (Kim and Greer 2009. With kind permission of Elsevier)

Figure 8.4 shows flow stresses at 10% strain, normalized by the ideal axial strength of Au on a log-log scale. A strong size effect is observed for both tension and compression, manifested by the flow-stress ratio decrease with the increasing dimensions of the tested specimens. There is no distinguishable difference between the tendencies of the flow stress-dimension relation between the compressed or the tensile-tested specimens. The change in yield strength (see Fig. 8.5a) follows the same trend as the one observed in Fig. 8.4 (for flow stress). However, there is a difference in flow stress in the Mo specimens during deformation by compression, as opposed to tension. Kim, and Greer claimed that the amount of strain hardening during tension in the BCC Mo, which is less pronounced than during compression, may be a consequence of: (i) the differences in the Peierls stress in twinning vs. anti-twinning deformations and (ii) the strong dependence of their CRSS on the non-glide, applied-stress tensor components.

The size effect on strengthening is probably associated with the smaller number of dislocation sources in small-sized specimens. In such cases, when the number of dislocation-emitting sources is small, effectively fewer dislocations are available

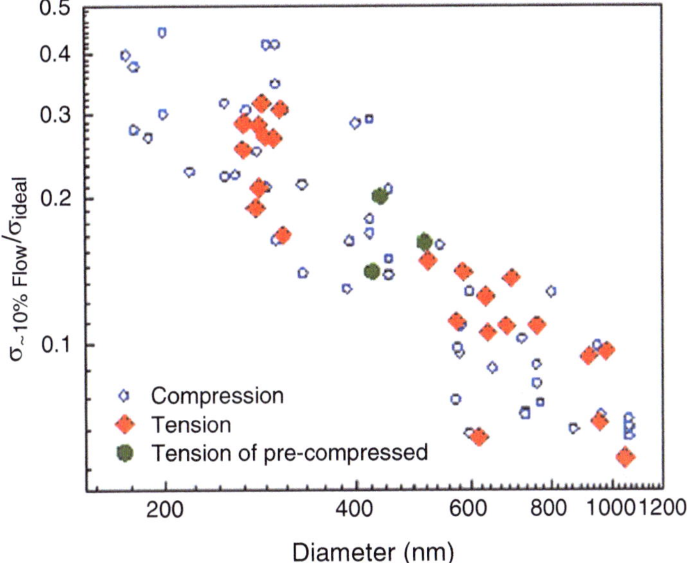

Fig. 8.4 Flow stresses in compression and tension for Au at ~10% strain normalized by the ideal strength of Au (Kim and Greer 2009. With kind permission of Elsevier)

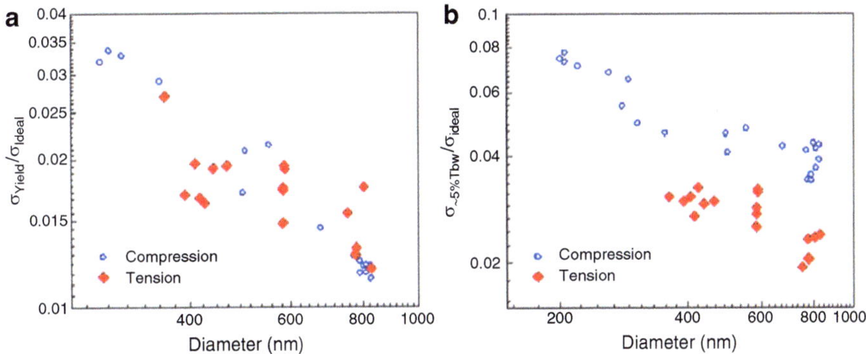

Fig. 8.5 (**a**) Yield strengths and (**b**) flow stresses at ~5% strain in compression and tension for Mo both normalized by the ideal strength of Mo (Kim and Greer 2009. With kind permission of Elsevier)

for plastic flow, resulting in the requirement of higher strength for deformation to proceed. The effect of size on the strength properties of materials has been recorded for various materials, single and polycrystalline alike, and affect properties such as yield strength, elongation (i.e., the strain to fracture), hardness, etc. An illuminating example of the generality of the size effect appears in Fig. 8.6, reproduced from the work of Uchic et al. for various metals investigated.

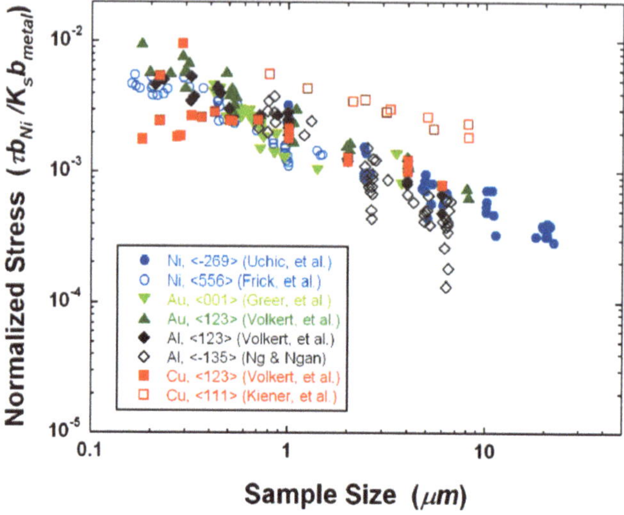

Fig. 8.6 Composite plot of published microcrystal flow stress data as a function of sample diameter for FCC metals. Data in the plot have been normalized by the shear modulus and Burgers vector. The scaling exponent n that best fits the data is approximately 0.6 (Uchic et al. Plasticity of micrometer-scale single-crystals in compression. A critical review (preprint), Air Force Research Laboratory, Materials and Manufacturing Directorate Wright-Patterson Air Force Base, OH 45433-7750, Air Force Materiel Command, United States Air Force, AFRL-RX-WP-TP-2008-4326. With kind permission of Dr. Uchic)

Based on various sources, certain test conditions require that normalized stress be plotted against sample size. In a regime in which size-dependent strengthening is being observed, the relation between flow stress, σ and sample diameter, d, can be empirically described by a power law:

$$\sigma = \mathrm{A}\mathrm{d}^{-\mathrm{n}} \tag{8.5}$$

Here, A is a constant and n is the power-law exponent. Generally, this relation holds for samples in the range of a few hundreds to a few tens of micrometers in diameter. The power-law exponent for FCC metals varies from study to study, ranging from 0.61 to 0.97 (Uchic et al.). As seen in Fig. 8.6, the strengthening data, as a function of specimen size, merge into a single band when the shear stress is normalized. Normalization is by the shear modulus and the relative ratio of the Burgers vector for each metal indicated. The scaling exponent taken from Eq. 8.7 farther below is 0.6, which is reasonable for all data indicated in Fig. 8.6. It is believed that the degree of size-dependent strengthening is related to the initial-dislocation density. It is also claimed that, for the FCC metals reported in Fig. 8.6, the relative importance of crystal orientation lessens as the sample diameter shrinks to the micrometer scale, since the stress-strain curves for both single-slip and multiple-slip orientations become qualitatively similar when size-affected strain hardening dominates plastic flow.

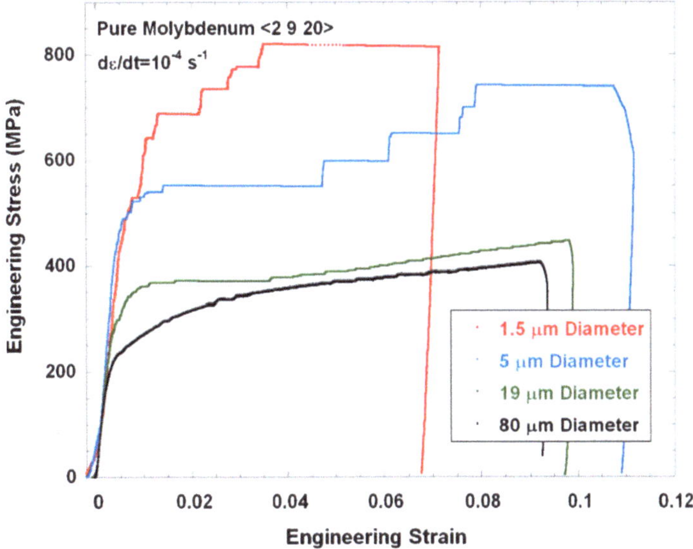

Fig. 8.7 Representative experimental flow curves for pure molybdenum <2 9 20>-oriented microcrystals that clearly demonstrate size-dependent strengthening effects (Uchic et al. Plasticity of micrometer-scale single-crystals in compression. A critical review (preprint), Air Force Research Laboratory, Materials and Manufacturing Directorate Wright-Patterson Air Force Base, OH 45433-7750, Air Force Materiel Command, United States Air Force, AFRL-RX-WP-TP-2008-4326. With kind permission of Dr. Uchic)

Size-dependent strengthening in BCC metals is seen in Fig. 8.7 for Mo single crystals, similar to what was observed in FCC metals. It was reported that microcrystals, presumed to be defect-free, display an incredibly high yield stress (9.3 GPa), which decreases dramatically when an initial-dislocation substructure is present in the microcrystal. It has also been reported (see, for example, Uchic et al.) that microcrystal deformation behavior is independent of size in the extreme cases, either when the material is dislocation-free or when the initial-dislocation densities are high. Also, at either extreme condition, the scatter in the flow stress is relatively small and, for high-density crystals, the flow curves vary smoothly and strain bursts are generally absent.

These observations seem to be logical in the light of the following facts: (a) in dislocation-free crystals, the strength approaches the theoretical strength (as calculated in Chap. 2 and as observed in whiskers) and (b) in initially-high dislocation-density specimens, the strength is high. The consequence of both extremes is that the contribution of the size to the overall high strength is relatively insignificant and its small strengthening effect becomes almost completely masked.

Unlike deformation in bulk crystals, usually associated with smoothly-varying flow curves, discrete bursts of strain activity are regularly observed in microcrystal flow curves. Examples of such strain bursts appear in Figs. 8.1 and 8.2. One may consider these observations in the following manner. Under stress, crystalline materials exhibit irreversible plastic deformation, caused by the motion of lattice dislocations. This also occurs in microcrystals, in which internal-dislocation avalanches lead to jumps in the stress-strain curves, known as 'strain bursts'. In macroscopic samples, plasticity appears as a smooth process. The phenomenology of these strain bursts closely resembles that of the macroscopic plastic instabilities of cases with stress-strain curves characterized by serrated yielding (Csikor et al.). In experiments performed on single-crystal samples, complications associated with grain-boundary strengthening or due to underlying substrates are eliminated and the net size effect on strength is obtained.

8.3.2 *Polycrystalline Materials*

The specimen used for deformation in polycrystalline materials shows both strong size-dependence and active grain-boundary hardening (Chen and Ngan). An example of the effects of specimen and grain sizes may be given by the deformation results for Ag polycrystalline materials. In Fig. 8.8, both these effects are illustrated. Ag wires with an initial diameter of ~52 μm were annealed in a vacuum furnace at a vacuum level of 10^{-5} torr for 0.5 h under different temperatures ranging from 170°C to 750°C in order to obtain different grain sizes. Figure 8.8a, b show the typical engineering stress (σ)-engineering strain (ε) curves of the Ag wires with various diameters and grain sizes. Figure 8.8a reveals that, at a given wire diameter of ~40 mm, as the grain size increases within the experimental range, both the tensile elongation and the ultimate-tensile strength decrease. Figure 8.8b shows that, at a constant grain size, the elongation of the wires decreases as their diameters decrease. The yield-stress increase is more pronounced for smaller grain sizes (3.5 μm grains) compared with the larger 40.6 μm grains. These wires were annealed at 1,023 K (750°C). The variation of 0.2% proof stress with grain size may be expressed by the H-P relation. As discussed earlier in Chap. 3, the H-P relation is given by:

$$\sigma_y = \sigma_0 + \frac{k_y}{\sqrt{d}} \tag{3.42}$$

However, another relation, based on the H-P relation, was suggested (Chen and Ngan) for the relation of the 0.2% proof stress to the grain size, given as:

$$\sigma_{0.2\%} = \sigma_1(t) + \sigma_2\left(\frac{t}{d}\right) + k_y d^{-1/2} \tag{8.6}$$

Equation 8.6 takes into account three parameters: (i) the grain size, d (as is expressed by the original H-P relation); (ii) the specimen diameter, t; and (iii) the specimen shape, expressed by the ratio t/d. This equation is also expressed in graphical form in Fig. 8.9.

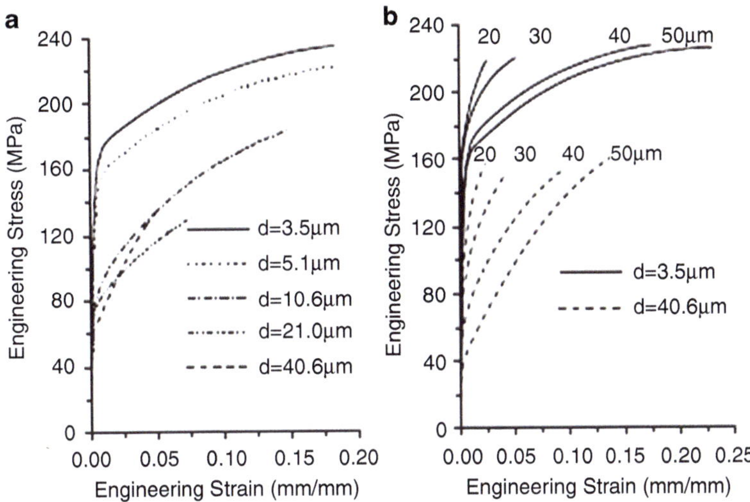

Fig. 8.8 Typical engineering stress (σ) vs. engineering strain (ε) behavior of Ag wires. (**a**) Grain sizes differ from 3.5 to 40.6 μm, with a constant diameter of 40 μm. (**b**) Specimen diameters differ from 50 to 20 μm, with grain sizes of 3.5 and 40.6 μm respectively (Chen and Ngan 2011. With kind permission of Elsevier)

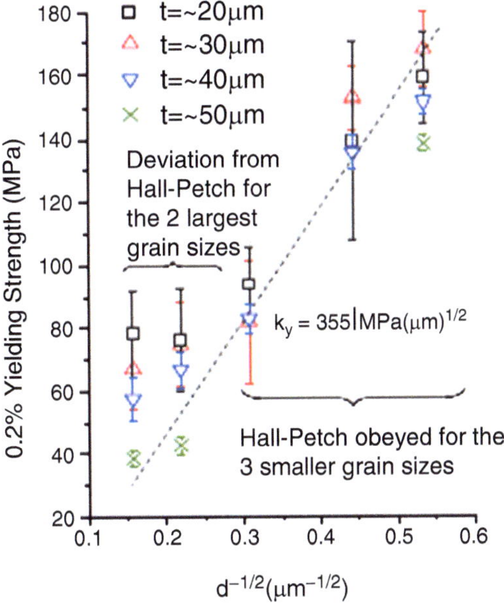

Fig. 8.9 The Hall–Petch relationship applied to the 0.2% yielding strength (specimen diameters: 50, 40, 30 and 20 μm). Error bars denote ±1 standard deviation (Chen and Ngan 2011. With kind permission of Elsevier)

Note that the relation holds for specimens with smaller size grains, but deviates from the extrapolated line for larger grain sizes. A common line drawn through the different specimen sizes (illustrated in Fig. 8.9) gives a slope of ~355 MPa (μm)$^{0.5}$. An analysis of the proof-stress data according to Eq. 8.6 is shown in Fig. 8.10,

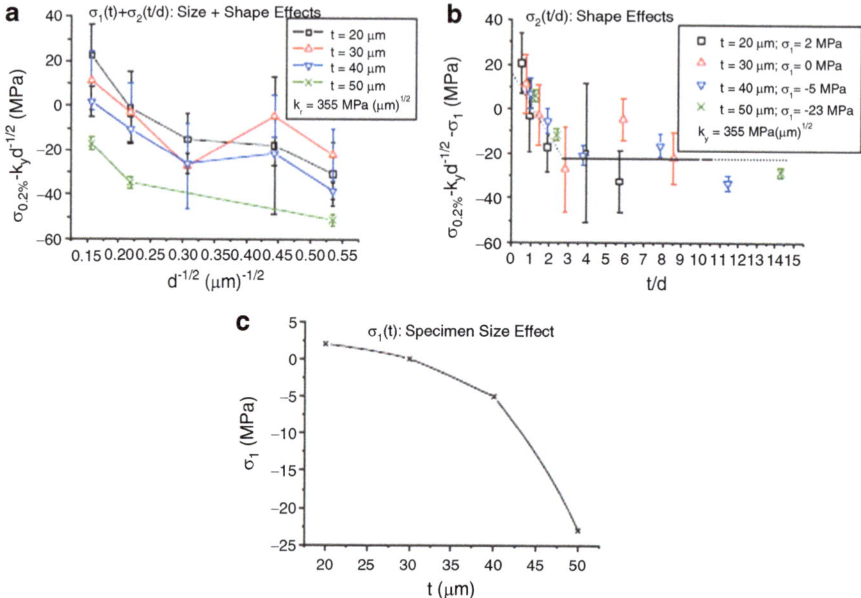

Fig. 8.10 Analysis of the proof strength data according to Eq. 8.6. (**a**) Plots of $\sigma_{0.2\%} - k_y d$ vs. $d^{1/2}$. (**b**) Plots of $\sigma_{0.2\%} - k_y d^{-1/2} - \sigma_1(t)$ vs. t/d, where the $\sigma_1(t)$ values are selected so as to cause the data for different t values to overlap. (**c**) Plot of $\sigma_1(t)$ vs. t (Chen and Ngan 2011. With kind permission of Elsevier)

emphasizing certain aspects of this relation. Thus, Fig. 8.10a, shows that the strength is, indeed, approximately expressed by Eq. 8.6. To this end, the plot demonstrates $\sigma_{0.2\%} - k_y d^{-1/2}$ vs. $d^{-1/2}$. This way, if Eq. 8.6 is correct, H-P grain-size dependence on strength is offset and any residual dependence on d and t should be due to specimen size, $\sigma_1(t)$, and shape, $\sigma_2(t/d)$, factors. In Fig. 8.10b, $\sigma_{0.2\%} - k_y d^{-1/2} - \sigma^1(t)$ is plotted vs. the shape factor, (t/d). The $\sigma_1(t)$ values are selected so as to cause the data for different t values to overlap more or less into a single band. In Fig. 8.10c, the $\sigma_1(t)$ values are plotted against specimen size, t, thus showing the size effect. The $\sigma_1(t)$ values, which made the plots in Fig. 8.10b, converge are those shown in Fig. 8.10c. Figure 8.10b shows that the shape effect is prominent only for t/d ≤ 3; in larger t/d ratios, the σ_2 (t/d) function tends to a constant value. The variation in the specimen-size effect, $\sigma_1(t)$, as shown in Fig. 8.10c, is −25 MPa over the range of t from 20 to 50 μm.

In conclusion, deformation in polycrystalline Ag tensile-tested microwires shows a significant specimen-size effect on strength, in addition to a grain-size effect. Careful analysis of these results also indicates the existence of a shape-dependent strengthening effect with decreased specimen size, according to the ratio of the specimen diameter to the grain size, provided that this ratio is less than 3.5 μm in diameter.

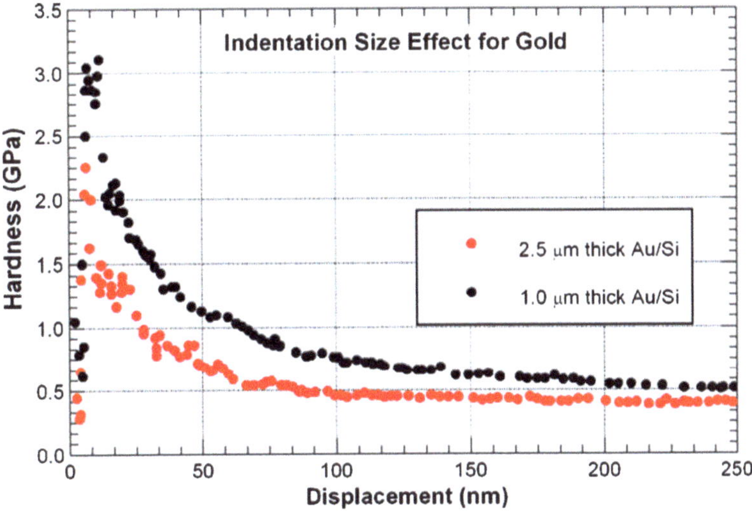

Fig. 8.11 Indentation size effect for gold thin films on silicon substrates showing that the hardness increases with decreasing indentation size (Nix et al. 2007. With kind permission of Elsevier)

8.3.3 Thin Films

Thin film materials of micron and submicron size are extensively used in industry for various technological applications. Thin films are produced by various metallurgical techniques, of which deposition is the most common. They may be tested by known methods of mechanical testing, such as tension-compression, bending and hardness. Various special nanoindenters are available on the market (see, for example, Micro Star Technologies, www.microstartech.com). Nanoindentation is a versatile technique for measuring mechanical properties at very small micron and submicron scales and, in addition to hardness, the elastic modulus and fracture toughness may be obtained with good accuracies (Pharr). Some of these thin film tests are presented below. A large number of nanoindentation experiments have been reported in the literature (Nix et al. and inherent references), indicating a strong inverse relation between hardness and indentation size (i.e., the depth of penetration). In the reported experiments, plastic deformation is not uniform and, therefore, considerable strain gradients arise. Figure 8.11 illustrates the commonly observed results of indentation on Au thin-film specimens on a silicon substrate. The variation of hardness with the change in film thickness, as indicated, is film-thickness-dependent; a shift in hardness values to higher levels is seen for thinner films.

The nanoindentation depth of epitaxial Au and Mo thin films is shown in Figs. 8.12 and 8.13. In both, the displacements are on a nanometer scale. Indentation was performed by means of a Berkovich diamond indenter with a rounded tip of about 49 nm. The solid line corresponds to the elastic-loading line, as predicted by the Hertz contact theory for a diamond sphere contacting a Mo surface. Note that two limiting cases of the Hertz application must be fulfilled (Yan and

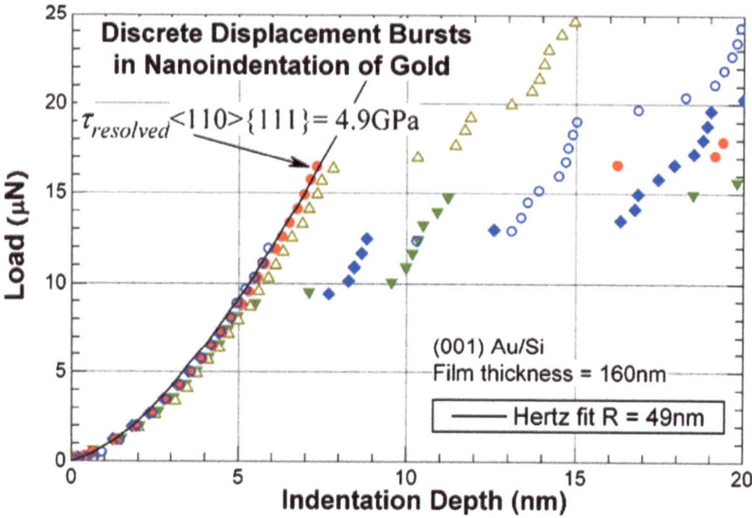

Fig. 8.12 Discrete displacement bursts associated with nanoindentation of an epitaxial film of Au. The displacement bursts are preceded by perfectly elastic loading (Nix et al. 2007. With kind permission of Elsevier)

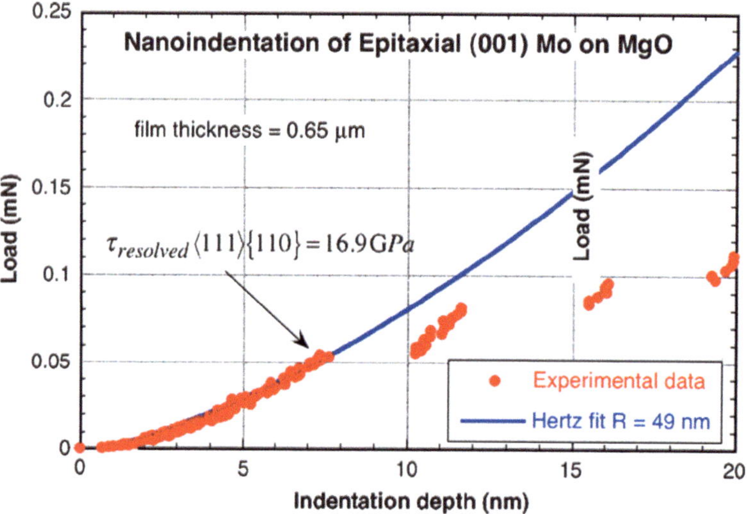

Fig. 8.13 Discrete displacement bursts associated with nanoindentation of an epitaxial film of Mo. The displacement burst is preceded by perfectly elastic loading (Nix et al. 2007. With kind permission of Elsevier)

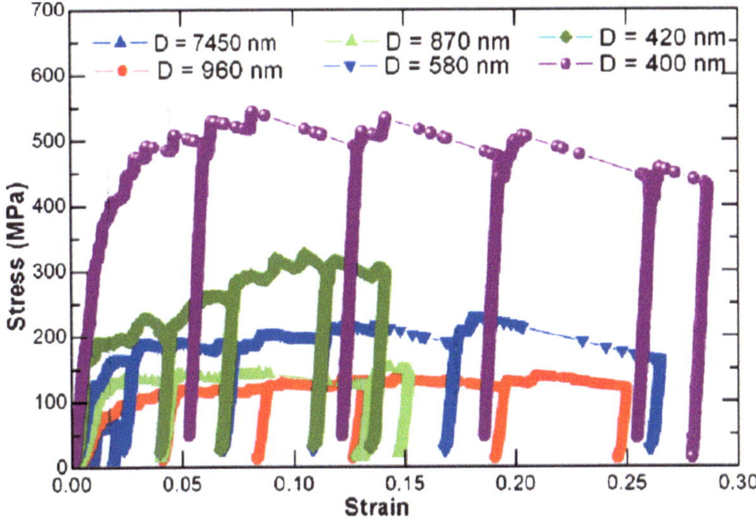

Fig. 8.14 Stress–strain relations for <001> gold pillars of different sizes showing the strong size effect for this kind of deformation (Nix et al. 2007. With kind permission of Elsevier)

Fischer), namely (i) the contact between elastic bodies should be frictionless and (ii) the significant dimensions of the contact area should be much smaller than the dimensions and radii of the curvatures of the contacting bodies.

The load-displacement data in the illustrated curves follow the Hertz loading curve up to a critical load, at which point a displacement burst occurs. Displacement bursts occur at shear stresses approaching the theoretical shear strength and are thought to coincide with the nucleation of dislocations. Using Hertz contact theory, the shear stress on the expected slip system at the point of maximum stress and at critical load was calculated. The results of these calculations are 4.9 and 16.9 GPa, as indicated in the illustrations for Au and Mo films, respectively. These values correspond to the expected, theoretical shear strength of the crystals or the critical stress at which dislocation nucleation may occur. The necessity of dislocation nucleation indicates that nanometer-scale indentation is associated with dislocation-starved crystals, requiring the nucleation of dislocations in order to initiate plasticity.

Compression is another testing method able to show the size effect. An example of deformation by gold micropillar hardening is shown in Fig. 8.14. The stress-strain curves shown are for a range of pillar sizes. These experiments were performed using multiple cycles of loading/unloading before reaching the maximum load, in order to ensure that the unloading data is, indeed, elastic. A FIB fabrication approach (following Uchic et al.) was used to create submicron pillars of <001>-oriented single crystal gold. The stress-strain curve is not smooth; rather it is characterized by discrete strain bursts, as mentioned earlier. Clearly, the flow stress of these gold pillars increases markedly with decreasing pillar diameter and, thus, size-dependent strengthening occurs.

Earlier (in Sect. 8.2.1), dislocation starvation was considered as one of the possible mechanisms introducing size-induced strengthening. This may explain the great strength observed in uniaxial compression. Unlike the case in bulk materials, dislocation motion leads to dislocation multiplication by double cross slip and similar processes, invariably leading to softening before strain hardening occurs, due to the elastic interaction of the dislocations. Dislocations in small crystals can only travel much smaller distances before annihilation at the free surface, thereby, reducing the probability of multiplication processes that might have led to initial softening before the hardening process. Dislocations leaving small-sized crystals, before they have a chance to multiply, leads to dislocation starvation. Very high stresses are required to nucleate new dislocations, leading to the observed high strengths in micron- and submicron-sized specimens.

As previously stated, sophisticated methods have been developed for testing small-sized specimens. The miniaturization of some FCC single crystals deformed by tension has been thoroughly discussed by Dehm. He presents an overview of recent progress in tensile testing on a small length scale, in the micron and submicron range, summarizing plasticity and the dislocation processes involved in deformation, focusing on tensile straining. Here, further consideration is given to the thermal straining of thin-films by using the differences in the thermal-expansion coefficients of the thin-films and the substrate materials. Figure 8.15 presents a technique for producing tensile specimens (see Dehm's paper). This technique was used for Cu single crystals. The tested sample is shown before and after it was gripped; it is shown before loading in Fig. 8.16, together with the W gripper. A 3 μm-thick stress-strain curve and a 15 μm-long Cu tensile specimen with a tensile axis along <−234> are both shown in Fig. 8.17. Note the stress drops in the displacement-controlled experiments. Points labeled (a–f) correspond to the simultaneously recorded SEM images presented in Fig. 8.18. At ~22% strain, the sample was unloaded and reloaded. In a series of SEM images, Fig. 8.18 shows the formation of distinct glide steps, first in the center of the sample (Fig. 8.18b) and finally along the whole gauge length (Fig. 8.18c–e). The corresponding positions along the stress-strain curve are indicated by arrows (Fig. 8.17b). Figure 8.18f shows the sample and the gripper after unloading.

Another means of straining specimens is by thermal straining. Wafer curvature experiments done on thin films on substrates provide stress-strain data by using the differences in the thermal-expansion coefficients of the thin-films and substrates. The average biaxial-film stress, σ, is calculated based on Stoney's equation, given as:

$$\sigma = \frac{Mt^2}{6h}K \tag{8.7}$$

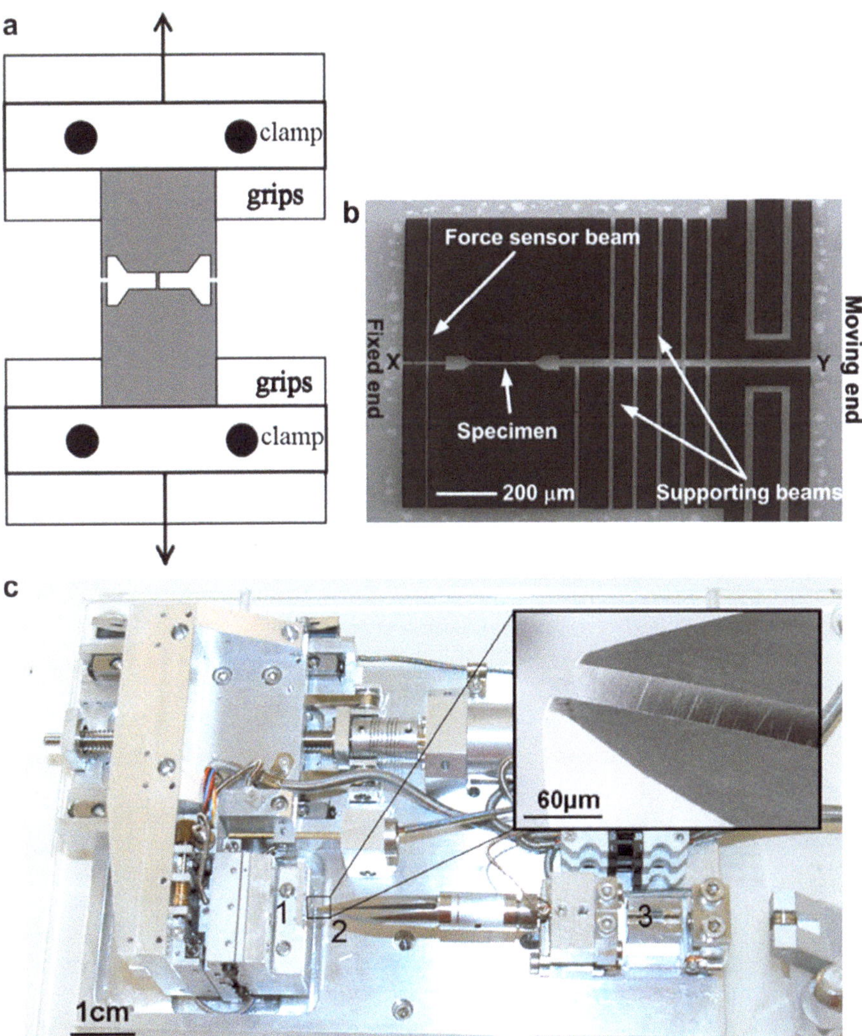

Fig. 8.15 (**a**) Schematic of a tensile sample made by lithography techniques consisting of a several hundred microns long freestanding metal film on an mm-sized Si support. The Si support can be attached to a tensile apparatus and after cutting the Si frame the free-standing metal film can bet strained (**b**) Micro-tensile testing chip for in situ experiments in electron microscopes. (**c**) Photograph of a commercial micro-tensile stage for use in a SEM. The sample is mounted on a piezo-driven stage (*1*), clamped by a tweezers (*2*, see also *inset* for magnified view on the *tip* of the tweezers), and load measured by a pre-strained wire (*3*) connected with the tweezers. (**d**, **e**) Fabrication of a miniaturized tensile sample from a bulk Cu single crystal. (**d**) A small needle is fabricated by cutting and electro-polishing, (**e**) and finally shaped by FIB milling. The *dog-bone* shaped sample can be placed in a W micro-gripper which is also structured by FIB milling. The W micro-gripper is attached to a microindenter which can operate in forward (compression) and backward (tension) loading mode. (**f**) Tensile samples with a pull hole at one end for manipulation with an actuated hook; The free-standing thin metal film is obtained by lithography and preferential etching (Dehm 2009. With kind permission of Elsevier)

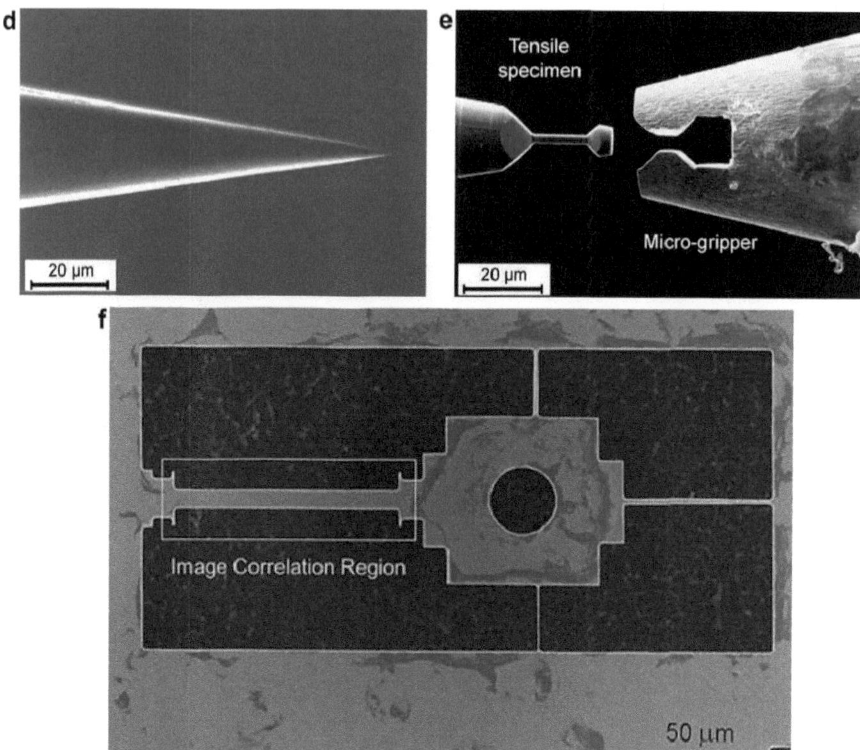

Fig. 8.15 (continued)

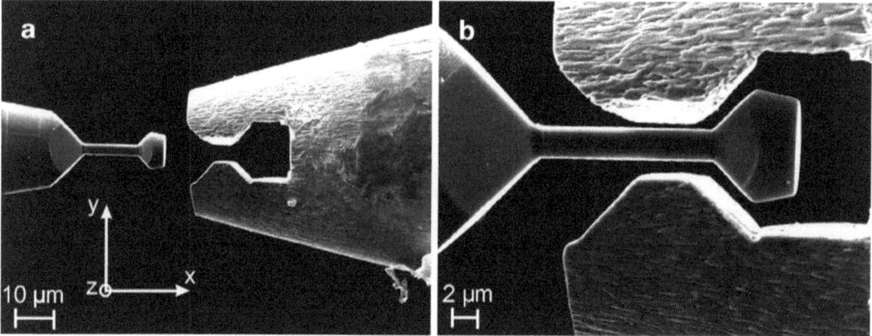

Fig. 8.16 (**a**) SEM image showing a single-crystal copper tension sample and the corresponding tungsten sample gripper before the test at a low magnification. (**b**) Sample and gripper aligned prior to loading. The loading direction is along <−234> (Kiener et al. 2008. With kind permission of Elsevier)

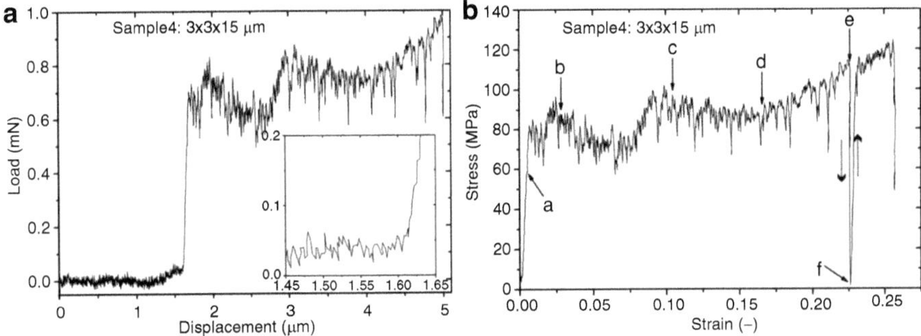

Fig. 8.17 (**a**) Load–displacement curve measured for the first loading of the sample depicted in Fig. 8.16. The *inset* shows details of the measured curve just before full contact between the sample and the gripper is established. (**b**) Stress–strain curve calculated from the load–displacement data. A second loading step (indicated by *arrows*) performed for this sample is included in the graph but not shown in (**a**) (Kiener et al. 2008. With kind permission of Elsevier)

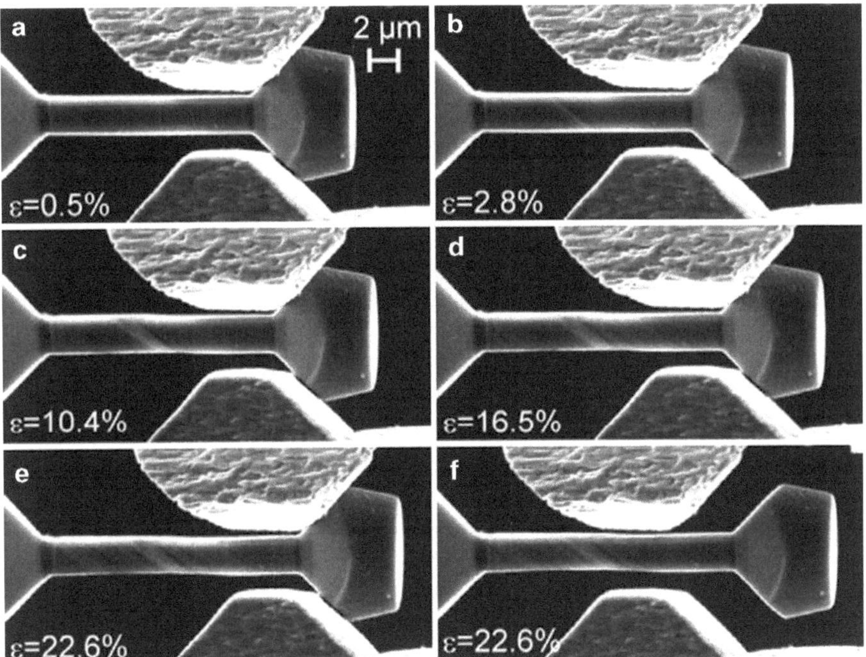

Fig. 8.18 SEM images of the sample shown in Fig. 8.16 taken during in situ tensile testing (**a–e**) and after unloading (**f**). The corresponding positions along the stress–strain curve are indicated by arrows in Fig. 8.17b (Dehm 2009. With kind permission of Elsevier)

K is the curvature of the thin-film/substrate system, h is film thickness, t is substrate thickness and M is the biaxial modulus of the substrate. $M = E/(1-\nu)$, where E and ν are the Young's modulus and Poisson's ratio, respectively. Basically, the above equation, for the thin-film case, which is much thinner than the substrate, may be more commonly presented as:

$$\sigma_f = \frac{E_s t_s{}^2}{6(1-\nu)t_f}\left(\frac{1}{R} - \frac{1}{R_0}\right) \tag{8.8}$$

The equivalence of these two relations is clear: $\sigma = \sigma_f$ and $M = E_s/(1-\nu)$ and, as indicated earlier, $t = t_s$, $h = t_f$ and $K = (1/R - 1/R_0)$. Here, R_0 = substrate curvature before film deposition and R is the substrate curvature measured after deposition and/or during the heating process. The thin-film strain is given by the product of the thermal-coefficient difference, $\Delta\alpha$ (between the thin film and the substrate) and the applied temperature change, ΔT. (Recall that linear-thermal expansion, resulting from a temperature increase, dT, may be expressed as $\frac{dl}{l} = \alpha dT$ with dl as the change in length, l, resulting from temperature change, α, is a function of temperature). It is essential to determine the initial curvature of the substrates prior to film deposition, so that it may be subtracted from the curvature of the thin film/substrate system. One method for measuring these curvature changes during thermal straining is by laser technology. Thin films may be either polycrystalline or single-crystal-like. The latter is usually obtained by epitaxial film growth. For this to occur, usually single-crystal substrates, such as Si or single-crystal oxides of Al, Mg, etc. are used. (If the deposited film is, as often observed, amorphous, than thermal treatment can induce crystallization to form polycrystalline film).

Thermally strained Al and Cu thin-films are examples illustrated in Fig. 8.19, showing the results of deformation in terms of biaxial stress vs. temperature and biaxial stress vs. strain. The epitaxial Al thin-films had a (100) orientation and were grown on (110) Si or on (0001) α-Al_2O_3 substrates. The (111) Cu thin-films were also grown on (0001) α-Al_2O_3 substrates. In the calculated stress-strain curves of the cooling portion of the thermal cycle shown in Fig. 8.19a, the temperature-dependence of the thermal-expansion coefficients was taken into account. The origin of the stress-strain curves in Fig. 8.19b was selected corresponding to the zero-stress value of the thermal-straining curves upon cooling. Since thermal straining is very sensitive to small strains, the biaxial 0.2% yield strength may be determined for the 600 nm thin-films shown in Fig. 8.19b, providing the values of 84 MPa for Al and 107 MPa for Cu, respectively. In the label for Fig. 8.19, the variation of the thermal expansions with temperature for Al, Cu and the α-Al_2O_3 substrate are given. Based on the slopes in the elastic regimes of the curves in Fig. 8.19, the biaxial-elastic modulus can be evaluated for the (111)-oriented thin-films: ${M^{Al}}_{111} = 87$ GPa and ${M^{Cu}}_{111} = 200$ GPa. These values are lower than the theoretic ones, which are ${M^{Al}}_{111} = 114$ GPa and ${M^{Cu}}_{111} = 260$ GPa, respectively. This may be the consequence of the high temperatures used (Fig. 8.19a) during which an ~20% reduction in the elastic constants occurs. The effect of the thin-film

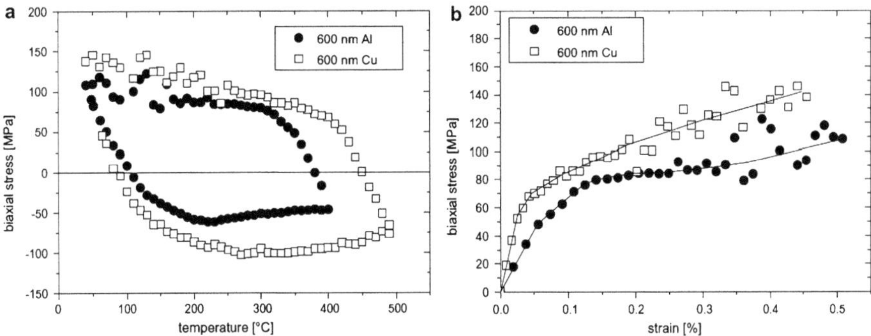

Fig. 8.19 (**a**) Thermal (biaxial) stress evolution of a 600 nm thick (111) oriented Al film up to 400°C and of a 600 nm thick (111) Cu film cycled between room temperature and 500°C. The scatter in the data upon cooling is a result of unintentional vibrations of the samples caused by flooding the furnace with an increasing flow of nitrogen to reduce the temperature. In contrast, a small and constant nitrogen flow is maintained upon heating (compare heating/cooling cycles). (**b**) Calculated stress–strain curves based on the cooling portions of the thermal cycles in (**a**). The following temperature dependent thermal expansion coefficients, α, were used to deduce strain values, where T is the temperature in °C: $\alpha(\text{Al}) = [23.64 + 0.008(T-27) + 0.000025\,(T-27)^2] \times 10^{-6}$; $\alpha(\text{Cu}) = [16.85 + 0.005\,(T-27) + 0.0000034\,(T-27)^2] \times 10^{-6}$, and $\alpha(\text{Al2O3}) = [7.0 + 0.0062\,(T+273) - 0.0000015\,(T+273)^2] \times 10^{-6}$ (Dehm 2009. With kind permission of Elsevier)

dimensions in terms of its thickness, h, on the biaxial stress may be seen in Fig. 8.20. The biaxial stress, σ, at the end of the thermal cycle at 40°C represents the size effect (see Fig. 8.19a). Note that the plot is on a log-log scale. On the abscissa, the inverse of film thickness is clearly indicated, thus, the curves show the size-dependent stress of the thin-films, i.e., the smaller is the film thickness the stronger it is.

The stress increase is described by a power law; for Al it is $\sigma \sim h^{-05}$ and for Cu it is $\sigma \sim h^{-0.7}$. The size effect is a consequence of the thin-film/substrate interface constraining dislocation motion. The Nix–Freund model yields a relation for flow stress, σ_{Nix}, which depends on film thickness, h, the elastic properties of the film (index: f), the substrate material (index: s) and on the geometry of the slip system, given as:

$$\sigma_{Nix} = \frac{\sin\phi}{\cos\phi\cos\psi}\frac{b}{4\pi\,(1-\nu)\,h}\left[\frac{2G_f G_s}{G_f+G_s}\ln\left(\frac{\beta_s}{b}h\right)\right] \tag{8.9}$$

where ϕ is the angle between the film normal and the inclined $11\bar{1}$ glide planes for a (111) film plane, ψ is the angle between the film normal and the Burgers vector, ν is the Poisson's ratio of the thin-film material, h is film thickness, G_f and G_s are the shear moduli of the thin-film and substrate, respectively, and b_s is a numerical constant. The experimental power-law fit with $\sigma \sim h^{-\alpha}$ having α 0.5 and 0.7 for Al and Cu thin-films, respectively, is somewhat smaller than for the Nix-Freund model, that predicts α 0.7 and 0.8 for these thin-films. This model assumes that

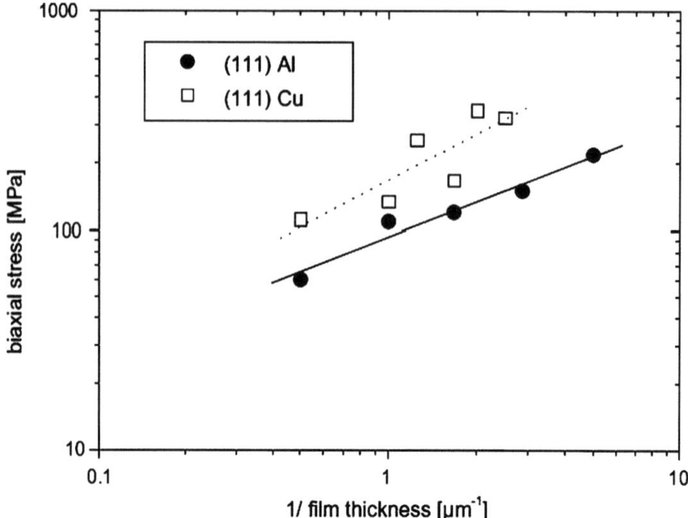

Fig. 8.20 Biaxial stresses at the end of a thermal cycle at 40°C for film thicknesses, h, between 0.2 and 2 μm. The stresses increase with $\sigma \sim h^{-0.5}$ for Al and $\sigma \sim h^{-0.7}$ for Cu as indicated by the *lines* in the graph (Dehm 2009. With kind permission of Elsevier)

the strengthening of thinner films has been rationalized in terms of threading dislocations, whose motion is confined to the metal layer, because of image forces generated by a stiff substrate. The plastic yielding of the layer corresponds to the motion of the first threading dislocations and, trailing in their wake, is a segment of interfacial dislocation. This motion is possible when the increase in interfacial-dislocation energy, due to the length increase and the work of the image forces, equals the film-strain energy reduction. Balancing these two terms leads to the Nix-Freund relation, which establishes that yield stress varies as 1/h.

TEM images of thermal-straining experiments reveal the dislocation structures of the single-crystalline Al and Cu thin-films on Si and α-Al_2O_3 substrates (considered above). It was found that dislocation nucleation occurs predominantly from dislocation sources at the thin-film/substrate interface, as indicated in Fig. 8.21a. Due to the lattice mismatch between the epitaxial films and their substrates, misfit-dislocation networks are already present at or near the interface. These dislocations act as sources for the formation of new dislocations. When dislocation sources in the interior of the thin-films are operating, dislocation pile-ups occur at the thin-film/substrate interface, exerting back stress on the dislocation source. For Cu films on α-Al_2O_3, it was observed that the pile-up reaches a constant number of dislocations, since, after each dislocation emission from the source, the leading dislocation of the pile-up is incorporated into the pre-existing dislocation network at the interface (as seen in Fig. 8.21c). A more illustrative presentation of the role of dislocation pile-ups is shown Fig. 8.22 for Cu thin-film deposited on sapphire (i.e.,

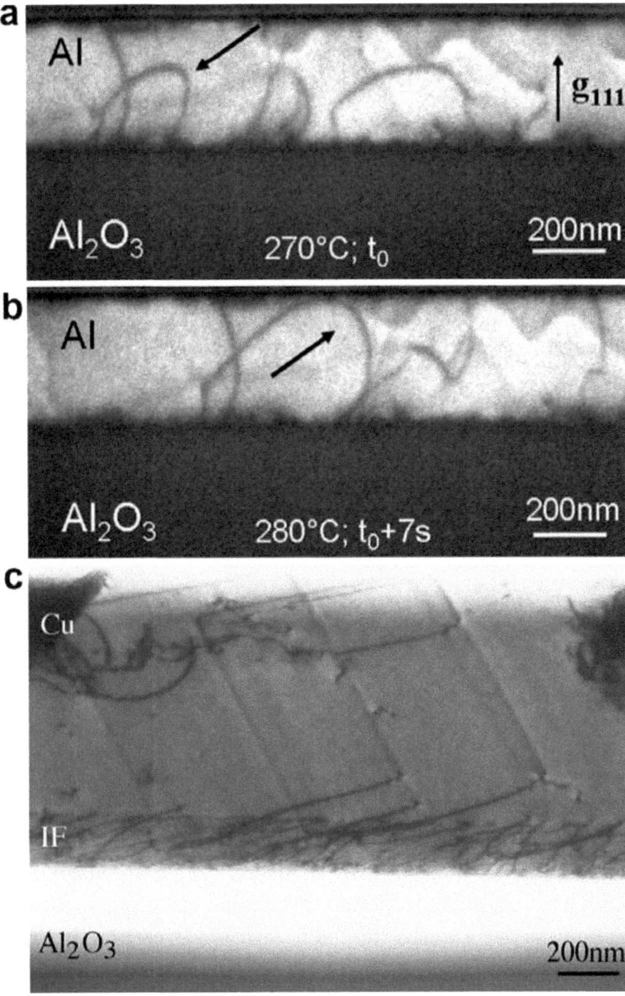

Fig. 8.21 In situ TEM images revealing deformation mechanisms in thin films: (**a**, **b**) Emission of dislocation loops from the Al/α-Al_2O_3 interface during thermal straining; (**c**) Pile-up of dislocations nucleated in the interior of a *Cu* film from a single-ended spiral source at the Cu/α-Al_2O_3 interface. *IF* denotes the interface, which is imaged under an inclined view. The region of *intense bright contrast* directly below the IF is an artifact of the TV-rate camera due its limited dynamical range (Dehm 2009. With kind permission of Elsevier)

α-Al_2O_3). Here, a common relaxation mechanism of the Cu layer is visible at the end of the thermal cycle. Dislocation sources within the film nucleate dislocations, as seen in Fig. 8.22a. The dislocation segments that are emitted by the source escape via the film surface at one end (at the top) or are blocked at the sapphire interface on the other Fig. 8.22b. This results in the formation of pile-ups (Fig. 8.22b), where five dislocations stack against the interface and note that interfacial dislocations

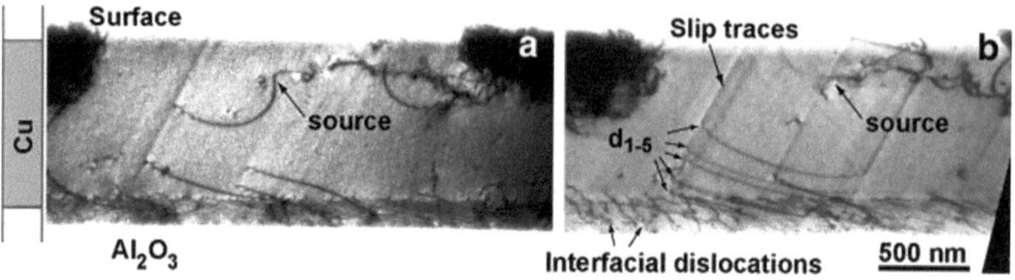

Fig. 8.22 Repulsive dislocation-interface interaction in a 900 nm thick Cu film deposited on sapphire and observed in cross-section. Still pictures taken at 208°C after in situ cycles up to 450°C. Figure (**a**) has been taken at the beginning of the relaxation at room temperature, figure (**b**) several minutes after. Dislocations (d1–5) emitted by the source pile-up against the interface, materialized by interfacial dislocations (Legros et al. 2009. With kind permission of Wiley and Sons)

are also visible. Occasionally, some dislocations, piling-up against the interfacial dislocation network, can become incorporated into the network after long waiting periods. During heating, some of the dislocations in the network are also directly emitted into the strip thickness and swept to the free surface.

8.3.4 *Free-Standing Films*

Further information on thin-film properties and their mechanical behavior, without the substrate influence, may be obtained by investigating free-standing, i.e., self-sustained films. A technique developed by Espinosa et al. for such research is performed by membrane-deflection experiments [henceforth: MDE]. Studies on FCC metals, such as Au, Al and Cu used this unique technique to provide important data on yield stress, the Young's modulus and fracture strength. The specimens tested are polycrystalline materials. The load application in MDE is illustrated schematically in Fig. 8.23. This technique involves the stretching of a free-standing thin-film membrane in a fixed-configuration. The membrane is attached at both ends and spans a micromachined window beneath. A nanoindenter applies a line-load at the center of the span to achieve deflection. The geometry of the membranes is such that it contains tapered regions to eliminate boundary-failure effects. The result is direct tension, in the absence of strain gradients, of a gauge region. Membrane size was varied, so as to maintain the aspect ratio, namely length/width, of the gauge region. Gold membranes with various widths and thicknesses were tested (see Fig. 8.24). The Young's modulus was determined from the slopes in the elastic region, e.g., for the 0.3 μm-thick and 2.5 μm-wide specimen, it is 53–55 GPa. In the case of specimens having the other widths, their behavior was almost identical with a yield stress (the departure from linearity) of 170 MPa. This behavior is quite similar to that of the 0.5 μm-thick specimens, but with a more extended elastic zone and a larger yield stress of 220 MPa for the 2.5 μm membrane. The difference in behavior

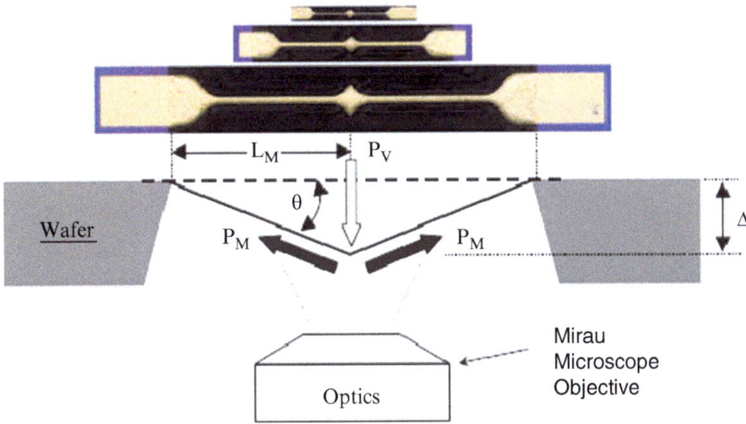

Fig. 8.23 Side view of the MDE test showing vertical load being applied by a nanoindenter, P_V, the membrane in-plane load, P_M, and the position of the Mirau microscope objective (Espinosa et al. 2004. With kind permission of Elsevier)

is attributed to the number of grains and, hence, to their sizes; thus, this is associated both with geometric constraints and the number of dislocation sources operating (Espinosa et al.). Different deformation behavior was observed in the 1.0 μm films. As in Fig. 8.24b, c, the drop in yield stress resulted from increasing the widths of the membranes. A significant change in mechanical behavior occurs as the membranes begin to deform plastically with decreasing stress-to-failure. Table 8.1 shows the results of this effect and the widths of the membranes on the yield stress.

Similar curves are obtained in other FCC metals, as illustrated by Cu and Al in Fig. 8.25. The elastic moduli derived from these curves are 125–129 and 65–70 MPa for Cu and Al, respectively. These values are close to those of bulk polycrystalline materials, being 121 and 70 GPa for Cu and Al, respectively. Note that Au does not usually oxidize and, hence, no oxide effects can exist; this is not necessarily the case for Cu and Al.

It was indicated (Espinosa et al.) that for submicron-thick Al films this layer has no effect on the elastic properties; however, it is expected to play a role in plastic deformation by acting as a barrier to dislocations reaching the surface. All three specimens of Cu (Fig. 8.25a) exhibit strain hardening until failure. The yield stress in Cu and Al membranes show strong thickness effects. For Cu, the yield stress more than doubles, from ~160 to ~345 MPa, when thickness is decreased from 1 to 0.2 μm. Al exhibits a similar behavior, however the 0.2 μm membranes show no evidence of plastic deformation and fail in a completely brittle manner.

The rate-of-deflection tests of the free-standing films measured were evaluated by the relation (Espinosa et al.):

$$\dot{\varepsilon} = \frac{d}{dt}\left(\frac{\Delta L(t)}{L_M}\right) = \frac{\Delta t}{L_M\sqrt{\Delta(t)^2 + L^2{}_M}}\left(\frac{d\Delta}{dt}\right) \tag{8.10}$$

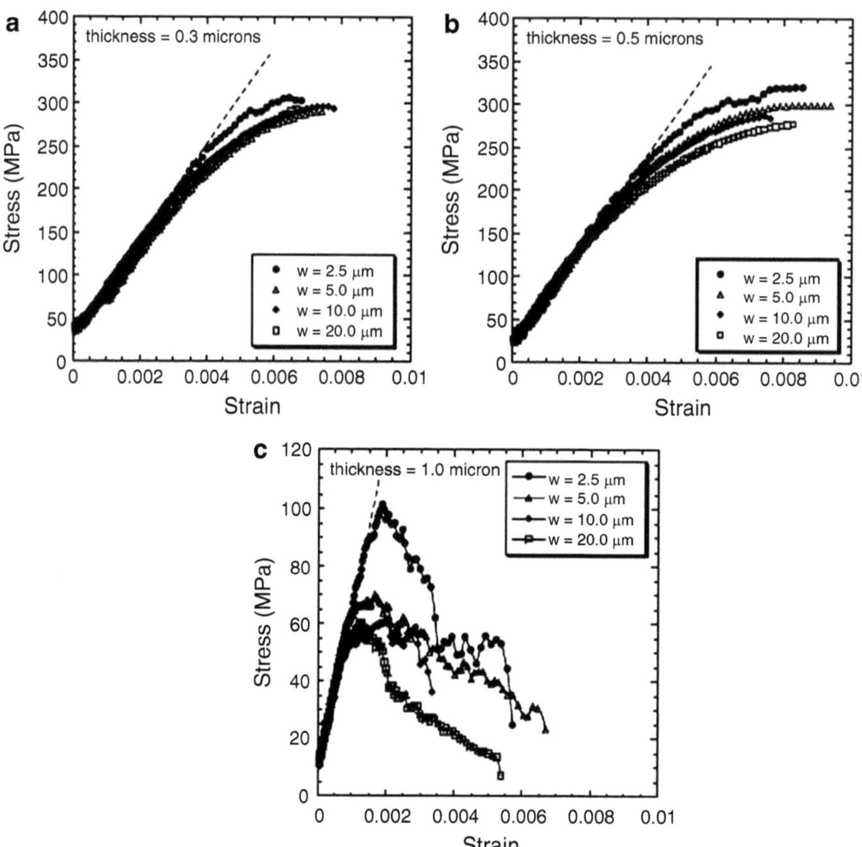

Fig. 8.24 Stress-strain plots comparing membrane width (2.5, 5.0, 10.0, and 20.0 μm) for (**a**) Au films 0.3 μm; (**b**) 0.5 μm; (**c**) 1.0 μm thick specimens. The slope of the dashed line in each plot is between 53 and 55 GPa (Espinosa et al. 2004. With kind permission of Elsevier)

Table 8.1 Yield stresses for each combination of thickness and width with corresponding strain rate at onset of yielding (From Espinosa et al. 2004. With kind permission of Elsevier)

Thickness (μm)	Width (μm)	2.5	5	10	20
0.3	σ_y (MPa)	220	170	170	170
	$\dot{\varepsilon}\,(1\times10^{-5}\ \mathrm{s}^{-1})$	5.8	3.5	1.7	0.8
0.5	σ_y (MPa)	220	170	170	140
	$\dot{\varepsilon}\,(1\times10^{-5}\ \mathrm{s}^{-1})$	5.8	3.5	1.5	0.7
1.0	σ_y (MPa)	90	65	55	55
	$\dot{\varepsilon}\,(1\times10^{-5}\ \mathrm{s}^{-1})$	4.5	2.2	0.9	0.3

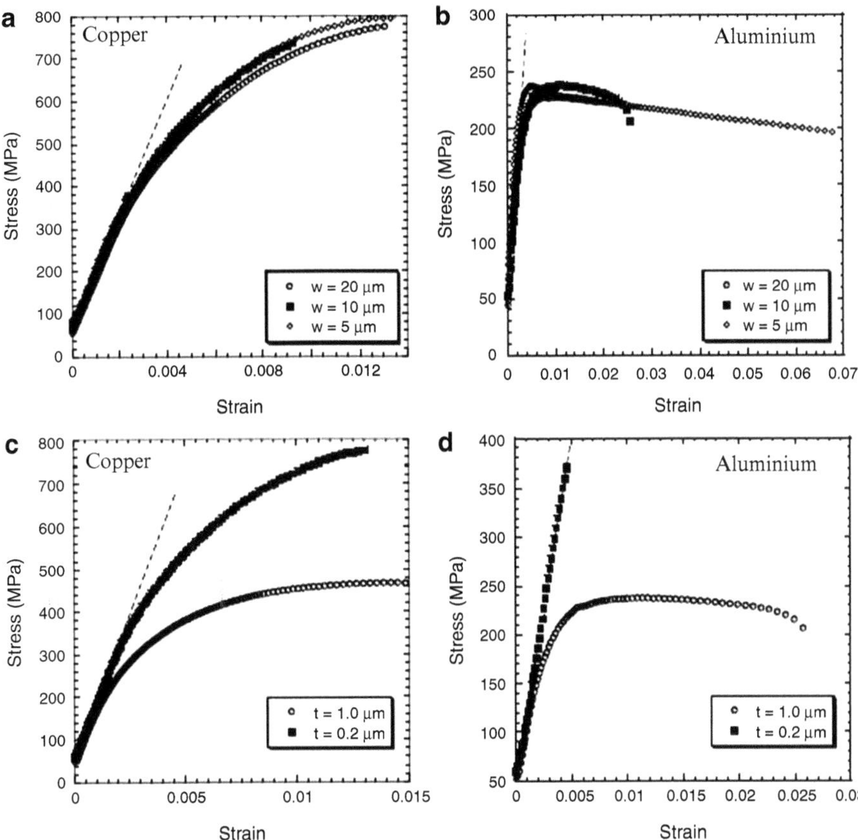

Fig. 8.25 Stress-strain plots for Cu 0.2 and Al 1 μm membranes showing effects of specimen widths and thickness: (**a** and **b**) width; (**c** and **d**) thickness (Espinosa et al. 2004. With kind permission of Elsevier)

where L_M is the membrane half-length, $\Delta L(t)$ is the change in membrane length and $(d\Delta/dt)$ is the vertical-displacement rate. The $(\Delta L(t)/L_M)$ term is obtained from the geometric relations between L_M and the length of the deflected membrane at time t (Fig. 8.23). The strain rate is not a constant value, but increases with vertical deflection.

The failure of the Au membrane is ductile in nature. In the TEM imaging of cross-sections, the gold appears to have undergone large, localized-plastic deformation before fracture. Figure 8.26 is a SEM image showing fracture in a gold membrane. Multiple-deformation bands are seen, some perpendicular to the tensile direction, denoted "a" and others parallel to the fracture surface and denoted "b". Note the transition regions between the two directions, denoted "c". Another feature observable in this top-view of the membrane is the reduction in film width in the

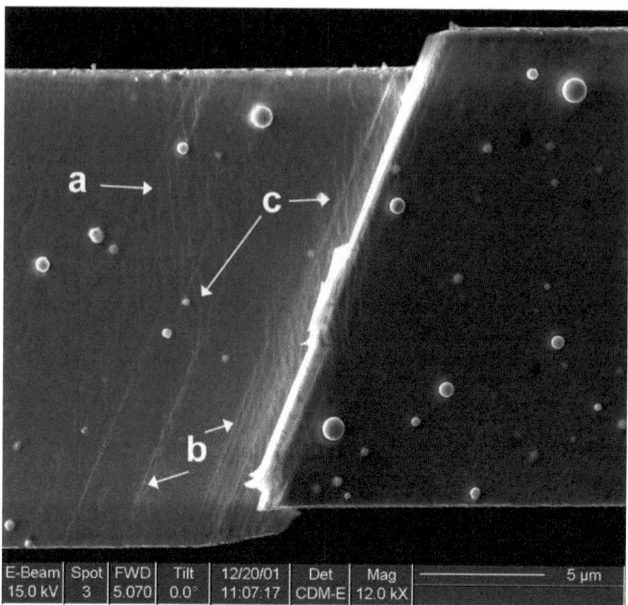

Fig. 8.26 SEM image of fracture for a Au film with thickness of 1.0 μm. Note that the fracture surface is not perpendicular to the tensile direction. In the illustration "*a*" denotes deformation bands perpendicular to the tensile direction, "*b*" denotes those parallel to the fracture surface and "*c*" denotes regions of transitions between the two (Espinosa et al. 2004. With kind permission of Elsevier)

neighborhood of the fracture plane, consistent with film necking. Failure of this sort is believed to result from maximum plastic dissipation along these directions, due to the plate-like nature of the specimens.

The change in deformation behavior is observed in a SEM image during the stress-strain analysis, as thickness was reduced from 1.0 to 0.5 μm. The membranes failed in a brittle manner (see Fig. 8.27 below). No deformation bands are observed for membranes of this thickness or thinner, although some ductility has occurred. These findings are consistent with the measured stress-strain curves. In Fig. 8.28, a series of interferometric images and their corresponding stress-strain curves are shown for different time intervals during MDE. At $t = 0$, the membrane is in an unloaded state. At $t = 1$, the membrane has been deformed through its linear-elastic region and has just begun to deform plastically. Uniform-plastic deformation continues through $t = 2$. By $t = 3$, discontinuity appears in the fringes, consistent with shear localization and a discrete jog is recorded in the stress-strain curve. A second shear localization and corresponding jog appear at $t = 4$. The membrane finally fractures in frame $t = 5$. These images are taken from the bottom side of the membrane, shown in the SEM image in Fig. 8.27. Measured stress-strain data allow for the identification of: Young's modulus of the thin-film, the onset of plastic deformation and the fracture stress.

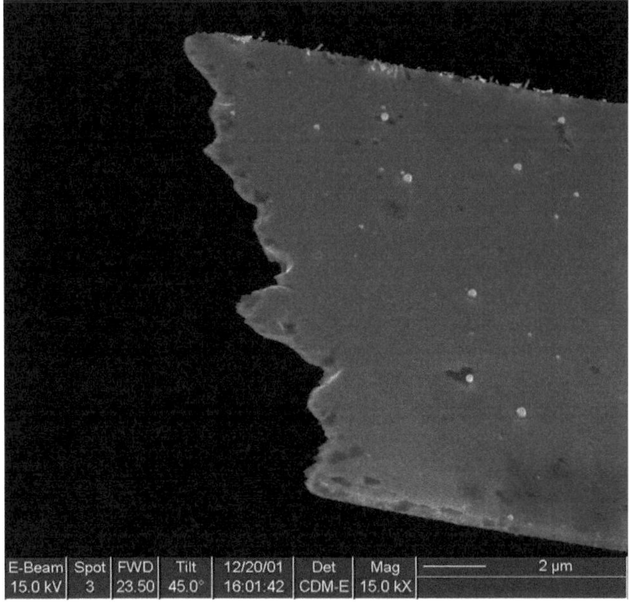

Fig. 8.27 SEM image of fracture for a Au film with thickness of 0.5 μm. Note the absence of deformation bands and zig-zag fracture surface (Espinosa et al. 2004. With kind permission of Elsevier)

It was revealed in the self-sustaining thin-film experiments that film width and thickness have an effect on the yield stress. Dimensional parameters exhibit size effects, but thickness, in particular, has a greater influence when it is decreased from 1.0 to 0.5 μm.

Several features are apparent from the experimental data. First of all, the deformation behavior in small-sized materials (represented here by free-standing thin-films) is the same as in bulk materials. No different deformation mechanism is observed, but rather the kind of mechanism exhibited during the testing of micron and submicron materials requires sophisticated techniques, in order to enable the characterizations of the responses to acting forces. Thus, multiple striations or deformation bands are seen near failure regions, as observed in bulk materials. Failure can occur either by necking down, when thin-film are ductile (Fig. 8.26) or even as brittle failure (Fig. 8.27). Furthermore, work hardening (Fig. 8.25) or works softening (Fig. 8.24) occur with noticeable shear observed before failure (Fig. 8.28).

8.3.5 Whiskers

Several techniques are available for producing whiskers (or filaments) of various materials. To review the subject of the growth and properties of whiskers, one may

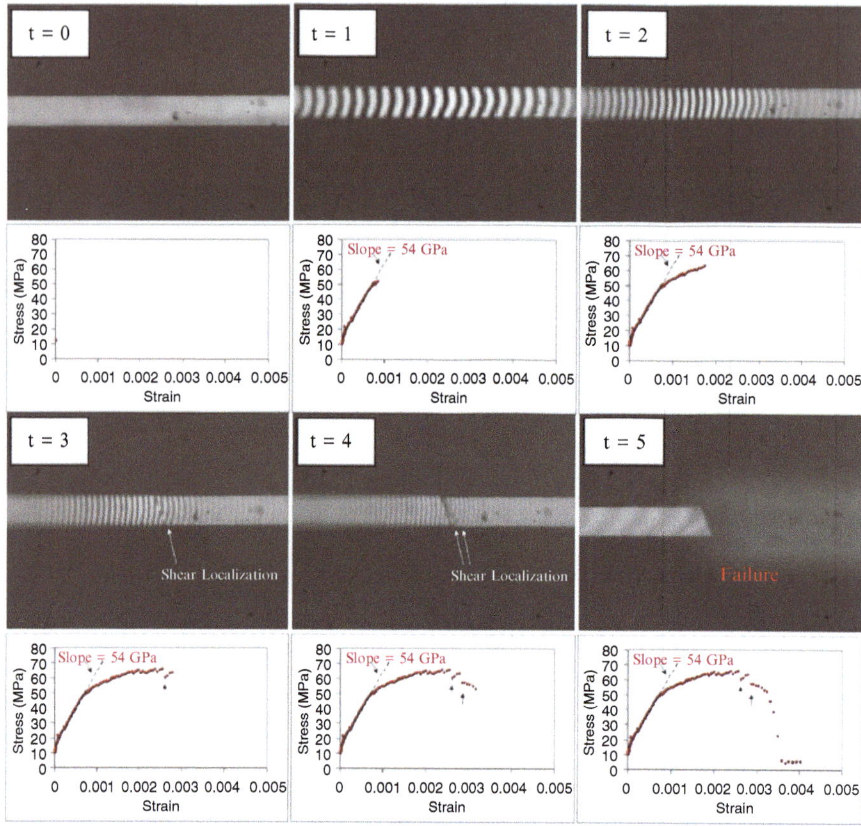

Fig. 8.28 Series of optical interferometric images and the corresponding stress–strain curves for different instances of time during MDE testing (Espinosa et al. 2004. With kind permission of Elsevier)

consult Brenner. Most whiskers exhibit very high strengths, approaching theoretical values. Crystals, in the form of thin whiskers (or filaments), have strengths often between 100 and 1,000 times greater than those measured in bulk materials. It is not known precisely whether the high strength of whiskers is due to their very small size or to their structural or surface perfection. It is true that both confined dimensions (i.e., small sizes) and crystal perfections (each one separately) increase whisker strength, but it is most likely that a combination of these strengthening factors act in real materials. Dislocation density and orientation are also expected to be involved in the strength of whiskers. Webb et al. have studied the operation of screw dislocations in whiskers and in their growth. There are, however, reports (Gotoh) that in Cu whiskers, whose diameters range from several hundreds of Å to 100 μm, that there are only few dislocations. These results are based on electron and fine-focus,

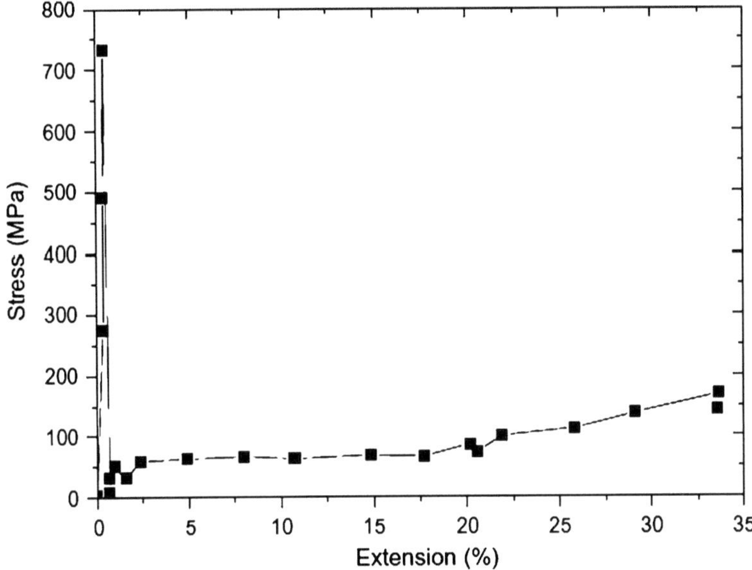

Fig. 8.29 Stress-strain curve of a [111] oriented Cu whisker. After yielding the stress drops by a factor ~10 and remains constant up to ~20% strain before hardening sets in (Redrawn from Brenner's work. Dehm 2009. With kind permission of Elsevier)

X-ray-diffraction microscopes. Yoshida et al. have studied stress-strain curves of 150 Cu whiskers and found a sharp yield point in thinner whiskers (Fig. 8.29). The upper-yield stress of some thin whiskers reaches the theoretical strength characterizing perfect crystals, but decreases as thickness increases. Single-crystal whisker studies were conducted on [100], [110] and [111] orientations and the [111] orientation had higher yield stress, flow stress and strain hardening levels (Brenner). The yield drop observed in Fig. 8.29 is followed by a relatively constant easy-glide stage, up to ~15–20% extension, after which strain hardening sets in. A high-yield stress of 225 kg/mm^2 was obtained in a Cu whisker with <100> orientation, as reported by Yoshida et al. for a specimen thickness of 12.5 μm. This stress value, in terms of resolved shear stress, is about G/40, where G refers to the shear modulus of Cu. The theoretical stress-level is ~G/30.

Easy glide following the yield drop is characterized by Lüders-band formations. These seem to develop from glide bands nucleated somewhere in the whisker, with the slip steps of the band resolved microscopically in SEM during the in-situ tensile test. The specimens in such tests may be obtained by FIB fabrication. These Lüders bands propagate along the easy glide regime. It has been found by Yoshida et al. that: (i) the slip bands nucleate midway along the gauge length; (ii) in specimens thinner than ~10 μm, only one slip band is nucleated, which develops into Lüders bands; and (iii) in whiskers >10 μm, Lüders bands sometimes propagate from several places. There is a tendency to lower stress with larger whisker diameters.

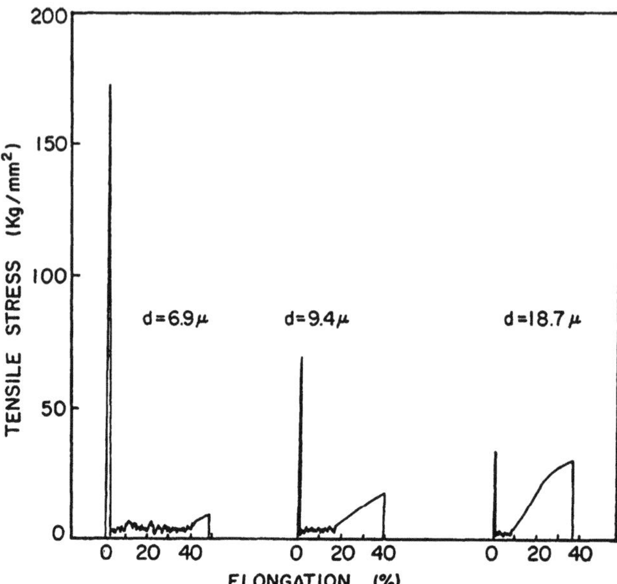

Fig. 8.30 Variation of the tensile stress-strain curve with thickness. Specimen axis is the <111> orientations for all values of the diameter. Note the Lüders bands in all specimens but the extent decreases with increased diameter (Yoshida et al. 1968. With kind permission of JPSJ)

As thickness increases, the upper yield stress, the flow stress in the easy-glide region and the Lüders bands (strain) decrease. These may be seen in Figs. 8.30, 8.31, and 8.32. The work-hardening rate, however, increases with increasing whisker diameter or thickness and, in cases of single-crystal whiskers, irrespective of their orientation (illustrated in Fig. 8.30). Flow stress in the easy-glide region is whisker-size-dependent.

Shear-stress variation may be seen in Fig. 8.32 for two kinds of Cu specimens: a (100) whisker and a Cu FIB specimen with (−234) orientation. The yield-stress pattern behaves in the same way as the flow stress, decreasing with the increasing diameter of the whisker (as illustrated for Cu in Fig. 8.33). Since stage-I work hardening is very shallow, almost constant (as seen in Fig. 8.29), the work hardening occurring after easy glide refers to stage-II hardening. The reason for the increased strain-hardening rate with whisker size is apparently related to the fact that, with increased size, the dislocation density increases (there is less escape from the crystal, because of the longer path to the surface) and the probability of dislocation-dislocation interactions between different slip systems increases. Easy glide is terminated when the Lüders bands have propagated through the entire length of the whisker and, thus, the Lüders strain is equal to the end of the easy-glide region. An interesting illustration in Fig. 8.34 shows the Lüders-band propagation in a whisker of <111> orientation.

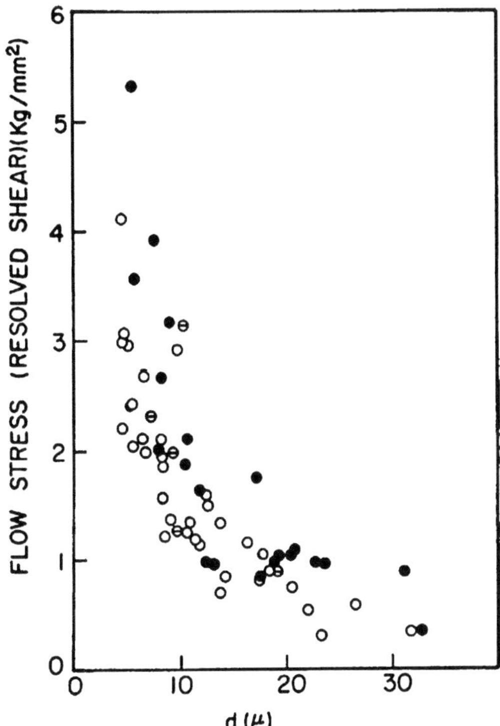

Fig. 8.31 Thickness-dependent flow stress (i.e., the resolved shear stress) in the easy glide region. The stress is the upper limit of the serrations observed in the stress-strain curve. The meaning of the symbols is • ⊖ ∘ < 100>, <110> and <111> orientations, respectively (Yoshida et al. 1968. With kind permission of JPSJ)

So far, the deformation aspects of whiskers have been considered for upper-yield stress, flow stress and when expressed in terms of resolved shear stress. Again, work hardening in the easy-glide region (stage I) is very shallow, almost non-existent, but during the second stage, an increased hardening rate is observed following the termination of Lüders strain. These are general concepts that were previously discussed in Chap. 1, representing the conventional, mechanical behavior characterization of materials. At this point, a few words on twinning are in order, since twinning also makes a significant contribution to overall deformation.

8.3.6 Twinning

In Chap. 3, Sect. 3.5, where twinning was presented in regard to bulk materials, two types of twins were mentioned: annealing twins and mechanical twins. In bulk materials, mechanical twins are also known as 'deformation twins' and are most common in HCP metals, whereas annealing twins., also called 'growth twins', are more frequent in worked, heat-treated FCC metals. Below, a few examples of observed twins in small-sized specimens are discussed.

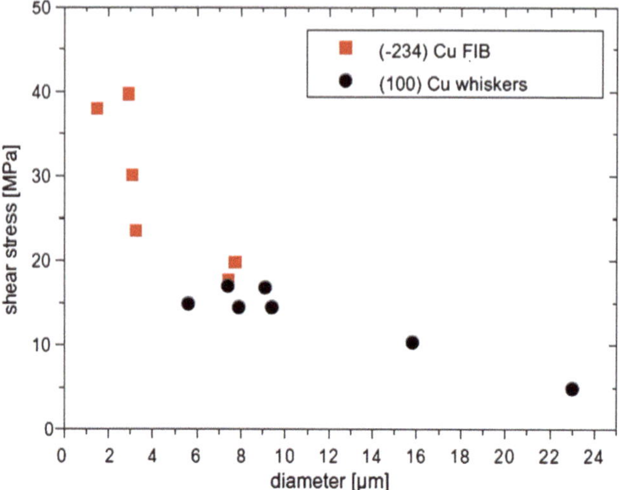

Fig. 8.32 Shear stress values in the easy-glide region of Cu whiskers and FIB shaped tensile samples (Dehm 2009. With kind permission of Elsevier)

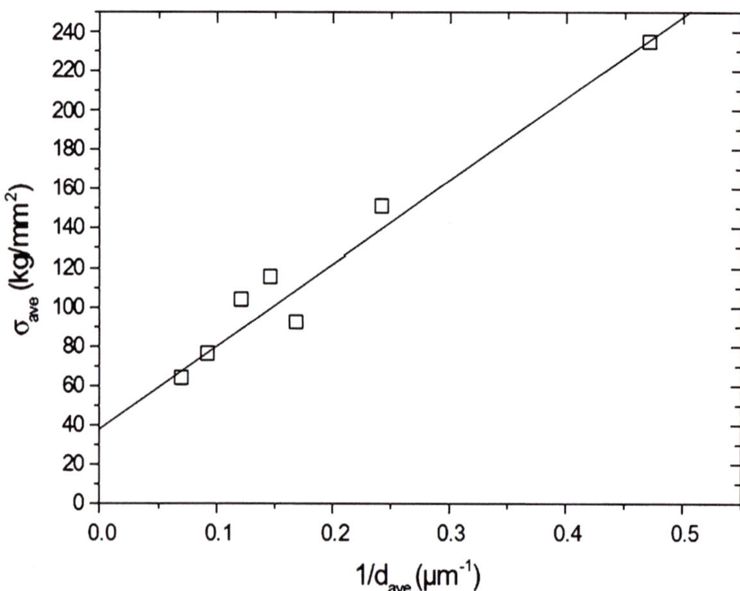

Fig. 8.33 Yield stress values of Cu whiskers versus their diameter (Dehm 2009. With kind permission of Elsevier)

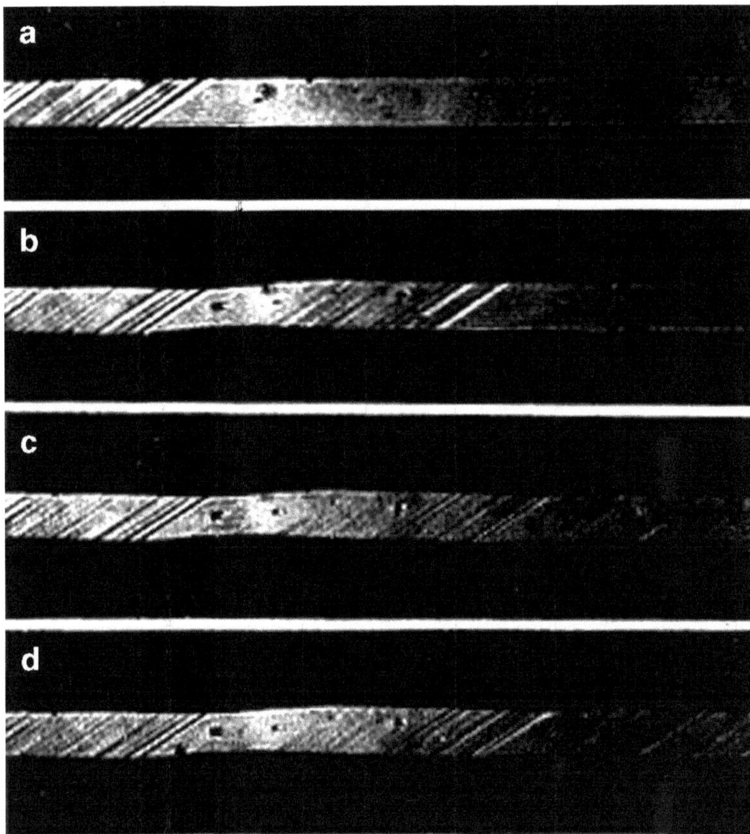

Fig. 8.34 Propagation of a Lüders band with increasing strain of Cu whisker with <111> orientation. Whisker diameter is 15 mm: (**a**) tensile strain e = 10.0%, (**b**) e = 10.2%, (**c**) 10.4%, (**d**) 10.55% (Yoshida et al. 1968. With kind permission of JPSJ)

Mechanical twinning and slip may be considered to be competing deformation processes, often occurring simultaneously, side by side, during a plastic-deformation process. Mechanical twining in thin-films is a sort of internal-stress relaxation mechanism, arising, for example, from phase transformation, lattice mismatch or thermal mismatch between substrate and film. Gold films grown epitaxially on NaCl-substrate single crystals were studied in recent work by Dehm et al. Following film growth, all the samples were spin-coated with a μm polyimide precursor, cured at 350°C. This was followed by dissolving the NaCl substrate, to produce an Au/polyimide system. An example of a cross-sectional TEM image of Au gold film is shown in Fig. 8.35. An increase in twin density with decreasing film thickness is both observed by plan-view TEM studies and predicted by theoretical considerations (see Fig. 8.36). The curve was evaluated by a quantitative analysis of

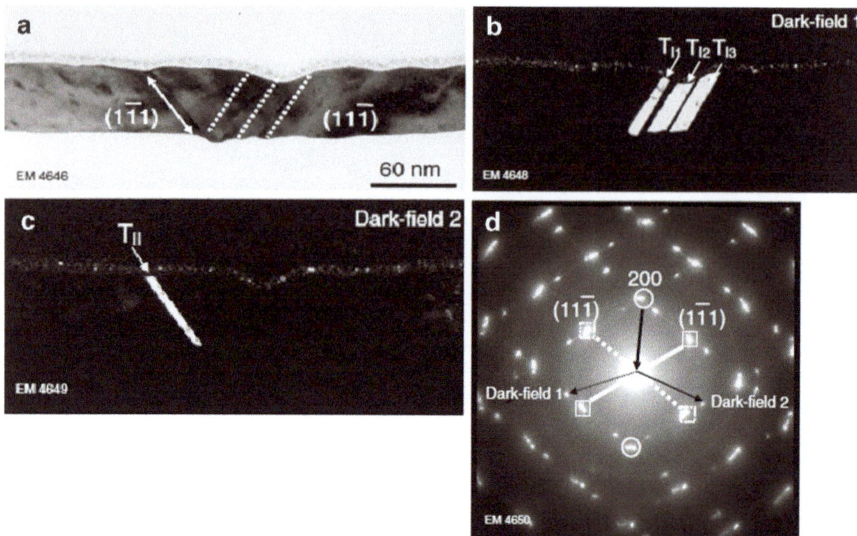

Fig. 8.35 (**a**) Cross-sectional TEM image recorded in the [011] zone axis with three twins on the (11$\bar{1}$) plane and one twin on the (1$\bar{1}$1) plane. Dark-field images with the (**b**) (11$\bar{1}$)and (**c**) (1$\bar{1}$1) twins imaged with (200) reflections, labelled as *Dark-field 1* and *2* in the diffraction pattern (**d**). The diffraction pattern is recorded along the [011] zone axis (Dehm et al. 2007. With kind permission of Elsevier)

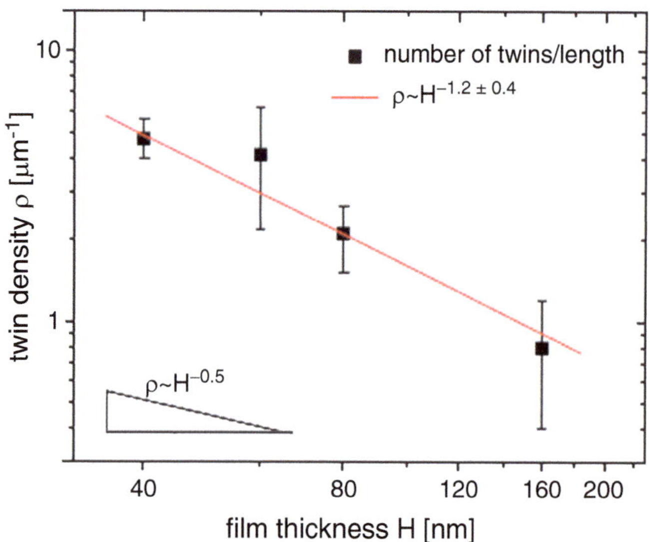

Fig. 8.36 Twin density vs. film thickness obtained from TEM plan-view images by the linear intercept method (Dehm et al. 2007. With kind permission of Elsevier)

Table 8.2 Available twin systems for the (100)-oriented Au films

Twin system	Twin normal $\underline{n}$	Twin direction $\underline{t}$	Twin trace in (100) plane
I	$[11\bar{1}]$	$[2\bar{1}1]$	$[01\bar{1}]$
II	$[1\bar{1}1]$	$[21\bar{1}]$	$[0\bar{1}1]$
III	$[111]$	$[2\bar{1}\bar{1}]$	$[011]$
IV	$[1\bar{1}\bar{1}]$	$[211]$	$[0\bar{1}\bar{1}]$

the twin density, carried out by applying the linear intercept method to the plan-view TEM images and counting all the intersects along <011> directions with twins.

As discussed above (in Chap. 3, Sect. 3.5) regarding FCC metals and alloys, twinning stress is directly related to SF energy. The twinning shear stress, τ_T, may be expressed (following Meyers et al.) as:

$$\frac{\tau_T}{G} = \frac{\gamma}{Gb} + \frac{\alpha b}{l} \tag{8.11}$$

In the above relation, l is the source size and γ is the SF energy. According to this relation, an increase in twinning stress is predicted with increasing SF energy and decreasing source size. Another relation, in terms of the Burgers vector of partial dislocations (recall that partial dislocations are associated with SF energies; see Chap. 2) is:

$$\frac{\tau_T}{G} = \frac{\lambda}{2b_p G} \tag{2.12}$$

Mechanical twinning readily occurs in materials with low SF energy. Furthermore, in cases where it is difficult to activate perfect dislocations (i.e., where there is a propensity for dislocations to split into partial dislocations), deformation by twinning occurs. Au single crystals, although they usually deform by dislocation glide, also have low-SF energy and, therefore, may possibly deform by means of mechanical twinning. Moreover, it has been observed that partial dislocations in FCC metals, such as Au, are readily formed when their grain size or film thickness are in the submicron (nanometer) range. The low SF energy of Au in the submicron range and its tendency to form partial dislocations promote twinning in Au films with small dimensions. Moreover, the Burgers vector for partial dislocations is the same at (a/6) <211>.

A total of four twin systems have been observed in Au on a NaCl substrate, as listed in Table 8.2. TEM studies indicate that, of these four twin systems, only two are active (see Fig. 8.35a, b), i.e., one for each <011> orthogonal direction. One of the rare cases observed, in which the second possible twin system along the <011> direction may occur is seen in Fig. 8.35c.

These twins are on average 1,250 $\pm$ 625 nm apart, while, for the 40 nm-thick films, an average twin spacing of only 208 $\pm$ 34 nm occurs. The widths of the twins,

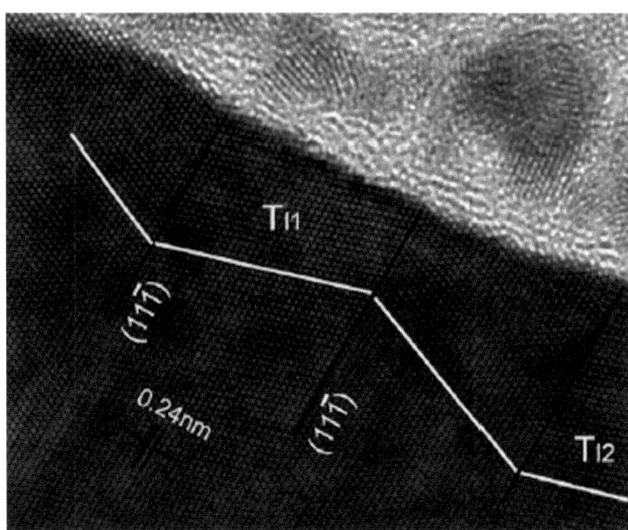

Fig. 8.37 High-resolution TEM image of twinning in a 60 nm thick Au film near the surface (originally the interface to NaCl) recorded in the [011] zone axis. *TI1* and *TI2* indicate the twins resolved in the dark-field image of Fig. 8.35b (Dehm et al. 2007. With kind permission of Elsevier)

measured from a conventional cross-section (Fig. 8.35) and high-resolution TEM images, recorded in the <011> zone axis (Fig. 8.37) are in the range of 3–15 nm.

Two illuminating examples are provided in Fig. 8.38, helping to determine if dislocation glide or twinning mechanisms are involved in Au. A 40 nm-thick Au film is shown in plan-view (Fig. 8.38a) and in a cross-sectional TEM image combined with an inclined plan-view (Fig. 8.38b). In this way, a three-dimensional impression of mechanical twinning was obtained for a 160 nm-thick Au film. In addition to the orthogonal-twin network, dislocations are also observed by TEM in Fig. 8.38b. Each of the two orthogonal <011> directions laying on the (100) surface belong to two twin systems, making a total of four possible twin systems (as indicated in Table 8.2).

For further details on the analysis of twin formation in thin gold films, the work of Dehm et al. may be consulted.

Growth twins have been observed in various FCC metals and their formation is associated with their SF energy. Equations 8.11 and 8.12 express such relations. In Fig. 8.39, a plot is given for some FCC metals in accordance with Eq. 8.12, relating twinning stress to SF energy. Note the high twinning stress presumably required (in the extrapolated line) for Al, which has a high-SF energy of ~150 mJ/m^2. This is, of course, associated with decreased grain size, expected to render twinning more difficult. On the effect of grain size, see more below regarding the H-P

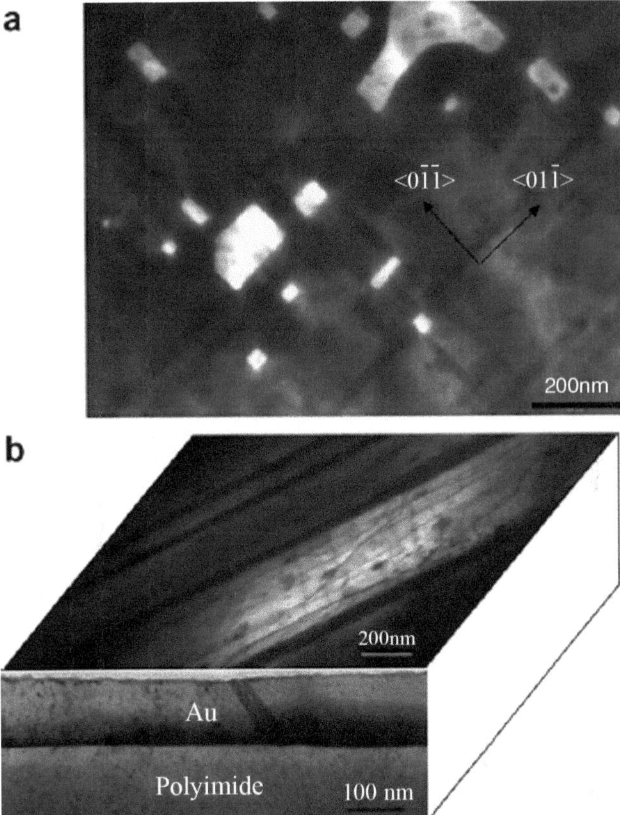

Fig. 8.38 (**a**) Plan-view TEM image of a 40 nm thick Au film revealing twins along two orthogonal <011> directions. Additionally, pores are visible which stem from cuboidal surface facets of the NaCl template. (**b**) Cross-sectional TEM image artificially combined with an inclined plan-view TEM micrograph in order to obtain a three-dimensional impression of mechanical twinning in the 160 nm thick Au film (Dehm et al. 2007. With kind permission of Elsevier)

relation in submicron materials. Aluminum, which has an SF energy higher than Cu, cannot twin either at high-shock pressures or by deformation close to 0 K. Nevertheless, molecular-dynamic simulations predict mechanical twinning in the deformation of nanocrystalline Al. Actual twinning in Al has been observed following molecular-dynamic simulation, as shown in Fig. 8.40. Figure 8.40b shows twins and SFs. The magnified, outlined region is shown in Fig. 8.40c and d. Twinning is also observed in stainless steel, as indicated in Fig. 8.41.

The suggested, critical nucleus size, r_c, for twinning as a function of applied stress, was given by Meyer et al.

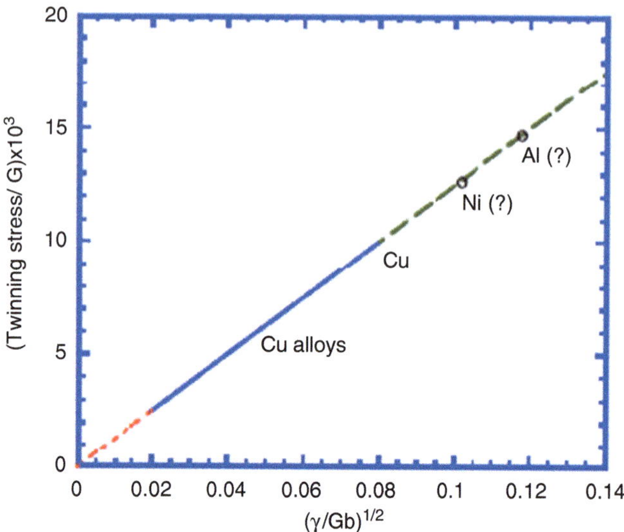

Fig. 8.39 Effect of stacking fault energy on twinning stress (Meyers et al. 2006. With kind permission of Elsevier)

The relation given below takes into account the twin-boundary energy, γ_{TB}, which is directly related to the SF energy:

$$r_c = \frac{5\pi G \gamma_{TB}}{4\tau'_T} \tag{8.13}$$

Here, τ'_T is the local-twinning shear stress. The critical radius is plotted against the normalized shear stress for several FCC elements (Fig. 8.42).

Strength and ductility may be increased by growth (annealing) twins. By the rolling-deformation process of Cu and its annealing, nanoscale-ultrafine grains in the range of 80–200 nm were fabricated. The heavily cold-worked Cu has a high dislocation density. The annealing resulted in a mixture of coarse- and fine-grained microstructures. Growth twins were observed in the larger grains. It is believed that the very good combination of strength and ductility is a consequence of: (i) multi-axial stress states in the confined grains; (ii) the presence of growth twins in the larger grains; and (iii) the preferential accommodation of strain in the larger grains. Strength may be increased by introducing a high density of growth twins on a nanoscale. The strength of the Cu is increased ~10 times compared to conventional, coarse grained (CG) Cu, as a result of the presence of nanoscale growth twins. The reason for such an increase in strength is connected with the effective blockage of dislocation motion by numerous coherent twin boundaries (see Fig. 8.43).

Further studies on NC materials, such as Al, with polycrystalline structures have indicated that the rate of plastic deformation is higher for a given applied-stress

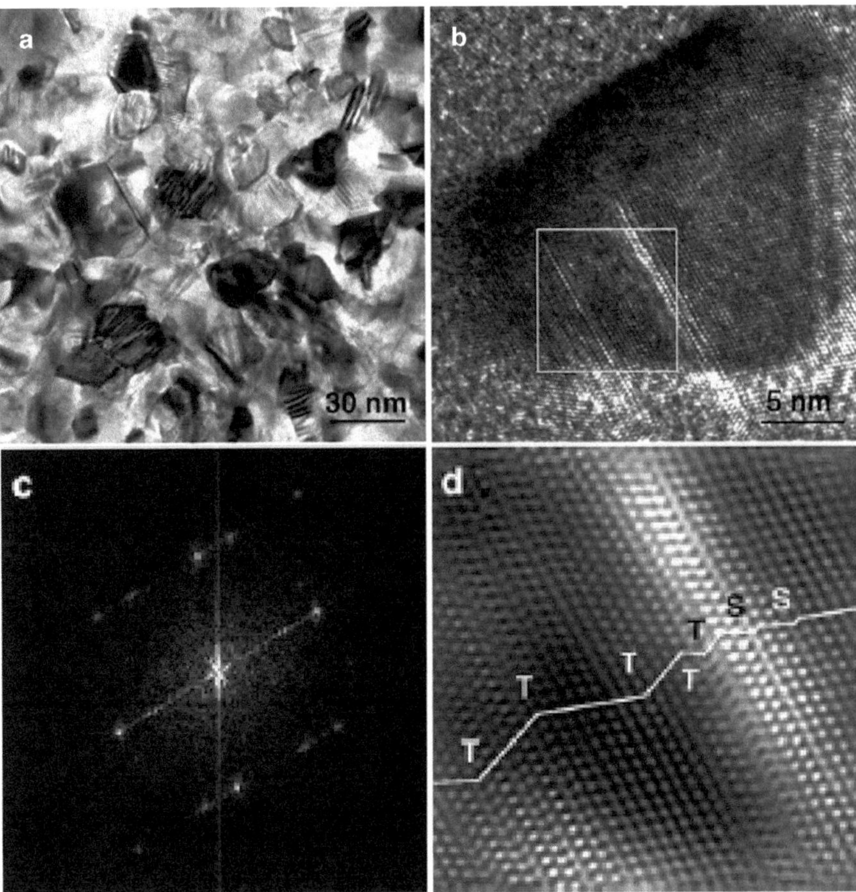

Fig. 8.40 TEM micrographs of nanocrystalline aluminum deformed by grinding: (**a**) overall view of nanograins with planar features within grains; (**b**) multiple deformation twins and stacking faults; (**c**) blowup of *box* in (**d**) showing atomic resolution and zig-zag due to twinning (Meyers et al. 2006. With kind permission of Elsevier)

level for grains with grown-in twin boundaries, due to twin migration. Structures with mean grain diameters of 12 nm and deformation strain-time curves are shown in Figs. 8.44 and 8.45, respectively. The deformation curves for the different Al samples have an applied uniaxial tensile stress of 1.2 GPa. The strain-time curves indicate that the presence of pre-twins in the nano sized grains facilitate plastic deformation.

At this point, it is reasonable to consider the effects of NC grains in terms of the well-known H-P relation, in order to get a more complete picture of the mechanical behavior in materials of small dimensions.

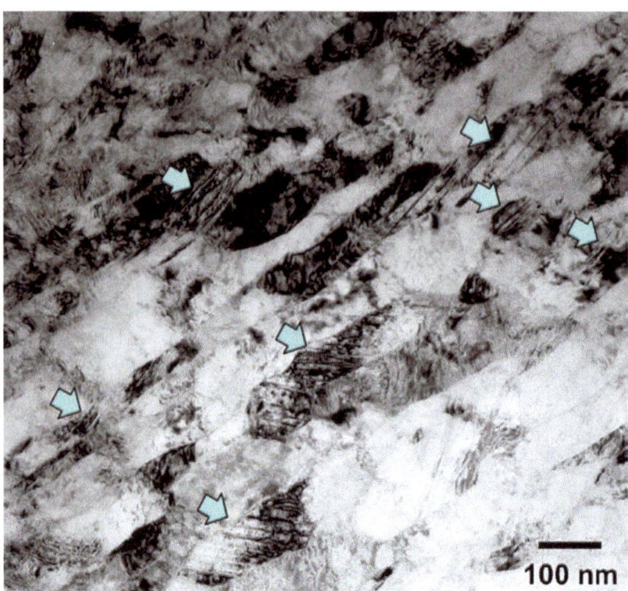

Fig. 8.41 Nanocrystalline AISI 305 SS produced within shear localization region in high-strain rate deformation (Meyers et al. 2006. With kind permission of Elsevier)

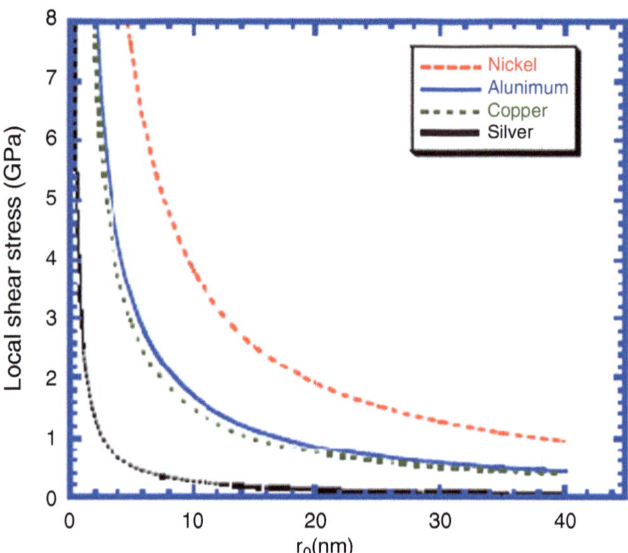

Fig. 8.42 Critical shear stress required for the activation of a twin nucleus as a function of size; Notice that nucleus size is in the nanocrystalline domain for stresses on the order of the flow stress of nanocrystalline metals (Meyers et al. 2006. With kind permission of Elsevier)

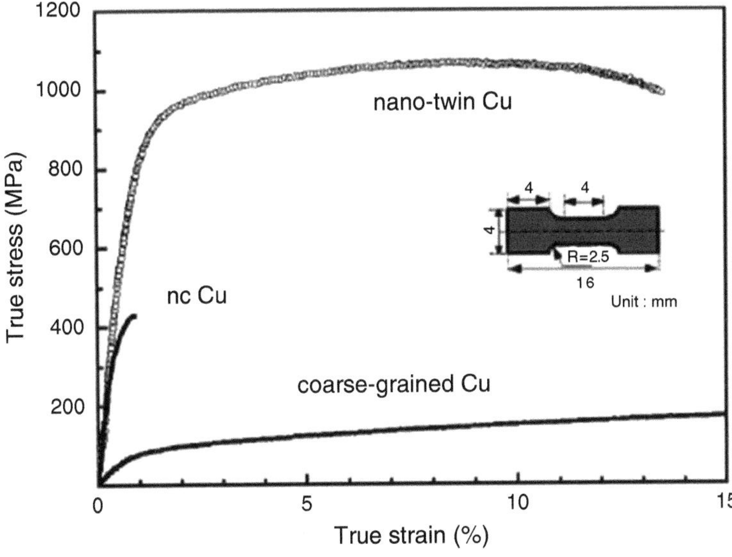

Fig. 8.43 Effect of nanotwin density on the mechanical response of copper (Meyers et al. 2006. With kind permission of Elsevier)

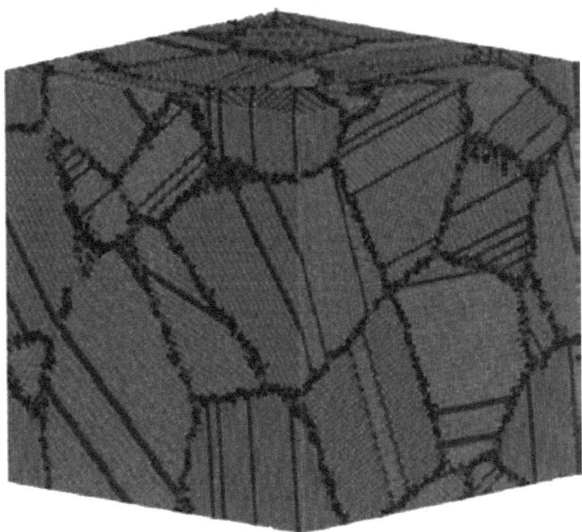

Fig. 8.44 The twinned nc-Al sample with an average grain diameter of 12 nm (Meyers et al. 2006. With kind permission of Elsevier)

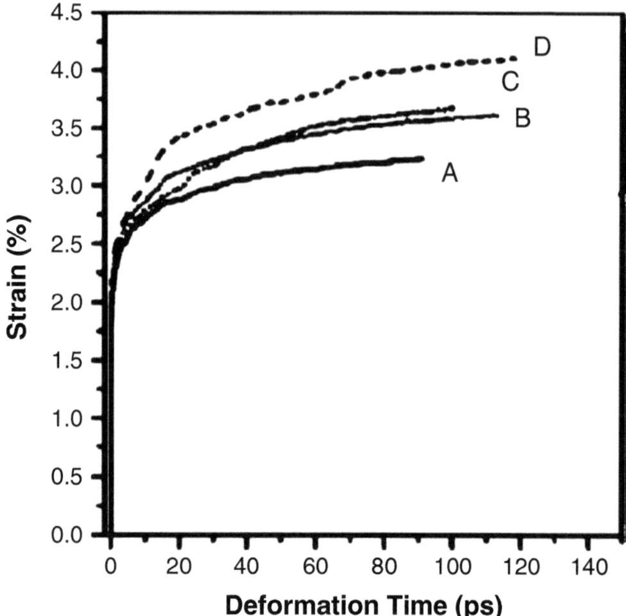

Fig. 8.45 Deformation curves for the 12-nm grain size nc samples, (*A*) Al with defect-free grains; (*B*) Al samples with pre-twin boundaries are rotated into another {111} family of planes; (*C*) Al sample with twins/grain reduced to two and (*D*) Al sample with twin boundaries, at an applied uniaxial tensile stress of 1.2 GPa [320] (Meyers et al. 2006. With kind permission of Elsevier)

8.3.7 *The Hall-Petch Relation (H-P) in Materials of Small Dimensions*

Whether the H-P holds for specimens of small dimensions has been a debatable problem for some time and still is an unresolved issue. The main controversy is pivoted around the slope direction of some known mechanical property vs. grain size. Not only experimental curves were constructed (which might sometimes be misinterpreted), but molecular-dynamic calculations were also performed to gain a better understanding of the subject. As discussed in an earlier chapter:

$$y = y_0 + \frac{k}{\sqrt{d}} \tag{3.42}$$

where y_0 is the friction stress and k is a material dependent constant.

However, as the grain size is decreased to the submicrometer range, the k value tends to decrease, and when the grain size is decreased to less than 100 nm, the k value often becomes negative. The latter fact is called the ‘inverse H-P relation’ and calculations have been done to determine if the inverse H-P effect is real or

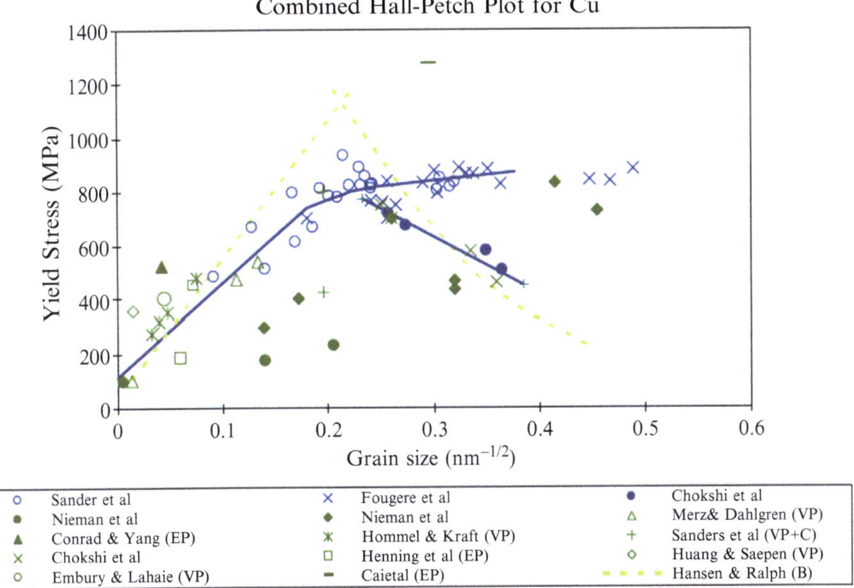

Fig. 8.46 Compiled yield stress versus grain size plot for Cu from various sources ranging from coarse to nanograin size. The plots show different trend as the grain size falls below a critical size (Meyers et al. 2006. With kind permission of Elsevier)

an artifact. The experimental curves in Fig. 8.46 show the H-P plot for Cu, taken from different sources. One can observe an ambiguity in the trend of the plot as the grain size falls to a value below ~25 nm. While some results predict a plateau, others show a decrease. The H-P trend for different NC samples, crystallized from amorphous solids, is plotted in Fig. 8.47a. Here two distinct trends are seen: (i) the formation of a plateau and (ii) a decrease in σ_y (or hardness) as d is decreased below 25 nm. Note that, in Fig. 8.47b, the curve extends down to the amorphous state, as the lower limit, where the hardness value is ~6 GPa for Ni and Ni-P alloys. The explanation for this observation must be related to some change in mechanism, setting in at a critical grain size. Chokshi et al. were the first to report this negative H-P effect in NC Pd and Cu, as seen in Fig. 8.48a. In comparison, Weertman's H-P relation is indicated in Fig. 8.48b, as presented by Meyer et al. The mechanism, responsible for the negative slope in the H-P relation, according to Chokshi et al., is a result of rapid-diffusional creep at room temperature. Specifically, Coble creep was considered to be the operating mechanism. The strain rate, $\dot{\varepsilon}$, for this creep is given as:

$$\dot{\varepsilon} = \frac{150\Omega\delta D_{gb}\sigma}{\pi kTd^3} \tag{8.14}$$

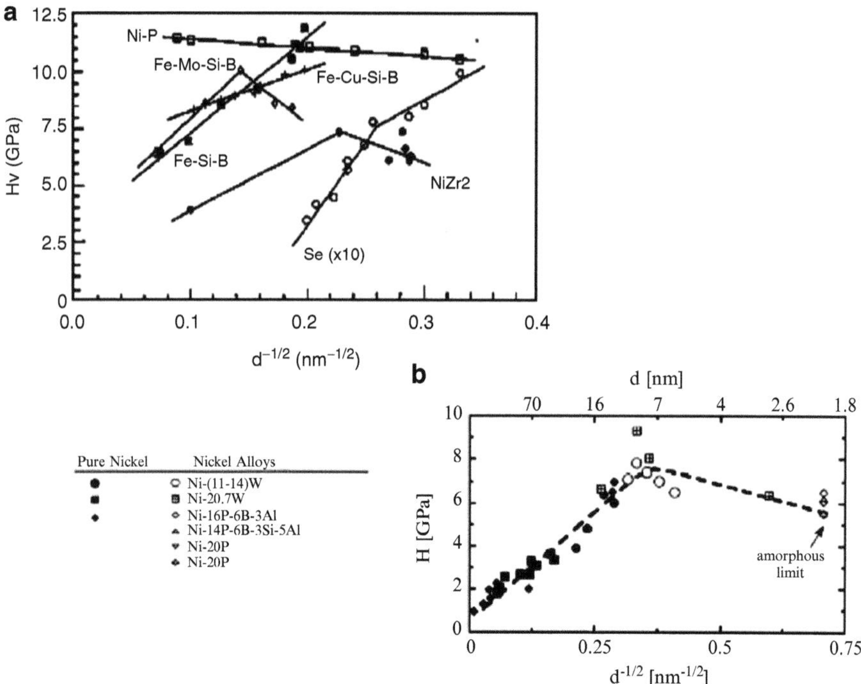

Fig. 8.47 (**a**) Hall–Petch plots for different nanocrystalline samples crystallized from amorphous solids; (**b**) hardness as a function of grain size for pure Ni and Ni–P alloy going all the way to amorphous limit (Meyers et al. 2006. With kind permission of Elsevier)

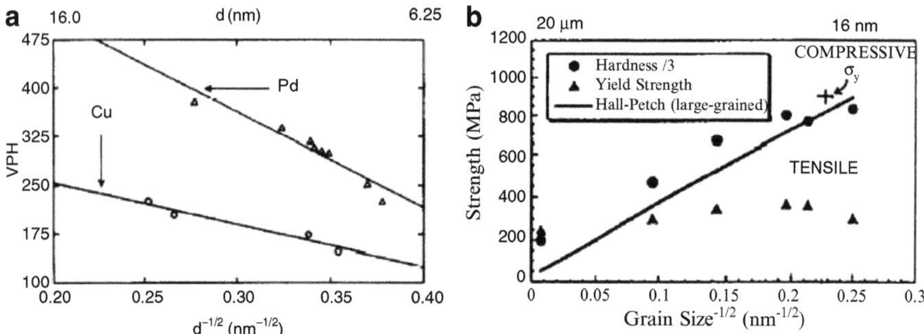

Fig. 8.48 (**a**) Inverse Hall Petch trend for Cu and Pd, (**b**) Positive Hall–Petch slope with higher values for compressive (from hardness measurements) than for tensile strengths (Meyers et al. 2006. With kind permission of Elsevier)

As usual, Ω is the atomic volume, δ is grain-boundary width, D_{gb} is grain-boundary diffusion, k is the Boltzmann's constant, T is absolute temperature, and σ is the stress, respectively. Using stress values of 100 MPa and 1,000 MPa at 300 K and the values of

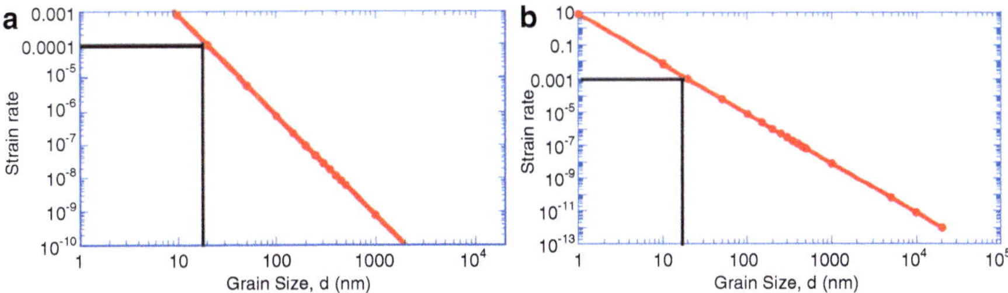

Fig. 8.49 Log–log plot of strain rate versus grain size for stresses of (**a**) 100 MPa and (**b**) 1,000 MPa for Coble creep as used by Chokshi et al. (Meyers et al. 2006. With kind permission of Elsevier)

$$\Omega = 1.3 \times 10^{-29}\mathrm{m}^3; \delta = 1\ \mathrm{nm}; \mathrm{D_{gb}} = 3 \times 10^{-9}\exp\left(-62,000/\mathrm{RT}\right)\mathrm{m}^2\mathrm{s}^{-1}$$

plots for the strain rate as a function of grain size were obtained as shown in Fig. 8.49. Weertman, who presents his results in terms of tensile tests, obtains a positive H-P slope (Fig. 8.48b) and considers the observation of a negative slope as being the result of the presence of flaws. Furthermore, he suggests that the negative slope obtained by Chokshi et al. was due to the use of a single sample, subjected to repeat anneals in order to change the grain size. Thus, the observed negative slope was merely a heat-treatment artifact and not the result of real change in the mechanism.

Deviations from the H-P relation at grain sizes in the micron and submicron ranges have been recorded for several metals, as shown in Fig. 8.50. There is a deviation from the conventional H-P curve in the NC range and there is also a significant decrease in the slope for small grain sizes. However, the nature of the curves at grain sizes below ~10–15 nm is not known from these graphs. Researchers have debated the existence of the negative H-P effect, but there is still insufficient information to validate the existence of this effect, despite the large amount of experimental work done to clarify this matter. Kadau et al. performed a molecular-dynamic study of NC Al undergoing tensile loading. An inverse H-P effect was found showing a linear relation between flow stress and the inverse of the density, leading researchers to conclude that the deformation mechanism for small grains is related to the surface-to-volume ratio. The negative H-P slope vs. 1/d may be seen in Fig. 8.51. They also report that their data scales well with $d^{-1/2}$, the scaling expected in H-P relation, as seen in the inset in Fig. 8.51.

There is extensive literature on the inverse H-P effect indicating that, at a certain grain sizes, there may also be a negative slope. Despite the ongoing controversy, there is a consensus that the H-P relationship breaks down with a decrease in slope to the 1 μm–100 nm range. Evidence that the conventional H-P relationship is not obeyed may be seen for several metals in Fig. 8.50. The yield strength of NC materials has been measured and there are scattered reports of a plateau or a negative slope in the H-P relationship for grains <20 nm. Figures 8.46, 8.47, and 8.48

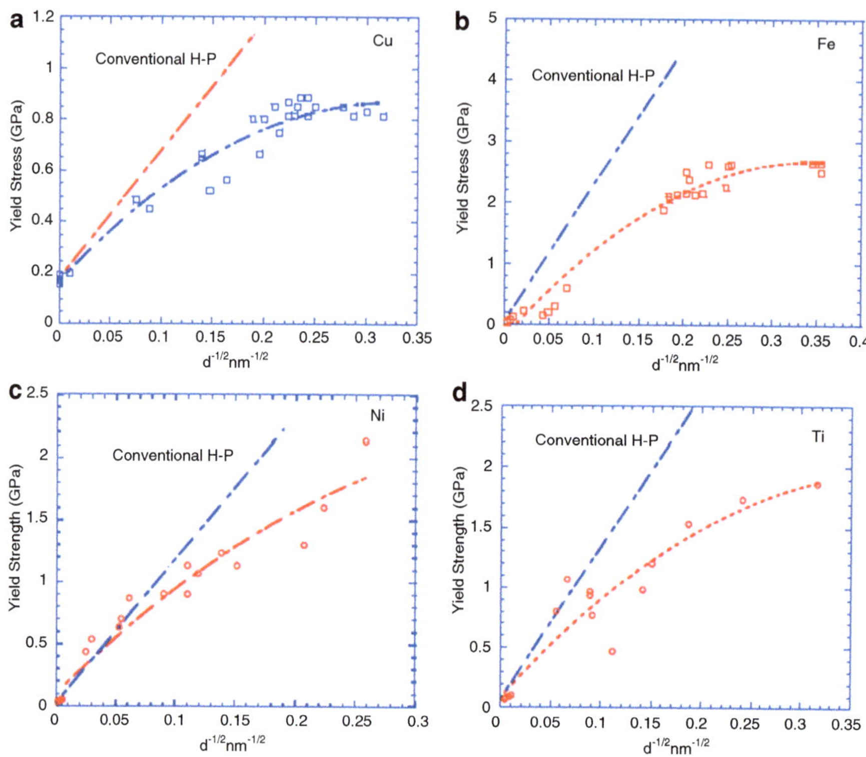

Fig. 8.50 Plots showing the trend of yield stress with grain size for different metals as compared to the conventional Hall–Petch response: (**a**) copper, (**b**) iron, (**c**) nickel and (**d**) titanium (Meyers et al. 2006. With kind permission of Elsevier)

provide experimental data supporting these reports. Further support for the possible existence of the negative H-P slope may be obtained from Fig. 8.49, showing negative strain-rate dependence on grain size. In addition to the GBS concept, which rules the deviation from common H-P relationship in larger grain sizes, there is also the concept of partial-dislocation involvement in the observed phenomena at the nanometer scale. This mechanism was put forward by Froseth et al.; they recently used molecular-dynamic calculations to describe the emission of partial dislocations from grain boundaries. Additional molecular-dynamic calculations were performed by Van Swygenhoven et al. and by TEM investigations, showing low dislocation density after considerable plastic deformation.

Grain boundaries act as both as sources and sinks for dislocation formation and dislocation disappearance at boundaries. When grain sizes are reduced to the nanometer range, the mean free paths of the dislocations generated at the grain boundaries are greatly limited. These dislocations can run freely until they meet the next grain boundary, which acts as sink. Thus, dislocation density remains low throughout the deformation process and work hardening is insignificant.

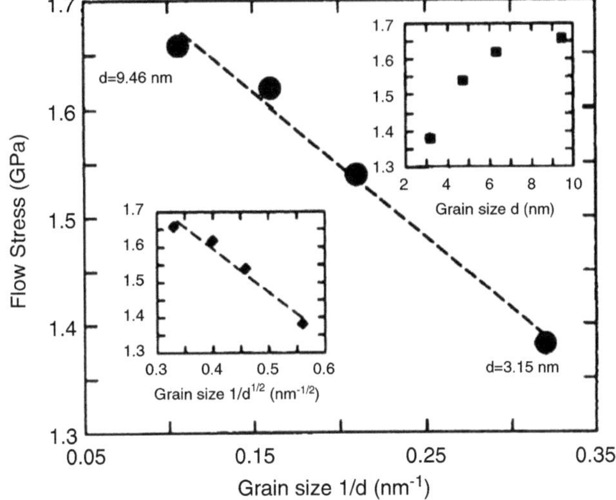

Fig. 8.51 Negative Hall–Petch slope computed by Kadau et al. using molecular dynamics for aluminum (Meyers et al. 2006. With kind permission of Elsevier)

A consequence of the limitation of the mean free paths by small grain size is that dislocation reactions, cross slips and dislocation multiplications are effectively prohibited. This may explain inverse H-P.

H-P strengthening is limited by the size of the dislocations. Once the grain size reaches ~10 nm (3.9×10^{-7} in.), grain boundaries start to slide, as indicated above. The most probable behavior is that yield strength will plateau below a critical grain size. The real trend remains to be determined along with the knowledge of whether it varies for different materials. An extensive summary on the various aspects of deformation in micron and submicron/nanometer materials is found in the comprehensive review work done by Meyers et al.

8.3.8 Superplasticity

Superplasticity is a known feature of ultra fine grained (UFG) materials. This topic has received broad treatment in the literature and certainly deserves a few words with some good examples within the framework of this chapter. In superplasticity, high strength is coupled with extended deformation before breaking, often up to ~200% or even beyond it during tension. Metals and ceramics can show superplastic behavior. The prerequisites for superplasticity are having fine grain-sized materials and the dispersion of fine, stable particles, able to pin the grain boundaries. However, not only grain size is an essential factor in the superplastic behavior of materials; these materials must exhibit appropriate temperature and strain rates, as well. Strain-rate sensitivity is an important feature of superplastic behavior. Of the many concepts explaining superplasticity, one relates to grain-boundaries sliding past each other. When testing includes high temperatures, recovery and dynamic recrystallization may be responsible for superplastic appearance, even in CG materials (Sharma et al.). The ability to achieve superplastic behavior, as observed for example in the high tensile ductility of polycrystalline materials, is not only of scientific, but also of technical interest. One important application is in the forming industry, where, in the multipass forming (rolling, drawing, etc.), a reduction in the pass number is possible. Since temperature plays an important role in superplasticity and the homologous temperature is often cited as the test temperature, one can understand that, for lower melting-temperature materials, these tests may be performed at room or near-room temperatures, whereas, for high melting-point materials, these tests are performed at higher temperatures. In the following, a few superplastic cases will be considered.

In the recent work of Figueiredo and Langdon, an Mg AZ31 alloy (Mg-3% Al-1% Zn) processed by equal-channel angular-processing [henceforth: ECAP] resulted in a refined grain size of 2.2 μm. The specimens pulled by tension in the 623–723 K range, at strain rates 10^{-3} to $10^{-4}s^{-1}$, showed ductility in the range of 500–1,000% before failure. Their experiments showed a strain-rate sensitivity of ~0.5 and activation energy close to the value for grain-boundary diffusion. Calculations showed that the experimental results are in good agreement with a model for superplasticity based on GBS. Strain-rate sensitivity may be expressed by a relation similar to Eq. 8.14 as:

$$\dot{\varepsilon} = \frac{AD_{gb}Gb}{kT}\left(\frac{b}{d_s}\right)^P\left(\frac{\sigma}{G}\right)^n \tag{8.15}$$

A is a dimensionless constant, D_{gb} is the grain-boundary diffusion coefficient, d_s is the spatial grain size, p is the exponent of the inverse grain size, n is the stress exponent and the meaning of the other symbols are as usual. It is now well-established that the superplastic-flow mechanism is based on the movement of dislocations along grain boundaries with some limited intragranular slip acting as an accommodation process. The values in the above relation are $A \approx 10$, $p = 2$ and $n = 2$. The inverse of the stress exponent is equivalent to the strain-rate sensitivity, m, obtained by the slope of a logarithmic plot of σ/G (normalized stress by G) versus strain rate (see Eq. 1.31). Since atomic diffusion occurs in the grain boundaries during this process, the well-known relation for the diffusion coefficient in the grain boundaries is recalled:

$$D_{gb} = \delta D_0 \exp\left(-\frac{Q_{gb}}{kT}\right) \tag{8.16}$$

δ is the grain boundary width, D_0 is the pre-exponential or frequency factor in grain boundaries and Q_{gb} is the activation energy for diffusion in grain boundaries.

Substituting the value for grain-boundary diffusion into Eq. 8.15, the strain rate, in terms of the activation energy in the grain boundaries may be obtained as:

$$\dot{\varepsilon} = A\left(\frac{D_0\delta G}{kT}\right)\left(\frac{b}{d_s}\right)^p\left(\frac{\sigma}{G}\right)^n \exp\left(-\frac{Q_{gb}}{kT}\right) \tag{8.17}$$

Much lower elongation than indicated above may be obtained in a similar alloy of Mg AZ31, also produced by ECAP. Material structures vary according to the number of processing passes by ECAP fabrication and, after seven such passes, a homogeneous structure is obtained having a grain size of 9 μm. Figure 8.52 illustrates optical-microscope images showing grain-size variation with the number of passes after ECAP at 623 K. Grain-size variation with the number of ECAP passes is summarized in Fig. 8.53. Figure 8.54 shows the mechanical properties of the Mg AZ31 alloy before and after ECAP at 623 K for different passes. The uniform elongation of the Mg AZ31 alloy increases continuously and strength decreases after ECAP. In particular, the elongation increases from 28% of the as-extruded alloy to 50% after 2-pass ECAP, after which it reaches a stable value. X-ray diffraction experiments indicate that with higher ECAP-pass numbers, the intensity of the $\{10\bar{1}1\}<10\bar{1}2>$ increases, indicating that the main deformation is by twinning, while, with additional passes, the main deformation mode is by dislocation slip. It is known that twinning in hexagonal structures, such as Mg, occurs in the $(10\bar{1}2)[\bar{1}011]$ system. Clearly, the experimental conditions dictate the outcomes of superplastic behavior, even in the same alloy. Furthermore, not all similar alloys show the same elongation, even when the experiments are performed by the same researchers. This may be seen in the next two Figs. 8.55 and 8.56, performed by Langdon's associates, although the alloy had a different composition (i.e., Mg-0.6%Zr). Whereas, in Langdon's recent paper (on Mg-3% Al-1% Zn), an elongation of 500–1,000% before failure was reported, in Figs. 8.55 and 8.56 below, only an ~450% elongation was observed at a strain rate of $3.3 \times 10^{-4}\ s^{-1}$. A strain-rate increase, as is known, increases the stress, but this occurs at the expense of elongation, not unlike in conventional stress-strain curves. The variation in stress-strain (in terms of the percentage of elongation) is seen in Fig. 8.55, which illustrates the effect of strain rate on the stress-strain relation in an Mg alloy prepared under the experimental conditions indicated. The effect of strain rate, specifically on elongation under different experimental conditions on the same alloy (Fig. 8.56), shows the importance of ECAP, which is universally used in superplastic-fabrication technology. ECAP can refine the structure of a material to an average grain size of ~2.2 μm and below that. In this figure the effect of fabrication indicated. (i) relates to a cast alloy with the process of ECAP (symbol triangle), (ii) relates only to extrusion of the cast alloy (circles), while (iii) includes all aspects of fabrication, namely the extruded cast alloy which has been in addition processed by ECAP. (iii) provides the highest elongation of the alloy at a given strain rate.

High-temperature superplasticity, for practical purposes, should be kept at a level at which the potential contribution of creep phenomena during use is avoided. Therefore, low-temperature superplasticity is of great interest. The Mg alloys

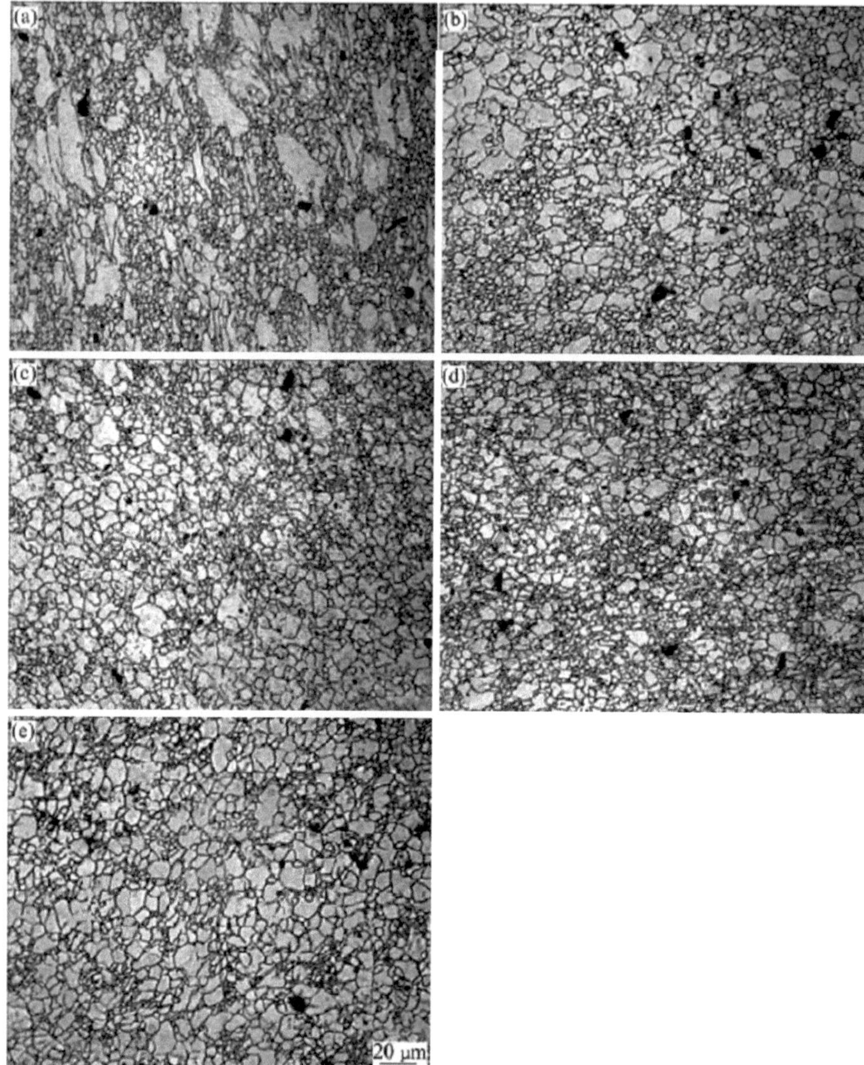

Fig. 8.52 Optical microstructures on cross sections of AZ31 Mg alloys after different passes of ECAP: (**a**) One pass; (**b**) Three passes; (**c**) Five passes; (**d**) Seven passes; (**e**) Eight passes (Xiao-ming and Tao-tao 2009. With kind permission of Elsevier)

considered thus far were handled at around 573 K or somewhat above it. Mg alloys showing high-superplastic behavior at lower temperatures are of particular interest. Remember that Mg alloys are among the lightest structural materials and improving their mechanical behavior will enhance their wider use. Commercially-cast Mg alloys usually have an elongation of less than 20% at room temperature and not above ~40% at ~300°C. The ductility of Mg at room temperature may be improved by creating a grain structure which converts the alloy to a superplastic

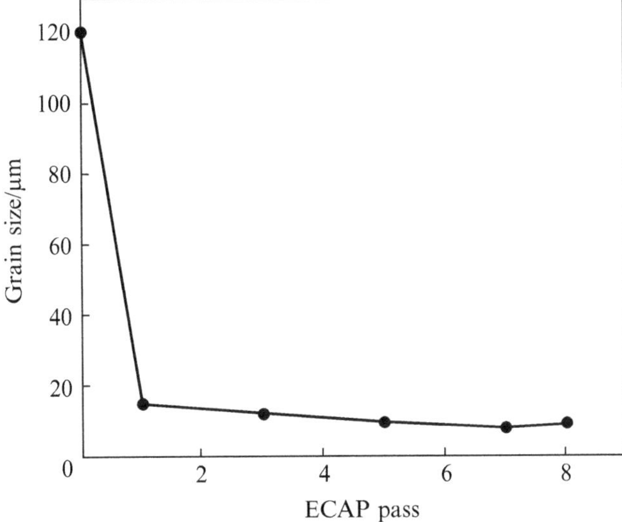

Fig. 8.53 Grain size evolution during ECAP (Xiao-ming and Tao-tao 2009. With kind permission of Elsevier)

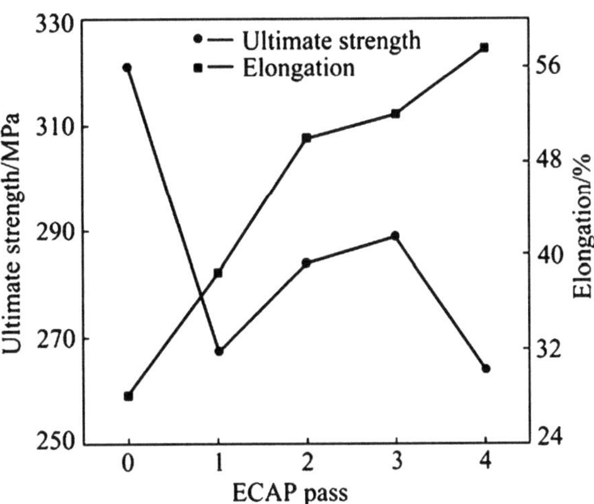

Fig. 8.54 Mechanical properties of AZ31 Mg alloy as a function of ECAP passes (Xiao-ming and Tao-tao 2009. With kind permission of Elsevier)

material. Thus, the low-temperature superplasticity of Mg is of great practical value and is considered below for the 273–673 K range. Two Mg cast alloys, ZK291 (Mg-6wt%Zn-0.5wt%Zr) and AZ91 (Mg-9wt%Al-1wt%Zn-0.2wt%Mn), with grain sizes of 20 μm and 15 μm, respectively, were fabricated by ECAP, resulting in mean rain sizes of ~1 and 0.8 μm for ZK60 and AZ91, respectively.

In Fig. 8.57, superplasticity tests of these alloys are presented as stress-strain relations. These tests were carried out at 200–430°C at a fixed strain rate of 3×10^{-3} s^{-1}.

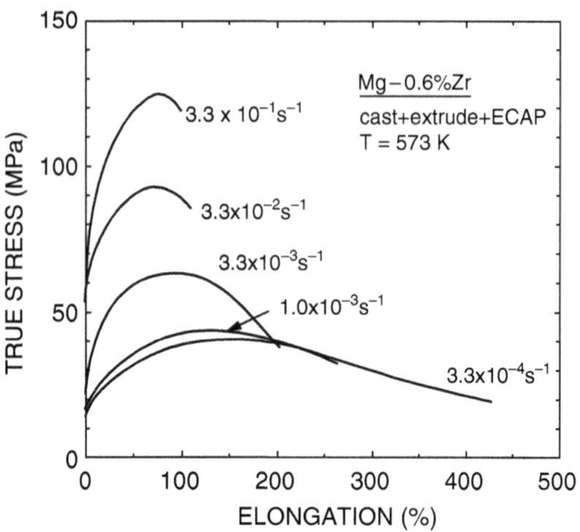

Fig. 8.55 True stress versus elongation curves at 573 K for a Mg–0.6% Zr alloy at different strain rates after extrusion and ECAP (Horita et al. 2002. With kind permission of Elsevier)

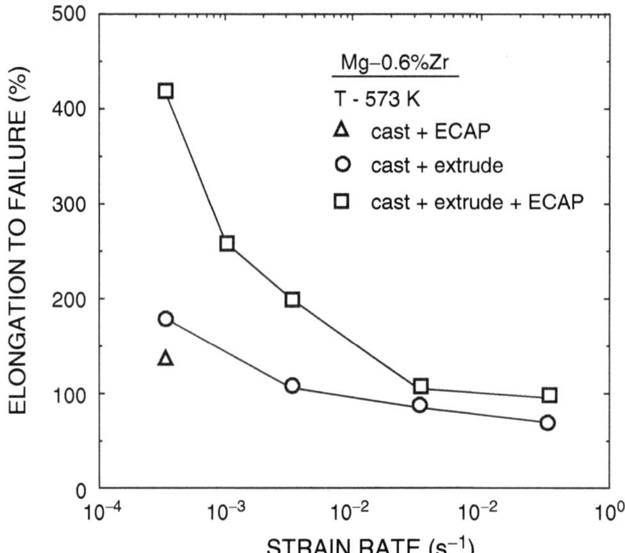

Fig. 8.56 Elongation to failure versus strain rate at 573 K for specimens fabricated (i) by ECAP of the cast alloy, (ii) by extrusion of the cast alloy and (iii) by extrusion and ECAP of the cast alloy (Horita et al. 2002. With kind permission of Elsevier)

An increase in ductility and a decrease in flow stress are observed between 200°C and 260°C. In Fig. 8.57a, the highest ductility was obtained at 260°C (ZK291 alloy), whereas it occurred at a somewhat higher temperature in AZ91, i.e., at 300°C.

The effect of temperature on elongation-to-failure is presented more clearly in Fig. 8.58 for both alloys, indicating that the maximum elongations are ~800% and ~570% for the ZK291 and AZ91 alloys, respectively. The decrease in ductility at

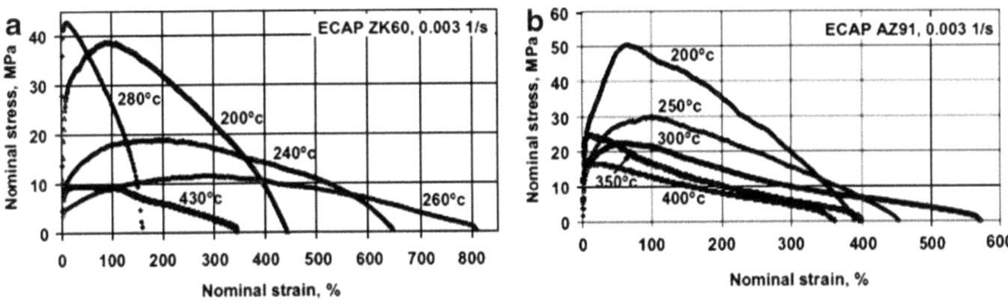

Fig. 8.57 Nominal stress-strain relations at strain rate of $3 \times 10^{-3}\mathrm{s}^{-1}$ for microcrystalline ZK60 (**a**) and AZ91 (**b**) Mg alloys (Chuvil'deev et al. 2004. With kind permission of Elsevier)

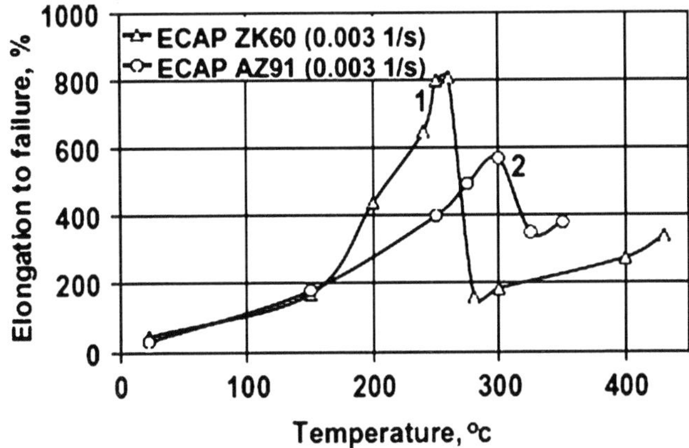

Fig. 8.58 Elongation to failure as a function of deformation temperature for microcrystalline ZK60 (*1*) and AZ91 (*2*) alloys at a strain rate of $3 \times 10^{-3}\ \mathrm{s}^{-1}$ (Chuvil'deev et al. 2004. With kind permission of Elsevier)

higher temperatures has probably occurred because of grain growth. The elongation-to-failure and the ultimate tensile strength, as functions of the strain rate at the optimal superplasticity temperature (260°C for ZK60 and 300°C for AZ91), are presented in Fig. 8.59. With a decreasing strain rate, the values for elongation increase, reaching 960% and 570%, respectively, for the ZK60 and AZ91 alloys at $\dot{\varepsilon} = 6.5 \times 10^{-4}\ \mathrm{s}^{-1}$ and $\dot{\varepsilon} = 3 \times 10^{-3}\ \mathrm{s}^{-1}$, respectively.

Prediction of the optimum temperature at which the maximum tensile elongation is achieved may be determined by internal friction measurements. The temperature at which the internal friction is a maximum seems to coincide with the temperature of maximum elongation, which can be used to evaluate the optimum temperature for superplasticity (Chuvil'deev et al.). This relation, between the internal friction peak and optimal superplasticity in terms of elongation, is illustrated in Fig. 8.60.

Now, some experimental results for superplasticity in FCC metals, such as Au and other commercially significant alloys, will be presented. Previously,

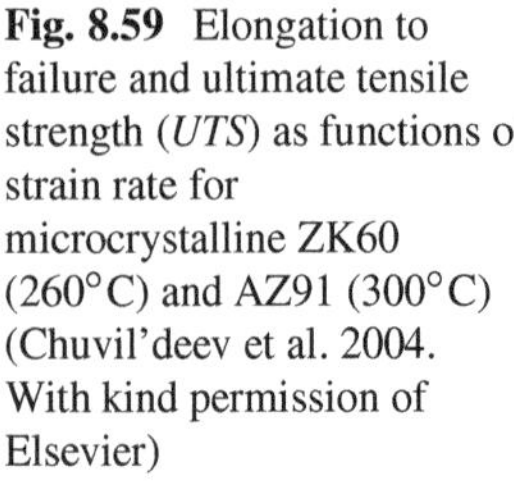

Fig. 8.59 Elongation to failure and ultimate tensile strength (*UTS*) as functions of strain rate for microcrystalline ZK60 (260° C) and AZ91 (300° C) (Chuvil'deev et al. 2004. With kind permission of Elsevier)

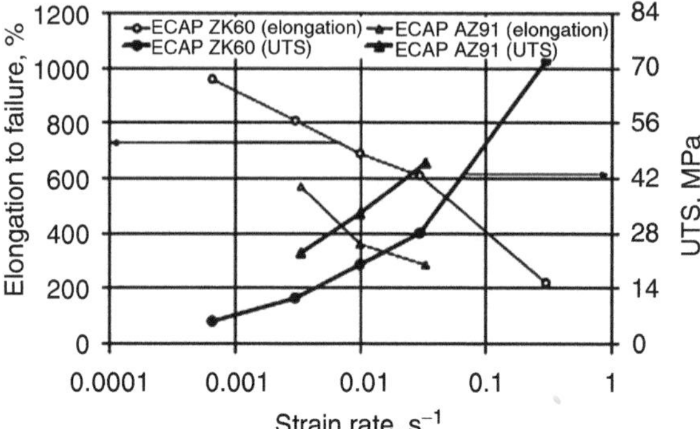

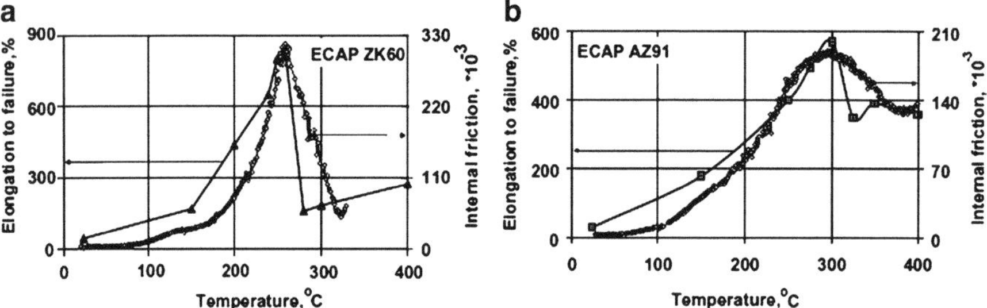

Fig. 8.60 Comparison of elongation to failure as a function of deformation temperature and temperature dependence of internal friction in microcrystalline ZK60 (**a**) and AZ91 (**b**) alloys (Chuvil'deev et al. 2004. With kind permission of Elsevier)

superplasticity was considered in the micron grain-size range. Here, one of the materials of interest is the Ti-6Al-4V alloys, that show superplastic behavior on the nanometer scale at ~300 nm. An overall indication of superplasticity may be seen in Fig. 8.61, showing the tensile extension of test specimens. This photograph shows various deformations exhibited by two types of Ti-6Al-4V alloys at different temperatures and strain rates. The extent of the deformation exhibited by conventional, superplastic Ti-6Al-4V is much lower than that shown by the submicron-grained Ti-6Al-4V, especially at the lower strain rates of 0.001 and 0.0001 s^{-1}.

The grain sizes compared in the experiments were 3 and 0.3 μm. The superplastic performance of these two materials is evaluated on the basis of the tensile-test data. Figure 8.62 shows the flow curves for Ti-6Al-4V, tested at two different temperatures and at various strain rates (see inset). Under identical strain rate and temperature conditions, the two materials exhibited significantly different strength values (Fig. 8.62). The 3 μm alloy tested at 775°C and the high strain rate shows appreciable strain hardening, according to Fig. 8.62a. At lower strain rates, the ductility exhibited by the fine-grained Ti-6Al-4V is exceptionally high, having a 700% elongation, as seen in Fig. 8.62b.

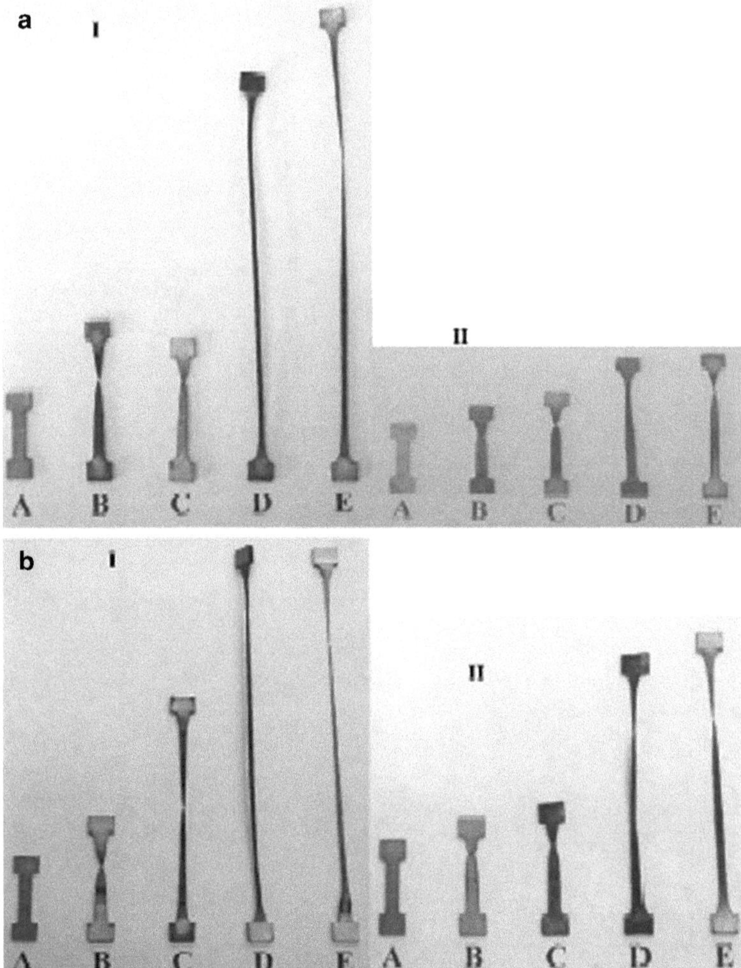

Fig. 8.61 (**a**) Ti-6Al-4V Tensile specimens with (*I*) grain size of 0.3 μm and (*II*) grain size of 3 μm tested at 775°C and strain rates of (*B*) 0.1 s^{-1}, (*C*) 0.01 s^{-1}, (*D*) 0.001 s^{-1} and (*E*) 0.0001 s^{-1} in comparison to untested specimen (*A*); (**b**) Ti-6Al-4V tensile specimen with (*I*) grain size of 0.3 μm and (*II*) grain size of 3 μm tested at 875°C and the same strain rates as in (**a**) and compared to the untested specimen (*A*) (Patankar et al. 2002. With kind permission of Elsevier)

Note the scale differences for stress and strain between the 3 and 0.3 μm tensile-tested specimens. Clearly, the increase in ductility is at the expense of stress. Although the strain-rate effect is visible (Fig. 8.62), a more comprehensive illustration of its effect appears in Fig. 8.63.

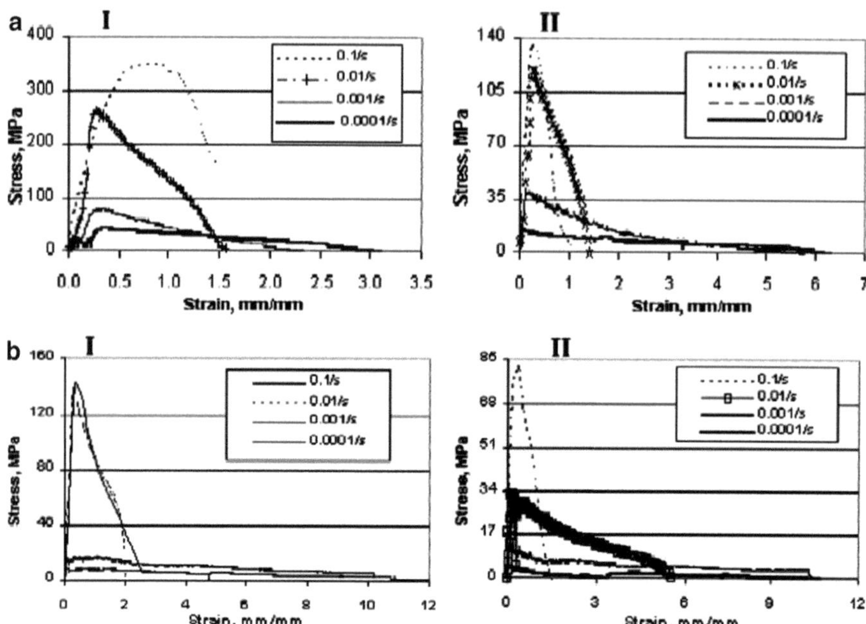

Fig. 8.62 Flow curves of Ti-6Al-4V with (**a**) grain size of 3 μm tested at (*I*) 775°C and (*II*) 875°C, and (**b**) grain size of 0.3 mm tested at (*I*) 775°C and (*II*) 875°C (Patankar et al. 2002. With kind permission of Elsevier)

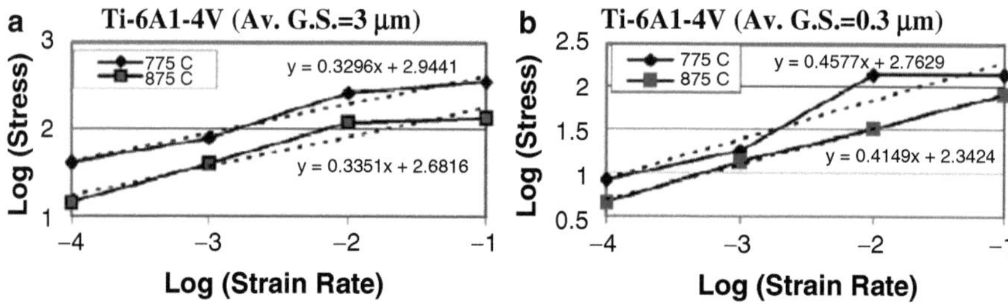

Fig. 8.63 Variation of yield strength as a function of strain rate in (**a**) Ti-6Al-4V with a grain size of 3 μm and (**b**) Ti-6Al-4V with a grain size of 0.3 μm (Patankar et al. 2002. With kind permission of Elsevier)

The strain-rate sensitivity may be determined from the slope of such curves; it is a very important characteristic of superplastic materials. As discussed in an earlier chapter, the strain-rate sensitivity, m, is given by:

$$\sigma = k\dot{\varepsilon}^m \tag{8.18}$$

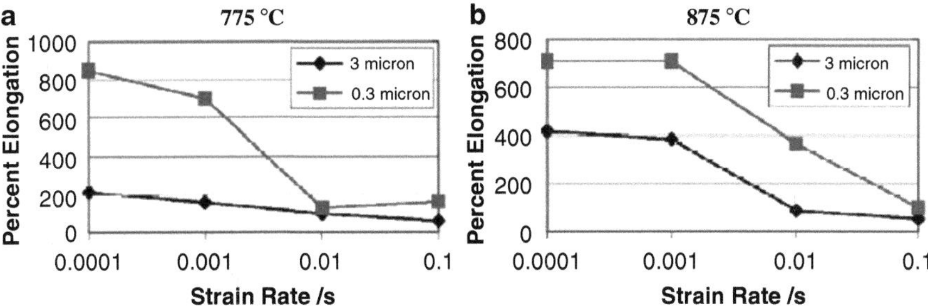

Fig. 8.64 Variation of ductility as a function of strain rate in Ti-6Al-4V with different grain sizes tested at (**a**) 775°C and (**b**) 875° (Patankar et al. 2002. With kind permission of Elsevier)

In Ti-6Al-4V, m = 0.3 for the 3 μm grain-sized material, as obtained from the logarithmic expression in Eq. 8.18. The 0.3 μm grain-sized Ti-6Al-4V shows a higher strain-rate sensitivity, as seen from the slope in Fig. 8.63b. Elongation, as a function of strain rate, is shown in Fig. 8.64 for both grain sizes at the experimental temperatures. Note the strong effect of the strain rate on the superplastic behavior. In the case of Ti-6Al-4V at a strain rate of $10^{-4}s^{-1}$ for a grain size of 0.3 μm at a temperature of 775°C, an elongation >800% was achieved.

In Fig. 8.64, the combined effect of all three factors: (i) grain size; (ii) strain rate; and (iii) temperature, which determine superplastic behavior are shown. A summary of the experimental data showing the effect of these three factors in the Ti-6Al-4V alloy is presented in Table 8.3. The reason for the different behaviors of the two grain sizes may be stated as follows:

(a) Generally, in conventional materials, strength properties vary inversely with elongation. This is, indeed, observed in the Table 8.3 for both grain sizes; as the stress (at 50% strain) increases, elongation decreases.
(b) The large number of grain boundaries in small grain-sized materials probably induces GBS, representing some sort of contribution to overall elongation, which means reduced strength and vice versa. There is less GBS (in larger grain-sized alloys) when the number of boundaries is smaller.
(c) Grain-boundary diffusion (which promotes GBS) is more pronounced in small grain-sized alloys. It was also observed in the Ti-6Al-4V alloy that forming operations can be performed at lower temperatures with a grain size of 0.3 μm and the results also indicate that the deformation along both rolling directions (longitudinal and transverse) are uniform. By extrapolating these results to other materials, this behavior may serve as a general pattern.

It was mentioned above that the effect of dislocation in grain boundaries is considered to be a mechanism for superplasticity. In the Mg AZ31 alloy, twinning, in addition to dislocation, has also been shown to act as a mechanism for

Table 8.3 High temperature mechanical properties of Ti-6Al-4V sheet with different grain size (Patankar et al. 2002. With kind permission of Elsevier)

Temperature, °C	Grain size, μm	Strain rate, s^{-1}	Flow stress (MPa) at 50% of strain	Elongation, %
775	0.3	10^{-4}	9	850
		10^{-3}	17	700
		10^{-2}	140	170
		10^{-1}	140	190
	3	10^{-4}	45	200
		10^{-3}	80	190
		10^{-2}	250	170
		10^{-1}	350	80
875	0.3	10^{-4}	5	720
		10^{-3}	12	720
		10^{-2}	30	380
		10^{-1}	82	110
	3	10^{-4}	15	420
		10^{-3}	38	390
		10^{-2}	122	90
		10^{-1}	138	80

superplastic behavior. The following is a discussion on twinning as the dominant deformation mechanism inducing superplasticity in defect-free materials. Usually, nanomaterials, in the form of nanowires (NWs) are expected to show superplasticity induced by a twinning mechanism.

Au defect-free NWs show superplasticity when tested by tensile deformation. High-resolution electron microscopy indicates that, after initial superplastic deformation, NWs show superplastic deformation induced by coherent-twin propagation with the consequence of complete crystal reorientation from <110> to <100>. In a snapshot of an Au single-crystal NW, tensile deformation was recorded until final fracture (reproduced in Fig. 8.65).

Fabricating defect-free materials may be possible by reducing their sizes below certain critical values. Defect-free metal NWs are ideal, model systems for exploring tensile properties at the nanoscale level, since they are free of complex interactions and have well-defined geometries and high-aspect ratios. Atomistic simulations have predicted strong, but ductile, behavior for single-crystalline metal NWs by twinning deformation, but only a few experimental studies have been reported on the processes in NWs (Jong-Hyun Seo et al. and inherent references). In Fig. 8.66, the progress of deformation, as observed by TEM, is indicated.

Similar observations were made for other metals, even for iron NWs (Jianwei Zhu and Daning Shi), in which crystal reorientation takes place involving a martensitic phase transformation. The TEM images in Figure 8.66 illustrate how the atomic crystal-structure extension seen in Fig. 8.65 changes during the tensile-loading process. During an initial stage of plastic deformation, two nanoscale twin nuclei are displayed in Fig. 8.66a, b, each composed of two SFs generated by the successive gliding of partial dislocations across the NW on neighboring

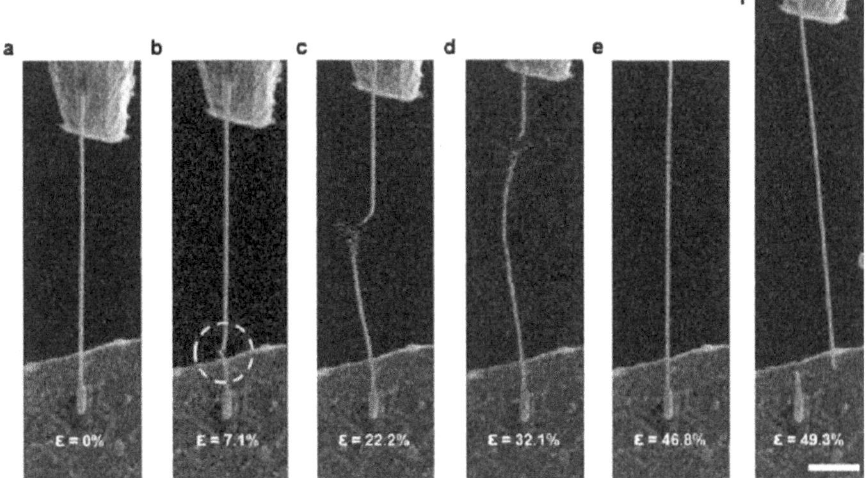

Fig. 8.65 Snapshots on the tensile deformation of a <110> Au NW. (**a–f**) SEM images captured from Supporting Information Movie S1 that shows the entire tensile process until a final fracture of a Au NW. ε represents total strain in each tensile stage. A *yellow circle* in (**b**) highlights an arch-shaped feature formed at an initial stage of plastic deformation. The scale bar is 1 μm (After Seo et al. 2011)

{111} crystal planes. The initially-nucleated twin is extended by the coherent-twin propagation process to an ~1 μm length, showing a clear twin boundary (Fig. 8.66d). The two twins in Fig. 8.66a are combined into one by twin propagation. Selected area electron diffraction [henceforth: SAED] patterns of the original and the twinned regions indicated in Figs. 8.65e–f, showing that the twin formation reorients the lattice of the NW from <110> to <100>. The tensile process of the load variation of a cantilever is indicated in terms of a stress-strain relation in Fig. 8.67. The strain was measured by SEM video images. Actually, the progress of the tensile deformation seen in Fig. 8.65 is depicted in Fig. 8.67 for specimens of different dimensions in terms of the stress-strain relation.

The three stages in Fig. 8.67 represent: (i) the superplastic and superstrong region (region 1); (ii) the superplastic in regions 2 and 3, where the tensile process of a now-reoriented <100> occurs; leading to (iii) final fracture via necking. The atomistic picture of the tensile-deformation dynamics is made clear by the information in the TEM observations. In Fig. 8.67, region 1 indicates the start of deformation of a <110>-oriented NW. The yield strength of ~1.54 GPa is considerably higher than the bulk value of Au (55–200 MPa) and is quite close to the theoretical value of 1.8 GPa. Thus, the NW shows a yield having super strength. Furthermore, Young's modulus, derived from the slope, is 72 GPa, also close to the theoretical value of 78 GPa, whereas the value of the bulk Au is only in the range of 53–55 GPa. It is believed that the super strength is a consequence of the Au NW being defect-free and, therefore, very close to the theoretical value.

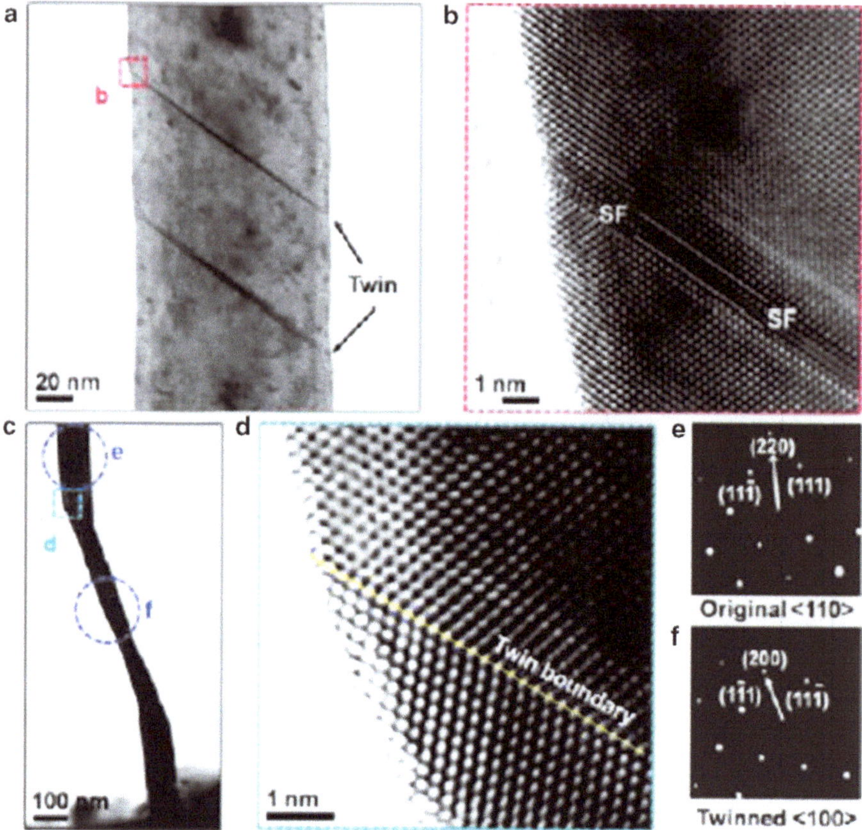

Fig. 8.66 TEM images of a Au NW during the tensile deformation. (**a**) Two nanotwins initially nucleated in the plastic deformation. (**b**) HRTEM image magnified from the *magenta square* in (**a**). The twin is composed of two stacking faults (*SFs*), as clearly seen between two *dotted lines* in (**b**). (**c**)TEM micrograph showing a twin extended to 1 μm length by twin propagation (supporting information in Fig. 8.66). (**d**) HRTEM image magnified from the *cyan square* in (**c**), showing a clear twin boundary along the *yellow line*. (**e, f**) SAED patterns acquired from two *blue dotted circles* in (**c**) with a 180 nm aperture, the NW lattice is changed from an original <110> (**e**) to a twinned <100> (**f**) (After Seo et al. 2011)

Region 2 starts with the sudden generation of nanotwins, decreasing the stress and continuing with subsequent superplastic elongation by twin propagation. A strain of 41% was measured from the SEM images. This value is the same as the geometric elongation resulting from the lattice reorientation of the NW from <110> to <100>. The stress-strain curve indicates that the flow stress required for the nucleation and propagation of a twin in the <110> Au NW is ~200 MPa, much lower than the yield strength for the first nucleation of a partial dislocation at point P1 in Fig. 8.67. Thus, twin expansion is energetically preferred over the nucleation of additional partial dislocations at other locations.

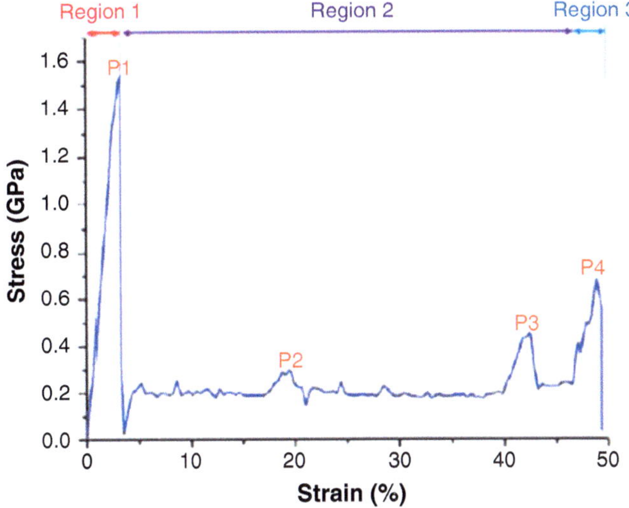

Fig. 8.67 The stress_strain curve of a <110> Au NW for the whole tensile process. *Region 1* represents elastic deformation, *Region 2* represents the plastic deformation of a <110> Au NW, and *Region 3* represents the deformation of a reoriented <100> Au NW (After Seo et al. 2011)

The mechanism of the twin propagation may be explained as follows. At the yield point, P1, a partial dislocation nucleates and, as expected in FCC crystals, it glides across the wire on some {111} plane and is annihilated upon reaching the opposite side (of the path of its motion). A second partial dislocation nucleates next to the first, gliding on an adjacent {111} plane and annihilates in the same way as the previous partial dislocation.

By such repeated motion, twins propagate coherently along the NW axis to the end, inducing the observed superplasticity. Twin propagation converts the initial <110> NW, having a rhombic cross-section and {111} side facets, to a <100> NW, having a rectangular cross-section and {100} side facets. After the Au NW reorientation from <110> to <100> is completed, the final stage (see region 3, in Fig. 8.67) shows the elastic deformation of a now <100> NW and its final fracture via necking by continued tensile loading. The observed superstrength of a <110> Au NW is ascribed to the difficulty in its initial-dislocation nucleation in a perfectly crystalline Au NW.

This subsection may be concluded by summing up the observations discussed above:

1. The deformation of materials in the nanometer or micron/submicron range, either in the form of single crystals or polycrystalline, superfine-grained structures, occurs in about the same manner as in bulk materials. The difference is not of kind, but rather of degree, as dictated by the sizes involved.
2. Because of the small dimensions, either of the nanoscale single crystals or of the UFG polycrystalline structures (micron/submicron or nanometer), their properties may be quite different from those of their bulk counterparts.

3. Specifically-high superplastic strength, large elongation and the consequent behavior stemming from these properties characterize such materials. Superstength, as in bulk specimen behavior, occurs at the expense of elongation and vice versa. High strength is usually characterized by lower strain, whereas high strain occurs at the expense of strength.
4. When the amount of extension (in terms of elongation) is high, strain hardening does not reach appreciable values or, in many cases, does not exist.
5. The three major characteristics that control the superplasticity level are: grain size (or specimen size in single crystals), strain rate and temperature. These three factors determine the quality and magnitude of suprerplasticity.
6. In gross observations (macroscale) of experimental results, GBS is considered to be the major mechanism of superplasticity. However, in atomic dimensions (i.e., at a microscale level) dislocation motion, interactions and splitting into partials are responsible for superplasticity and, to some extent, the mechanism of deformation may also be controlled by twinning in defect-free crystals.

8.4 Time-Dependent Deformation (Creep)

In Chap. 5, the subject of creep was broadly described. There are, however, different effects due to the superfine grains in polycrystalline materials yet to be dealt with. This is the subject of this section. First of all, it was observed in bulk materials that yield-stress variation is an inverse function of grain size according to the H-P relation. Beyond a certain, small grain size the H-P line develops a plateau and even becomes a line with a negative slope (i.e., the strength properties decrease). This observation is called the 'inverse H-P effect' (illustrated in Figs. 8.46, 8.47, and 8.49). The deviation from the usual linear strengthening appears in Fig. 8.50. One of the explanations considered above was that the dislocation pile-up model does not hold, because of the very fine grains in the nanometer/submicron size, which cannot support dislocations, due to the small spacing between them at the level of the grain size itself. The volume fraction of the boundaries greatly increases in these nanometer/submicron-scaled materials and, thus, different deformation mechanisms may begin to operate in both static and time-dependent deformations. The change from the normal to the inverse H-P at some critical grain size provides experimental evidence of the preference for grain-boundary mechanisms over dislocation-based plasticity.

Chapter 5 stated that three parameters determine deformation: (a) temperature; (b) applied stress; and (c) grain size; however, strain rate also has a significant effect. Diffusion-controlled creep for the NH creep is expressed by the relation:

$$\dot{\varepsilon} = 16\alpha \frac{D_V C_V{}^0}{\pi d^2} \frac{\sigma V}{kT} = \frac{16\alpha D_S \sigma V}{\pi d^2 kT} \tag{5.44}$$

and for Coble creep as:

$$\frac{d\varepsilon_{gb}}{dt} = \dot{\varepsilon} = A_{CO}\frac{\sigma\Omega D_{gb}\delta}{l^3kT} \tag{5.63}$$

Equations 5.44 and 5.63 are often given as:

$$\dot{\varepsilon} = 14\frac{\sigma\Omega D_L}{kT}\frac{1}{d^2} \tag{5.44a}$$

and:

$$\frac{d\varepsilon_{gb}}{dt} = \dot{\varepsilon} = 14\pi\frac{\sigma\Omega D_{gb}\delta}{kT}\frac{1}{d^3} \tag{5.63a}$$

where V (the atomic volume) is replaced by Ω and l is replaced by d. It can be seen from Eqs. 5.63a or 5.63 that the strain rate is inversely proportional to $1/d^3$, whereas, in the NH relations, the strain rate varies as $1/d^{\,2}$. This means that Coble creep is preferable to the NH creep as grain size becomes smaller, because it varies as d^{-3}. In creep at low stresses, when diffusion-controlled creep occurs, deformation may be considered as being affected by both relations; therefore, the sum of Eqs. 5.44a and 5.63a must be taken into account:

$$\dot{\varepsilon} = \frac{\sigma\Omega}{kT}\left(A_{NH}\frac{D}{d^2} + A_{CO}\frac{D_{gb}}{d^3}\right) \tag{8.19}$$

where A_{NH} and A_{CO} stand for NH and Coble creeps, respectively. D is often found in the literature as D_L or D_S for lattice diffusion (in Eq. 5.44, D_S stands for self-diffusion). This relation expresses the transport mechanism occurring in and along the grain in the boundaries. Since diffusion requires vacancies, the grain boundary cannot merely be considered the location where diffusion takes place, but also as the source of vacancy generation (also acting as a sink for vacancies). Equation 8.19 may also be given as:

$$\dot{\varepsilon} = 14\frac{\sigma\Omega}{kTd^2}D_L\left(1 + \frac{\pi\delta}{d}\frac{D_{gb}}{D_L}\right) \tag{8.20}$$

This is obtained from summing Eqs. 5.44a and 5.63a. Equation 8.19, as suggested by Raj and Ashby, was quite recently used by Millett et al., for the atomistic simulation of diffusional creep. Since creep deformation in materials with supersmall-sized grains may be accommodated in grain boundaries with considerable GBS, events are controlled by grain-boundary diffusion. The deformation rate for GBS is given by Lüthy et al. as:

$$\dot{\varepsilon} = \frac{2x10^5 D_{gb}Gb}{kT}\left(\frac{b}{d}\right)^3\left(\frac{\sigma_e}{G}\right)^2 \tag{8.21}$$

This relation has often been used for many alloys and metals, for example by Chua et al. to determine the creep behavior in nanostructure Mg alloys at ambient and low temperatures. In a more general form, Eq. 8.21 for the strain rate,$\dot{\varepsilon}$, was given by Bird, Dorn, and Mukherjee as:

$$\dot{\varepsilon} = \frac{AD_{gb}Gb}{kT}\left(\frac{b}{d}\right)^p\left(\frac{\sigma_e}{G}\right)^n \tag{8.21a}$$

where A is a dimensionless constant, G, b, k and T have their usual meanings, p is the inverse grain-size exponent and n is the stress exponent. This equation is similar to the original Coble creep, when the exponents are $p = 3$, $n = 1$. Coble creep involves vacancy flow along the grain boundaries and, with the indicated exponents, the expression is:

$$\dot{\varepsilon}_{CO} = \frac{A_{CO}D_{gb}Gb}{kT}\left(\frac{b}{d}\right)^3\left(\frac{\sigma_e}{G}\right) \tag{8.21b}$$

NH creep is among the established, diffusional creep mechanisms in CG materials, which, unlike Coble creep, involves vacancy flow through the lattice, given when $p = 2$ and $n = 1$ by:

$$\dot{\varepsilon}_{NH} = \frac{A_{NH}D_{L}Gb}{kT}\left(\frac{b}{d}\right)^2\left(\frac{\sigma_e}{G}\right) \tag{8.21c}$$

$A_{CO} = 33$ and $A_{NH} = 28$.

As mentioned above, Coble creep predominates at nanoscale grain sizes. This occurs in diffusional creep under the conditions shown below, where the possibility of either Coble or NH creep may occur. Of these two independent mechanisms, the faster reaction controls the overall creep behavior. Thus, for small nanoscale materials:

$$\dot{\varepsilon}_{CO} > \dot{\varepsilon}_{NH} \tag{8.22}$$

Writing the values of the respective strain rates for Eq. 8.22 from Eq. 8.21b and Eq. 8.21c, one obtains:

$$\left(\frac{b}{d}\right)\left(\frac{D_{gb}}{D_L}\right) > \frac{A_{NH}}{A_{CO}} \tag{8.23}$$

The condition in Eq. 8.23 occurs at low temperatures with very small grain sizes. In nanometer-sized materials, grain growth occurs at relatively lower temperatures, namely, at small homologous temperatures. Therefore, creep studies are often performed at lower temperatures (sometimes even at room temperatures). An example of the low temperature, -actually room temperature- creep behavior of NC Ni is shown in Fig. 8.68. Although the applied stress is close to the yield strength, the

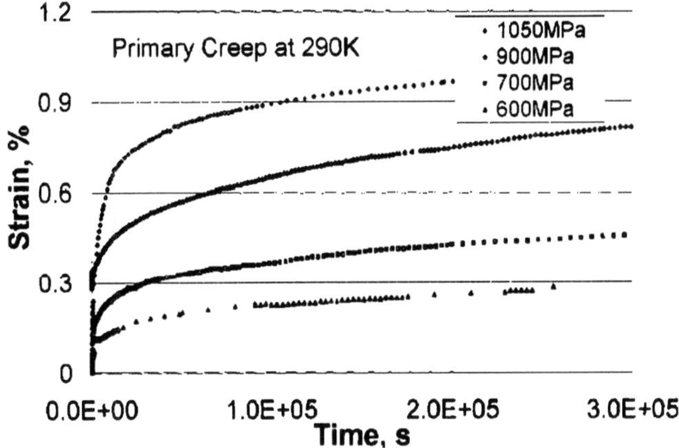

Fig. 8.68 Primary creep of nanocrystalline nickel at room temperature (Yin et al. 2001. With kind permission of Elsevier)

creep rate is very small in NC Ni at room temperature. Here, a prolonged primary-creep rate is observed and at 600 and 700 MPa, it lasts more than 200 h and only decreased at higher stress levels. The minimum strain rate at room temperature was achieved when the load was decreased from 1,050 to 800 MPa, as seen in Fig. 8.69b. A prolonged incubation period of more than 60 h was observed. The NC Ni exhibited accelerated-creep deformation at 373 K, where the creep rate at 700 MPa reached a minimum value 8.8×10^{-8} s^{-1}, which is about two orders of magnitude larger than that (1.1×10^{-9} s^{-1}) at room temperature (Fig. 8.69a).

The structure of the as-received Ni had equiaxed grains with an average size of ~30 nm, as shown in the TEM microstructure in Fig. 8.70. The specimens from the room-temperature creep tests were investigated by TEM and did not show any grain growth. From plots of the stress dependence of creep deformation, a plot of strain rate versus stress was made, based on constant load and step-load tests (shown in Fig. 8.71). From such curves, the stress exponent was evaluated as 1.1 for room-temperature creep and 6.5 for creep at 373 K. The grains were stable up to 353 K, with no evidence of grain growth.

Nano-nickel is one of the NC metals that have been subjected to extensive scientific investigations; therefore, further consideration will also be given to this element. In Fig. 8.72, the static-creep behavior of 20 nm NC Ni at the given stress levels is shown for comparison. These tests were carried out at room temperature for about 40,000 s and the creep rates were obtained from the slopes of these curves. Creep-strain rates were plotted as a function of applied stresses for three different grain sizes, as shown in Fig. 8.73.

The results indicate that grain size has a strong influence on creep behavior as the magnitudes and slopes of the curves change with grain size. By fitting the experimental data to relations such as Eq. 8.21, the dashed lines in Fig. 8.73 were obtained.

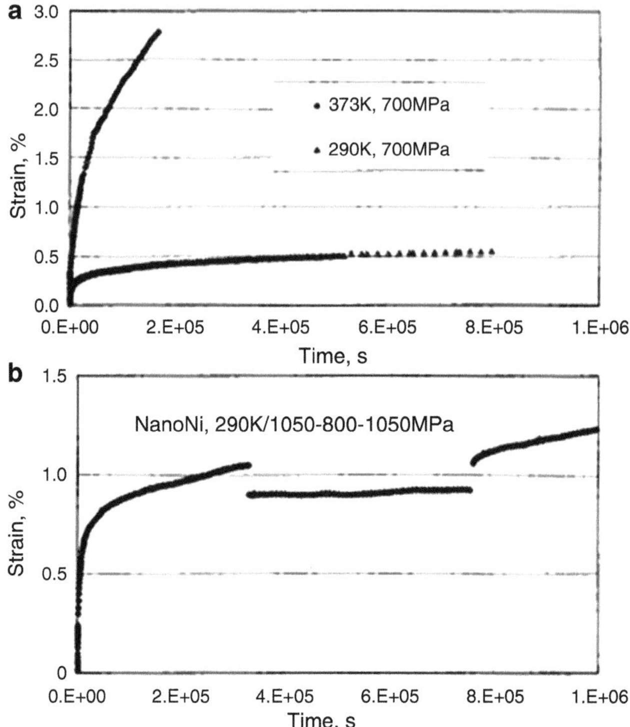

Fig. 8.69 Tensile creep curves of nanocrystalline nickel, (**a**) 700 MPa, (**b**) Step-load (Yin et al. 2001. With kind permission of Elsevier)

It was found that the model of GBS controlled by grain-boundary diffusion fits the data the best; however, the GBS model underestimates the creep rate for smaller grain-sized specimens (as seen for the 6 nm grain-sized data). This result indicates that the room-temperature creep of NC Ni is controlled by the GBS mechanism and that the contribution of the total creep strain, due to diffusion along the intercrystalline components becomes significant only at smaller grain sizes, i.e., $d < 20$ nm.

Figure 8.73 also shows that the experimental data for 40 nm indicate a transition between deformation mechanisms. It appears that either power-law creep or dislocation creep become the predominant deformation mechanism at high-stress levels. The slope is found to be about 5.3 using the least-square fitting method. Furthermore (in Fig. 8.73), the experimental results prove that: (i) smaller grain size is associated with a faster creep rate; (ii) for grain sizes >6 nm with smaller stresses, the effect of grain size seems to be consistent with the GBS model, controlled by grain-boundary diffusion; and (iii) the behavior of grains >6 nm at higher stresses is attributed to the contribution of dislocation creep.

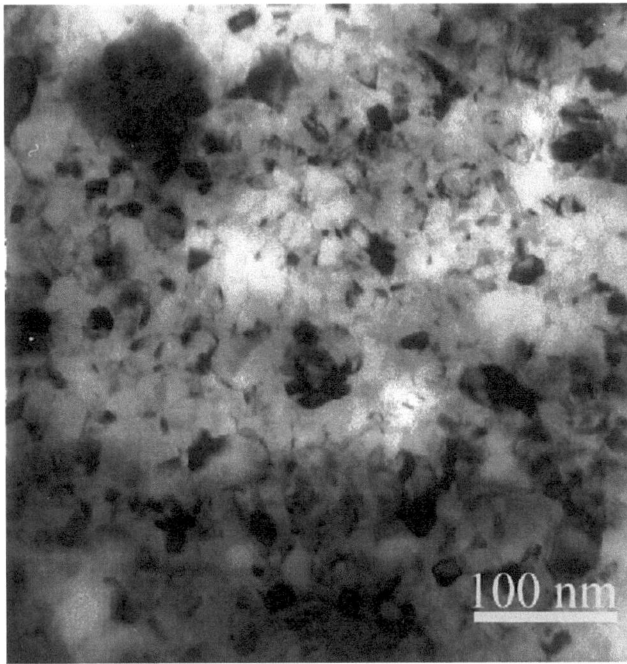

Fig. 8.70 Microstructure of as received nanocrystalline nickel (Yin et al. 2001.With kind permission of Elsevier)

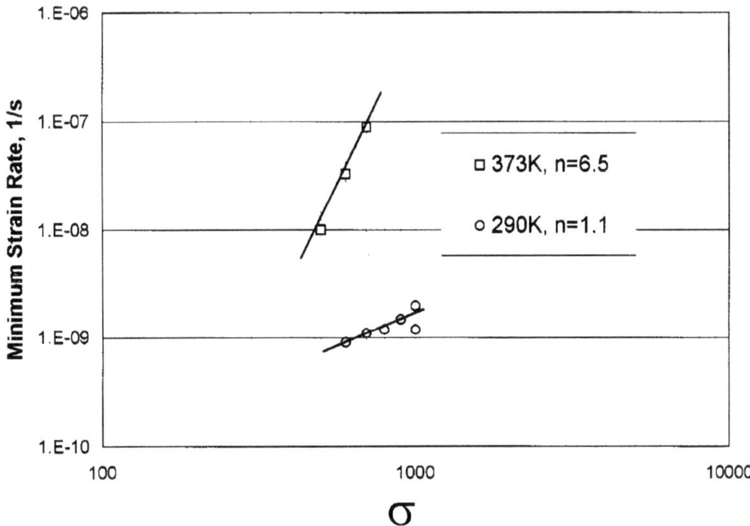

Fig. 8.71 Minimum strain rate vs. stress (Yin et al. 2001. With kind permission of Elsevier)

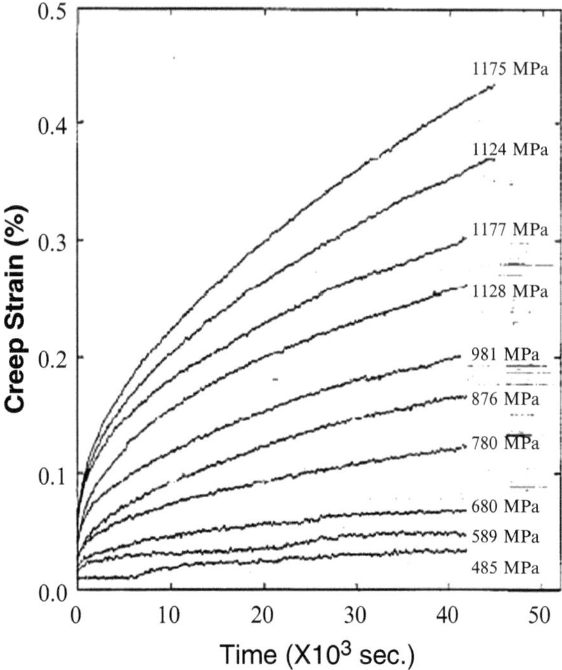

Fig. 8.72 The room temperature static creep curves of nanocrystalline (20 nm) nickel electrodeposits at different stress levels (Wang et al. 1997. With kind permission of Elsevier)

The effect of coarser grains than those considered above is compared with the NC Ni-P alloy (Deng et al.). The stress dependence of creep is actually indicated in both coarse and NC grains. In evidence, stress-jump curves are provided in Fig. 8.74 for NC 28 nm and for coarser 257 nm samples. Each stress-jump was carried out after a steady stage had been reached for a sufficient time under the preceding stress. The creep rate of the coarser samples is relatively smaller. The duration of these experiments was adjusted so as to obtain an accurate steady-state creep rate. From the log-strain rate–log-stress plots, the stress exponents for the two cases were evaluated (Fig. 8.74). The stress exponents for the 28 and 257 nm samples are 1.2 and 2.5, respectively. Their different stress exponents suggest that their creep mechanisms are different. Recall that the combined effect of stress and temperature for steady-state creep rate is expressed either by Eq. 5.3 or by Eq. 5.3a, rewritten as:

$$\dot{\varepsilon} = B\sigma^n \exp - \left(\frac{Q}{kT}\right) \tag{5.3}$$

or:

$$\dot{\varepsilon}_s = A\sigma^n \exp - \left(\frac{Q_c}{RT}\right) \tag{5.3a}$$

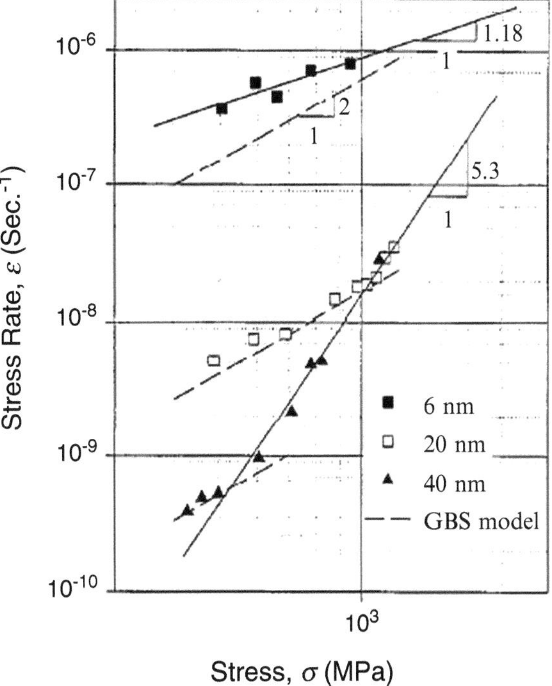

Fig. 8.73 Effect of grain size on the steady creep rate of nanocrystalline pure nickel electrodeposits as a function of applied stress (Wang et al. 1997. With kind permission of Elsevier)

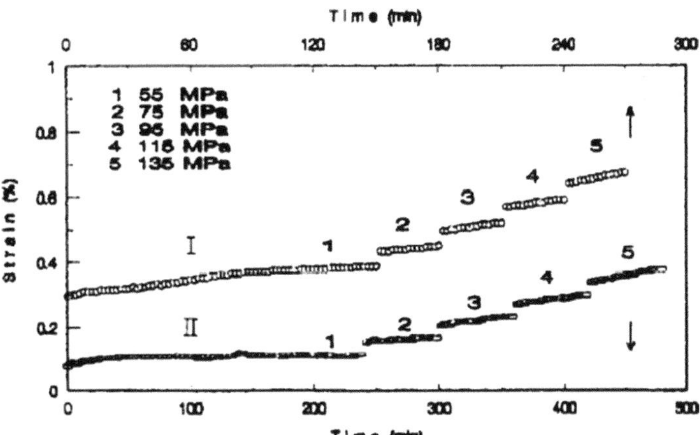

Fig. 8.74 Typical stress jump curves at 573 K for 28 nm samples (*curve I*) and for 257 nm samples (*curve II*) (Deng et al. 1995. With kind permission of Elsevier)

associated with the dislocation-creep mechanism. As a matter of fact, in all the strain-rate equations indicated earlier in this section, the diffusion coefficients may be replaced by the well-known Arrhenius relation for diffusion. By using a relation

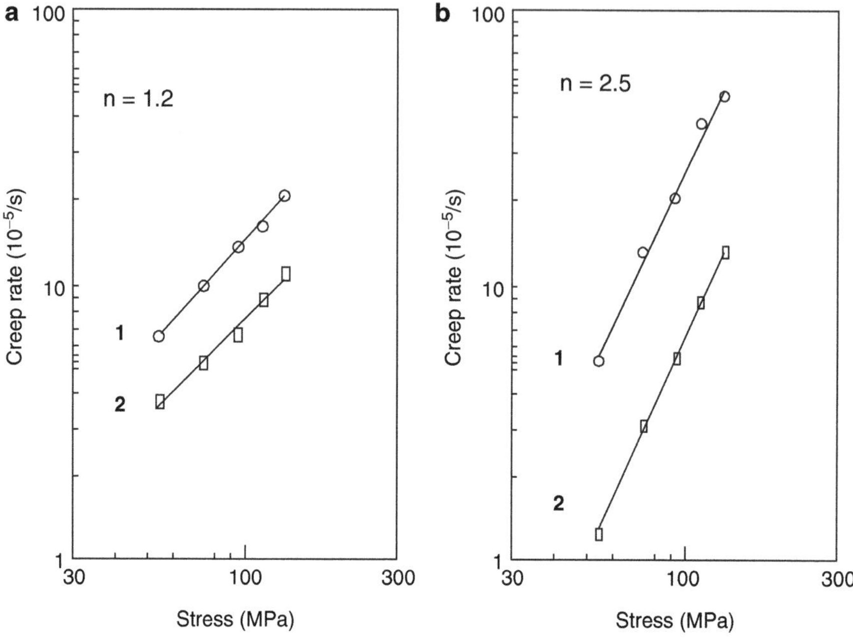

Fig. 8.75 Steady state creep rate vs. tensile stress for (**a**) 28 nm sample for 1 at 573 K and 2 at 543 K; (**b**) for a 257 nm ample and the meaning of 1 and is the same as in (**a**) (Deng et al. 1995. With kind permission of Elsevier)

such as Eq. 5.3 (used by Deng et al.) along with the exponents seen in Fig. 8.75, the diffusion coefficients may be related to the strain rate by:

$$\dot{\varepsilon} = A_1 \sigma^{1.2} \exp\left(-\frac{0.71eV}{kT}\right)$$

and:

$$\dot{\varepsilon} = A_2 \sigma^{2.5} \exp\left(-\frac{1.1eV}{kT}\right)$$

The creep mechanisms in both cases (for fine and coarse grain sizes) are based on the stress exponents and the magnitude of the activation energies, suggesting that, in the 28 nm grain-sized samples, grain-boundary diffusion is dominant. This, is based on the stress exponent which is close to unity and the low activation energy, apparently representing grain-boundary diffusion. A stress exponent of 1 may be characteristic of either NH or Coble creep, but the low activation energy favors grain-boundary diffusion, which controls creep. The 2.5 stress exponent of the CG samples, favor neither lattice nor grain-boundary diffusion. For interface reactions controlled by a GBS mechanism in very small-sized nanocrystallites, the exponent

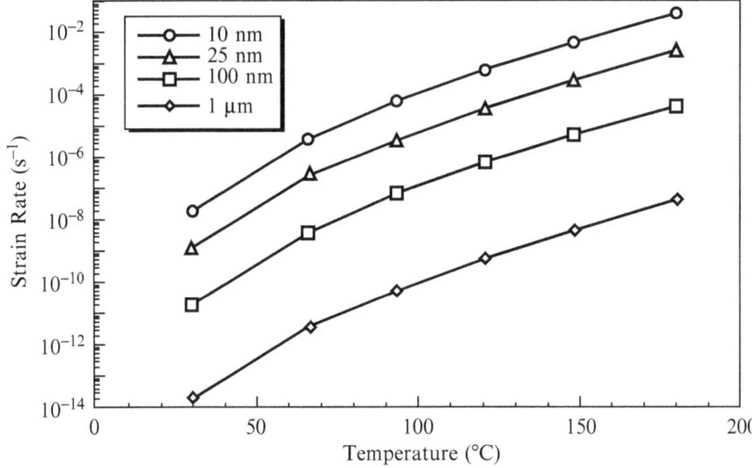

Fig. 8.76 Creep rate of Cu predicted by the equation for Coble creep as a function of temperature for various grain sizes. The creep rates are calculated for a stress of 50 MPa and are based on grain boundary diffusion coefficients for coarse-grain copper (Sanders et al. 1997. With kind permission of Elsevier)

is 2, which is not the case in the samples in Fig. 8.75b, with a grain size of 257 nm. Since dislocations can exist in samples with such a large grain size, dislocation creep is possible. The stress exponent for the dislocation-creep mechanism is in the range of 3–5, which is larger than the observed value. The stress exponent of 2.5 (being between 2 and 3) may indicate that the creep rate of the CG samples is possibly controlled by a combined mechanism, involving grain-boundary diffusion and dislocation creep, as well (Meyers et al.).

Additional information on the effect of grain size on creep may found in Fig. 8.76. Here, these curves were calculated for a stress of 50 MPa by using the values of grain-boundary diffusivity for CG Cu. At the relatively low temperature of 180°C for the grain size of 25 nm, a strain rate of $\sim 10^{-3}$/s is predicted for 50 MPa stress, but the experimental value is by orders of magnitude smaller, at 10^{-7}/s. The experimental values for all the other cases shown (Fig. 8.76) are also less than those predicted by orders of magnitude. The calculated creep curves (shown in the figure) assume grain-boundary diffusion for different grain sizes. These curves are based on the equation for Coble creep. Sanders concluded that the prevalence of low-energy grain boundaries, together with the inhibition of dislocation activity caused by small grain sizes, is responsible for low strain rates and higher than predicted or expected creep resistance. The lower than predicted creep rates in Cu (and Pd) for Coble creep at low temperatures is believed to be the result of the high fraction of low-energy grain boundaries together with the limitation of dislocation activity by the small grain sizes. At moderate temperatures, the NC-creep rates are comparable or slower than the creep rates of similar, but CG, Cu.

There are indications in the literature that diffusional creep may become significant at nanometer grain sizes, even at room temperature. However, disagreement still exists over the properties of the grain boundaries for sizes less than 1 μm and for nanometer grain sizes. As mentioned earlier, Deng et al. reported an increased creep rate (3.7 times higher) for Ni-20% P nanocrystalline 28 nm samples compared to the relatively coarser specimen (257 nm) at 280°C. (Nanocrystalline materials generally refer to the class of materials whose average grain sizes are less than 100 nm). Again, this difference was attributed to two different mechanisms (see above). However, the study of the creep properties of the pure NC metals, Pd and Cu, carried out by Nieman et al. led to the conclusion that creep is not enhanced in UFG materials at room temperature. Furthermore, there are several studies in the literature that report a decrease in creep resistance due to Coble creep, in which the creep rate is proportional to d^{-3}. Nevertheless, conflicting results indicate a creep resistance much higher than the Coble prediction. This may be due to the contamination of the grain boundaries by impurities that act as barriers to GBS.

Concluding this section on creep in small-sized specimens, it may be stated that the system behaves differently from the observations discussed in Chap. 5 only below a certain grain-size value. Only when grain sizes are in the low range of the NC materials the creep behavior was found to be different. Existing models, such as grain-boundary diffusion creep and GBS models are insufficient to describe the creep behavior. Perhaps the processing history in regard to defects such as impurities and excess free volume in alloys, changes material-diffusivity rates, leading to different creep behaviors in NC alloys. Faster creep rates, in comparison with their macroscopic-sized counterparts, are frequently reported. A faster creep rate is the result of GBS, which is dominated by grain-boundary diffusion. The diffusion-creep rate of a polycrystalline material may be enhanced by reducing its grain size. Thus, NC materials are expected to exhibit much greater creep when GBS and diffusional creep contribute. The effect of the stress exponent in the creep mechanism is an important index for visualizing the mechanism; therefore, very extensive work has been focused on evaluating the magnitude of the exponent. Stain rate, stress level and temperature are known to affect the mechanism of creep in materials having a certain grain size. Note that, while in macrosized specimens grain boundaries strengthen, weakening may occur in nanoscaled materials without the known grain-boundary hardening, because GBS combined with grain-boundary diffusion occur. Also note that the inverse H-P, beyond a critical grain size, was explained as being a consequence of diffusional creep accompanied by GBS. However, other experimental evidence shows that, for pure Cu with grain sizes as small as 10 nm, hardness still follows the classic H-P relation and the rate sensitivity value is one order of magnitude higher than that of CG Cu. The flow-stress activation volume for NC (=nm) Cu with d = 10 nm is about $8b^3$. This experimental evidence indicates that grain-boundary diffusion-related activities are much enhanced by grain refinement in the nanometer scale, though they are not yet the dominating mechanism in the plastic deformation of NC Cu samples, even with d values as small as 10 nm (Fig. 8.77). As such, the measured hardness of NC Cu with grain sizes as small as 10 nm still follows the H-P relation.

Dao et al. used Fig. 8.77, which indicates that the hardness of NC Cu with an average grain size of 10 nm may be as high as 3 GPa, corresponding to a yield

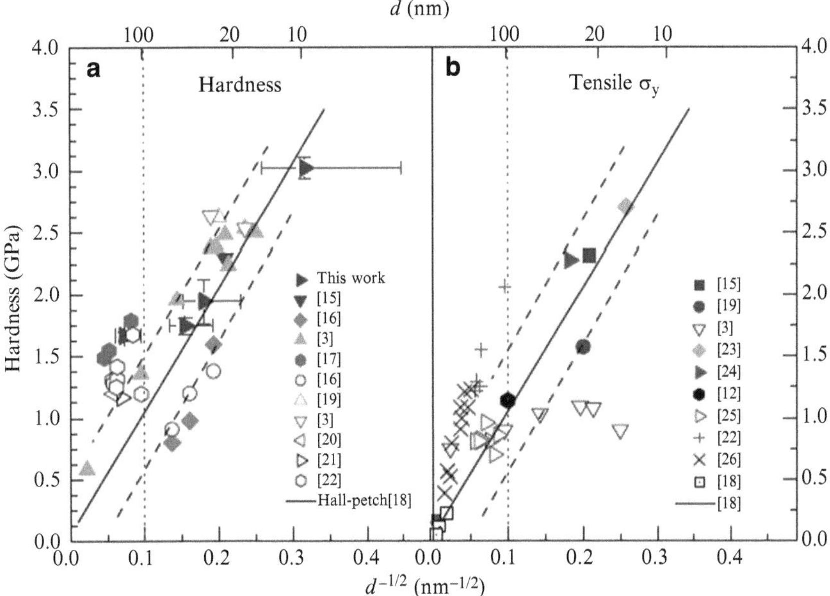

Fig. 8.77 (**a**) Variation of hardness with $d^{-1/2}$ for various Cu samples. Literature data on hardness (*solid symbols*) and yield strength (multiplied by 3) from compression tests (*empty symbols*) are also included. (**b**) Variation of tensile yield strength with $d^{-1/2}$ for various Cu samples reported in the literature. The *straight lines* represent the H–P relation extrapolated from the coarse grained (CG)-Cu (Chen et al. 2006. With kind permission of Elsevier)

strength of $\sigma_y \approx 1$ GPa, which is more than one order of magnitude higher than that of CG Cu ($\sigma_y \approx 50$ MPa). Furthermore, Fig. 8.78 presents the variation in the strain sensitivity, m as a function of d for NC Cu. Here, an increase in m values with a decreasing d from the macro to the submicron scale is seen. The m value increases from about 0.02 at d = 100 nm to 0.06 at d = 10 nm. Such a transition in the m-d variation tendency may imply a change in the plastic-deformation mechanism. Although the m value of Cu is enhanced by one order of magnitude when d is reduced to 10 nm, it is still much smaller than expected for a plastic-deformation process controlled by GBS or Coble creep. In addition, the strain rate of GBS in polycrystalline materials is related to d by a Lüthy relation:

$$\dot{\varepsilon}_{GBS} = 2 \times 10^5 \frac{D_{gb}Gb}{kT}\left(\frac{b}{d}\right)^3\left(\frac{\sigma_e}{G}\right)^2 \tag{8.24}$$

By using parameters taken from the literature, the estimated strain rate of $\dot{\varepsilon}_{GBS} \sim 10^{-5}\text{s}^{-1}$ at room temperature, even at d = 10 nm, is much smaller than the measured rate of 10^{-3} s^{-1}. This means that the observed plastic deformation in the NC Cu sample is unlikely to be dominated by GBS. Similarly, the grain-boundary diffusion-related Coble-creep rate may be estimated as $\sim 10^{-7}\text{s}^{-1}$ at RT for Cu with

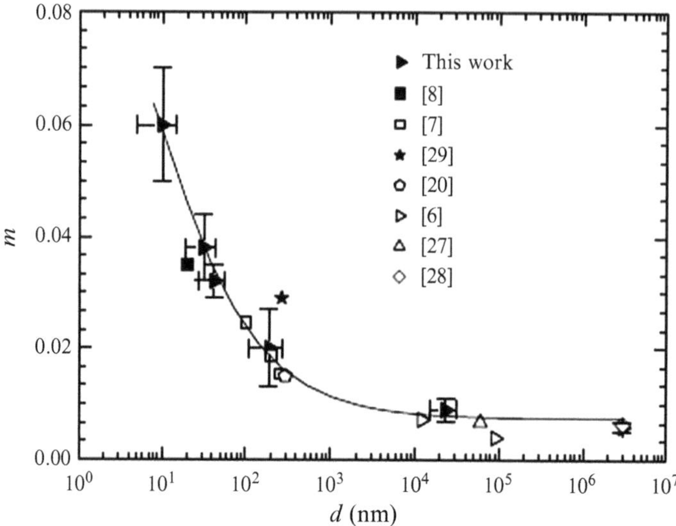

Fig. 8.78 The variation of m as a function of *d* for Cu using experimental data from the literature and from the present work (Chen et al. 2006. With kind permission of Elsevier)

d = 10 nm, which is about 3–4 orders of magnitude below the measured rate. Hence, Coble creep can also be ruled out; it does not dominate plastic deformation in NC Cu. These analyses prove that the plastic deformation for Cu, with d as small as 10 nm at RT, is not yet dominated by grain-boundary diffusion or Coble creep.

Recent developments on the path to obtaining a quantitative understanding of NC mechanical behavior consider the role of partial/perfect-dislocations or deformation-twin mediated processes as they interact with grain boundaries. The crystal plasticity of pure Cu indicates the role of nanoscale twins in the strengthening mechanism. Taking this approach, it is possible to extend classic dislocation theories about grain-boundary hardening to grain sizes on the order of 5–30 nm, but below this grain size, it is uncertain whether these models are valid. Improved strength is also found when many twins are present, which may be introduced into materials with low SF energy. Partial dislocations are involved in SF generation. Twins can lead to strengthening independently of grain size, with strengthening being determined by twin spacing, instead of grain-boundary spacing. Despite recent efforts devoted to studying the creep and superplastic behavior of NC materials, the data obtained by different investigators are either inconsistent or clearly contradictory, as mentioned above.

There are two possible reasons for the differences reported for creep in nm-sized materials. Firstly, mechanical properties in NC materials are very sensitive to their initial microstructures. This sensitivity is the result of the non-equilibrium nature of NC materials. Due to this nature, two grades of NC material (same composition and structure), prepared by different procedures but having a similar grain size, may exhibit different mechanical properties. Secondly, it is very difficult to produce NC materials free of porosity and other flaws. Because of this difficulty, the dependence of strain rate on stress, grain size and temperature in NC materials is, at present, still unclear. The reader may get more details on the strengthening of NC materials by twinning and dislocations/partial dislocations from the research of Dao et al. and Morris.

Alloying to produce solid solutions, complex-crystal structures or mixed-phase microstructures offers a possibility of improving the stability of nanostructures while also improving overall mechanical behavior.

8.5 Fatigue Behavior

8.5.1 Introduction

Previously in this chapter, when dealing with the mechanical behavior of small-sized systems in the micrometer-nanometer range, it was indicated that the strength properties are different than when in their macroscale dimensions. The results of creep from the various tests showed contrary observations and it was also indicated that creep-resistant behavior is size-dependent, even on the small scale. Ultrafine-grained structures (in the micron and submicron ranges) appeared to behave similarly during creep as their macroscale counterparts and the difference was outstanding only in the very low nanoscale size. Since micro-sized specimens became of technological interest for various micro- and nanosystems, such as for micro-electro-mechanical systems [henceforth: MEMS], serving as interconnects and small functional or structural components in microelectronics, micromachines or in biomedical applications, e.g., implants in biological/bio-implantable materials (Ritchie et al.), the understanding of the fatigue behavior of such components has become of special interest. This will be discussed below.

Various fabrication methods exist for obtaining small-sized devices. Connolley et al. categorized these methods into three types: (i) those in which the component is built up layer by layer (e.g., in photolithography, LIGA, i.e., Lithographie, Galvanoformung und Abformung); (ii) those in which the component is machined by removing material from a solid piece (e.g., in electrostatic discharge machining, laser cutting, electropolishing, FIB machining); and (iii) those in which the component is fabricated by manipulating a certain form, such as wire or foil. It is important to understand the fabrication process, because it may influence the mechanical behavior of the finished component (see Connolley et al.).

8.5.2 Fatigue in Micron-/Submicron-Sized Materials

Silicon-based MEMS material is of special interest for integrated circuits and there is the advantage of having available fabrication techniques in the microelectronics

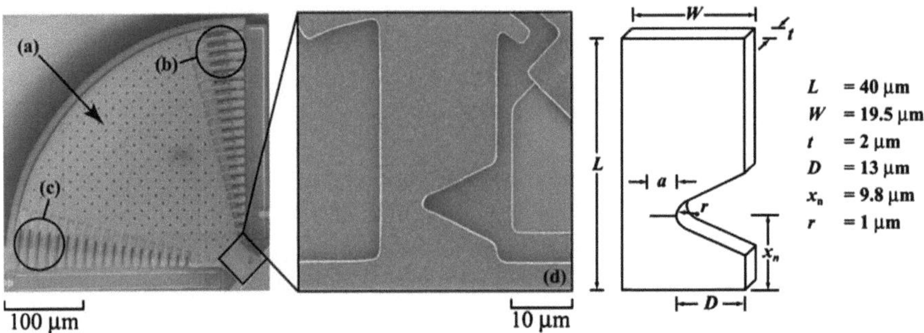

Fig. 8.79 SEM of the micron-scale fatigue characterization structure containing a notched cantilever-beam specimen. The (*a*) mass, (*b*) comb drive actuator, (*c*) capacitive displacement sensor, and (*d*) notched cantilever-beam specimen are shown. The nominal dimensions of the specimen are as indicated in the schematic (Muhlstein et al. 2004. With kind permission of Elsevier)

industry. The long-term performance of such systems is of major concern, since premature-fatigue failure must be minimized and, therefore, accumulated knowledge and experimental data are valuable. The micromechanical, resonant-fatigue characterization structure has been widely used to evaluate the fatigue behavior of thin-film structures of single crystal and polycrystalline Si for MEMS applications, as described in several publications (for example, Muhlstein et al. or Van Arsdell and Brown). The structure shown in Fig. 8.79 was originally developed by Van Arsdell and used to evaluate fatigue-crack growth behavior in Si thin-films under resonant loading conditions. This structure is square in overall shape (~300 μm on a side) and is analogous to a specimen tested by means of an electromechanical load frame and capacitive-displacement transducer, as found in conventional, mechanical testing systems. The specimen is a notched-cantilever beam, attached to a large, perforated plate that serves as a resonant mass. The mass and the beam are electrostatically forced to resonate and the resulting motion is measured capacitively. On opposite sides of this resonant mass are interdigitated 'fingers', commonly referred to as 'comb drives'. One side of these drives is for electrostatic actuation; the other side provides capacitive motion sensing. The resonance generates fully-reversed, constant-amplitude, sinusoidal stresses at the notch, i.e., a load ratio, R (the ratio of minimum-to-maximum load) of -1. The structure is designed for testing at ~40 kHz and is excited at a fixed frequency, updated periodically during the test. For further details on this structure, the reader is referred to the original work of Van Arsdell and the other cited references.

An S-N plot is obtained by the resonance method as shown in Fig. 8.80. The relation between angular rotation and maximum principal stress, determined using the finite-element model, was used to calculate the applied-stress amplitude, ranging from ~2.2 to 4.4 GPa. A similar technique was also used for single crystals (see, for example, Baumert et al.). An example of the S-N curve for single crystals is shown in Fig. 8.81, tested at a relative humidity [henceforth: RH] of 50%. The reason for

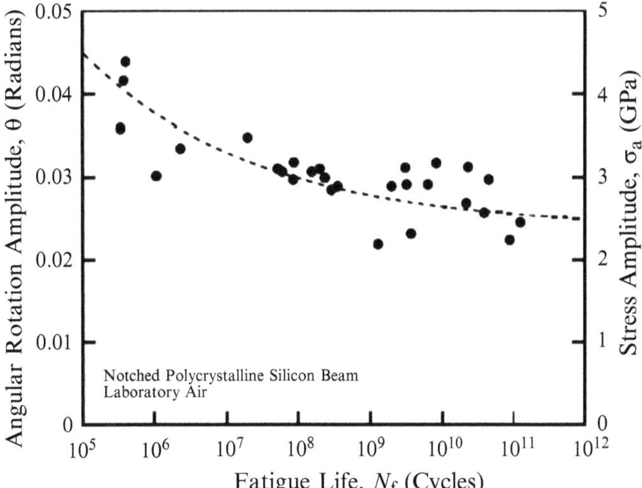

Fig. 8.80 Typical stress-life (S-N) fatigue behavior of ~2 μm thick, polycrystalline silicon obtained using the micron-scale fatigue characterization structure under fully reversed, tension – compression loading at 40 kHz in moist room air (Muhlstein et al. 2004. With kind permission of Elsevier)

testing in a humid atmosphere is an outcome of the experimental findings stated by Baumert et al.: "From the many experiments performed on Si, it is believed that Si thin films do not fatigue in ultra-high vacuum, while they do in the presence of oxygen and/or water." Fatigue lives are much longer in medium vacuum than in laboratory air and are significantly affected by the partial pressure of water in air. A different technique for evaluating fatigue in the LCF of single-crystal silicon thin-films is described by Liu et al.

An example of the ultrafine-grain Ti, produced by severe plastic deformation via ECAP technique is displayed in Fig. 8.82. Pure Ti is important in medical applications, owing to its low weight, excellent corrosion resistance and high biocompatibility. Note that the UFG Ti produced by ECAP exhibits both an increased fatigue-life limit and increased fatigue life under constant-load testing compared with those in the CG state (i.e., grains with 9 or 32 μm), when plotted on the S-N diagram (Wöhler diagram). The UFG Ti has a size of 0.3 μm, while the CG reference material has 25 μm (Fig. 8.82). The fatigue limits are, respectively, 380 + 5 MPa and 238 MPa for the UFG and the 9 μm samples. Figures 8.83 (after ECAP) and 8.84 show these microstructures before and after fatigue. The SAED pattern seen in Fig. 8.83 was recorded from a region having a 3 μm diameter. The SAED pattern consists of rings of diffraction spots, demonstrating that the grain boundaries have high angles of miss-orientation. The average grain size, 300 nm, was evaluated from the TEM micrographs. Nevertheless, numerous low-angle grain boundaries are also frequently observed, as evidenced by the dark-field image.

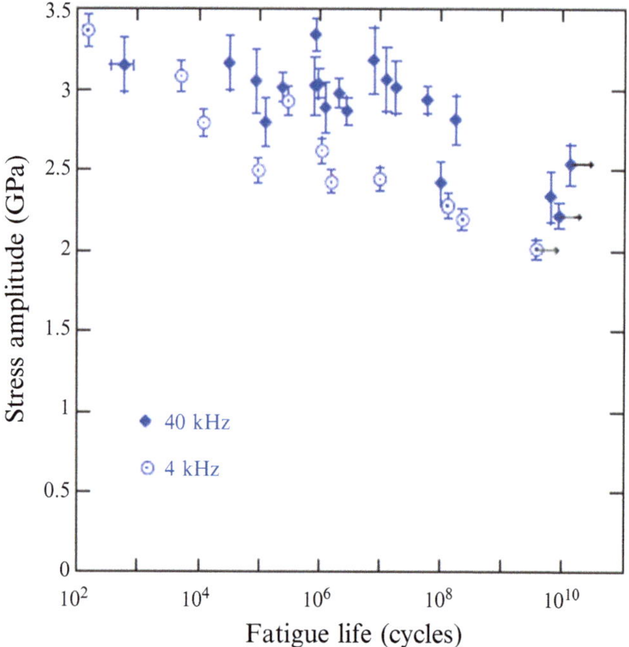

Fig. 8.81 S–N curves at 30°C, 50% RH, for both 4 and 40 kHz resonators. The *arrows* indicate run out tests (Baumert et al. 2010. With kind permission of Elsevier)

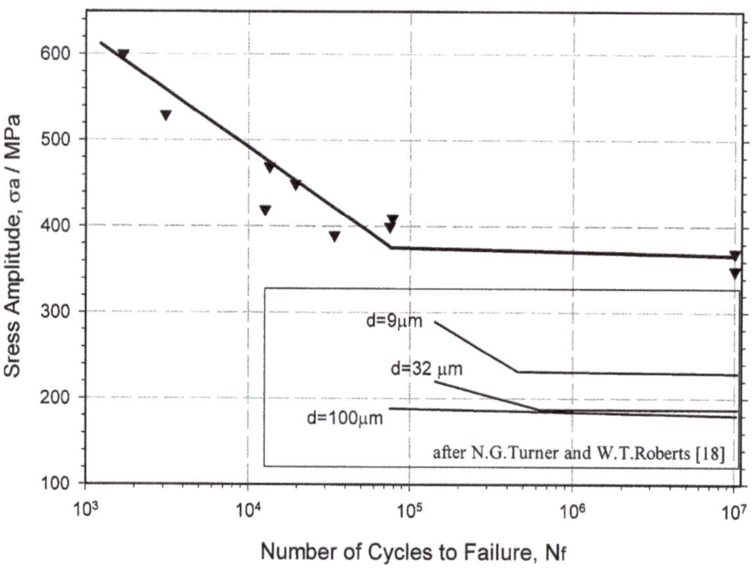

Fig. 8.82 S–N diagram for ECAP Ti (Vinogradov et al. 2001. With kind permission of Elsevier)

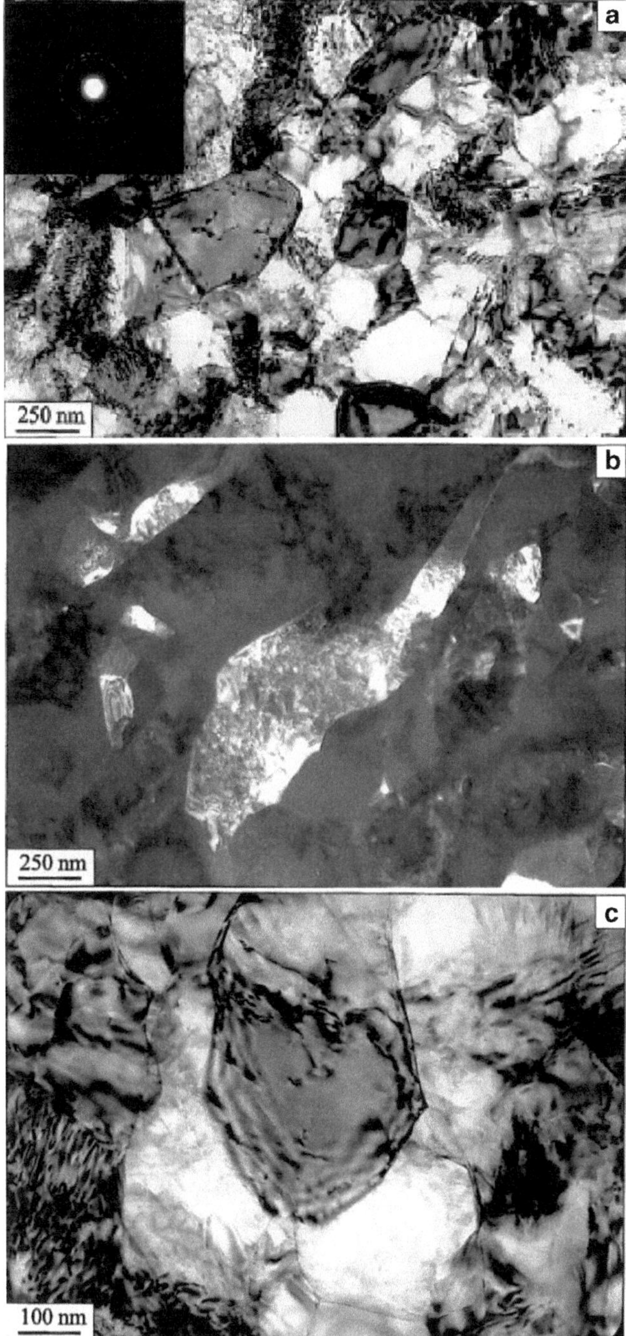

Fig. 8.83 Mictostructure of ECA-processed Titanium. (**a**) Light-field image TEM photo and selected area electron diffraction pattern; (**b**) dark-field image; (**c**) individual grain shape (Vinogradov et al. 2001. With kind permission of Elsevier)

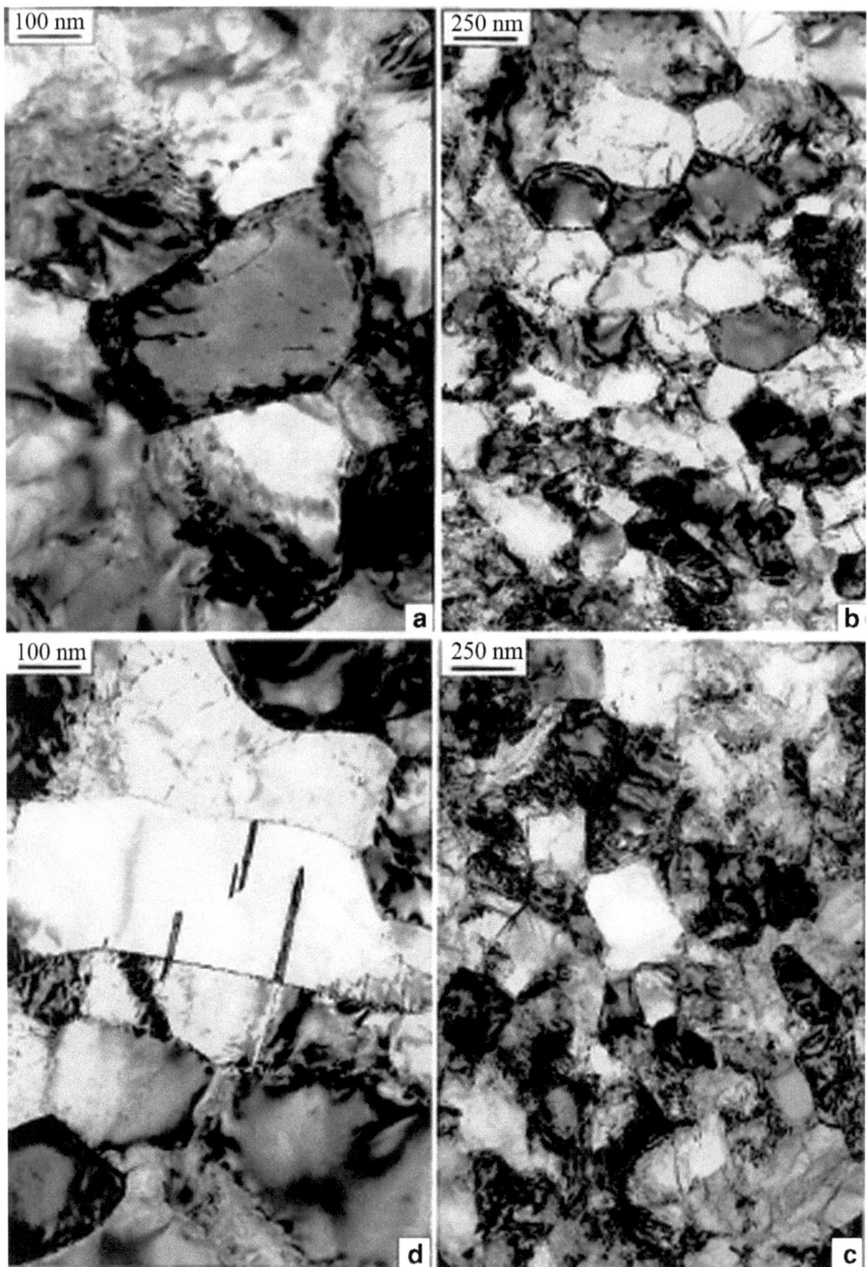

Fig. 8.84 Structure of ECAP Ti after fatigue under constant plastic strain amplitudes to failure: (**a** and **b**) $\varepsilon_{pl} = 5 \times 10^{-3}$; (**c** and **d**) $\varepsilon_{pl} = 7.5 \times 10^{-4}$ (Vinogradov et al. 2001. With kind permission of Elsevier)

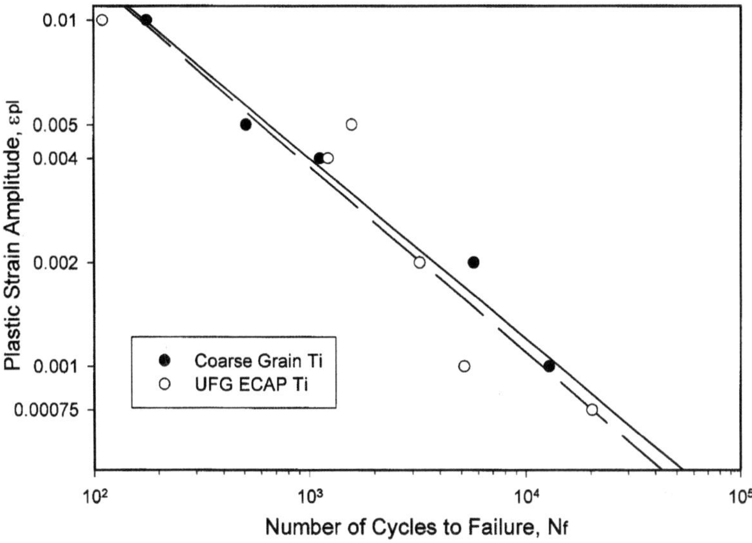

Fig. 8.85 Coffin–Manson plot for ECAP UFG Ti and its conventional coarse-grain counterpart (Vinogradov et al. 2001. With kind permission of Elsevier)

The estimated, average-dislocation density is high and close to the maximum resolvable values of 10^{13}–10^{14} m^{-2}. No gross changes in the character of dislocation distribution are evident when compared to different grain sizes. The TEM does not show significant differences in the fine structure before and after fatigue.

Figure 8.82 is clearly based on Eq. 6.8 from Chap. 6, given as:

$$\sigma_a = \sigma'_f (2N_f)^b \tag{6.8}$$

Titanium does not demonstrate any reduction in its fatigue performance under constant, plastic-strain cyclic testing as is evidenced by the nearly identical Coffin-Manson behavior in fine- and CG states (see Fig. 8.85). These curves are based on the Manson-Coffin equation (see Chap. 6) for the plastic component:

$$\frac{\Delta\varepsilon_p}{2} = \varepsilon'_f (2N_f)^c \tag{6.19}$$

which is a part of the total strain in Eq. 6.7 shown below:

$$\varepsilon = \varepsilon_a = \varepsilon_e + \varepsilon_p = \frac{\sigma}{E} + \varepsilon_p = \frac{\sigma'_f (2N_f)^b}{E} + \varepsilon'_f (2N_f)^c \tag{6.7}$$

No cyclic softening or degradation were noticed during strain-controlled experiments on Ti, contrary to materials with wavy slip, such as Cu subjected to ECAP.

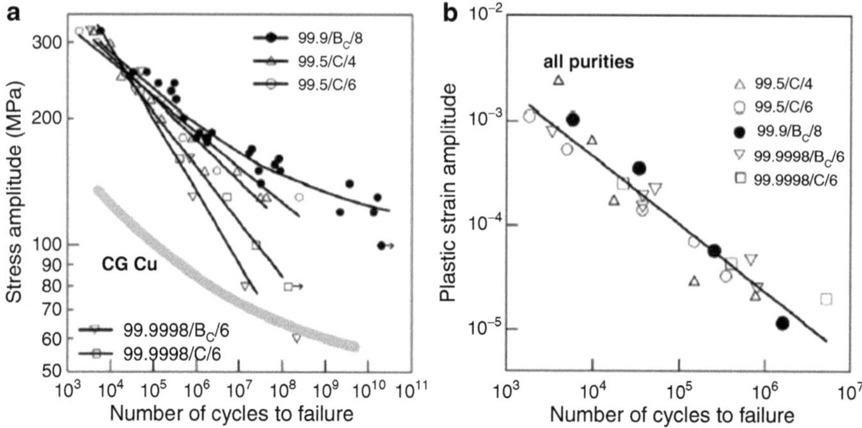

Fig. 8.86 Fatigue life curves of UFG copper of different purities and of CG copper. (**a**) Wöhler (S–N) representation, showing strong dependence of fatigue life of UFG copper on purity, in particular in the HCF and UHCF (ultra high cyclic fatigue) ranges. (**b**) Coffin-Manson representation, almost collapse of data of UFG copper of different purities. Specimen designations refer to purity/route/number of ECAP passes (Mughrabi and Höppel 2010. With kind permission of Elsevier)

Unlike copper, with its many symmetric slip modes, Ti tries to slip only on prismatic planes, unless the grain is very unfavorably oriented. This may help to explain why Ti is less susceptible to softening, since its substructure (developed during ECAP) cannot be relaxed by cross-slip. Recall that, in the FCC Cu, there are 12 slip systems, compared to only 3 slip systems found in Ti. Thus, in Ti dislocation, hardening occurs during fatigue.

Grain refinement is known as an effective means for enhancing the strength of crystalline materials. The H-P relation, discussed in this chapter and in one of the earlier chapters, is an indication of the grain-size-dependence of strength. In general, the fatigue strength of UFG materials is considerably enhanced by grain refinement, in particular, in the HCF regime. Numerous examples may be listed on this grain-size effect, though only some are presented below. Ultrafine-grained Cu is compared with CG Cu in Fig. 8.85. In Fig. 8.86a, the plot follows Eq. 6.18, while Fig. 8.86b gives the Coffin-Manson relation, expressed graphically, following Eq. 6.19 (Fig. 8.86).

Relation (6.18) was given earlier in the chapter on fatigue rewritten below

$$\sigma_a = \frac{\Delta\sigma}{2} = \frac{\Delta(E\varepsilon_e)}{2} = \frac{E\Delta\varepsilon_e}{2} = \sigma_f'(2N_f)^b \tag{6.18}$$

In Fig. 8.87, examples of surface shear bands are seen, either as patches (Fig. 8.87a) or in more extended form (Fig. 8.87b). An example of very pronounced, fatigue-induced grain coarsening, that occurs in some areas due to dynamic

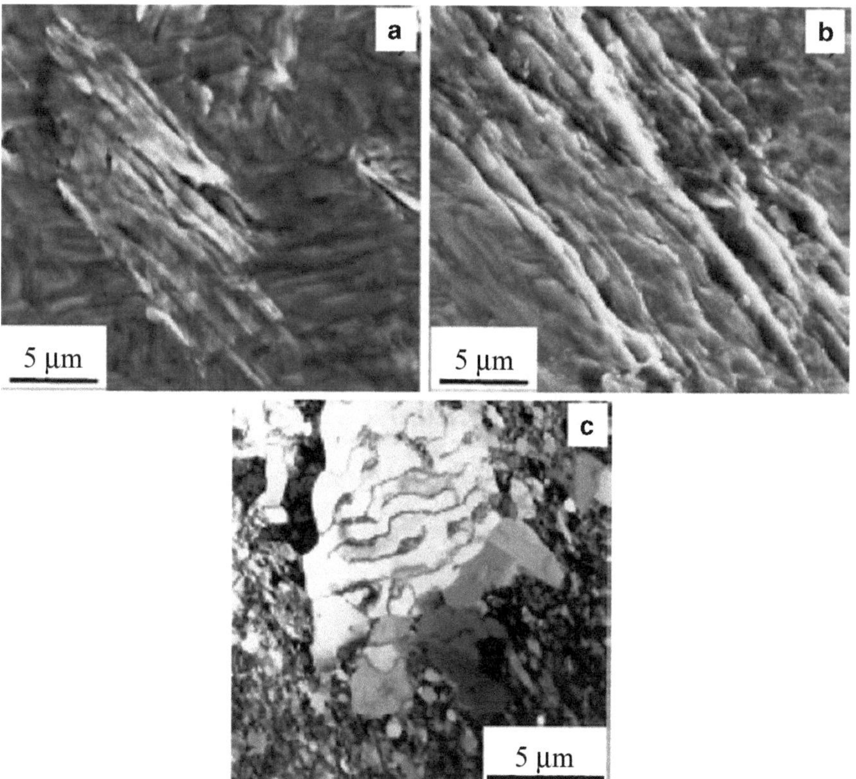

Fig. 8.87 Cyclic strain localization and shear banding in UFG copper fatigued at RT. Stress axis horizontal. Surface observations. (**a**) Shear bands in "patches". (**b**) Extended shear bands. (**c**) TEM observation of locally coarsened grain/subgrain microstructure (Mughrabi and Höppel 2010. With kind permission of Elsevier)

recrystallization at the low homologous temperature of ~0.2 T_m, during strain-controlled fatigue, is seen, while, in other areas, the original UFG microstructure was retained (Fig. 8.87c).

Increased strength during ECAP grain refinement, results in the reduction of ductility, with the consequent lowering of the LCF resistance of UFG materials. In the case of UFG copper, a low-temperature annealing treatment (~2 h at170°C) improves ductility at the expense of some strength loss. ECAP grain refinement has been shown to improve the LCF fatigue resistance. A bimodal grain structure develops (as indicated in Fig. 8.88a). After fatigue, typical dislocation patterns/walls are found inside the larger grains (Fig. 8.88b) as a consequence of the annealing. Such dislocation patterns are typical of fatigued CG materials.

The effect of the fabrication process for obtaining fine-grained structures may be seen in Fig. 8.89 for an Al alloy, i.e., alloy 2124. The role of fine grain size on the behavior of materials under strain-controlled conditions is clear. In the total

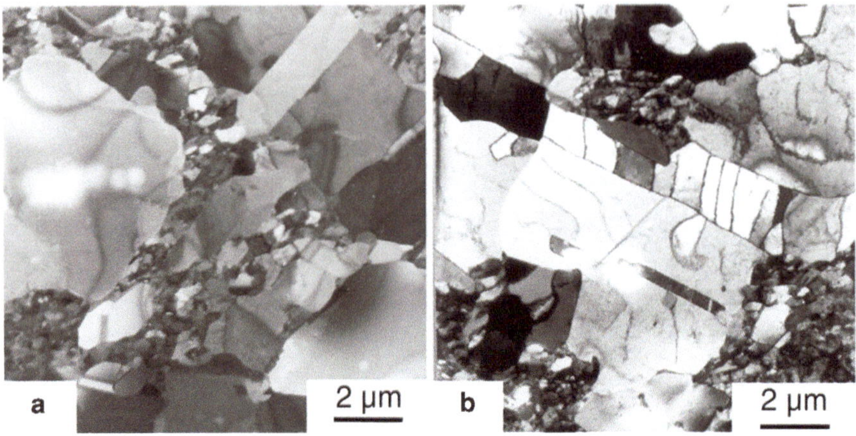

Fig. 8.88 Fatigue of high-purity UFG copper containing a bimodal grain structure, obtained by annealing for 2 h at 170°C. (**a**) Prior to fatigue. (**b**) After RT – fatigue at $\Delta\varepsilon_{pl}/2 = 10^{-3}$. The fatigue-induced dislocation substructure inside the larger grains should be noted (Mughrabi and Höppel 2010. With kind permission of Elsevier)

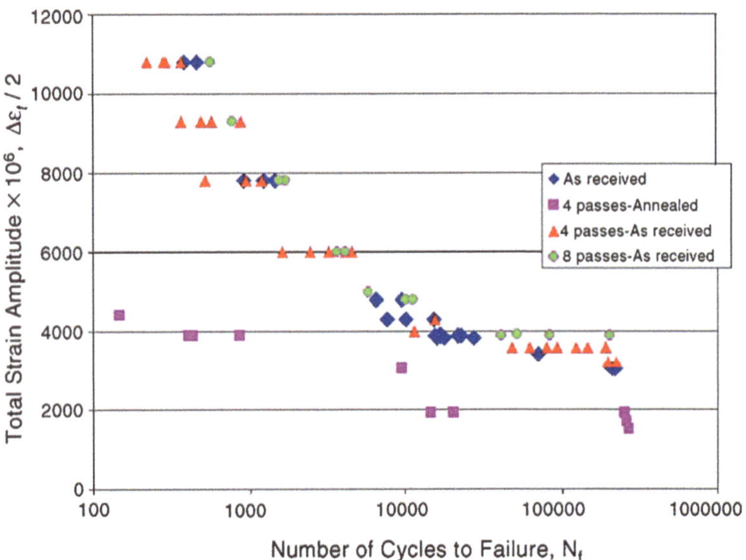

Fig. 8.89 Strain-life fatigue results (Lapovok et al. 2006. With kind permission of Elsevier)

strain-fatigue life diagram shown in Fig. 8.89, LCF and HCF are incorporated. The as-received CG alloy is compared with ECAP 8-pass processed results, which exhibits higher resistance over the whole range of fatigue lives investigated. No cracks have developed in these samples after the maximum number of passes (eight passes) with an applied back pressure of 150–75 MPa.

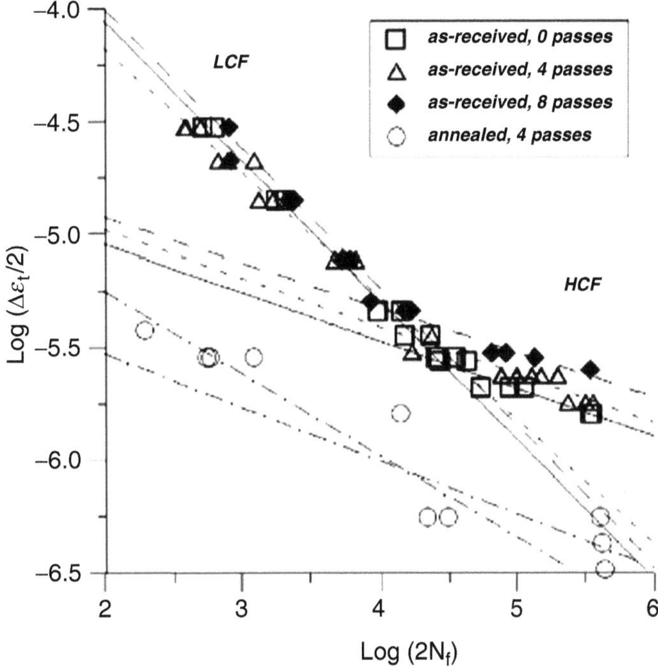

Fig. 8.90 Fit of the fatigue data by Eq. 6.7b (Lapovok et al. 2006. With kind permission of Elsevier)

A significant drop in LCF life range is observed for the 4-pass annealed samples compared to the as-received and the as-received 4-pass ECAP samples, respectively. The considerable improvement in the fatigue properties of the 8-pass ECAP samples is probably the result of the use of back pressure during ECAP. As a result, there is an increase in the fatigue-strength coefficient, σ'_f, due to the formation of an UFG structure by severe plastic deformation. The experimental data were fitted to Eq. 6.7b and the results are presented in Fig. 8.90. These plots are on a log-log scale of the total-strain amplitude, ε_t, vs. the number of strain reversals, $2N_f$.

$$\frac{\Delta\varepsilon}{2} = \frac{\Delta\varepsilon_e}{2} + \frac{\Delta\varepsilon_p}{2} = \frac{\sigma}{E} + \frac{\Delta\varepsilon_p}{2} = \frac{\sigma'_f\left(2N_f\right)^b}{E} + \varepsilon'_f\left(2N_f\right)^c \tag{6.7b}$$

Based on these plots, various quantities may be established, i.e., the fatigue-ductility coefficient, ε'_f, the fatigue-ductility exponent, c (by the Coffin-Manson law), the ratio of the fatigue-strength coefficient to Young's modulus, σ'_f/E, and the fatigue-strength exponent, b (in Basquin's law).

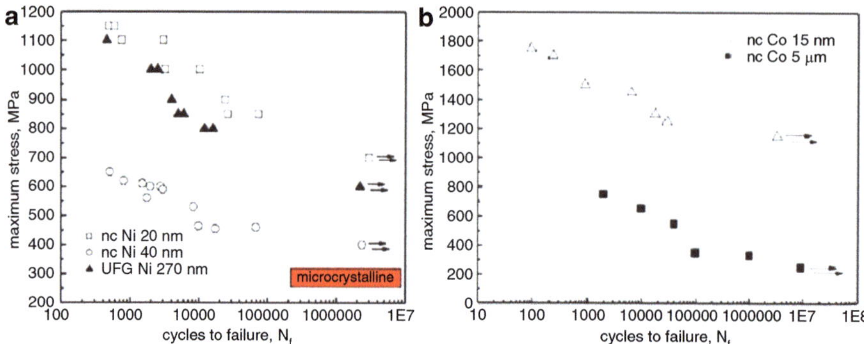

Fig. 8.91 S–N behavior (**a**) of pure *Ni* in the *nc* and *UFG* regimes, (**b**) of pure electrodeposited *Co* in the *nc* and *mc* regimes (Cavaliere 2009. With kind permission of Elsevier)

8.5.3 The Fatigue of Nanocrystalline (NC) Materials

Pure, small grain-sized Ni has been under extensive investigation for various grain sizes. As indicated earlier in this chapter, it is generally observed that strength values increase when decreasing the mean grain size to below 100 nm. The decrease in strength, with decreased grain size to below ~20–10 nm mean grain size (as observed in inverse H-P behavior), indicates a change in the deformation mechanism. The breakdown of the H-P relation is believed to indicate a transition in the deformation modes of metals, as the grain size in the NC range decreases to very low levels. Recent investigation suggests that dislocation-accommodated boundary sliding is the main deformation process governing the whole deformation in NC metals below ~10 nm. Small-sized grains lead to the suppression of early-crack nucleation and propagation, associated with reduced flaw size and the difficulty of existing stress concentration at the tip to exceed the natural strength of the material. It is usually acknowledged that a deformation mechanism at the 100–600 nm range occurs in the accepted manner (as in the bulk materials), while in the range of ~50–100 nm, grain-size dislocations form and are both emitted and annihilated at grain boundaries. In the ~10–50 nm range, partial dislocations and twinning are the dominant deformation features. Below ~10 nm, GBS is the major mechanism for deformation. The interest in NC materials is not only a consequence of understanding the mechanisms resisting deformation damage, but also due to their potential use in modern industries. In particular, fatigue properties are of interest under cyclic loading, since it has been confirmed that the fatigue limit of NC metals increases with decreasing grain size and crack-initiation susceptibility decreases. In Fig. 8.91, the S-N behavior of Ni and Co is shown. Here, the effect of grain refinement on the S-N plots indicates a significant improvement in the endurance limit over that of pure microcrystalline pure (MC) Ni.

Nanocrystalline Ni (20 nm) has a slightly higher tensile-stress range compared to the UFG crystalline Ni. Both NC and UFG conditions have a significantly higher

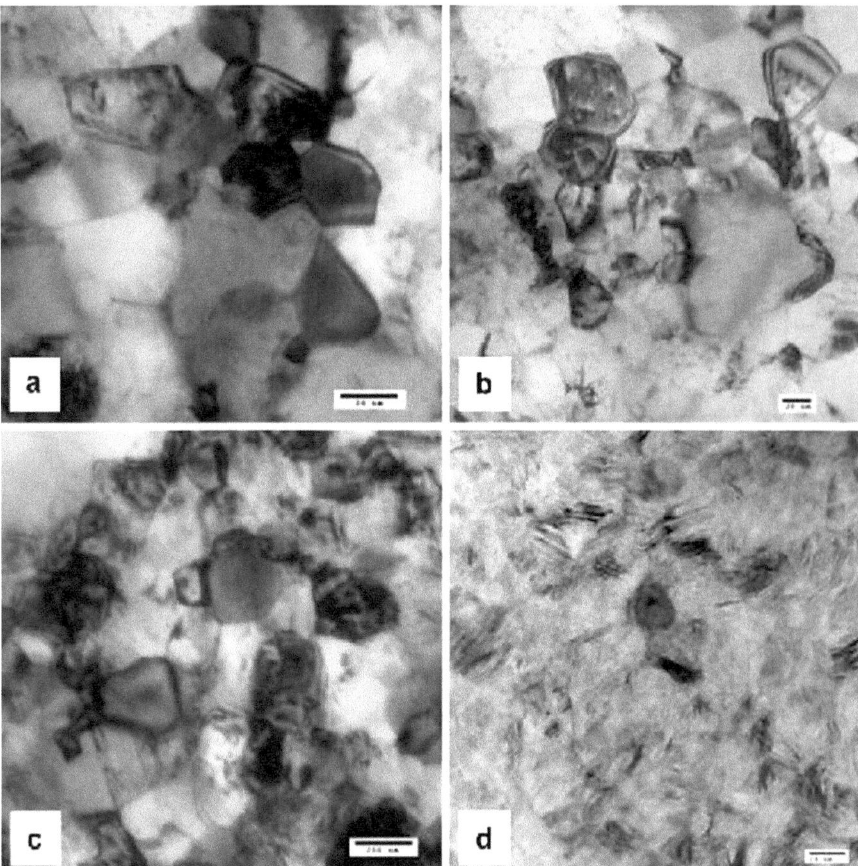

Fig. 8.92 TEM micrographs showing the grain size and distribution of the electrodeposited nanocrystalline Ni 20 nm (**a**) and 40 nm (**b**), for comparison the microstructure of the ultra-fine grain electrodeposited Ni is reported (**c**), TEM image showing the microstructure of the electrodeposited nanocrystalline cobalt, the mean grain size was in the range 15–20 nm; and a microstructure characterized by an high density of stacking faults and nano-twins (**d**) (Cavaliere 2009. With kind permission of Elsevier)

fatigue-endurance limit than the MC metal (marked in red). Usually, the ductility of materials is reduced as grain size decreases and this phenomenon is accompanied by a reduction of plasticity, as was evidenced by the strong reduction in necking. It has been demonstrated that the grain-size reduction in the UFG and NC regimes leads to an increase in the fatigue limit of materials. The S-N curve of Co (shown in Fig. 8.91b) also indicates that the NC material has a much higher endurance limit than that of the MC sample. The TEM micrographs of the electrodeposited Ni (presented in Fig. 8.92) show mean grain-size dimensions of 20 and 40 nm for the NC Ni, but 270 nm for the UFG material. The TEM micrographs of the electrodeposited, pure Co show mean grain-size dimensions of 10–20 nm (Fig. 8.92d).

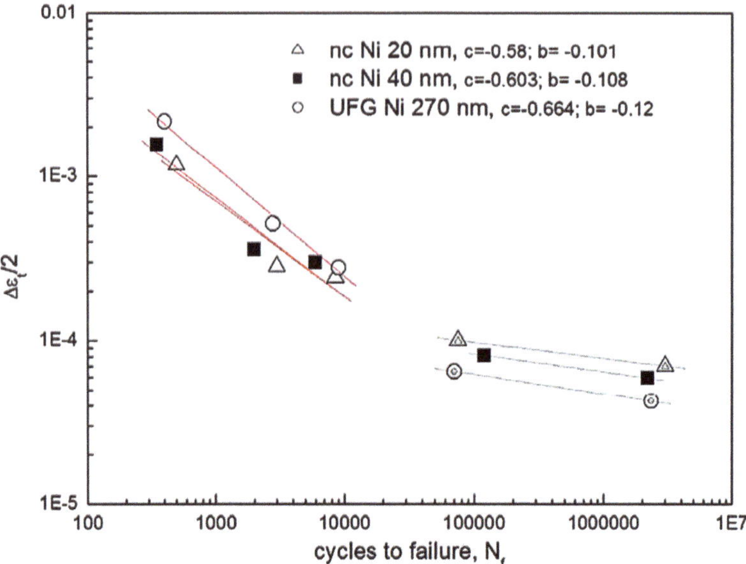

Fig. 8.93 Total strain range as a function of cycles to failure for the electrodeposited *nc* and *UFG Ni* (Cavaliere 2009. With kind permission of Elsevier)

The strain amplitude, as a function of the number of cycles, is expressed as the Coffin-Manson plot according to Eq. 6.7b and is presented in Fig. 8.93. The Paris law is given as:

$$\frac{da}{dN} = C(\Delta K)^m \tag{6.51}$$

and expresses crack growth in Ni and Co.

The mean grain size of a material has a strong influence on crack initiation and growth. The fatigue-crack rate, as a function of ΔK at a load ratio of R = 0.25, is seen in Fig. 8.94. For all the investigated stress levels, this material is less sensitive to crack initiation as the mean grain size decreases. However, the resistance to crack growth decreases with grain refinement.

It is evident that the resistance to fatigue-crack growth is substantially lower in NC Ni 20 nm, relative to UFG Ni, and an intermediate behavior is found in NC Ni 40 nm. This same behavior pattern has also been observed in the case of Co.

Additional experiments on pure Ni (shown in Fig. 8.95) confirm the results presented in Fig. 8.91. In Fig. 8.95, the effect of grain refinement on the S-N plots indicates a significant improvement in the endurance limit compared with the pure MC Ni. Nanocrystalline Ni has a slightly higher tensile stress range compared to UFG crystalline Ni (~300 nm).

Figure 8.96 shows the variation in crack length as a function of the number of fatigue cycles for NC, UFG, and MC Ni, where constant cyclic load experiments

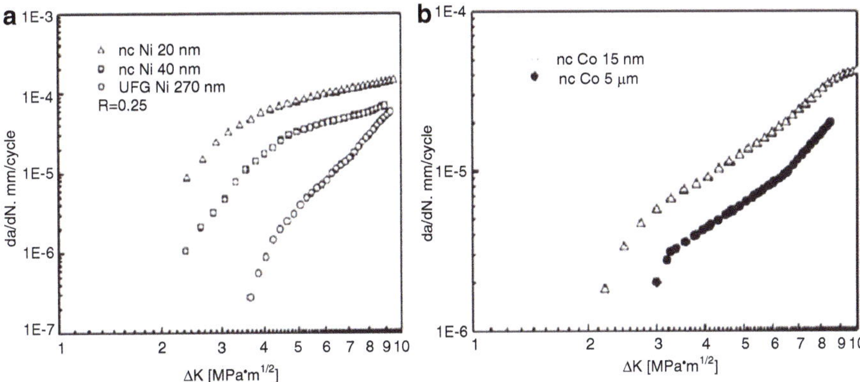

Fig. 8.94 (**a**) Crack growth rate as a function of ΔK for the electrodeposited pure Ni in the nc and UFG regimes and (**b**) fatigue crack growth behavior for electrodeposited pure Co with nanocrystalline and microcrystalline grain size (Cavaliere 2009. With kind permission of Elsevier)

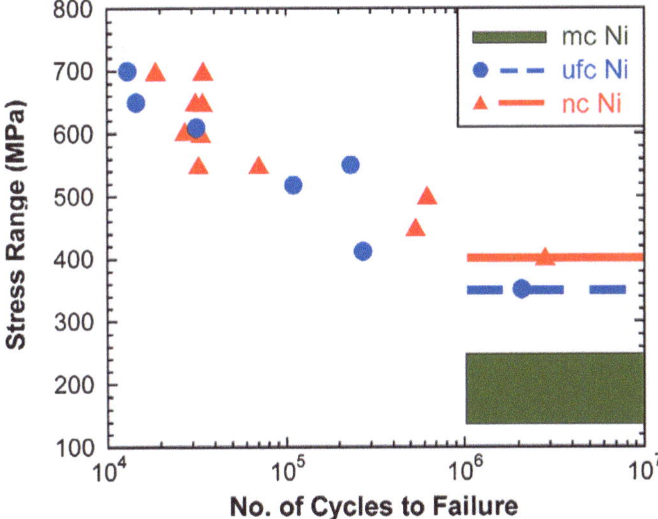

Fig. 8.95 S-N fatigue response of electrodeposited, fully dense nanocrystalline Ni (grain size 20–40 nm) compared to that of a similarly produced ultrafine crystalline Ni (grain size approximately 300 nm) at $R = 0$ and frequency of 1 Hz in laboratory air environment. Also sown for comparison are values of the range of endurance limit for microcrystalline pure Ni (from the literature) (Kumar et al. 2003. With kind permission of Elsevier)

were conducted with an initial stress-intensity factor range of 11.5 MPa $m^{1/2}$. It is clearly seen that the fatigue-crack length increases at a much faster rate with fatigue cycling in the NC Ni than in the UFG crystalline or MC Ni under identical loading conditions.

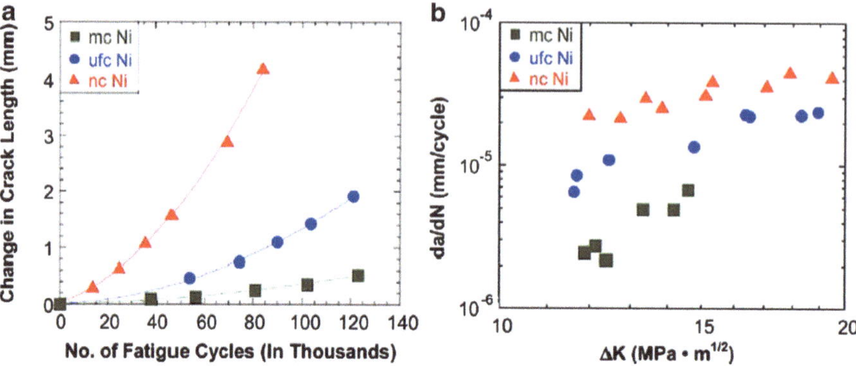

Fig. 8.96 A comparison of the fatigue crack growth of pure Ni as a function of grain size. (**a**) Increase in crack length with fatigue cycles for the nc, ufc and mc Ni at a common initial $\Delta K = 11.5$ MPa m$^{1/2}$, at $R = 0.3$ and n = 10 Hz. (**b**) Constant amplitude fatigue crack growth response under the same test conditions after Hanlon et al. (Kumar et al. 2003. With kind permission of Elsevier)

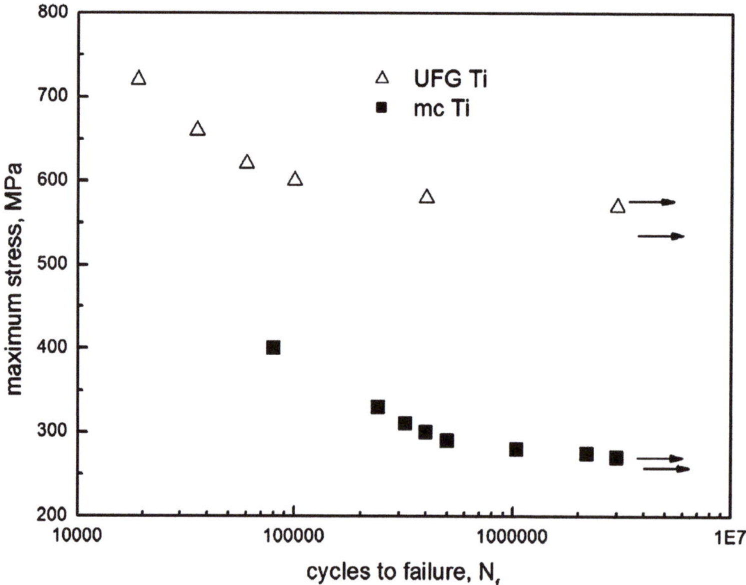

Fig. 8.97 S–N curves of the two studied Ti revealing the increase in fatigue behavior due to the severe plastic deformation and consequent grain refinement (Cavaliere 2009. With kind permission of Elsevier)

This variation in fatigue-crack growth rate (FCGR), da/dN with ΔK in the Paris regime, is plotted in Fig. 8.96b, showing that fatigue-crack growth occurs up to four times faster in NC Ni than in UFG crystalline Ni. Among these three materials, the MC Ni showed the slowest FCGR at the given ΔK, where the crack-growth rate was several times smaller than that of UFG crystalline Ni.

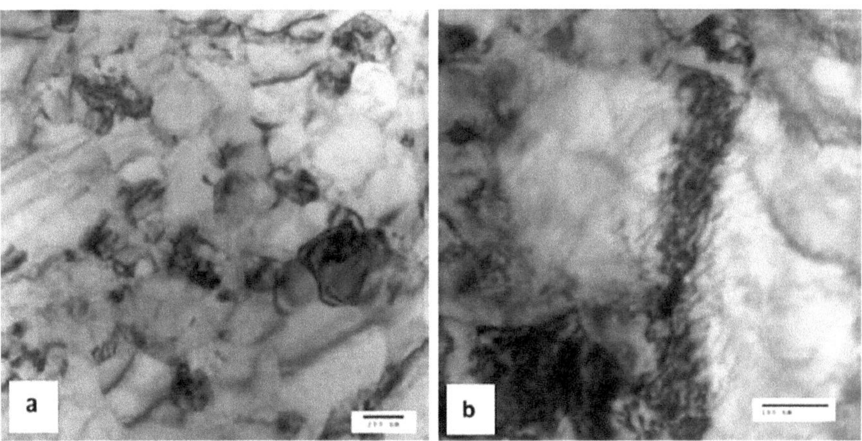

Fig. 8.98 TEM micrograph of the pure ECAP Ti (**a**), TEM micrograph of ECAP Ti after fatigue loading (600 MPa maximum stress) (**b**) (Cavaliere 2009. With kind permission of Elsevier)

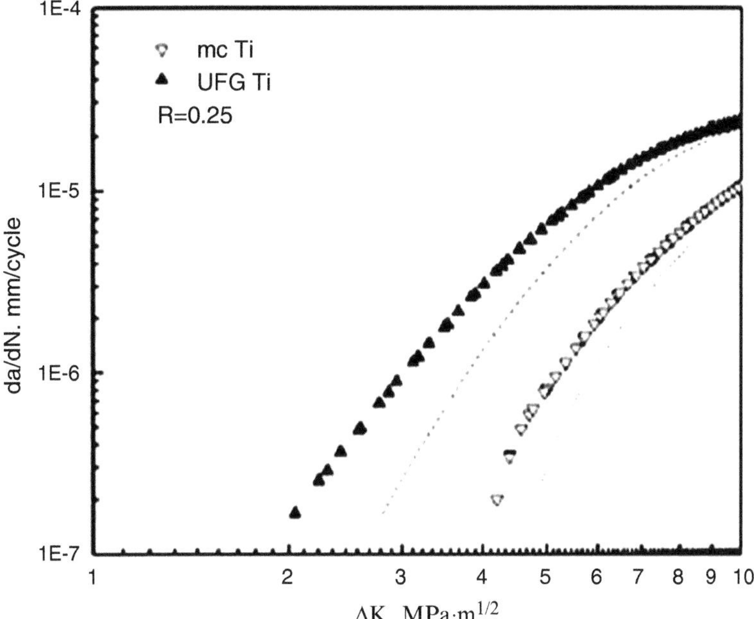

Fig. 8.99 Crack growth behavior as a function of ΔK for the ECAP and microcrystalline pure Ti (Cavaliere 2009. With kind permission of Elsevier)

In Fig. 8.97, S-N curves of pure Ti, produced by 8-pass ECAP, are illustrated. The UFG crystalline sample is compared with MC Ti and the results are, again, in line with the experimentally observed improvement in fatigue properties in many metals. The TEM micrographs of these fatigue-tested specimens are shown in Fig. 8.98, before and after the fatigue deformation. Their crack-growth behavior is presented in Fig. 8.99. The TEM revealed a mean grain size of 300 nm with a relatively high density of dislocations characterizing the grain boundaries and a relatively dislocation-free state in the grain interiors. Also, twin formation is observed, but having a very low density. A strong increase in the fatigue limit and in the number of cycles-to-failure is observed with respect to the as-received materials. The yield stresses of the UFG and MC Ti are, respectively, 720 and 420 MPa. The crack-propagation curves observed in Fig. 8.99 for pure MC and ECAP UFG Ti are fully characterized from threshold to final failure. It may be observed in Fig. 8.99 that the grain refinement, due to the severe plastic deformation, produces a decrease in the ΔK threshold and an increase in the crack-propagation rate. The primary mechanism responsible for accelerated FCGR, observed with decreasing grain size, is thought to be associated with the reduction in crack-path deflection due to grain refinement. In CG materials, periodic deflections in the fatigue cracks at grain boundaries during crystallographic fracture may lead to a relatively tortuous crack path, as compared to the straight crack path in UFG Ti, resulting in increased crack propagation in UFG Ti.

Summarizing this section, it may be stated that the grain refinement of materials to the NC and UFG ranges may significantly affect both the total life of these materials and crack growth when fatigue deformation is stress-controlled (see Figs. 8.93, 8.95b or 8.98). During stress-controlled tests, grain refinement via ECAP leads to an increase in fatigue properties. An increase in crack-propagation rate is observed in many pure metals, such as Ni, Ti, Al, etc. Their sensitivity to cyclic hardening increases with the decreased mean grain size of the material.

8.6 Fracture

8.6.1 Introduction

It has been indicated in this chapter that the conventional concept of "smaller is stronger" has initiated a great interest in applying materials of small dimensions (in the micron-submicron/nano range) for industrial purposes. Therefore, efforts are underway to manufacture, characterize and analyze the behavior of materials in these small ranges and to understand their performances in the elastic and plastic ranges at these small scales. The main feature of small-dimensioned material structures is the extremely small grain size, in other words, the consequent large area of grain boundaries per unit volume, possibly effecting plastic deformation by changing dislocation mechanisms. The possibilities of: limiting dislocation motion, affecting dislocation multiplication and potentially changing dislocation-nucleation capability are all factors that require consideration and experimental verification. At the beginning of this chapter, the suggested dislocation mechanisms were considered in order to better understand deformation behavior in materials with small dimensions. The final stage of material deformation is fracture initiation by some failure mechanism. Below, fracture will be considered in small-sized materials.

8.6.2 The Characteristics

Recalling the elements of fracture (described in Chap. 7), this phenomenon is associated with the initiation of cracks, their propagation, coalescence and final failure, when the material is separated into two or more parts or is sometimes completely disintegrated. This process, in small-sized materials, is the same as in macroscale materials, differing only in the mechanism causing the events leading to fracture. Clearly, coalescence before failure is related to microcracks or nanocracks, depending on the size of the specimens, which form or nucleate by the coalescence of micro- or nanovoids. Any of the three dislocation mechanisms (discussed above in Sect. 8.2.1) have been advocated by researchers who studied materials with small dimensions, single crystal or UFG polycrystalline. However, the role of boundary sliding (in polycrystalline material) cannot be overlooked in cases of samples below the 10–20 nm range. Furthermore, the contribution of twinning is another characteristic feature involving material resistance to deformation leading to fracture. In the following several examples illustrate the behavior of small-dimensioned materials. Ni (which has been widely investigated) is a good example for explaining deformation leading to fracture. Nanocrystalline Ni was obtained by electrodeposition from two different sources. The mean grain size from the first source is $\sim$30 nm, while, from the second source, it was $\sim$40 nm. High-resolution images are shown in Fig. 8.100. A well-defined crystalline structure is seen in Fig. 8.100a, b, without dislocation in the grain interior. Growth twins are prevalent in this material, as shown in a grain in Fig. 8.100c.

The progress of the deformation was documented continuously in real-time in order to demonstrate the presence of dislocation activity during plastic deformation. The image on the screen was sent to a video camera, connected to a viewing monitor, as well as to a computer capable of capturing freeze-frame images at various stages of the deformation process. These freeze-frame images are shown in Fig. 8.101. Three grains are seen in the figure the boundaries of which are identified by the white arrows and the crack tip is denoted by A (Fig. 8.101a). This image was obtained after one displacement pulse, when the specimen was under a load. The crack was there prior to the application of the displacement pulse, since the specimen had been under a load which had already produced the crack. The white arrows in Fig. 8.101a indicate the presence of grain-boundary microcracks and triple-junction voids.

After four displacement pulses (Fig. 8.101b), the crack appears to proceed part way through grain 1 and the grain boundary cracks and a triple junction void appears to have grown. A second crack (crack B) arrives in this region from the top right corner (in Fig. 8.101b.). This same location (in Fig. 8.101c) is shown after 12 pulses, where the progression of crack A through grain 1, as well as through most of grain 2, is seen. The progression of crack B, from the top of the image down into grain 3,

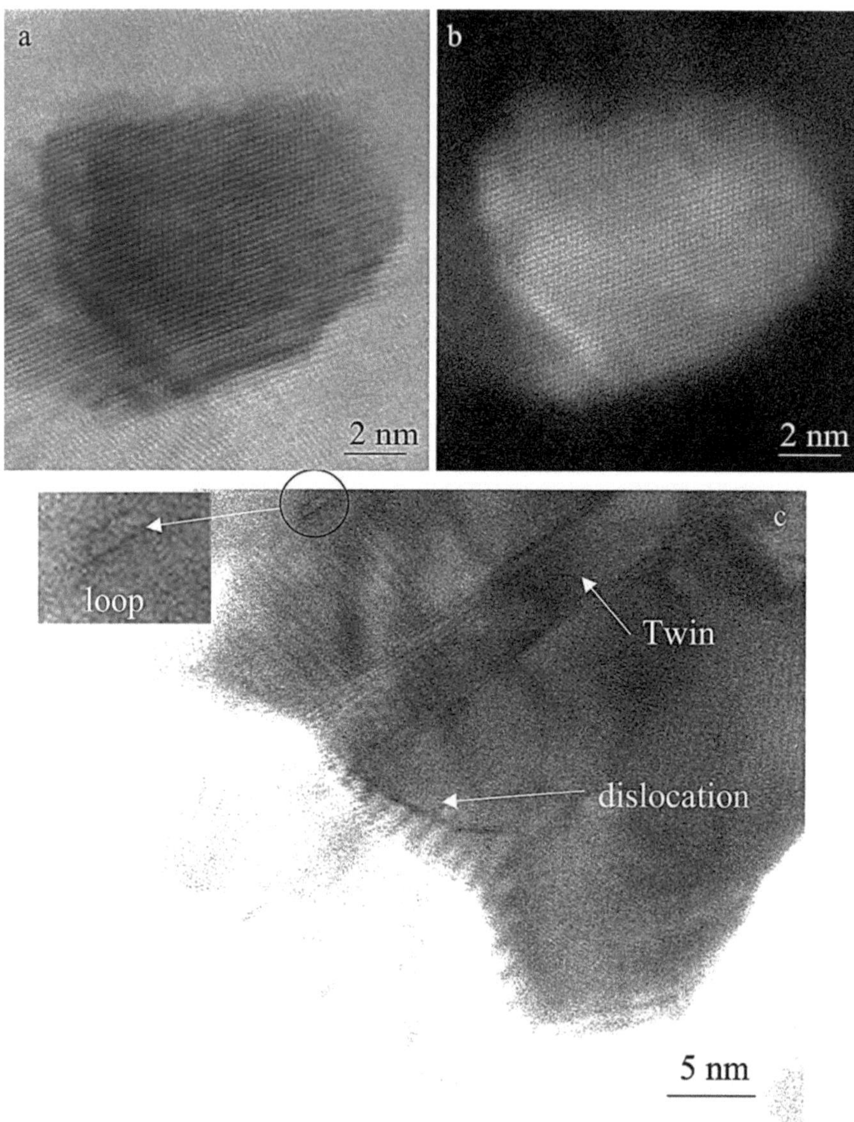

Fig. 8.100 High resolution STEM images of the nc-Ni from source 1 (**a**, **b**) bright field-dark field pair confirming a crystalline structure in the grain interior that extends all the way to the grain boundary with no indication of dislocations within the grains, and (**c**) another grain that contains a twin, a single dislocation at the boundary and a small dislocation loop within the grain (Kumar et al. 2003. With kind permission of Elsevier)

is also apparent, with only a segment of grain 3 left intact. An array of dislocations appears to have been emitted from crack B; at a later time (see Fig. 8.101d), this array is still visible, but it has a different configuration, and the inset in Fig. 8.101d shows the array at this stage at a higher magnification.

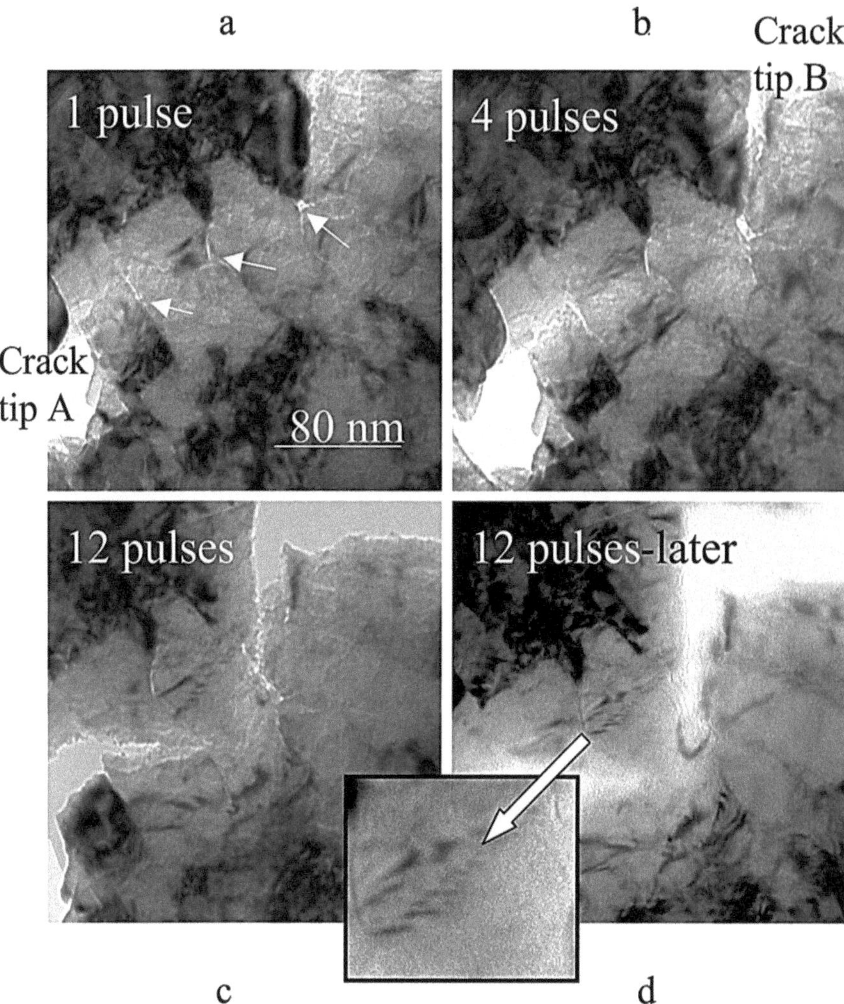

Fig. 8.101 A sequence of "freeze-frame" images captured during an in situ deformation test in the TEM of a microtensile specimen from source 2. Images (**a–d**) show the microstructural evolution and progression of damage with an increase in the applied displacement pulses. The presence of grain boundary cracks and triple-junction voids (indicated by *white arrows* in (**a**)), their growth, and dislocation emission from *crack tip B* in (**b–d**) in an attempt to relax the stress at the crack tip as a consequence of the applied displacement can all be seen. The magnified *inset* in (**d**) highlights the dislocation activity (Kumar et al. 2003. With kind permission of Elsevier)

In Ni, the room-temperature fracture surface consists of dimples, as seen in Fig. 8.102. The fracture surface corresponding to the thicker section of the tested specimen illustrates dimpled rupture. In the SEM fractographs (Fig. 8.102a), the average size of the dimples is considerably larger than the average grain size.

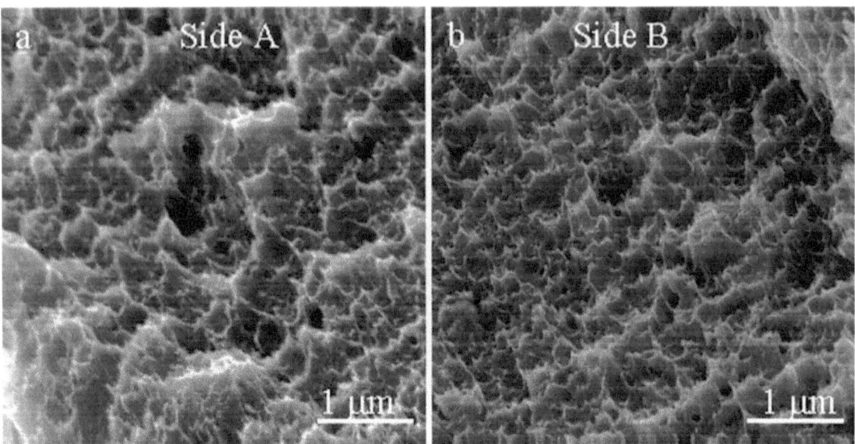

Fig. 8.102 A pair of SEM images (**a, b**) obtained from mating fracture surfaces of an in situ deformed tensile specimen of nc-Ni from source 2. The images were obtained from matching locations on the two surfaces and clearly confirm dimpled rupture on both surfaces (Kumar et al. 2003. With kind permission of Elsevier)

The TEM in situ observations (shown in Fig. 8.101) indicate extensive dislocation activity at the crack tips, demonstrating the dominant role of dislocation plasticity during the deformation of NC Ni.

Earlier in this chapter, strengthening was indicated by the H-P relation and a change in slope was indicated below a small grain size by the inverse H-P plots. Thus, according to the classic H-P concept, whenever strengthening occurs, resulting in higher strength, this usually also leads to higher fracture stress. However, the mechanisms responsible for the observed behavior below a grain size of ~100 nm are still not well established. There is a general consensus that plastic deformation occurs by dislocation motion and that dislocation interaction is probably the primary deformation mode, at least down to 25–30 nm (see Fig. 8.46, where a decrease in σ_y or hardness is indicated when d is decreased below 25 nm). At this stage, there appears to be no clear evidence of substantial dislocation activity. At grain sizes of ~10–15 nm, alternate mechanisms have been suggested involving GBS operating in conjunction with dislocation processes. Yet, at grain sizes of 5–10 nm, dislocation activity apparently ceases completely to be operative. The possibility that twinning may be an active deformation mode at very low grain sizes is suggested. This is a reasonable possibility for FCC materials, such as Al or Ni, having high SF energy. Growth twins are present in NC Ni (as indicated in Fig. 8.100c). The possibility of deformation twins in NC Ni has been observed by video-tape recording. It is known that twinning requires higher critical stresses, which can be obtained in NC metals. Figure 8.103 indicates the likelihood of the presence of twins in NC Ni.

Triple points play an important role in crack formation leading to fracture. Voids may readily form at such locations and in grain boundaries. Voids, before their coalescence into cavities to form sub-cracks, must be formed either before or during deformation. Even if they are not present a priori, voids may form in the wake

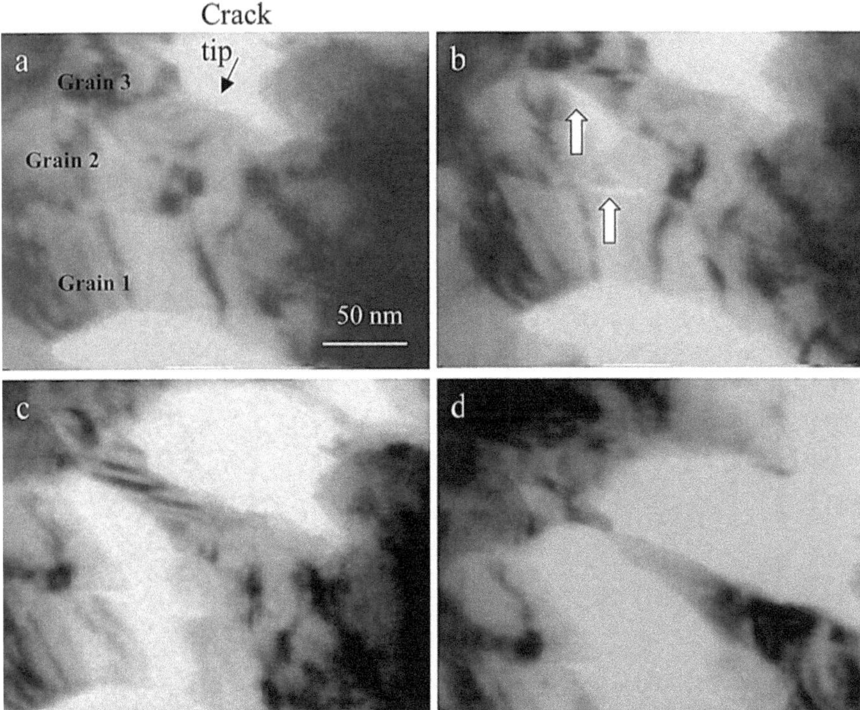

Fig. 8.103 A sequence of images extracted from a video-tape recording of the progression of deformation an in situ straining experiment. In image (**a**), the *crack tip* and a ligament composed of three grains can be seen. This ligament has already undergone plastic deformation and the three grains are elongated. Further deformation results in grain boundary separation (**b**) and subsequent transgranular failure of *grains 1* and *2* (**c**). The ligament left intact in (**c**) continues to deform plastically, developing parallel bands of alternate *black-and-white* contrast that could perhaps be a consequence of twinning. Eventually, the ligament necks down to a chisel point and ruptures (**d**) (Kumar et al. 2003. With kind permission of Elsevier)

of deformation: (i) during dislocation emission from grain boundaries; (ii) when created by GBS, leaving wedges at triple points; and (iii) by GBS inducing cavity formation on ledges. Several schemes have been suggested for the mechanism of deformation leading to fracture via crack formation, two of which are provided below.

Figure 8.104 reproduces the scheme of Kumar et al. A schematic illustration shows how deformation evolves in NC Ni and leads to (a) dislocation emission from grain boundaries, void formation/growth at grain boundaries and triple junctions, GBS; (b) the formation of the partially unconstrained ligaments that deform plastically; (c) the generation of cavities at grain-boundary ledges; and (d) the interaction of these various features to produce final dimple size. The interaction of these various features, shown in the schematic figure, produces the eventual fracture morphology in the nanocrystal.

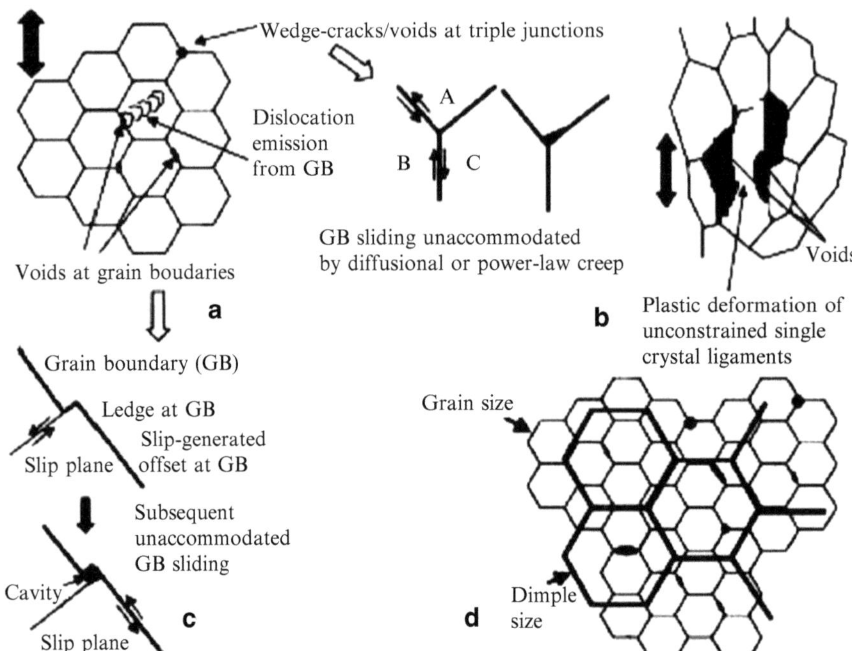

Fig. 8.104 Schematic illustration showing how deformation evolves in the nanocrystalline nickel and leads to (**a**) dislocation emission from grain boundaries, void formation/growth at grain boundaries and triple junctions, *GBS*; (**b**) the formation of partially unconstrained ligaments that deform plastically; (**c**) generation of cavities at grain-boundary ledges and (**d**) the interaction of these various features to produce final dimple size [331] (Meyers et al. 2006, after Kumar et al. With kind permission of Elsevier)

Crystal plasticity, in macroscaled specimens is based on dislocation motion; however, the dimensions of the individual grains in a NC material may be small enough so that dislocations do not behave in the classic manner. An indication of this may be obtained from Fig. 8.105.

These in situ experiments provide direct evidence for significant dislocation activity in NC Ni. Dislocations emitted at grain boundaries are found in Fig. 8.101, but GBS-induced void/crack formation at triple points or along grain boundaries are also observed. The formation and the growth of voids enable partial relaxation and further deformation of the NC material (here Ni) providing significant local plasticity and the formation of dimples, as observed at the fractured surface. The contribution of twinning deformation is a possible aspect of the overall deformation in NC materials.

One can state that the fracture surface of NC metals is quite complex. By bending a Ni foil and fracturing it, for example, river markings that correspond to the steps separating the regions of ductile fracture may be observed (see Fig. 8.106). Again, river patterns are features of fracture in macroscopic specimens. Their presence in bulk specimens indicates that, when fracture is brittle on a macroscopic scale,

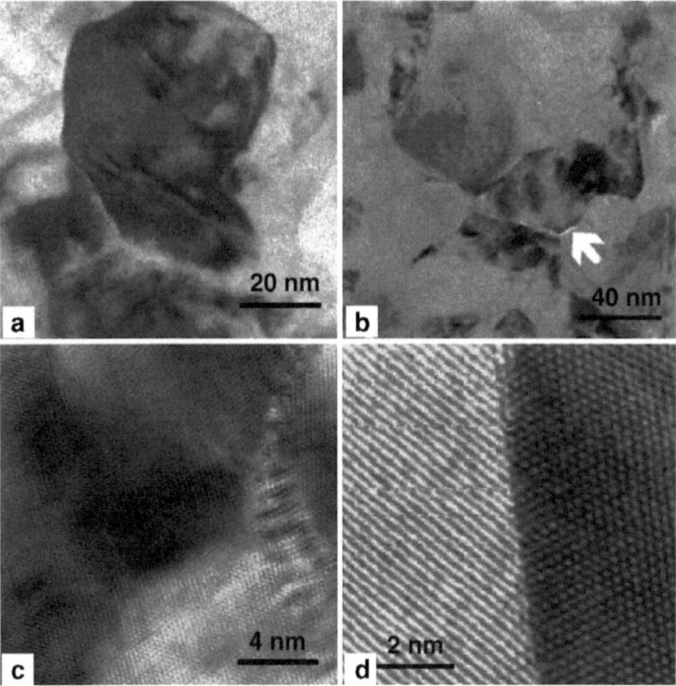

Fig. 8.105 Microstructure of nc-Ni after 4% plastic strain in compression: (**a**) bright-field TEM image showing a few dislocations within a grain; (**b**) a possible crack at a grain-boundary triple junction; (**c**) substantially dislocation-free grains and possibly a low-angle grain boundary and (**d**) clean grain boundaries with no evidence of residual dislocation debris left after the deformation (Meyers et al. 2006. With kind permission of Elsevier)

there is some localized-plastic flow in the vicinity of the fractured surface. On closer examination, the flat surfaces, usually representing cleavage planes, consist of dimples (Fig. 8.106c). This may indicate that localization between regions of ductile tearing is an inherent characteristic of fracture in NC metals. The resemblance of river patterns in NC materials to what is a common feature of fracturing in brittle macroscale samples, suggests that the stress concentrations at the roots of microcracks may cause the plastic flow ahead of them.

When a microcrack propagates rapidly, then there is no time for it to initiate plastic deformation ahead of itself, while a slow-moving crack has sufficient time for plastic flow to occur. Thus, one may conclude, as mentioned above, that increased fracture strength in ultrafine grained materials not only resists crack propagation, but makes it more difficult for cracks to nucleate within them. As such, grain boundaries in nanostructure materials are not significantly different from those found in conventional polycrystals.

A different approach to crack formation and growth (to macrocrack-size) in NC materials is related to deformation by GBS (illustrated in Fig. 8.107). In general, NC

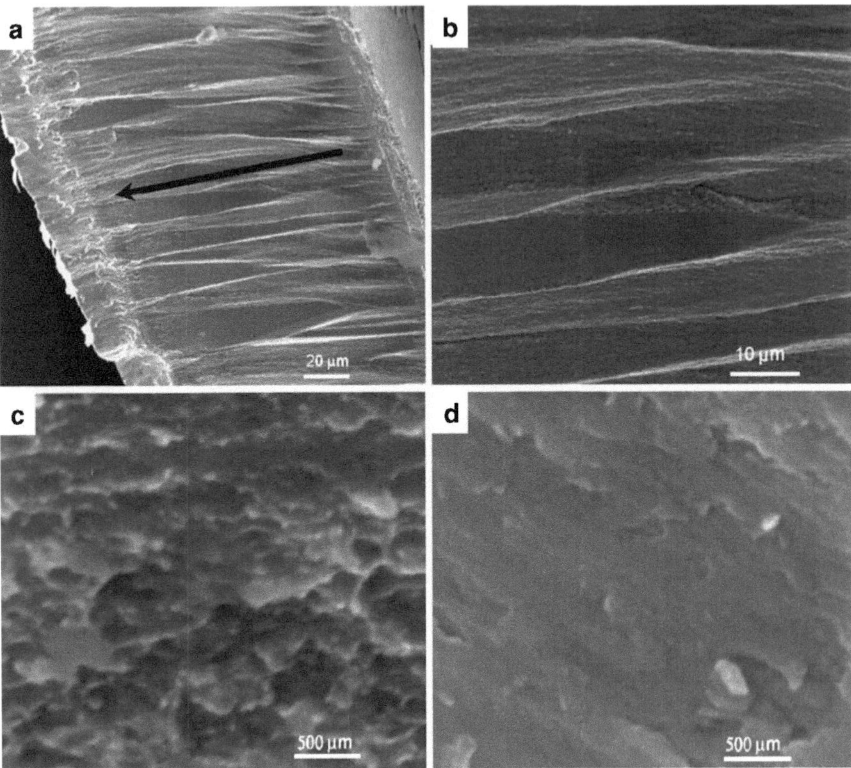

Fig. 8.106 Stepped fracture surface containing smooth steps produced by bending nanocrystalline Ni foil until fracture: (**a**) river pattern with plateaus and steps; (**b**) closer view of (**a**); (**c**) dimpled fracture in plateaus and (**d**) shear surface in steps (Meyers et al. 2006. With kind permission of Elsevier)

materials may fracture either due to the convergence of pre-existing cracks/voids or as a result of the generation of new nanocracks during plastic deformation. In Fig. 8.106, the case of an NC material without pre-existing cracks and voids is considered. This model considers GBS in NC materials, potentially leading to the formation of grain-boundary nanocracks during plastic deformation. As the number of grain boundaries containing nanocracks increases, and when the applied stress and strain reach some critical values, nanocracks grow by convergence forming a macrocrack, resulting in the catastrophic fracture of the NC solid.

Understanding the contribution of GBS to deformation leading-to-fracture is of importance, since the basic processes of plastic deformation by some dislocation mechanism is affected in small-sized specimens. The distribution of dislocation sources is progressively changed as the total volume of the sample decreases. Dislocation multiplication is restricted, relative to the case in macroscopic crystals. Furthermore, with size reduction, dislocations escaping from the crystal restricted multiplication, lending more support to the dislocation starvation mechanism. The

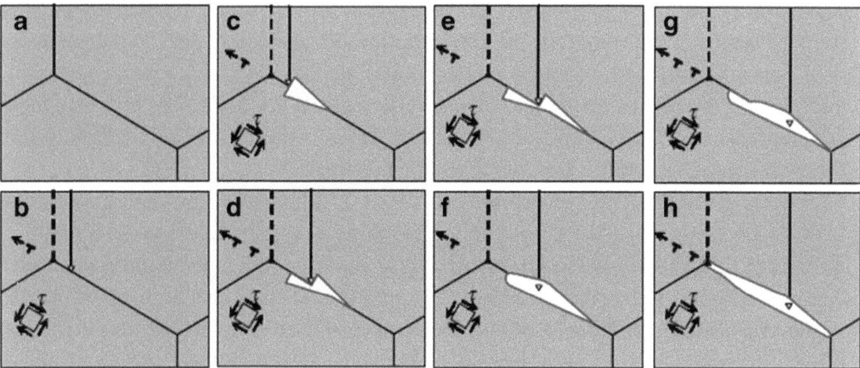

Fig. 8.107 Formation of a disclination dipole and grain boundary nanocrack as a result of GBS in a deformed nanocrystalline material. (**a**) Initial state of the grain boundary. (**b**) Disclination* dipole is generated in the grain boundary due to grain boundary sliding. Simultaneously, a dislocation forms in a triple junction. Then the dislocation is emitted into the grain interior. (**c**) Nanocrack generates at one of the dipole disclinations. (**d**, **e**) Owing to grain boundary sliding, the nanocrack changes its shape and elongates. At the same time, the second dislocation forms in the triple junction. Then the dislocation is also emitted into the grain interior. (**f**) Nanocrack is smoothen up due to diffusion. (**g**) Subsequent grain boundary sliding leads to the shear of the crack surfaces and crack elongation to the triple junction. (**h**) Crack growth to the *left*, along the rest of the grain boundary, owing to grain boundary diffusion (Meyers et al. 2006. With pemission of Dr. Ivan Archakov, Staff Editor of Rev. Adv. Mater. Sci. *A disclination is a line defect in which rotational symmetry is violated)

decreased dislocation density means that plastic deformation now requires high stress levels for the repeated generation of dislocations, which explains the high stresses observed in NC materials and their resistance to fracture.

8.7 Epilogue

It is now well-documented that materials with micron-submicron/nanometer dimensions do not behave as expected according to the general behavior of bulk macroscopic dimensioned materials. The mechanical properties of materials with such small dimensions cannot be derived simply from direct extrapolation from tests performed on macroscopic specimens. Furthermore, the methods, the equipment and the regular mechanical tests are not suitable for performing experiments on micrometer- and nanometer-scaled samples. It was indicated earlier that several new, sophisticated methods have been developed in tandem with the production of high-technological equipment and complete systems, currently used in a variety of applications (see, for example, Sharpe). The excellent performance of these novel systems permit the close investigation of micron-submicron/nanometer materials and enable a direct understanding of their behaviors. Many new and appropriate methods now exist in order to investigate various mechanical properties, such as: elastic, plastic, time-dependent deformations, and dynamic properties such as fatigue, fracture, etc. Very thin micron-submicron/nanometer films and structures, NWs, nanorods and nanotubes of various materials are now objects of experiments which have led to the development of several novel test methods. The accurate measurement of the mechanical properties of these objects promotes new understanding of their responses to acting stresses while being strained.

Despite disagreements regarding the exact mechanisms associated with the observed phenomena of these materials, one thing is clear, materials having small dimensions (e.g., UFG), or, more specifically in the NC range, may exhibit increased strength/hardness, improved toughness, improved ductility and high resistance to fatigue. Admittedly, different results have been recorded for experiments performed on the same materials, though they share the same purity, composition and grain-size, the data obtained by different investigators remain inconsistent. This may be due to the fact that the mechanical properties of NC materials are very sensitive to their initial microstructures, as they follow different preparation procedures (but having the same purity, composition and grain size). Another explanation is that it is very difficult to produce large bulk NC materials free from porosity and other flaws. Clearly this second reason (i.e., indicating the presence of pores) may be responsible for the difference between two grades of the same NC material. Microstructures strongly influence the mechanical behaviors of materials, especially having small dimensions. The challenge now is to reproduce microstructures of the same small-dimensioned materials and then to obtain the same experimental results, despite different production techniques.

Despite the present lack of a general agreement regarding all the observed and often contradictory experimental results, this does not prevent contemporary engineers, researchers and inventors from placing small-dimensioned materials, especially nanomaterials, into immediate use in industry. The work ahead is demanding, time is short and the technological potential is limitless; yet, new progressive methods and techniques must be sought. Students of materials science face great challenges while pursuing their careers in an age of endless possibilities.

Reference

D. Hull, D. Bacon, *Introduction to Dislocations* (Elsevier Butterworth-Heinemann, Oxford, 2004)

Further References

E.K. Baumert, P.-O. Theillet, O.N. Pierron, Acta Mater. **58**, 2854 (2010)
S.S. Brenner, Science **128**, 56 (1958)
P. Cavaliere, Int. J. Fatigue **31**, 1476 (2009)
X.X. Chen, A.H.W. Ngan, Scr. Mater. **64**, 717 (2011)
J. Chen, L. Lu, K. Lu, Scr. Mater. **54**, 1913 (2006)

A.H. Chokshi, A. Rosen, J. Karch, H. Gleiter, Scr. Mater. **23**, 1679 (1989)
B.W. Chua, L. Lu, M.O. Lai, Mater. Res. Bull. **41**, 2102 (2006)
V.N. Chuvil'deev, T.G. Nieh, M.Yu. Gryaznov, A.N. Sysoev, V.I. Kopylov, Scr. Mater. **50**, 861 (2004)
R.L. Coble, J. Appl. Phys. **34**, 1679 (1963)
F.F. Csikor, Ch Motz, D. Weygand, M. Zaiser, S. Zapperi, Science **318**, 251 (2007)
M. Dao, L. Lu, R.J. Asaro, J.T.M. De Hosson, E. Ma, Acta Mater. **55**, 4041 (2007)
G. Dehm, S.H. Oh, P. Gruber, M. Legros, F.D. Fischer, Acta Mater. **55**, 6659 (2007).
G. Dehm, Prog. Mater. Sci. **54**, 664 (2009)
J. Deng, D.L. Wang, Q.P. Kong, J.P. Shui, Scr. Metall. Mater. **32**, 349 (1995)
D.M. Dimiduk, M.D. Uchic, T.A. Parthasarathy, Acta Mater. **53**, 4065 (2005)
H.D. Espinosa, B.C. Prorok, B. Peng, J. Mech. Phys. Solids **52**, 667 (2004)
R.B. Figueiredo, T.G. Langdon, J. Mater. Sci. **43**, 7366 (2008)
G.E. Fougere, J.R. Weertman, R.W. Siegel, S. Kim, Scr. Metall. Mater. **26**, 1879 (1992)
L.B. Freund, J. Appl. Mech. **54**, 553 (1987)
L.H. Friedman, D.C. Chrzan, Philos. Mag. A **77**, 1985 (1998)
A.G. Froseth, P.M. Derlet, H. Van Swygenhoven, Acta Mater. **52**, 5863 (2004)
Y. Gotoh, Jpn. J. Appl. Phys. **11**, 1403 (1972)
J.R. Greer, W.D. Nix, Phys. Rev. B **73**, 245410 (2006)
J.R. Greer, W.C. Oliver, W.D. Nix, Acta Mater. **53**, 1821 (2005)
S. Han, C. Lim, C. Kim, S. Kim, Metall. Mater. Trans. A **36A**, 467 (2005)
T. Hanlon, Y.-N. Kwon, S. Suresh, Scr. Mater. **49**, 675 (2003)
Z. Horita, K. Matsubara, K. Makii, T.G. Langdon, Scr. Mater. **47**, 255 (2002)
H.-K. Liu, B.J. Lee, P.-P. Liu, Sens. Actuators A **140**, 257 (2007)
K. Kadau, T.C. Germann, P.S. Lomdahl, B.L. Holian, D. Kadau, P. Entel, M. Kreth, F. Westerhoff, D. Wolf, Metall. Mater. Trans. **35A**, 2719 (2004)
D. Kiener, W. Grosinger, G. Dehm, R. Pippan, Acta Mater. **56**, 580 (2008)
Ju-Y Kim, J.R. Greer, Acta Mater. **57**, 5245 (2009)
Ju-Y Kim, J.R. Greer, Acta Mater. **58**, 2355 (2010)
K.S. Kumar, H. Van Swygenhoven, S. Suresh, Acta Mater. **51**, 5743 (2003)
R. Lapovok, C. Loader, F.H. Dalla Torre, S.L. Semiatin, Mater. Sci. Eng. A **425**, 36 (2006)
M. Legros, M. Cabie, D.S. Gianola, Microsc. Res. Tech. **72**, 270 (2009)
K. Lu, W.D. Wei, J.T. Wang, Scr. Metall. Mater. **24**, 2319 (1990)
H. Lüthy, R.A. White, O.D. Sherby, Mater. Sci. Eng. **39**, 211 (1979)
M.A. Meyers, A. Mishra, D.J. Benson, Prog. Mater. Sci. **51**, 427 (2006)
P.C. Millett, T. Desai, V. Yamakov, D. Wolf, Acta Mater. **56**, 3688 (2008)
Y. Miyahara, K. Matsubara, Z. Horita, T.G. Langdon, Metall. Mater. Trans. **36A**, 1705 (2005)
D.G. Morris, Rev. Metal. **46**, 173 (2010)
H. Mughrabi, H.W. Höppel, Int. J. Fatigue **32**, 1413 (2010)
C.L. Muhlstein, R.T. Howe, R.O. Ritchie, Mech. Mater. **36**, 13 (2004)
G.W. Nieman, J.R. Weertman, R.W. Siegel, J. Mater. Res. **6**, 10121 (1991)
W.D. Nix, Metall. Trans. A **20**, 2217 (1989)
W.D. Nix, J.R. Greer, G. Feng, E.T. Lilleodden, Thin Solid Films **515**, 3152 (2007)
A. Ovidko, Rev. Adv. Mater. Sci. **10**, 89 (2005)
S.N. Patankar, J.P. Escobedo, D.P. Field, G. Salishchev, R.M. Galeyev, O.R. Valiakhmetov, F.H. (Sam) Froes, J. Alloys Compd. **345**, 221 (2002)
G.M. Pharr, Mater. Sci. Eng. **A253**, 151 (1998)
R. Raj, M.F. Ashby, Trans. Metall. Soc. AIME **2**, 1113 (1971)
R.O. Rittchie, J.J. Kruzic, C.L. Muhlstein, R.K. Nalla, E.A. Stach, Int. J. Fract. **128**, 1 (2004)
P.G. Sanders, M. Rittner, E. Kiedaisch, J.R. Weertman, H. Kung, Y. Lu, Nanostruct. Mater. **9**, 433 (1997)
J.-H. Seo, Y. Yoo, Na-Y Park, S.-W. Yoon, H. Lee, S. Han, S.-W. Lee, T.-Y. Seong, S.-C. Lee, K.-B. Lee, P.-R. Cha, H.S. Park, B. Kim, J.-P. Ahn, Nano Lett. **11**, 3499 (2011)
G. Sharma, R. Kishore, M. Sundararaman, R.V. Ramanujan, Mater. Sci. Eng. A **419**, 44 (2006)

W.N. Sharpe Jr., Strain **44**, 20 (2008)
M.D. Uchic, D.M. Dimiduk, J.N. Florando, W.D. Nix, Science **305**, 986 (2004)
W.W. Van Arsdell, S.B. Brown, J. Microelectromech. Syst. **8**, 319 (1999)
H. Van Swygenhoven, P.M. Derlet, A. Froseth, Acta Mater. **52**, 2259 (2004)
A.Yu. Vinogradov, V.V. Stolyarov, S. Hashimoto, R.Z. Valiev, Mater. Sci. Eng. **A318**, 163 (2001)
C.A. Volkert, E.T. Lilleodden, Philos. Mag. **86**, 5567 (2006a)
B. von Blanckenhagen, P. Gumbsch, E. Arzt, Philos. Mag. Lett. **83**, 1 (2003)
N. Wang, Z. Wang, K.T. Aust, U. Erb, Mater. Sci. Eng. **A237**, 150 (1997)
Y. Wang, M. Chen, F. Zhou, E. Ma, Nature **419**, 912 (2002)
W.W. Webb, R.D. Dragsdorf, W.D. Forgeng, Phys. Rev. **108**, 498 (1957)
J.R. Weertman, Mater. Sci. Eng. A **166**, 161 (1993)
F.E.N.G. Xiao-ming, A.I. Tao-tao, Trans. Nonferr. Met. Soc. China **19**, 293 (2009)
W. Yan, F.D. Fischer, Arch. Appl. Mech. **70**, 255 (2000)
W.M. Yin, S.H. Whang, R. Mirshams, C.H. Xiao, Mater. Sci. Eng. **A301**, 18 (2001)
K. Yoshida, Y. Gotoh, M. Yamamoto, Jpn. J. Appl. Phys. **24**, 1099 (1968)
J. Zhu, D. Shi, J. Phys. D: Appl. Phys. **44**, 055404 (2011)

About the Author

Joshua Pelleg received his B.S. in Chemical Engineering Technion Institute of Technology, Haifa, Israel, M.S. in Metallurgy, Illinois Institute of Technology, Chicago, IL, USA and Ph.D. Metallurgy, University of Wisconsin, Madison, WI. He is with Ben Gurion University of the Negev, Materials Engineering Department, Beer Sheva, Israel since 1970. Recipient of the Sam Ayrton Chair in Metallurgy. He has taught ever since the subjects of Mechanical Properties of Materials, Diffusion in Solids and Defects in Solids. Prior to arriving to BGU, Pelleg was assistant and then Associate Professor in the Department of Materials and Metallurgy, University of Kansas, Lawrence, KS, USA. Professor Pelleg was Visiting Professor in: Metallurgy Department, Iowa State University, Institute for Atomic Research, US Atomic Energy Commission, Ames, IA, USA in Department of Metallurgy, McGill University, Montreal, QC, Canada, Tokyo Institute of Technology, Applied Electronics Dept., Nagatsuta Campus, Yokohama, Japan and in Curtin University, Department of Physics, Perth, Australia. Among his non-academic research and industrial experience one can note: Chief Metallurgist in Urdan Metallurgical Works LTD., Netanya, Israel, Research Engineer in International Harvester, Manufacturing Research, Chicago IL., Associate Research Officer, National Research Council of Canada, Structures and Materials, National Aeronautical Establishment, Ottawa, ON, Physics Senior Research Scientist, Nuclear Research Center, Beer Sheva, Israel, Materials Science Division, Argonne National Labs, Argonne, IL, USA., Atomic Energy of Canada, Chalk River, Ont. Canada,Visiting Scientist, CSIR, National Accelerator Centre, Van de Graaf Group Faure, South Africa, Bell Laboratories, Murray Hill, NJ, USA, GTE Laboratories, Waltham, MA, USA. His main research interests are diffusion in solids, thin film deposition and properties (mostly by sputtering) and characterization of thin films among them various silicides.

J. Pelleg, *Mechanical Properties of Materials*, Solid Mechanics and Its Applications 190, DOI 10.1007/978-94-007-4342-7,

Index

J. Pelleg, *Mechanical Properties of Materials*, Solid Mechanics and Its Applications 190, DOI 10.1007/978-94-007-4342-7,

D

If you have any concerns about our products,
you can contact us on
ProductSafety@springernature.com

In case Publisher is established outside the EU,
the EU authorized representative is:
Springer Nature Customer Service Center GmbH
Europaplatz 3, 69115 Heidelberg, Germany

Printed by Libri Plureos GmbH
in Hamburg, Germany